High Performance Computing in Science and Engineering '13

Wolfgang E. Nagel • Dietmar H. Kröner
Michael M. Resch
Editors

High Performance Computing in Science and Engineering '13

Transactions of the High Performance Computing Center, Stuttgart (HLRS) 2013

Springer

Editors
Wolfgang E. Nagel
Zentrum für Informationsdienste
und Hochleistungsrechnen (ZIH)
Technische Universität Dresden
Dresden
Germany

Dietmar H. Kröner
Abteilung für Angewandte Mathematik
Universität Freiburg
Freiburg
Germany

Michael M. Resch
Höchstleistungsrechenzentrum
Stuttgart (HLRS)
Universität Stuttgart
Stuttgart
Germany

Front cover figure: Visualisation of turbulent structures in the draft tube of a Francis turbine with pressure isosurfaces coloured by eddy viscosity ratio. The red colour indicates regions with a higher amount of turbulence. Details can be found in "Flow Simulation of a Francis Turbine using the SAS Turbulence Model", by T. Krappel, A. Ruprecht and S. Riedelbauch, Institute of Fluid Mechanics and Hydraulic Machinery, University of Stuttgart.

ISBN 978-3-319-02164-5 ISBN 978-3-319-02165-2 (eBook)
DOI 10.1007/978-3-319-02165-2
Springer Cham Heidelberg New York Dordrecht London

Library of Congress Control Number: 2013956623

Mathematics Subject Classification (2010): 65Cxx, 65C99, 68U20

Printed on acid-free paper

Springer is part of Springer Science+Business Media (www.springer.com)

Preface

The Cray supercomputer HERMIT—installed in October 2011—is still the working horse at HLRS. It is a large Cray XE-6 with a peak speed of more than 1 PFLOP/s, based on the AMD Interlagos chip with overall more than 113,000 cores, integrated into 38 water-cooled cabinets. Additionally, the system is tightly integrated with external servers for pre- and postprocessing to support complex workflows. The system entered the Top500 list already in November 2011 and was—at that time—ranked number 12 on a worldwide level, achieving a Linpack value of 831,4 TFLOP/s. In the list published in November 2013, this system still is ranked on spot 39. Having Hermit as an infrastructure, and together with the new research buildings for VISUS, SimTech, and HLRS, the Universität Stuttgart is well positioned to become one of the leading science nodes for simulation technology in Germany, as well as abroad.

In 2014, the second delivery phase will follow, and the final Cray system will then have a peak performance of roughly 5 PFLOP/s. The plan is to have a Tier-0 HPC system within the GCS operating at any time within the 5-year period.

The HLRS participates in the European project PRACE (Partnership for Advances Computing in Europe) as part of the GCS, extending its reach to all European member countries, and has provided hundreds of millions of cpu hours to the European user community. Additionally, HLRS participates with partners in Germany in two Exascale Software Initiatives on European Level, namely TEXT and CRESTA where the challenges on the efficient use of current and future computing systems are investigated.

While the GCS has successfully addressed the high end computing needs, it was clear from the very beginning that an additional layer of support is required to maintain the longevity of the center, via a network of competence centers across Germany. This gap is addressed by the Gauß–Allianz (GA), in which regional and local centers teamed up to create the necessary infrastructure, knowledge, and the required methods and tools. The mission of the Allianz is to coordinate the HPC-related activities of its members. By providing versatile computing architectures and by combining the expertise of the participating centers, the necessary ecosystem for computational science has been created. Strengthening the research and increasing the visibility to compete at the international level are the further goals of the Gauß–Allianz. To disseminate information about its activities, the Gauß–Allianz is publishing a flyer (GA-Infobrief, http://www.gauss-allianz.de/infobrief), issued several times a year.

A number of projects of the third BMBF HPC-call have started as early as October 2013. This call was directed towards proposals that enable and support petascale applications on more than 100,000 processors, as they are also currently available at HLRS. While the projects of the first funding round started in early 2009 and the second one in April 2011, the follow-up call had been delayed again by more than 18 months. Nevertheless, all experts and administration authorities continue to acknowledge the strong need for such a funding program, given that the main issue identified in nearly all applications is that of *scalability*. The strategic funding plan involves more than 10 million Euros, with the hope of follow-up calls over the next 5 years, for projects that develop scalable algorithms, methods, and tools to support massively parallel systems. This can be seen as a very large investment. Nevertheless, in relation to the investment in computing hardware within Germany over this 5-year period, the investment in software is still comparatively small amounting to less than 20 % of the hardware investment. Furthermore, the investment in software will produce the "brains" that will be needed to use the newly developed innovative methods and tools, to accomplish technological breakthroughs in scientific as well as industrial fields of applications.

It is widely known that the long-term target is aimed not only at Petascale but also at Exascale systems. We do not only need competitive hardware but also excellent software and methods to address—and solve—the most demanding problems in science and engineering. The success of this approach is of significant importance for our community and will also greatly influence the development of new technologies and industrial products. Beyond being important, the success of this approach will finally determine whether Germany will be an accepted partner alongside the leading technology and research nations.

It is, therefore, a pleasure to announce that the German Research Foundation (DFG) has started the additional Priority Program 1648 "Software for Exascale Computing (SPPEXA)" in the field of HPC in January 2013. The funding is available for 6 years, with about 4 million Euros per year, to support fundamental and basic research questions in several specific areas related to HPC.

Since 1996, the HLRS has supported the scientific community as part of its official mission. Just as in the past years, the major results of the last 12 months were presented at the 15th annual Results and Review Workshop on High Performance Computing in Science and Engineering, which was held from September 30 to October 1, 2013, at the Universität Stuttgart. The workshop proceedings contain the written versions of the research work presented. The papers were selected from all projects running at the HLRS and the SSC Karlsruhe during a 1-year period between October 2012 and September 2013. Overall, a number of 48 papers were chosen from Physics, Solid State Physics, Chemistry, Reactive Flow, Computational Fluid Dynamics (CFD), Transport, Climate, and numerous other fields. The largest number of contributions originated from the CFD field, just as in many previous years, with 15 papers. Even though such a small collection cannot entirely represent an area this vast, the selected papers demonstrate the state of the art in high performance computing in Germany. The authors were encouraged to emphasize the computational techniques used in solving the problems examined. This is an often forgotten aspect and was the major focus of the workshop proceedings. Nevertheless, the importance of the newly computed scientific results for the specific disciplines is impressive.

We gratefully acknowledge the continuing support of the federal state of Baden-Württemberg in promoting and supporting high performance computing. Grateful acknowledgments are also due to German Research Foundation (Deutsche Forschungsgemeinschaft (DFG) and the Germany Ministry for Research and Education (BMBF), as many projects pursued on the HLRS and SSC computing machines could not have been carried out without its support. Also, we thank Springer Verlag for publishing this volume and, thus, helping to position the national activities in an international framework. We hope that this series of publications contributes to the global promotion of high performance scientific computing.

Dresden, Germany — Wolfgang E. Nagel
Freiburg, Germany — Dietmar Kröner
Stuttgart, Germany — Michael Resch
November 2013

Contents

Part I
Physics

Peter Nielaba

In this section, ten physics projects are described, which achieved important scientific results by using the HPC resources of the HLRS/SCC. Fascinating new results are being presented in the following pages for quantum systems (quarks, correlated electrons, and Bose systems), lithium niobate systems, soft matter systems (colloids, colloid-polymer mixtures), subsurface structures and stars.

The studies in the soft matter systems have focused on dynamical properties, spinodal decomposition, nucleation and flow in porous media.

J. Zhou and F. Schmid from the University of Mainz in their project *CCAC* studied a "Dissipative Particle Dynamics" (DPD) model for the dynamics of a charged colloid. A mesoscopic colloid model has been developed in which a spherical colloid is represented by many interacting sites on its surface. The model is based on DPD, with full considerations of the hydrodynamic and electrostatic interactions. The authors have investigated the electrophoretic mobility of a charged colloid in an external electric field, and the influence of salt concentration and colloid charge. The simulations have been done on HERMIT.

A. Statt, A. Winkler, P. Virnau and K. Binder from the University of Mainz in their project *colloid* have investigated effects of confinement on colloid-polymer mixtures enclosed in slits between repulsive walls or in spheres by computer simulations on HERMIT. Static properties have been studied by large-scale Monte Carlo simulation, including "ensemble switch" methods to estimate excess free energies due to confining walls. The kinetics of the phase separations has been studied by a combination of Molecular Dynamics and a multi-particle collision method. Due to the large scales of the structures that form via spinodal decomposition, large systems had to be chosen, containing e.g. 236,859 colloidal particles, about a million soft spheres representing the polymers, and 52 million effective solvent particles. Using the domain decomposition scheme and message passing (MPI), 4096 cores could

P. Nielaba
Fachbereich Physik, Universität Konstanz, 78457 Konstanz Germany
e-mail: peter.nielaba@uni-konstanz.de

be used in parallel to study such systems, with almost no loss in efficiency, in comparison to simulations without solvent particles.

Surface excess free energies due to the walls have been obtained, varying the range of the colloid-wall soft repulsive potential. This information has been used to clarify the wetting properties of the walls and the shift of the phase transitions due to finite film thickness (or finite sphere radius, respectively). It has been shown that a quantitative understanding of the "capillary condensation"-like shift of the phase transition has been reached.

D. Roehm, K. Kratzer and A. Arnold from the University of Stuttgart studied in their project *HYNUC* the heterogeneous and homogeneous crystallization of soft spheres in suspension. Using GPU-accelerated Lattice Boltzmann and Forward Flux Sampling, in combination with the ESPResSo package, the authors investigated the problem of crystallization of charged macromolecules like proteins or colloidal particles. The effects of hydrodynamic correlations on the heterogeneous nucleation in Yukawa-type colloids between two parallel plates has been studied with 16.384 particles, where a typical production run on a single 9 core node with MPI parallelization and attached GPU took up to 6 wall time hours on the Nehalem cluster. In case of homogeneous nucleation, a Forward Flux Sampling method has been used. By using the Nehalem cluster, it was possible to to study the effect of hydrodynamic interaction on the velocity of crystal growth.

J. Harting, F. Janoschek, B. Kaoui, T. Krüger, and F. Toschi from the Universities of Stuttgart, Eindhoven, London and Rome studied in their project *icpsusp* simplified models for coarse-grained blood flow simulations. On a microscopic scale the authors implemented a combined Lattice Boltzmann and immersed boundary method to investigate the behaviour of a single deformable viscous vesicle as a model system for a red blood cell, and the authors found that the transition from tank-treading to tumbling can be obtained by changing the viscosity contrast and the confinement or other relevant dimensionless parameters. Additional research results have been obtained using a more coarse grained model for blood flow, including a validation with experimental data of blood flow in cylindrical tubes. The computations have been done on the XC2 at the SCC.

Studies of quantum mechanical properties have been done for quarks, correlated electrons, Bosons, and for lithium niobate systems.

S. Krieg and the Wuppertal-Budapest collaboration in their project *HighPQCD* aim at a high precision calculation of the equation of state of Quantum Chromodynamics (QCD) including the (dynamical) effects of the charm quark. The authors use a lattice discretized version of QCD, and most of the finite temperature ensembles have been generated on less scalable architectures, the zero temperature renormalization runs, however, require a scalable architecture, and the authors used HERMIT with up to 256 nodes to generate these essential ensembles.

A. Moreno, J. M. P. Carmelo,and A. Muramatsu from the University of Stuttgart studied in their project *CorrFer* unconventional fractionalization of strongly correlated electrons. The authors investigated the effect of an injected spin-1/2 fermion into a strongly correlated 1D system, in particular the t-J model. The time evolution of a corresponding wave packet has been computed by a time dependent density matrix renormalization group (t-DMRG) method, using 600 DMRG vectors. The authors conclude, that charge and spin fractionalization occurs beyond the predictions of the Luttinger liquid theory. A comparison with results from Bethe-Ansatz allowed to identify charge and spin excitations that split into components at high and low energies. Computations have been done at the HLRS, utilizing large memory processors.

A.U.J. Lode, K. Sakmann, R.A. Doganov, J. Grond, O.E. Alon, A.I. Streltsov and L.S. Cederbaum from the University of Heidelberg studied in their project *MCTDHB* many-body non-equilibrium quantum dynamics of Bose–Einstein condensates by their method termed multiconfigurational time-dependent Hartree for bosons (MCTDHB). The authors have focused on two applications: (a) benchmarking the MCTDHB package with the harmonic-interaction model, and (b) the many-body decay by tunneling of a Bose–Einstein condensate to open space. The computations with about 400,000 core hours per month have been done on HERMIT and the Nehalem cluster.

A. Riefer, M. Rohrmüller, M. Landmann, S. Sanna, E. Rauls, N.J. Vollmers, R. Hölscher, M. Witte, Y. Li, U. Gerstmann, A. Schindlmayr and W.G. Schmidt from the University of Paderborn have studied in their project *MolArch1* the lithium niobate dielectric function and the second-order polarizability tensor. The authors have computed the frequency-dependent dielectric function and the second-order polarizability tensor of ferroelectric $LiNbO_3$. Density functional theory methods have been used (VASP package) within the generalized gradient approximation to obtain the structural and ground-state electronic properties of ferroelectric stoichiometric and congruent lithium niobate systems, the quasiparticle band structure has been calculated within the GW approximation for the exchange-correlation self-energy, and for the linear dielectric function of stoichiometric lithium niobate, the Bethe-Salpeter equation for coupled electron-hole excitations has been solved. The authors conclude, that the linear and nonlinear optical response calculated for a Li site vacancy model as well as a Nb site vacancy model are in closer agreement with available experimental data than the spectra calculated for stoichiometric crystals. Compared to stoichiometric lithium niobate, the congruent lithium niobate simulations lead to a redshift of the spectral features, wash out part of the fine structure, and reduce the nonlinearities. The computations have been done on HERMIT, linear scaling up to 1024 cores has been achieved.

On different length scales compared to the quantum and soft matter systems described above, the project *AEFWT* has focused on the recovery of subsurface images from seismic data, and the project *sotedimp* on the treatment of turbulent plasma and particle transport in the heliosphere.

T. Bohlen, S. Butzer, L. Groos, S. Heider, S. Jetschny, A. Kurzmann, A. Przebindowska and M. Schäfer from the Karlsruhe Institute of Technology have applied and developed in their project *AEFWT* a new seismic inversion and imaging technique that uses the full information content of the seismic recordings. Each echo or reflection from geological discontinuities in the earth is used to reveal the earth structure. For an estimation of subsurface parameters the authors exploit the full information contained in the seismic waveforms ("acoustic and elastic full waveform tomography”), and find a subsurface model by iteratively minimizing the misfit between observed and modeled data by using a conjugate gradient method. The 2D code has been applied to real streamer data of the North Sea, where sedimentary layers could be reconstructed. The computations have been done on HERMIT.

S. Lange and F. Spanier from the University of Würzburg studied in their project *sotedimp* turbulent plasma systems combined with the transport of high-energy particles. The modeling of such a system is achieved by a hybrid code (“GISMO”), which simulates the turbulent background plasma via a magnetohydrodynamic approach and the particle transport via a kinetic particle Ansatz. The authors results confirm theories stating that the turbulence cascade is anisotropic, and they show the importance of amplified wave modes to the particle scattering.

A Dissipative-Particle-Dynamics Model for Simulating Dynamics of Charged Colloids

Jiajia Zhou and Friederike Schmid

Abstract A mesoscopic colloid model is developed in which a spherical colloid is represented by many interacting sites on its surface. The hydrodynamic interactions with thermal fluctuations are taken accounts in full using Dissipative Particle Dynamics, and the electrostatic interactions are simulated using Particle–Particle–Particle Mesh method. This new model is applied to investigate the electrophoretic mobility of a charged colloid under an external electric field, and the influence of salt concentration and colloid charge are systematically studied. The simulation results show good agreement with predictions from the electrokinetic theory.

1 Introduction

Colloidal dispersions have numerous applications in different fields such as chemistry, biology, medicine, and engineering [1, 2]. In an aqueous solution, colloidal particles are often charged, either by ionization or dissociation of a surface group, or preferential adsorption of ions from the solution. A good understanding of the dynamics of charged colloids is important from the fundamental physics point of view. Furthermore, such an understanding may also provide insights to improve the material properties of colloidal dispersions. Theoretic studies of charged colloids are difficult because of the complexity of the system and various different interactions among the colloid, solvents, and microions.

Molecular simulations can shed light on the dynamic phenomena of charged colloids in a well-defined model system. Such studies are numerically challenging because two different types of long-range interactions are involved: the electrostatic

J. Zhou · F. Schmid (✉)
Institut für Physik, Johannes Gutenberg-Universität Mainz, Staudingerweg 7, 55099 Mainz, Germany
e-mail: zhou@uni-mainz.de; friederike.schmid@uni-mainz.de

W.E. Nagel et al. (eds.), *High Performance Computing in Science and Engineering '13*, DOI 10.1007/978-3-319-02165-2_1,

and the hydrodynamic interactions. In recent years, a number of coarse-grained simulation methods have been developed to address this class of problem. The general idea is to couple the explicit charges with a mesoscopic model for Navier–Stokes fluids. One of the examples is the coupling scheme developed by Ahlrichs and Dünweg [3], which combines a Lattice–Boltzmann (LB) approach for the fluid and a continuum Molecular Dynamics model for the polymer chains. This method has been applied to study the polyelectrolyte electrophoresis and successfully explained the maximum mobility in the oligomer range for flexible chains [4–6]. Besides the Lattice–Boltzmann method [7–12], there are a few choices of the fluid model in the literature, such as the Direct Numerical Simulation (DNS) [13–16], Multi-Particle Collision Dynamics (MPCD) [17, 18], and Dissipative Particle Dynamics (DPD) [19–22]. In this paper, we choose the DPD approach. DPD is a coarse-grained simulation method which is Galilean invariant and conserves momentum. Since it is a particle-based method, microions can be introduced in a straightforward manner. A recent comparative study [23] indicated that the electrostatic interaction is the most expensive part in terms of the computational cost. Therefore, for intermediate or high salt concentrations, different methods for modelling the fluid becomes comparable.

For colloidal particles, one requirement of the simulation model is the realization of no-slip boundary condition on the colloid surface. In this work, we present such a colloid model based on the Dissipative Particle Dynamics, with full considerations of the hydrodynamic and electrostatic interactions. We apply this model to study the dynamics of a charged colloidal particle under static electric fields. The remainder of this article is organized as follows: in Sect. 2, we introduce the simulation model and describe relevant parameters for the system. We present the simulation results of electrophoretic mobility in Sect. 3. Finally, we conclude in Sect. 4 with a brief summary.

2 Simulation Model

Our simulation system consists of three parts: the solvent, the colloidal particle and the microions. All simulations were carried out using the open source package ESPResSo [24].

2.1 *Fluids*

The fluids are simulated using Dissipative Particle Dynamics (DPD), an established method for mesoscale fluid simulations. For two fluid beads i and j, we denote their relative displacement as $\mathbf{r}_{ij} = \mathbf{r}_i - \mathbf{r}_j$, and their relative velocity $\mathbf{v}_{ij} = \mathbf{v}_i - \mathbf{v}_j$. The distance between two beads is denoted by $r_{ij} = |\mathbf{r}_{ij}|$ and the unit vector is

$\hat{\mathbf{r}}_{ij} = \mathbf{r}_{ij}/r_{ij}$. The basic DPD equations involve the pair interaction between fluid beads. The force exerted by bead j on bead i is given by

$$\mathbf{F}_{ij}^{\mathrm{DPD}} = \mathbf{F}_{ij}^{\mathrm{D}} + \mathbf{F}_{ij}^{\mathrm{R}}. \tag{1}$$

The dissipative part $\mathbf{F}_{ij}^{\mathrm{D}}$ is proportional to the relative velocity between two fluid beads,

$$\mathbf{F}_{ij}^{\mathrm{D}} = -\gamma^{\mathrm{DPD}}\,\omega^{\mathrm{D}}(r_{ij})(\mathbf{v}_{ij}\cdot\hat{\mathbf{r}}_{ij})\hat{\mathbf{r}}_{ij}, \tag{2}$$

with a friction coefficient γ^{DPD}. The weight function $\omega^{\mathrm{D}}(r_{ij})$ is a monotonically decreasing function of r_{ij}, and vanishes at a given cutoff $r_{\mathrm{c}}^{\mathrm{DPD}}$,

$$\omega^{\mathrm{D}}(r) = \begin{cases} \left(1 - \dfrac{r}{r_{\mathrm{c}}}\right)^2 & \text{if } r \leq r_{\mathrm{c}}^{\mathrm{DPD}}, \\ 0 & \text{if } r > r_{\mathrm{c}}^{\mathrm{DPD}}. \end{cases} \tag{3}$$

The cutoff radius $r_{\mathrm{c}}^{\mathrm{DPD}}$ characterizes the finite range of the interaction.

The random force $\mathbf{F}_{ij}^{\mathrm{R}}$ has the form

$$\mathbf{F}_{ij}^{\mathrm{R}} = \sqrt{2k_BT\gamma^{\mathrm{DPD}}\,\omega^{\mathrm{D}}(r_{ij})}\,\xi_{ij}\hat{\mathbf{r}}_{ij}, \tag{4}$$

where $\xi_{ij} = \xi_{ji}$ are symmetric, but otherwise uncorrelated random functions with zero mean and variance $\langle\xi_{ij}(t)\xi_{ij}(t')\rangle = \delta(t-t')$ (here $\delta(t)$ is Dirac's delta function). The fluctuation-dissipation theorem relates the magnitude of the stochastic contribution to the dissipative part, to ensure the correct equilibrium statistics. The pair forces between two beads satisfy Newton's third law, $\mathbf{F}_{ij} = -\mathbf{F}_{ji}$, hence the momentum is conserved. The momentum-conservation feature of DPD leads to the correct long-time hydrodynamic behavior (i.e. Navier–Stokes equations).

In the following, physical quantities will be reported in a model unit system of σ (length), m (mass), ε (energy), e(charge) and a derived time unit $\tau = \sigma\sqrt{m/\varepsilon}$. We use the fluid density $\rho = 3.0\,\sigma^{-3}$. The friction coefficient is $\gamma^{\mathrm{DPD}} = 5.0\,m/\tau$ and the cutoff for DPD is $r_{\mathrm{c}}^{\mathrm{DPD}} = 1.0\,\sigma$. To measure the shear viscosity, we implement the method in [25] to simulate the Poiseuille and Couette flows in a thin channel geometry. The viscosity of the fluid with our parameter setting is $\eta_s = 1.23 \pm 0.01\,m/(\sigma\tau)$.

2.2 *Microions*

Microions (either counterions to balance the colloid charge or the dissolved electrolytes) are introduced as the same DPD beads as the fluid, but carry charges and have exclusive interactions (to other charged beads but not to the fluid beads). We only consider the monovalent case where microions carry a single elementary charge $\pm e$. The exclusive interaction is necessary to prevent the collapse of charged system. A modified, pure repulsive Lennard–Jones interaction is used [26],

$$V(r) = \begin{cases} 4\varepsilon\left[\left(\dfrac{\sigma}{r-r_0}\right)^{12} - \left(\dfrac{\sigma}{r-r_0}\right)^{6} + \dfrac{1}{4}\right] & \text{if } r - r_0 \le r_c^{\mathrm{LJ}}, \\ 0 & \text{if } r - r_0 > r_c^{\mathrm{LJ}}, \end{cases} \tag{5}$$

where r is the distance between two charged beads. The cutoff radius is set at the potential minimum $r_c^{\mathrm{LJ}} = \sqrt[6]{2}\,\sigma$. The microions have a size of $1.0\,\sigma$ ($r_0 = 0$). Charged microions also interact by Coulomb interactions, and we compute the electrostatic interactions using Particle–Particle–Particle Mesh (P3M) method [27–29]. The Bjerrum length $l_B = e^2/(4\pi\epsilon_m k_B T)$ of the fluid is set to $1.0\,\sigma$ and the temperature is $k_B T = 1.0\,\varepsilon$.

One useful quantity is the diffusion constant of the microion D_{I}. We measure the mean-square displacement of the microion, then use a linear regression at late times to obtain the diffusion constant,

$$\lim_{t\to\infty} \langle (\mathbf{r}(t) - \mathbf{r}(0))^2 \rangle = 6 D_{\mathrm{I}} t. \tag{6}$$

The diffusion constant depends on the salt concentration ρ_s. We perform simulations with different salt concentrations, $\rho_s = 0.003125 - 0.2\,\sigma^{-3}$, in a simulation box $L = 30\,\sigma$. The simulation results are compared with the empirical Kohlrausch law [30], which states that microion's diffusion constant depends linearly on the square root of the salt concentration $\sqrt{\rho_s}$,

$$D_{\mathrm{I}} = A - B\sqrt{\rho_s}, \tag{7}$$

where A and B are two fitting parameters. Figure 1 shows the simulation results and a fit to Kohlrausch law.

2.3 Colloid

The colloidal particle is represented by a large sphere which has modified Lennard–Jones type conservative interaction to the fluid beads. The interaction has a similar form of (5), but with a larger radius $R = r_0 + \sigma = 3.0\,\sigma$.

To implement the boundary condition at the surface, a set of DPD interaction sites is distributed evenly on the surface $R = 3.0\,\sigma$, and the number of the sites is N_s. The position of the interacting sites is fixed with respect to the colloid center. These sites interact with the solvent beads through the DPD dissipative and stochastic interactions, with the friction constant γ^{CS} and the same cutoff r_c^{DPD}. The total force

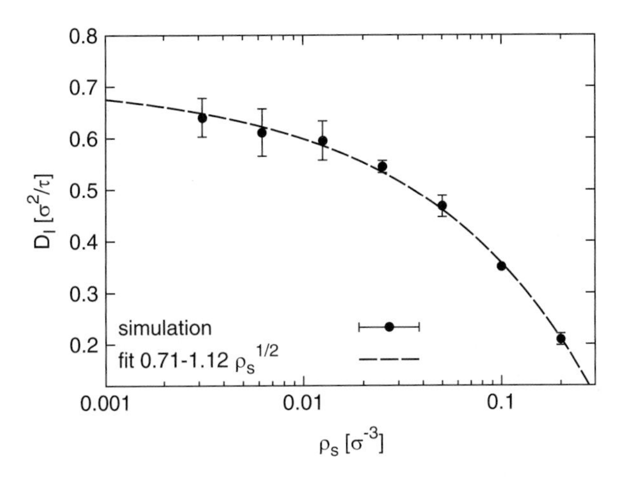

Fig. 1 The diffusion constant D_I of microions as a function of salt concentration ρ_s. *Symbols* are the simulation results. Error bars are obtained by averaging five independent simulation runs. The *curve* is a fit to Kohlrausch law with fitting parameters $A = 0.71$ and $B = 1.12$

exerted on the colloid is given by the sum over all DPD interactions due to the surface sites, plus the conservative excluded volume interaction,

$$\mathbf{F}^\text{C} = \sum_{i=1}^{N_s} \mathbf{F}_i^\text{DPD}(\mathbf{r}_i) + \mathbf{F}^\text{LJ}. \tag{8}$$

Here $\mathbf{r}_i$ denotes the position of i-th surface sites. Similarly, the torque exerted on the colloid can be written as

$$\mathbf{T}^\text{C} = \sum_{i=1}^{N_s} \mathbf{F}_i^\text{DPD}(\mathbf{r}_i) \times (\mathbf{r}_i - \mathbf{r}_\text{cm}), \tag{9}$$

where $\mathbf{r}_\text{cm}$ is the position vector of the colloid's center-of-mass. Note that the excluded volume interaction does not contribute to the torque because the associated force points towards the colloid center. The total force and torque are then used to update the position and velocity of the colloid in a time step using the Velocity–Verlet algorithm. The mass of the colloidal particle is $M = 100\,m$ and the moment of inertia is $I = 360\,m\sigma^2$, corresponding to a uniformly distributed mass. Figure 2 shows a representative snapshot of a single charged colloidal particle with counterions.

As a benchmark for our colloid model, we have performed simulations of an uncharged colloid in a simulation box $L = 60\,\sigma$ and measured the autocorrelation functions. Two functions are obtained from the simulations: the translational and rotational velocity autocorrelation functions

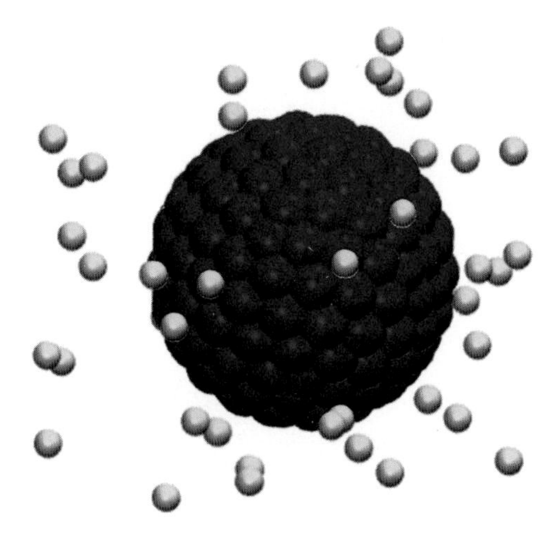

Fig. 2 Snapshot of a colloidal particle in a salt-free solution. The surface sites are represented by the *dark beads*, and the *light beads* are counterions. For clarity, solvent beads are not shown here

$$C_v(t) = \frac{\langle \mathbf{v}(0) \cdot \mathbf{v}(t) \rangle}{\langle \mathbf{v}^2 \rangle}, \tag{10}$$

$$C_\omega(t) = \frac{\langle \boldsymbol{\omega}(0) \cdot \boldsymbol{\omega}(t) \rangle}{\langle \boldsymbol{\omega}^2 \rangle}, \tag{11}$$

where $\mathbf{v}(t)$ and $\boldsymbol{\omega}(t)$ are the translational velocity and rotational velocity of the colloid at time t, respectively. Figure 3 shows the simulation results for $\gamma^{CS} = 10.0\,m/\tau$ in log-log plots.

For short time lags, both autocorrelation functions show exponential relaxation. The decay rate can be calculated using the Enskog dense-gas kinetic theory [31–34]

$$\lim_{t \to 0} C_v(t) = \exp(-\zeta^v_{\mathrm{ENS}} t), \tag{12}$$

$$\lim_{t \to 0} C_\omega(t) = \exp(-\zeta^\omega_{\mathrm{ENS}} t), \tag{13}$$

where the Enskog friction coefficients are

$$\zeta^v_{\mathrm{ENS}} = \frac{8}{3} \left(\frac{2\pi k_B T m M}{m + M} \right)^{1/2} \rho R^2 \frac{2}{M}, \tag{14}$$

$$\zeta^\omega_{\mathrm{ENS}} = \frac{8}{3} \left(\frac{2\pi k_B T m M}{m + M} \right)^{1/2} \rho R^2 \frac{5}{2M}, \tag{15}$$

where m and M are the mass for the fluid bead and the colloid, respectively, and ρ is the solvent density. Equations (12) and (13) are plotted as solid curves in Fig. 3 and show reasonable agreement with the simulation data when $t < 0.1\,\tau$.

For long time lags, hydrodynamic effects set in and lead to a slow relaxation for autocorrelation functions [35]. This so-called long-time tail is the manifestation

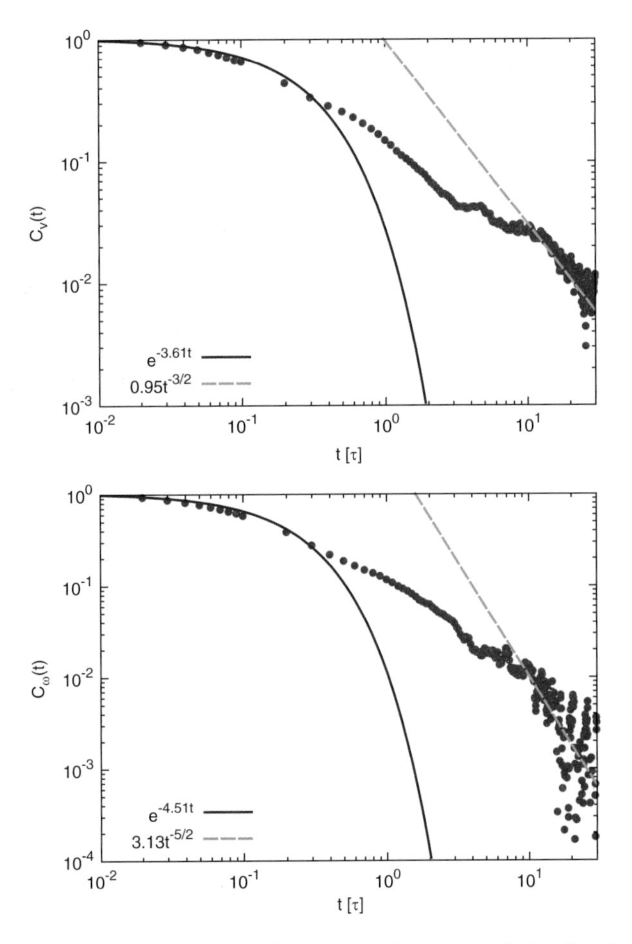

Fig. 3 Translational (*top*) and rotational (*bottom*) velocity autocorrelation functions. The measurement is performed for a single uncharged colloid with radius $R = 3.0\sigma$ in a salt-free solution. The temperature is $k_B T = 1.0\,\varepsilon$ and surface-fluid DPD parameter $\gamma^{\mathrm{CS}} = 10.0\,m/\tau$

of momentum conservation, as the momentum must be transported away from the colloid in a diffusive manner. Mode-coupling theory predicts an algebraic behavior at long times [36]

$$\lim_{t\to\infty} \langle \mathbf{v}(0)\cdot\mathbf{v}(t)\rangle = \frac{k_B T}{12 m\rho[\pi(\nu + D)]^{3/2}}\, t^{-3/2}, \tag{16}$$

$$\lim_{t\to\infty} \langle \boldsymbol{\omega}(0)\cdot\boldsymbol{\omega}(t)\rangle = \frac{\pi k_B T}{m\rho[4\pi(\nu + D)]^{5/2}}\, t^{-5/2}, \tag{17}$$

where $\nu = \eta_s/\rho$ is the kinematic viscosity and D is the diffusion constant of the colloid, which is much smaller than ν and can be neglected. The results are plotted

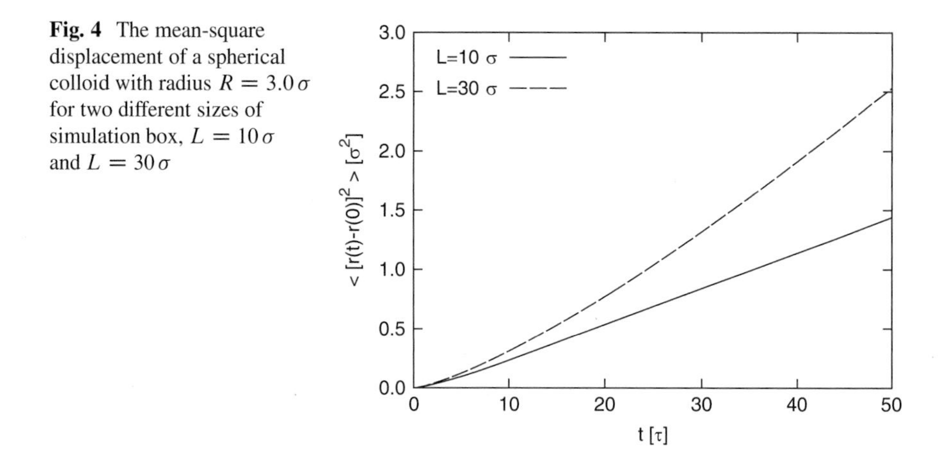

Fig. 4 The mean-square displacement of a spherical colloid with radius $R = 3.0\,\sigma$ for two different sizes of simulation box, $L = 10\,\sigma$ and $L = 30\,\sigma$

as dashed lines in Fig. 3. The data are consistent with the theoretical prediction for $t > 10\,\tau$, but the rotational autocorrelation function exhibits large fluctuations for large times. This is mainly due to the fact that the statistics for long-time values becomes very bad, and very long simulations are required to obtain accurate values.

The diffusion constant of the colloid can be calculated from the velocity autocorrelation function using the Green–Kubo relation

$$D = \frac{1}{3}\int_0^{\infty} \mathrm{d}t \, \langle \mathbf{v}(0) \cdot \mathbf{v}(t) \rangle. \tag{18}$$

Alternatively, the diffusion constant can be obtained from the mean-square displacement, similar to (6). Due to the periodic boundary condition implemented in simulations, the diffusion constant for a colloid in a finite simulation box depends on the box size. Figure 4 demonstrates the finite-size effect by plotting the mean-square displacement as a function of time for two different simulation boxes $L = 10\,\sigma$ and $L = 30\,\sigma$.

The diffusion constant increases with increasing box size. For small simulation box, the long-wavelength hydrodynamic modes are suppressed due to the coupling between the colloid and its periodic images. An analytic expression for the diffusion constant in terms of an expansion of powers of $1/L$ was derived by Hasimoto [37]

$$D = \frac{k_B T}{6\pi \eta_s}\left(\frac{1}{R} - \frac{2.837}{L} + \frac{4.19R^2}{L^3} + \cdots\right). \tag{19}$$

In Fig. 5, simulation results of the diffusion constant are plotted in terms of $1/L$, the reciprocal of the box size, and the curve is the prediction from (19). The simulation results and the hydrodynamic theory agree well.

Recently, studies of flow over superhydrophobic surfaces demonstrate that no-slip boundary condition is not always appropriate [38,39]. A more general boundary

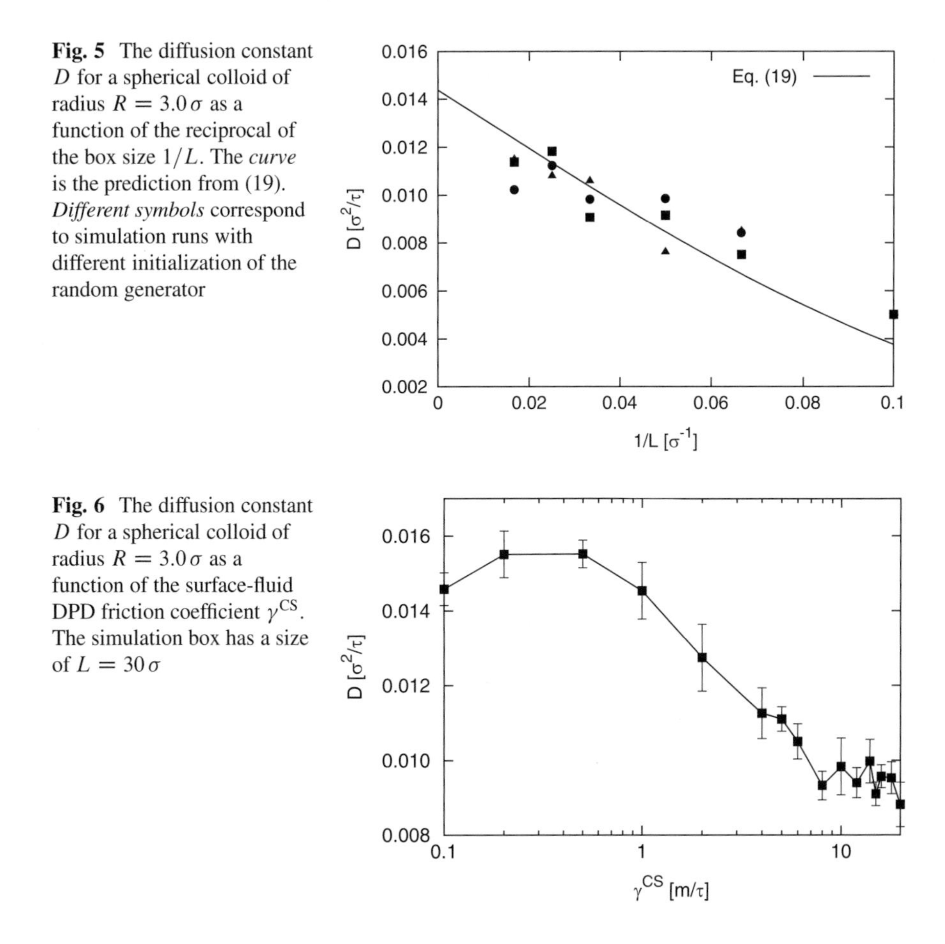

Fig. 5 The diffusion constant D for a spherical colloid of radius $R = 3.0\,\sigma$ as a function of the reciprocal of the box size $1/L$. The *curve* is the prediction from (19). *Different symbols* correspond to simulation runs with different initialization of the random generator

Fig. 6 The diffusion constant D for a spherical colloid of radius $R = 3.0\,\sigma$ as a function of the surface-fluid DPD friction coefficient γ^{CS}. The simulation box has a size of $L = 30\,\sigma$

condition is the Navier boundary condition, where finite slip over the surface is allowed. One advantage of our colloid model is the ability to adjust the boundary condition from no-slip to full-slip by changing the surface-fluid DPD friction γ^{CS}. Figure 6 illustrates the change of the diffusion constant by varying γ^{CS} in a simulation box $L = 30\,\sigma$. This freedom provides opportunities to study the effect of hydrodynamic slip on the colloidal dynamics [40, 41].

3 Electrophoretic Mobility

In this section, we apply our DPD-based colloid model to investigate the electrophoretic mobility of a charged colloid, and compare simulation results with predictions from electrokinetic theories [1, 42, 43].

Charged colloids in an aqueous solution are surrounded by counterions and dissolved salt ions. Counterions accumulate around the colloid surface due to the

Coulomb attraction between opposite charges and form an electric double layer (EDL). In equilibrium, the counterion cloud has spherical symmetry for a spherical colloid. The thickness of the EDL is characterized by the Debye screening length

$$l_D = \kappa^{-1} = \left[4\pi l_B \sum_i z_i^2 \rho_i(\infty) \right]^{-\frac{1}{2}}, \tag{20}$$

where the summation runs over different ion species; z_i and $\rho_i(\infty)$ are the valence and the bulk concentration for i-th ion, respectively. When an external electric field is applied to the suspension, the colloid (assumed to be positively charged) starts to move in the direction of the field, while the counterion cloud (negatively charged) is deformed and elongated in the opposite direction of the field. A steady state is reached when the electric driving force is equal to the hydrodynamic friction acting on the colloid. When the field strength is small, the final velocity of the colloid $\mathbf{u}$ depends linearly on the applied electric filed $\mathbf{E}$, and the proportionality defines the electrophoretic mobility μ:

$$\mathbf{u} = \mu \mathbf{E}. \tag{21}$$

The electrophoretic mobility is in general a second-order tensor, but is reduced to a scalar for spherical colloids. Due to the complexity of the system and the coupling between the hydrodynamic and electrostatic interactions, analytic solutions for the mobility only exist for limiting cases.

In the literature, the electrophoretic mobility is often expressed in terms of a ζ-potential, defined as the electrostatic potential at the shear plane. In the limit of $\kappa R \ll 1$ where the thickness of electric double layer is much larger than the colloidal size, the mobility is given by the Hückel formula [44]

$$\mu_{\mathrm{H}} = \frac{2\epsilon_m \zeta}{3\eta_s}, \tag{22}$$

where ϵ_m is the fluid permittivity and η_s is the shear viscosity. In the opposite limit when the Debye screen length is much thinner in comparison to the colloid size ($\kappa R \gg 1$), the famous Smoluchowski's formula states [45]

$$\mu_{\mathrm{S}} = \frac{\epsilon_m \zeta}{\eta_s} = \frac{3}{2} \mu_{\mathrm{H}}. \tag{23}$$

For more general cases of intermediate values of κR, one has to rely on numerical methods to solve the electrokinetic equations [42, 43]. In this work, we compute the electrophoretic mobility using the software MPEK.[1]

[1]http://reghanhill.research.mcgill.ca/research/mpek.html.

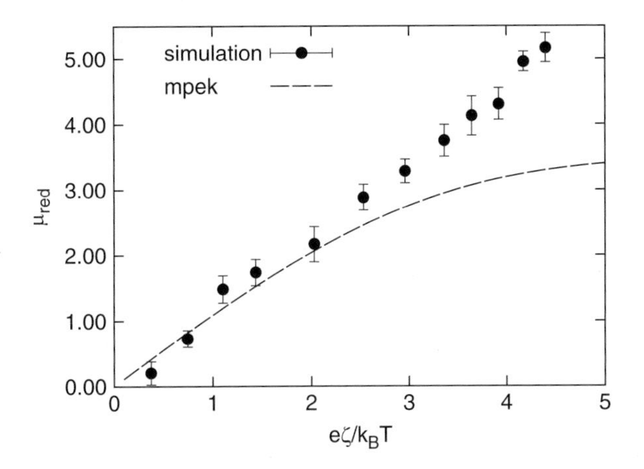

Fig. 7 The reduced mobility as a function of ζ-potential for a spherical colloid of size $R = 3.0\,\sigma$ and salt density $\rho_s = 0.05\,\sigma^{-3}$

In simulations, we use the colloid charge as a controlling parameter. To convert the colloid charge to the zeta-potential, one need to solve the Poisson–Boltzmann equation. For a spherical particle, numerical tables for the solution to Poisson–Boltzmann equation were given by Loeb et al. [46]. An analytic expression for the relationship between the ζ-potential and the surface charge density σ was derived by Ohshima et al. [47, 48]

$$\sigma = \frac{2\epsilon_m \kappa k_B T}{e} \sinh\left(\frac{e\zeta}{2k_B T}\right) \left[1 + \frac{1}{\kappa R} \frac{2}{\cosh^2(e\zeta/4k_B T)} + \frac{1}{(\kappa R)^2} \frac{8 \ln[\cosh(e\zeta/4k_B T)]}{\sinh^2(e\zeta/2k_B T)}\right]^{1/2}. \tag{24}$$

We use the above equation to compute the zeta-potential from known surface charge density.

Figure 7 plots the electrophoretic mobility as a function of the dimensionless zeta-potential $e\zeta/k_B T$. The simulations are performed for a charged spherical colloid of radius $R = 3.0\,\sigma$, and the solution has a salt concentration $\rho_s = 0.05\,\sigma^{-3}$. We follow the standard to use the reduced mobility, which is a dimensionless number and is defined as

$$\mu_{\text{red}} = \frac{6\pi\eta_s l_B}{e}\mu. \tag{25}$$

The simulation results agree well with the prediction from the electrokinetic theory at small zeta potentials. At large zeta potential, the steric effects of the microions may play a role [41, 49, 50], resulting in an increase of the mobility.

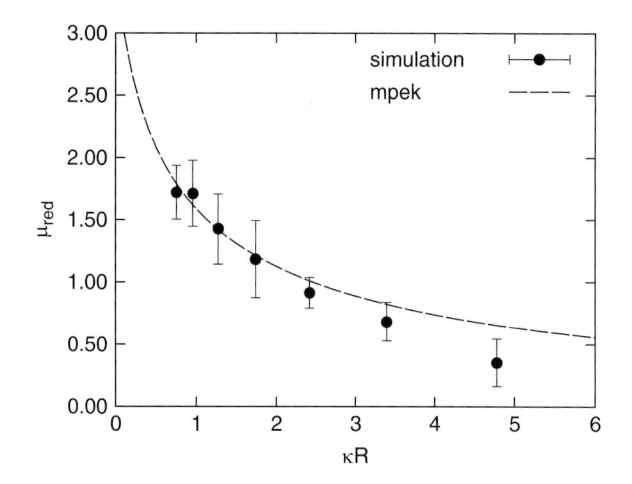

Fig. 8 The reduced mobility as a function of κR for a spherical colloid of size $R = 3.0\,\sigma$ and charge $Q = 10\,e$

Figure 8 plots the reduced electrophoretic mobility as a function of κR. In simulations, the colloid has a fixed radius $R = 3.0\,\sigma$, and the salt concentration is varied to change the value of κR. The charge on the colloid is small, $Q = 10\,e$, to ensure that the zeta-potential is also small. The simulation results and the theoretic predictions agree well, except at very large salt concentration (large κR value). One possible explanation to the discrepancy is the change of microion's diffusion constant with respect to the salt concentration (see Fig. 1), which is not taken into account in the electrokinetic theory.

4 Summary

We have developed a mesoscopic colloid model based on the Dissipative Particle Dynamics. We have taken accounts in full for the hydrodynamic interaction with thermal fluctuations, using Dissipative Particle Dynamics, and the electrostatic interactions, using Particle–Particle–Particle Mesh method. We applied this new colloid model to investigate the electrophoretic mobility of a charged colloid under a static external field. The simulation results show good agreement with the predictions from electrokinetic theories. Furthermore, this model has been applied successfully to study the dynamic and dielectric response of a charged colloid to alternating electric fields [51–53].

Acknowledgements We are grateful to Prof. Reghan Hill for providing the computer program MPEK. We thank the HLRS Stuttgart for a generous grant of computer time on HERMIT. This work is funded by the Deutsche Forschungsgemeinschaft (DFG) through the SFB-TR6 program "Physics of Colloidal Dispersions in External Fields".

References

1. W.B. Russel, D.A. Saville, W. Schowalter, *Colloidal Dispersions* (Cambridge University Press, Cambridge, 1989)
2. J. Dhont, *An Introduction to Dynamics of Colloids* (Elsevier, Amsterdam, 1996)
3. P. Ahlrichs, B. Dünweg, J. Chem. Phys. **111**, 8225 (1999)
4. K. Grass, U. Böhme, U. Scheler, H. Cottet, C. Holm, Phys. Rev. Lett. **100**, 096104 (2008)
5. K. Grass, C. Holm, Soft Matter **5**, 2079 (2009)
6. K. Grass, C. Holm, Faraday Discuss. **144**, 57 (2010)
7. V. Lobaskin, B. Dünweg, New J. Phys. **6**, 54 (2004)
8. V. Lobaskin, B. Dünweg, C. Holm, J. Phys. Condens. Matter **16**, S4063 (2004)
9. V. Lobaskin, B. Dünweg, M. Medebach, T. Palberg, C. Holm, Phys. Rev. Lett. **98**, 176105 (2007)
10. A. Chatterji, J. Horbach, J. Chem. Phys. **122**, 184903 (2005)
11. A. Chatterji, J. Horbach, J. Chem. Phys. **126**, 064907 (2007)
12. G. Giupponi, I. Pagonabarraga, Phys. Rev. Lett. **106**, 248304 (2011)
13. H. Tanaka, T. Araki, Phys. Rev. Lett. **85**, 1338 (2000)
14. Y. Nakayama, R. Yamamoto, Phys. Rev. E **71**, 036707 (2005)
15. K. Kim, Y. Nakayama, R. Yamamoto, Phys. Rev. Lett. **96**, 208302 (2006)
16. Y. Nakayama, K. Kim, R. Yamamoto, Eur. Phys. J. E **26**, 361 (2008)
17. A. Malevanets, R. Kapral, J. Chem. Phys. **110**, 8605 (1999)
18. G. Gompper, T. Ihle, D.M. Kroll, R.G. Winkler, Adv. Polym. Sci. **221**, 1 (2009)
19. P.J. Hoogerbrugge, J.M.V.A. Koelman, Europhys. Lett. **19**, 155 (1992)
20. J.M.V.A. Koelman, P.J. Hoogerbrugge, Europhys. Lett. **21**, 363 (1993)
21. P. Español, P.B. Warren, Europhys. Lett. **30**, 191 (1995)
22. R.D. Groot, P.B. Warren, J. Chem. Phys. **107**, 4423 (1997)
23. J. Smiatek, M. Sega, C. Holm, U.D. Schiller, F. Schmid, J. Chem. Phys. **130**, 244702 (2009)
24. H. Limbach, A. Arnold, B. Mann, C. Holm, Comput. Phys. Commun. **174**, 704 (2006)
25. J. Smiatek, M. Allen, F. Schmid, Eur. Phys. J. E **26**, 115 (2008)
26. J.D. Weeks, D. Chandler, H.C. Andersen, J. Chem. Phys. **54**, 5237 (1971)
27. R. Hockney, J. Eastwood, *Computer Simulation Using Particles* (Adam Hilger, Bristol, 1988)
28. M. Deserno, C. Holm, J. Chem. Phys. **109**, 7678 (1998)
29. M. Deserno, C. Holm, J. Chem. Phys. **109**, 7694 (1998)
30. M.R. Wright, *An Introduction to Aqueous Electrolyte Solutions* (Wiley, Chichester, 2007)
31. G. Subramanian, H. Davis, Phys. Rev. A **11**, 1430 (1975)
32. J. Hynes, Annu. Rev. Phys. Chem. **28**, 301 (1977)
33. J.T. Padding, A. Wysocki, H. Löwen, A.A. Louis, J. Phys. Condens. Matter **17**, S3393 (2005)
34. J.K. Whitmer, E. Luijten, J. Phys. Condens. Matter **22**, 104106 (2010)
35. B.J. Alder, T.E. Wainwright, Phys. Rev. A **1**, 18 (1970)
36. J.P. Hansen, I.R. McDonald, *Theory of Simple Liquids*, 3rd edn. (Academic, London, 2006)
37. H. Hasimoto, J. Fluid Mech. **5**, 317 (1959)
38. L. Bocquet, J.L. Barrat, Soft Matter **3**, 685 (2007)
39. O.I. Vinogradova, A.V. Belyaev, J. Phys. Condens. Matter **23**, 184104 (2011)
40. J.W. Swan, A.S. Khair, J. Fluid Mech. **606**, 115 (2008)
41. A.S. Khair, T.M. Squires, Phys. Fluids **21**, 042001 (2009)
42. R.W. O'Brien, L.R. White, J. Chem. Soc. Faraday Trans. 2 **74**, 1607 (1978)
43. R.J. Hill, D.A. Saville, W.B. Russel, J. Colloid Interface Sci. **258**, 56 (2003)
44. E. Hückel, Phys. Z. **25**, 204 (1924)
45. M.v. Smoluchowski, Z. Phys. Chem. **92**, 129 (1917)
46. A.L. Loeb, J.T.G. Overbeek, P.H. Wiersema, *The Electrical Double Layer Around a Spherical Colloid Particle* (MIT, Massachusetts, 1961)
47. H. Ohshima, T. Healy, L. White, J. Colloid Interface Sci. **90**, 17 (1982)

48. H. Ohshima, *Theory of Colloid and Interfacial Electric Phenomena* (Academic, Amsterdam, 2006)
49. J. López-García, M. Aranda-Rascón, J. Horno, J. Colloid Interface Sci. **316**, 196 (2007)
50. J. López-García, M. Aranda-Rascón, J. Horno, J. Colloid Interface Sci. **323**, 146 (2008)
51. J. Zhou, F. Schmid, J. Phys. Condens. Matter **24**, 464112 (2012)
52. J. Zhou, F. Schmid, Eur. Phys. J. E **36**, 33 (2013)
53. J. Zhou, R. Schmitz, B. Dünweg, F. Schmid, J. Chem. Phys. **139**, 024901 (2013)

Phase Separation of Colloid Polymer Mixtures Under Confinement

Antonia Statt, Alexander Winkler, Peter Virnau, and Kurt Binder

Abstract Colloid polymer mixtures exhibit vapor-liquid like and liquid-solid like phase transitions in bulk suspensions, and are well-suited model systems to explore confinement effects on these phase transitions. Static aspects of these phenomena are studied by large-scale Monte Carlo simulations, including novel "ensemble switch" methods to estimate excess free energies due to confining walls. The kinetics of phase separation is investigated by a Molecular Dynamics method, where hydrodynamic effects due to the solvent are included via the multiparticle collision dynamics method.

1 Introduction

Colloidal suspensions are of great interest for various applications, both in the chemical and food industries (paints, coatings, ink, cosmetic creams, tooth paste, diary products, etc.), and at the same time some colloidal suspensions are model systems for the study of cooperative phenomena in condensed matter. Here, we are only concerned with colloidal suspensions containing spherical particles of uniform size (in the range from 0.1 μm to a few μm in diameter) in good solvent (grafting surfactants or electrical charges prevent particle aggregation). Adding (smaller) polymers to the solution creates the well-known entropic depletion attraction ("Asakura–Oosawa (AO) model" [1]) among these spheres. The range and strength of this attraction can be tuned via the size (or molecular weight, respectively) and concentration of the polymers over a wide range, and different types of phase diagrams are found [2, 3], containing vapor-like phases (poor in colloids), liquid-like phases (rich in colloids, but the arrangement of the latter is disordered) and

A. Statt · A. Winkler · P. Virnau · K. Binder (✉)
Institut für Physik, Johannes Gutenberg-Universität,
55099 Mainz, Staudinger Weg 7, Germany
e-mail: kurt.binder@uni-mainz.de

W.E. Nagel et al. (eds.), *High Performance Computing in Science and Engineering '13*, DOI 10.1007/978-3-319-02165-2_2,

solid-like phases (typically the arrangement of the colloidal particles corresponds to a face-centered cubic crystal).

Due to the large size and slow dynamics of the colloidal particles these systems are particulary well-suited for experimental studies of phase separation, including interfacial phenomena and kinetics under confinement [4, 5]. When condensed matter systems are confined such that one (or more) linear dimensions of the system exceed the particle size only by one or two orders of magnitude, a delicate interplay of finite size effects, surface effects, and interfacial phenomena stabilized by the surface can be expected [6]. As is well-known, statistical fluctuations are enhanced in reduced dimensionality, analytic theories (mostly of mean-field type) are a bad guide for the understanding of such phenomena [7]. So computer simulation is the method of choice [6, 8], but at the same time challenging since the system typically develops inhomogeneous, and slowly relaxing structures with characteristic linear dimensions on the scale of the finite linear dimension(s) itself. Even using advanced simulation techniques [6–12] the use of supercomputers is mandatory in order to obtain valid results. The results described below have been obtained at the massively parallel supercomputer HERMIT of the HLRS, after preliminary studies at the computers of the ZDV Mainz and the JUROPA computer of the JSC available through NIC Jülich.

This report is arranged as follows. In the next section the model is specified and the simulation methodology is indicated. Then a survey of typical results on static properties for planar (Sect. 3) and spherical (Sect. 4) confinements is given. Section 5 then contains some representative results on the kinetics of phase separation, while Sect. 6 summarizes our conclusions.

2 Models and Methods

For the study of the vapor-liquid like phase transitions, polymers are included explicitly, while for liquid-solid transitions their effect is included only implicity via a suitable effective potential, that can be exactly derived from the AO model, by integrating out the polymer degrees of freedom, if the size ratio $q = \sigma_p/\sigma_c < 0.154$ (we denote the diameter of the polymers by σ_p, and of the colloids by σ_c, respectively). In the standard AO model [1–3], colloids are treated as hard spheres, polymers as soft spheres, which can overlap each other with zero energy cost (while overlap of polymers and colloids is strictly forbidden). For liquid-vapor phase separation, $q = 0.8$ has been chosen throughout.

While this original version of the AO model is convenient when only static properties are addressed, it is not useful when one is concerned with dynamic properties as well. To be able to study the latter, a soft version of the AO model has been developed [13], where hard-core repulsion is replaced by a truncated and shifted Lennard–Jones potential

$$U^{\alpha\beta}(r) = 4\varepsilon_{\alpha\beta}[(\sigma_{\alpha\beta}/r)^{12} - (\sigma_{\alpha\beta}/r)^{6} + 1/4], \tag{1}$$

for $r < r_c = 2^{1/6}\sigma_{\alpha\beta}$ while $U^{\alpha\beta}(r > r_c) \equiv 0$. Here $\alpha\beta$ stands for colloid-colloid (cc) or colloid-polymer (cp) pairs, and we choose $\sigma_{cc} = 1$ (unit of length) while $\sigma_{cp} = 0.9$, and energy parameters $\varepsilon_{cc} = \varepsilon_{cp} = 1.0$ (unit of temperature, $k_B \equiv 1$). The polymer-polymer interaction is [13]

$$U^{pp}(r) = 8\varepsilon_{pp}[1 - 10(r/r_c)^3 + 15(r/r_c)^4 - 6(r/r_c)^5], \qquad (2)$$

with $\varepsilon_{pp} = 0.0625$, $r_c = 2^{1/6}\sigma_{pp}$, $\sigma_{pp} = 0.8$, $U^{pp}(r > r_c) \equiv 0$. The choice of these parameters was dictated by computational efficiency [13].

In order to deal with confinement, the effect of the confining walls also needs to be specified. Choosing a rectangular box geometry, $L_x \times L_y \times L_z$, with periodic boundary conditions in x and y directions, but walls placed at $z_w = \pm L_Z/2$, we choose a potential similar to (1) to describe the effect of the walls,

$$U^{\alpha}_{\text{wall}}(z) = 4\varepsilon_{w\alpha}\left[\left(\frac{\sigma_{w\alpha}}{z - z_w}\right)^{12} - \left(\frac{\sigma_{w\alpha}}{z - z_w}\right)^{6} + \frac{1}{4}\right], \qquad (3)$$

where $\alpha = (c, p)$, $z_w = -L_z/2$, $z_w < z < z_w + 2^{1/6}\sigma_{w\alpha}$, and similarly for the wall at $z_w = +L_z/2$. We choose $\varepsilon_{wp} = \varepsilon_{wc} = 1$, $\sigma_{wp}/\sigma_{wc} = 0.8$, while σ_{wc} is varied. Note that $\sigma_{wc} = 0.5$ and $\sigma_{wp} = 0.4$ correspond to the natural case in which colloids of radius 0.5 and polymers of radius 0.4 as defined by the interparticle interactions just touch the wall. For the original AO model, instead simple hard wall potentials are used, $U_{wc}(z) = \infty$ if $z - z_w < \sigma_{wc}$ but $U_{wc}(z) = 0$ if $z - z_w > \sigma_{wc}$, instead of (3). Of course, these wall potentials can straightforwardly be used also for confinement by a spherical wall.

Of course, for our study it is crucial that these wall potentials provide conditions where the wetting properties of the walls can be varied. Figure 1 therefore presents snapshot pictures of the AO model confined by planar walls, where σ_{wc} is varied from 0.5 to 0.8. One can see that for $\sigma_{wc} = 0.5$ the depletion attraction of colloids to the walls still dominates, while for $\sigma_{wc} = 0.8$ the situation is opposite, polymers now are attracted to the walls. So a relatively small variation of σ_{wc} suffices to change conditions from complete wetting (contact angle $\theta = 0$, $\cos\theta = 1$) to complete drying (contact angle $\theta = 180°$, $\cos\theta = -1$). In the range $0.6 \le \sigma_{wc} \le 0.7$ several cases of partial wetting or drying are shown, the case $\sigma_{wc} = 0.65$ clearly is close to the "neutral walls", i.e. $\theta = 90°$. However, the snapshot pictures also reveal that the interfaces between the coexisting polymer-rich and colloid-rich phases are very rough, there are strong statistical fluctuations, and when one performs a coarse-graining (right part of Fig. 1) one can see that the finite width of the interfacial region would make quantitative estimation of the contact angle rather ambiguous. This example is chosen such that the colloid packing fraction $\eta_c = (\pi\sigma_{cc}^3/6)N_c/V$ (N_c is the number of colloidal particles in the simulation box of volume V) is in the center of the two-phase coexistence region, at the chosen value of the polymer reservoir packing fraction, $\eta_p^r = (\pi\sigma_{pp}^3/6)\exp(\mu_p/k_BT)$, see Fig. 2 (A. Statt, Diploma Thesis (Johannes Gutenberg Universität Mainz, 2012, unpublished)).A closer look even reveals that the colloidal particles near the walls

Fig. 1 Snapshot pictures (*left part*) of colloid-polymer mixtures at a colloid packing fraction $\eta_c = 0.18$ and a polymer reservoir packing fraction $\eta_p^r = 1.50$ for a system with linear dimensions $L_x = L_y = 16$, $L_z = 64$, varying $\sigma_{wc} = 0.5, 0.6, 0.65, 0.7$ and 0.8, respectively (*from top to bottom*). Colloids are shown as *yellow spheres*, polymers as *black spheres*. The *right part* shows corresponding coarse-grained colloid density distributions, in the form of a contour diagram

show a pronounced "layering effect", Fig. 3, which makes the study of interfaces near the hard walls particularly difficult. In such small simulation boxes, it clearly is impossible to distinguish a moderate surface enrichment of colloidal particles at a wall from a wetting layer (that may emerge only in the thermodynamic limit $L_z \to \infty$, of course.)

Thus for characterizing precisely the wetting properties of the walls, a different approach has been adopted, based on the use of Young's equation,

$$\gamma_{wg} - \gamma_{w\ell} = \gamma_{\ell g} \cos(\theta), \tag{4}$$

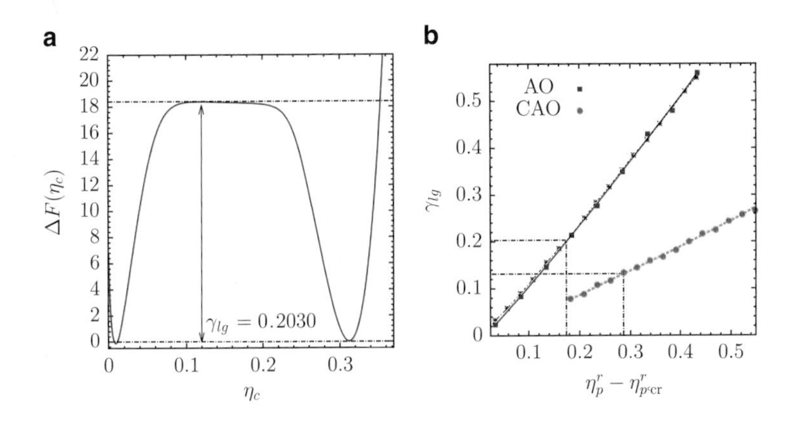

Fig. 2 (**a**) Free energy difference $\Delta F(\eta_c)$ relative to the free energy at phase coexistence, for the AO model at a polymer reservoir packing fraction 0.94, choosing a simulation box $L_x = L_y = 6.735$ and $L_z = 14.6$. From the hump in the free energy (which is due to two parallel interfaces of area $L_x \times L_y$) the interfacial tension of the "gas-liquid" interface can be estimated, $\gamma_{\ell g} = 0.203$. (**b**) Interfacial tension plotted versus the distance from the critical point, $\eta_p^r - \eta_{p,r}^r$, both for the standard AO model and the continuous AO model (CAO, defined by (1), (2)). Here $\eta_{p,cr}^r = 0.7670 \pm 0.0002$ (AO) and $\eta_{p,cr}^r = 1.285 \pm 0.0010$ was estimated from a finite size scaling analysis (A. Statt, Diploma Thesis (Johannes Gutenberg Universität Mainz, 2012, unpublished)). The *horizontal* and *vertical straight lines* mark the conditions for which ensemble switch simulations were performed (A. Statt, Diploma Thesis (Johannes Gutenberg Universität Mainz, 2012, unpublished))

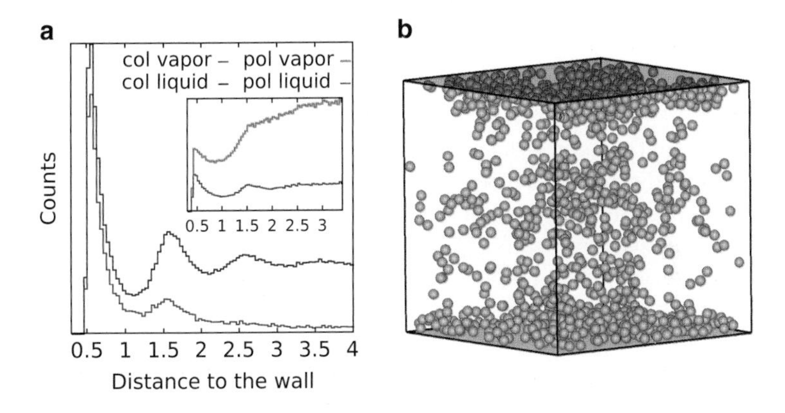

Fig. 3 (**a**) Colloid "density" profile of the vapor phase at $\eta_c = 0.0436$ and of the liquid phase at $\eta_c = 0.2773$, for the CAO model at $\eta_p^r = 1.57116$ and $\sigma_{wc} = 0.5$, using (3) obtained in a box where $L_x = L_y = L_z = 24$. (**b**) Snapshot of typical "vapor" configuration containing 1,100 colloids. Polymers are not shown

which requires to estimate the excess free energies due to the wall of the vapor-like phase (γ_{wg}) and the liquid-like phase (γ_{wl}) at conditions of phase coexistence, cf. Fig. 2a. The liquid-vapor interfacial tension $\gamma_{\ell g}$ has already been obtained (Fig. 2a,b) using Successive Umbrella Sampling [14]. The task of obtaining γ_{wg} and γ_{wl}

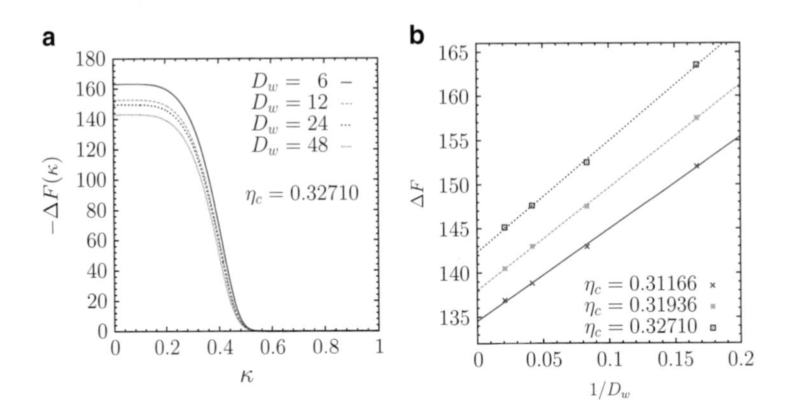

Fig. 4 (**a**) A typical example of the negative free energy $-\Delta F(\kappa)$ as a function of the mixing parameter κ, for different wall distances D_w as indicated. A box with linear dimensions $L_x = L_y = 6.375$ is used, and the colloid packing fraction is $\eta_c = 0.3271$. The wall interaction range was set to $\sigma_{wc} = 0.5$ (**b**) Each ΔF is plotted versus $1/D_w$ and extrapolated to $D_w \to \infty$

can be solved by the so called "ensemble switch method" [9]. In this method, a generalized Hamiltonian $\mathcal{H}(\{\mathbf{r}_i\})$ is constructed from a system with periodic boundary conditions in all three space directions $\{\mathcal{H}_1(\{\mathbf{r}_i\})\}$ and a corresponding system with two parallel, equivalent walls, $\mathcal{H}_2(\{\mathbf{r}_i\})$. The linear dimensions and the particle numbers of both systems are precisely the same, $\{\mathbf{r}_i\}$ denoting the positions of all the particles. The generalized Hamiltonian then is defined as

$$\mathcal{H}(\{\mathbf{r}_i\}) = (1-\kappa)\mathcal{H}_1(\{\mathbf{r}_i\}) + \kappa\mathcal{H}_2(\{\mathbf{r}_i\}), \tag{5}$$

where $\kappa \in \{0, 1\}$ denotes the "mixing percentage" between both systems. In every Monte Carlo move (at $\kappa = \text{const}$) in the canonical ensemble (involving a displacement of a randomly chosen particle), the standard acceptance probability of the move is calculated from this extended Hamiltonian as usually done. In addition to these moves, one also considers moves where κ changes (discretizing the set of κ-values into e.g. 100 discrete κ_i, and considering only moves to neighboring values $\kappa_i \to \kappa_{i-1}$ or κ_{i+1}). Sampling then the probability $P(i)$ that the system resides in a state with κ_i, using the Wang–Landau algorithm [15], one finds the free energy difference between two neighboring states as $k_B T(\ln[P(i)] - \ln[P(i-1)])$, and sampling the full range from $\kappa = 0$ to $\kappa = 1$ the free energy difference due to the two walls can be constructed (Fig. 4). Since there is a significant dependence on the distance D_w between both walls, an extrapolation against $1/D_w \to \infty$ is necessary (A. Statt, Diploma Thesis (Johannes Gutenberg Universität Mainz, 2012, unpublished)). Carrying out this procedure both for the vapor-like and liquid-like phase, as described already in [9], one can obtain rather accurate results on the variation of the contact angle θ with σ_{wc}, both for the AO model and the continuous AO model (CAO model).

Finally, we discuss the simulation of dynamic aspects of this model system. Then the solvent (which does not affect the static properties of the system, on the scale of colloidal particles the solvent particles act like point-like ideal gas particles) needs to be included via the multiparticle collision dynamics (MPC) method. It is this aspect which could not have been done at all without access to a massively parallel supercomputer such as HERMIT, since due to the large scales of the structures that form via spinodal decomposition [10, 11] one needs to study very large systems, containing e.g. 236,859 colloidal particles, about a million soft spheres representing the polymers, and 52 million effective solvent particles. Using the domain decomposition scheme and the message passing interface (MPI), 4,096 cores could be used in parallel to study such systems, with almost no loss in efficiency, in comparison to simulations without solvent particles (A. Winkler, Dissertation (Johannes Gutenberg Universität Mainz, 2012, unpublished)). More details about this algorithm were already given in a previous report [16], and hence are omitted here.

We only mention that in the domain decomposition algorithm for MPC developed by Sutmann et al. [17] the system is divided into an array of subsystems, where each of these subsystems has to communicate at the borders with the neighboring subsystems. This is achieved by adding "halo layers" to the borders that contain all the necessary information from the neighboring domain to perform the MD integration time step with the Velocity Verlet algorithm. The communication occurs via MPI. Since the interaction potentials are of short range, these "halo layers" are reasonably small. The collision step of the solvent according to the MPC algorithm is implemented as described in [17]. When a MPC cell is shared by multiple cores, the needed information (positions and velocities of the particles) is sent in x, y, and z-directions in a serial manner. After the collision is performed, the resulting velocities are sent back in reversed order. In comparison to standard Molecular Dynamics simulations of phase separation of highly asymmetric mixtures, where uneven load of processors dealing with different regions of the system creates problems, the combined MD/MPC algorithm does not suffer from this problem, due to the homogeneous distribution of the large number of solvent particles, the density in all subdomains is approximately the same. These considerations explain, why for this problem one can use up to 4,096 processors (necessary to deal with the 52 million solvent particles) with very little performance loss, as already demonstrated by the scaling plot published in [16]. This massive parallelization was indispensable for the present study, since we needed to go to very large times, up to 30,000 MD time units (with a step of the MD integrator as small as $\delta t = 0.002$). Thus, it was the MD part of this project where somewhat more than 90 % of the computing resources needed to be invested.

3 Static Properties for Planar Confinement

As described in Sect. 2, we can obtain all three interfacial free energies that enter Young's equation for the contact angle θ from independent simulations. Figure 5 shows then the resulting variation of the contact angle with the range of the colloid-wall interaction σ_{wc}, including both data for the AO and the CAO model. It is seen

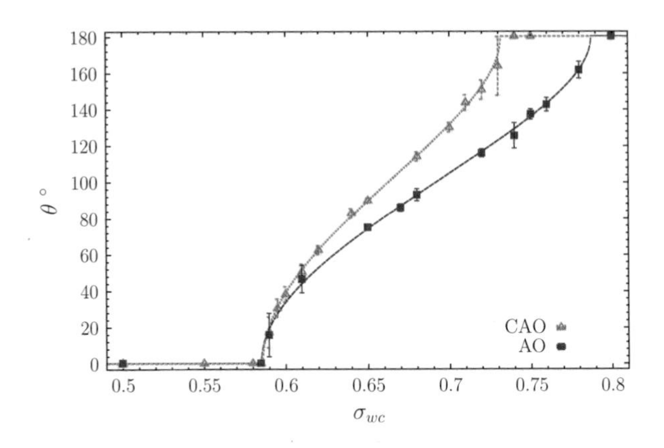

Fig. 5 Contact angle θ plotted versus the range of the colloid wall interaction, as calculated from Young's equation (4), both for the standard AO model and its continuous version (CAO)

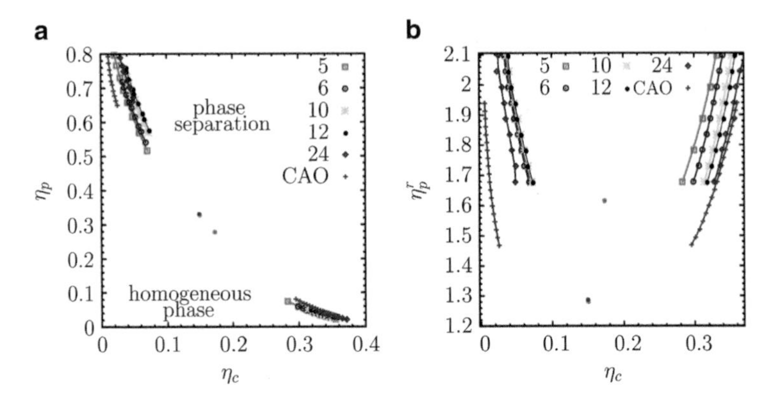

Fig. 6 Coexistence curves of the continuous AO model in the bulk (denoted as CAO) and of $L_x \times L_x \times D_w$ system with two walls a distance D_w apart, at which the wall potentials with $\sigma_{wc} = 0.5$ act. Case (**a**) shows the phase diagrams in the (η_p, η_c) plane and case (**b**) the phase diagrams of the same systems in the (η_p^r, η_c) plane. System sizes were $L_x = 6.735$, $D_w = 6, 10, 12$ and 24 or $L_x = 10$, $D_w = 5$, respectively. Critical points (at $\eta_p^r \approx 1.285$, $\eta_c \approx 0.15$ and $\eta_p^r \approx 1.62$, $\eta_c \approx 0.17$) are only shown for the limiting cases of the bulk and $D_w = 5$, respectively (extracted from a finite size scaling analysis, while the phase boundaries shown as points connected by lines are just taken from the minima of $F(\eta_c)$, cf. Fig. 2a)

that one can cover the full range from complete wetting to complete drying, as expected from Fig. 1. Considering then systems confined between planar walls for specific choices of σ_{wc}, we then can judge whether the system (in the limit $D_w \to \infty$) would correspond to completely wet, partially wet or dry, or completely dry conditions of the walls.

As an example, Fig. 6 presents coexistence curves of the continuous AO model, both plotted in the η_p^r and η_c-plane and in the $\eta_p - \eta_c$ plane (η_p is the packing fraction of the colloids, $\eta_p = (\pi\sigma_{pp}^3/6)N_p/V$, N_p being the number of polymers in the simulation box). One can clearly recognize that confinement enhances the compatibility of the mixture somewhat, and also causes (for the chosen wall potential) a shift of the critical point towards larger colloid packing fractions.

4 Static Properties for Spherical Confinement

The model of Sect. 2 can be straightforwardly applied to confinement in spheres of radius R; in (3) one simply needs to reinterpret z as the normal distance of a point from the sphere surface. Details of the results obtained can be found in [12], so we here restrict attention only to some salient features.

One important point to note is that a system confined in a sphere of finite radius means there are also only a finite number of particles in the system, unlike the case of planar confinement where a thermodynamic limit still exists, taking $L_x \to \infty$ (although in practice finite size effects associated with the finite size of L_x are not of interest, when one takes periodic boundary conditions in x and y directions, as done in Sect. 3, and for many purposes these finite size effects can be disregarded). One consequence of finiteness is that in spherical geometry the condensation process from the vapor-like to the liquid-like phase of the colloid-polymer mixture depends on the ensemble of statistical mechanics in which it is studied, and the results from one ensemble cannot be translated into the conjugate ensemble via a Legendre transformation (although this translation is true in the thermodynamic limit, $R \to \infty$). As an example, Fig. 7 compares the "condensation isotherms", i.e., the thermal average $\langle\eta_c\rangle$ versus the (scaled) colloid chemical potential $\beta\mu = \mu/k_BT$, as obtained in the μVT ensemble, to the result for the thermal average of the chemical potential $\beta\langle\mu\rangle$, versus η_c, as obtained in the canonical N_cVT ensemble. In the thermodynamic limit, of course, the results are identical: the Legendre transformation simply means an interchange of abscissa and ordinate. For finite R, however, the isotherm in the grand canonical ensemble is monotonic, the maximum slope scales like R^3, and so the isotherm smoothly converges to the singular isotherm that results in the thermodynamic limit (the vertical part in Fig. 7a reflects two-phase coexistence). In the canonical ensemble, the isotherms for any finite R exhibit loops (Fig. 8b), and hence for finite R there is no simple relation between the isotherms in the two ensembles. While the loops in Fig. 8b, look like van der Waals loops, the interpretation is completely different: while the metastable and unstable part of a van der Waals loop (predicted for the thermodynamic limit!) is a mean-field artefact [18], the loops seen in Fig. 7b are results pertaining to full thermal equilibrium, and simply reflect the interfacial effects of two-phase coexistence within a sphere of finite radius. Two-phase coexistence within the sphere then exists not only for the chemical potential corresponding to a Maxwell construction (horizontal broken lines in Fig. 7b), but for a whole range of chemical potentials (corresponding to the whole loops). Unlike

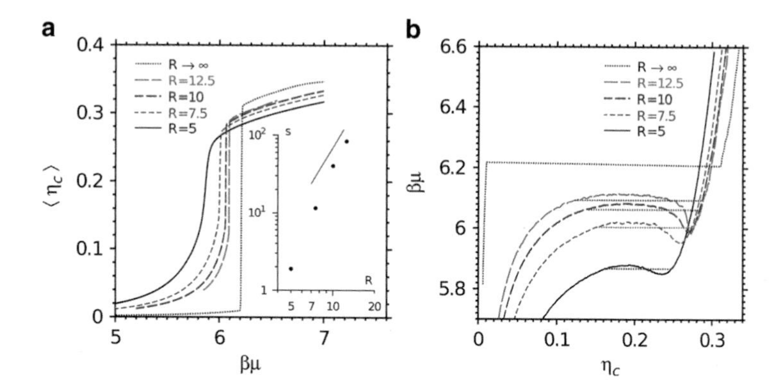

Fig. 7 "Condensation isotherms" for colloid-polymer mixtures confined to spheres of finite radius R, plotting either (**a**) the average packing fraction $\langle \eta_c \rangle$, sampled in the grand canonical ensemble, versus the chemical potential, or (**b**) the average chemical potential $\beta \langle \mu \rangle$, sampled in the canonical ensemble, versus η_c. Several choices of R are included, as indicated. *Insert* in (**a**) shows a log-log plot of the maximum slope s of the isotherm versus R, to illustrate the relation $s \propto R^3$ (*straight line*)

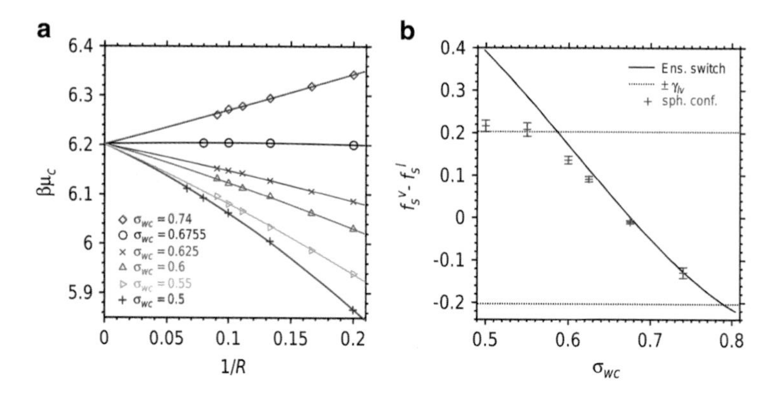

Fig. 8 (**a**) Chemical potential $\mu_c(R)$ at phase coexistence for $\eta_p^r = 0.94$ in the AO model plotted versus 1/R for six values of σ_{wc}, as indicated. *Curves* show fits to $\mu_c(R) = \mu_c(\infty) + a/R + b/R^2$, to test (6). (**b**) Difference of wall free energies $\gamma_{wg} - \gamma_{w\ell}$ plotted versus σ_{wc}, for the AO model at $\eta_p^r = 0.94$. *Full curve* corresponds to the data of Fig. 5, while *symbols* with error bars are from the coefficient a of the fit in (**a**). Note that for complete wetting (4) gets replaced by $\gamma_{wg} - \gamma_{w\ell} = \gamma_{\ell g}$, since $|\cos\theta| \leq 1$ (*upper horizontal straight line*), while for complete drying $\gamma_{wg} - \gamma_{w\ell} = -\gamma_{\ell g}$ (*lower horizontal straight line*)

the thermodynamic limit, there is no horizontal part of the isotherm in the canonical ensemble present for spherical confinement (and the Maxwell construction has no physical significance here!).

The chemical potential where the maximum slope occurs in Fig. 7 (which is roughly equivalent to the chemical potentials corresponding to the horizontal straight lines in Fig. 7b) are systematically shifted relative to the chemical potential

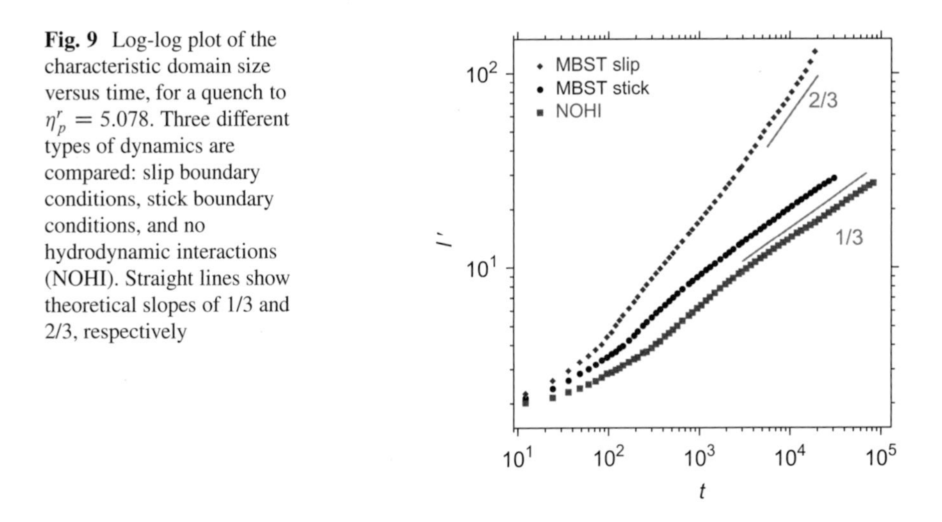

Fig. 9 Log-log plot of the characteristic domain size versus time, for a quench to $\eta_p^r = 5.078$. Three different types of dynamics are compared: slip boundary conditions, stick boundary conditions, and no hydrodynamic interactions (NOHI). Straight lines show theoretical slopes of 1/3 and 2/3, respectively

where phase coexistence occurs in the bulk. This is an expression of "capillary condensation", and theory predicts that [12]

$$\mu_{\text{coex}}(R \to \infty) - \mu_{\text{coex}}(R) = \frac{3}{R}\frac{\gamma_{w\ell} - \gamma_{wg}}{\rho_c^{(\ell)} - \rho_c^{(g)}} + o\left(\frac{1}{R^2}\right) \tag{6}$$

Equation (6) is nothing but the familiar Kelvin equation for capillary condensation adapted to spherical geometry. Using the data for $\gamma_{w\ell} - \gamma_{wg}$, which were already used to construct the contact angle (Fig. 5), we have been able to demonstrate the validity of (6) explicitly, see Fig. 8.

5 Quasi-two Dimensional Spinodal Decomposition in Ultrathin Films

Some preliminary results on phase separation kinetics for films of width $D_w = 5$ were already presented in our previous project report [16], and extensive results for both $D = 1.5, 5$, and 10 were published in [10, 11]. Here we just present, as an example, results for the case of the ultrathin film ($D = 1.5$), where we choose the parallel linear dimension L_x particularly large, $L_x = 512$. Figure 9 shows that for a deep quench one clearly finds the growth law $\ell_d(t) \propto t^{2/3}$ in the case of slip boundary conditions of the fluid at the walls (then one expects hydrodynamic interactions to be effective). In the case of stick boundary conditions, however, hydrodynamic interactions are to a large extent screened: then one observes ordinary diffusive growth, $\ell_d(t) \propto t^{1/3}$, similar to the case when hydrodynamic effects are completely "turned off" (fluid particles then act simply as a Maxwell–Boltzmann thermostat, without local momentum conservation). Figure 9 demonstrates the wide range of times that is accessible with our approach, and this fact has been crucial for the success of this work, yielding a clear confirmation of the pertinent theoretical concepts.

6 Concluding Remarks

In this work, effects of confinement on colloid-polymer mixtures enclosed in slits between repulsive walls or in spheres have been elucidated by large scale Monte Carlo and Molecular Dynamics simulations. The bulk phase diagrams describing liquid-vapor type phase separation into colloid-rich and polymer-rich phases have been obtained in earlier work, but in the present study surface excess free energies due to the walls have been obtained, varying the range of the colloid-wall soft repulsive potential. This information has been used to clarify the wetting properties of the walls and the shift of the phase transitions due to finite film thickness (or finite sphere radius, respectively). It has been shown that a quantitative understanding of the "capillary condensation"-like shift of the phase transition has been reached. Extending the study to the kinetics of phase separation, the growth of domain coarsening and the corresponding effects of hydrodynamic interactions could be clarified.

So far, only liquid-vapor type phase transitions in these colloid-polymer mixtures have been studied. However, for other parameters also fluid-solid phase transitions occur [2, 3], and first exploratory studies of fluid-solid coexistence in confined geometry have begun [19]. In future stages of the present project, very large systems of this type will also be studied, applying again domain decomposition-based use of massively parallel codes. This generalization is not straightforward, since for crystal-fluid interfaces the anisotropy of the interfacial free energy presents a significant complication. In a first step, we shall analyze the consequences for homogeneous nucleation of crystals from the fluid.

Acknowledgements This work was supported in part by the Deutsche Forschungsgemeinschaft (DFG) under projects TR6/A5 and SPP 1296 VI 237/4-3. We are grateful to the NIC Jülich where exploratory runs, mainly relating to static aspects, were made the JUROPA supercomputer of the Jülich Supercomputer Centre. The simulations on the kinetics of phase separation were carried out at HERMIT (HLRS). Helpful interactions with G. Gompper, R.G. Winkler, C. Huang and G. Sutmann are acknowledged.

References

1. S. Asakura, F. Oosawa, J. Polym. Sci. **33**, 183 (1958)
2. S.M. Ilett, A. Orrock, W.C.K. Poon, P.N. Pusey, Phys. Rev. E **51**, 1344 (1995)
3. W.C.K. Poon, J. Phys. Condens. Matter **14**, R859 (2002)

4. E.A.G. Jamie, G.J. Davies, M.D. Howe, R.P.A. Dullens, D.G.A.L. Aarts, J. Phys. Condens. Matter **20**, 494231 (2008)
5. E.A.G. Jamie, R.P.A. Dullens, D.G.A.L. Aarts, J. Chem. Phys. **137**, 204902 (2012)
6. K. Binder, J. Horbach, R. Vink, A. De Virgiliis, Soft Matter **4**, 1555 (2008)
7. K. Binder, P. Virnau, D. Wilms, A. Winkler, Eur. Phys. J. Spec. Top. **197**, 227 (2011)
8. D. Wilms, A. Winkler, P. Virnau, K. Binder, Phys. Rev. Lett. **105**, 45701 (2010)
9. A. Statt, A. Winkler, P. Virnau, K. Binder, J. Phys. Condens. Matter **24**, 464122 (2012)
10. A. Winkler, P. Virnau, K. Binder, R.G. Winkler, G. Gompper, J. Chem. Phys. **138**, 054901 (2013)
11. A. Winkler, P. Virnau, K. Binder, R.G. Winkler, G. Gompper, Europhys. Lett. **100**, 16003 (2012)
12. A. Winkler, A. Statt, P. Virnau, K. Binder, Phys. Rev. E **87**(Article ID 032307) (2013)
13. J. Zausch, P. Virnau, K. Binder, J. Horbach, R.L.C. Vink, J. Chem. Phys. **130**, 064906 (2009)
14. M. Müller, P. Virnau, J. Chem. Phys. **120**, 10925 (2004)
15. F. Wang, D.P. Landau, Phys. Rev. Lett. **86**, 2050 (2001)
16. A. Winkler, P. Virnau, K. Binder, in *High-Performance Computing in Science and Engineering '12*, ed. by W.E. Nagel et al. (Springer, Berlin, 2013), pp. 29–38
17. G. Sutmann, R.G. Winkler, G. Gompper, Simulating hydrodynamics of complex fluids: multi-particle collision dynamics coupled to molecular dynamics on massively parallel computers. doi:10.1016/j.cpc.2013.10.004
18. K. Binder, B.J. Block, P. Virnau, A. Tröster, Am. J. Phys. **80**, 1099 (2012)
19. D. Deb, A. Winkler, P. Virnau, K. Binder, J. Chem. Phys. **136**, 134710 (2012)

Heterogeneous and Homogeneous Crystallization of Soft Spheres in Suspension

Dominic Roehm, Kai Kratzer, and Axel Arnold

Abstract Nucleation, i.e., the onset of a phase transition like crystal growth, is a rare event with waiting times in the order of days. Yet, it is an event on the molecular scale, and therefore difficult to study, both experimentally and by computer simulations. Our interest is in the role of long range interactions in nucleation, in particular electrostatic and hydrodynamic interactions mediated by solvent molecules.

In order to model the solvent, we use a lattice fluid that is propagated by the fluctuating Lattice Boltzmann (LB) method. Our implementation uses a graphics card (GPU) to propagate the solvent and is coupled to the Molecular Dynamics (MD) simulation package ESPResSo. Using this code, we study the heterogeneous crystallization in Yukawa-like colloidal systems. Our simulations allow to observe the growth of a crystal in a channel with and without hydrodynamic interactions, and indicate that hydrodynamic interactions slow down the crystallization.

Additionally, we present results on the homogeneous crystallization of Yukawa particles. While heterogeneous nucleation can be observed directly in simulations, homogeneous nucleation requires special sampling techniques. We use our own Forward Flux Sampling implementation, the Flexible Rare Event Sampling Harness Systems (FRESHS). FRESHS can control popular MD simulation packages as back-end, making it a versatile tool to study rare events. Our simulations confirm previous results at higher supersaturations, which show that the nucleation mechanism involves two steps, namely the formation of a metastable bcc phase and the transformation to a stable fcc phase.

D. Roehm (✉) · K. Kratzer · A. Arnold
Institute for Computational Physics, Allmandring 3, 70569 Stuttgart, Germany
e-mail: dominic.roehm@icp.uni-stuttgart.de; kratzer@icp.uni-stuttgart.de; arnolda@icp.uni-stuttgart.de

W.E. Nagel et al. (eds.), *High Performance Computing in Science and Engineering '13*, DOI 10.1007/978-3-319-02165-2_3,

1 Introduction

Crystals play a role in many applications, from photonic structures to protein structure determination. In the latter case the proteins, charged macromolecules in suspension, need to be crystallized. This requires the careful addition of agents such as crowding polymers or multivalent salts. The optimal amounts differ from protein to protein and need to be determined by trial and error at the moment. In order to allow for systematic studies, a theory of nucleation, i.e. the onset of crystal growth, in charged macromolecular systems is necessary. In our project, we concentrate on the influence of long range interactions on nucleation, namely electrostatic and hydrodynamic interactions. In order to improve our understanding, we use computer simulations with coarse-grained interactions, which allows us to study different contributions separately.

However, nucleation is a rare event and therefore difficult to access, both experimentally and in computer simulations. The waiting time between events is orders of magnitude longer than the time the actual event takes, which make direct observation unfeasible. The problems become even larger, if hydrodynamic interactions are to be taken into account, which are mediated by the embedding solvent. The necessary number of solvent molecules is too large to perform direct simulations, so that one needs to reduce the number of degrees of freedom spent on the solvent.

A possible solution is the Lattice–Boltzmann (LB) method, which is able to deliver a proper solution of the Navier–Stokes equation on a simpler lattice fluid. LB allows for taking into account thermal fluctuations, which are quite pronounced on the scale of proteins or colloidal particles. The LB method is much faster than computing solvent molecules explicitly, but still consumes more computation time than the dynamics of the embedded particles, which can be overcome by off-loading to GPUs.

In the following we will give a brief introduction to our implementation of the LB solver on a GPU, which is attached to our Molecular Dynamics (MD) software ESPResSo (Extensible Simulation Package for Research on Soft matter [1, 2]). With the help of our GPU accelerated LB solver, we investigate the effects of hydrodynamic correlations on the crystallization in colloid systems. In our simulations, we observe the growth of a colloidal crystal from a wall, including hydrodynamic interactions. The simulations, which previously would have been considered computationally unfeasible, can be performed in a reasonable amount of time using the GPU equipped Nehalem cluster and our optimized software.

Nevertheless, these simulations where only possible due to the inclusion of walls, which greatly enhance the rate of nucleation. In a homogeneous system, spontaneous nucleation is much more rare, and will not happen in a brute force computer simulation. This can be overcome by special simulation techniques, such as Forward Flux Sampling (FFS). We introduce our Flexible Rare Event Sampling Harness System (FRESHS) environment, which executes FFS simulations using a client-server model. The FRESHS server manages multiple clients running different established MD codes like ESPResSo, GROMACS [3] or LAMMPS [4].

Using FRESHS and our automatic interface placement methods for FFS simulations, we present some results on the nucleation process in colloidal suspensions.

2 Molecular Dynamics and the Lattice Boltzmann Method

The microscopic details of the solvent are irrelevant to the motion of the solute particles. Therefore, any approach reproducing the macroscopic Navier–Stokes equations can be used to obtain correct hydrodynamics. The Lattice–Boltzmann (LB) method is such an approach, propagating discrete particle populations to approximate the behavior described by the Navier–Stokes equations [5] in the limit of low Mach and Reynolds numbers. Computationally it is very simple, representing the phase-space particle populations discretized in space, time and velocity. The populations stream between the discrete spatial positions, the nodes, while at the same time the velocities at each node relax towards equilibrium. The first approach by Bhatnagar, Gross and Krook (BGK) uses a single relaxation time [6], alternatively one can use different relaxation times for independent hydrodynamic modes, which is known as the multiple relaxation time (MRT) approach [7]. Complex flow geometries can be easily implemented by imposing no-slip boundary conditions on individual nodes of the discretized particle populations [8]. Streaming and collision are completely node-local, therefore the LB method is easy to implement and exhibits ideal parallel scaling.

In order to study macromolecules, cells or other small objects in solution, the LB method can be coupled to embedded particles. LB-particle couplings have first been realized using moving boundaries by Ladd [8]. Dünweg et al. [9] brought forward another approach, namely to couple the particles to the fluid through a frictional drag. This allows to use a much coarser discretization for the LB populations, which saves computation time, and since the coupling is shape-agnostic, it can couple to arbitrarily shaped particles, even with soft interactions such as polymer coils. The price for this is that the coupling is only correct in the far field.

Whenever such very small objects are immersed in a liquid, thermal fluctuations play an important role and need to be included in the LB scheme, known as fluctuating LB. The theory to do this consistently has been formulated recently by several groups [10]. Computationally, the fluctuating LB method has the same ideal scaling as the pure LB approach, but requires more computational effort per grid point, since it utilizes the MRT scheme and a Gaussian random number per degree of freedom.

Due to its algorithmic simplicity and broad application spectrum, the LB method was one of the first algorithms to be ported to Graphics Processing Units (GPUs). In principle, GPUs are small, massively parallel computers, which adhere to the SIMD (single instruction, multiple data) paradigm. This is particularly suitable for streaming lattice-based methods such as LB. Consequently, already in 2003, Li et al. presented an implementation of the BGK model using 8-bit 2D textures [11]. With the appearance of general purpose programming interfaces for GPUs, such

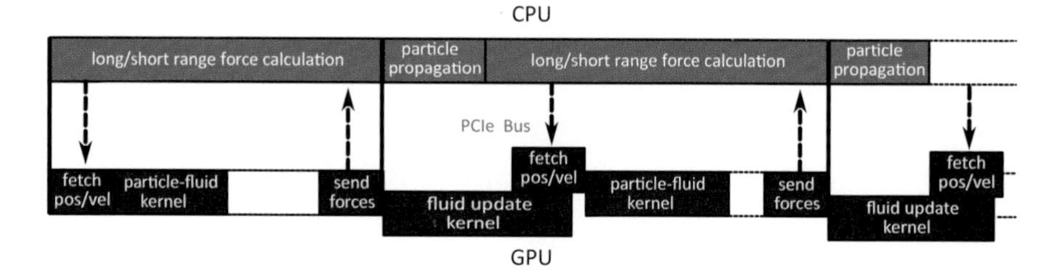

Fig. 1 Illustration of the concurrent CPU and GPU propagation. *Solid lines* represent points, where streams of kernels and memory transfers are invoked, *dotted lines* represent the memory transfers of particle positions and velocities to the GPU and forces back. The fetching of the positions is always scheduled at the beginning of the force calculation. However, it actually takes place only after the fluid update. By this, the time consuming fluid update can happen simultaneously with the time consuming force calculation on the CPU

as NVIDIA CUDA [12, 13], a number of LB implementations on GPUs of different flavors have been reported, for example the MRT model [14] or multicomponent fluids [15]. The highly scalable waLBerla LB framework can make use of multiple GPUs [16].

Our code implements the fluctuating LB equations and the Dünweg particle coupling, making it ideal for dynamic simulations of nano-scale objects in solution, for example colloidal suspensions or polymers. We focussed on the optimization of the overall simulation, by interweaving the fluid computation on the GPU with the conservative force calculation for the embedded particles on conventional CPUs. The code is included in the open-source package ESPResSo starting from version 3.0.0, and can be downloaded from http://www.espressomd.org.

The MD code ESPResSo, to which our GPU LB code couples, is parallelized using the Message Passing Interface (MPI) [17]. This means, that the particle data is distributed among the available compute cores using a domain decomposition. However, only one of these cores can communicate with the GPU. Since ESPResSo uses a master-worker scheme, we naturally use the master node to communicate with the GPU. During each time step, it collects the particle positions and velocities from all workers via MPI, and sends the fluid drag forces back, after they have been computed by the GPU.

Both the propagation of the fluid and the force calculation are rather time consuming operations. Therefore, one would like to overlap these computations on the CPU and GPU. This is possible by using GPU streams (Fig. 1). Several operations on the GPU, such as kernel calls or memory transfers, can be combined into a stream, which will be executed by the GPU without blocking the CPU. Our implementation starts the propagation of the fluid simultaneously with the propagation of the particles. Typically, the particle propagation is much faster, after which the CPU immediately starts to calculate the conservative forces. As soon as the GPU is done with the fluid update and the CPU with the particle propagation, the GPU starts to calculate the fluid forces on the particles. Only at the end of the calculation of the conservative forces and the particle-fluid interaction, a barrier is needed, as only now the forces from the GPU can be transferred back to the CPU. Since this is the only synchronization barrier, fluid update and force calculation happen in parallel.

In order to compute the interactions with the particles, the GPU needs the current positions of the particles and has to return the forces. This communication has to pass the rather slow PCIe bus. However, we can overlap the transfer of the particle positions with the fluid update calculation, since CUDA allows to asynchronously copy data from the host computer to the GPU. Therefore, the sending of the positions and velocities happens potentially in parallel both with the force calculation on the CPU and the fluid update on the GPU. Only the transfer of the forces is synchronous since it serves as barrier, as discussed before. To improve the communication via the PCIe bus, we use pinned memory in the CPU RAM, which on recent GPUs allows to use fast DMA for the transfer [18].

Benchmarks

In this section, we compare the performance of our LB GPU code to the CPU-based MPI-parallelized version, which are both part of the ESPResSo package and offer the same features. For the benchmarks, we used servers with two Intel Xeon E5620 quad core processors with 2.4 GHz, 12 GB RAM and two NVIDIA Tesla M2050, which come with 448 CUDA cores running at 1.1 GHz and DDR5 VRAM at 1.5 GHz. The CPU code was compiled using the GNU Compiler Collection gcc version 4.5.3 and executed on all eight cores of the two Intel CPUs. We used the NVIDIA CUDA toolkit 3.2 to compile our code for compute architecture 2.0, and performed the tests with the Linux NVIDIA driver version 260.19.26.

Figure 2 shows the performance of our implementation when simulating a pure thermalized fluid, measured in Mega Lattice Updates per second (MLUps). The unit MLUps denotes the number of lattice nodes that can be updated by the LB process during a second, and therefore is a system size independent unit measuring the performance of an implementation. We chose the same parameters for both code, although the performance is independent of the parameters. In dimensionless units [10], the values were: fluid density 1.0, viscosity 3.0, temperature 1.0, lattice spacing 1.0 and time step 0.01. The most notable result is that the GPU code is faster than the CPU code running on the eight cores of two CPUs by a factor of about 50. This speedup is much larger than the estimated factor of 10–20 from a simple comparison of the GPU and CPU peak flop performances and is due to the fact that most memory access on the GPU can be handled in the ultrafast registers in contrast to the CPUs. The speedup factor is practically independent of the system size, even for small meshes, and both codes show a nearly ideal scaling over a wide range of mesh sizes.

A particular focus of our implementation lies on the efficient coupling of the LB fluid to embedded MD particles. In order to illustrate the advantages and disadvantages of such a coupling, we compare three different scenarios: using the GPU-based

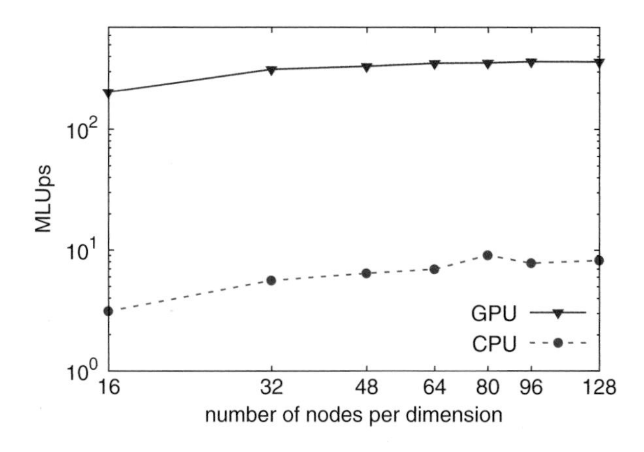

Fig. 2 Performance of the CPU and GPU LB codes in mega lattice updates per second (MLUPS) for a pure thermalized fluid in a cubic box, as a function of the mesh size. The *blue continuous line* gives the execution time of the LB GPU code running on the GPU, the *red dashed line* the time on 8 CPU cores

fluctuating LB solver, the CPU-based fluctuating LB solver, and a conventional simulation, in which hydrodynamic interactions are ignored and the thermalization is provided by a Langevin thermostat. The latter is much faster to compute. For static properties, the underlying dynamics is irrelevant, and therefore simple thermostats such as the Langevin are widely employed in molecular simulations. However, when studying dynamic processes like nucleation, this may or may not be sufficient, which we want to find out. In our example, the particles interact via a Weeks–Chandler–Andersen (WCA) [34] potential, which is a Lennard–Jones potential cut off at its minimum. This rather hard interaction mimics solid objects like colloidal particles. We were using particle number densities of 0.3 and 0.7, which are common for dense suspensions. These particles were coupled with friction constant 1.0 to a fluid of density 1.0, viscosity 3.0 and temperature 1.0. The lattice spacing was equal to the diameter of the WCA particles, which is a common choice.

Figure 3 shows the resulting execution times. Due to the constant density, the Lennard–Jones interaction, the Langevin thermostat and the LB solver scale proportional to the simulation box volume. Despite the similar scaling of the simulations, the execution times per time step are fairly different. There is a large difference between the execution times of the simulations using the LB CPU code compared to the simulations using the LB GPU code or the Langevin thermostat. For the more common, dilute system, the CPU code is almost a factor of four slower than the Langevin simulation, which explains why one usually avoids to include hydrodynamics if possible. Even for the dense system, the LB doubles the simulation time. The GPU code instead costs only about 20–30 % additional time, which make it much more practical to include hydrodynamics. The reason why the overhead caused by the LB GPU code is so small, is the that the execution of the fluid update and force calculation on the GPU happens simultaneously. In systems

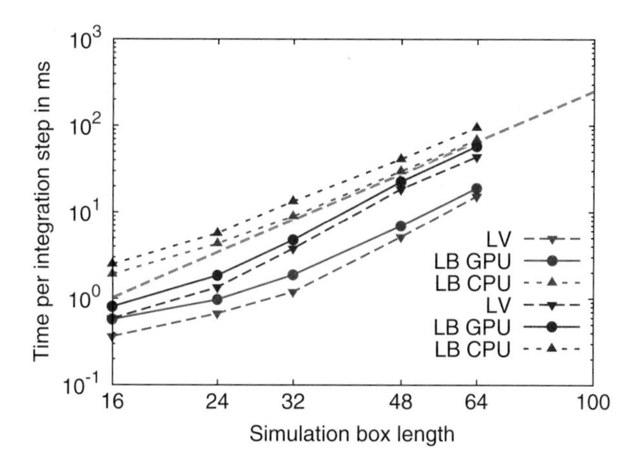

Fig. 3 Execution time per time step of a system of WCA particles, as a function of the system size. *Blue lines* give the execution time at particle density of 0.7, *red lines* at density 0.3. Including hydrodynamics through the CPU solver increases the computation time compared to the bare Langevin thermostat by a factor of 2–4, depending on the density, while the GPU only costs about 20–30 % extra. The *gray dashed line* represents the ideal scaling proportional to the volume of the simulation box

with more expensive interactions, for example electrostatics, this overhead can decrease even to a couple of percent, essentially allowing to include hydrodynamics without using additional computation time.

3 Rare Event Sampling

The second important numerical ingredient in our simulations is the sampling of rare events, in our case homogeneous nucleation. The technique that we use is Forward Flux Sampling (FFS), which is used to compute the transition rate k_{AB} from an initial state A to a final state B. The transition rate k_{AB} can be computed as $k_{AB} = \Phi P_B$ [19], where Φ is the flux of trajectories leaving the initial state, and P_B is the probability that a trajectory that leaves this state reaches the final state. The initial and final states are defined in terms of an order parameter λ, such that if $\lambda < \lambda_A$, the system is in the initial state, and if $\lambda > \lambda_B$, it is in the final state. In our case, this order parameter is the average bond order parameter $\overline{q}_6$, which we will discuss later. The intermediate region is divided into a series of n interfaces, such that $\lambda_i < \lambda_{i+1}$, $\lambda_0 = \lambda_A$ and $\lambda_n = \lambda_B$ (see Fig. 4). This subdivision is useful, since the probability P_B can be written as [19]

$$P_B = \prod_{i=0}^{n-1} p_i, \tag{1}$$

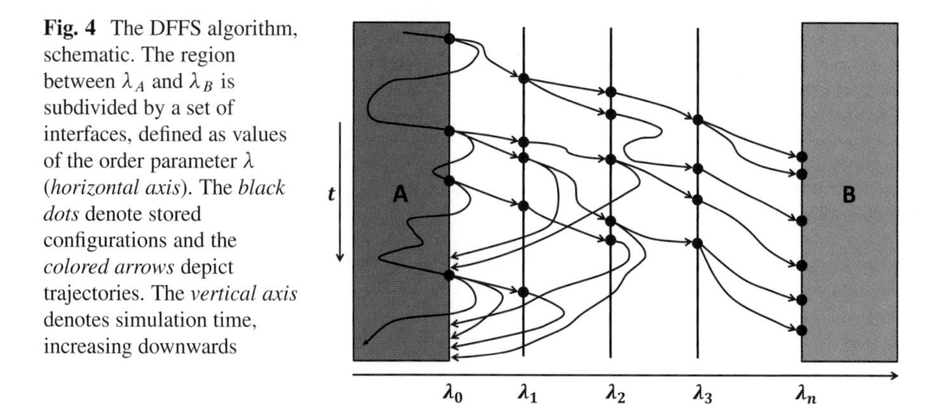

Fig. 4 The DFFS algorithm, schematic. The region between λ_A and λ_B is subdivided by a set of interfaces, defined as values of the order parameter λ (*horizontal axis*). The *black dots* denote stored configurations and the *colored arrows* depict trajectories. The *vertical axis* denotes simulation time, increasing downwards

where p_i is the transition probability from λ_i to λ_{i+1}. Typically, P_B is extremely small, leading to the low transition rate and making it very hard to sample. The individual contributions p_i are much larger, and therefore can be sampled by conventional simulation techniques. For example, if $P_B = 10^{-10}$, this means that one would have to fire around 10^{10} direct simulations to record at least one successful transition. On the other hand, with just $n = 10$ interfaces, the probabilities p_i are of the order 0.1 each, so that one can obtain reasonable statistics with already 100 runs per interface, or 1,000 simulations in total. By tracking back, the algorithm also generates transition trajectories.

While several variants of FFS exist [20–22], we focus here on the direct FFS algorithm (DFFS) [21–23]. The DFFS algorithm has two calculation stages. First, the flux Φ across the interface λ_0 is computed by simulating a system in the initial state and monitoring the trajectory crossings per time on λ_0 in the direction of increasing λ. When these crossings happen, the configuration of the system is stored; this simulation thus generates not only a measurement of Φ but also a collection of N_0 configurations corresponding to states of the system at the moments of crossing λ_0. In the second stage of the algorithm, the probabilities p_i are computed step-wise (see Fig. 4 for illustration). To compute p_1, one chooses configurations at random from the set which was stored at the previous interface λ_0 and uses them to start new trajectories, which are continued until they either reach the next interface λ_1 ("success") or return to λ_0 ("failure"). For successful trajectories, the final configuration at λ_1 is stored in a new set. After M_0 trial trajectories have been launched, p_1 is computed by dividing the number of successes by M_0. One then repeats the same procedure for the next interface, using again the configurations at the previous interface as starting points for M_1 trajectories that are continued until they reach λ_2 or return to λ_0, and so on, until the border of the final state is reached and one has a complete set of estimated probabilities p_i. Transition trajectories from the initial to the final state can then be extracted from the collection of successful trajectories between interfaces [20, 21].

FFS is a technique that is not specific to a particular type of simulation dynamics or code. Therefore, we have implemented a generic FFS server infrastructure, the

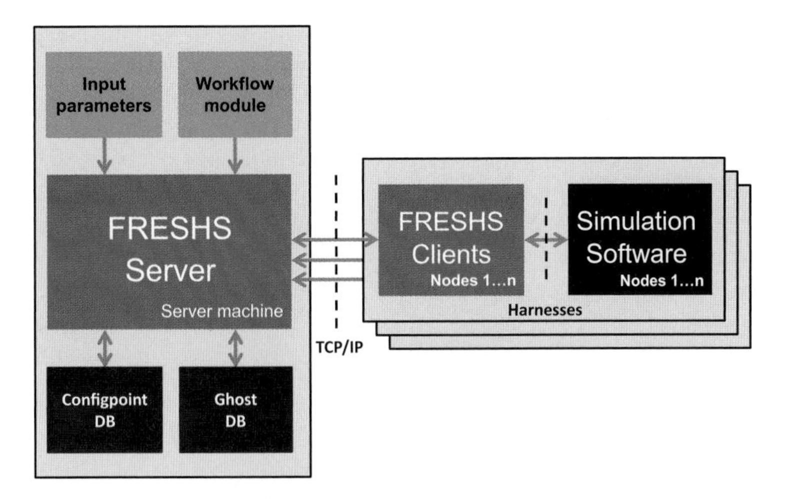

Fig. 5 Schematic structure of FRESHS. The server reads the input parameters from a configuration file, loads the work flow module corresponding to the desired sampling method and stores information in a database. The clients connect to the server via a socket layer. Clients can be on the same machine as the server or on different calculation nodes. Each client instance communicates with its own instance of the MD simulation program via the corresponding harness scripts. The simulation software can itself be parallel and distributed over several nodes

Flexible Rare Event Sampling Harness System (FRESHS). In FRESHS, the work flow of the rare event simulation is controlled by the server. Calculation clients consist of the simulation software and harness scripts that connect to the server. Thereby, the client simulations themselves can be parallel. If we perform a FFS run with 100 clients using 4 nodes with 8 CPUs each, we have 3,200 CPUs working for us in parallel (cf. Fig. 5).

The crucial input to the FFS method is the placement of the interfaces. It is by no means optimal to place these equally spaced, and manual placing of the interfaces can easily lead to conditions where no FFS trial trajectories succeed in reaching the next interface. In this case, one has to start over with a finer set of interfaces to overcome the bottleneck region. Even worse, if the simulation continues, but with a very low transition probability on one of the interfaces, it is easily possible to compute wrong transition rates. Therefore we have derived an automatic, optimal interface placement, which has the potential to greatly improve the feasibility and computational efficiency of FFS simulations and is capable of placing interfaces in their optimal locations, on-the-fly [24]. Thereby, the interface placement is based on the minimization of the statistical error via the principle of constant flux [25] and the minimization of the computational cost.

There are two methods to place the interfaces, which are both based on firing a small number of trial runs. The "trial interface method" uses an interface which is placed between the last known interface λ_i and the final state B. This interface position is shifted according to the outcome of the trial runs, which have the possibility to either reach this trial interface or to fall back to the initial state A. In the "exploring scouts method", the calculation steps of the trial runs is limited and no trial interface is placed. Each trial run is aborted if either the maximum number of steps is reached or the trial run reaches A or B. During the trial run, the order parameter fluctuations are monitored and at the end of the run, the maximum of the order parameter is reported. From this information, the new interface location λ_{i+1} is derived.

As a measurement for the constant flux rule and hence for the accuracy of the interface placement, the quantity

$$f_i = \frac{\sum_{j=0}^{i-1} \log p_j}{\sum_{j=0}^{n-1} \log p_j}, \tag{2}$$

was introduced by Borrero and Escobedo [25]. At constant flux, i.e. $p_j = p =$ const, we have $f_i = i/n$, so that f_i should be linear when plotted against the interface index i. For further details of the placement methods refer to [24].

4 Effects of Hydrodynamic Correlations in Crystallization of Yukawa-Type Colloids

We use the GPU-accelerated LB solver in order to study the influence of hydrodynamically induced correlations on the crystallization in a soft sphere system with Yukawa-like interactions. These particles interact via a potential of the form

$$U_{\text{Yukawa}}(r) = \epsilon \frac{\exp(-\kappa(r/\sigma - 1))}{r/\sigma}. \tag{3}$$

The Yukawa potential is a screened Coulomb potential, suitable for modeling charged particles whose electrostatic interactions are screened by surrounding ionic atmospheres. The screening length κ^{-1} describes the range of electrostatic interactions under screening by the surrounding salt, ϵ describes the interaction strength and σ the apparent diameter of the particles. This is usually a good approximation to the interaction of likely charged colloidal particles. Unlike experiments, computer simulations allow to switch on or off particular interactions, so that we can determine the influence of hydrodynamic interactions easily.

Most nucleation studies are performed in the bulk, and special sampling methods are necessary to compute the evolution of the crystal growth efficiently. However, including a solvent into these simulations is difficult, in particular, since usually the isothermal-isobaric ensemble is used, which cannot be coupled to the Lattice Boltzmann method. Therefore, we studied the crystallization confined between two planar walls, which can be simulated directly with reasonable effort in the constant volume isothermal ensemble. We prepare our systems as an undercooled

Table 1 Values of q_6 and $\overline{q}_6$ for BCC, FCC structures and the undercooled liquid established by Lechner and Dellago [29]

	q_6	$\overline{q}_6$
BCC	0.440526	0.408018
FCC	0.507298	0.491385
HCP	0.445384	0.421810
LIQ	0.360012	0.161962

liquid and let the system crystallize. The influence of hydrodynamic correlations on the crystallization process was evaluated by employing both a fluctuating lattice Boltzmann method [10] that includes hydrodynamic interactions, and Langevin dynamics [26] that ignores hydrodynamic interactions.

In order to track the growing crystal it is necessary to distinguish between particles which are part of the solid phase and the liquid phase. The state-of-the-art algorithm is the Steinhardt order parameter [27]:

$$q_l(i) = \sqrt{\frac{4\pi}{2l+1}\sum_{m=-l}^{l}|q_{lm}(i)|^2}, \tag{4}$$

where q_{lm} are spherical harmonics. Depending on the choice of l, the Steinhardt order parameter is sensitive to various crystal symmetries. Investigations by Moroni et al. [28] showed that especially q_4 and q_6 are good choices to determine whether cubic or hexagonal structures are present in the system. In the following, we will focus on FCC and BCC crystal structures, which the q_6 order parameter distinguishes well from a liquid. The strong fluctuations of the plain Steinhardt order parameter can be dampened at the cost of spatial resolution by averaging over the first shell of neighbors [29]:

$$\overline{q}_l(i) = \sqrt{\frac{4\pi}{2l+1}\sum_{m=-l}^{l}|\overline{q}_{lm}(i)|^2}, \tag{5}$$

with

$$\overline{q}_{lm}(i) = \frac{1}{\tilde{N}_b(i)}\sum_{k=0}^{\tilde{N}_b(i)} q_{lm}(k). \tag{6}$$

For our measurements we always used the $\overline{q}_6$ order parameter. The literature values for the $\overline{q}_6$ order parameter in the relevant phases are given in Table 1.

Since our main goal is the investigation of hydrodynamic correlations in soft sphere systems, we started by reproducing parts of the known phase diagram of Yukawa-type colloidal systems in bulk [30]. To this aim, we set up simulation of 4,096 particles in a three dimensional periodic system without walls. After a short warm-up simulation, the production runs were each executed on four cores using

Fig. 6 Snapshot of the colloidal crystal growing from the walls, *color* refers to the $\overline{q}_6$ parameter (*blue solid, red fluid*). Single crystal layers can be seen, starting from both walls

up to 24 h of wall time, which results in 50–100 million integration steps. These simulations allowed us to gauge our simulations in order to perform simulations reasonably close to the phase transition border. This is important in order to avoid early agglomeration, which leads to many defects.

As we want to compare the same system with and without hydrodynamic interactions, we had to deal with the question what properties of the system have to be matched. Therefore a vast number of production runs have been executed to measure the tracer diffusion of the particles in the periodic system. These runs were needed to sample a certain range of friction coefficients. As hydrodynamic interactions are subject to finite size effects, these runs require the same large simulation box as the final system. We therefore used 16,384 particles in a box of size 64×16×16. In the following, if results are reported for systems with and without hydrodynamic interactions, the friction coefficients were always chosen to produce the same long time diffusion coefficient in bulk liquid. Due to the interactions between the particles, there is no simple law connecting these two quantities, which made this matching procedure necessary.

Using this gauging, we studied the heterogeneous crystal growth from a flat wall. This was done in a system of 16,384 particles in a simulation box of size $66 \times 16 \times 16$ confined by two planar walls located at $x = 0.5$ and $x = 65.5$ (Fig. 6). In order to track the progress of the crystal front, we record the density profile and the average $\overline{q}_6$ bond-order parameter across the simulation box, which we computed on the fly to reduce the amount of output data. Usual production runs use a single 8 core node with MPI parallelization and the attached GPU, and take up to 6 wall time hours on the Nehalem cluster. To sample sufficient statistics dozen successful runs are needed, since under unfortunate conditions, even heterogeneous nucleation can take rather long to set in. Furthermore, runs have to be performed for different hydrodynamic interaction strengths.

Our simulation parameters were chosen to form a BCC-solid close to the solid-fluid transition line. The undercooled fluid confined between two planar walls then grows into a crystal with BCC structure starting from a HCP wall layer. In Fig. 7 the peak positions of the measured average $\overline{q}_6$ bond-order parameter and the density are shown. Confined fluids show a strong layering near walls, where regions of high density interleave with regions of very low density. As the order parameter can only be detected in the regions where there are particles, we report only the peak positions. As one can clearly see, the average $\overline{q}_6$ gives much smoother results. Therefore, $\overline{q}_6$ is well suited to detect the actual position of the front, and all our

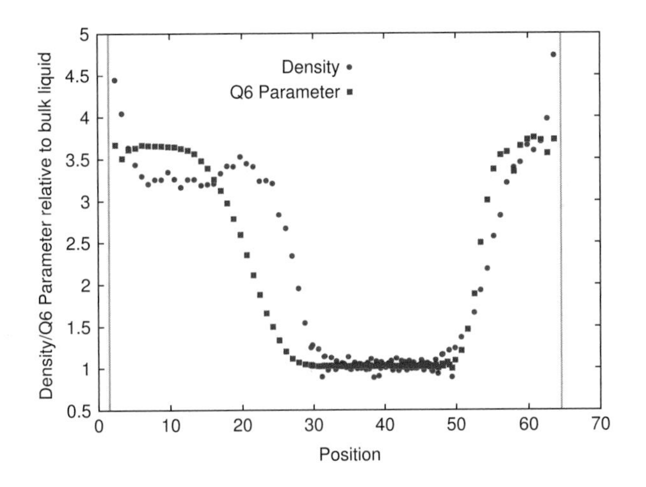

Fig. 7 Values of the $\overline{q}_6$ order parameter and the density in x-direction of the simulation box. The location of the crystal front can be detected more precisely by analyzing the $\overline{q}_6$ profile compared to the density profile. A higher concentrating is observed in the vicinity of the growing crystal. Based on the shape of the profile a fitting procedure is used to evaluate the actual position of the crystal front

following investigations are based on the $\overline{q}_6$ profiles. Note that in the figure one can clearly distinguish the BCC solid and the undercooled fluid. The values in the fluid are somewhat higher than usually reported, since we only report peak values, rather than the average. To determine the position of the crystal, we performed a fit of the $\overline{q}_6$ data points with a function of shape

$$\frac{q_{bcc} + q_{liq}}{2} - \frac{q_{bcc} - q_{liq}}{2} \arctan[w \cdot (x - d)], \tag{7}$$

where x is the x-position of the front in the simulation box, d is the estimated front position and w is the width of the front region. q_{bcc} and q_{liq} denote the peak values of $\overline{q}_6$ in the BCC solid and bulk liquid, respectively.

Performing fits at different time steps, we can determine the time dependent front position $d(t)$, as reported in Fig. 8. There is a broad region, in which the crystal grows very uniformly with a constant velocity. In this region, the velocity of the front can be fitted with a linear function of type:

$$g(x) \propto u \cdot r_x, \tag{8}$$

where u is the velocity of front. As one can clearly see in the reported example, the growth velocity is significantly reduced when hydrodynamic interactions are taken into account. Further simulations are necessary to quantify this effect. Our impression so far is, that there is significant influence of hydrodynamic correlations on the crystallization at least in Yukawa-type colloidal suspensions, even

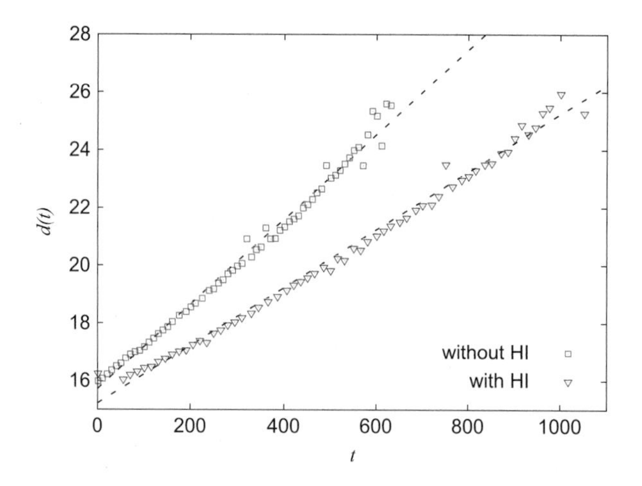

Fig. 8 An example of the position of the crystal front as a function of time. *Squares* are used for results without and *triangles* are used for results with HIs. Both systems show a linear trend, as illustrated by the *black dashed lines*. The slopes of the linear fits are the front growth velocities

in parameter regions where hydrodynamic interactions would usually have been considered negligible. This would have strong implications for the often drawn analogy between metal melts and colloidal suspensions, which is based on the assumption that hydrodynamic interactions, that are only present in the colloidal model system, do not play a role.

5 Homogeneous Nucleation of Yukawa/WCA Macromolecules

A second problem of our interest is homogeneous nucleation in the same Yukawa system. Unlike heterogeneous nucleation, this however cannot be studied directly, and requires special rare event sampling techniques. Here, we use Forward Flux Sampling [21] in our FRESHS implementation. In addition to the Yukawa interaction given by (3), we add a Weeks–Chandler–Andersen (WCA) [34] short range hard core repulsion given by

$$U_{\mathrm{WCA}}(r) = \begin{cases} 4\left(\left(\frac{\sigma}{r}\right)^{12} - \left(\frac{\sigma}{r}\right)^{6} + \frac{1}{4}\right) & r < \sigma^{\frac{1}{6}} \\ 0 & \text{else.} \end{cases} \tag{9}$$

The repulsive WCA potential is used to characterize the excluded volume of the particles, which is important at the higher pressures necessary for spontaneous homogeneous nucleation (note that the energy scale for the WCA potential is set to $k_B T = 1$ in our simulation units). In this work, the parameters of the Yukawa potential are the inverse screening length $\kappa = 5$ and the value of the repulsive potential at contact $\epsilon = 8$. Despite the existence of important previous work [31,32], the nucleation mechanism in these systems is still not clarified, to which rare event sampling simulations can contribute by providing theoreticians with nucleation rates and transition paths [31]. However, because the Yukawa interaction requires a larger cutoff than the often studied Lennard–Jones interaction, simulations of these systems are computationally expensive, especially at important conditions, e.g. for low salt concentrations, which means that the number of trajectories which can be simulated using FFS is limited. This makes setting up of a standard FFS simulation difficult, particularly under interesting conditions, e.g. close to coexistence where the transition rate is low [33]. Here, our automatic interface placement for FFS becomes important, as we will demonstrate now.

We performed molecular dynamics (MD) simulations of 4,096 WCA-Yukawa particles in a cubic box with periodic boundary conditions in the NPT ensemble at constant pressure $P = 38$ (LJ units) with a Langevin thermostat. For these simulations, we used the software package ESPResSo [1, 2] in combination with the DFFS algorithm, implemented in our Flexible Rare Event Sampling Harness System FRESHS.

Note that FFS requires stochastic dynamics: here this is provided by the Langevin thermostat, not the LB solver, since we want to connect to previous studies [31]. Adding hydrodynamic interactions is planned for a later stage. The system is initially prepared in the liquid phase, which is undercooled (and therefore metastable). We are interested in the transition to the stable FCC crystal phase. Our order parameter λ is the size of the largest cluster of solid particles, where particles are classified as solid or liquid based on the local $\overline{q}_6$ order parameter, as described above. The boundaries of the initial and final states were fixed such that the system is in the initial state if less than 0.5 % of the particles are in the largest solid cluster and in the final state if more than 90 % of the system's particles are in the largest solid cluster. This corresponds to $\lambda_A = 15$ and $\lambda_B = 3700$.

In our DFFS simulations, we compared three methods for interface placement: (1) placing the interfaces manually using a logarithmic scheme, (2) the trial interface method and (3) the exploring scouts method. All our DFFS simulations used $N_0 = 80$ configurations at the first interface and $M = 50$ trial runs per interface.

For nucleation problems, where simulations are computationally expensive, manual interface placement in FFS is very demanding. Our problem has a steep free energy barrier and so placing interfaces evenly between λ_A and λ_B results in poor success rates for early interfaces. Therefore, as a manual choice, we placed more interfaces at the beginning, which means a closer distribution between the early interfaces. Even with this choice of interfaces Fig. 10a shows that this results in success probabilities that are far from equal (inset) and many of the p_i values are very low: this happens in steep regions of the free energy landscape, where the guessed number of interfaces was insufficient. A low success probability stands for the fact that much computational effort will be wasted on non-successful trajectories. Beyond that, for later interfaces, the conditional probabilities are close

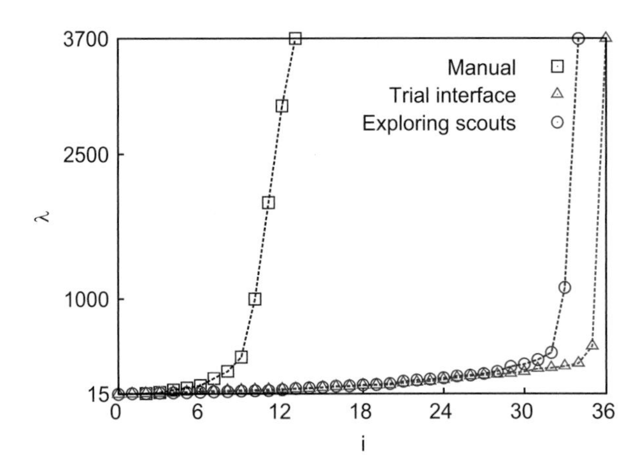

Fig. 9 Interface positions λ_i (plotted as a function of interface index i) for the manual interface placement (*squares*), the trial interface method (*triangles*) and the exploring scouts method (*circles*)

to 1. In this region of the free energy landscape, the crystal grows spontaneously and no interfaces would have been necessary. This results in an extra computational overhead in storing configurations and additional start up times of the MD engine. The fact that the manual interface placement is not optimal is also visible in the highly non-linear form of the function f_i when plotted against the interface index i (main plot in Fig. 10a and (2)).

Figures 9 and 10 also show the results of the automatic interface placement methods. For both methods, the trial interface and the exploring scouts method, we set $M_{\text{trial}} = 8$. For the trial interface method, we used an acceptance range of the probability from $p_{\min} = 0.3$ to $p_{\max} = 0.6$ and the initial position for λ_{i+1} was set at $\lambda_i + 0.1(\lambda_B - \lambda_A)$. For the exploring scouts method, we used the destination probability $p_{\text{des}} = 0.45$ and a maximum number of $m_{\max} = 10{,}000$ time steps. Figure 9 shows that both these methods produce similar interface numbers and positions, which are different from those of the manual placement. The automatic methods place the interfaces closer to the A state. There are no interfaces located at λ values greater than 1120. This suggests that the nucleation barrier is located closer to λ_A than to λ_B in our case. When the system has overcome the barrier, the conditional probability is always larger than the destination value and thus no further interfaces must be placed. However, without the pre-knowledge, there would be no way to guess this when placing the interfaces manually. Figure 10b, c show that both automatic interface placement methods perform well, the success probabilities p_i are much more uniform, with no p_i values close to zero which is shown in the insets. The functions f_i are also much more linear for the automatic interface placement methods than for the manual interface placement which can be seen in the main plots.

A great advantage of the automatic methods is that setting up a FFS simulation becomes a lot easier and less time-consuming than placing the interfaces manually.

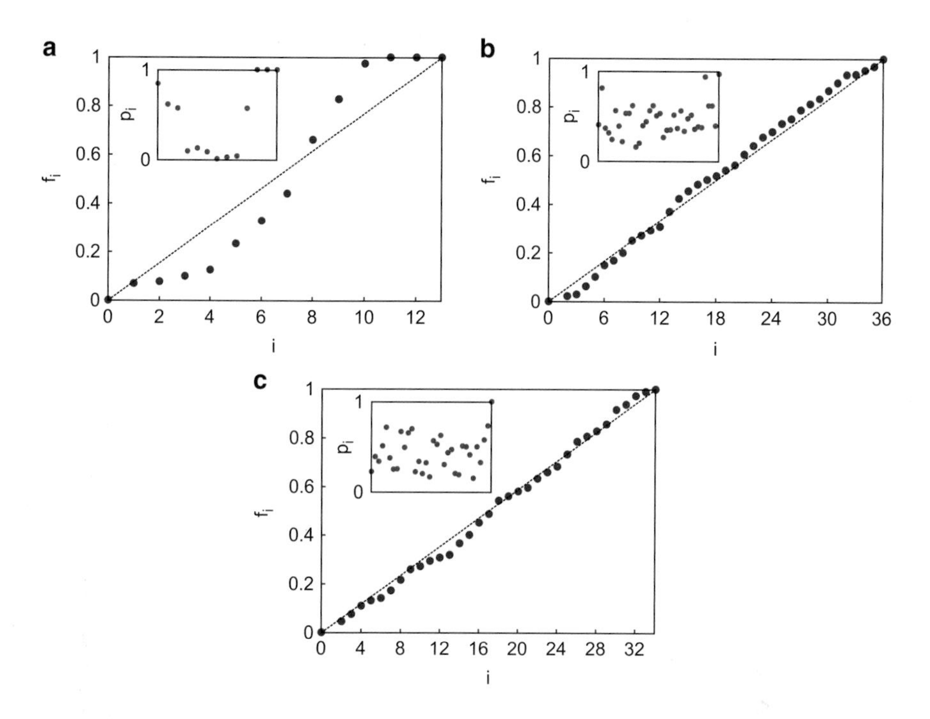

Fig. 10 Optimization criteria for the interface sets generated manually (**a**), with the trial interface method (**b**) and with the exploring scouts method (**c**). The main plots show the function f_i (2), plotted against the interface index i, from our simulations (*symbols*)—the *dashed lines* show the optimal case. The *insets* show the conditional probabilities p_i for each interface

Table 2 Rate constant k_{AB} (in $\sigma^{-3}\tau^{-1}$ with the simulation time unit τ) for DFFS simulations using manual interface placement, the trial interface method and the exploring scouts method

Method	k_{AB}
Manual placing	$5 \times 10^{-12\pm3}$
Trial interface	$6 \times 10^{-14\pm1}$
Exploring scouts	$2 \times 10^{-14\pm1}$

The error bars in k_{AB} were determined by repeated simulations

In addition, the resulting calculations are more efficient with the optimized interface collections. Table 2 shows that the error bars, which were computed by repeated FFS calculations, are larger for the manual interface placement. Furthermore, the computational cost of the FFS calculation was about a factor of 10 lower for the automatically placed interfaces than for those that were placed manually (see Table 3). For this simulation, the exploring scouts method required about 25 % fewer simulation steps than the trial interface method. Table 3 also shows estimates for the variance in the obtained rate constant, and the resulting computational efficiency. The estimated computational efficiency is three orders of magnitude higher if we

Table 3 Computational cost, variance in the rate constant and resulting computational efficiency, for DFFS simulations using manual interface placement, the trial interface method and the exploring scouts method

Method	Cost $\mathscr{C}$	Variance $\mathscr{V}$	Efficiency $\mathscr{E}$
Manual placing	4×10^7	18652	10^{-12}
Trial interface	4×10^6	188	10^{-9}
Exploring scouts	3×10^6	251	10^{-9}

The cost was measured in simulation time steps including the cost of exploratory trial runs for the automatic methods

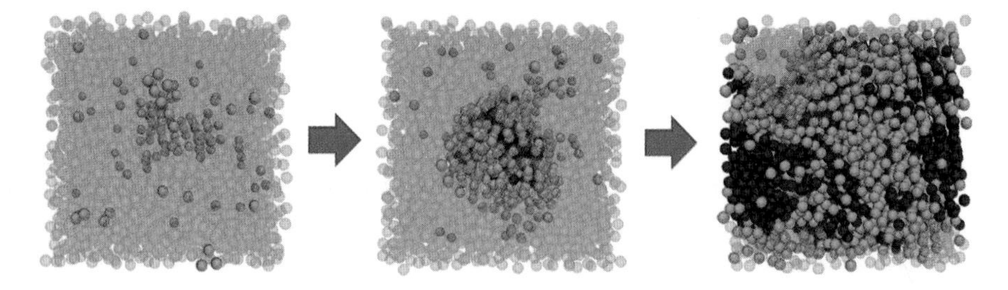

Fig. 11 Snapshots on different interfaces of the crystallization process of a FFS run, obtained via backtracing successful trajectories. The *golden particles* are detected as BCC-like, the *red particles* are FCC-like. Liquid particles are shown as *transparent blue*. Note that the surrounding particles which have not enough bonds are not shown as solid-like. *Left*: BCC-like particles where the crystal will form, *middle*: critical nucleus forms, *right*: spontaneous growth

use the automatic interface placement methods, compared to the manual interface collection.

FFS does not only provide the transition rate for this crystallization process, but also successful crystallization trajectories, which can be obtained by backtracking successful pathways. Figure 11 shows snapshots of such a successful trace exemplarily. We find that the crystallization in this case is a two-step process: First, we observe particles crystallizing in a bcc-like structure in the candidate domain for a critical nucleus. Then, a stable fcc-like structured nucleus forms, and if the critical cluster size is overcome, the crystal grows spontaneously.

6 Conclusion

Using two frontier techniques in computational physics, GPU-accelerated Lattice Boltzmann and Forward Flux Sampling, we tackle the problem of crystallization of charged macromolecules like proteins or colloidal particles. Depending on the density and interactions of the system, using the lattice Boltzmann GPU code for the computation of the hydrodynamics on a NVIDIA C1060 device accelerates the simulation by a factor of $\approx$20, so that our simulations take a day instead of a month. This enables us to run several thousand simulations in feasible time, and by that for the first time to study the effect of hydrodynamic interaction on the velocity of crystal growth. Only a larger GPU cluster such as the Nehalem cluster at HLRS makes such studies possible.

Forward Flux Sampling in turn parallelizes trivially, so that one can use large numbers of processors in parallel. On the other hand, FFS is used in computationally hard systems, so that using thousands of processors simultaneously is necessary in order to obtain results in reasonable time.

References

1. H.J. Limbach, A. Arnold, B.A. Mann, C. Holm, Comp. Phys. Commun. **174**(9), 704 (2006)
2. A. Arnold, O. Lenz, S. Kesselheim, R. Weeber, F. Fahrenberger, D. Roehm, P. Košovan, C. Holm, in *Meshfree Methods for Partial Differential Equations VI*, ed. by M. Griebel, M.A. Schweitzer. Lecture Notes in Computational Science and Engineering, vol. 89 (Springer, Berlin, 2013), pp. 1–23, http://www.springer.com/mathematics/computational+science+%26+engineering/book/978-3-642-32978-4
3. B. Hess, C. Kutzner, D. van der Spoel, E. Lindahl, J. Chem. Theory Comput. **4**(3), 435 (2008)
4. S.J. Plimpton, J. Comput. Phys. **117**, 1 (1995)
5. S. Succi, *The Lattice Boltzmann Equation for Fluid Dynamics and Beyond* (Oxford University Press, New York, 2001)
6. P.L. Bhatnagar, E.P. Gross, M. Krook, Phys. Rev. **94**(3), 511 (1954)
7. D. d'Humieres, Philos. Trans. R. Soc. Lond. A Math. Phys. Eng. Sci. **360**(1792), 437 (2002)
8. A.J.C. Ladd, J. Fluid Mech. **271**, 285 (1994)
9. P. Ahlrichs, B. Dünweg, Int. J. Mod. Phys. C **9**(8), 1429 (1998)
10. B. Dünweg, A.J.C. Ladd, in *Advanced Computer Simulation Approaches for Soft Matter Sciences III*. Advances in Polymer Science, vol. 221 (Springer, Berlin, 2009), pp. 89–166. doi:10.1007/12_2008_4
11. W. Li, X. Wei, A. Kaufman, Vis. Comput. **19**(7), 444 (2003)
12. NVIDIA Corporation, *Getting Started, NVIDIA CUDA Development Tools 3.2 Installation and Verification on Linux* (NVIDIA Corporation, Santa Clara, 2010)
13. NVIDIA Corporation, *NVIDIA CUDA C Programming Guide Version 3.2* (NVIDIA Corporation, Santa Clara, 2010)
14. J. Myre, S. Walsh, D. Lilja, M. Saar, Concurr. Comput. **23**(4), 332 (2011)
15. M.A. Safi, M. Ashrafizaadeh, A.A. Ashrafizaadeh, in *International Conference on Fluid Mechanics, Heat Transfer, and Thermodynamics*, vol. 73 (World Academy of Science, Engineering and Technology, Las Cruces, 2011), pp. 875–882
16. C. Feichtinger, S. Donath, H. Köstler, J. Götz, U. Rüde, J. Comput. Sci. **2**, 105–112 (2011)
17. MPI Consortium. The Message Passing Interface (MPI) Standard (2004), http://www.mcs.anl.gov/research/projects/mpi/Homepage
18. NVIDIA Corporation, *NVIDIA CUDA Reference Manual Version 3.2* (NVIDIA Corporation, Santa Clara, 2010)
19. T.S. van Erp, D. Moroni, P.G. Bolhuis, J. Chem. Phys. **118**, 7762 (2003)
20. R.J. Allen, D. Frenkel, P.R. ten Wolde, J. Chem. Phys. **124**, 024102 (2006)
21. R.J. Allen, C. Valeriani, P.R. ten Wolde, J. Phys. Condens. Matter **21**(46), 463102 (2009)
22. F.A. Escobedo, E.E. Borrero, J.C. Araque, J. Phys. Condens. Matter **21(33)**, 333101 (2009)
23. R.J. Allen, P.B. Warren, P.R. ten Wolde, Phys. Rev. Lett. **94**, 018104 (2005)
24. K. Kratzer, A. Arnold, R.J. Allen, J. Chem. Phys. **138**(16), 164112 (2013)
25. E.E. Borrero, F.A. Escobedo, J. Chem. Phys. **129**(2), 024115 (2008)
26. E. Hinch, J. Fluid Mech. **72**, 499 (1975)

27. P. Steinhardt, D. Nelson, M. Ronchetti, Phys. Rev. B **28**(2), 784 (1983)
28. D. Moroni, P. Ten Wolde, P. Bolhuis, Phys. Rev. Lett. **94**(23), 235703 (2005)
29. W. Lechner, C. Dellago, arXiv preprint arXiv:0806.3345 (2008)
30. S. Hamaguchi, R. Farouki, D. Dubin, J. Chem. Phys. **105**, 7641 (1996)
31. S. Auer, D. Frenkel, J. Phys. Condens. Matter **14**(33), 7667 (2002)
32. E. Sanz, C. Valeriani, D. Frenkel, M. Dijkstra, Phys. Rev. Lett. **99**, 055501 (2007)
33. F.E. Azhar, M. Baus, J.P. Ryckaert, E.J. Meijer, J. Chem. Phys. **112**(11), 5121 (2000)
34. J.D. Weeks, D. Chandler, H.C. Andersen, J. Chem. Phys. **54**, 5237 (1971)

Simplified Models for Coarse-Grained Hemodynamics Simulations

J. Harting, F. Janoschek, B. Kaoui, T. Krüger, and F. Toschi

Abstract Human blood can be approximated as a dense suspension of red blood cells in plasma. Here, we present two models we recently developed to investigate blood flow on different scales: in the first part of the paper we concentrate on describing individual cells or model systems such as vesicles with high resolution in order to understand the underlying fundamental properties of bulk hemodynamics. Here, we combine a lattice Boltzmann solver for the plasma with an immersed boundary algorithm to describe the cell or vesicle membranes. This method allows a detailed study of individual particles in complex hydrodynamic situations. Further, this model can be used to provide parameters for a more coarse-grained approach: in that second approach we simplify much further than existing particulate models. We find the essential ingredients for a minimalist description that still recovers hemorheology. These ingredients include again a lattice Boltzmann method describing hydrodynamic long range interactions mediated by the plasma between cells.

J. Harting (✉) · F. Janoschek
Department of Applied Physics, Eindhoven University of Technology, Den Dolech 2, 5600MB Eindhoven, The Netherlands

Institute for Computational Physics, University of Stuttgart, Allmandring 3, 70569 Stuttgart, Germany
e-mail: jens@icp.uni-stuttgart.de

B. Kaoui
Department of Applied Physics, Eindhoven University of Technology, Den Dolech 2, 5600MB Eindhoven, The Netherlands

T. Krüger
Centre for Computational Science, University College London, 20 Gordon Street, London WC1H 0AJ, UK

F. Toschi
Department of Applied Physics, Eindhoven University of Technology, Den Dolech 2, 5600MB Eindhoven, The Netherlands

CNR-IAC, Via dei Taurini 19, 00185 Rome, Italy

W.E. Nagel et al. (eds.), *High Performance Computing in Science and Engineering '13*, DOI 10.1007/978-3-319-02165-2_4,

The cells themselves are simplified as rigid ellipsoidal particles, where we describe the more complex short-range behavior by anisotropic model potentials. Recent results on the behaviour of single viscous red blood cells and vesicles in confined flow situations are shown alongside with results from the validation of our simplified model involving thousands or even millions of cells.

1 Introduction

Human blood can be approximated as a suspension of deformable red blood cells (RBCs, erythrocytes) in a Newtonian blood plasma. Since the other constituents like leukocytes and thrombocytes appear only in small numbers, they can be neglected [1]. Typical volume concentrations for RBCs are 40–50 % under physiological conditions. In the absence of external stresses, erythrocytes assume the shape of biconcave discs of approximately 8 μm diameter [2]. An understanding of their effect on the rheology and the clotting behavior of blood is necessary for the study of pathological deviations in the body and the design of microfluidic devices for improved blood analysis.

Typical models to simulate blood flow either restrict themselves to a continuous description at larger scales [3] or resolve individual cells with high resolution. The latter often include a Navier–Stokes solver for the plasma (classical CFD, lattice Boltzmann (LB), or particle based methods) combined with an elaborate model of deformable cell membranes [4–8]. Somewhere inbetween are models which coarse-grain the cell details and describe them as rigid particles [9–11].

In this contribution we follow two routes to contribute to the understanding of blood flow: on the microscale, we investigate vesicles as a model system for red blood cells. Vesicles are closed membranes made of phospholipid molecules. Next to being considered a good model system for living cells, they are also used for micro-encapsulation of active materials in drug delivery. The thickness of the vesicle membrane ($\sim$5 nm) is negligibly small compared to the vesicle size (typically $\sim$10 μm for giant unilamellar vesicles). Therefore, the membranes are considered as two-dimensional incompressible Newtonian liquids. As a consequence, the membrane area cannot undergo extension or compression, which implies local and global conservation of the vesicle area, as is the case for RBCs [12]. These vesicles are modeled using a combined algorithm involving an implementation of the immersed boundary method (IBM) for the membrane dynamics which is coupled to a lattice Boltzmann solver for the inner and outer fluids [13, 14].

For larger scales we developed a coarse-grained blood model. It aims at a minimal resolution of red blood cells which allows for a simple and highly efficient but still particulate description of blood as a suspension. The ultimate goal is to perform large-scale simulations that allow to study the flow in realistic geometries but also to link bulk properties, for example the effective viscosity, to phenomena at the level of single erythrocytes. Only a computationally efficient description allows the reliable accumulation of statistical properties in time-dependent flows which is necessary for this task. The main idea of this model is to distinguish between the long-range hydrodynamic coupling of cells and the short-range interactions that are related to the complex mechanics, electrostatics, and the chemistry of the membranes. The short-range behavior of RBCs is described on a phenomenological level by means of anisotropic model potentials [11]. Long-range hydrodynamic interactions are accounted for by means of an LB method. Our models are well suited for the implementation of complex boundary conditions and an efficient parallelization on parallel supercomputers. Both are necessary for the study of realistic systems such as branching vessels and the accumulation of statistically relevant data in bulk flow situations.

2 Simulation Methods

We apply a Bhatnagar–Gross–Krook (BGK) lattice Boltzmann method for modeling the blood plasma [15]. The single particle distribution function $n_r(\mathbf{x}, t)$ resembles the fluid traveling with one of $r = 1, \ldots, M$ discrete velocities $\mathbf{c}_r$ at the lattice position $\mathbf{x}$ and discrete time t. In the cases relevant to this paper we restrict ourselves to $M = 9$ (2D) and $M = 19$ (3D). The evolution of the single particle distribution function in time is determined by the lattice Boltzmann equation

$$n_r(\mathbf{x} + \mathbf{c}_r, t + 1) - n_r(\mathbf{x}, t) = -\frac{1}{\tau}\left(n_r(\mathbf{x}, t) - n_r^{\mathrm{eq}}(\varrho(\mathbf{x}, t), \mathbf{u}(\mathbf{x}, t))\right), \qquad (1)$$

where the right hand side is the BGK-collision term with a single relaxation time τ. The equilibrium distribution function $n_r^{\mathrm{eq}}(\varrho, \mathbf{u})$ is an expansion of the Maxwell–Boltzmann distribution. $\varrho(\mathbf{x}, t) = \sum_r n_r(\mathbf{x}, t)$ and $\varrho(\mathbf{x}, t)\mathbf{u}(\mathbf{x}, t) = \sum_r n_r(\mathbf{x}, t)\mathbf{c}_r$ can be identified as density and momentum. In the limit of small velocities and lattice spacings the Navier–Stokes equations are recovered with a kinematic viscosity of $\nu = (2\tau - 1)/6$. Here, we keep $\tau = 1$ if not mentioned otherwise.

The immersed boundary method (IBM) is used as a front tracking approach [16] to describe the deformable vesicle membrane. Within the IBM, the interface is considered sharp (zero thickness) and is represented by a cluster of marker points (nodes) which constitute a moving Lagrangian mesh. This mesh is immersed in a fixed Eulerian lattice representing the fluid. To consider correct dynamics, a bi-directional coupling of the lattice fluid and the moving Lagrangian mesh has to be taken into account: the interface is moving along with the ambient fluid velocity. A deformation of the interface generally leads to stresses reacting back onto the fluid via local forces. As an example, interfacial stresses are caused by local bending of a fluid-fluid interface (surface tension) or shearing of a capsule membrane (shear deformation). The two-way coupling is accomplished in two main steps: (1) velocity interpolation and Lagrangian node advection and (2) force spreading (reaction).

Interpolation: First, the fluid flow field is computed using LB. As a result, the velocity $\mathbf{u}$ is known at each lattice site $\mathbf{x}$. Afterwards, the fluid velocity is interpolated to obtain its value at the position $\mathbf{r}_i$ of each membrane node i which equals the velocity $\dot{\mathbf{r}}_i$ of that node. Finally, the interface is advected by updating the position of each membrane node using an Euler scheme, $\mathbf{r}_i(t + \Delta t) = \mathbf{r}_i(t) + \dot{\mathbf{r}}_i(t)\Delta t$.

Reaction: By advecting the interface, it is generally deformed. The new shape of the interface is not necessarily its equilibrium shape which would minimize its energy. Therefore, each node exerts a reaction force (which is computed from the known deformation state and the constitutive interface properties) on its surrounding fluid. The reaction force has to be taken into account by the Navier–Stokes solver as a local acceleration in the next time step and is being computed according to

$$\mathbf{f} = \left[\kappa_\mathrm{B}\left(\frac{\partial^2 c}{\partial s^2} + \frac{c^3}{2}\right) - c\,\zeta\right]\mathbf{n} + \frac{\partial \zeta}{\partial s}\mathbf{t}, \tag{2}$$

where $\mathbf{n}$ and $\mathbf{t}$ are the normal and tangential unit vectors, respectively. The membrane is characterized by the membrane bending modulus κ_B and the local effective tension ζ. The local curvature is c, and s is the arclength coordinate. Our algorithm allows for different internal and external viscosities. Their ratio Λ is called viscosity contrast.

For a coarse-grained description of the hydrodynamic interaction of cells and blood plasma, a method similar to the one by Aidun et al. modeling rigid moving particles of finite size is applied [17, 18]. Starting point is the mid-link bounce-back boundary condition: the confining geometry is discretized on the LB lattice and all internal nodes are turned into fluid-less wall nodes. A freely moving particle i is modeled by such moving walls and is defined by its continuous position $\mathbf{r}_i$. In contrast to the biconcave equilibrium shape of physiological RBCs we choose a simplified ellipsoidal geometry that is defined by two distinct half-axes $R_\parallel$ and $R_\perp$ parallel and perpendicular to the unit vector $\hat{\mathbf{o}}_i$ which points along the direction of the axis of rotational symmetry of each particle i. Since the cell-fluid interaction volumes are rigid we need to allow them to overlap in order to account for the deformability of real erythrocytes.

In order to account for the complex behavior of real RBCs at small distances we add phenomenological pair potentials between cells. The idea is to develop simple model potentials and adjust their free parameters in order to match the pair interactions of cells. In a very similar way, the well-known Lennard–Jones potential is applied in classical molecular dynamics simulations to model atomic interactions. A fit can be achieved by comparison of simulation results and experimental data from the literature, especially regarding blood rheology. For the moment, we concentrate on high shear rates $\dot{\gamma} > 10\,\mathrm{s}^{-1}$ where aggregation can be neglected [19]. Therefore, the model potential has to account only for deformation effects. As a simple way to describe elastic deformability, we use the repulsive branch of a Hookean spring potential

$$\phi(r_{ij}) = \begin{cases} \varepsilon\left(1 - r_{ij}/\sigma\right)^2 & r_{ij} < \sigma \\ 0 & r_{ij} \geq \sigma \end{cases} \tag{3}$$

for the scalar displacement r_{ij} of two cells i and j. With respect to the disc-shape of RBCs, we follow the approach of Berne and Pechukas [20] and choose the energy and range parameters

$$\varepsilon(\hat{\mathbf{o}}_i, \hat{\mathbf{o}}_j) = \frac{\bar{\varepsilon}}{\sqrt{1 - \chi^2\left(\hat{\mathbf{o}}_i \times \hat{\mathbf{o}}_j\right)^2}} \tag{4}$$

and

$$\sigma(\hat{\mathbf{o}}_i, \hat{\mathbf{o}}_j, \hat{\mathbf{r}}_{ij}) = \frac{\bar{\sigma}}{\sqrt{1 - \frac{\chi}{2}\left[\frac{\left(\hat{\mathbf{r}}_{ij}\times\hat{\mathbf{o}}_i + \hat{\mathbf{r}}_{ij}\times\hat{\mathbf{o}}_j\right)^2}{1+\chi\hat{\mathbf{o}}_i\times\hat{\mathbf{o}}_j} + \frac{\left(\hat{\mathbf{r}}_{ij}\times\hat{\mathbf{o}}_i - \hat{\mathbf{r}}_{ij}\times\hat{\mathbf{o}}_j\right)^2}{1-\chi\hat{\mathbf{o}}_i\times\hat{\mathbf{o}}_j}\right]}}, \tag{5}$$

in which $\times$ denotes the scalar product as functions of the orientations $\hat{\mathbf{o}}_i$ and $\hat{\mathbf{o}}_j$ of the cells and their normalized center displacement $\hat{\mathbf{r}}_{ij}$. We achieve an anisotropic potential with a zero-energy surface that is approximately that of ellipsoidal discs. Their half-axes parallel $\sigma_\parallel$ and perpendicular $\sigma_\perp$ to the symmetry axis enter (4) and (5) via $\bar{\sigma} = 2\sigma_\perp$ and $\chi = (\sigma_\parallel^2 - \sigma_\perp^2)/(\sigma_\parallel^2 + \sigma_\perp^2)$, whereas $\bar{\varepsilon}$ determines the potential strength. For modeling the cell-wall interaction we assume a sphere with radius $\sigma_{\mathrm{w}} = 1/2$ at every lattice node on the surface of a vessel wall and implement similar forces as for the cell-cell interaction. See Fig. 1 for a two-dimensional cut as an outline of the full model [11].

Based on experimental measurements of RBC geometries[2], the half-axes of the simplified volume defining the cell-cell interaction in our model are set to $\sigma'_\perp = 4\,\mu\mathrm{m}$ and $\sigma'_\parallel = 4/3\,\mu\mathrm{m}$. The spatial resolution quantified by the physical distance corresponding to one lattice spacing is set to $\delta x = 2/3\,\mu\mathrm{m}$ [11]. Primed variables are used to distinguish quantities given in physical units from the same unprimed variable measured in lattice units. Supposing that ν matches the kinematic plasma viscosity of $\nu' = 1.09 \times 10^{-6}\,\mathrm{m}^2\,\mathrm{s}^{-1}$, the time discretization is determined as $\delta t = 6.80 \times 10^{-8}\,\mathrm{s}$. $\delta m = 3.05 \times 10^{-16}\,\mathrm{kg}$ is chosen arbitrarily.

The simulations presented in this report utilize several simulation codes. The vesicle work was performed using a 2D implementation of the combined lattice Boltzmann and immersed boundary method. Here, many very long simulations were required to cover the full parameter space of interest. The coarse-grained blood model utilizes our 3D implementation "LB3D". "LB3D" combines massively parallel versions of a multicomponent lattice Boltzmann solver and a molecular dynamics algorithm for suspended particles. Recently, also a 3D immersed boundary module was added to extend the 2D vesicle work presented here to 3D. Since we already reported on the performance of "LB3D" in several places, the reader is referred to the available literature [11,21–25].

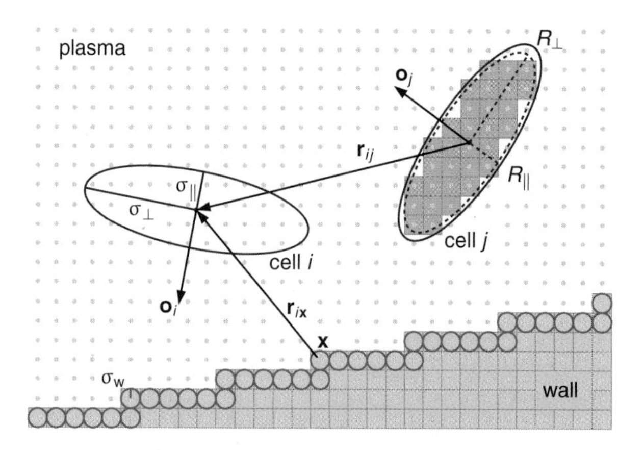

Fig. 1 2D cut as an outline of the simplified particular model. Shown are two cells with their axes of rotational symmetry depicted by $\hat{\mathbf{o}}_{i/j}$. The volumes defined by the cell-cell interaction are approximately ellipsoidal with half-axes $\sigma_{\perp/\parallel}$. The smaller ellipsoidal volumes (half-axes $R_{\perp/\parallel}$) of the cell-plasma interaction are discretized on the underlying lattice. The cell-wall potential assumes spheres with radius σ_w on all surface wall nodes $\mathbf{x}$ (taken from [11])

3 High Resolution Simulations of Single Vesicles and Red Blood Cells

We investigate the interplay between the confinement and the viscosity contrast (ratio between internal and external fluid viscosities) on the dynamics of viscous vesicles and RBCs. This is relevant, for example, to understand the reproducibility and the accuracy of blood viscosity measurements, where the interplay of rheometer wall effects and the dynamical state of each individual RBC have a significant impact. While confinement can lead to an overestimation of the measured effective viscosity of blood as mentioned in [26], the transition of the dynamical states of vesicles and RBCs induced by varying the viscosity contrast was found to minimize the effective viscosity [27]. It is known that an unconfined vesicle or RBC subjected to shear undergoes either a steady liquid-like motion called *tank-treading* (the vesicle main axis assumes a steady inclination angle with the flow direction while the membrane undergoes a tank-treading-like motion) or unsteady solid-like motion called *tumbling* (the vesicle flips as a solid particle) [28–31]. The transition between dynamical states depends on dimensionless control parameters: (1) the swelling degree Δ, (2) the viscosity contrast Λ, (3) the capillary number Ca (ratio between viscous and membrane bending forces), and (4) the degree of confinement χ [13, 14, 32–35].

Figure 2 demonstrates the influence of the viscosity contrast on the dynamical behaviour of a (stiff, almost not deformable) vesicle: if the inner and outer viscosities are identical ($\Lambda = 1$), the vesicle tank-treads (Fig. 2a). The tank-treading motion of the membrane generates a rotational flow inside the vesicle. By increasing

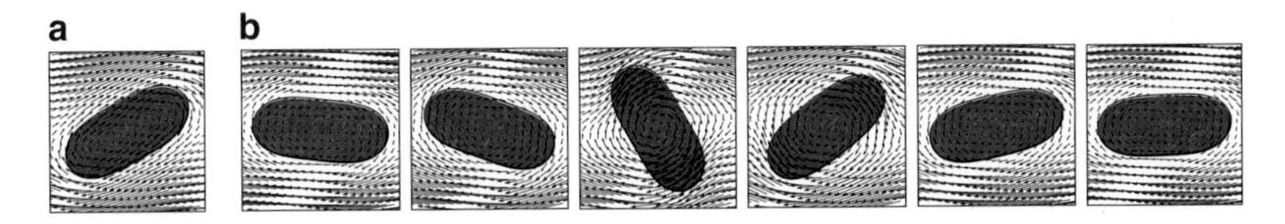

Fig. 2 Snapshots showing the dynamics of a viscous vesicle under shear flow. (**a**) Tank-treading: the vesicle assumes a steady inclination angle with the flow direction while its membrane undergoes a tank-treading motion. (**b**) Tumbling: the vesicle rotates as a rigid elongated particle (from [36])

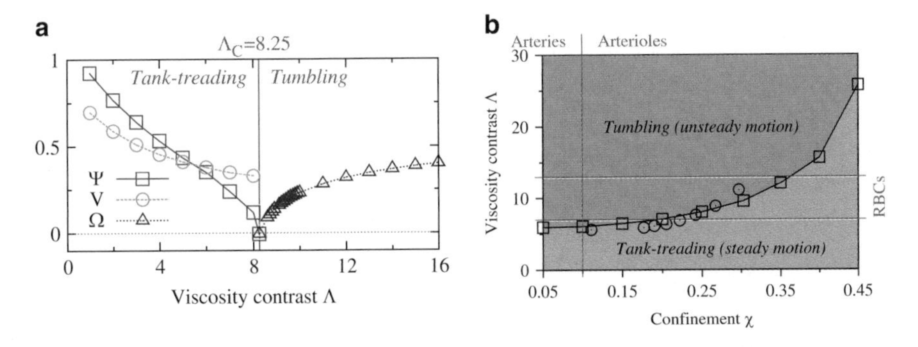

Fig. 3 (**a**) The transition of the vesicle dynamical state induced by increasing the viscosity contrast. *Left panel*: the inclination angle Ψ and the tank-treading velocity V decrease with Λ. *Right panel*: the tumbling frequency Ω increases with Λ. (All quantities are normalized, see text.) (**b**) χ-Λ phase-diagram of the dynamical states of confined viscous vesicles. The confinement shifts or even inhibits the transition to tumbling. *Circle symbols* are data from [29] (from [14])

Λ above a certain threshold, we induce a transition from tank-treading to tumbling. This motion is depicted in Fig. 2b for $\Lambda = 15$.

We compute the physical quantities characterizing each vesicle dynamical state when varying Λ, while keeping the confinement fixed at $\chi = 0.26$. The tank-treading motion is characterized by the steady inclination angle Ψ and the membrane tank-treading velocity V, while the tumbling motion is characterized by the tumbling frequency Ω. Figure 3a shows the variation of these quantities when increasing Λ from 1 to 16. For convenience, Ψ is normalized by $\pi/6$, V and Ω are normalized, respectively, by the rotational velocity $(\gamma R_0/2)$ and frequency $(\gamma/4\pi)$ of a rigid cylinder rotating in unbounded shear flow. At lower Λ, the vesicle performs the tank-treading motion. By increasing Λ, both Ψ and V decrease. Ψ decreases until it vanishes at a critical value $\Lambda_C = 8.25$ above which tumbling takes over. In the tumbling regime, Ω increases with Λ until the limit of a rigid particle is reached.

In Fig. 3b, the critical viscosity contrast Λ_C is plotted against the confinement χ. This gives the χ-Λ phase-diagram of the dynamical states of a confined viscous vesicle subjected to shear flow. It can be seen that the transition threshold Λ_C is pushed up by increasing confinement. The same trend has been observed by

Beaucourt et al. [29] for $0.1 \leq \chi \leq 0.3$. A comparison of the present data obtained by IBM/LBM (square symbols), with phase-field simulations reported in [29] (circle symbols) reveals a good qualitative and quantitative agreement. We find that the transition between the tank-treading and the tumbling motion can also be induced by varying the confinement χ alone, without changing the viscosity contrast Λ. For example, by taking a tumbling vesicle with $\Lambda > \Lambda_C$ at lower confinement, we can force it to tank-tread by increasing the confinement above a certain threshold χ_C. We expect the same confinement-induced transition (beside the shear-induced transition) to happen for RBCs, for which $7 \leq \Lambda \leq 13$, during their displacement from arteries to arterioles.

4 Mesoscale Simulation of Hemodynamics

We study the dependence of the viscosity of a suspension of RBCs in plasma using a Couette setup, where at the side planes a constant shear velocity is imposed by an adaption of the Lees–Edwards shear boundary condition to the LB method [37]. The apparent viscosity is calculated from the average shear rate and shear stress [11]. Recently, an alternative method of viscosity measurement based on Kolmogorov flow was demonstrated which allows the employment of more simple periodic boundary conditions [38]. The effect of the stiffness parameter of the cell-cell potential $\bar{\varepsilon}$ is studied. The viscosity as a function of the shear rate increases with increasing $\bar{\varepsilon}$. For very stiff cells this dependence on the shear rate decreases considerably which is in asymptotic consistency with the experimental results of Chien [19] who measured the apparent shear viscosity of a suspension of artificially hardened RBCs and found a significantly increased yet mostly constant viscosity. At a cell-fluid volume concentration of 43 %, we find best agreement for $\bar{\varepsilon}' = 1.47 \times 10^{-15}$ J and use this parametrization for all following investigations [11].

We now confine our model suspension in a cylindrical channel of diameter D and a length of 43 μm with periodic boundaries. Both cells and plasma are steadily driven by a volume force equivalent to a pressure gradient $\mathrm{d}P/\mathrm{d}z$. We arbitrarily choose $\bar{\varepsilon}'_{\mathrm{w}} = 1.47 \times 10^{-16}$ J for the strength of the cell-wall interaction as a value that reliably prevents cells from penetrating the vessel wall. The pseudo-shear rate, which is defined via the volume flow rate Q, is $\bar{v}' = 4Q'/(\pi D'^3) = 61\,\mathrm{s}^{-1}$. A preferential alignment of the cells largely perpendicular to the velocity gradient is visible. We compare radial velocity profiles for different flow velocities, a cell-fluid volume concentration of $\Phi = 42\,\%$, and $D' = 63$ μm. While for high velocities the result looks parabolic in the central region, there is increasing blunting of the profile when the flow rate is reduced. The blunting can be understood as a consequence of the shear-thinning behavior of the model and is qualitatively consistent with experimental data from the literature [1]. Generally, apparent slip is visible close to the vessel wall. It is due to a cell depletion layer that to some extent can be controlled via $\bar{\varepsilon}_{\mathrm{w}}$. Thus, the observations can partly be described by an existing model assuming a homogeneous core region with high hematocrit and consequently high viscosity and a cell-depleted boundary layer close to the vessel wall [11, 39].

As it is well known, the formation of this cell depletion layer influences the flow resistance of vessels which can be expressed in terms of their apparent viscosity [39]. By inserting the measured volume flow rate Q through a cylindrical vessel into the theoretical expression for a Newtonian fluid

$$Q = \frac{\pi D^4}{128\mu} \frac{\mathrm{d}P}{\mathrm{d}z} \tag{6}$$

and solving for the dynamic viscosity μ, the respective apparent viscosity μ_{app} can be determined. Except for Q, all known quantities in (6) are constant and set as parameters of our simulation code. Only the flow rate undergoes stochastic fluctuations and—at the beginning of each simulation—shows a strong time dependence as the system relaxes from an arbitrary initial condition to a macroscopically steady state. Thus, typical simulations last for up to $\sim 10^7$ time steps corresponding to ~ 1 s of physical time. Q is averaged over typically ~ 0.1 s and its statistical error serves to estimate the accuracy of the resulting viscosity. The main dependency of μ_{app} is on the hematocrit Φ. A comparison with experimental data in the case of Couette flow was published earlier [11]. However, μ_{app} depends also on the vessel diameter D. This is known as Fåhræus–Lindqvist effect [39]. Pries et al. combine a large set of experimental studies for $\bar{v}' > 50\,\mathrm{s}^{-1}$ and provide an empirical fit [40]. The resulting expression for μ_{app} is plotted as a function of Φ for two discrete values of D as lines in Fig. 4a. The symbols stand for simulation results at the same diameters D and pseudo-shear rates of $(62 \pm 1)\,\mathrm{s}^{-1}$. In consistency with the still significant shear-thinning that experiments but also our simulations exhibit above $\dot{\gamma}' = 50\,\mathrm{s}^{-1}$ [11, 19], our results show a clear dependency on the pseudo-shear rate, which cannot be covered by the curves by Pries et al. that were obtained from averaging over different $\bar{v}' > 50\,\mathrm{s}^{-1}$ [40]. Nevertheless, a comparison confirms the agreement concerning the order of magnitude of our results and the observation that the apparent viscosity increases with D. This can be explained by the decreasing relative influence of the cell depletion layer. While the effect is captured realistically for low volume concentrations $\Phi \lesssim 0.3$, the dependence of μ_{app} on D becomes less clear for higher Φ. We explain this discrepancy by the fact that the thickness of the cell-depleted layer at high Φ is determined by a balance of the short-range interactions of cells in the core acting towards a decrease of the depletion layer thickness and the short-range interactions of cells and the vessel wall acting towards its increase.

While the results for the straight channel studied first could still mostly be reproduced by a homogenous fluid with a specially tuned shear rate or position dependent viscosity [3, 41], this is not possible at the scale of capillaries. Here, our model allows to account for clearly particulate effects like clogging or local changes of flow rate and pressure [22]. Such effects can lead to a distinct unsteadiness of the local cell volume concentration which is present also in human microvascular networks [39].

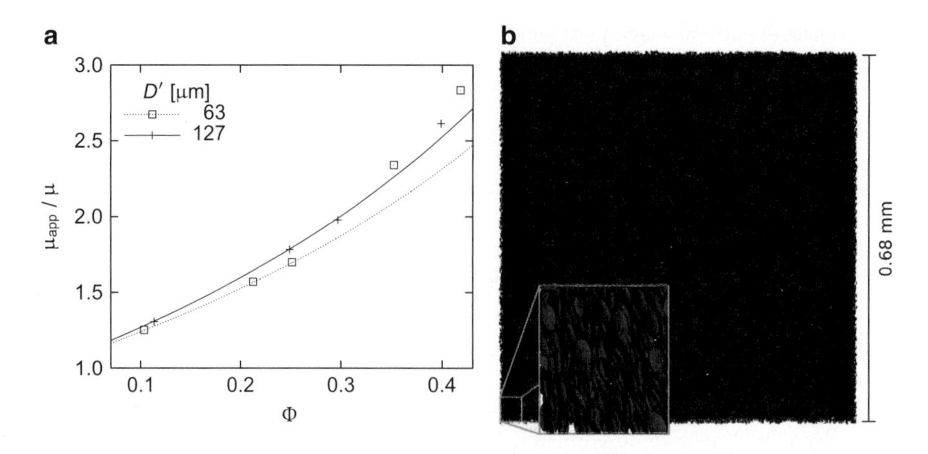

Fig. 4 (**a**) Dependence of the apparent viscosity μ_{app} in a cylindrical vessel with diameter D on the volume concentration Φ. The *lines* show empirical results by Pries et al. [40] while the *symbols* represent our simulations for $\bar{\dot{\gamma}}' = (62 \pm 1)\,\mathrm{s}^{-1}$. For $\Phi \lesssim 0.3$ there is good consistency of simulation results and experimental data. (**b**) Schematic view of one of the square side planes of a benchmark system containing $1024^2 \times 2048$ lattice sites and more than 4×10^6 RBCs. The simulated volume resembles $0.68^2 \times 1.37\,\mathrm{mm}^3$ of blood (from [22])

Despite the simplifications of the model, parallel supercomputers are necessary to simulate realistic vessel networks or large bulk systems. This makes the scalability of the code crucial. Compared to deformable particle models our method not only has a lower overall number of computations at a given resolution but is also easier to parallelize efficiently because each RBC has only six degrees of freedom. Figure 4b shows a snapshot from a simulation of a suspension consisting of $1024^2 \times 2048$ lattice sites and 4.1×10^6 cells which was performed on 1,024 CPUs of the XC2 at SSC Karlsruhe.

5 Conclusions

We demonstrated two possible simulation approaches to improve our understanding of blood flow: on a microscopic scale we implemented a combined lattice Boltzmann and immersed boundary method to investigate the behaviour of a single deformable viscous vesicle as a model system for a red blood cell. We presented that the transition from tank-treading to tumbling can be obtained by changing the relevant dimensionless control parameters and in particular the viscosity contrast and the confinement. In the second part of this report we presented recent results obtained using a more coarse-grained model for blood flow including studies of the impact of some model parameters and a validation with experimental data of blood flow in cylindrical tubes.

Acknowledgements We thank the Scientific Supercomputing Center Karlsruhe for providing the computing time and technical support for the presented work.

References

1. H.L. Goldsmith, R. Skalak, Hemodynamics. Annu. Rev. Fluid Mech. **7**, 213–247 (1975)
2. E. Evans, Y.C. Fung, Improved measurements of the erythrocyte geometry. Microvasc. Res. **4**, 335–347 (1972)
3. J. Boyd, J.M. Buick, S. Green, Analysis of the Casson and Carreau-Yasuda non-Newtonian blood models in steady and oscillatory flows using the lattice Boltzmann method. Phys. Fluids **19**, 093103 (2007)
4. H. Noguchi, G. Gompper, Shape transitions of fluid vesicles and red blood cells in capillary flows. Proc. Natl. Acad. Sci. USA **102**, 14159–14164 (2005)
5. M.M. Dupin, I. Halliday, C.M. Care, L. Alboul, L.L. Munn, Modeling the flow of dense suspensions of deformable particles in three dimensions. Phys. Rev. E **75**, 066707 (2007)
6. J. Wu, C.K. Aidun, Simulating 3D deformable particle suspensions using lattice boltzmann method with discrete external boundary force. Int. J. Numer. Methods Fluids **62**(7), 765–783 (2010)
7. T. Krüger, F. Varnik, D. Raabe, Efficient and accurate simulations of deformable particles immersed in a fluid using a combined immersed boundary lattice Boltzmann finite element method. Comput. Math. Appl. **61**, 3485–3505 (2011)
8. T. Krüger, F. Varnik, D. Raabe, Particle stress in suspensions of soft objects. Philos. Trans. R. Soc. Lond. A **369**, 2414–2421 (2011)
9. C. Sun, C. Migliorini, L.L. Munn, Red blood cells initiate leukocyte rolling in postcapillary expansions: a lattice boltzmann analysis. Biophys. J. **85**(1), 208–222 (2003)
10. T. Hyakutake, T. Matsumoto, S. Yanase, Lattice Boltzmann simulation of blood cell behavior at microvascular bifurcations. Math. Comput. Simul. **72**(2–6), 134–140 (2006)
11. F. Janoschek, F. Toschi, J. Harting, Simplified particulate model for coarse-grained hemodynamics simulations. Phys. Rev. E **82**, 056710 (2010)
12. T.M. Fischer, Is the surface area of the red cell membrane skeleton locally conserved? Biophys. J. **61**, 298 (1992)
13. B. Kaoui, J. Harting, C. Misbah, Two-dimensional vesicle dynamics under shear flow: effect of confinement. Phys. Rev. E **83**, 066319 (2011)
14. B. Kaoui, T. Krüger, J. Harting, How does confinement affect the dynamics of viscous vesicles and red blood cells? Soft Matter **8**, 9246 (2012)
15. S. Succi, *The Lattice Boltzmann Equation for Fluid Dynamics and Beyond*. Numerical Mathematics and Scientific Computation (Oxford University Press, Oxford, 2001)
16. C. Peskin, The immersed boundary method. Acta Numer. **11**, 479 (2002)
17. C.K. Aidun, Y. Lu, E.-J. Ding, Direct analysis of particulate suspensions with inertia using the discrete Boltzmann equation. J. Fluid Mech. **373**, 287–311 (1998)
18. N.-Q. Nguyen, A.J.C. Ladd, Lubrication corrections for lattice-Boltzmann simulations of particle suspensions. Phys. Rev. E **66**, 046708 (2002)
19. S. Chien, Shear dependence of effective cell volume as a determinant of blood viscosity. Science **168**, 977–979 (1970)
20. B.J. Berne, P. Pechukas, Gaussian model potentials for molecular interactions. J. Chem. Phys. **56**, 4213–4216 (1972)
21. J. Harting, J. Chin, M. Venturoli, P.V. Coveney, Large-scale lattice boltzmann simulations of complex fluids: advances through the advent of computational grids. Philos. Trans. R. Soc. Lond. A **363**, 1895–1915 (2005)
22. F. Janoschek, F. Toschi, J. Harting, Simulations of blood flow in plain cylindrical and constricted vessels with single cell resolution. Macromol. Theory Simul. **20**, 562 (2011)

23. F. Günther, F. Janoschek, S. Frijters, J. Harting, Lattice boltzmann simulations of anisotropic particles at liquid interfaces. Comput. Fluids **80**, 184 (2013)
24. J. Harting, T. Zauner, A. Narvaez, R. Hilfer, Flow in porous media and driven colloidal suspensions, in *High Performance Computing in Science and Engineering '08*, ed. by W. Nagel, D. Kröner, M. Resch (Springer, Berlin, 2008)
25. S. Schmieschek, A. Narváez Salazar, J. Harting, Multi relaxation time lattice boltzmann simulations of multiple component fluid flows in porous media, in *High Performance Computing in Science and Engineering '12*, ed. by W. Nagel, D. Kröner, M. Resch (Springer, Berlin, 2013), p. 39
26. A.M. Forsyth, J.D. Wan, P.D. Owrutsky, M. Abkarian, H.A. Stone, Multiscale approach to link red blood cell dynamics, shear viscosity, and atp release. Proc. Natl. Acad. Sci. USA **108**, 10986 (2011)
27. V. Vitkova, M.-A. Mader, B. Polack, C. Misbah, T. Podgorski, Micro-macro link in rheology of erythrocyte and vesicle suspensions. Biophys. J. **95**, L33 (2008)
28. S. Keller, R. Skalak, Motion of a tank-treading ellipsoidal particle in a shear flow. J. Fluid Mech. **120**, 27 (1982)
29. J. Beaucourt, F. Rioual, T. Seon, T. Biben, C. Misbah, Steady to unsteady dynamics of a vesicle in a flow. Phys. Rev. E **69**, 011906 (2004)
30. V. Kantsler, V. Steinberg, Transition to tumbling and two regimes of tumbling motion of a vesicle in shear flow. Phys. Rev. Lett. **96**, 036001 (2006)
31. G.B. Jeffery, The motion of ellipsoidal particles immersed in a viscous fluid. Proc. R. Soc. Lond. A **102**, 161 (1922)
32. H. Noguchi, G. Gompper, Swinging and tumbling of fluid vesicles in shear flow. Phys. Rev. Lett. **98**, 128103 (2007)
33. V.V. Lebedev, K.S. Turitsyn, S.S. Vergeles, Dynamics of nearly spherical vesicles in an external flow. Phys. Rev. Lett. **99**, 218101 (2007)
34. B. Kaoui, A. Farutin, C.C. Misbah, Vesicles under simple shear flow: elucidating the role of relevant control parameters. Phys. Rev. E **80**, 061905 (2009)
35. J. Deschamps, V. Kantsler, V. Steinberg, Phase diagram of single vesicle dynamical states in shear flow. Phys. Rev. Lett. **102**, 118105 (2009)
36. T. Krüger, S. Frijters, F. Günther, B. Kaoui, J. Harting, Numerical simulations of complex fluid-fluid interface dynamics. Eur. Phys. J. Spec. Topics **222**, 177 (2013)
37. A.J. Wagner, J.M. Yeomans, Phase separation under shear in two-dimensional binary fluids. Phys. Rev. E **59**, 4366–4373 (1999)
38. F. Janoschek, F. Mancini, J. Harting, F. Toschi, Rotational behavior of red blood cells in suspension—a mesoscale simulation study. Philos. Trans. R. Soc. Lond. A **369**(1944), 2337–2344 (2011)
39. A.S. Popel, P.C. Johnson, Microcirculation and hemorheology. Annu. Rev. Fluid Mech. **37**(1), 43–69 (2005)
40. A.R. Pries, D. Neuhaus, P. Gaehtgens, Blood viscosity in tube flow: dependence on diameter and hematocrit. Am. J. Physiol. Heart Circ. Physiol. **263**(6), H1770–1778 (1992)
41. T.W. Secomb, Mechanics of red blood cells and blood flow in narrow tubes, in *Modeling and Simulation of Capsules and Biological Cells*, ed. by C. Pozrikidis (Chapman & Hall, London, 2003), pp. 163–196

Report: Thermodynamics with 2 + 1 + 1 Dynamical Quark Flavors

Stefan Krieg

Introduction

The aim of our project is to compute the charmed equation of state for Quantum Chromodynamics (for details, see [1]). We are using the lattice discretized version of Quantum Chromodynamics, called lattice QCD, which allows simulations of the theory through importance sampling methods. Our results are important input quantities for phenomenological calculations and are required to understand experiments aiming to generate a new state of matter, called Quark-Gluon-Plasma, such as the upcoming FAIR at GSI, Darmstadt.

The present status of the field is marked by our paper on the $N_f = 2 + 1$[1] equation of state [2]. In the time since the publication of the aforementioned work, the hotQCD collaboration have improved the precision of their results. It was found that some discrepancies between our and their results still remain (see e.g. [3]). It is the aim of this work to provide a high precision calculation of the equation of state of QCD including the (dynamical) effects of the charm quark, in order to remedy the above situation.

Our simulations are performed using so-called staggered fermions. In the continuum limit, i.e. at vanishing lattice spacing a, one staggered Dirac operator implements four flavors of mass degenerate fermions. At finite lattice spacing, however, discretization effects induce an interaction between these would be flavors

[1]This refers to dynamical up/down and strange quarks—including a dynamical charm quark is what was proposed here.

S. Krieg (✉)
Bergische Universität Wuppertal, Fachbereich C - Physik, Gaußstraße 20, 42119 Wuppertal, Germany

Forschungszentrum Jülich GmbH, IAS/JSC, 52425 Jülich, Germany
e-mail: krieg@uni-wuppertal.de

W.E. Nagel et al. (eds.), *High Performance Computing in Science and Engineering '13*, DOI 10.1007/978-3-319-02165-2_5,

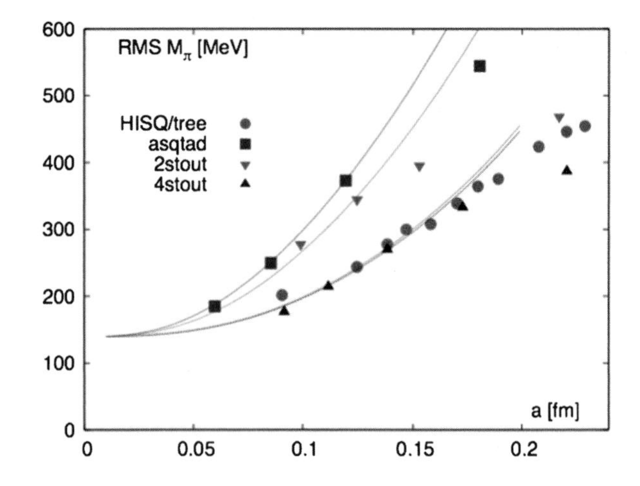

Fig. 1 RMS pion mass for different staggered fermion actions, in the continuum limit

lifting the degeneracy. The "flavors" are, consequentially, renamed to "tastes", and the interactions are referred to as "taste-breaking" effects. Even though the tastes are not degenerate, in simulations one takes the fourth root of the staggered fermion determinant to implement a single flavor. This procedure is not proven to be correct—however, practical evidence suggests that is does not induce errors visible with present day statistics.

Taste-breaking is most severely felt at low pion masses and large lattice spacing, as the pion sector is distorted through the taste-breaking artifacts: there is one would-be Goldstone boson, and 15 additional heavier "pions", which results in an RMS pion mass larger than the mass of the would-be Goldstone boson. This effect is depicted in Fig. 1 for different staggered type fermion actions. As can be seen for this figure, the previously used twice stout smeared action ("2stout") has a larger RMS pion mass and thus taste-breaking effects than the HISQ/tree action. If, however, the number of smearing steps is increased to four, with slightly smaller smearing strength ("4stout"), the RMS pion mass measured agrees with that of the HISQ/tree action. In order to have an improved pion sector, we, therefore, opted to switch to this new action and to restart our production runs.

Status

The status of our production is summarized in Table 1 and Fig. 2. We generated most finite temperature ensembles on less scalable architectures, such as GPU clusters, available to the collaboration at Wuppertal and Budapest universities. Our (zero temperature) renormalization runs require a scalable architecture, and we, as proposed, used HERMIT to generate these essential ensembles. Our present and upcoming finite temperature simulations on finer, thus, larger lattices also benefit from scalability.

Table 1 Production status with the new "4stout" action

Data set	Status
$N_t = 8$	Ready
$N_t = 10$	Production
$N_t = 12$	Thermalizing

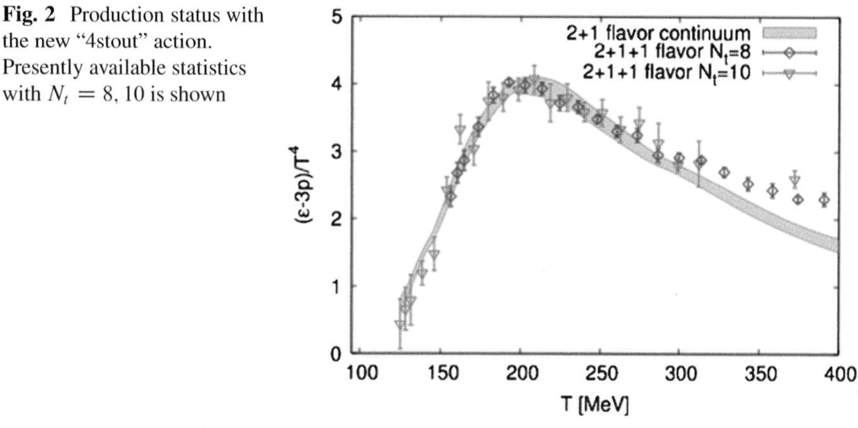

Fig. 2 Production status with the new "4stout" action. Presently available statistics with $N_t = 8, 10$ is shown

Results

In order to be able to reach very fine lattice spacings, or, equivalently, large temperatures at $N_t = 12$, we had to extend our line-of-constant-physics (LCP) beyond the range of lattice spacings available to us. Our previous strategy was the following ($m_c = 11.85\, m_s$):

1. Simulate a "reasonable" rectangle of up/down and strange quark mass values and to measure M_π / f_π and M_K / f_π.
2. Interpolate to the quark mass values where above ratios take their physical values. This gives m_{ud} and m_s.
3. Extract the lattice spacing by interpolating $a f_\pi$ to the above quark mass point.

The results of this procedure are shown in Figs. 3 and 4, for the data points with $\beta < 3.9$. The remaining points have been generated with the procedure described in the following. At very fine lattice spacing, the zero temperature runs at physical mass parameters would require enormous lattices and would also likely face issues related to the freezing of the topological charge at $a \approx 0.5$ fm. We, therefore, adapted our strategy:

1. For every $\beta < 3.9$: Simulate $N_f = 3+1$ flavors in the flavor-symmetric point [5] ($\bar{m} = (2m_{ud,phys} + m_{s,phys})/3$), using the quark mass parameters found above.
2. Measure f_{PS} and M_{PS}/f_{PS}.
3. Perform a continuum extrapolation.

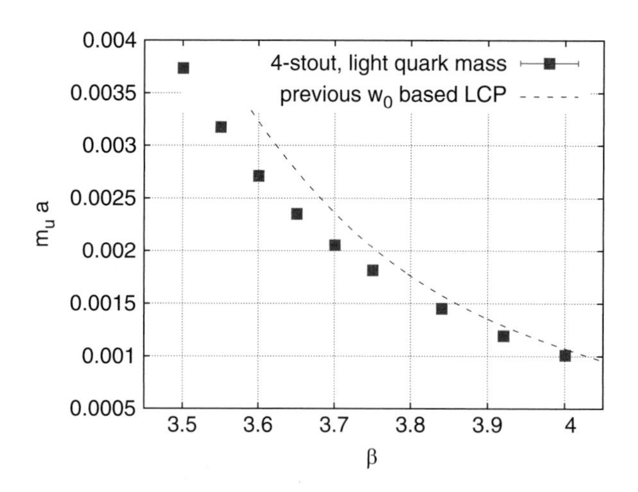

Fig. 3 LCP for our new 4stout action. Shown are the new data points for the light quark mass parameter. The *dashed line* indicates a first iteration using w_0 [4]. We prefer to use the ratio M_π/f_π as it is directly related to the pion sector

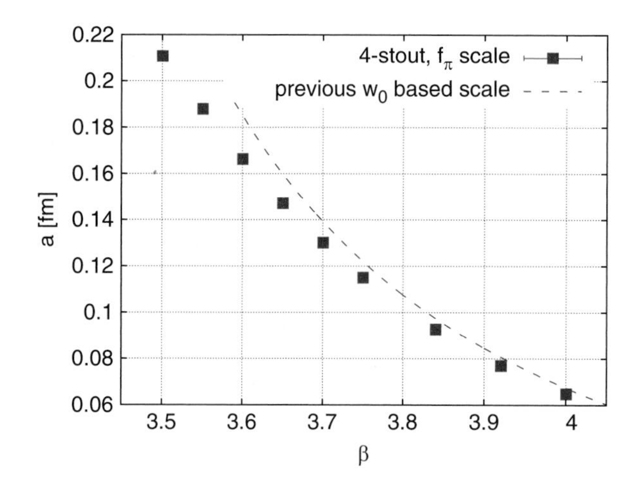

Fig. 4 LCP for our new 4stout action. Shown are the new data points for the lattice spacing. The *dashed line* indicates a first iteration using w_0 [4]. We prefer to use the pion decay constant as it is directly related to the pion sector

We now can estimate the expected values for f_{PS} and M_{PS}/f_{PS} at the target $\beta > 3.9$, see Figs. 5 and 6. Simulating several $\bar{m}$ values at these β we can find the value where the expected M_{PS}/f_{PS} is reproduced. This gives $\bar{m}(\beta)$, while the lattice spacing will be set through f_{PS}. The procedure is shown for one β in Figs. 7 and 8. It gives the LCP entries at $\beta > 3.9$ as shown above in Figs. 3 and 4.

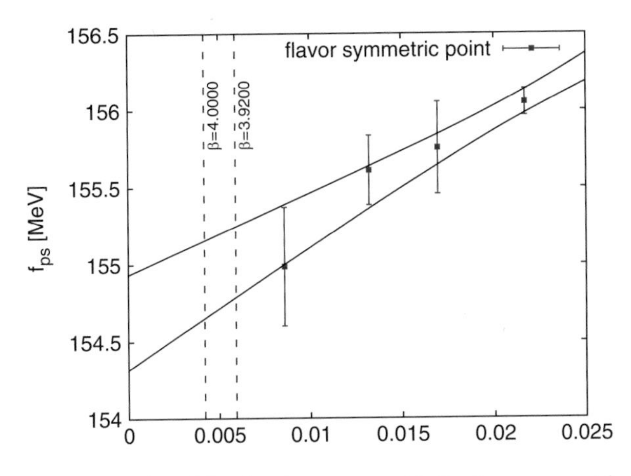

Fig. 5 Continuum extrapolation of the pion decay constant in the flavor-symmetric point (see text) using our new 4stout action. The target beta values are indicated by *dashed lines*. This extrapolation provides expected values for the pion decay constant at these β values

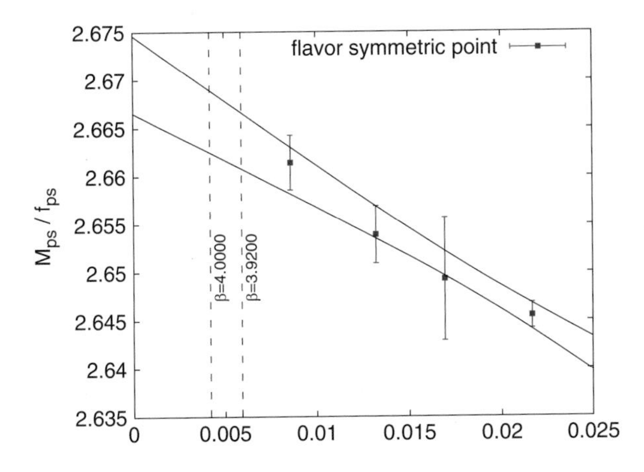

Fig. 6 Continuum extrapolation of the ration M_{PS}/f_{PS} in the flavor-symmetric point (see text) using our new 4stout action. The target beta values are indicated by *dashed lines*. This extrapolation provides expected values for the ratio at these β values

Additional Results

The reduced taste breaking allowed us to determine a further observable, which is of particular interest of the heavy ion community. While the equation of state is an input to the relativistic hydrodynamic calculations that reproduce the observed flow characteristics, there is an observable that can and is being directly measured at LHC: the non-gaussianity of the net charge and strangeness fluctuations at freeze-out temperature. We concentrated all our efforts on a single temperature, so far, and

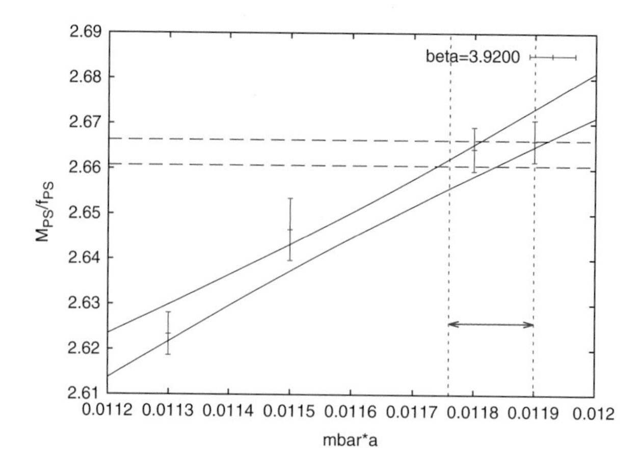

Fig. 7 Calculation of the (1σ region of) bare mass parameter for our new 4stout action. The *blue band* marks the expected value for the ration M_{PS}/f_{PS} at this β value, taken from the continuum extrapolation as shown in Fig. 6

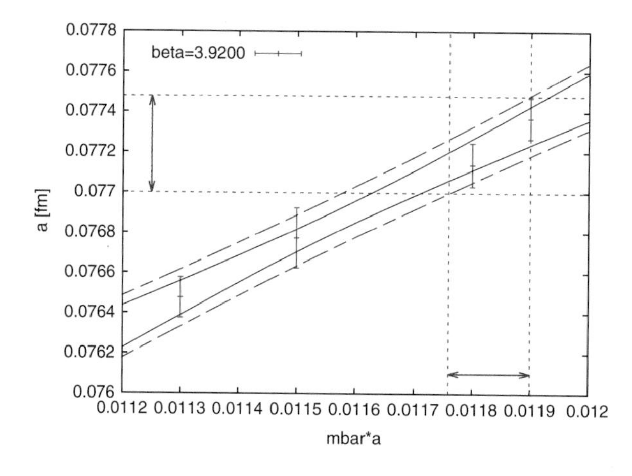

Fig. 8 Calculation of the lattice spacing at $\beta = 3.92$ for our new 4stout action. The correct value for $a\bar{m}$ has been extracted as shown in Fig. 7. The *blue band* marks the region defined by the measured f_π divided by the expected value from the continuum extrapolation (including errors)

show a continuum extrapolation in Fig. 9. This of course will need to be extended to a couple of more temperatures. The temperature dependence could be then used to measure the freeze-out temperature in heavy-ion collisions.

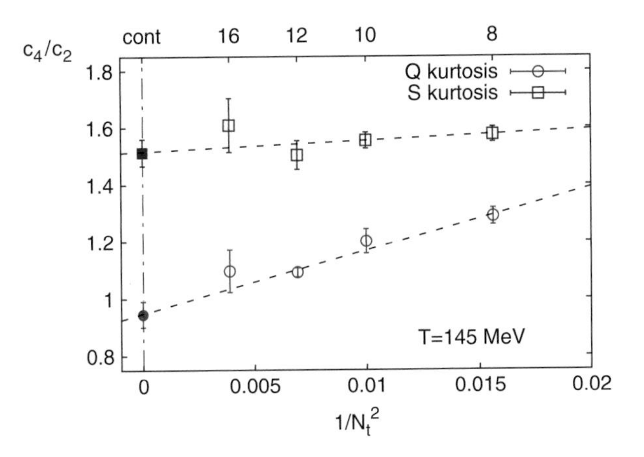

Fig. 9 Continuum limit of the c_4/c_2 (kurtosis × variance) at $T = 145\,\text{MeV}$ temperature for the net charge and strangeness. These data were taken using our new 4stout action

No of nodes	Gflop/node $N_s = 32$	Gflop/node $N_s = 48$
1	16.3	15.4
2	16.8	16.0
4	16.5	16.2
8	16.3	16.3
16	16.3	16.3
32	16.8	16.0
64	17.1	16.5
128	19.2	16.5
256	16.3	16.0

Production Specifics and Performance

Most of our production is done using modest partition sizes, as we found these to be most efficient for our implementation.

Performance

Our code shows nice scaling properties on HERMIT. For our scaling analysis below, we used two lattices ($N_s = 32$ and 48) and several partition sizes up to 256 nodes. We timed the most time consuming part of the code: the fermion matrix multiplication. The results are summarized in Fig. 9.

Production

Given the nice scaling properties of our code, we were able to run at the sweet spot for queue throughput, which we found to be located at a job size of 64 nodes. Larger job sizes proved to have a scheduling probability sufficiently low that benefits in the runtime due to the larger number of cores were compensated and the overall production throughput decreased. We, therefore, opted to stay at jobs sizes with 64 nodes.

Outlook

At the present level of ensemble generation, we believe we will be able to publish within 2013. HERMIT has proved to be an essential tool to be able to achieve this goal.

References

1. S. Borsanyi, G. Endrodi, Z. Fodor, S.D. Katz, S. Krieg, et al., PoS LATTICE2011, 201 (2011)
2. S. Borsanyi, G. Endrodi, Z. Fodor, A. Jakovac, S.D. Katz, et al., J. High Energy Phys. **1011**, 077 (2010). doi:10.1007/JHEP11(2010)077
3. P. Petreczky, PoS LATTICE2012, 069 (2012)
4. S. Borsanyi, S. Durr, Z. Fodor, C. Hoelbling, S.D. Katz, et al., J. High Energy Phys. **1209**, 010 (2012). doi:10.1007/JHEP09(2012)010
5. R. Horsley, et al., Phys. Rev. **D86**, 114511 (2012). doi:10.1103/PhysRevD.86.114511

Unconventional Fractionalization of Strongly Correlated Electrons

A. Moreno, J.M.P. Carmelo, and A. Muramatsu

Abstract While in condensed matter systems the constituents are well known, namely electrons, neutrons and protons, their interplay may give rise to unexpected states of matter. In this contribution we concentrate on strongly correlated electrons in one dimension driven out of equilibrium. This requires in principle, the solution of Schrödinger's equation dealing with a space of states, whose dimension increases exponentially with the number of electrons. Implementing an algorithm that requires only polynomially increasing computational resources, namely the time-dependent density matrix renormalization group (t-DMRG), we show that an electron injected into the system, fractionalizes in several portions, some of them carrying charge but no spin, and others carrying the spin and partial charge, in spite of the electron being an elementary particle in isolation. The characterization of such a fractionalization of charge and spin was made possible by the access to HPC platforms with large memory processors.

A. Moreno
Institut für Theoretische Physik III, Universität Stuttgart, Pfaffenwaldring 57, 70550 Stuttgart, Germany
e-mail: moreno@itp3.uni-stuttgart.de

J.M.P. Carmelo
Center of Physics, University of Minho, Campus Gualtar, 4710-057 Braga, Portugal

Institut für Theoretische Physik III, Universität Stuttgart, Pfaffenwaldring 57, 70550 Stuttgart, Germany

Beijing Computational Science Research Center, 100084 Beijing, China
e-mail: carmelo@fisica.uminho.pt

A. Muramatsu (✉)
Institut für Theoretische Physik III, Universität Stuttgart, Pfaffenwaldring 57, 70550 Stuttgart, Germany

Beijing Computational Science Research Center, 100084 Beijing, China
e-mail: mu@theo3.physik.uni-stuttgart.de

W.E. Nagel et al. (eds.), *High Performance Computing in Science and Engineering '13*, DOI 10.1007/978-3-319-02165-2_6,

1 Introduction

Condensed matter systems are composed of very well known particles, namely electrons, protons and neutrons. Furthermore, their physical properties are determined solely by the Coulomb interaction. In spite of this apparent simplicity, condensed matter systems display an enormous spectrum of physical phenomena, like superconductivity, superfluidity, and magnetism. In recent years, new phenomena like supersolidity [1] and superconductivity in iron compounds [2], and new states of matter like non-abelian states [3], topological insulators [4], or spin liquids [5], together with electronic systems in one dimension [6] became the center of attention in the discussion of exotic forms of matter [7].

In this contribution we concentrate on one-dimensional (1D) quantum systems, where recent experimental advances allowed to access phenomena like spin-charge separation and charge fractionalization [6]. At low energies these systems are well described by the Luttinger Liquid (LL) theory [8] that predicts two independent excitations carrying either only charge (holons) or only spin (spinons) and propagating with different velocities, and hence, spin-charge separation. Experimental evidences of its existence have been observed in quasi-1D organic conductors [9], semiconductor quantum wires [10], and quantum chains on semiconductor surfaces [11]. The LL theory also predicts the fractionalization of injected charge into two chiral modes (left- and right-going) [12], a phenomenon recently confirmed experimentally [13]. Along the experimental advances also theoretical progress was recently achieved pertaining extensions beyond the LL limit by incorporating nonlinearity of the dispersion, leading to qualitative changes in the spectral function [14–17] and relaxation processes of 1D electronic systems [18].

Here we review work where we showed that fractionalization of charge and spin beyond the forms described by LL theory takes place when a spin-1/2 fermion is injected into a strongly correlated 1D system, namely the t-J model [19]. By studying the time evolution of the injected wavepacket at different wavevectors k, using time-dependent density matrix renormalization group (t-DMRG) [20, 21] different regimes were obtained. When k is close to the Fermi wavevector k_F, the known features from LL theory like spin-charge separation and fractionalization of charge into two chiral modes result. On increasing k, a further fractionalization of charge and spin appears, in forms that depend on the strength of the exchange interaction J or the density n. Their dynamics can be understood at the supersymmetric (SUSY) point $J = 2t$ in terms of charge and spin excitations of the Bethe–Ansatz solution [22, 23]. For the region of the phase diagram [24, 25], where the ground state corresponds to a repulsive LL, two qualitatively different regimes are identified: one regime with $v_s > v_c$ and another where $v_s < v_c$. Here $v_{c(s)}$ is the velocity of the excitations mainly carrying charge (spin). For $v_s > v_c$ and $k > k_F$ the spin excitation starts to carry a fraction of charge that increases with k while v_c corresponds to a wavepacket carrying only charge. For $v_s < v_c$ and $k > k_F$ the situation is reversed and the fastest charge excitation carries a fraction of spin that increases with k while the wavepacket with v_s carries almost no charge, i.e. in this case spin fractionalizes.

2 Model and Algorithms

The Hamiltonian of the 1D t-J model is as follows,

$$H = -t \sum_{i,\sigma} \left(\tilde{c}_{i,\sigma}^{\dagger} \tilde{c}_{i+1,\sigma} + \text{h.c.} \right) + J \sum_{i} \left(\mathbf{S}_i \cdot \mathbf{S}_{i+1} - \frac{1}{4} n_i n_{i+1} \right), \tag{1}$$

where the operator $\tilde{c}_{i,\sigma}^{\dagger}$ ($\tilde{c}_{i,\sigma}$) creates (annihilates) a fermion with spin $\sigma = \uparrow, \downarrow$ on the site i. They are not canonical fermionic operators since they act on a restricted Hilbert space without double occupancy. $\mathbf{S}_i = \tilde{c}_{i,\alpha}^{\dagger} \boldsymbol{\sigma}_{\alpha\beta} \tilde{c}_{i,\beta}$ is the spin operator and $n_i = \tilde{c}_{i,\sigma}^{\dagger} \tilde{c}_{i,\sigma}$ is the density operator.

We study the time evolution of a wavepacket, corresponding to a fermion with spin up injected into the ground state, by means of t-DMRG [20, 21], that extends the DMRG method [26, 27] originally developed for the ground-state of quasi-one dimensional systems in equilibrium, to the solution of the time dependent Schrödinger's equation. A review of the original method and later advances can be found in review articles by Schollwöck [28, 29].

The state of a gaussian wavepacket $|\psi\rangle$ centered at x_0, with width Δ_x and average momentum k_0, is created by the operator $\psi_{\uparrow}^{\dagger}$ applied onto the ground state $|G\rangle$:

$$|\psi\rangle \equiv \psi_{\uparrow}^{\dagger} |G\rangle = \sum_{i} \varphi_i \tilde{c}_{i\uparrow}^{\dagger} |G\rangle, \tag{2}$$

with

$$\varphi_i = A e^{-(x_i - x_0)^2 / 2\Delta_x} e^{i k_0 x_i}. \tag{3}$$

A is fixed by normalization. The time evolved state $|\psi(\tau)\rangle$ by the Hamiltonian (1) determines the spin (s) and charge (c) density relative to the ground state as a function of time τ measured in units of $1/t$ ($\hbar = 1$),

$$\rho_{\alpha}(x_i, \tau) \equiv \langle \psi(\tau) | n_{i\alpha} | \psi(\tau) \rangle - \langle G | n_{i\alpha} | G \rangle, \tag{4}$$

where $\alpha = s, c$, $n_{ic} = n_{i\uparrow} + n_{i\downarrow}$, and $n_{is} = n_{i\uparrow} - n_{i\downarrow}$. Most of the numerical results were carried out on systems with $L = 160$ lattice sites, using 600 DMRG vectors (this translates into errors of the order of 10^{-4} in the spin and charge density up to times of $50/t$) and $\Delta_x = 5$ lattice sites (which corresponds to a width $\Delta_k \sim 0.06\pi$ in momentum space).

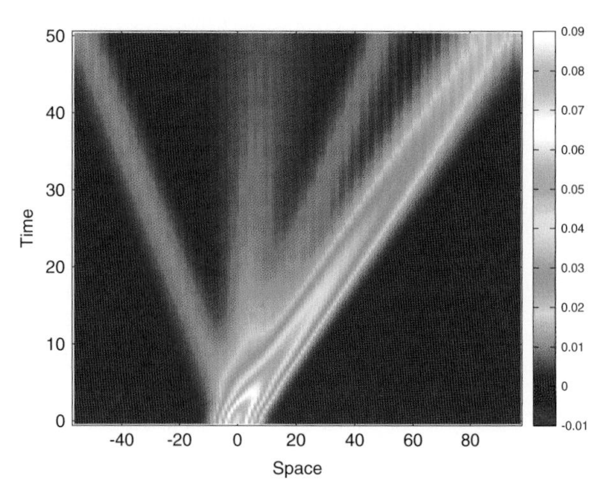

Fig. 1 Time evolution of $\rho_c(x_i, \tau)$ for a wavepacket initially at $x = 0$, with momentum $k = 0.7\pi$, at density $n = 0.6$, and $J = 2t$. Charge fractionalizes into four wavepackets, one to the left (P_1) and the rest (P_2, P_3, P_4) to the right. P_1 and P_3 have the same charge and speed but opposite velocities

3 Results

We discuss first the time evolution of a wavepacket at the SUSY point $J = 2t$, since here we will be able to identify the different portions in which the wavepacket splits on the basis of the Bethe–Ansatz solution.

Figure 1 shows the time evolution of $\rho_c(x_i, \tau)$ for a density of $n = 0.6$ ($n \equiv N/L$, with N the number of fermions). The momentum of the injected fermion is $k = 0.7\pi$, i.e. midway between $k_F = 0.3\pi$ and the zone boundary. The charge (i.e. ρ_c) splits into four fractions, one portion traveling to the left and the rest doing so to the right. A splitting into chiral modes is expected in the frame of LL theory [12], where for an injected right-going fermion, a splitting $Q_\alpha^{(\pm)} = (1 \pm K_\alpha)/2$ (where K_α is the so-called LL parameter and "+" ("-") corresponds to the right (left) propagating part) is predicted. The amount of charge (i.e. the integral of the wavepacket over its extension) corresponding to the portion denoted P_1 is $Q_c^{(-)} \sim 0.1$ which agrees well with the prediction of LL theory, since for the parameters in this case, $K_c \sim 0.8$ [25]. However, at long enough times, a further splitting of the right-going charge is observed (wavepackets P_i with $i = 2, 3, 4$), beyond the prediction of the LL theory.

Figure 2 displays both $\rho_c(x_i, \tau)$ (full line) and $\rho_s(x_i, \tau)$ (dashed line) for different values of the initial momentum of the injected wavepacket. The arrows indicate the direction of motion of each packet. As opposed to ρ_c, ρ_s does not split. In an SU(2) invariant system $K_s = 1$ [8], and hence $Q_s^{(-)} = 0$, i.e. no left propagating part is expected for the spin density. Moreover, part of the charge (P_4) is accompanying the spin, such that spin-charge separation does not appear to be complete. The amount of charge accompanying the spin increases as the momentum of the injected fermion approaches the zone boundary. These results make already evident that injecting a fermion at a finite distance from the Fermi energy leads to fractionalization of charge beyond the expectations from the LL theory.

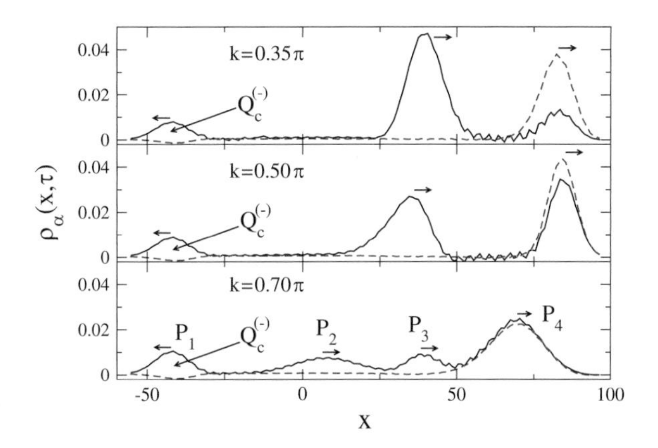

Fig. 2 Charge ($\rho_c(x_i, \tau)$, *full line*) and spin ($\rho_s(x_i, \tau)$, *dashed line*) densities for $J = 2t, n = 0.6$, at time $\tau = 40$, for different values of the momentum of the injected fermion

In order to understand the new forms of fractionalization that go beyond the LL frame, we consider the excitations corresponding to one-particle addition processes, whose energies can be obtained from the Bethe–Ansatz solution [16, 17, 30]. When adding an electron with momentum k, the single particle excitation energy is given by $\omega(k) = -\epsilon_c(q_c) - \epsilon_s(q_s)$, where $\epsilon_c(q_c)$ and $\epsilon_s(q_s)$ are the dispersion relations of the excitations for charge and spin, respectively, and the momenta are related to the momentum of the incoming particle as follows: $k = \pm 2k_F - q_c - q_s$, where $q_c \in [-q_{Fc}, q_{Fc}]$, and $q_s \in [-q_{Fs}, q_{Fs}]$, with $q_{Fc} = \pi - 2k_F$ and $q_{Fs} = \pi - k_F$ the pseudo-Fermi momenta for the excitations for charge and spin, respectively [23].

Figure 3 displays the velocities obtained from t-DMRG for the different wavepackets (symbols) compared to those obtained from Bethe–Ansatz (full lines), as a function of the momentum of the injected fermion. The velocity of each P_i is extracted by measuring the position of the maximum of the packet at the most convenient time, i.e, at that time where we can resolve P_i and the spreading of one packet does not destroy the other packets. The wavepackets P_1 (triangles) and P_3 (squares) have opposite directions, but the same speed and charges $Q_c^{(-)} \simeq Q_{c3}^{(+)}$, where the charges for the (+) branch are labelled by an index corresponding to the respective wavepackets. The velocity of the wavepacket P_4 (circles) agrees almost perfectly with the one corresponding to spin excitations. Its determination is best since it is the fastest wavepacket, such that it can be easily discerned from the rest. The velocity of the remaining wavepacket, P_2 (diamonds), is more difficult to assess, since it overlaps at the beginning with other ones. Nevertheless, its velocity closely follows the one of charge excitations. The wavepackets just described deliver a direct visualization of the excitations appearing in the Bethe–Ansatz solution, where only two different kind of particles are involved: the c and s pseudoparticles with their associated bands. The excitation associated with spin involves one hole in the c band with fixed momentum q_{Fc} and one hole in the s band with momentum q_s, where $q_s = \pm 2k_F - q_{Fc} - k$ [23]. In fact, the velocities of P_1 and P_3 correspond to

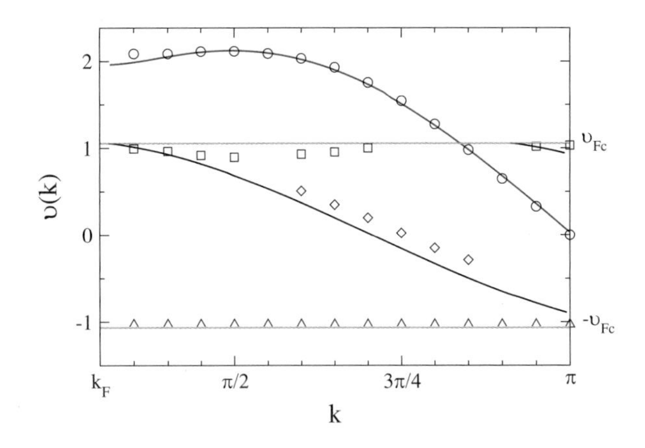

Fig. 3 *Full lines*: derivatives $v_\alpha(k) = \partial\epsilon_\alpha(k)/\partial k$ of the dispersions obtained by the Bethe–Ansatz solution. The *symbols* correspond to the velocities of the different wavepackets identified in Fig. 2: *triangles* (P_1), *diamonds* (P_2), *squares* (P_3) and *circles* (P_4). The *orange horizontal lines* stand for the Fermi velocity $v_{Fc} = \partial\epsilon_c(q_{Fc})/\partial q$ given by the Bethe–Ansatz solution

the group velocity at both pseudo-Fermi momenta $\pm q_{Fc}$. Furthermore, as shown in Fig. 3, the velocity of those fractions is independent of the momentum of the injected fermion, in agreement with the picture given by Bethe–Ansatz. The dispersion of the hole in the s-band gives rise to the velocity displayed by the red line in Fig. 3. Similarly, the c line (black line in Fig. 3) involves one hole in the s band with fixed momentum q_{Fs} and one hole in the c band with momentum q_c determined in terms of k by $q_c = \pm 2k_F - q_{Fs} - k$. Using the same argument as for the s line we can associate the P_2 packet (diamonds) with the c pseudoparticle. However, in this case we cannot observe wavepackets associated with spin and velocities corresponding to the group velocity at the pseudo-Fermi momenta $\pm q_{Fs}$. We understand this as due to the fact that $K_s = 1$, by analogy to what we observe in the $K_c = 1$ case. On the SUSY point this case is reached in the limit of vanishing density, where the system can be described by a Fermi gas. Hence, fractionalization is absent in this limit.

Next we depart from the SUSY point and examine how fractionalization takes place in the region of the phase diagram where the ground state corresponds to a LL with $K_c < 1$. As shown in Fig. 4, essentially the same features are observed as at the SUSY point both for $J > 2t$ and $J < 2t$. In all the cases shown in Fig. 4, where the velocity of spin excitations (v_s) remains higher than that of charge excitations (v_c) in most parts of the Brillouin zone, spin does not fractionalize, as opposed to charge, so that the interpretation derived from Bethe–Ansatz remains valid over an extended region of the phase diagram: charge splits into four portions of which one travels with the spin wavepacket, and two have the same speed but opposite group velocity which does not depend on the momentum of the injected fermion. It is tempting to assign those excitations to states at a pseudo-Fermi surface for charge excitations.

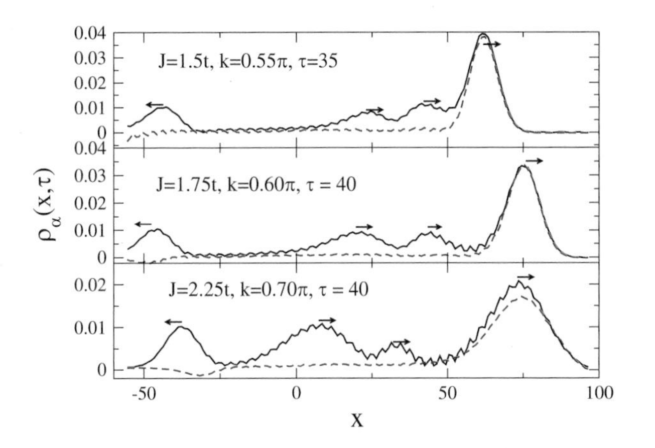

Fig. 4 Fractionalized wavepackets for different values of J/t away from the SUSY point, at a density $n = 0.6$. As in the SUSY case, charge fractionalizes into four pieces, while spin does not, and carries an appreciable amount of charge

In summary, we have shown through the time evolution of an injected spin-full fermion onto the t-J model, that charge and spin fractionalization occurs beyond the predictions of the Luttinger liquid theory. A comparison with results from Bethe–Ansatz allowed to identify charge and spin excitations that split into components at high and low energies. The components at high energy reveal the dispersion ϵ_c and ϵ_s of charge and spin excitations, respectively. The components at low energy have a velocity that does not depend on the momentum of the injected fermion and are very well described by states at the pseudo-Fermi momenta of the charge excitation. This picture can be extended to a wide region in the phase diagram of the t-J model as long as the ground state corresponds to $K_c < 1$, and $v_s > v_c$. In this region fractionalization is observed only in the charge channel. However, for $v_c > v_s$, a region that develops for J/t below ~ 1.5, the spin density shows fractionalization [19]. All over, the fastest excitation is accompanied by the complementary one, such that spin-charge separation is for them only partial. The other fractions present an almost complete spin-charge separation.

Acknowledgements A. Moreno and A. Muramatsu acknowledge support by the DFG through SFB/TRR 21. A. Muramatsu and J.M.P. Carmelo thank the hospitality and support of the Beijing Computational Science Research Center, where part of the work was done. J.M.P. Carmelo thanks the hospitality of the Institut für Theoretische Physik III, Universität Stuttgart, and the financial support by the Portuguese FCT both in the framework of the Strategic Project PEST-C/FIS/UI607/2011 and under SFRH/BSAB/1177/2011 , the German transregional collaborative research center SFB/TRR21, and Max Planck Institute for Solid State Research. A.M. thanks the KITP, Santa Barbara, for hospitality. This research was supported in part by the National Science Foundation under Grant No. NSF PHY11-25915. We are very grateful to HLRS (Stuttgart) and NIC (Jülich) for providing the necessary supercomputer resources.

References

1. S. Balibar, The enigma of supersolidity. Nature **464**, 176–182 (2010)
2. I. Mazin, Superconductivity gets an iron boost. Nature **464**, 183–186 (2010)
3. A. Stern, Non-Abelian states of matter. Nature **464**, 187–193 (2010)
4. C. Moore, The birth of topological insulators. Nature **464**, 194–198 (2010)
5. L. Balents, Spin liquids in frustrated magnets. Nature **464**, 199–208 (2010)
6. V.V. Deshpande, M. Bockrath, L.I. Glazman, A. Yacoby, Electron liquids and solids in one dimension. Nature **464**, 209–216 (2010)
7. D. Csontos, Exotic matter. Nature **464**, 175 (2010)
8. T. Giamarchi, *Quantum Physics in One Dimension* (Clarendon, Oxford, 2004)
9. T. Lorenz, M. Hofmann, M. Gruninger, A. Freimuth, G.S. Uhrig, M. Dumm, M. Dressel, Nature **418**, 614 (2002)
10. O.M. Auslaender, H. Steinberg, A. Yacoby, Y. Tserkovnyak, B.I. Halperin, K.W. Baldwin, L.N. Pfeiffer, K.W. West, Science **308**, 88 (2005)
11. C. Blumenstein, J. Schäfer, S. Mietke, A. Dollinger, M. Lochner, X.Y. Cui, L. Patthey, R. Matzdorf, R. Claessen, Nat. Phys. **7**, 776 (2011)
12. K.-V. Pham, M. Gabay, P. Lederer, Phys. Rev. B **61**, 16397 (2000)
13. H. Steinberg, G. Barak, A. Yacoby, L.N. Pfeiffer, K.W. West, B.I. Halperin, K.L. Hur, Nat. Phys. **4**, 116 (2008)
14. A. Imambekov, L.I. Glazman, Science **323**, 228 (2009)
15. A. Shashi, L.I. Glazman, J.-S. Caux, A. Imambekov, Phys. Rev. B **84**, 045408 (2011)
16. J.M.P. Carmelo, K. Penc, D. Bozi, Nucl. Phys. B **725**, 421 (2005)
17. J.M.P. Carmelo, K. Penc, D. Bozi, Nucl. Phys. B **737**, 351 (2006)
18. G. Barak, H. Steinberg, L.N. Pfeiffer, K.W. West, L. Glazman, F. von Oppen, A. Yacoby, Nat. Phys. **6**, 489 (2010)
19. A. Moreno, A. Muramatsu, J.M.P. Carmelo, Phys. Rev. B **87**, 075101 (2013)
20. S.R. White, A.E. Feiguin, Phys. Rev. Lett **93**, 076401 (2004)
21. A.J. Daley, C. Kollath, U. Schollwöck, G. Vidal, J. Stat. Mech. Theor. Exp. P04005 (2004)
22. P.A. Bares, G. Blatter, M. Ogata, Phys. Rev. B **44**, 130 (1991)
23. P.A. Bares, J.M.P. Carmelo, J. Ferrer, P. Horsch, Phys. Rev. B **46**, 14624 (1992)
24. M. Ogata, M. Luchini, S. Sorella, F. Assaad, Phys. Rev. Lett **66**, 2388 (1991)
25. A. Moreno, A. Muramatsu, S.R. Manmana, Phys. Rev. B **83**, 205113 (2011)
26. S.R. White, Phys. Rev. Lett **69**, 2863 (1992)
27. S.R. White, Phys. Rev. B **48**, 10345 (1993)
28. U. Schollwöck, Rev. Mod. Phys. **77**, 259 (2005)
29. U. Schollwöck, Ann. Phys. **326**, 96 (2011)
30. J.M.P. Carmelo, L.M. Martelo, K. Penc, Nucl. Phys. B **737**, 237 (2006)

Numerically-Exact Schrödinger Dynamics of Closed and Open Many-Boson Systems with the MCTDHB Package

Axel U.J. Lode, Kaspar Sakmann, Rostislav A. Doganov, Julian Grond, Ofir E. Alon, Alexej I. Streltsov, and Lorenz S. Cederbaum

Abstract The quantum many-body dynamics of indistinguishable, interacting particles are described by the time-dependent many-body Schrödinger equation (TDSE). The TDSE constitutes a difficult problem and is not solvable analytically in most cases. The present review article expedites and benchmarks the capabilities of a novel theoretical method, the multiconfigurational time-dependent Hartree method for bosons (MCTDHB) that is designed for solving the TDSE of interacting bosonic particles. The MCTDHB package is a software implementation that solves the equations of motion of MCTDHB numerically. It is assessed with a benchmark versus an analytically treatable model of trapped interacting bosons (a closed system) that the MCTDHB is capable of solving the TDSE for bosons numerically exactly, i.e., to any desired numerical precision. Furthermore, the structure and parallelization of the MCTDHB package as well as an application to the tunneling to open space dynamics [Proc. Natl. Acad. Sci. USA, **109**, 13521 (2012)] are discussed.

A.U.J. Lode (✉) · J. Grond · A.I. Streltsov · L.S. Cederbaum
Theoretische Chemie, Physikalisch-Chemisches Institut, Universität Heidelberg,
Im Neuenheimer Feld 229, 69120 Heidelberg, Germany
e-mail: Axel.Lode@pci.uni-heidelberg.de

K. Sakmann
Department of Physics, Stanford University, Stanford, CA 94305, USA

O.E. Alon
Department of Physics, University of Haifa at Oranim, Tivon 36006, Israel

R.A. Doganov
Department of Physics, Özyilmaz Group, National University of Singapore, S13-02-13,
2 Science Drive 3, 117542, Singapore

W.E. Nagel et al. (eds.), *High Performance Computing in Science and Engineering '13*, DOI 10.1007/978-3-319-02165-2_7,

1 Introduction and Theory

The first experiments with trapped Bose-Einstein condensates (BECs) made of ultracold atoms [1–3] have boosted an extremely vivid field in the intersection of atomic, molecular, optical and condensed-matter physics [4–7]. The present work reports on some of our recent theoretical activity in this lively field. Recently we developed a numerical method to tackle the time-dependent quantum mechanical many-body problem of interacting ultracold bosonic particles [8, 9]. The method is termed the multiconfigurational time-dependent Hartree for bosons (MCTDHB) method. In contrast to other methods used in the field, such as the Gross-Pitaevskii mean-field theory and the Bose-Hubbard model, it is in principle and in practice [10, 11] numerically-exact. Thus, one can obtain numerically-exact solutions if the number of orbitals, i.e., time-dependent variationally optimized basis functions, is large enough. For the quantum many-boson problem, the governing equation is the time-dependent many-body Schrödinger equation (TDSE),

$$i\partial_t \Psi(\mathbf{r}_1,\dots,\mathbf{r}_N,t) = \hat{H}\Psi(\mathbf{r}_1,\dots,\mathbf{r}_N,t). \tag{1}$$

Here, $\mathbf{r}_1,\dots,\mathbf{r}_N$ denote the coordinates of the N bosons, Ψ is the N-boson wavefunction, t is time and the Hamiltonian $\hat{H}$ is defined as follows:

$$\hat{H} = \sum_i^N \underbrace{(-\frac{1}{2m}\nabla^2_{\mathbf{r_i}} + \hat{V}(\mathbf{r_i}))}_{\hat{h}(\mathbf{r_i})} + \sum_{i<j}^N W(\mathbf{r_i}-\mathbf{r_j}). \tag{2}$$

Here, $\hat{h}$ is the one-body Hamiltonian containing the kinetic energy $-\frac{1}{2m}\nabla^2_{\mathbf{r_i}}$ and the potential energy operator $\hat{V}(\mathbf{r_i})$ for each boson and $\hat{W}$ is the interparticle interaction operator. To solve the TDSE, Eq. (1), as an initial value problem, means to give an initial many-body wavefunction $\Psi(t=0)$ and to find its evolution in time.

Because the standard methods such as configuration interaction (which employ time-independent basis functions) are not practical for particle numbers $N \gtrsim 5$ we developed the MCTDHB approach which are capable of accurate calculations with several thousands of atoms [12–14]. We tackle the TDSE, Eq. (1), with an ansatz that consists of expanding the wavefunction as a weighted sum of all possible configurations of N particles distributed among M time-dependent basis functions,

$$\Psi(t) = \sum_{\{\mathbf{n}\}} C_{\mathbf{n}}(t)|\mathbf{n},t\rangle = \Psi[\Phi_i(\mathbf{r},t),\dots,\Phi_M(\mathbf{r},t),\{C_{\mathbf{n}}\},t]. \tag{3}$$

Here, $\mathbf{n} = (n_1,\dots,n_M)$ denotes a single configuration which is weighted with the coefficient $C_{\mathbf{n}}$. The M time-dependent basis functions, $\Phi_1,\dots,\Phi_M$, are used to construct the time-dependent configuration (permanent) $|\mathbf{n},t\rangle$. The total number of possible configurations and the number of coefficients $C_{\mathbf{n}}(t)$ weighting them is given by the combinatorial relation $N_{\mathrm{conf}} = \binom{N+M-1}{N}$.

In order to derive the equations-of-motion (EOM) we formulate an action functional and use the principle of least action (often referred to as time-dependent or Dirac-Frenkel variational principle). The ansatz depends on the orbitals $\{\Phi_i(\mathbf{r},t), i = 1,\ldots,M\}$ and the coefficients $\{C_{\mathbf{n}}(t)\}$. These are the variational parameters as indicated in the second equality in Eq. (3). After somewhat lengthy but otherwise straightforward calculations (see [8, 9] for details) we obtain the following sets of M coupled non-linear integro-differential orbital and $\binom{N+M-1}{N}$ linear coefficient EOM (without loss of generality, here prescribed for contact inter-particle interaction):

$$i\partial_t|\Phi_j\rangle = \hat{P}\left[\hat{h}|\Phi_j\rangle + \lambda_0 \sum_{k,s,q,l=1}^{M} \{\boldsymbol{\rho}\}^{-1}_{jk}\rho_{ksql}\Phi_s^*\Phi_l|\Phi_q\rangle\right], \tag{4}$$

$$i\partial_t\mathscr{C}(t) = \mathscr{H}(t)\mathscr{C}(t). \tag{5}$$

Here, $\hat{P} = 1 - \sum_{k=1}^{M}|\Phi_k\rangle\langle\Phi_k|$ is a projector ensuring the orthogonality of the time derivative of an orbital to all the orbitals, ρ_{kq} and ρ_{kqsl} are the reduced one- and two-body density-matrix elements, $\mathscr{C}$ denotes the vector containing all the coefficients $\{C_{\mathbf{n}}\}$ and $\mathscr{H}$ is the Hamiltonian-matrix with the elements $\mathscr{H}_{\mathbf{nn}'} = \langle\mathbf{n},t|\hat{H}|\mathbf{n}',t\rangle$. To solve these EOM, i.e., Eqs. (4) and (5), we implemented a two step integration scheme in a Fortran application package termed the MCTDHB Package (current version is 2.3, 2013) [15]. The description of the MCTDHB Package is the topic of the following section.

2 Implementation and Parallelization

2.1 Algorithm

The problem set we aim to attack numerically [cf. Eqs. (4) and (5)] is actually twofold: The N_{conf} equations for the coefficients are linear in time, but the coefficients depend on the orbitals via $\mathscr{H}$. The M equations for the orbitals are nonlinear integro-differential equations because each of them contains the matrix elements ρ_{kqsl} and ρ_{kq} which are functions of the coefficients. Importantly, each of the latter contains integrals of different combinations of *all* the orbitals. Furthermore, for a general particle–particle interaction $\hat{W}(\mathbf{r}_i - \mathbf{r}_j)$, see Eq. (2), the integrals $W_{ksql} = \int\int \Phi_k^*(\mathbf{r},t)\Phi_s^*(\mathbf{r}',t)\hat{W}(\mathbf{r}-\mathbf{r}')\Phi_q(\mathbf{r},t)\Phi_l(\mathbf{r}',t)d\mathbf{r}d\mathbf{r}'$ have to be evaluated at each time step.

Technically, the set of the expansion coefficients is represented as a one-dimensional array of complex variables. The size of this array is equal to the total number of the configurations $N_{\text{conf}} = \binom{N+M-1}{N}$. Each element of this array contains one expansion coefficient. The orbital Φ_i is a one-dimensional complex array containing the values of Φ_i on a one-, two- or three-dimensional grid of the total size $N_{\text{dvr}} = N_x N_y N_z$. The two sets of EOM [see Eqs. (4) and (5)] constitute a system of coupled integro-differential equations, which we integrate using the following two-step scheme for a timestep $\Psi(0) \rightarrow \Psi(\tau)$.

1. Evaluate the matrix elements $h_{kq}(t = 0), W_{kqsl}(t = 0), \rho_{kq}(t = 0), \rho_{kqsl}(t = 0)$ with the initial condition $\Psi(0)$.
2. Propagate $\{C_{\mathbf{n}}(t = 0)\} \rightarrow \{C_{\mathbf{n}}(t = \frac{\tau}{2})\}$ using the short iterative Lanczos (SIL) method. Use the $\{C_{\mathbf{n}}(t = \frac{\tau}{2})\}$ and evaluate $\rho_{kqsl}(\frac{\tau}{2})$ and $\rho_{kq}(\frac{\tau}{2})$.
3. Obtain the next set of orbitals $\{\Phi_i(\mathbf{r}, t = \frac{\tau}{2}), i = 1, \ldots, M\}$ using the Adams-Bashfourth-Moulton predictor corrector integrator (ABM), the Bulirsch-Stoer algorithm (BS), the Runge-Kutta integrator of fifth or eighth order (RK5/8), or an implementation of a Gear-type second order backwards differentiation formula (ZVODE).
4. Propagate $\{\Phi_i(\mathbf{r}, t = 0), i = 1, \ldots, M\} \rightarrow \{\Phi_i^{'}(\mathbf{r}, \frac{\tau}{2}), i = 1, \ldots, M\}$ with ABM/BS/RK5/RK8/ZVODE using $\rho_{kq}(t = 0), \rho_{kqsl}(t = 0)$ for the determination of the accumulated error by comparing the $\{\Phi_i^{'}(\mathbf{r}, t = \frac{\tau}{2}), i = 1, \ldots, M\}$ with $\{\Phi_i(\mathbf{r}, t = \frac{\tau}{2}), i = 1, \ldots, M\}$ obtained in step 3. If the error is bigger than a predetermined threshold then decrease τ and restart at step 2.
5. In the next step, propagate $\{\Phi_i(\mathbf{r}, \frac{\tau}{2}), i = 1, \ldots, M\} \rightarrow \{\Phi_i(\mathbf{r}, \tau), i = 1, \ldots, M\}$ using ABM/BS/RK5/RK8/ZVODE and the matrix elements $\rho_{kqsl}(t = \frac{\tau}{2})$ and $\rho_{kq}(t = \frac{\tau}{2})$ obtained in step 2. From the orbitals at $t = \tau$ the integral quantities $h_{kq}(t = \tau), W_{kqsl}(t = \tau)$ are obtained.
6. With the $W_{kqsl}(t = \tau)$ and $h_{kq}(t = \tau)$ of step 5 and SIL propagate $\{C_{\mathbf{n}}(t = \frac{\tau}{2})\} \rightarrow \{C_{\mathbf{n}}(t = \tau)\}$.
7. Back-propagate $\{C_{\mathbf{n}}(t = \frac{\tau}{2})\} \rightarrow \{C_{\mathbf{n}}^{'}(t = 0)\}$ with SIL using $h_{kq}(t = \tau), W_{kqsl}(t = \tau)$ for the estimation of the error due to time discretization obtained from the comparison of $C_{\mathbf{n}}(t = 0)$ with $C_{\mathbf{n}}^{'}(t = 0)$. If the error is bigger than a predetermined threshold then decrease τ and restart at step 2.

2.2 *Implementation and Parallelization Details*

Our implementation of the described algorithm, the aforementioned MCTDHB Package, relies on a hybrid OpenMP[1] and MPI[2] parallelization. In the elementary integration step (see previous section) the computationally most demanding task is the evaluation of the right-hand sides of the EOM, Eqs. (4) and (5).

For the time propagation of the orbitals the ABM/BS/RK5/RK8/ZVODE needs to evaluate the right-hand side of the orbitals' EOM (4). For this purpose, we broadcast the orbitals (a complex array of the dimension $M \cdot N_{\mathrm{dvr}} = M \cdot N_x N_y N_z$) to at most a number of MPI processes proportional to M^4. The evaluation of the right-hand side of Eq. (4) is done using OpenMP directives. The distribution of the computational task is problem-size adaptive. If the problem is small, the communication overhead for the calculation of the action of the one-body part

[1]Open **M**ulti-**P**rocessing.

[2]**M**essage **P**assing **I**nterface.

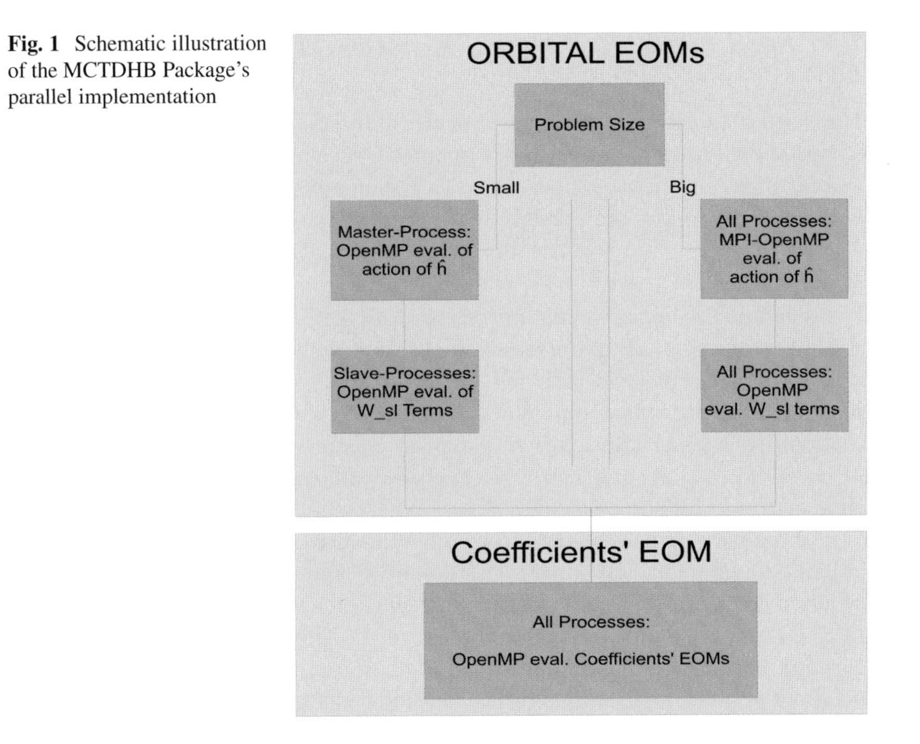

Fig. 1 Schematic illustration of the MCTDHB Package's parallel implementation

of the Hamiltonian, $\hat{h}$, is big. Hence, it is beneficial not to use a distributed memory parallelization with MPI and rely on shared memory parallelization with OpenMP instead. For bigger problems the evaluation of the action of the one-body Hamiltonian, $\hat{h}$, is also done in distributed memory. In both, small and big problem sets, the evaluation of the local interaction potentials, $\hat{W}_{sl}$, is MPI-parallelized, see Fig. 1. It is appropriate to state here, that the categorization of what *small* or *big* problem actually means does of course depend on the architecture. In the case of the NEC Nehalem machine Laki and the Cray XE6 machine Hermit the border was at a spatial grid size of the order of 10^3. Finally, after the evaluation of the right hand side, we collect the results using the collective MPI_GatherV operation.

The basic step for the time propagation of the coefficients EOM (5) using SIL is the successive application of the Hamiltonian $\hat{H}$ to the vector of expansion coefficients $C_{\mathbf{n}}$ utilizing a recently developed mapping algorithm for this operation (see [12,16]). We broadcast the vector of expansion coefficients—the complex array of dimension $\binom{N+M-1}{N}$—to all MPI processes. Each process computes the action of a set of Hamiltonian terms on the vector of expansion coefficients (see [16]). This evaluation is further parallelized using OpenMP directives. The required total action of the Hamiltonian is then obtained by a collective MPI_Reduce operation. For a schematic illustration of the MCTDHB Package's parallel implementation see Fig. 1.

2.3 *Applications' Scaling and Resources' Utilization*

The scaling of the program's performance is problem-dependent, i.e., different computational tasks require a different number of orbitals M, and are performed with a different number of particles N. The number of the non-linear integro-differential EOM for the orbitals (4) is M, where the size of each orbital is $N_{\mathrm{dvr}} = N_x N_y N_z$. The equation for the coefficients (4) is linear, but the number of coefficients, $N_{\mathrm{conf}} = \binom{N+M-1}{N}$, may be very large (several tens of millions).

For problems with dominating orbital part, i.e., a big number M or N_{dvr}, a number of up to M^4 MPI processes has a good scaling. For problems with a dominating coefficients part, the scaling is good up to an order of M^2 MPI processes. The necessary communication limits the scaling behavior in both cases. Considering a case with a few millions of coefficients (i.e., $N = 3{,}003$ and $M = 3 \Rightarrow N_{\mathrm{conf}} = 4{,}513{,}510$; see [12]), we still obtain a almost linear scaling with up to ten MPI processes.

The memory demands range from a few hundred MB to a few GB per MPI process, i.e., per NUMA[3]-node. The described knowledge and experience was acquired doing calculations, benchmarking and optimizations on the bwGRiD [17] facilities as well as our project on the Cray XE6 Hermit and the NEC-Nehalem cluster Laki. The total consumption of CPU-core hours of this test project was on the order of 400,000 h per month. The developments made during our project lead to important publications, see, e.g. [11, 13] which will be described in more details in Sects. 3 and 4.

For the systems we are researching, we use a Fortran compiler and the Intel Math Kernel Library (IMKL). To tackle highly non-equilibrium dynamics of many particles, with a bigger number of coefficients, a bigger grid, and more orbitals we have further increased the performance and the parallelization schemes of our MCTDHB Package, utilizing available performance analysis software (Scalasca, Tau, Threadspotter, Vampir, etc.) and, of course, the computational resources of the Cray XE6 Hermit and the NEC Nehalem cluster Laki at the HLRS in Stuttgart.

3 Application I: Closed System: Benchmarking the MCTDHB Package with the Harmonic-Interaction Model

When dealing with many-body methods, theoretically and numerically, it is essential to benchmark them against solvable many-body models. This is an important step in proving the usefulness and potential of the tools at hand to a larger community.

---

[3] **N**on-**U**niform **M**emory **A**ccess.

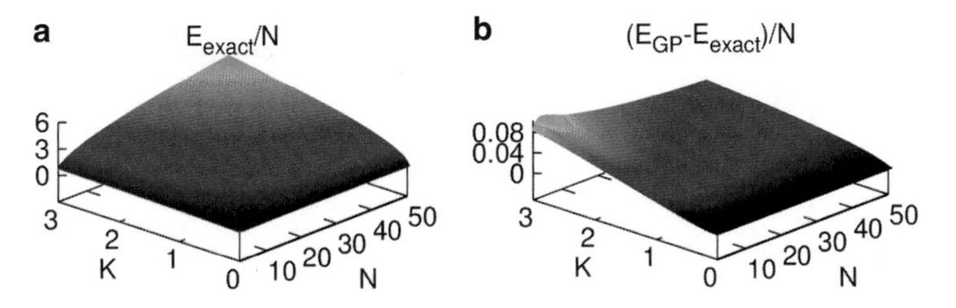

Fig. 2 Mean-field versus exact analytical many-body ground-state energies in the harmonic-interaction model (HIM). Here N is the number of bosons and K the inter-particle interaction (trap frequency ω, particle mass m, and $\hbar$ are all equal to 1). See [11], and references therein, for more details. All quantities are dimensionless

Within this project, we have taken the task of benchmarking the MCTDHB Package with the harmonic-interaction model (HIM) [18, 19] which, importantly for our needs, has also been extended to include time-dependency, see [11] and references therein for details.

The HIM is a system where identical bosons are trapped in a harmonic potential and all interact with a harmonic two-particle potential. Both attractive and repulsive bosons can be treated analytically-exactly within the HIM. The key point in being exactly solvable is the transformation to a set of center-of-mass and relative coordinates which renders the problem (in these coordinates) separable. Furthermore, both the mean-field and many-body ground-states can be obtained analytically, thereby providing a straightforward way to quantify the performance of the mean-field solution directly against the many-body one, see Fig. 2 for a visual representation.

We now move to benchmarking the MCTDHB Package. Since the MCTDHB treats identical bosons, the solution of the HIM problem is done numerically in the laboratory frame where all bosons (degrees of freedom) are coupled. The convergence of the ground-state energy of up to $N = 50$ interacting bosons is documented in Fig. 3.

The next and crucial step is to benchmark the MCTDHB Package for a time-dependent scenario. Here, within the HIM, we have considered a scenario where the many-boson system is driven by a time-dependent function that makes both the trap potential and the inter-particle interaction time-dependent quantities. It is possible to chose a time-dependent function such that only the center-of-mass of the system carries the system's time-dependency, which increases the numerical resources needed (recall that on the computer the many-boson Schrödinger equation is solved in the laboratory frame).

We would like to demonstrate here that the numerical time-dependent many-boson wavefunction converges to the exact one, and we do so by computing and checking the convergence of the eigenvalues (occupation numbers) of its reduced one-particle density matrix $\boldsymbol{\rho}$. The results of one such numerical experiment are collected in Fig. 4, where convergence is readily seen. This concludes our account of benchmarking the MCTDHB Package, for more details see [11].

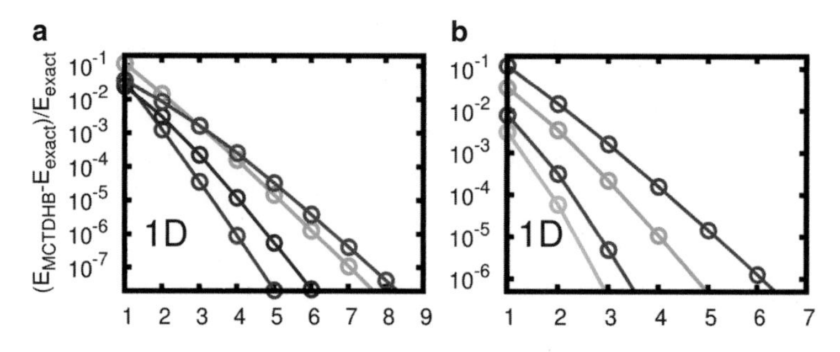

Fig. 3 Numerical convergence to the exact analytical ground-state energies in the harmonic-interaction model (HIM). Shown is the relative difference of the MCTDHB energy and the exact energy, $(E_{\text{MCTDHB}} - E_{\text{exact}})/E_{\text{exact}}$, for the one-dimensional HIM as a function of the number M of self-consistent orbitals. In panel (**a**) the *red line* is for $N = 2, K = 0.5$, the *blue line* for $N = 10, K = 0.5$, the *green line* for $N = 2, K = 2.0$ and the *magenta line* for $N = 10, K = 2.0$, respectively. The relative difference is decreasing exponentially with M. Panel (**b**) depicts the convergence with M for different particle numbers where the "mean-field" ratio $K/(N-1)$ is kept constant for $N = 2$ (*red line*), $N = 5$ (*green line*), $N = 10$ (*magenta line*) and $N = 50$ (*turquoise line*) particles. In this case the pace of the convergence improves when increasing the particle number. See [11], for more details. All quantities shown are dimensionless

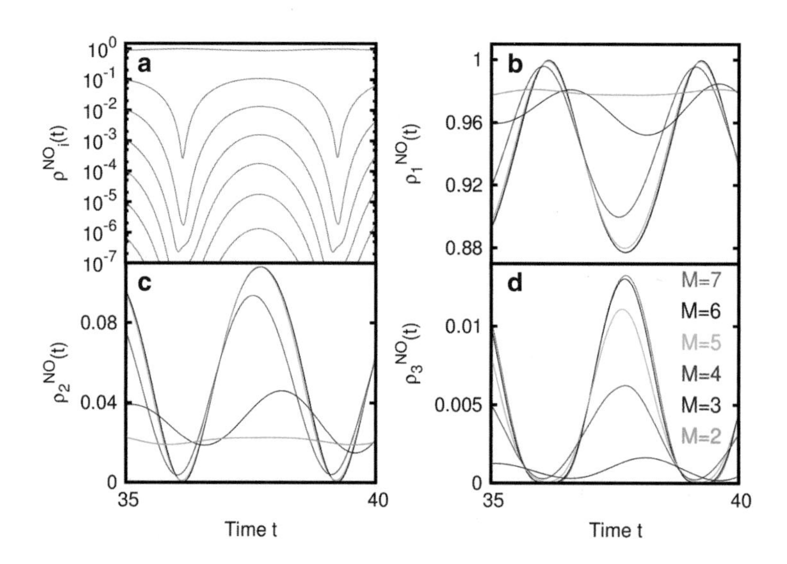

Fig. 4 Numerical convergence to the exact occupation numbers for a driven scenario in the harmonic-interaction model (HIM). Panel (**a**) shows all occupations of the converged $N = 10, K = 0.5$ time-dependent study with $M = 7$ time-adaptive orbitals. Panels (**b**), (**c**) and (**d**) show how the time-evolution of the first three occupation numbers $\rho_1^{NO}(t)$, $\rho_2^{NO}(t)$ and $\rho_3^{NO}(t)$ converges for increasing number of time-adaptive orbitals M. The *color code* indicated by the labels in panel (**d**) holds also for panels (**b**) and (**c**). See [11], for more details. All quantities shown are dimensionless

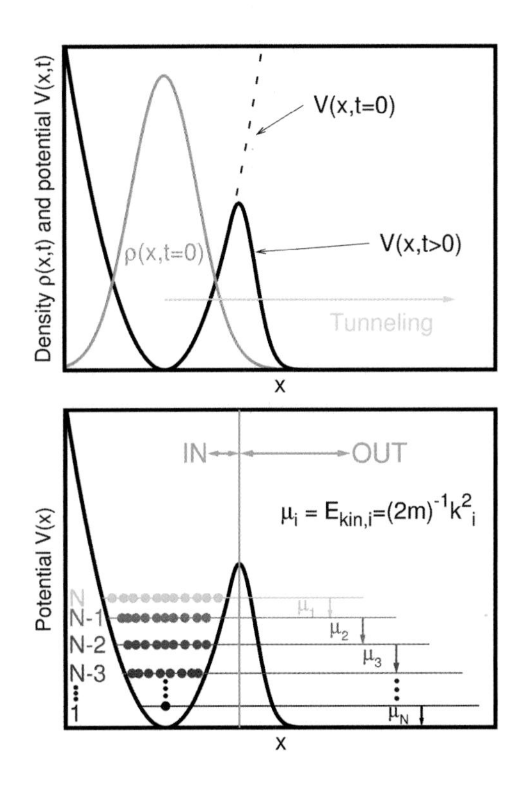

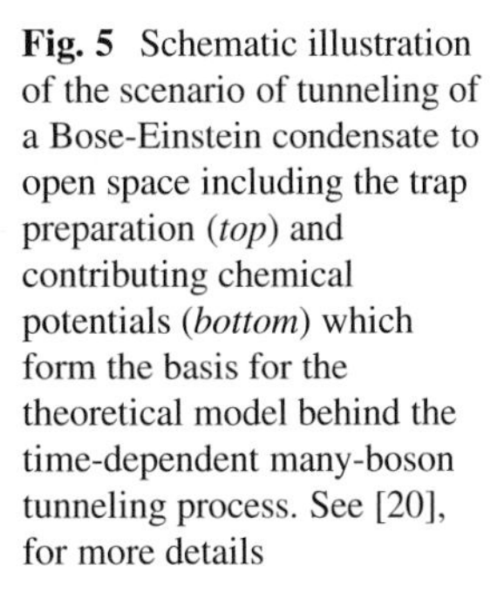
Fig. 5 Schematic illustration of the scenario of tunneling of a Bose-Einstein condensate to open space including the trap preparation (*top*) and contributing chemical potentials (*bottom*) which form the basis for the theoretical model behind the time-dependent many-boson tunneling process. See [20], for more details

4 Application II: Open System: Many-Body Decay by Tunneling to Open Space of a Bose-Einstein Condensate

As an application showing the great advantage that the Cray Hermit and NEC Laki machines have brought to our research we discuss the many-body decay by tunneling of a Bose-Einstein condensate to open space, which has recently been published in [20], also see the press release at, e.g., [21].

The scenario we consider is as the illustration in Fig. 5 (top) shows: We consider a weakly-interacting Bose-Einstein condensate prepared in the ground state of a one-dimensional harmonic potential. At $t > 0$ the trap suddenly opens up and the *interacting* particles are allowed to tunnel to open space. We actually looked in the problem for a much smaller system (number of particles and gird length) in the past [22], before we could benefit from Laki's and Hermit's computational capabilities. In turn we had to use non-unitary time-evolution in [22] (absorbing-boundary conditions) which made the analysis of the results (particularly of many-body properties) significantly more involved. Now, with the help of Hermit and Laki, we have taken a grid large enough to sustain and investigate this intricate non-equilibrium many-body process within the standard unitary time-evolution. It is appropriate here to state the extent of the quantum system's considered in our work [20]: we exactly described the many-body quantum dynamics of a system

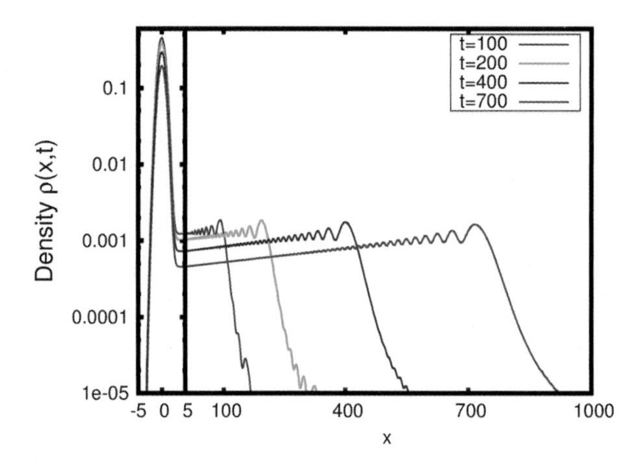

Fig. 6 Many-body tunneling to open space on an ultra-large grid can be described by a unitary time-evolution. See [20], for more details. All quantities shown are dimensionless

with a length of more than 7 mm (!) represented on a DVR grid of up to 131,072 points. See Fig. 6 which speaks for itself!

What happens as the many-body system tunnels to open space? Analyzing with the MCTDHB Package the numerically-exact many-body solution of the Schrödinger equation of the one-dimensional system tunneling to open space with repulsive interactions, we have managed to deduce the beautiful mechanism underlying the process: (1) The overall many-particle decay process is a quantum interference of simultaneously happening single-particle tunneling processes emerging from sources with different particle numbers, see Fig. 5 (bottom). (2) When the emitted particles leave the trapped and coherent source of bosons they lose their coherence with the source and between themselves, see Fig. 7. This concludes our brief account of the mechanism of many-body tunneling of a Bose-Einstein condensate to open space unraveled by the MCTDHB Package with the help of the Cray XE 6 Hermit and NEC Nehalem machines, for more details see [20].

5 Concluding Remarks and Outlook

The MCTDHB Package has performed very well on Hermit and on Laki. Its application led to breathtaking developments and research papers in the demanding field of many-body non-equilibrium quantum dynamics of Bose-Einstein condensates. Benefiting further from the high-performance computational facilities of the HLRS at Stuttgart, not less is expected to come. We list below three thrilling research directions which are on our agenda: (1) moving to two dimensions [23, 24] and eventually three dimensions [23] which is of great interest and a big challenge due to the large grids necessary, (2) moving to bosonic and other mixtures [25]

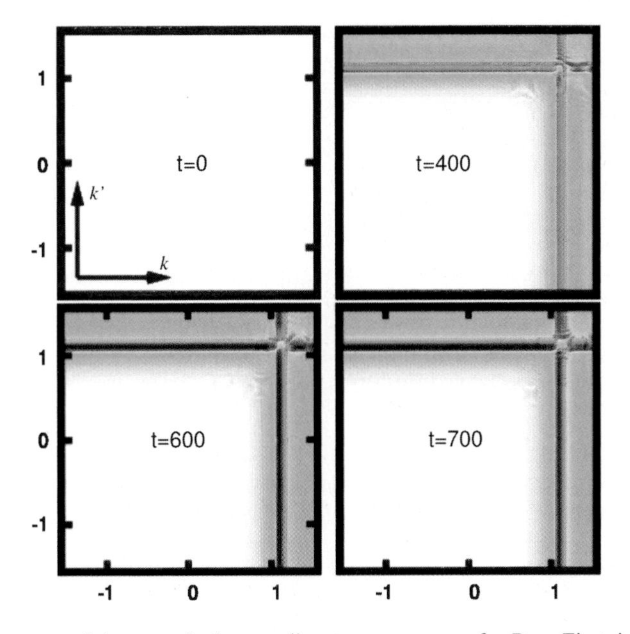

Fig. 7 Incoherence of the many-body tunneling to open space of a Bose-Einstein condensate. The first order correlation functions in momentum space $|g^{(1)}(k'|k;t)|^2$ for $N = 101$ weakly-interacting bosons (!) are plotted at $t = 0, 400, 600, 700$. Full coherence, $|g^{(1)}|^2 = 1$, is depicted as white, and total loss of coherence, $|g^{(1)}|^2 = 0$, is depicted as black. At $t = 0$ the system is almost totally coherent, i.e., $|g^{(1)}|^2 \approx 1$. At times $t > 0$, the system remains coherent everywhere in k-space apart from the region around the momentum $k = 1$, where the peaks in the momentum distributions appear. The loss of coherence, namely $|g^{(1)}|^2| \approx 0$, only in these regions allows one to conclude that the source (trapped) bosons remain coherent at all times while the emitted ones are incoherent. See [20], for more details. All quantities shown are dimensionless

which will lead to ever novel phenomena, and (3) studying excitation spectra of depleted and fragmented Bose-Einstein condensates [26] aiming at unraveling the microscopic properties of non-condensed systems, which are much more abundant as our extensive research has shown—with much help from the Cray XE6 cluster Hermit and the NEC Nehalem cluster Laki.

References

1. M.H. Anderson, J.R. Ensher, M.R. Matthews, C.E. Wieman, E.A. Cornell, Science **269**, 198 (1995)
2. C.C. Bradley, C.A. Sackett, J.J. Tollett, R.G. Hulet, Phys. Rev. Lett. **75**, 1687 (1995)
3. K.B. Davis, M.-O. Mewes, M.R. Andrews, N.J. van Druten, D.S. Durfee, D.M. Kurn, W. Ketterle, Phys. Rev. Lett. **75**, 3969 (1995)
4. F. Dalfovo, S. Giorgini, L.P. Pitaevskii, S. Stringari, Rev. Mod. Phys. **71**, 463 (1999)
5. A.J. Leggett, Rev. Mod. Phys. **73**, 307 (2001)

6. I. Bloch, J. Dalibard, W. Zwerger, Rev. Mod. Phys. **80**, 885 (2008)
7. A. Polkovnikov, K. Sengupta, A. Silva, M. Vengalattore, Rev. Mod. Phys. **83**, 863 (2011)
8. A.I. Streltsov, O.E. Alon, L.S. Cederbaum, Phys. Rev. Lett. **99**, 030402 (2007)
9. O.E. Alon, A.I. Streltsov, L.S. Cederbaum, Phys. Rev. A **77**, 033613 (2008)
10. K. Sakmann, A.I. Streltsov, O.E. Alon, L.S. Cederbaum, Phys. Rev. Lett. **103**, 220601 (2009)
11. A.U.J. Lode, K. Sakmann, O.E. Alon, L.S. Cederbaum, A.I. Streltsov, Phys. Rev. A **86**, 063606 (2012)
12. A.I. Streltsov, K. Sakmann, O.E. Alon, L.S. Cederbaum, Phys. Rev. A **83**, 043604 (2011)
13. K. Sakmann, A.I. Streltsov, O.E. Alon, L.S. Cederbaum, Phys. Rev. A **78**, 023615 (2008)
14. I. Březinová, A.U.J. Lode, A.I. Streltsov, O.E. Alon, L.S. Cederbaum, J. Burgdörfer, Phys. Rev. A **86**, 013630 (2012)
15. A.I. Streltsov, K. Sakmann, A.U.J. Lode, O.E. Alon, L.S. Cederbaum, *The Multiconfigurational Time-Dependent Hartree for Bosons Package*, Version 2.3 (Many-body Theory of Bosons Group, University of Heidelberg, Heidelberg, 2013); see the MCTDHB Homepage: http://mctdhb.org
16. A.I. Streltsov, O.E. Alon, L.S. Cederbaum, Phys. Rev. A **81**, 022124 (2010)
17. bwGRiD Homepage: http://www.bw-grid.de
18. L. Cohen, C. Lee, J. Math. Phys. **26**, 3105 (1985)
19. J. Yan, J. Stat. Phys. **113**, 623 (2003)
20. A.U.J. Lode, A.I. Streltsov, K. Sakmann, O.E. Alon, L.S. Cederbaum, Proc. Natl. Acad. Sci. USA **109**, 13521 (2012)
21. http://www.pro-physik.de/details/news/2717321/Vielteilchentunnelprozess_erstmals_exakt_beschrieben.html and http://www.innovations-report.de/html/berichte/physik_astronomie/mechanismus_vielteilchentunnelprozesses_erstmals_203395.html
22. A.U.J. Lode, I. Streltsov, O.E. Alon, H.-D. Meyer, L.S. Cederbaum, J. Phys. B **42**, 044018 (2009); A.U.J. Lode, I. Streltsov, O.E. Alon, H.-D. Meyer, L.S. Cederbaum, J. Phys. B **43**, 029802 (2010)
23. A.I. Streltsov, Phys. Rev. A **88**, 041602(R) (2013)
24. R.A. Doganov, S. Klaiman, O.E. Alon, A.I. Streltsov, L.S. Cederbaum, Phys. Rev. A **87**, 033631 (2013)
25. O.E. Alon, A.I. Streltsov, K. Sakmann, A.U.J. Lode, J. Grond, L.S. Cederbaum, Chem. Phys. **401**, 2 (2012)
26. J. Grond, A.I. Streltsov, L.S. Cederbaum, O.E. Alon, Phys. Rev. A **86**, 063607 (2012); J. Grond, A.I. Streltsov, A.U.J. Lode, K. Sakmann, L.S. Cederbaum, O.E. Alon, Phys. Rev. A **88**, 023606 (2013)

Lithium Niobate Dielectric Function and Second-Order Polarizability Tensor From Massively Parallel Ab Initio Calculations

A. Riefer, M. Rohrmüller, M. Landmann, S. Sanna, E. Rauls, N.J. Vollmers, R. Hölscher, M .Witte, Y. Li, U. Gerstmann, A. Schindlmayr, and W.G. Schmidt

Abstract The frequency-dependent dielectric function and the second-order polarizability tensor of ferroelectric $LiNbO_3$ are calculated from first principles. The calculations are based on the electronic structure obtained from density-functional theory. The subsequent application of the GW approximation to account for quasiparticle effects and the solution of the Bethe–Salpeter equation yield a dielectric function for the stoichiometric material that slightly overestimates the absorption onset and the oscillator strength in comparison with experimental measurements. Calculations at the level of the independent-particle approximation indicate that these deficiencies are at least partially related to the neglect of intrinsic defects typical for the congruent material. The second-order polarizability calculated within the independent-particle approximation predicts strong nonlinear coefficients for photon energies above 1.5 eV. The comparison with measured data suggests that self-energy effects improve the agreement between experiment and theory. The intrinsic defects of congruent samples reduce the optical nonlinearities, in particular for the 21 and 31 tensor components, further improving the agreement with measured data.

1 Introduction

The electro-optic, photorefractive, and nonlinear optical properties of ferroelectric lithium niobate [1, 2] ($LiNbO_3$, LN) are exploited in a large number of devices, such as optical modulators, acousto-optic devices, optical switches for gigahertz frequencies, Pockels cells, optical parametric oscillators, or Q-switching devices. The vast majority of actual applications and measurements employ congruent

A. Riefer · M. Rohrmüller · M. Landmann · S. Sanna · E. Rauls · N.J. Vollmers · R. Hölscher · M. Witte · Y. Li · U. Gerstmann · A. Schindlmayr · W.G. Schmidt (✉)
Lehrstuhl für Theoretische Physik, Universität Paderborn, 33095 Paderborn, Germany
e-mail: W.G.Schmidt@upb.de

W.E. Nagel et al. (eds.), *High Performance Computing in Science and Engineering '13*, DOI 10.1007/978-3-319-02165-2_8,

crystals grown by the Czochralski method. In fact, LN crystals are in general not stoichiometric (SLN) but congruent (CLN), i.e., Li deficient. Many physical properties, such as the Curie temperature, depend strongly on the existence of point defects related to doping or the Li deficiency of congruent material [3]. There are also indications that the optical properties of SLN and CLN differ notably [4, 5]. However, all existing first-principles studies of the linear [6, 7] and nonlinear [8, 9] optical properties are restricted to stoichiometric LN.

The present work [10] aims at a better understanding of the influence of nonstoichiometry on the LN optical response. Calculations for the ordinary and extraordinary dielectric function as well as for the energy dependence of the four nonvanishing components of the second-order polarizability tensor, which is responsible for second-harmonic generation (SHG), are presented.

2 Methodology

CLN is strongly Li deficient and exhibits a [Li]/[Nb] ratio of 0.94. At first sight, it might seem as if this could be well described by a supercell with 80 atoms, from which one Li atom is removed, resulting in a [Li]/[Nb] ratio of 0.9375. Such a supercell is easily constructed by a $2 \times 2 \times 2$ repetition of the primitive cell, which contains two formula units of $LiNbO_3$. However, a realistic simulation of the congruent material requires more than the mere adjustment of the Li concentration [11]. Several experimental investigations [12, 13] have excluded the presence of oxygen vacancies V_O, which are the hallmark of most oxides, but instead revealed the presence of a large amount of Nb_{Li}^{+4} antisite defects (Fig. 1b) [14]. These might be charge compensated either by Nb vacancies (Fig. 1d), where four V_{Nb}^{-5} compensate for five Nb_{Li}^{+4} antisites in the so-called *Nb site vacancy model*, or by Li vacancies (Fig. 1c), where four V_{Li}^{-} compensate for one Nb_{Li}^{+4} antisite within the *Li site vacancy model*. Nowadays, the Li site vacancy model, which we denote by CLN(Li) in the following, is widely accepted and used for the interpretation of the LN material properties. According to this model, we simulate CLN by a charge-neutral supercell containing 360 atoms, which we construct by a $3 \times 3 \times 4$ repetition of the primitive cell, with one Nb_{Li}^{+4} antisite and four Li vacancies V_{Li}^{-}. This corresponds to a [Li]/[Nb] ratio of 0.92. For comparison, however, we also perform calculations within the Nb site vacancy model, denoted by CLN(Nb). In this case a supercell of the same size containing five Nb_{Li}^{+4} antisites and four Nb vacancies V_{Nb}^{-5} is used, which again yields a [Li]/[Nb] ratio of 0.92.

The calculation of the optical response proceeds in three steps: First, we use density-functional theory (DFT) within the generalized gradient approximation (GGA) to obtain the structural and ground-state electronic properties of ferroelectric SLN and CLN. In detail, we employ the VASP implementation [15, 16] of the projector-augmented-wave method. The wave functions are expanded into plane waves up to a cutoff energy of 400 eV, while the mean-field effects of exchange and correlation are modeled with the PW91 functional [17]. A $6 \times 6 \times 6$ **k**-point mesh

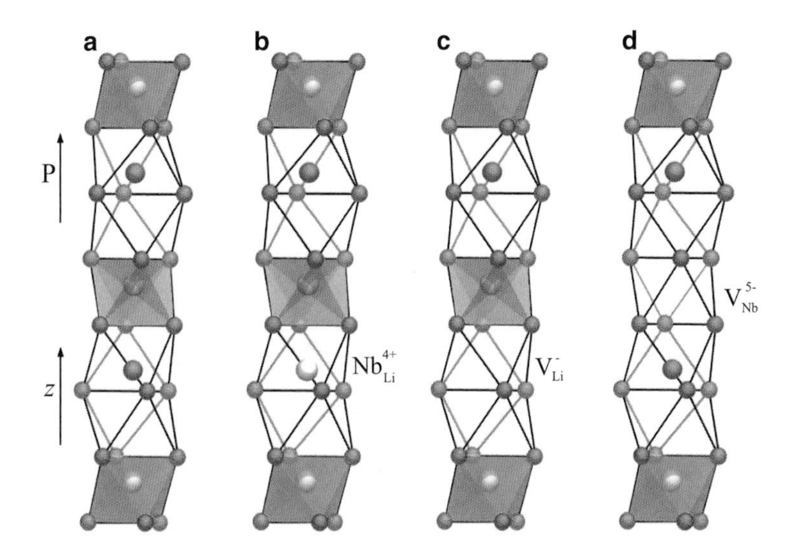

Fig. 1 Schematic representation of (**a**) bulk SLN and the defects considered in the CLN models: (**b**) Nb^{+4}_{Li} antisite, (**c**) lithium vacancy V^{-}_{Li}, (**d**) Niobium vacancy V^{-5}_{Nb}. Oxygen atoms are *red*, Li *black*, and Nb *white*. The Nb octahedra are *shaded*, the *z* axis and the polarization direction are indicated

is used to sample the Brillouin zone corresponding to the primitive cell of bulk SLN. To determine the structural properties of CLN, including the relaxation of the atoms in the vicinity of the defect sites, with high accuracy, we choose $4 \times 4 \times 4$ **k**-points to sample the much smaller Brillouin zone corresponding to the 360-atom supercell. To compute the electronic and optical properties of CLN we then change the sampling to $2 \times 2 \times 2$, which corresponds approximately to the same density of **k** points as the $6 \times 6 \times 6$ mesh used for SLN.

Second, the quasiparticle band structure is calculated within the *GW* approximation[18] (GWA) for the exchange-correlation self-energy. As usual in practical applications of this scheme, we obtain the self-energy corrections to the electronic eigenvalue spectrum from a perturbative solution of the quasiparticle equation, where the GGA exchange-correlation potential is replaced by the nonlocal and energy-dependent self-energy operator. We evaluate the latter in the standard non-self-consistent G_0W_0 approximation [19], using the implementation described in [20], from the convolution of the single-particle propagator G_0 and the dynamically screened Coulomb interaction W_0 in the random-phase approximation.

Third, to obtain the linear dielectric function of SLN, we solve the Bethe–Salpeter equation (BSE) for coupled electron-hole excitations [21–23], which incorporates the screened electron-hole attraction as well as the unscreened electron-hole exchange. Here we use the time-evolution method described in [24, 25] to obtain the polarizability.

Within the independent-particle approximation (IPA) or the independent-quasiparticle approximation (IQA), the linear dielectric function is given by

$$\varepsilon_{\alpha\beta}(\omega) = \delta_{\alpha\beta} + \frac{4\pi e^2}{\hbar m_{\mathrm{e}}^2 V \omega^2} \sum_{n,m} \sum_{\mathbf{k}} f_{nm}(\mathbf{k}) \frac{p_{nm}^{\alpha}(\mathbf{k}) p_{mn}^{\beta}(\mathbf{k})}{\omega_{mn}(\mathbf{k}) - \tilde{\omega}} \tag{1}$$

with the abbreviation $f_{nm}(\mathbf{k}) = f_{n\mathbf{k}} - f_{m\mathbf{k}}$, where $f_{n\mathbf{k}}$ and $f_{n\mathbf{k}}$ are the Fermi occupation factors for the Bloch states $|n\mathbf{k}\rangle$ and $|m\mathbf{k}\rangle$, respectively. Furthermore, V denotes the volume of the crystal, and $\tilde{\omega} = \omega + i\eta$ is the shorthand notation for the frequency ω with an additional small positive imaginary part η, which turns the electromagnetic field on adiabatically. The symbols

$$p_{nm}^{\alpha}(\mathbf{k}) = \langle n\mathbf{k}|\hat{p}^{\alpha}|m\mathbf{k}\rangle \tag{2}$$

represent the momentum matrix elements of the system, and $\hbar\omega_{mn}(\mathbf{k}) = \epsilon_{m\mathbf{k}} - \epsilon_{n\mathbf{k}}$ are the transition energies between the bands m and n at the point $\mathbf{k}$. The single-electron energies are taken from the DFT eigenvalue spectrum in the IPA and from the GWA quasiparticle band structure in the IQA.

Corresponding expressions for the coefficients of the second-order polarization tensor $\chi_{\alpha\beta\gamma}^{(2)}(\omega)$ were derived, e.g., in [26–28]. Here we calculate the nonlinear response as the sum of the two-band contribution

$$\chi_{\alpha\beta\gamma}^{(2),\mathrm{two}}(\omega) = -\frac{ie^3}{m_{\mathrm{e}}^3 \hbar^2 V} \sum_{n,m} \sum_{\mathbf{k}} \frac{p_{nm}^{\alpha}(\mathbf{k}) \{\Delta_{mn}^{\beta}(\mathbf{k}) p_{mn}^{\gamma}(\mathbf{k})\}}{[\omega_{mn}(\mathbf{k})]^4} \times f_{nm}(\mathbf{k}) \left(\frac{16}{\omega_{mn}(\mathbf{k}) - 2\tilde{\omega}} - \frac{1}{\omega_{mn}(\mathbf{k}) - \tilde{\omega}} \right) \tag{3}$$

and the three-band contribution

$$\chi_{\alpha\beta\gamma}^{(2),\mathrm{three}}(\omega) = -\frac{ie^3}{m_{\mathrm{e}}^3 \hbar^2 V} {\sum_{n,m,l}}' \sum_{\mathbf{k}} \tag{4}$$
$$\times \frac{p_{nm}^{\alpha}(\mathbf{k}) \{p_{ml}^{\beta}(\mathbf{k}) p_{ln}^{\gamma}(\mathbf{k})\}}{\omega_{ln}(\mathbf{k}) - \omega_{ml}(\mathbf{k})} \left(\frac{16 f_{nm}(\mathbf{k})}{[\omega_{mn}(\mathbf{k})]^3 [\omega_{mn}(\mathbf{k}) - 2\tilde{\omega}]} + \frac{f_{ml}(\mathbf{k})}{[\omega_{ml}(\mathbf{k})]^3 [\omega_{ml}(\mathbf{k}) - \tilde{\omega}]} + \frac{f_{ln}(\mathbf{k})}{[\omega_{ln}(\mathbf{k})]^3 [\omega_{ln}(\mathbf{k}) - \tilde{\omega}]} \right).$$

The prime at the sum symbol indicates that the summation is restricted to the index combinations $n \neq m \neq l$. As in the case of the linear dielectric function, the formulas can be evaluated either with independent-particle or independent-quasiparticle energies. To simplify the notation, we have defined

$$\{p_{ml}^{\beta}(\mathbf{k}) p_{ln}^{\gamma}(\mathbf{k})\} = \frac{1}{2} \left[p_{ml}^{\beta}(\mathbf{k}) p_{ln}^{\gamma}(\mathbf{k}) + p_{ml}^{\gamma}(\mathbf{k}) p_{ln}^{\beta}(\mathbf{k}) \right]. \tag{5}$$

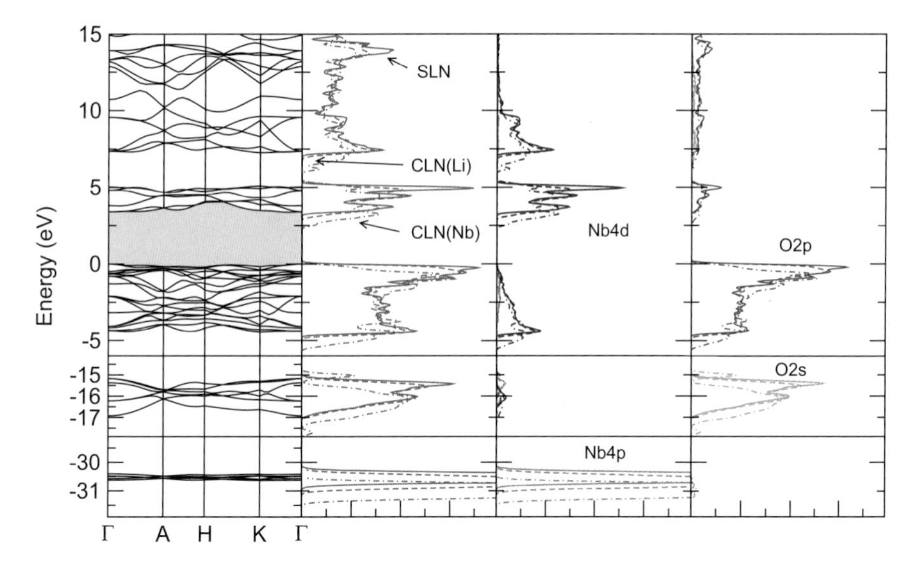

Fig. 2 Electronic band structure of SLN within DFT-GGA together with the densities of states of SLN (*solid line*), CLN(Li) (*dashed line*), and CLN(Nb) (*dash-dotted line*) in atomic units. The notation of the high-symmetry points follows [6]. Separately shown are the O $2s$, O $2p$, Nb $4p$, and Nb $4d$ projected densities of states

Finally, the matrix elements of the intraband transitions

$$\Delta_{mn}^{\beta}(\mathbf{k}) = p_{mm}^{\beta}(\mathbf{k}) - p_{nn}^{\beta}(\mathbf{k}) \tag{6}$$

appearing in (3) are obtained from

$$p_{nn}^{\beta}(\mathbf{k}) = \frac{m_e}{\hbar} \partial_{k_\beta} \epsilon_n(\mathbf{k}). \tag{7}$$

Due to symmetry reasons, the second-order polarizability tensor of ferroelectric LN has four independent nonvanishing components $\chi_{\alpha\beta\gamma}^{(2)}(\omega)$, namely the index combinations $211 = -222 = 112$, $223 = 131$, $311 = 322$, and 333. These are commonly labeled by the contracted indices $21 = -22 = 16$, $24 = 15$, $31 = 32$, and 33, respectively. These four components are studied in the present work. We follow [29] concerning the notation for the Cartesian axes.

Calculated SHG data are known to be rather sensitive with respect to numerical details, such as the number of **k** points used to sample the Brillouin zone and the number of electronic bands included in the calculation [30]. Concerning the Brillouin-zone sampling, we find a $6 \times 6 \times 6$ **k**-point mesh necessary to obtain converged spectra for SLN. The density of states (DOS) calculated at the DFT-GGA level is displayed in Fig. 2. It is characterized by a continuum of conduction bands and well-separated groups of valence bands. The latter are dominated by O $2p$ and

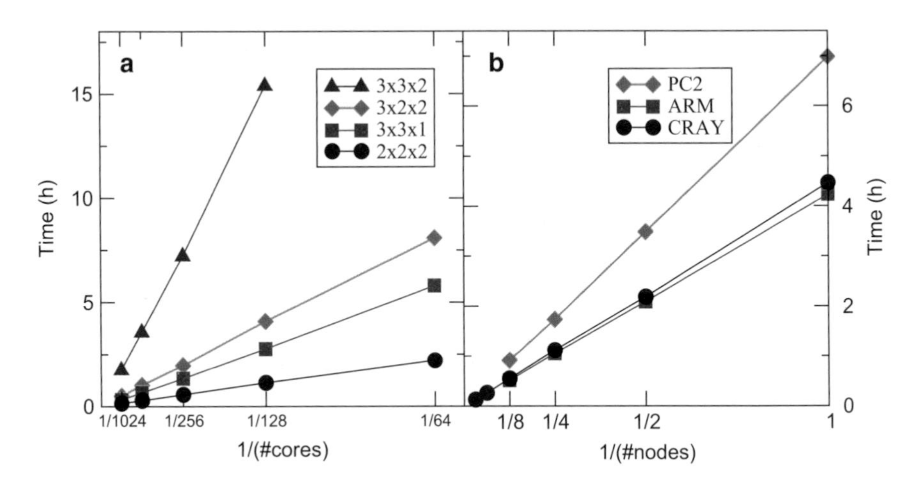

Fig. 3 Computation time of the three band contribution of the second harmonic generation tensor (4). Computation time on the HLRS CRAY XE6 vs number of cores is shown in (**a**). Thereby $2 \times 2 \times 2$, $3 \times 3 \times 1$, $3 \times 2 \times 2$, and $3 \times 3 \times 2$ repetitions of the primitive LN unit cell were probed. As can be seen, the scaling is linear up to 1,024 cores. A comparison between HLRS CRAY XE6 (CRAY), Arminius+ Cluster (ARM) and the Pling2 Cluster (PC2) in case of the a $2 \times 2 \times 2$ cell is shown in (**b**)

Nb $4d$ states in the energy range between 0 and -5 eV, by O $2s$ states between -15 and -17 eV , and by Nb $4p$ states around -30.5 eV. These DOS features also determine the convergence of the SHG calculations. We find that the SHG spectra in the energy range up to 6 eV only vary weakly if the number of conduction bands increases above a cutoff energy of about 20 eV. The results are essentially converged for a cutoff energy of about 25 eV, which is hence used in the following. The dependence of the SHG data on the number of valence bands included in the calculation is more complex, however. While the inclusion of the O $2p$, Nb $4d$, and O $2s$ states already yields relatively well converged 21 and 24 components of the second-order polarizability tensor, the 31 and 33 components exhibit an oscillatory behavior and require the additional inclusion of the Nb $4p$ states in the calculation. The transitions between the Nb $4p$ and Nb $4d$ states cause noticeable modifications in the intensity of the three band contribution (4) for these elements, as illustrated below in Fig. 5. On the other hand, the inclusion of states below -36 eV has no impact on the calculated spectra for photon energies up to 6 eV.

Turning to the numerics, it is clear from the expressions given above for the linear and nonlinear optical response that in particular the evaluation of the three-band contribution to the second harmonic generation tensor, (4), shows an unfavorable $\mathcal{O}^3$ scaling with system size. On the other hand, it can be easily parallelized. We choose a parallelization with respect to the energy $\hbar\omega$. Results for $2 \times 2 \times 2$, $3 \times 3 \times 1$, $3 \times 2 \times 2$, and $3 \times 3 \times 2$ repetitions of the LN unit cell are exemplarily shown in Fig. 3a. Obviously, the scaling with the number of cores is roughly linear up to 1,024 cores, allowing for highly efficient calculations. In Fig. 3b we compare the run time on the HLRS CRAY XE6, which presents the main computational resource used for the calculations in this project, with other architectures available to us: the Arminius+ Cluster (two Intel Xeon X5650, 2.67 GHz) and the Pling2 Cluster (two Intel Xeon E5540 Nehalem, 2.53 GHz) hosted by the Paderborn Center for Parallel Computing (PC^2).

3 Results

In Fig. 4 we show the calculated linear optical response of ferroelectric SLN according to the three levels of theory described above, i.e., within (1) the independent-particle approximation (IPA) based on the DFT eigenstates and eigenenergies, (2) within the independent-quasiparticle approximation (IQA), where the electronic self-energy is either obtained from the GWA or simply approximated by a scissors operator that widens the fundamental gap by 2.03 eV, and (3) from the solution of the Bethe–Salpeter equation (BSE). The spectra obtained within the IPA show two main features of the optical absorption around 5 and 8 eV. The inclusion of self-energy effects within the GWA leads to a nearly rigid blueshift of the spectra by about 2 eV. The GWA spectra shown here differ slightly from our previous results [6], where the self-energy was evaluated with a model dielectric function rather than the fully frequency-dependent random-phase approximation employed here. We find that the self-energy effects in the present calculations can be simulated very well by a simple scissors shift (Sci) of 2.03 eV, as illustrated in Fig. 4. The inclusion of the Coulomb correlation between electrons and holes by means of the BSE both repositions the spectrum on the energy axis and changes the line shapes. The first peak of the low-energy main feature of the optical absorption becomes more pronounced, and the entire structure is redshifted relative to the GWA. It is now centered at 5.5 and 5.6 eV for $\varepsilon_\perp$ or $\varepsilon_{||}$, respectively. The oscillator strength of the originally rather broad ($\sim$ 2 eV) high-energy main feature of the optical absorption is also redshifted and transferred into a single sharp peak at about 9.3 eV for $\varepsilon_\perp$ and $\varepsilon_{||}$. This peaks originates mainly from electronic transitions that involve the O $2p$ valence states and conduction states with energies of between 6 and 8 eV.

Comparison of the BSE calculations for SLN with experimental data [31, 32] shows qualitative agreement but clear deviations concerning some quantitative details. On the one hand, the calculations overestimate the onset of the optical absorption by about 0.2 eV. On the other hand, the double-peak structure of the first absorption maximum is far more pronounced in the calculations than observed experimentally. This may partially be related to temperature effects, which are neglected in the present calculations. Furthermore, deviations between experiments and our calculations can be expected due to the modified stoichiometry and structure of congruent LN compared to the stoichiometric material considered so far. In order to explore the differences between the optical response of CLN and SLN, we now turn to the former, using the CLN(Li) and CLN(Nb) models introduced above. Since realistic simulations require large supercells with 360 atoms, these calculations are

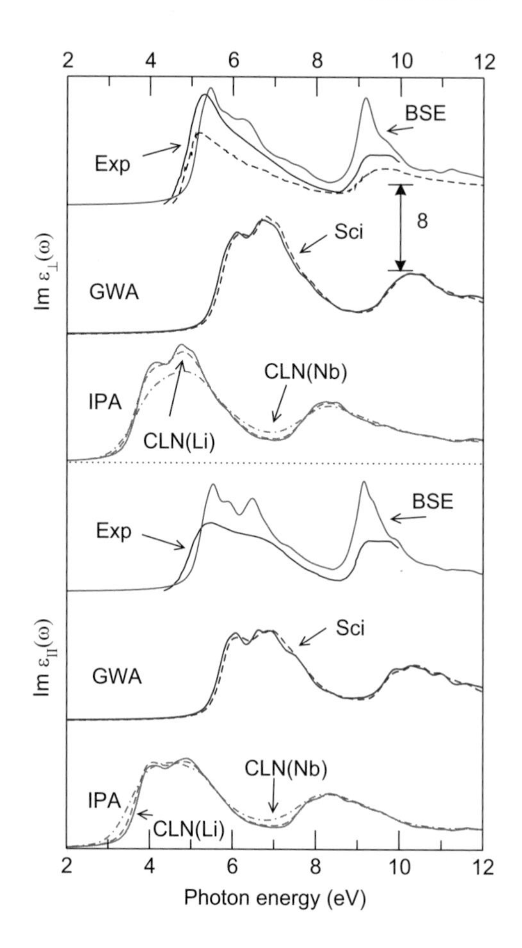

Fig. 4 Imaginary part of the ordinary (*above*) and extraordinary (*below*) dielectric function of ferroelectric SLN. In both cases we compare the independent-particle approximation (IPA) with results that include quasiparticle self-energy effects either evaluated within the GW approximation (GWA) or obtained from a scissors operator (Sci), as well as the solution of the Bethe–Salpeter equation (BSE), which includes electron-hole attraction and local-field effects. Also shown are IPA calculations for CLN together with experimental data (*solid line*: [31], *dashed line*: [32])

restricted to the IPA level of theory. As can be seen in Fig. 4, the differences in the linear optical response between the CLN(Li) model and SLN are rather small. Slightly more pronounced are the changes that arise if the CLN(Nb) structure for the congruent material is used. Irrespective of the model underlying the calculations, however, we observe the same trend: The onset of optical absorption is redshifted by up to 0.2 eV for CLN in comparison to SLN, especially for $\varepsilon_{||}$, and the double-peak structure of the first absorption maximum is washed out, most notably for $\varepsilon_{\perp}$. As a consequence, the results for CLN are closer to the experimental data for the linear optical response than the SLN calculations. However, it should be kept in mind that the differences in the optical response between CLN and SLN, which we have here studied in the IPA, may be affected by many-body effects and show a different behavior at the level of the BSE. In particular, large exciton binding energies for localized defect states (see, e.g. [33]) may further increase the optical-absorption redshift of CLN compared to stoichiometric samples.

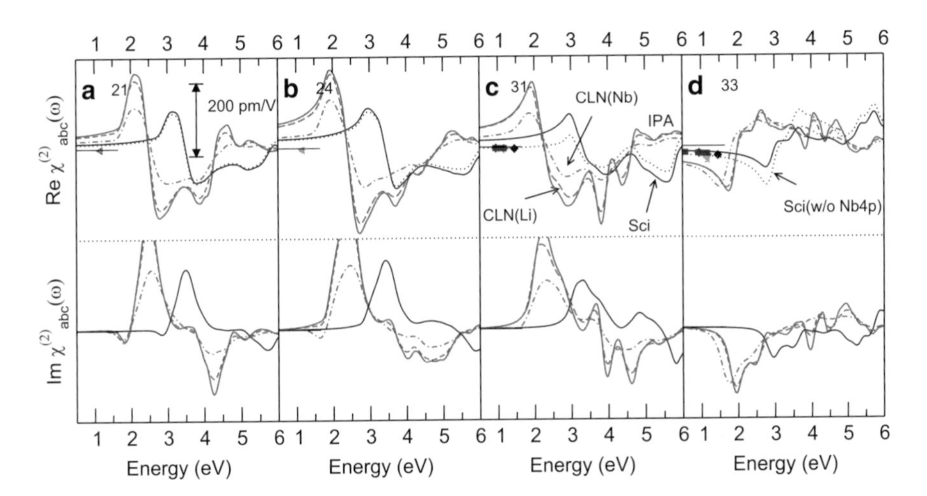

Fig. 5 Coefficients of the second-order polarizability tensor $\chi^{(2)}(\omega)$ for SLN and CLN calculated within the IPA and with a scissors shift (Sci), compared to experimental data from [34–42] (*diamonds, squares, crosses, plusses, left triangles, right triangles, down triangles, up triangles, stars*). We follow the notation of [43] concerning the orientation of the axes and apply the scaling described in [34]. Sci(w/o Nb $4p$) denotes calculations that do not include the Nb $4p$ states

In order to understand the smoothing of the double-peak feature for $\varepsilon_\perp$ in the range between 3.5 and 5.5 eV due the defects typical for CLN, we analyzed the joint density of states (JDOS) as well as the oscillator strengths of the relevant transitions. We find the JDOS to be very similar for SLN and CLN, apart from a reduction of the gap in the congruent material, which can already be identified in the DOS displayed in Fig. 2, and a rather uniform intensity reduction for energy differences above 4.5 eV. The fine structure of the first absorption maximum is found to be caused by the oscillator strengths rather than the JDOS, however. The fading of this fine structure for CLN is mainly related to the reduced transition matrix elements between the O $2p$ valence states and the Nb $4d$ conduction states. The redshift of the absorption onset, on the other hand, is caused by the modified DOS.

In Fig. 5 we present the 21, 24, 31, and 33 components of the second-order polarizability tensor for ferroelectric LN calculated within the IPA. Both the real and the imaginary part of $\chi^{(2)}(\omega)$ are shown up to 6 eV. The main features of the SHG spectra lie between 1.5 and 3.0 eV, with additional strong nonlinearities at higher energies in the range between 4 and 5 eV, in particular for the 21 and 31 components. The present results for the 33 component are very similar to earlier IPA calculations [9] that neglected intraband transitions and considered only the three-band contribution (4). We find that the inclusion of the two-band term (3) only leads to marginal changes in the case of LN and neither affects the line shapes nor the magnitude of the SHG coefficients substantially. The differences between the present calculations and the work by Cabuk [44] appear more pronounced, on the other hand. However, these are partially related to the different broadening parameters η. In fact, if we decrease the broadening from 0.15 eV, the value used in our calculations, to 0.05 eV, then the obtained line shapes are in good correspondence with those in [44]. In particular, the 33 component shows a very similar spectral dependence.

The SHG coefficients are strongly affected by self-energy effects. As demonstrated in Fig. 4, the energy and $\mathbf{k}$ dependence of the quasiparticle corrections is sufficiently small to allow for their approximation by a scissors operator. Therefore, we follow the same approach here to obtain the second-order polarizability at the IQA level of theory. The results are displayed in Fig. 5. We observe a blueshift in the SHG features by about 1 eV if self-energy effects are included, accompanied by a notable lowering of the intensity. The dependence of the SHG spectra on transitions involving the Nb $4p$ states is also illustrated in Fig. 5. Evidently, the inclusion of the Nb $4p$ states does not give rise to any new spectral signatures in the energy range considered here, but it modifies the intensities considerably, in particular for the 31 and 33 components.

Several experimental measurements of SHG data for LN have been reported in the literature [34–43]. The available data points are indicated by symbols in Fig. 5. We also include measurements of the modulus $|\chi^{(2)}(\omega)|$ and take these to represent the real part of the polarizability, because the imaginary part is very small in the energy region experimentally probed. The comparison with the calculated spectra at the IPA and IQA levels of theory in Fig. 5 shows that the former clearly overestimates the measured data, whereas the IQA calculations, which include electronic self-energy effects, describe the optical nonlinearities better, especially the 33 component. Nevertheless, even the IQA calculations for SLN still yield stronger nonlinearities than observed experimentally. In this respect, it is interesting to note that the CLN results at the IPA level of theory indicate that the congruent material gives rise to weaker SHG signals than predicted for SLN. This holds in particular for CLN(Nb). In addition, for the Nb site vacancy model the spectra are somewhat smoothed, and the features are redshifted. In the case of the 33 component this leads to an enhanced signal in the energy region experimentally probed. As for the linear response, the proper inclusion of defects in the congruent material thus also seems to improve the agreement between experiments and theory concerning $\chi^{(2)}(\omega)$.

A word of caution is necessary when comparing the experimental and theoretical data, however. First, all of the experimental values are clearly in the nonresonant region around 1 eV and, in particular for the 21, 24, and 31 components, of very low absolute magnitude. We remark that the intensity is, e.g., two orders of magnitude lower than the SHG signal of GaAs in this region. Second, the effects of the crystal local fields and the electron-hole interaction are not included in the present SHG calculations. Luppi et al. [45], for instance, found that $\chi^{(2)}(\omega)$ can decrease up to 30 % due to local-field effects, depending on the material and on the frequency range. Excitonic effects, on the other hand, are expected to lead to a rather uniform but noticeable increase of the SHG signal [28, 45]

4 Summary

Our work represents a first attempt towards a first-principles description of the optical response of congruent lithium niobate. The linear and nonlinear optical response calculated for the Li site vacancy model as well as the Nb site vacancy model are shown to be closer to the available experimental data than the spectra calculated for stoichiometric crystals: Compared to SLN, the CLN simulations lead to a redshift of the spectral features, wash out part of the fine structure, and reduce the nonlinearities. This holds in particular for CLN(Nb). It is clear, however, that further work is required to fully understand the influence of crystal imperfections on the optical response of lithium niobate. On the one hand, a variety of further defect configurations and defect complexes will have to be investigated, see, e.g., [46]. In this respect, the structure models adopted in this work are created by placing the defects at randomly chosen lattice sites and can be further refined by taking clustering phenomena into account [47]. On the other hand, we expect many-body effects, which were shown here to have a strong impact on the optical response of SLN, to give rise to even stronger modifications of the CLN signal.

Acknowledgements We gratefully acknowledge financial support from the DFG as well as supercomputer time provided by the HLRS Stuttgart and the Paderborn PC^2.

References

1. A. Räuber, Curr. Top. Mater. Sci. **1**, 481 (1978)
2. R.S. Weis, T.K. Gaylord, Appl. Phys. A **37**, 191 (1985)
3. L. Galambos, S.S. Orlov, L. Hesselink, Y. Furukawa, K. Kitamura, S. Takekawa, J. Cryst. Growth **229**, 228 (2001)
4. I.V. Kityk, M. Makowska-Janusik, M.D. Fontana, M. Aillerie, F. Abdi, J. Appl. Phys. **90**, 5542 (2001)
5. I.V. Kityk, M. Makowkska-Janusik, M.D. Fontana, M. Aillerie, F. Abdi, J. Phys. Chem. B **105**, 12242 (2001)
6. W.G. Schmidt, M. Albrecht, S. Wippermann, S. Blankenburg, E. Rauls, F. Fuchs, C. Rödl, J. Furthmüller, A. Hermann, Phys. Rev. B **77**, 035106 (2008)
7. W.Y. Ching, Z.Q. Gu, Y.N. Xu, Phys. Rev. B **50**, 1992 (1994)
8. A.M.M.H. Akkus, S. Cabuk, Int. J. Nanoelectron. Mater. **3**, 53 (2010)
9. A. Riefer, S. Sanna, A.V. Gavrilenko, W.G. Schmidt, IEEE Trans. Ultrason. Ferroelectr. Freq. Control **59**, 1929 (2012)
10. A. Riefer, S. Sanna, A. Schindlmayr, W.G. Schmidt, Phys. Rev. B **87**, 195208 (2013)
11. T. Volk, M. Wöhlecke, *Lithium Niobate: Defects, Photorefraction and Ferroelectric Switching*. Springer Series in Materials Science (Springer, Berlin, 2010) [Softcover reprint of hardcover 1st ed. 2009 edn.]
12. N. Zotov, H. Boysen, F. Frey, T. Metzger, E. Born, J. Phys. Chem. Solids **55**(2), 145 (1994)
13. A.P. Wilkinson, A.K. Cheetham, R.H. Jarman, J. Appl. Phys. **74**(5), 3080 (1993)
14. N. Iyi, K. Kitamura, F. Izumi, J. Yamamoto, T. Hayashi, H. Asano, S. Kimura, J. Solid State Chem. **101**(2), 340 (1992)
15. G. Kresse, J. Furthmüller, Comput. Mat. Sci. **6**, 15 (1996)

16. G. Kresse, D. Joubert, Phys. Rev. B **59**, 1758 (1999)
17. J.P. Perdew, J.A. Chevary, S.H. Vosko, K.A. Jackson, M.R. Pederson, D.J. Singh, C. Fiolhais, Phys. Rev. B **46**, 6671 (1992)
18. L. Hedin, Phys. Rev. **139**, A769 (1965)
19. M.S. Hybertsen, S.G. Louie, Phys. Rev. B **34**, 5390 (1986)
20. M. Shishkin, G. Kresse, Phys. Rev. B **74**, 035101 (2006)
21. S. Albrecht, L. Reining, R. DelSole, G. Onida, Phys. Rev. Lett. **80**, 4510 (1998)
22. L.X. Benedict, E.L. Shirley, R.B. Bohn, Phys. Rev. Lett. **80**, 4514 (1998)
23. M. Rohlfing, S.G. Louie, Phys. Rev. Lett. **83**, 856 (1999)
24. W.G. Schmidt, S. Glutsch, P.H. Hahn, F. Bechstedt, Phys. Rev. B **67**, 085307 (2003)
25. P.H. Hahn, W.G. Schmidt, F. Bechstedt, Phys. Rev. Lett. **88**, 016402 (2001)
26. D.E. Aspnes, Phys. Rev. B **6**, 4648 (1972)
27. J.L.P. Hughes, J.E. Sipe, Phys. Rev. B **53**, 10751 (1996)
28. R. Leitsmann, W.G. Schmidt, P.H. Hahn, F. Bechstedt, Phys. Rev. B **71**, 195209 (2005)
29. S. Sanna, W.G. Schmidt, Phys. Rev. B **81**, 214116 (2010)
30. M.Z. Huang, W.Y. Ching, Phys. Rev. B **47**, 9464 (1993)
31. E. Wiesendanger, G. Güntherodt, Solid State Commun. **14**, 303 (1974)
32. A.M. Mamedov, M.A. Osman, L.C. Hajieva, Appl. Phys. A **34**, 189 (1984)
33. P. Rinke, A. Schleife, E. Kioupakis, A. Janotti, C. Rödl, F. Bechstedt, M. Scheffler, C. Van de Walle, Phys. Rev. Lett **108**(12), 126404 (2012)
34. I. Shoji, T. Kondo, A. Kitamoto, M. Shirane, R. Ito, J. Opt. Soc. Am. B **14**, 2268 (1997)
35. M.M. Choy, R.L. Byer, Phys. Rev. B **14**, 1693 (1976)
36. G.D. Boyd, R.C. Miller, K. Nassau, W.L. Bond, A. Savage, Appl. Phys. Lett. **5**(11), 234 (1964)
37. D.A. Kleinman, R.C. Miller, Phys. Rev. **148**, 302 (1966)
38. R.C. Miller, A. Savage, Appl. Phys. Lett. **9**(4), 169 (1966)
39. J. Bjorkholm, IEEE J. Quantum Electron. **4**(11), 970 (1968)
40. W.F. Hagen, P.C. Magnante, J. Appl. Phys. **40**(1), 219 (1969)
41. R.C. Miller, W.A. Nordland, P.M. Bridenbaugh, J. Appl. Phys. **42**, 4145 (1971)
42. B.F. Levine, C.G. Bethea, Appl. Phys. Lett. **20**(8), 272 (1972)
43. D. Roberts, IEEE J. Quantum Electron. **28**(10), 2057 (1992)
44. S. Cabuk, Cent. Eur. J. Phys. **10**(1), 239 (2011)
45. E. Luppi, H. Hübener, V. Véniard, Phys. Rev. B **82**(23), 235201 (2010)
46. H.H. Nahm, C.H. Park, Phys. Rev. B **78**(18), 184108 (2008)
47. H. Xu, D. Lee, J. He, S. Sinnott, V. Gopalan, V. Dierolf, S. Phillpot, Phys. Rev. B **78**(17), 174103 (2008)

Acoustic and Elastic Full Waveform Tomography

A. Kurzmann, S. Butzer, S. Jetschny, A. Przebindowska, L. Groos, M. Schäfer, S. Heider, and T. Bohlen

Abstract For a better estimation of subsurface parameters we develop imaging methods that can exploit the richness of full seismic waveforms. Full waveform tomography (FWT) is a powerful imaging method and emerges as an important procedure in hydrocarbon exploration and underground construction. It is able to recover high-resolution subsurface images from recorded seismic data. In this work, we demonstrate the performance of our parallel time-domain implementations applied to 2D and 3D problems. The results of synthetic feasibility studies—representing applications with a realistic background—illustrate the ability of FWT to reconstruct structures at small scale lengths. We also show first results of an FWT application to real seismic data recorded by a marine survey. Optimal parallelization strategies, such as domain decomposition and shot parallelization, are used in seismic modeling to exploit the possibilities of the high-performance computer *HERMIT* and to allow efficient FWT computations. FWT implementations show a convincing scaling behaviour. However, a competing benchmark of seismic modeling exhibits some performance drawbacks. Further software improvements of the modeling implementations might be necessary.

A. Kurzmann (✉) · S. Butzer · L. Groos · M. Schäfer · S. Heider · T. Bohlen
Karlsruhe Institute of Technology, Karlsruhe, Germany
e-mail: andre.kurzmann@kit.edu

S. Jetschny
Petroleum Geo-Services ASA, Oslo, Norway

A. Przebindowska
Baker Hughes INTEQ GmbH, Celle, Germany

W.E. Nagel et al. (eds.), *High Performance Computing in Science and Engineering '13*, DOI 10.1007/978-3-319-02165-2_9,

1 Introduction

Living in a time where natural resources are scarce and precious, the number of underground constructions is increasing, and the storage of waste and other material in the earth is necessary, it is important to find accurate ways of mapping geological structures in the Earth's interior. A considerable amount of money and effort has been spent on the field of reflection seismology, the science of collecting echoes and transforming them into images of the subsurface. Conventional seismic imaging methods applied in reflection seismology utilize a small portion of the information of the echoes we obtain from the subsurface. Most methods analyze the arrival time of the echoes or specific signal amplitudes only.
In this project, we further apply and develop a new seismic inversion and imaging technique that uses the full information content of the seismic recordings. Each echo or reflection from geological discontinuities in the earth is used to unscrample the earth structure. For an improved estimation of subsurface parameters we thereby exploit the full information contained in the seismic waveforms. Full waveform tomography (FWT) is a cutting-edge inverse method that accounts for the full seismic waveform recorded over a broad range of frequencies and apertures. It iteratively retrieves multiparameter models of the subsurface by solving the full wave equations. It allows for a mapping of structures on spatial scales down to less than the seismic wavelength, hence providing a tremendous improvement of resolution compared to traveltime tomography based on ray-theory. Especially in exploration geophysics and earthquakes seismology the interest in FWT is increasing continuously.
The FWT can be applied in time- or frequency-domain [13]. Although first implementations of the FWT in the 1980s were conducted in the time-domain by [7, 12], the frequency-domain version of FWT developed in the 1990s by [8] has now emerged as an efficient imaging tool. In our work we concentrate on the implementation of the former method and its application to seismic problems. It comprises two- and three-dimensional modeling of acoustic and elastic wavefields in the time-domain. A main advantage of this method is the efficient parallelization by domain decomposition [1] and shot parallelization [4, 5] leading to a significant speedup on parallel computers.

2 Method and Implementation

Full waveform tomography aims to find the optimal subsurface model by iteratively minimizing the misfit between observed and modeled data. The implementation is based on the conjugate gradient method as described by [7, 12].

2.1 Seismic Modeling

Seismic modeling is the fundamental part of full waveform tomography and requires nearly all the computation time. In dependence of the field of application, the wave-propagation physics for an underlying subsurface model has to be described by an appropriate wave equation. On the one hand, this comprises the acoustic or elastic wave equations. On the other hand, they have to be solved for two-dimensional or three-dimensional subsurface models (corresponding applications are referred to as 2D or 3D FWT). The numerical implementation of the wave equations consists of a time-domain finite-difference (FD) time-stepping method in cartesian coordinates. In the elastic case the FD-scheme solves the velocity-stress formulation by utilizing particle velocities and stress components of the wavefield. However, the acoustic approximation is limited to the pressure formulation with only one wavefield component. Due to finite model sizes, the wave equations are expanded by perfectly matched layer terms (PML) to avoid artificial boundary reflections.

2.2 Full Waveform Tomography

The solution of the inverse problem comprises several steps shown in Fig. 1. The method is initialized by the choice of a 2D or 3D initial parameter model. Seismic velocities or mass density are assigned to the model m_0 at the first iteration. The initial model can be either assumed using a priori information or computed by conventional imaging methods. For each source of the acquisition geometry seismic modeling is applied, i.e. the wavefield is emitted by the source and forward-propagates across the medium. This wavefield has to be stored in memory with respect to the whole volume and propagation time. Synthetic seismic data is recorded at the receivers and the difference of observed and synthetic data is calculated—resulting in residuals. For each source the residual wavefield is back-propagated from the receivers to the source position. The cross-correlation of forward- and back-propagated wavefields yields shot-specific steepest-descent gradients δm_n. The computation of the final gradient is given by the summation of all shot-specific gradients. The result is the global gradient of the entire acquisition geometry. An optimized gradient $\widetilde{\delta m}_n$ is computed by subsequent preconditioning and application of the method of conjugate gradient. The update of the model parameter is the final step of a FWT iteration. The gradient d_n has to be scaled by an optimal step length μ_n to get a proper model update at iteration step n. The estimation of μ_n is performed at each iteration and requires additional modelings.

2.3 Parallelization and Performance

Apart from other factors the success of a tomography depends on a sufficient ray coverage of the model area. Thus, several shot and receiver positions are

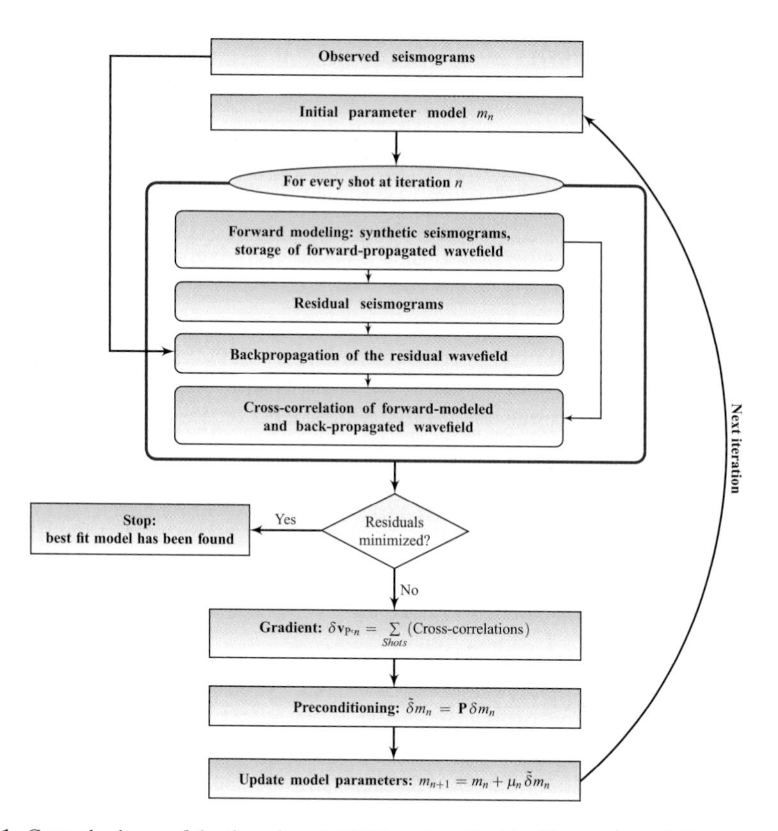

Fig. 1 General scheme of the time-domain FWT: as described by TARANTOLA [12], the model is improved by using an iterative gradient method

necessary. Reasonable numbers of shots may vary from 10 to 100. Due to the separate wavefield simulation of each shot, modelings require most of the entire computation time. This results in huge computational efforts, which can be handled by a heavy parallelization. Our FWT implementation offers two types of parallelization. The model area can be decomposed into subdomains, which are assigned to all available cores (Fig. 2). Additional padding layers with half the size of the spatial differential operator are located around the subdomains. At each time step these model boundaries are exchanged using the Message Passing Interface (MPI) to allow wave propagation across the internal boundaries. Due to increasing communication, 2D and 3D modelings cannot benefit from the decomposition of a model into a high number of very small subdomains. Hence, the domain decomposition should be restricted to the cores of one node (if there is enough memory available) and combined with shot parallelization. This second possibility allows the simultaneous modeling for several shots. These modelings can be distributed to all available nodes. Thus, they are performed independently and no inter-node network communication is necessary. A comparison of scaling

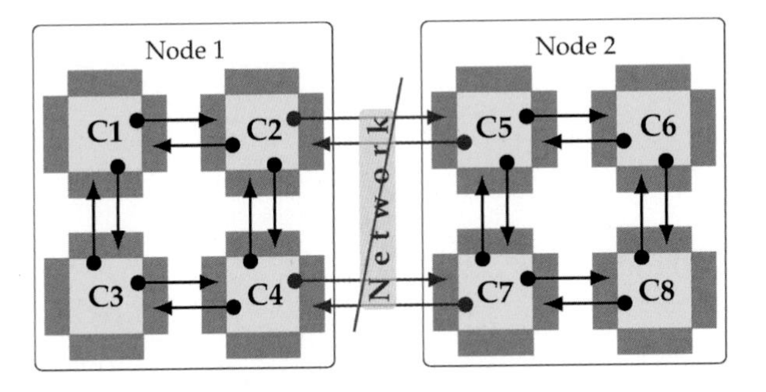

Fig. 2 Parallelization strategies on a cluster computer (2 nodes, 4 cores "C#" each). The model is divided into subdomains (*yellow*) with padding layers to exchange wavefields. Pure domain decomposition requires intra-node (*black*) and inter-node (*blue*) communication (*arrows*). Shot parallelization distributes simulations among the nodes and omits (*red line*) inter-node (*blue*) communication

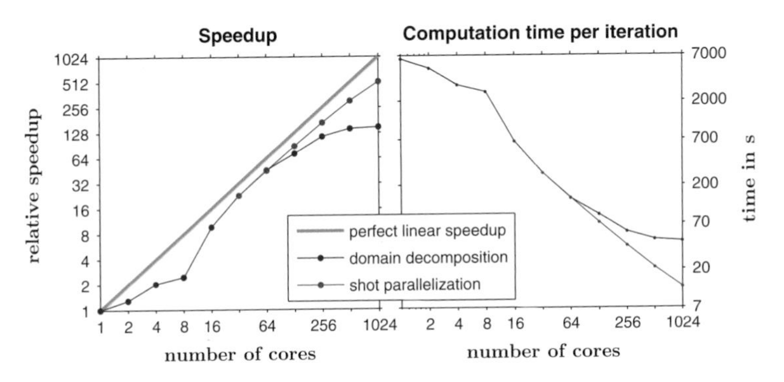

Fig. 3 Scaling behaviour of the 2D acoustic FWT code. Due to the combination with a 8×4 domain decomposition, the shot parallelizations takes effect at 64 and more cores

behaviour (Fig. 3) for an example with 64 sources and a total number of 32 nodes with 32 cores per node—performed on *HERMIT*—illustrates the advantages of shot parallelization. Altogether, nearly all the computation time is consumed by forward modeling. The remaining inversion algorithms are performed quickly. The feasibility of FWT is only given on high performance computers. In particular, even small-scale 3D FWTs require extensive computational resources (see examples in Sects. 4 and 5).

3 Performance Tests of 3D Seismic Modeling

As described in the previous chapter, the scaling behaviour of the inversion codes on large cluster computers is mainly influenced by the forward solver. It is therefore natural to perform benchmarking tests using the forward modeling code first.

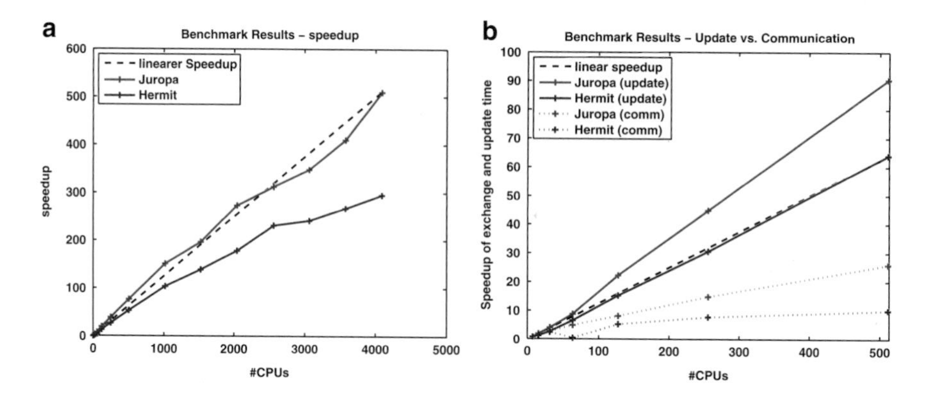

Fig. 4 Speedup benchmarking results of our SOFI3D Finite Difference forward modeling code: (**a**) speedup with respect to the total simulation time, (**b**) speedup with respect to the update (computation) and communication (exchange) times

Our finite difference seismic forward modeling code is parallelized using the classical domain decomposition method, i.e. the overall grid (model domain) is divided into sub-grids (sub-domains) and each processing element (PE) updates the elastic parameters on each sub-grid separately. Usually, a PE is a single core on a multi-core node within the cluster computer. At discrete time steps, elastic parameters are exchanged between neighboring sub-grids using the Message Passing Interface (MPI) in order to allow seismic waves propagating through the entire model domain. In general, the forward modeling can be, thus, divided into a computational part where mainly the local CPU computes updates on each sub-grid and a communication part, where sub-grids exchange information.

We have performed both a speedup and scale-up benchmark test using only our Finite Difference forward modeling code SOFI3D which is also part of the 3D FWT codes. As compiler flags, individual parameters optimal for each supercomputer are used. The speedup benchmark tests the scaling behavior if a wave simulation on a constant model domain is repeated while increasing the number of processing cores. In our case, the modeling domain consists of $512 \times 512 \times 512$ grid points. Within each sub-domain, a source position is defined. The total number of sources is, therefore, depending on the number of sub-grids and number of PEs, respectively. Ideally, if the number of cores is doubled, the total simulation time is halved. In this ideal scenario the ratio $\frac{NP_2}{NP_1}$ equals $\frac{t_1}{t_2}$ with NP_1 is the number of PEs use for the second simulation with a total runtime t_1 and NP_2 is the number of PEs use for the first simulation with a total runtime t_2, i.e. the speedup is 2 (see Fig. 4a, black dashed line). This is about the case for the SOFI3D speedup results gained from the benchmark performed on the *JUROPA* Supercomputer at the Juelich Supercomputing Centre (JSC) (Fig. 4a, red line). Up to 2,000 CPU cores the scaling is even better than linear which is due to caching effects when the sub-grid is small enough to be stored in fast memory of the PE (L2 or L3 cache) instead of the main memory.

The very same benchmark performed on the *HERMIT* Supercomputer at the HLRS looks a little bit different (Fig. 4a, blue line). Up to 1,000 cores (PEs), the speedup is close to be linear. If the number of cores exceeds 1,000, increasing the number of cores by the factor of 2 does not decrease the simulation time by the factor 2, as well, but only e.g. a factor of 1.6, which indicates a non-optimal scaling behavior. To investigate this effect in more details, we can plot the update (computation) and communication (exchange) times separately (see Fig. 4b). First it is noticeable that the computation times scale perfectly linear (solid blue line) or even over-linear (solid red line) while the communication times do not (dashed lines). Obviously the communication is the bottleneck of the forward modeling performance which might be due to the limited bandwidth of the communication layer which has to be shared by all users of the cluster computer or due to the fact that the communication has to be synchronized at each time step. As a consequence, all PEs have to wait if only one PE is delayed for various reasons. Also the MPI communication overhead increases with increasing number of PEs which affects the overall communication performance.

Second, it can be observed that both the computation and communication times scale less with increasing number of cores in case of the *HERMIT* Supercomputer. The computation does not benefit from any caching effects which might be due to a different hardware architecture of the CPU cores and the scaling of the communication times is by comparison worse than the results gained from the *JUROPA* Supercomputer. After all, each simulation on *HERMIT* is speedup when increasing the number of cores and only a fraction of the increased cluster performance is consumed by the communication overhead.

For the scale-up benchmark, the modeling domain for each processing element (CPU core) is constant ($128 \times 128 \times 128$ grid points). With increasing number of CPU cores, the overall modeling domain increases as well. Ideally, the total simulation runtime is constant, i.e. the simulation time on a grid twice the size with twice the number of CPU cores is the constant. According to Fig. 5a, this is the case for both the computation timing results on *JUROPA* and *HERMIT*. The effect that computational effort perfectly scales with increasing number of PEs has been already shown by the speedup and proves that the numerical problem itself can be perfectly parallelized. However, please note that for visibility reasons a constant time of 1,100 s has been subtracted from the *HERMIT* update (computations) times (solid blue line). Actually, each simulation takes more than twice as long on the *HERMIT* supercomputer than on the *JUROPA* supercomputer. If we now focus on Fig. 5b and only the communication times, we—again—see that efforts in order to exchange information between neighboring sub-grid increase with increasing number of increase. This growing communication overhead is especially noticeable on the *HERMIT* supercomputer (solid blue line).

Possible ways to address the communication bottleneck will involve performance toolkits to get insights into the MPI communication. We also plan to implement an overlap of computation and communication. This means, that parts of the sub-grid that will be exchanged with neighboring sub-domains will be updated first. Than communication will be initialized and while most of the communication takes place,

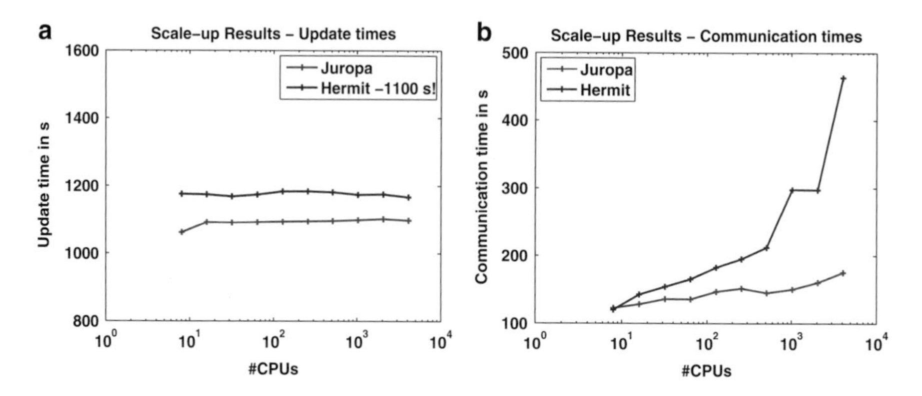

Fig. 5 Scale-up benchmarking results of our SOFI3D Finite Difference forward modeling code: (**a**) scale-up with respect to only the update (computation) times; (**b**) scale-up with respect to only the communication (exchange) times

the remaining and in most cases the majority of the remaining sub-domain will be updated. We believe that this can significantly mask the communication overhead even with increasing number of PEs because computation times are usually larger than computational times.

4 Subproject 1: Acoustic FWT

4.1 Further Development and Synthetic Application

In practice, the general structure of the subsurface is unknown. The occurrence of a simple 1D or 2D geology allows a meaningful application of a 2D FWT. However, a 2D FWT yields an inaccurate reconstruction of the velocity model by projecting events in the seismic data, which arise from significant 3D heterogeneities, to the 2D model. Under these circumstances, the application of a 3D FWT is mandatory. We present the preliminary results of a 3D acoustic FWT in the time domain applied to a 3D geology in a marine shelf environment—representing a typical scenario of hydrocarbon exploration. This synthetic reflection experiment demonstrates the characteristics and difficulties arising from a 3D tomography obtained from a off-shore measurement. The 3D model is based on a section of the 2D Marmousi-II model [6]. The true velocity model is characterized by a water layer on top and a complex geology with fault structures which are embedded into layered sediments (Fig. 6). The model size amounts to $5600 \times 2400 \times 2400$ m. The reflection acquisition geometry is located at the sea surface and consists of 24 explosive sources along three lines—emitting compressional waves into the sea and the subsurface—as well as a total number of 1,778 hydrophones along seven

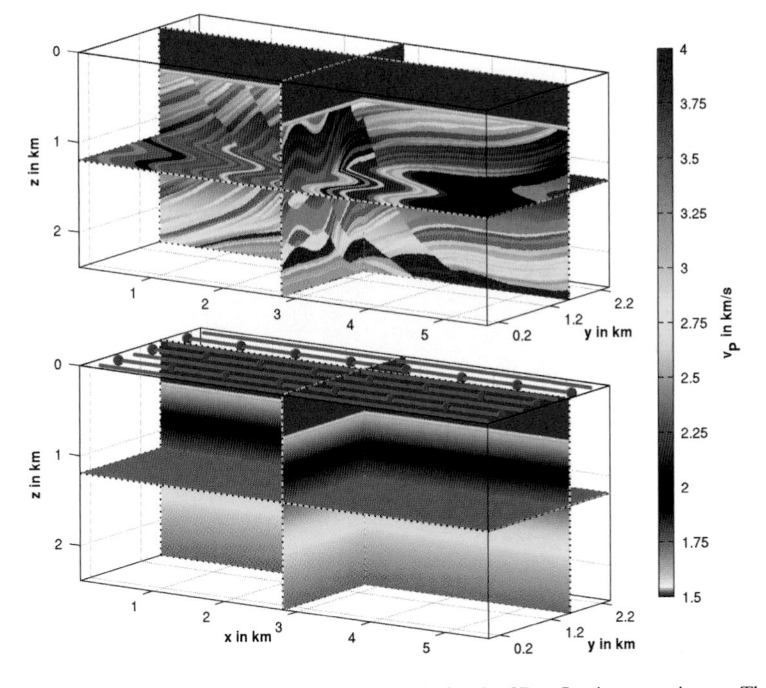

Fig. 6 True (*top*) and initial (*bottom*) velocity models for the 3D reflection experiment. The true model consists of a complex 3D geology and a water layer on top. The acquisition geometry is located at the sea surface and consists of 24 explosive sources (*red markers*) and 1,778 hydrophones (*green lines*)

lines—recording reflected seismic data. The large receiver array is used to ensure an adequate illumination of the subsurface.

For waveform tomography, we have to choose an assumption of the subsurface—the initial velocity model. A meaningful choice is given by a linear 1D velocity gradient (Fig. 6). Apart from the fact that seismic velocities increase with depth, it does not require additional a priori information. It contains the water layer of the true model, which is excluded from the model update. In practice, a 1D velocity model can be obtained by auxiliary boreholes measurements.

The results of the synthetic experiment are illustrated by representative cross sections across the 3D model (Fig. 7). First of all, the 3D FWT sufficiently recovers complex 3D structures of the subsurface model. In particular, upper structures at small scale lengths—such as the layer containing hydrocarbons—are reconstructed very well. It is well-known that both the lateral and vertical resolution of a seismic reflection experiment decreases with increasing depth. This resembles the limitations of the given acquisition geometry. In contrast to a real seismic survey with more than 100 source locations, the sources are spatially undersampled which reduces the illumination of the subsurface and causes some artificial noise in the resulting velocity model. A more appropriate acquisition geometry is necessary.

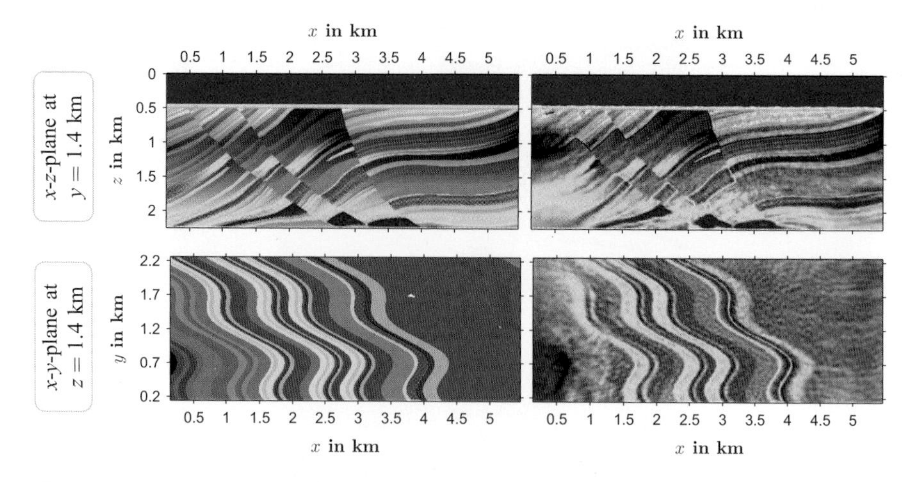

Fig. 7 Exemplary vertical (*top*) and lateral (*bottom*) cross sections of true model (*left column*) and inversion result (*right column*). All planes are located centrally (compare Fig. 6)

Table 1 Computational efforts of FWT on the supercomputer *HERMIT* for applications discussed in Sects. 4, 5.2 and 6

	2D acoustic	3D acoustic	2D elastic	3D elastic
Discretization and efforts of finite-difference modeling				
Grid size in grid points	1200 × 400	560 × 240 × 240	468 × 664	300 × 200 × 350
No. of time steps	5,000	4,000	3,000	3,300
No. of sources	62	24	76	10
Total no. of simulations	18,840	33,720	40,000	2,044
Resource consumption and computational performance				
No. of cores	160	3 072	96	800
Overall memory usage	99 GiB	1.9 TiB	20 GiB	760 MiB
No. of FWT iterations	120	281	250	73
Total computation time	2.9 h	29 h	24 h	42 h
Total core hours	466	88,000	2,300	33,600

However, the computational efforts of the FWT are proportional to the number of sources.

Table 1 makes clear that a 3D time-domain FWT is highly demanding. It summarizes the resource consumption on *HERMIT* of a 3D FWT performed to obtain the result in Fig. 7. The 3D time-domain method is characterized by high computational efforts. In particular, the sophisticated combination of different parallelization strategies (see Sect. 2.3) and the consideration of several seismic sources in modeling allows the involvement of a huge number of CPU cores. Although only 666 out of 4,000 3D snapshots of the forward-propagating wavefield (compare Sect. 2.2) are required for appropriate model improvements, their storage is realized at the expense of an extensive memory consumption.

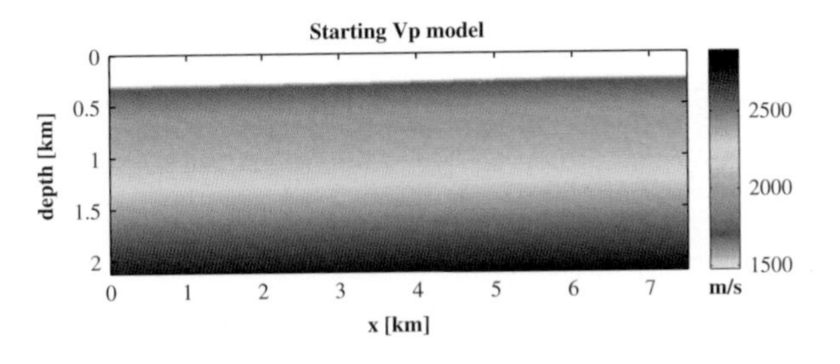

Fig. 8 Starting P-wave velocity model from the traveltime tomography

4.2 Acoustic FWT of Field Data

We apply the 2D acoustic FWT to seismic streamer data acquired in the North Sea [9]. The survey gathered over 30 km of 2D seismic data. The acquisition geometry is composed of a 160-channel, 4,000 m streamer towed at the depth of 10 m. The source is an airgun array, the original shot spacing is 25 m. From the available data, we selected a subregion that extends over 7.5 km, we use 62 shots spaced every 200 m (every fourth shot). Because of the moving streamer geometry, the maximum offset is varying from 975 to 4,000 m.

Prior to the waveform inversion, a specific preprocessing was applied to the raw field data. This data preparation is a fundamental step for FWT. Main objectives are to fit observed and synthetic seismograms, to improve signal-to-noise ration, and to reduce the non-linearity of the inversion.

Marine seismic data are always band limited and lack very low frequencies. Therefore a good starting model, that contains long-wavelength features of the subsurface, is necessary to fill in a gap of low frequencies before the inversion of available frequencies. The starting velocity model for the waveform tomography was generated with the refraction traveltime tomography (Fig. 8).

To reduce the high complexity of the inverse problem, we use the multi-stage inversion approach. The inversion starts at low frequencies and higher frequency content is gradually added. Furthermore, a dynamic time windowing is applied, i.e. for each frequency range, muting windows of different length are applied to the data. It means that early arrivals are inverted before the late arrivals information is included. In this way, the inversion is proceeding from the shallow to the deeper parts of the model. To correct for the amplitude loss with depth due to geometrical spreading and to enhance deeper parts of the model, the gradient scaling with depth is applied. Source wavelet estimation is part of the inversion algorithm and is based on solving a linear least-squares inversion problem.

Figure 9 shows the reconstructed velocity model from waveform tomography for frequencies 3–11 Hz. To check the validity of the inverted model the real data is compared with the synthetic seismograms generated for the inversion result. Although there are some amplitude discrepancies, the synthetic data are comparable to the field data (Fig. 10).

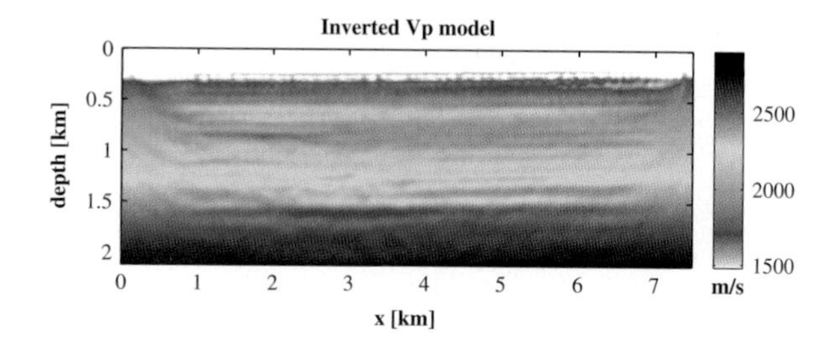

Fig. 9 Waveform inversion results after 90 iterations for the frequency range from 3 to 11 Hz

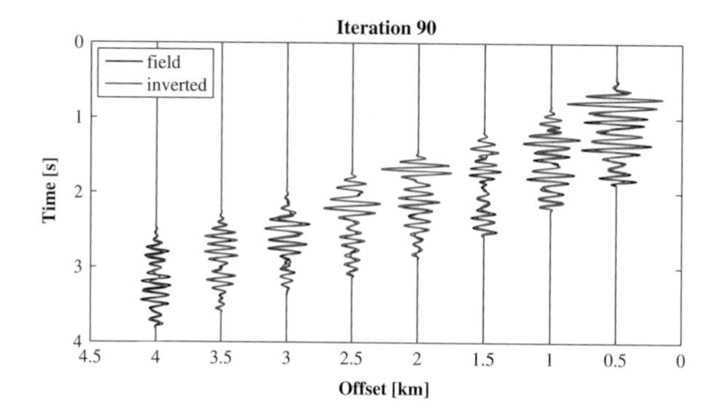

Fig. 10 Comparison of synthetic waveform (*red*) obtained by using the final full waveform inversion result and observed waveform (*black*)

The 2D acoustic FWT has been successfully applied to the real marine-streamer data. Our inversion result demonstrates that the waveform inversion provides more detailed image of the subsurface structures than the traveltime tomography. Furthermore, in case of the applicability of a 2D FWT it is characterized by a quite low resource consumption—compared to a 3D FWT (see Table 1).

5 Subproject 2: Elastic FWT

5.1 2D Elastic FWT of Surface Waves

We applied a 2D elastic FWT code which was originally developed by Köhn [3] to shallow seismic surface waves. Shallow seismic surface waves can be easily excited by a hammer blow and have a high sensitivity to the S-wave velocity v_s in the first few meters of the subsurface.

In shallow seismic field data the effects of viscoelastic damping are significant. We normally observe quality factors Q between 10 and 50. Therefore, we investigate to which degree we have to consider viscoelasticity during the FWT. We run

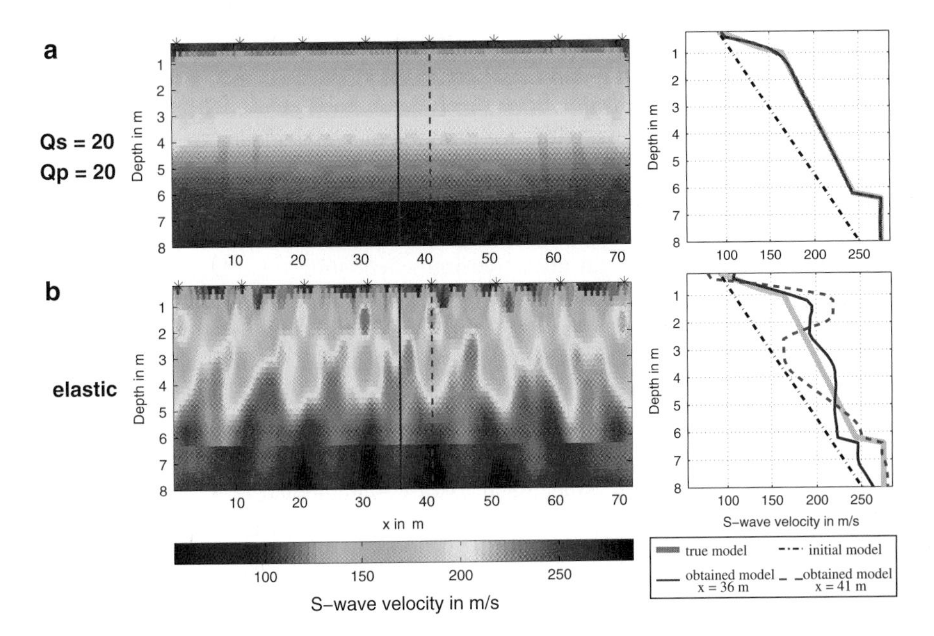

Fig. 11 Results of reconstruction tests for the investigation of the influence of attenuation. On the *left* the obtained S-wave velocity models are plotted. The source positions are marked by the *red stars*. On the *right* vertical velocity profiles at x=36 m (*solid blue line*) and x=41 m (*dashed red line*) are shown together with the true model (*thick grey line*) and the initial model (*dash-dotted black line*). The locations of the profiles are marked by the *blue* and *red lines* in the models

reconstruction tests for simulated viscoelastic observations with a Q factor of 20 using both elastic as well as viscoelastic forward modeling with different Q factors. We do not invert for Q factors but use them as a priori known fixed parameters in the inversion. We only invert for the S-wave velocity. Figure 11a displays the reference result where we used viscoelastic forward modeling in the inversion with the correct Q factor of 20. The S-wave velocity model is reconstructed almost perfectly. In contrast, if we use purely elastic wave propagation in the inversion the reconstruction of the S-wave velocity model is no longer possible. There are lots of artefacts in the S-wave velocity model (Fig. 11b) that partly compensate the differences between viscoelastic and elastic wave propagation. If we use slightly wrong Q factors (between approximately 15 and 30) in the forward modeling of the inversion results the results are similar to the results obtained by an inversion using the correct Q factor. If we use Q factors that stronger deviate from the Q factor of the observed data the inversion result becomes worse.

5.2 3D Elastic FWT

Today, most elastic FWT applications are performed in 2D. 3D FWTs are computationally much more expensive and mainly accomplished in the acoustic approximation. In exchange, detailed images of 3D structures can be achieved. Furthermore, in highly heterogenous 3D medium, the 2D approximation is not valid anymore. Thus, with increasing computational power, 3D FWT will become more and more important. In this section, we describe our implementation of 3D elastic FWT and show some synthetic results (see [2]).

Our implementation is based on a time-frequency approach [11]. This is a combination of forward modeling in the time domain and inversion in the frequency domain and offers an attractive implementation regarding storage costs and runtime. Time domain FWT requires the storage of wavefields of the whole time series for gradient calculations. By contrast, in the frequency domain only few discrete frequencies can be sufficient for the gradient calculations, known as single frequency inversion (e.g. [10]). As the wavefield needs to be stored only for these frequencies (ω), storage costs decrease dramatically. Using discrete Fourier transformations, the wavefields can be transformed from time into frequency domain on the fly without using much extra computational time. The gradients are then calculated as multiplications of forward and conjugate backpropagated wavefields in the frequency domain. This approach additionally enables to increase the frequency during inversion without further effort. Starting from sufficiently low frequencies is extremely important to decrease the ambiguity of the inverse problem (e.g. [10]).

The computational and overall performance of the code is shown for two synthetic inversion tests, which are plotted in Fig. 12. Figure 12a shows the acquisition geometry, used in both tests, which consists of 12 sources and 416 receivers in transmission geometry. This geometry provides a good wave path coverage. We used homogeneous starting models with $v_p = 6{,}300\,\mathrm{m/s}$, $v_s = 3{,}500\,\mathrm{m/s}$ and $\rho = 2{,}800\,\mathrm{kg/m^3}$ and inverted for v_p and v_s. We show results for v_s only. The inverted P-wave velocity models are much smoother due to their larger wavelengths and require higher inversion frequencies.

As a first test a resolution study was performed, using a checkerboard with $\pm 5\,\%$ velocity variation and an edge length of 20 m. Frequencies were increased from 140 to 240 Hz. The result (Fig. 12b) illustrates the good recovery of the alternating cubes (excepting boundaries where wave path coverage is lower). The sharp contrast between the cubes cannot be resolved due to the lack of high frequencies.

In a second inversion test, we inverted data of a random medium model, described by an exponential autocorrelation function in space. The real shear wave velocity model is plotted in Fig. 12c. The model represents a crystalline rock environment and its differently sized 3D structures can be used to test the resolution of 3D FWT on different scales. Frequencies were increased from 140 to 320 Hz during inversion. The S-velocity result is shown in Fig. 12d. The inversion method could successfully recover the differently sized 3D random medium structures down to the scale of a wavelength. Higher frequencies would be required to reconstruct smaller features. Similar to the resolution test, we find that the resolution of the result decreases towards model boundaries, where wavepath coverage is less. Obviously, the quality strongly depends on acquisition geometry, the corresponding wave path coverage and the frequency range used during the inversion.

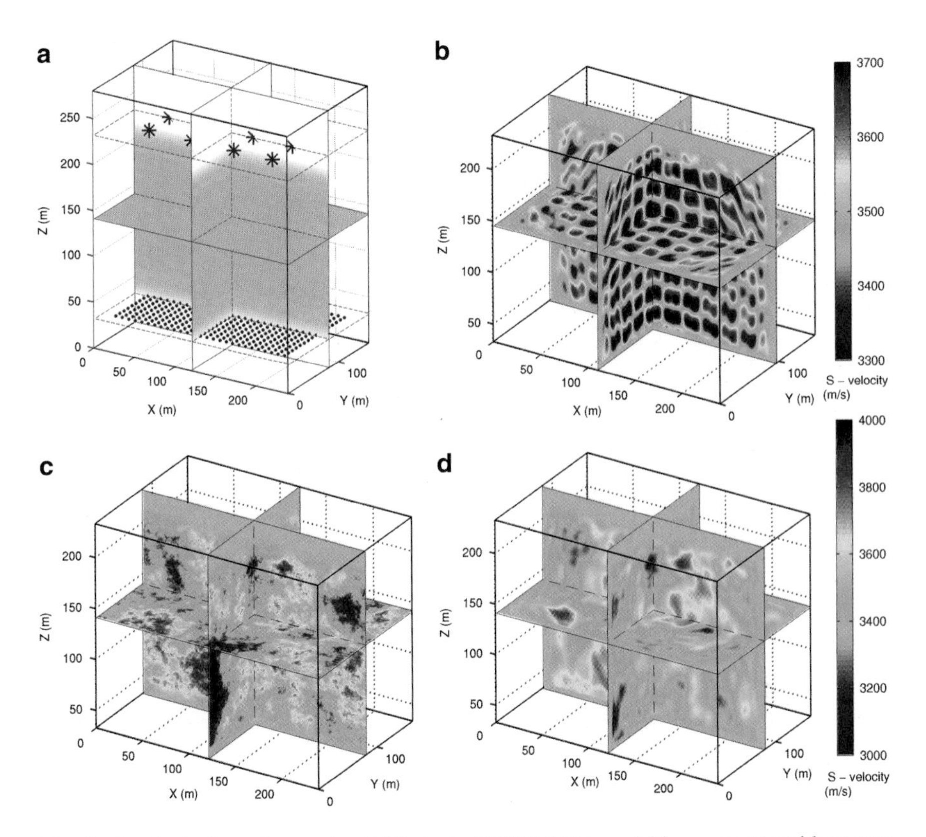

Fig. 12 S-velocity inversion results of 3D elastic FWT: (**a**) 3D acquisition geometry with sources (*) and receivers (+), (**b**) inversion result checkerboard (±5 % variation), (**c**) real random medium model8 and (**d**) random medium inversion result

Table 1 shows the runtime of the synthetic test. Overall, 73 iterations were performed, with 28 forward modelings in each iteration. A domain decomposition with 800 CPUs was employed and the inversion took 42 h. By contrast, a 2D inversion of the random medium model is a small sized problem and requires only 10.5 h on eight cores (Table 1), which is a factor of 400 less compared to 3D. Still, a 3D FWT is able to image 3D structures in the subsurface and can be essential in case of stronger 3D heterogeneities.

6 Subproject 3: Elastic FWT in a Tunnel Environment

Besides the application on shallow seismic surface waves, the FWT is also used for inverting body waves recorded in an underground environment.

Our survey area is an old silver mine in Freiberg, Germany. A block of crystalline rock (Freiberger Greygneiss) of about 50 m × 80 m is surrounded by three

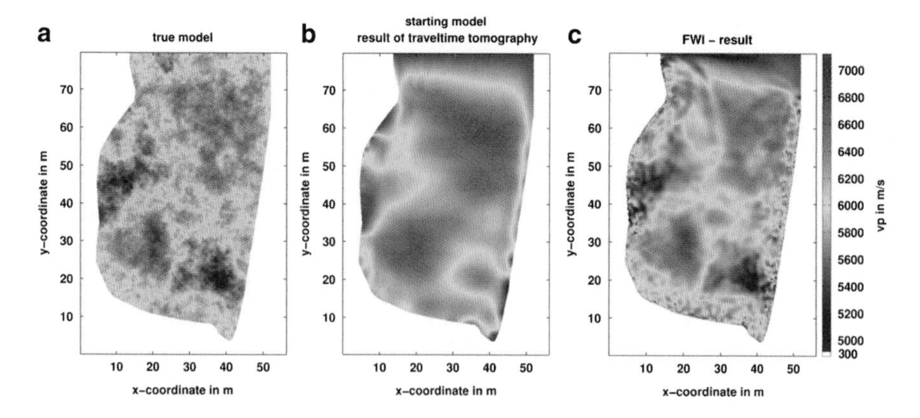

Fig. 13 (**a**) True model: A Random media model with a velocity perturbation which simulates a crystalline rock. (**b**) Starting model: The starting v_p velocity model for the inversion which was derived from a first arrival travel time tomography of the true model. (**c**) FWT-result: Inverted v_p velocity model using the first arrival P-waves only

galleries. The German Research Centre for Geosciences (GFZ), Potsdam, has set up a laboratory (GFZ-ULab) and the necessary infrastructure to perform seismic measurements belowground. Thirty 3-component receivers and 76 shots positions were used for the measurement. The source was a magnetostrictive vibrator which generates sweeps of 300 to 3,000 Hz.

Before inverting for the real data set, some synthetic test on the preprocessing of the data were calculated. The field data had shown that the average v_p velocity is roughly in the range of 6,000 m/s. We used this value as the mean v_p for the synthetic randomly distributed velocity model with an exponential autocorrelation function with a correlation length of 10 m and a Gaussian distribution of $\sigma = 5\,\%$. The v_s velocity model is calculated by the $v_s \approx \frac{v_p}{\sqrt{3}}$ criteria. The density model is kept constant at 2,550 kg/m^3 which is typical for crystalline host rocks. The model is illustrated in Fig. 13a and has been discretized on a 468 × 664 Cartesian grid.

We applied a 2D elastic full waveform tomography (FWT) code which was originally developed by [3] only to the first P-wave arrivals. As a source wavelet, we have chosen a half period of a sin^3 with a frequency of 1,000 Hz. In the next step the autocorrelation of the sweep signal, a Klauder wavelet, is used to get closer to the field data. The starting model was gained from a travel time tomography of the random velocity model is shown in Fig. 13b.

Figure 13c shows the FWT-result for the v_p velocity model. First of all, it is noticeable that the inversion corrects for the too high velocity of the starting model. In the inverted model, the large scale structures could also be reconstructed very well. Even inside these structures, many detailed velocity distributions are similar to the true model. The typical structure size of objects which cannot be reconstructed by the FWT is approximately 2–3 m and is in the range of the smallest wavelength. Nevertheless, some artifacts at the receiver positions and high velocity contrasts, close to the gallery walls can be observed. In the upper most part, the inversion did not update the model because of zero ray coverage.

For the inversion 250 iterations with an average of 160 forward modelings in each iteration were performed. On 96 cores considering 3,000 time steps for every forward modeling the inversion took about 24 h.

7 Summary

Full waveform tomography (FWT) aims to find an optimal subsurface model by iteratively minimizing the misfit between observed and modeled data. We use a conjugate gradient method. Depending on the seismic problem the seismic wavefield can be approximated with different wave equations. We apply 2D and 3D (visco-)acoustic or (visco-)elastic wavefield simulations which are numerically implemented using finite differences. In this report we presented some applications of FWT to synthetic and real data.

2D applications of real sized problems and 3D applications are computationally expensive and require the use of high performance computations. All codes are parallelised with MPI routines using domain decomposition. Furthermore shot parallelisation is implemented in the acoustic FWT codes, which additionally speeds up the inversion. We tested the computational performance of the forward modeling codes by performing scale up and speedup tests. Hereby it was found, that the speedup is about 1.6 for the *HERMIT* supercomputer, but the same test shows a speedup of 2 for the JUROPA supercomputer. This difference is mainly caused by the longer communication times observed for *HERMIT*. Further software development will be necessary to improve this speedup. The code scales well with increasing grid size.

The main area of the acoustic formulation is the marine environment, The inversion of synthetic streamer data was successfully performed in 2D and was extended to 3D synthetic application, where 3D structures of the subsurface are successfully recovered. The 2D code is already applied to real streamer data of the North Sea, where sedimentary layers can be reconstructed. Hereby, the choice of a good starting model and specific preprocessing is necessary.

A main focus of 2D elastic FWT is the inversion of shallow seismic surface waves. This gives insights into the shear wave velocity in the first meters of the subsurface. In synthetic studies, the influence of different parameters, like attenuation was tested. Good results were achieved when using attenuation as a passive parameter in inversion. Additionally, the 2D code will be applied to real data of crystalline rock recorded in an underground lab. Synthetic tests already show the potential of elastic FWT for this problem. The elastic code was also extended to 3D. The 3D FWT performs well in the first small-scale synthetic applications and will be applied to larger models in future work.

References

1. T. Bohlen, Parallel 3-D viscoelastic finite difference seismic modeling. Comput. Geosci. **28**, 887–899 (2002)
2. S. Butzer, A. Kurzmann, T. Bohlen, 3D elastic full-waveform inversion of small-scale heterogeneities in transmission geometry. Geophys. Prospect. **61**(6), 1238–1251 (2013)
3. D. Köhn, Time domain 2D elastic full waveform tomography, Dissertation, Christian-Albrechts-Universität zu Kiel, 2011
4. A. Kurzmann, Applications of 2D and 3D full waveform tomography in acoustic and viscoacoustic complex media, Dissertation, Karlsruhe Institute of Technology, 2012
5. A. Kurzmann, D. Köhn, A. Przebindowska, N. Nguyen, T. Bohlen, 2D acoustic full waveform tomography: performance and optimization, in *Extended Abstracts of the 71st EAGE Conference and Technical Exhibition*, Amsterdam, 2009
6. G.S. Martin, R. Wiley, K.J. Marfurt, Marmousi2—an elastic upgrade for Marmousi. Leading Edge **25**, 156–166 (2006)
7. P. Mora, Nonlinear two-dimensional elastic inversion of multioffset seismic data. Geophysics **52**, 1211–1228 (1987)
8. R.G. Pratt, Seismic waveform inversion in the frequency domain, part 1: theory and verification in a physical scale model. Geophysics **64**, 888–901 (1999)
9. A. Przebindowska, Acoustic full waveform inversion of marine reflection seismic data, Dissertation, Karlsruhe Institute of Technology, 2013
10. L. Sirgue, R.G. Pratt, Efficient waveform inversion and imaging: a strategy for selecting temporal frequencies. Geophysics **69**, 231–248 (2004)
11. L. Sirgue, J.T. Etgen, U. Albertin, 3D frequency domain waveform inversion using time domain finite difference methods, in *Extended Abstracts of the 70th EAGE Conference and Technical Exhibition*, Rome, 2008
12. A. Tarantola, Inversion of seismic reflection data in the acoustic approximation. Geophysics **49**, 1259–1266 (1984)
13. J. Virieux, S. Operto, An overview of full-waveform inversion in exploration geophysics. Geophysics **74**, WCC1–WCC26 (2009)

Particle Simulation in Turbulent Plasmas with Amplified Wavemodes

Sebastian Lange and Felix Spanier

Abstract The dynamic of the heliosphere is governed by plasma turbulence. Furthermore, charged high-energy particles are generated by coronal mass ejections and shock acceleration. In order to understand these complex dynamics and be able to predict the transport of high-energy particles in future, numerical simulations are the method of choice. We present a hybrid simulation code, which is capable of modelling such a system. According to the theories by Goldreich and Sridhar, the turbulence cascade is anisotropic. This is validated by our simulation results. Furthermore, we show the importance of amplified wave modes to the particle scattering. The limits of quasilinear theory of particle transport are discussed and tools for the interpretation of resonance patterns are presented.

1 Introduction

A large spectrum of astrophysical phenomena are connected to plasma turbulence. Especially the evolution of a turbulent magnetised plasma, i.e. our heliosphere is not understood, yet. Numerical simulations poses the unique opportunity to investigate the mechanisms which governs the dynamic of the interplanetary medium, since analytical or experimental approaches can only cover fractions of the jigsaw.

The simulation of plasma turbulence is today one of the major applications in supercomputing. In particular collision dominated plasmas can be described as an magnetised fluid which is governed by the magnetohydrodynamic (MHD) equations.

The investigated system is the solar wind, a plasmaflow, emitted by the Sun, streaming outwards with speeds of 400 up to 1,000 km/s. This plasma is fully

S. Lange · F. Spanier (✉)
Institut für Theoretische Physik und Astrophysik, Universität Würzburg, Emil-Fischer-Str. 31, 97074 Würzburg, Germany
e-mail: felix@fspanier.de

W.E. Nagel et al. (eds.), *High Performance Computing in Science and Engineering '13*, DOI 10.1007/978-3-319-02165-2_10,

ionized and embedded in the magnetic field of the Sun. The solar wind does not only contain thermal particles, but also particles accelerated to energies well above thermal energies. These solar energetic particles (SEP) have typical energies of 10–100 MeV and in extreme cases of up to 1 GeV.

The aim of our simulations is a description of this turbulent system which includes the transport of high-energy particles. Especially a fundamental understanding of particle scattering is necessary to predict the impact of SEP events. The modelling of such a system is achieved by a hybrid code, which simulates the turbulent background plasma via a magnetohydrodynamic approach and the particle transport via a kinetic particle ansatz.

2 Mathematical Description

2.1 Magnetohydrodynamic

The set of equations which is solved in our simulations are the incompressible MHD equations

$$\frac{\partial \boldsymbol{u}}{\partial t} = \boldsymbol{b} \cdot \nabla \boldsymbol{b} - \boldsymbol{u} \cdot \nabla \boldsymbol{u} - \nabla P + \nu \nabla^{2h} \boldsymbol{u} \tag{1}$$

$$\frac{\partial \boldsymbol{b}}{\partial t} = \boldsymbol{b} \cdot \nabla \boldsymbol{u} - \boldsymbol{u} \cdot \nabla \boldsymbol{b} + \nu \nabla^{2h} \boldsymbol{b} \tag{2}$$

with the magnetic field $\boldsymbol{b} \equiv \boldsymbol{B}/\sqrt{4\pi\rho}$ with constant mass density ρ and the fluid velocity $\boldsymbol{u}$. The total pressure is denoted by P and describes both, thermal and magnetic pressure with $P = p + B^2/(8\pi)$. Viscous and Ohmic dissipation are given by the generalised resistivity ν, which causes wavenumber-diffusion. We consider here also hyperdiffusivity, which occurs for $h > 1$. Especially for the fast solar wind, which we are interested in, this fluid can be considered as incompressible. This leads together with the solenoidality condition for the magnetic field to the boundary conditions

$$\nabla \cdot \boldsymbol{u} = 0 \tag{3}$$

$$\nabla \cdot \boldsymbol{b} = 0 \tag{4}$$

Using these boundary conditions, it is possible to find a closure for the MHD equations. The pressure P may be derived by taking the divergence of the MHD equations. This in turn yields [1]

$$\nabla^2 P = \nabla \boldsymbol{b} : \nabla \boldsymbol{b} - \nabla \boldsymbol{u} : \nabla \boldsymbol{u}. \tag{5}$$

The solution for incompressible fluid problems can be achieved by the spectral method.

In the incompressible regime of a magnetised plasma the MHD-turbulence consists of only two types of waves, which propagate along the parallel direction—the so-called pseudo- and shear Alfvén waves. First ones are the incompressible limit of slow magnetosonic waves and play a minor role within anisotropic turbulence [1]. The pseudo Alfvén waves polarisation vector is in the plane spanned by the wavevector $\boldsymbol{k}$ and $\boldsymbol{B}_0$. The shear waves are transversal modes with a polarisation vector perpendicular to the $\boldsymbol{k}$ - $\boldsymbol{B}_0$ plane. They are circularly polarised for parallel propagating waves. Both species exhibit the dispersion relation $\omega^2 = (v_A k_\parallel)^2$.

Since the model consists only of these two wave types it is suitable to use a description with Alfvénic waves moving either forwards or backwards. This is achieved by introducing the Elsässer variables [2]

$$\begin{aligned} \boldsymbol{w}^- &= \boldsymbol{v} + \boldsymbol{b} - v_A \boldsymbol{e}_\parallel \\ \boldsymbol{w}^+ &= \boldsymbol{v} - \boldsymbol{b} + v_A \boldsymbol{e}_\parallel, \end{aligned} \tag{6}$$

and transforming (2) into a suitable form of

$$\begin{aligned} (\partial_t - v_A k_z)\,\tilde{w}_\alpha^\mp &= \frac{i}{2}\frac{k_\alpha k_\beta k_\gamma}{k^2}\left(\widetilde{w_\beta^+ w_\gamma^-} + \widetilde{w_\beta^- w_\gamma^+}\right) \\ &\quad - i k_\beta \widetilde{w_\alpha^\mp w_\beta^\pm} - \frac{\nu}{2} k^{2h} \tilde{w}_\alpha^\mp \\ k_\alpha \tilde{w}_\alpha^\pm &= 0 \end{aligned} \tag{7}$$

where the tilde-notation represents quantities in Fourier space. The components of the wavevector are written as k_α.

To resemble the excitation of waves through nonthermal particle distributions, particles act resonantly with the plasma waves. These interactions cause a streaming instability, which has the highest growth rate in the background field direction. Consequently we assume that only purely parallel-propagating Alfvén waves are modified.

The Alfvén wave generation mechanism by nonthermal particles will not be investigated in detail. The driving mechanism which is assumed for our simulations is the streaming instability. The estimate of the wave growth rate is described in [3]. The streaming instability is caused by energetic proton scattering off interplanetary Alfvén waves. During the scattering process the particle changes its pitch angle cosine by $\Delta\mu$ while its momentum in the wave frame remains constant. Thus the particles' energy in the plasma frame is changed by $v_A p \Delta\mu$ and consequently also the Alfvén wave energy due to energy conservation.

The actual growth rate is

$$\Gamma(k) = \frac{\pi \omega_{cp}}{2 n_p v_A} \int \mathrm{d}^3 p \; v\mu |k| \; \delta\left(|k| - \frac{\omega_{cp}}{\gamma v_p}\right) f \tag{8}$$

with proton cyclotron frequency ω_{cp}, proton speed v_p, the Lorentz factor γ, the proton number density n_p, as well as μ the pitch angle cosine and f the proton distribution function. We will use peaks at $k = 2\pi \cdot 8$ and $k = 2\pi \cdot 24$ which represent proton energies of $E \approx 64$ MeV and $E \approx 7$ MeV respectively. Using the resonance condition this leads to a length scale of

$$L_{\text{scale}} = \frac{2\pi n}{eB_0} \gamma m_p c v \approx 10^8\,\text{cm}. \tag{9}$$

2.2 *Particle Transport*

The second aim of our simulations is the kinetic description of high-energy particles, namely the SEPs, which are influenced by the plasma turbulence with amplified wave modes significantly. Those charged particles are governed basically by the relativistic Lorentz force

$$\frac{\mathrm{d}}{\mathrm{d}t}\,\gamma\,\boldsymbol{v} = \frac{q}{mc}\left[c\boldsymbol{E}(\boldsymbol{x},t) + \boldsymbol{v} \times \boldsymbol{B}(\boldsymbol{x},t)\right], \tag{10}$$

which acts on the particles at position $\boldsymbol{x}$ through the electric $\boldsymbol{E}$ and magnetic fields $\boldsymbol{B}$ of the plasma. The particle velocities are denoted by $\boldsymbol{v}$, c is the speed of light and γ represents the Lorentz factor. A charged particle within a magnetic field will perform a gyromotion with the frequency

$$\Omega = \frac{ZeB}{\gamma mc} \quad \text{and Larmor radius} \quad r_L = \frac{v}{\Omega}. \tag{11}$$

In analytical or experimental approaches it is impossible to trace individual particle. Thus, a statistical description of particle transport is necessary to compare the different results to our simulations.

The general statistical approach of particle transport by the relativistic Vlasov equation. It contains the generalised forces g_{X_σ} represent the effects of the electromagnetic fields to the expectation value $F_T = \langle f_T \rangle$ of a particle distribution f_T of species T and are basically the time derivatives of the denoted variables, e.g. $g_{X_\sigma} = \dot{X}_\sigma$. This equation is in general analytically unsolvable.

A common approach to describe particle transport analytically is the QLT which was first suggested by Jokipii [4] in the context of energetic charged particle transport in turbulent magnetic fields. Its core is the assumption of unperturbed particle orbits. This implies the fluctuation amplitudes to be small, leading to a quasilinear system. The Vlasov equation would then simplify to

$$\frac{\partial F_T}{\partial t} + v\mu \frac{\partial F_T}{\partial Z} - \Omega \frac{\partial F_T}{\partial \phi} = S_T(X_\sigma, t)$$

$$+\frac{1}{p^2}\frac{\partial}{\partial X_\sigma}\left(p^2\frac{\partial F_T}{\partial \widehat{X_\sigma}}\underbrace{\int_0^t \mathrm{d}s\,\langle g_{X_\sigma} g_{\widehat{X_\sigma}}(\widehat{X_\sigma}, s)\rangle}_{D_{X_\sigma \widehat{X_\sigma}}}\right), \tag{12}$$

where ϕ is the polar angle and μ is the pitch angle cosine

$$\mu = \cos(\alpha) = \frac{\boldsymbol{v}\cdot\boldsymbol{B}}{|\boldsymbol{v}||\boldsymbol{B}|}. \tag{13}$$

Further, the method of characteristics was applied and the hat-symbol represents quantities along the characteristics [5]. This equation is known as the Fokker–Planck equation with the Fokker–Planck coefficients $D_{X_\sigma \widehat{X_\sigma}}$. One of the most interesting variables is the pitch angle diffusion coefficient $D_{\mu\mu}$. It describes the pitch angle scattering of the particle and is consequently connected to the scattering mean free path, which can be evaluated by the observable angle distribution and Monte Carlo simulations [6]. Here, scattering means a resonant wave–particle interaction of nth order which fulfills the condition

$$k_\parallel v_\parallel - \omega + n\,\Omega = 0, \quad n \in \mathbb{Z} \tag{14}$$

[7], where ω is the wave frequency and $k_\parallel$ its parallel wavenumber. Ω is again the particle gyrofrequency and $v_\parallel$ its parallel velocity component. Whether a particle with $v_\parallel$ interacts resonantly with a wave with $k_\parallel$ is determined by the polarisation [8]. Because our MHD-model consists of pseudo and shear Alfvénwaves only, certain values of n can be connected to the different types of waves. The Cherenkov resonance, n=0, is generated only by waves with compressive magnetic field ($\delta\mathbf{B}\cdot\mathbf{B}_0 \neq 0$), i.e., pseudo Alfvén waves. Purely parallel waves can contribute only to the $n = \pm 1$ resonance, and resonances with $|n| > 1$ occur only for waves with non-vanishing perpendicular wavenumber, i.e., with obliques Alfvén waves.

For further reading and a detailed insight of the derivation of (12) we would like to refer to [9]. A serious problem of the QLT is the limited applicability. The approximation of small fluctuations only holds for weak turbulent systems, where $\delta B/B_0 \ll 1$. This is important for the local field which acts on the individual particles. For instance, a strong turbulent region would change effectively the direction of the mean magnetic field and consequently the gyromotion of the particles. Hence the assumption of unperturbed orbits would be invalid. Another problem of the QLT is the inadequate description of particles propagating perpendicular to the mean magnetic field, $\mu \approx 0$. However, regarding (14) a very small parallel particle velocity will generate a singularity for $v_\parallel \to 0$. One aspect of our simulations concentrate on the applicability of the QLT to describe solar energetic particle transport.

The pitch angle scattering coefficient can be calculated by different methods. In order to compare it to the QLT, we use the definition

$$D_{\mu\mu} = \lim_{\Delta t \to \infty} \frac{(\Delta\mu)^2}{2\,\Delta t} \overset{t \gg t_0}{\approx} \frac{(\Delta\mu)^2}{2\,\Delta t}, \tag{15}$$

where the time interval $\Delta t = t - t_0$ compared to an initial state at t_0 is assumed to be large, i.e. the time evolution t has to be sufficient to develop resonant interactions. Instead of using the change of the pitch angle cosine $\Delta\mu = \mu_e - \mu_0$, the definition with the change of the pitch angle $\Delta\alpha = \alpha_e - \alpha_0$ itself leads to

$$D_{\alpha\alpha} = \frac{(\Delta\alpha)^2}{2\,\Delta t} = \frac{D_{\mu\mu}}{1-\mu^2}, \tag{16}$$

which represents the scattering frequency.

3 The GISMO Code

The hybrid-code GISMO was developed to simulate particle transport in turbulent plasmas. The application of the heliosphere and SEP event is only a fraction of its capabilities. In order to model both, MHD and kinetic particle movement, the code divides into two main parts: the pseudospectral GISMO-MHD and the kinetic GISMO-Particles. In the following we present a brief description of these codes.

3.1 GISMO-*MHD*

The MHD part of our code is a fully parallelised incompressible pseudospectral MHD-Code. Its heart is the calculation of the incompressible MHD-equations (2). The general flow diagram is given in Fig. 1.

The main advantage of using the approach of pseudospectral methods is obviously the transformation of PDEs into to ODEs. Unfortunately for the case of the MHD equations as nonlinear equations it is necessary to calculate the nonlinearities in real space. This requires two Fourier transforms per time step, which are in best case $\mathcal{O}(N \log N)$ where N is the number of grid points. This effort is only reasonable as pseudospectral methods allow an easy implementation of the incompressibility condition.

Using the scheme described above it is possible to simulate the MHD equations with much less dissipation compared to methods involving differencing schemes. This provides a more physical energy spectrum with an inertial range over many wavenumbers. Nevertheless the k-space resolution has to be reduced due to aliasing errors in the worst case by $2/3$ of all k-values. This is why large grids are necessary.

In Fig. 1 the schematic of the code is shown. The MHD-simulation runs mostly in Fourier space. After setting the initial conditions which in our case consist of the divergence free turbulent magnetic- and velocityfields, the code checks the stability

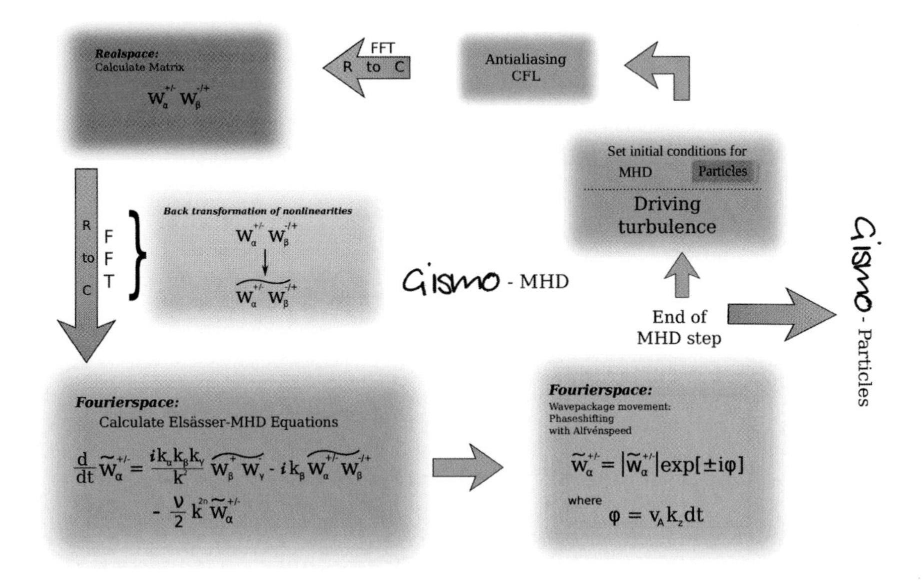

Fig. 1 Schematic flow diagram of the GISMO-MHD code

of the system and removes aliasing errors which in spectral methods is almost the same. Then a transformation into realspace is executed to calculate the nonlinear terms. Back in wavenumber space the time-evolution of the MHD-equations are solved by Runge–Kutta fourth order method. Depending on the type of turbulence additional energy is injected or not. Afterwards the timestep is finished and the whole algorithm repeats.

The output of the simulations which is mainly the mentioned physical fields can be done at arbitrary timesteps and is stored in a HDF5 datatype. The MHD-code supplies full ability to restart at any timestep from these quantities and it is possible to change the setup during this e.g. to injected peakstructures to the spectrum. Another small output is the energy of the magnetic and velocity fields as well as crosshelicity. For analytical reasons and due to its datasize these are stored in binary format.

3.2 GISMO-*Particles*

A suitable numerical approach for solving (10) for gyrating particles is the implicit scheme of the *Boris-push*. The basic idea has been given by Boris [10] where the iterations of the Lorentz force are separated in two partial steps. First, the particles are accelerated by the electric field within a half time step. Second, the gyromotion of the particles is calculated, which is caused by the magnetic field. After that, the electric fields acts again for another half time step to complete the iteration.

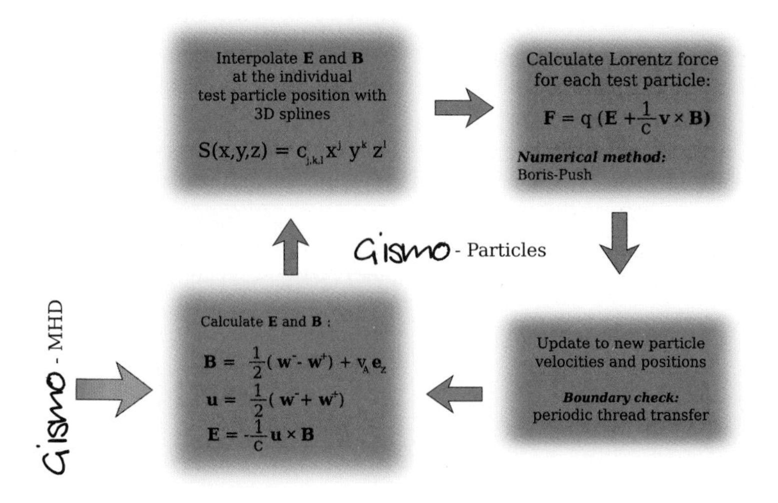

Fig. 2 Schematic flow diagram of the GISMO-Particles code

This approach leads to a discretisation of the Lorentz force (for the detailed set of equations see [11]).

The advantage of the Boris-push is the very high numerical stability. The particles are assumed to undergo gyromotions, hence the particle orbits themselves are stable for an arbitrary time discretisation. Even in the limit of $\Delta t^{\mathrm{num}} \gg \Omega^{-1}$ the particle orbit is stable, but converges to an adiabatic drift motion. The limitation of this method is the correct resolution of the Larmor radius r_L. If the timestep is chosen too large, this would lead to a big deviation from the analytical r_L. GISMO-PARTICLES measures the deviation from r_L and adapt it to the preferred value. To specify, in our simulations an accuracy of the order of $|r_L - r_{\mathrm{measured}}|/r_L \approx 10^{-5}$ was used.

A limitation to the method of the Boris-push are ultrarelativistic particle speeds. In this case the conservation of energy would be violated, since the ideal ohmic law is not fulfilled anymore. Beyond Lorentz factors of $\gamma \approx 10^3$ fictitious forces start to act and this approach is not applicable furthermore [12].

The flow diagram of GISMO-Particles is illustrated in Fig. 2. Both parts of GISMO are calculated for each step. After iterating the Elsässer MHD-fields $w^{\pm}$, they are transformed into the physical electric and magnetic fields which are transferred to GISMO-PARTICLES. Then the Boris-push will be performed. Each particle will respond to its local fields, which are calculated by an averaging method via three-dimensional splines [13, 14]. Periodic spatial boundary conditions were used, thus the number of particles remained constant in each simulation.

3.3 Resources

For this project simulations with 256^3 and 512^3 grid points have been performed. There are various physical setups for the background simulations, which require typically $\approx 4 \cdot 10^4$ timesteps to converge. Every simulation with 256^3 grid points needed approximately 3,500 CPU-h and the 512^3 runs ca. 32,000 CPU-h, accordingly. In the converged stage the peak driver has been used, which is running for 10^3 timesteps. The decay of the amplified wave modes is dependent on the used simulation setup and the total energy input of the peak. Typically, this stage requires additional 10^4 timesteps. Each peak simulation required roughly 2,000 CPU-h for the 256^3 simulations, i.e. 17,000 CPU-h for the 512^3.

After the MHD-part was simulated, the test-particles were injected at different stages of the peak evolution. A typical timescale for resonant interaction is around 20 gyro-periods. Depending on the sensitivity of the Larmorradius (see Sect. 3.2) and the simulation setup, a typical simulation time was 1,500 CPU-h for 256^3 grid points and respectively 14,000 CPU-h for 512^3.

4 Scientific Results

In this section we present selected results from our simulations. We focus on the particle transport simulations, after a brief overview of the peaked turbulence results. A detailed insight is given by Lange and Spanier [15] for the turbulence evolution with amplified wave modes and [16] for particle pitch angle scattering in such turbulence.

4.1 Simulation Setup

We focus on the environmental conditions in the solar corona at a distance of three solar radii. The magnetic background field in this region is approximately $B_0 = 0.174$ G which yields assuming a particle density of $10^5\,\text{cm}^{-3}$, an Alfvén speed of $v_A = 1.2 \cdot 10^8\,\text{cm}\,\text{s}^{-1}$ [17]. This region is of special interest because particle acceleration by CME driven shocks is strongest there.

The outer length scale of the simulated system is $L_{\text{scale}} = 3.4 \cdot 10^8$ cm, which was estimated by using the wave growth rate from [3], given by (8).

A remark on the notation: the wavenumber is in general defined as $k = (2\pi j)/L$ where j stands for the numerical grid position. For simplicity we used the normalised wavenumbers $k' = kL = (2\pi \cdot j)$ throughout.

On this scale the magnetic background field is assumed to be homogeneous. The spatial resolution is 256^3 grid points, resulting in 128^3 points in k-space of which $|\boldsymbol{k}'| = 2\pi[0 \cdots 42]$ wave modes are active modes that remain unaffected by (anti)aliasing. The resistivity parameter was set to $\nu = 1$ in numerical units and the hyperdiffusivity coefficient to $h = 2$.

The turbulence has been simulated using an anisotropic driving mechanism. Energy is, therefore, constantly injected into the simulation volume at the first five numerical wavenumbers in perpendicular ($k'_\perp = 2\pi[0\cdots4]$) and 15 in parallel direction ($k'_\parallel = 2\pi[0\cdots14]$). The reference frame is the solar wind frame.

The anisotropy was chosen for two reasons. First, to mimic the preferential direction of the solar wind, where particles streaming radially away from the Sun form the Parker spiral. Consequently, these particles can deposit their energy in a parallel direction on different scales. This is mainly valid in the vicinity of the Sun, which we are interested in. Second, a slab-component of SW turbulence is observed also at small scales in the parallel direction. To ensure turbulence evolution up to high parallel wavenumbers, the driving range was extended along the parallel axis. This is necessary because the parallel evolution is much weaker than the perpendicular [18, 19]. Even though this is primarily a technical aspect to ensure the extent of the spectrum to higher $k_\parallel$, it is still in line with observations. An isotropic driver would not yield sufficiently turbulent modes at high $k_\parallel$.

The turbulence driving is performed by allocating an amplitude with a phase to the Elsässer-fields within the Fourier space. The amplitude follows a power-law of $|\boldsymbol{k}|^{-2.5}$ and is initialised using a normal distribution. The phase was randomly chosen between zero and 2π. These settings are divergence free and Hermitian symmetric. After this initialisation the values were scaled to the desired scenario, which in our case is a $\delta B/B_0$ ratio of roughly 10^{-2}. In this inertial range energy is injected every 0.03 s (10 MHD-steps), which leads to a saturated turbulence—an equilibrium between dissipation and injection.

Within the saturated turbulence a Gaussian distributed energy peak with purely parallel wavenumber $\boldsymbol{k} = k\boldsymbol{e}_\parallel$ was injected. For this purpose two different wave modes were chosen. To investigate the physics of an SEP-event a wavenumber of $k_\parallel = 1.5\cdot10^{-7}\,\text{cm}^{-1}$ is used. This corresponds to a numerical wavenumber of $2\pi\cdot8$, which is still within the driving range of the turbulence. To represent injection at smaller scales, a peak is set at $k_\parallel = 4.4\cdot10^{-7}\,\text{cm}^{-1}$. With $k'_\parallel = 2\pi\cdot24$ this wave mode is not in the driving range of the background turbulence. The injection of energy in these modes was done gradually over a certain time interval.

The grid resolution is especially important for the turbulence evolution. More grid points means also a larger wavenumber space with less antialiasing and consequently a higher number of active modes. For this purpose another simulation setup was used with 512^3 grid points.

To summarise, these four simulations with excited wave modes at $k'_\parallel = 2\pi\cdot8$ and $k'_\parallel = 2\pi\cdot24$ both within turbulent fields governed by a strong and a weak B_0, are the starting point for the test particle simulations and the calculations of the scattering coefficient $D_{\mu\mu}$.

All of the test particle simulations by GISMO-PARTICLES have been performed with 10^5 protons with an initial uniform distribution in μ and ϕ and random positions $\boldsymbol{x}$. This amount has proven to be sufficient in test simulations for good statistics. These initial conditions aim to a complete coverage of the phase space in μ for the test particles. Thus, we are not interested in the development of special distribution functions.

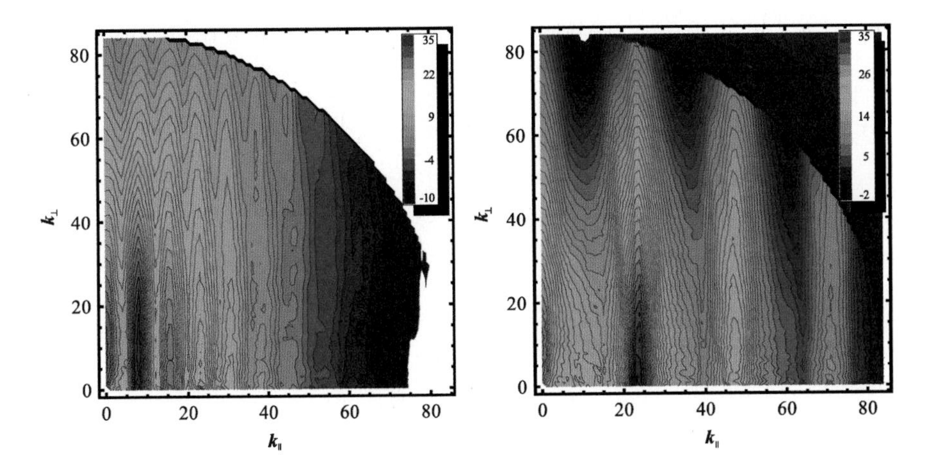

Fig. 3 Two-dimensional magnetic energy spectra of the decay stage for both peaks in the simulation with $B_0 = 0.174\,\text{G}$ and higher resolution with a grid of 512^3 cells. The *left figure* shows the state for the peak at $k'_\parallel = 2\pi \cdot 8$ at $t = 14.45\,\text{s}$. The *right figure* shows the $k'_\parallel = 2\pi \cdot 24$ peak at $t = 3.4\,\text{s}$. The larger Fourier space grid reveals the higher harmonics of the $k'_\parallel = 2\pi \cdot 24$ peak, which are not observable in the simulations with lower resolution 256^3

A constant absolute value of the momentum was chosen that particles with a certain parallel velocity component fulfill the resonance condition (14). Consequently, a resonant value of μ

$$\mu_R = \frac{\omega - n\,\Omega}{k_\parallel\, v} = \frac{\omega - n\,\Omega}{L_{\text{scale}}^{-1}\, k'_\parallel\, v} \tag{17}$$

must be within the interval $[-1; 1]$ for a given particle speed v. The resonant μ_R for particles with speed $v = 1.21 \cdot 10^{10}\,\text{cm}\,\text{s}^{-1}$ is $\mu_R = 0.86$ for the $k'_\parallel = 2\pi \cdot 8$ and $\mu_R = 0.29$ for the $k'_\parallel = 2\pi \cdot 24$ peak.

To investigate the particle scattering in different stages of the turbulence evolution, multiple simulations were performed. The most promising match between QLT and the simulation results is expected to be in stages with small $\delta B/B_0$. According to this, the particle simulations were performed in the driving phase of the peaks as well as in the decaying phase. A simulation at maximum driven peaks would simply lead to random scattering where no reasonable prediction can be made by QLT.

4.2 Selected Results

In Fig. 3 the evolution of the turbulence with amplified wave modes is shown by two dimensional energy spectra within the decay stage of the peaks for the 512^3 simulations. Two remarkable developments can be observed. First, despite the peak

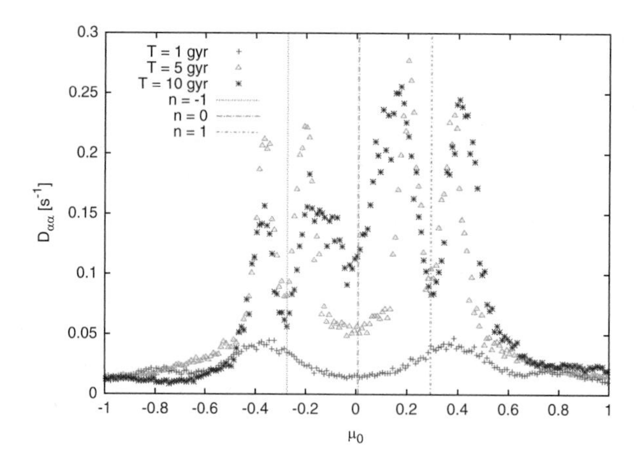

Fig. 4 Time evolution of the pitch angle scattering coefficient $D_{\alpha\alpha}$. The used setup is the low B_0^l, 256^3 gridsize, peak position $k'_{\|} = 2\pi \cdot 24$, driven stage. The *vertical lines* represent the predicted positions of the resonances according to (17)

is driven in parallel direction only, the energy has spread towards perpendicular wavenumbers. This evolution is fast compared to the parallel development, since the energy structure is strongly anisotropic. Second, the generation of higher harmonics can be observed. This mechanism—most probable caused by wave steepening or n-wave-interactions—lead to an effective energy transport mechanism towards high parallel wavenumbers. Furthermore, these modes can interact resonantly with high-energy particles and lead to scattering.

The pitch angle scattering coefficient $D_{\alpha\alpha}$ was calculated for different stages of the evolution. In Fig. 4 we show $D_{\alpha\alpha}$ for the first simulation setup within the 256^3 grid with the peak position $k'_{\|} = 2\pi \cdot 24$ at the driven stage. We observe a convergence between five and ten gyration periods. The resonant interaction with the $k'_{\|} = 2\pi \cdot 24$ mode is located at $\mu_R = 0.29$, whereas Fig. 4 shows two maxima at $\mu = 0.2$ and $\mu = 0.4$ which seem to move away from each other during the time development.

The real resonant structure is revealed by the scatter plot in Fig. 5. Indeed, the resonance is centered at the predicted position, but tilted because the particles orbits are not unperturbed as assumed by QLT. The finite influence of the waves to a particle will not only generate interactions with a single μ, but also within a small interval in μ. Therefore, the mean value of the pitch angle will interact resonantly. The tilt spreads the particles to a wide $\Delta\mu$ range, which results in splitting of the maximum of the pitch angle diffusion coefficient into two maxima at both sides of the resonant pitch angle $\mu_R = 0.29$. This is because the calculation of $D_{\alpha\alpha}$ in the QLT with (16) is not dependent on the sign of $\Delta\mu$. Consequently, the scattering coefficient is mapped due to the square value of $\Delta\mu$ to two different maxima. This also explains the movement of the maxima between five and ten gyration periods in Fig. 4, because $\Delta\mu$ increases with time. The smaller resonance at $n = -1$ i.e.

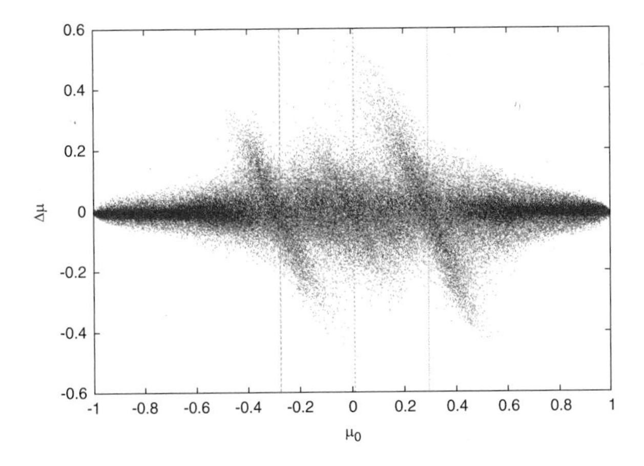

Fig. 5 Scatter plot for the 256^3 gridsize, peak position $k'_{\|} = 2\pi \cdot 24$, driven stage, $t = 10$ gyration periods. Each *dot* represents the total change of μ of an individual particle. Three resonance patterns are visible and centered at $\mu_R = -0.27, 0, 0.29$

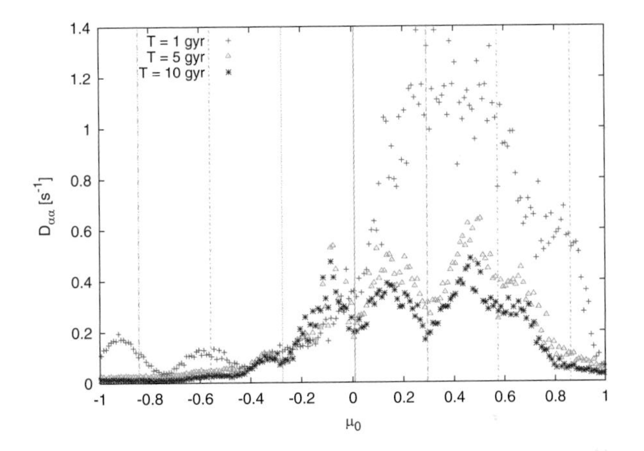

Fig. 6 Time evolution of the pitch angle scattering coefficient $D_{\alpha\alpha}$. The used setup is the 256^3 gridsize, peak position $k'_{\|} = 2\pi \cdot 24$, decay stage. The structure became more complex due to higher order resonances. Again, the *vertical lines* represent the predicted positions of the resonances according to (17)

$\mu_R = -0.27$ is caused by the different polarisations of the peaked mode. That means resonances with $\mu < 0$ are caused by left-handed circular polarised parts of the peaked mode and $\mu > 0$ by right-handed. Furthermore, in the scatter plot the Cherenkov resonance $n = 0$ is visible. It is hardly observable in the $D_{\alpha\alpha}$ plot, due to the dominant $|n| = 1$ resonances and their tilt.

The test particle simulation with the decay phase of the peaked mode at $k'_{\|} = 2\pi \cdot 24$ shows resonant interactions beyond the fundamental resonance. For example the maximum located near $\mu = 0.6$ in Fig. 6 represents the $n = 2$ interaction. Note

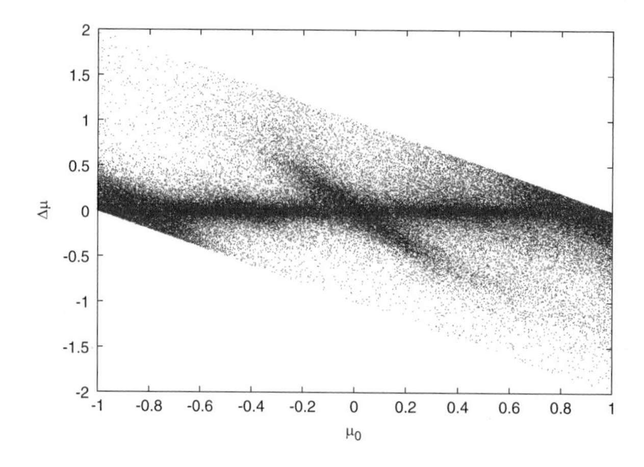

Fig. 7 Scatter plots for the 512^3 gridsize, peak position $k'_{\parallel} = 2\pi \cdot 8$, decay stage, $t = 25$ gyration periods. The higher resolution of the grid causes slightly stronger scattering, due to the higher amount of active modes. The resonance patterns are comparable to the 256^3 grid

that the tilt in the corresponding scatter plot (not shown here) causes again the split of the maximum in $D_{\alpha\alpha}$. In this consequence, the interpretation of the resonance pattern is more difficult, since the resonances overlaps each other.

At last we present the results of the pitch angle scattering results for the higher resolution with 512^3 grid points and an amplified wave mode at peak position $k'_{\parallel} = 2\pi \cdot 8$. In Fig. 7 the scatter plot is shown, where the resonances at the predicted positions, $\mu_R = 0.86$ for $n = 1$ and $\mu_R = -0.84$ for $n = -1$, are clearly visible. The higher number of active modes due to the higher resolution causes a stronger ballistic scattering background. Consequently, the tilt is much more prominent and many particles reach the maximum change from their initial μ_0, which is indicated by the clear straight lines which delimitate $\Delta\mu$. Because of this strong scattering, the coefficient $D_{\alpha\alpha}$ would appear very unstructured. Therefore, it is not shown here.

5 Conclusion

Our turbulence simulations with amplified wave modes show a strong anisotropic evolution of energy. This means, the energy transport by turbulent cascading is fast towards high perpendicular wavenumbers. However, an efficient energy transport to high parallel wave modes can be achieved by the generation of higher harmonics. These can be caused by high-energy protons which generate streaming instabilities and hence cause wave growth in parallel direction.

The pitch angle scattering of charged particles in such a turbulent state is influenced by the amplified wave modes significantly. However, the pitch angle scattering coefficient $D_{\alpha\alpha}$, which is calculated by a quasilinear approach, is not easy to interpret. The main reason is the finite distortion of the particle orbits, which is not described by QLT. A valuable tool to understand the underlying resonant structure is the scatter plot. Clear interactions at the predicted resonant μ_R (17) can be observed in this picture.

References

1. J. Maron, P. Goldreich, Astrophys. J. **554**, 1175 (2001). doi:10.1086/321413
2. W.M. Elsässer, Phys. Rev. **79**, 183 (1950). doi: 10.1103/PhysRev.79.183, http://link.aps.org/doi/10.1103/PhysRev.79.183
3. R. Vainio, Astron. Astrophys. **406**, 735 (2003). doi:10.1051/0004-6361:20030822
4. J.R. Jokipii, Astrophys. J. **146**, 480 (1966). doi:10.1086/148912
5. R. Schlickeiser, *Cosmic Ray Astrophysics* (Springer, Berlin, 2002)
6. N. Agueda, R. Vainio, D. Lario, B. Sanahuja, Adv. Space Res. **44**, 794 (2009). doi:10.1016/j.asr.2009.05.023
7. R. Schlickeiser, Astrophys. J. **336**, 243 (1989). doi:10.1086/167009
8. C.F. Kennel, F. Engelmann, Phys. Fluids **9**, 2377 (1966). doi:10.1063/1.1761629
9. R. Schlickeiser, *Cosmic Ray Astrophysics* (Springer, Berlin, 2002)
10. J.P. Boris, in *Proceedings of the Fourth Conference on Numerical Simulation Plasmas* (Naval Res. Lab., Washington, DC, 1970), pp. 3–67
11. C.K. Birdsall, A.B. Langdon, *Plasma Physics via Computer Simulation*, 1st edn. (Taylor and Francis, New York, 2005)
12. J. Vay, Phys. Plasmas **15**(5), 056701 (2008). doi:10.1063/1.2837054
13. F. Spanier, M. Wisniewski, Astrophys. Space Sci. Trans. **7**, 21 (2011). doi:10.5194/astra-7-21-2011
14. M. Wisniewski, F. Spanier, R. Kissmann, Astrophys. J. **750**, 150 (2012). doi:10.1088/0004-637X/750/2/150
15. S. Lange, F. Spanier, Astron. Astrophys. **546**, A51 (2012). doi:10.1051/0004-6361/201219579
16. S. Lange, F. Spanier, M. Battarbee, R. Vainio, T. Laitinen, Astron. Astrophys. **553**, A129 (2013). doi:10.1051/0004-6361/201220804
17. R. Vainio, T. Laitinen, H. Fichtner, Astron. Astrophys. **407**, 713 (2003). doi:10.1051/0004-6361:20030914
18. P. Goldreich, S. Sridhar, Astrophys. J. **438**, 763 (1995). doi:10.1086/175121
19. P. Goldreich, S. Sridhar, Astrophys. J. **485**, 680 (1997). doi:10.1086/304442

Part II
Solid State Physics

Holger Fehske

The following chapter reveals that solid state physics based research has profited substantially from the new leading edge supercomputer technology now being available at the High Performance Computing Center Stuttgart. The three contributions selected use primarily first-principle ab initio molecular dynamics and density functional techniques to obtain largely unbiased results for electronic, structural, electrochemical and vibronic properties of several fascinating systems, ranging from three-dimensional ionic materials with long-range interactions, quasi two-dimensional metal-oxide metal interfaces and ferroelectric surfaces, to finite colloidal semiconductor nanoclusters or quantum dots.

In their Stuttgart–Bielefeld–Warwick teamwork J. Roth et al. determine successfully different crystal structures of silica, magnesia and aluminia. Methodically the Wolf summation for long-range interactions combined with the Tangney–Scandalo model for polarisable oxygen and the Streitz–Mintmire model for variable charges at metal oxide-metal interfaces were implemented in the molecular dynamics (MD) package `IMD` used by the authors. The parallelization of the minimisation procedure that gives the chemical potential will be a challenge for continuing investigations. From a physical point of view, the study of cracks leads to interesting new effects as flexoelectricity in cubic periclase MgO and antiflexoelectricity in α-aluminia.

The Paderborn University theory group headed by W. G. Schmidt addresses the effect of ferroelectric poling on the water adsorption lithium niobate surfaces. To this end the adsorption energy, site and configuration are determined and the bonding between water and the surface is explored in dependence on temperature and pressure. In notable agreement with experiments, the adsorption configuration and adsorbate mobility are found to be strongly polarisation dependent, and the water adsorption occurs mainly non-dissociatively, independent of the coverage.

H. Fehske
Institut für Physik, Lehrstuhl Komplexe Quantensysteme, Ernst-Moritz-Arndt-Universität Greifswald, Felix-Hausdorff-Str. 6, 17489 Greifswald, Germany
e-mail: fehske@physik.uni-greifswald.de

The density functional calculations (DFT) performed in this study are based on the `VASP` and `QuantumEspresso` simulation packages, where a substantial speedup could be achieved using the `ScaLAPACK` library.

G. Bester and P. Hang from the Stuttgart MPI for Solid State Research continued their ab initio DFT investigations of vibrational properties of nanostructures, now with a focus on confinement effects for group III–V and II–VI atoms and thermodynamic properties of colloidal quantum dots. Furthermore the electron–phonon interaction and the vibron–vibron coupling in (semiconductor) silicon clusters are analyzed by means of MD simulations. Particularly the vibrational cooling process of selected Si–Si and Si–H stretching modes is investigated. The nanoclusters studied were constructed by cutting a sphere where geometry relaxation is accomplished by the Broyden–Fletcher–Goldfarb–Shano procedure for the optimization of the ionic positions. The `CPMD` code package used by the authors in the numerical work admits of a scalable DFT implementation within a hybrid `MPI–OpenMP` programming model.

Summing up it could be assessed that in the field of computational solid state physics a variety of interesting activities have been undertaken. Without any doubt the presented projects were of high scientific quality and beyond that demonstrate the large impact of supercomputing in this area.

Molecular Dynamics Simulations with Long-Range Interactions

Johannes Roth, Philipp Beck, Peter Brommer, Andreas Chatzopoulos, Franz Gähler Stephen Hocker, Siegfried Schmauder, and Hans-Rainer Trebin

Abstract The Wolf summation (Wolf et al., J. Chem. Phys. **110**, 8254 (1999)), an order $O(N)$ method for the calculation of long-range interactions, has been adapted successfully to the simulation of metal oxides. We present the combination of the method with the Tangney–Scandolo model for polarizable oxygen and the Streitz–Mintmire model for variable charges at metal oxide–metal interfaces. The methods have been implemented successfully in our molecular dynamics package IMD and applied to the structure of several metal oxides. The new methods allow for the simulation of cracks in oxides and the study of the flexoelectricity effect.

J. Roth (✉) · P. Beck · A. Chatzopoulos · H.-R. Trebin
Institut für Funktionelle Materalien und Quantentechnologien, Universität Stuttgart, Stuttgart, Germany
e-mail: johannes@itap.physik.uni-stuttgart.de; trebin@itap.uni-stuttgart.de; beck@itap.physik.uni-stuttgart.de; andreas@itap.physik.uni-stuttgart.de

P. Brommer
Department of Physics, University of Warwick, Coventry CV4 7AL, UK
e-mail: P.Brommer@warwick.ac.uk

F. Gähler
Fakultät für Mathematik, Universität Bielefeld, Bielefeld, Germany
e-mail: gaehler@math.uni-bielefeld.de

S. Hocker · S. Schmauder
Institut für Materialprüfung, Werkstoffkunde und Festigkeitslehre, Universität Stuttgart, Stuttgart, Germany
e-mail: stephen.hocker@iwmf.uni-stuttgart.de; siegfired.schmauder@iwmf.uni-stuttgart.de

W.E. Nagel et al. (eds.), *High Performance Computing in Science and Engineering '13*, DOI 10.1007/978-3-319-02165-2_11,

1 Introduction

Computer simulations with long-range interactions are challenging for large systems, since the basic method, Ewald summation [7], is of order $O(N^{3/2})$ at best. In the Ewald method the charges are smeared out with a Gaussian, thus becoming short-ranged. To get the correct interaction, the smeared out part has to be computed in Fourier space where it is also short-ranged. Adapting free parameters results in the $O(N^{3/2})$ complexity. The drawback is that the method only works for periodic boundary conditions. Surfaces for example cannot be simulated with the standard method.

Wolf et al. [23] showed that it is possible under certain special conditions to avoid the reciprocal part. The order of the algorithm is reduced $O(N)$ and simulations are possible for more general boundary conditions without a reciprocal space part. The Wolf summation is especially suitable for compensated systems, i.e. structures where the number of positive and negative charges is the equal and the system is neutral as in metal oxides for example. The Wolf summation leads to a performance gain of 3 orders of magnitude in molecular dynamics (MD) simulations.

In the case of oxides it is not sufficient to treat the oxygen atoms as point charges, since they are polarizable. The Tangney–Scandolo (TS) model takes care of this property by assigning an induced dipole to each oxygen atom. In the simulations the dipole strengths are computed self-consistently between each molecular dynamics (MD) simulation step. Since the TS model was developed for the Ewald summation and SiO_2, we first had to re-optimize the interaction model parameters [5]. This was carried out with our force-matching code `potfit` [4]. New results could then be obtained for magnesia [1]. The importance of the polarizability of the oxygen atoms was demonstrated also in [2]. Since the Wolf summation permits open boundaries, the simulation for cracks in alumina became possible [11]. This lead to the discovery of (anti-)flexoelectricity: α-alumina Al_2O_3 cannot have a permanent polarization even if it is deformed homogeneously since it is inversion symmetric. A spontaneous polarization, however, is observed in front of a crack since the deformation is inhomogeneous. Thus the crack generates a deformation gradient which leads to polarization even in the case of inversion symmetry. Only a short account of this effect is given below since it requires a more detailed study.

The third step in our sequence of methods is the combination of metal oxides and metals. Up to now *fixed* charges have been preset in the process of modeling the interaction with `potfit` by fitting the interaction to quantum mechanical ab initio data. Since metals are conducting and metal oxides are insulating, a combination of the two across an interface will lead to charge exchange and interface layers. The atoms in the metal will be neutral, while the atoms in the metal oxide will be charged. Thus we have to introduce variable charges and we must determine their size. Streitz and Mintmire [19] have developed a model (SM) which permits the determination of the charges by minimizing the chemical potential. We have implemented the SM model in IMD and applied it successfully to aluminum-alumina interfaces. New interactions have been determined again by `potfit`, dynamical simulations, however, have not yet been successful since the interfaces turned out to be unstable in this model.

A further challenge is the determination of the chemical potential. Since this is a global observable, the minimization has to be carried out for the whole sample which requires the application of massively parallel minimization packages. Up to now a simple conjugate gradient algorithm is used which works well but is not easily parallelizable. However, it should be possible to precompile the charges from small samples since the chemical potential is constant in the bulk far from the surface and to apply the calculated charges in the large samples. Since it varies only slowly during simulation it should be possible to adjust it iteratively like we do for the induced dipoles. The Wolf summation and the TS model on the other hand are trivially parallelizable since they fit perfectly in the standard interaction scheme which uses cell decomposition [15, 20].

2 Description of the Models

In the following a short account of the different models is given as they have been implemented in `IMD`. The molecular dynamics simulation package `IMD` itself has been described in [15, 20] and in several HLRS reports [17]. The code can be obtained from the webpage imd.itap.physik.uni-stuttgart.de, together with a detailed userguide.

2.1 The Wolf Summation for Long-Range Interactions

Ewald summation [7] is the method of choice for long-range interactions, but it is limited to small systems. For larger systems lattice- and tree-based methods exist which can lead to a complexity of nearly $O(N)$. The drawback is that they do not fit very well into the simulation code `IMD` and are difficult to parallelized. For small systems they are typically slower than the Ewald summation due to a large pre-factor in the complexity. Methods which involve Fourier transform are limited to periodic boundary conditions and cannot be applied to crack simulations for example.

The Wolf summation [23] is especially suitable for ionic melts and solids since these systems are compensated: the ions screen their opposite charges across rather short distances already. Wolf et al. have evaluated the Ewald method and found conditions under which the sum in reciprocal space and thus the Fourier transform part is negligible. The drawback is a rather large cut-off radius which is about twice as large as without electrostatic contributions. Neighbor lists can be quite big, but on today's machines that is not a problem. The gain of the method is a short-ranged potential which can be simulated together with all the other short-range interactions. It therefore leads to the full flexibility in the choice of boundary conditions.

2.2 The Tangney–Scandolo Model for Polarizable Oxygen

Tangney and Scandolo [21] devised a method for the simulation of silica SiO_2. The non-electrostatic parts of the interaction are modeled by a parameterized Morse-stretch potential. The ions get a fixed charge while the oxygen atoms possess an additional dipole moment. The polarizability of the oxygen ions is fixed and their dipole moment is calculated from the local field intensity. The field is computed in an iterative way from the monopoles and dipoles. The electrical field is determined by the ionic charges and dipole moments of the previous time step, in the original TS model by Ewald summation. The starting value of the iteration is predicted from the previous time steps to increase the efficiency.

After extending `IMD` for the simulation of monopoles with the Wolf summation we split up the generally non-convergent sums for energies and forces of the dipoles into a direct and a reciprocal space part. We could show that this second part can be neglected as it was the case for the monopoles. The real space part has to be cut off smoothly such that the forces vary continuously. For the conservation of energy it was found that the cutoff for monopoles and dipoles has to be chosen consistently. The final choice of the cutoff radius is a compromise between precision and efficiency.

The original TS calculations could be reproduced exactly with our implementation of the Ewald summation. With the Wolf summation the simulation time could be boosted by three orders of magnitude. The next step was to optimize the potential parameters for the different behavior of the Wolf summation. This was carried out with our `potfit` code and was applied to silica, alumina and magnesia MgO [1].

2.3 Details of the Implementation in `IMD` of the Wolf Summation Combined with the TS Model

Smooth Cut-Off

For MD simulations with limited-range interactions the potentials and their first derivatives must go to zero continuously at a cut-off radius r_c; otherwise, atoms crossing this threshold might get unphysical kicks. For the Morse-stretch pair potential this is generally not problematic, as it decays with r_{ij} fast enough. Following Wolf [23] the potential $U_{\mathrm{MS}}(r_{ij})$ is replaced by

$$\tilde{U}_{\mathrm{MS}}(r_{ij}) = U_{\mathrm{MS}}(r_{ij}) - U_{\mathrm{MS}}(r_c) - (r_{ij} - r_c)U'_{\mathrm{MS}}(r_c), \tag{1}$$

where a prime denotes a derivative with respect to r.

The other functions used in the TS model have a general r dependence of the form $r^{-n}, n \in \{1, 2, 3\}$. Especially the Coulomb energy with its r^{-1} dependency cannot simply be cut off without a treatment as in (1), for otherwise the energy of the system would fluctuate strongly with r_c, without convergence to the proper value. But even with a smooth cut-off (1), for which the Coulomb energy converges, a rather large cut-off radius would be required to make shifting of the potential negligible. Fortunately, the Wolf direct summation method [23] includes a weak exponential damping of the Coulomb potential by $\operatorname{erfc}(\kappa r)$. Such a damped potential can be cut off smoothly at a much smaller radius r_c without affecting the result. All integer powers of r^{-1} are treated in a way to conserve the differential relationship between the functions, i.e. the damped functions are

$$r^{-1} \rightarrow r^{-1} \operatorname{erfc}(\kappa r) =: f_{-1}(r), \tag{2}$$

$$\begin{aligned} r^{-2} = -\frac{d(r^{-1})}{dr} &\rightarrow -\frac{d(r^{-1}\operatorname{erfc}(\kappa r))}{dr} \\ &= r^{-2}\operatorname{erfc}(\kappa r) - \frac{2\kappa \exp(-\kappa^2 r^2)}{\sqrt{\pi} r} =: f_{-2}(r). \end{aligned} \tag{3}$$

This procedure is also required to conserve the energy during an MD simulation, as discussed in more detail in Sect. 2.3.

The damped potentials are then shifted to zero and zero derivative at the cut-off radius, as in (1). This allows for limited-range MD simulations with a standard MD code. The computational effort of such a simulation scales linearly in the number of particles (as the number of interactions that need to be evaluated per particle does not increase with the number of particles), but scales roughly with $O(r_c^3)$.

Energy Conservation

In MD simulations the energy is conserved, if the forces on the particles are exactly equal to the gradient of the potential energy with respect to the atomic coordinates. Otherwise, the energy might oscillate or even drift off if not controlled by a thermostat. In standard MD simulations, the requirement is usually automatically fulfilled: The forces are calculated as the derivative of the potential, which depends directly on the atomic positions. In the TS model, there is also an indirect dependence, as the potential is also a function of the dipole moments:

$$U = U(\{\boldsymbol{r}_i\}, \{\boldsymbol{p}_i(\{\boldsymbol{r}_j\})\}). \tag{4}$$

This would in principle lead to an extra contribution to the derivative of the potential,

$$\frac{dU}{d\{\boldsymbol{r}_i\}} = \frac{\partial U}{\partial\{\boldsymbol{r}_i\}} + \frac{\partial U}{\partial\{\boldsymbol{p}_i\}}\frac{\partial\{\boldsymbol{p}_i\}}{\partial\{\boldsymbol{r}_j\}}, \tag{5}$$

which would be practically impossible to determined effectively. Luckily, if the dipole moments are iterated until convergence is reached, we are at an extremal value of the potential energy, with $\partial U/\partial\{\boldsymbol{p}_i\} = 0$, and so this part need not be evaluated. Imperfections in convergence may lead to a drift in the energy, however, as was already observed by Tangney and Scandolo [21].

When applying the Wolf formalism to the TS potential, another issue arises concerning the conservation of energy. It can most easily be explained with a simple one-dimensional example. Given are two oppositely charged point charges $\pm q$ at a mutual distance r. If the negatively charged one is polarizable with polarizability α, it will get a dipole moment $p = \alpha q/(kr^2)$, with $k = 4\pi\epsilon_0$. This leads to a total interaction energy

$$U = \underbrace{-2 \cdot \frac{1}{2}\frac{1}{k}\frac{q^2}{r}}_{q-q} \underbrace{-2 \cdot \frac{1}{2}\frac{q}{k}\frac{p}{r^2}}_{q-p} + \underbrace{\frac{1}{2}\frac{p^2}{\alpha}}_{\text{dipole}}, \tag{6}$$

from which it follows that

$$\frac{\partial U}{\partial p} = -\frac{1}{k}\frac{q}{r^2} + \underbrace{\frac{p}{\alpha}}_{=\frac{1}{k}\frac{q}{r^2}} = 0. \tag{7}$$

Here, $q-q$ denotes the Coulomb interaction between charges, $q-p$ the interactions between charge and dipole, and the last term is the dipole energy. When we now damp and cut off the interactions, we replace the r^{-1}, r^{-2} functions by their damped and smoothed counterparts $\tilde{f}_{-1}(r), \tilde{f}_{-2}(r)$. If energy conservation is to be maintained, the differential relation between the $\tilde{f}_{-n}$ must be the same as for the r^{-n}:

$$\frac{d\tilde{f}_{-1}(r)}{dr} = -\tilde{f}_{-2}(r). \tag{8}$$

As a consequence, the first two derivatives of the smoothed damped Coulomb potential must be zero at r_c.

In MD simulation it is computationally advantageous to represent pair potential functions internally as functions of r^2, and their derivative as $f^{\hat{\prime}} := r^{-1}df/dr$. In this way, Wolf summation can be applied to dipolar interactions in the TS potential model.

2.4 The Streitz–Mintmire Model for Variable Charges

At interfaces of metals and metal oxides charges have to be variable at least in an transition region. The charge of the metal atoms will depend on the local oxygen concentration and vary from neutral to the maximal charge in the oxide. There are nearly no models for large scale simulations. It is not clear if mirror charges or interface diffusion has to be taken into account. In the model of Streitz and

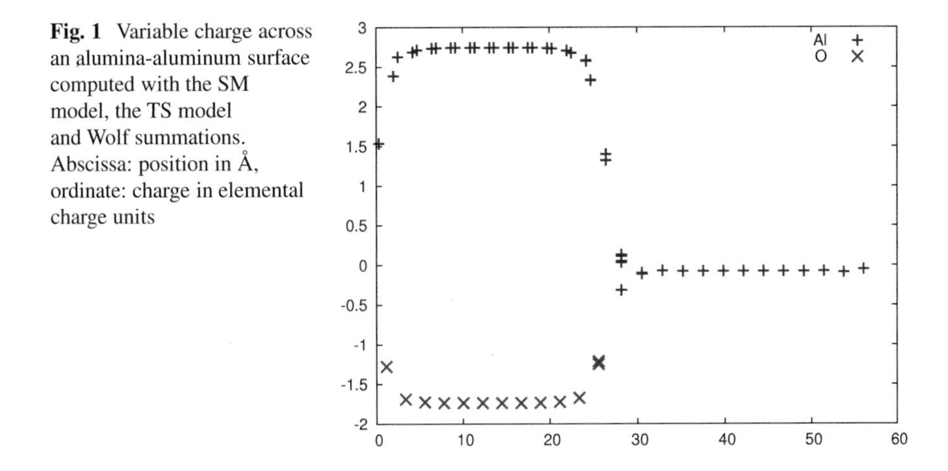

Fig. 1 Variable charge across an alumina-aluminum surface computed with the SM model, the TS model and Wolf summations. Abscissa: position in Å, ordinate: charge in elemental charge units

Mintmire [19] long-range interactions are taken into account adequately. The total energy is composed of an embedded atom potential and an electrostatic contribution which contains the charges to second order. The charge of each atom is obtained by minimizing the electrostatic energy under the condition of charge neutrality. The optimized charges minimized a quadratic expression which leads to a linear system of equations. Streitz and Mintmire [19] have inverted the matrices directly which is not suitable for large systems and massively parallel computations. We have implemented a conjugate gradient method which solves the equations iteratively. The efficiency can be increased by taken the values of the previous step as starting conditions since the optimized charges change only little from time step to time step.

Again we apply the Wolf summation at all steps: for the determination and minimization of the charge matrix as well as for the calculation of the long-range energies and forces. Figure 1 shows the variable charges across a alumina aluminum surface.

A weak point of the SM model is the necessity to minimize the global chemical potential which involves global communication. An improvement has been studied by Elsener et al. [6] for the case of single oxygen ions in a metal matrix. There the global chemical potential can be replaced by a local chemical potential. A similar approach could be advisable for interfaces, since in the bulk far from the interface the charges should attain the values of the pure oxide or metal. Consequently, the proper values can be precompiled from smaller samples and the actual values will change only little during simulation. Thus the can be adjusted iteratively.

3 Results

A short overview is given of the validation of the new models in simulations with `IMD` for a number of structures by studying structural and thermodynamic properties. Crack simulations and first results for (anti-)flexoelectricity will also be presented.

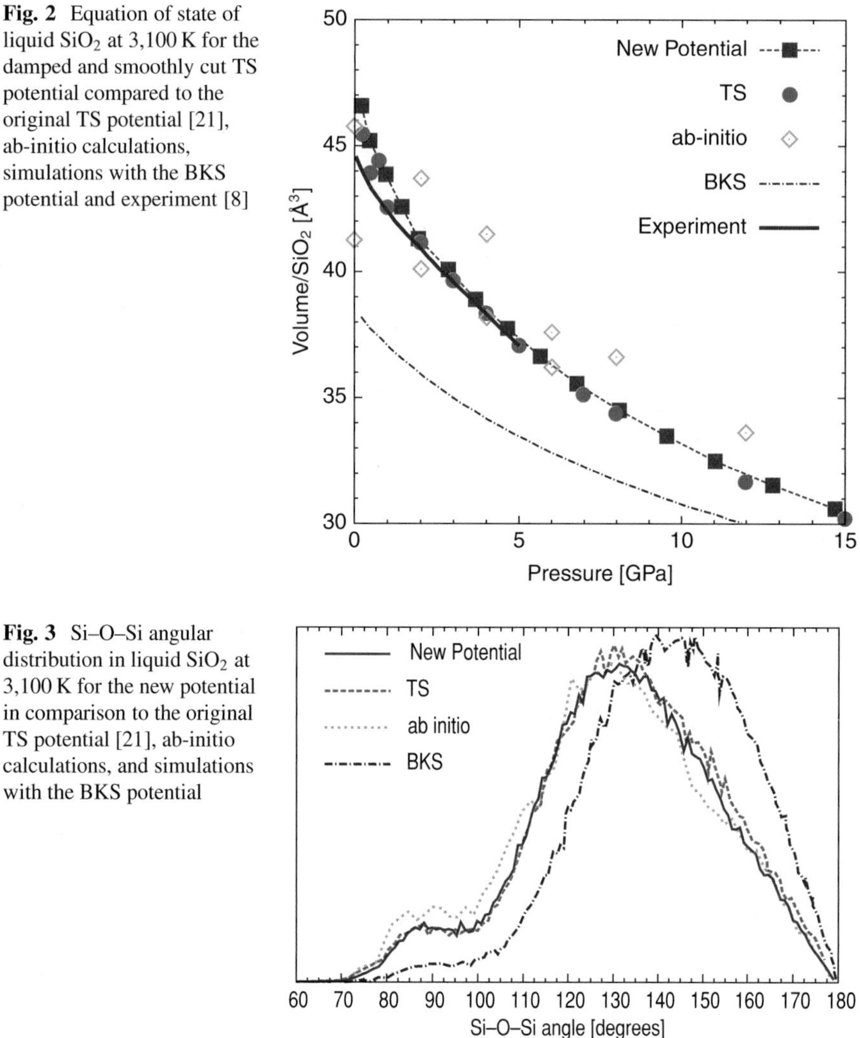

Fig. 2 Equation of state of liquid SiO_2 at 3,100 K for the damped and smoothly cut TS potential compared to the original TS potential [21], ab-initio calculations, simulations with the BKS potential and experiment [8]

Fig. 3 Si–O–Si angular distribution in liquid SiO_2 at 3,100 K for the new potential in comparison to the original TS potential [21], ab-initio calculations, and simulations with the BKS potential

3.1 Simulation of Oxides

Validation of the Wolf Summation

A comparisons of the Ewald and Wolf summation of silica shows that the differences are marginal. As a first step we have assumed the original TS parameters for SiO_2. Figures 2–4 clearly show a good agreement with ab-initio data and the original TS model with Ewald summation. For comparison simulations with the pair potential of Beest et al. [22] are given which is still widely used.

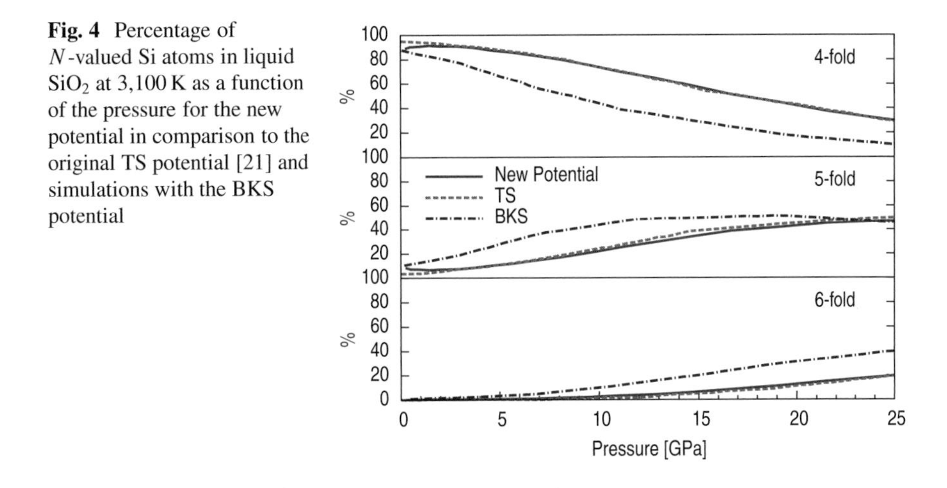

Fig. 4 Percentage of N-valued Si atoms in liquid SiO_2 at 3,100 K as a function of the pressure for the new potential in comparison to the original TS potential [21] and simulations with the BKS potential

The original potential has also been applied to other structure variants of SiO_2, namely quartz, coesite, and cristobalite [5], where it also lead to good agreement with experimental data and the results of Tangney and Scandolo. As a rule of thumb, the real space cut-off radius required for the Wolf summation should be about five times the largest nearest neighbor distance of opposite charges in the system. Still we obtain a speedup of more than two orders of magnitude compared to the original TS model with Ewald summation.

Silica and Magnesia

After validating the TS model with Wolf summation for the original interactions new interactions have been fitted with `potfit` for silica and magnesia [1]. For both structures the parameters of the Morse-stretch potential, the polarizabilities and the charges were obtained. The potential was validated by determining thermodynamic, microstructural and vibrational properties of silica. They are in excellent agreement with ab initio data. As a test, the new parameters were applied to amorphous silica which was not used for the fitting procedure. The vibrational density of state could be reproduced very well. A further test was the application to low-pressure α-quartz which yields a hard case for the transferability of the interaction. We get reasonable results although the interaction was not optimized for such crystal structures.

A further test was the application of the TS model to magnesia MgO. Details and interaction parameters can be found in [1]. The interaction parameters were fitted with `potfit` to a database of small reference structures for which the forces, stresses, and energies were computed with `VASP` [13, 14]. The new interactions are validated in the same way for liquid magnesia. Again an excellent agreement was found if compared to a study with ab initio data [12]. Although the potential was fitted only up to 15 GPa, it yielded good values up to 160 GPa. The vibrational density of state is also in good agreement with experiment [3] and ab initio calculations [9].

Not surprisingly the scaling with system size is linear for both materials, silica and magnesia.

The Necessity of Polarizability

The question is how important the polarizability is for the simulation of the oxides. A systematic comparison has been carried out for silica, magnesia and alumina [2] for two types of force fields which only differ in the polarizable term; the non-electrostatic and Coulomb term have the same form for both force fields. The influence on the radial distribution functions is negligible, the influence on the bond angle distribution function is stronger. The interactions without polarizability overestimate the lower angles and underestimate higher angles. Interactions with polarizability yield the correct behavior for increasing temperatures and are in better agreement with ab-initio data. For alumina the cohesive energy was also calculated. It was found that it is strongly overestimated without polarizability. The equation of state for silica and magnesia is underestimated without polarizability as compared to ab initio calculations. In summary we have clearly shown that the interactions with polarizable oxygen atoms yield much better results than without polarization.

3.2 Crack Simulations of Alumina

Interactions for α-alumina Al_2O_3 have been determined in pretty much the same way as for silica and magnesia described in Sect. 3.1. Special care was taken for a correct behavior of surface relaxation and surface energies since crack simulations were intended. A good agreement with ab initio calculations and experimental data was obtained. The new interaction underestimates the energy of different twins, but qualitative agreement is achieved. Details about the crack studies can be found in [11]. Cracks do not move in close packed planes. Cracks which are not placed not in the cleavage plane tend to be deflected into The most important result is the observation of spontaneous polarization of the sample in front of the crack. This is especially remarkable since α-alumina is inversion symmetrical which forbids polarization. Since effects from the boundary conditions can be excluded by applying periodic boundary conditions the only remaining influence is an inhomogeneous deformation which leads to (anti-)flexoelectricity as described in the next section.

3.3 Flexoelectricity

If an ionic crystalline structure without inversion symmetry is deformed it will show piezoelectricity since the deformation tensor ϵ_{jk} will couple to the polarization vector P_i via a third order tensor d_{ijk}:

$$P_i = d_{ijk}\epsilon_{jk} \tag{9}$$

This leads to a free energy f_{piezo} equal to

$$f_{\text{piezo}} = \frac{1}{2}E_i P_i = \frac{1}{2}E_i d_{ijk}\epsilon_{jk} \tag{10}$$

where summation over equal indices is assumed. Since E_i is an polar vector while ϵ_{jk} and d_{ijk} are inversion invariant, the energy must vanish for polar structures since f_{piezo} would change its sign. This is the case for periclase and α-alumina.

If the deformation is inhomogeneous, however, it could couple to P_i through its gradient ∂_i via a fourth order tensor μ_{ijkl}:

$$P_i = \mu_{ijkl}\partial_j \epsilon_{kl} \tag{11}$$

This is called *flexoelectricity*. The free energy f_{flexo} is equal to

$$f_{\text{flexo}} = \frac{1}{2}E_i P_i = \frac{1}{2}E_i \mu_{ijk}\partial_j \epsilon_{kl} \tag{12}$$

Since the gradient ∂_j is also a polar vector, f_{flexo} will not change its sign on inversion and therefore the flexoelectric effect will be present for all crystal structures. In the case of polar crystals it will be a second order correction to piezoelectricity.

The coupling of the inhomogeneous deformation to polarization was the reason for the spontaneous polarization of the sample in the crack simulations on α-alumina. Since the crack geometry is difficult to describe and analyse we invented a geometry where flexoelectricity is more easy to study: A rectangular plate was deformed in such a way that it formed a trapezoidal wedge. With this geometry two of the three independent constants of μ_{ijkl} in the cubic case could be determined for periclase MgO. The constants are μ_{11}=2× 10^{-13} C/cm and $\mu_{12} = -12\times10^{-13}$ C/cm.

If two wedges with opposite deformation are fit together, a Néel wall is formed as shown in Fig. 5.

Things become much more complicated for α-alumina. First of all, ten independent tensor components μ_{ijkl} are expected since this structure is hexagonal. Second, as can bee seen in Fig. 6, the material becomes antiflexoelectric! The domains are not as clearly separated as for periclase. Up to now we have not fully understood what is happening.

In summary we have found that it is possible to simulate flexo- and antiflexoelectricity with classical MD simulations if the potential allows for induced dipoles. The investigations of the observed structures and the determination of materials parameters through adequate deformation geometries are still at the very beginning.

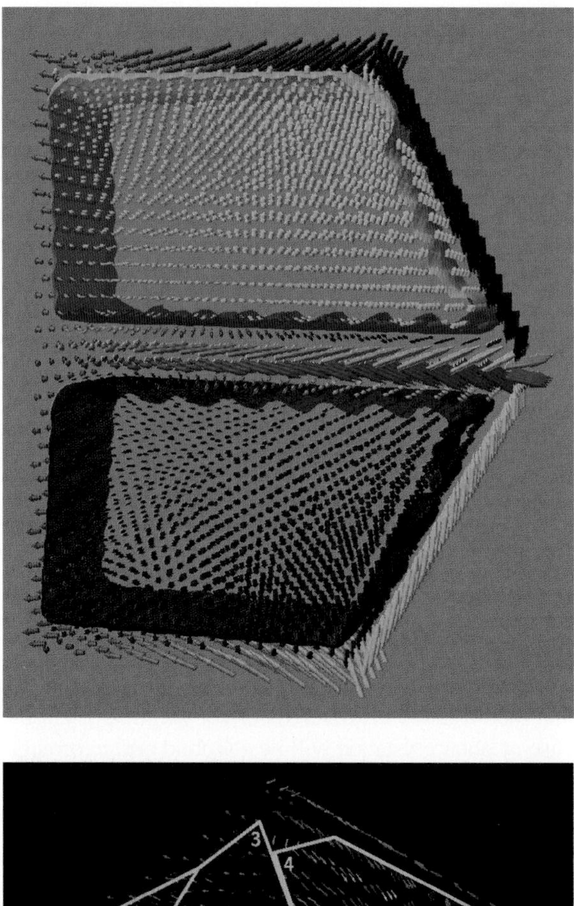

Fig. 5 Two cubic MgO periclase blocks deformed as indicated by the *boundaries*. The polarization in the *green part points* to the *upper right corner*, the polarization in the *lower part points* to the *lower right corner*. The different polarizations are adjusted in the center by an Néel wall. Visualization using `MegaMol` [10, 18]

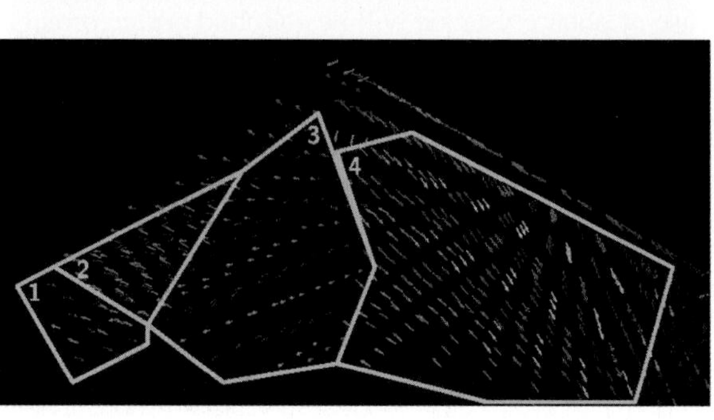

Fig. 6 Two hexagonal α-alumina Al_2O_3 blocks deformed as indicated by the *boundaries*. The *arrows* indicate the local polarization. As nearby arrows point in opposite directions, we find that the sample splits into antiflexoelectric domains. Visualization using `MegaMol` [10, 18]

4 Performance and Benchmarks

General benchmark data for `IMD` have been given by Stadler et al. [20]. The data demonstrate that `IMD` scales almost linearly in weak scaling (same number of atoms per processor) and fairly well for strong scaling (total number of atoms constant, thus communication load growing). This behavior is still valid as a systematic study on the Blue Gene/L clearly shows (see the previous HLRS report [17]). New data especially for the simulation of laser ablation have been reported last year (see [16]). Further benchmarks especially for long-range interactions depend on the algorithms only and are therefore given in the original publications (See e.g. [5]). Results for production simulations on the HLRS Hermite are not yet available since our project did not include access to this machine up to now. Other benchmark data for XE6, XK6 and XT5 machines obtained by Cray have been reported last year [16].

5 Summary

We have reported the extension of our molecular dynamics simulation package `IMD` to the effective simulation of ionic materials with long-range interactions including a model of polarizable oxygen. This part is fully parallelized with `MPI`. `IMD` has been extended further to variable charges and to the computation of these charges at metal oxide–metal interfaces by the minimization of the chemical potential. The latter part still has to be parallelized. As indicated there are good reasons that the chemical potential can be precalculated and iteratively adjusted during simulations. The new program capacities have been applied successfully to the study of several crystal structures of silica SiO_2, magnesia MgO and alumina Al_2O_3. The study of cracks in crystalline α-alumina lead to the discovery of flexoelectricity in cubic periclase MgO and antiflexoelectricity in hexagonal α-alumina. The new effects are still under investigation and final results can not yet been given.

References

1. P. Beck, P. Brommer, J. Roth, H.-R. Trebin, *Ab initio* based polarizable force field generation and application to liquid silica and magnesia. J. Chem. Phys. **135**, 234512 (2011)
2. P. Beck, P. Brommer, J. Roth, H.-R. Trebin, Influence of polarizability on metal oxide propeties studied by molecular dynamics simulations. J. Condens. Matter **24**, 485401 (2012)
3. A. Bosak, M. Krisch, Phonon density of states probed by inelastic x-ray scattering. Phys. Rev. B **72**, 224305 (2005)
4. P. Brommer, F. Gähler, Potfit: effective potentials from ab-initio data. Model. Simul. Mater. Sci. Eng. **15**, 295–304 (2007)
5. P. Brommer, P. Beck, A. Chatzopoulos, F. Gähler, J. Roth, H.-R. Trebin, Direct Wolf summation of a polarizable force field for silica. J. Chem. Phys. **132**, 194109 (2010)
6. A. Elsener, O. Politano, P.M. Derlet, H. Van Swygenhoven, Model. Simul. Mater. Sci. Eng. **16**, 025006 (2008)
7. P.P. Ewald, Die Berechnung optischer und elektrostatischer Gitterpotentiale. Ann. Phys. (Leipzig) **64**, 253–287 (1921)
8. G.A. Gaetani, P.D. Asimov, E.M. Stolper, Determination of the partial molar volume of SiO_2 in silicate liquids aat elevated pressures and temperatures: a new experimental approach. Geochim. Cosmochim. Acta **62**, 2499–2508 (1998)
9. S. Ghose, M. Krisch, A.R. Oganov, A. Beraud, A. Bosak, R. Gulve, R. Seelaboyina, H. Yang, S.K. Saxena, Lattice dynamics of MgO at high pressure: theory and experiment. Phys. Rev. Lett. **96**, 035507 (2006)
10. S. Grottel, P. Beck, C. Müller, G. Reina, J. Roth, H.-R. Trebin, T. Ertl, Visualization of electrostatic dipoles in molecular dyanmics of metal oxides. IEEE Trans. Vis. Comput. Graph. **18**, 2061–2068 (2012)

11. S. Hocker, P. Beck, S. Schmauder, J. Roth, H.-R. Trebin, Simulation of crack propagation in alumina with *ab initio* based polarizable force field. J. Chem. Phys. **136**, 084707 (2012)
12. B.B. Karki, D. Bhattarai, L. Stixrude, First-principles calculations of the structural, dynamical, and electronic properties of liquid MgO. Phys. Rev. B **73**, 174208 (2006)
13. G. Kresse, J. Furthmüller, Efficient iterative schemes for *ab initio* total-energy calculations using a plane-wave basis set. Phys. Rev. B **54**, 11169–11186 (1996)
14. G. Kresse, Hafner, *Ab initio* molecular dynamics for liquid metals. J. Phys. Rev. B **47**, 558–561 (1993)
15. J. Roth, F. Gähler, H.-R. Trebin, A molecular dynamics run with 5.180.116.000 particles. Int. J. Mod. Phys. C **11**, 317–322 (2000)
16. J. Roth, J. Karlin, M. Sartison, A. Krauš, H.-R. Trebin, Molecular dyanmics simulations of laser ablation in metals: parameter dependence, extended models and double pulses, in *High Performance Computing in Science and Engineering '12*, ed. by W.E. Nagel, D.B. Kröner, M.M. Resch (Springer, Heidelberg, 2013), pp. 105–107
17. J. Roth, C. Trichet, H.-R. Trebin, S. Sonntag, Laser ablation of metals, in *High Performance Computing in Science and Engineering '10*, ed. by W.E. Nagel, D.B. Kröner, M.M. Resch (Springer, Heidelberg, 2011), pp. 159–168
18. K. Scharnowski, M. Krone, F. Sadlo, P. Beck, J. Roth, H.-R. Trebin, T. Ertl, 2012 IEEE visualization contest winner: visualizing polarization domains in barium titanate. IEEE Comput. Graph. Appl. **13**(5), 9–17 (2013)
19. F.H. Streitz, J.W. Mintmire, Electrostatic potentials for metal-oxide surfaces and interfaces. Phys. Rev. B **50**, 11996–12003 (1994)
20. J. Stadler, R. Mikulla, H.-R. Trebin, IMD: A software package for molecular dynamics studies on parallel computers. Int. J. Mod. Phys. C **8**, 1131–1140 (1997)
21. P. Tangney, S. Scandalo, An *ab initio* parametrized interatomic force field for silica. J. Chem. Phys. **117**, 8898–8904 (2002).
22. B.W.H. Van Beest, G.J. Kramer, R.A. van Santen, Force fields for silica and aluminophospaheres based on ab initio calculations. Phys. Rev. Lett. **64**, 1955–1958 (1990)
23. D. Wolf, P. Keblinski, S.R. Philipot, J. Eggebrecht, Exact method for the simulation of Coulombic sysztems by spherically truncated, pairwise r^{-1} summation. J. Chem. Phys. **110**, 8254–8282 (1999)

Polarization Dependent Water Adsorption on the Lithium Niobate Z-Cut Surfaces

S. Sanna, A. Riefer, M. Rohrmüller, M. Landmann, E. Rauls, N.J. Vollmers, R. Hölscher, M. Witte, Y. Li, U. Gerstmann, and W.G. Schmidt

Abstract The effect of ferroelectric poling on the water adsorption characteristics of lithium niobate Z-cut surfaces is investigated by ab initio calculations. Thereby we model the adsorption of H_2O monomers, small water clusters and water thin films. The adsorption configuration and energy are determined as a function of the surface coverage on both the positive and negative $LiNbO_3(0001)$ surfaces. Thereby polarization-dependent adsorption energies, geometries and equilibrium coverages are found. The different affinity of water to the two surfaces is explained in terms of different bonding scenarios as well as the electrostatic interactions between the substrate and the polar molecules. Surface phase diagrams for the Z-cuts in equilibrium with water are predicted from atomistic thermodynamics.

1 Introduction

Lithium niobate ($LiNbO_3$, LN) is one of the most important optic materials, being the equivalent in the field of non-linear optics and optoelectronics to silicon in electronics [1]. Recently the (microscopic) surface and interface properties have become important [2]. In particular, the polarization domains of ferroelectric oxide surfaces can be manipulated by an external electric field, in order to tailor the surface reactivity for specific applications. Indeed, polarization-dependent physical and chemical surface phenomena have been reported. Surface conductivity [3], threshold energy for photoelectron emission [4], thermally stimulated electron emission [5] and etching rate in acid solutions [6, 7] have been shown to be very different for differently polarized domains. Polarization-dependent adsorption of particle and molecules, either directly on the ferroelectric surface [8, 9] or on metal

S. Sanna · A. Riefer · M. Rohrmüller · M. Landmann · E. Rauls · N.J. Vollmers · R. Hölscher · M. Witte · Y. Li · U. Gerstmann · W.G. Schmidt (✉)
Lehrstuhl für Theoretische Physik, Universität Paderborn, 33095 Paderborn, Germany
e-mail: W.G.Schmidt@upb.de

W.E. Nagel et al. (eds.), *High Performance Computing in Science and Engineering '13*, DOI 10.1007/978-3-319-02165-2_12,

and semiconducting thin films deposited in a ferroelectric support [10], have been demonstrated too. In addition, photochemical deposition reactions can be combined with the local control of the ferroelectric polarization to drive the assembly of surface nanostructures [11]. Thus, domain engineering opens the possibility for the realization of molecular detectors and other devices at nanoscale level. In fact, it has been suggested that molecular adsorption may stabilize opposite poling directions in ferroelectric thin films, allowing for the realization of ferroelectric chemical sensors [12].

Water molecules have a prominent position among the common adsorbates, because of their role in natural phenomena such as catalysis, electrochemistry, corrosion and because of a variety of applications, including hydrogen production, fuel cells and biological sensors. Furthermore water will be present and influence the performance of the LN-based devices, unless they operate in ultra high vacuum (UHV). The functionality of devices might thus depend on the relative humidity. Indeed, water temperature programmed desorption (TPD) measurements at the positive and negative surface of $LiNbO_3$ indicate that the molecule-surface interaction are both coverage and polarization dependent. Another reason to study the water-LN interface is related to the fact that up to now high-resolution atomic force images of the LN surface could only be obtained in liquid environment [13].

Here we study the coverage dependent adsorption of water at the positive and negative LN(0001) surface [14] by means of atomistic simulations in the framework of the density functional theory (DFT). Adsorption energy, site and configuration are determined and the bonding between water and surface is analyzed and discussed. Surface thermodynamics is used to predict the ground state of water covered LN surfaces in dependence of temperature and pressure.

2 Methodology

Total energy density functional calculations have been performed within the PW91 formulation of the generalized gradient approximation (GGA) [15] as implemented in the VASP simulation package [16]. This approach allows for the accurate treatment of hydrogen bonds and water structures [17] and leads to reliable structures and energies for both LN bulk and surfaces [14, 18]. PAW potentials [19] with projectors up to $l = 3$ for Nb, $l = 2$ for Li, O and $l = 1$ for H, have been used for the calculations. The electronic wave functions are expanded into plane waves up to a kinetic energy of 400 eV.

Our work is based on the surface models proposed in [18, 20], which are in agreement with the experimental observation. According to these models, the positive surface is –Nb–O_3–Li_2 terminated, with one of the two top Li atoms relaxing down to the lower laying oxygen layer and the other above it, as represented in Fig. 1 (left). The negative surface is –Li–O terminated, instead (see Fig. 1 (right)). These models are used as basis for the investigation of the H_2O adsorption. Thereby we use slabs consisting of 18 atomic layers within a 2×2 periodicity (124 atoms

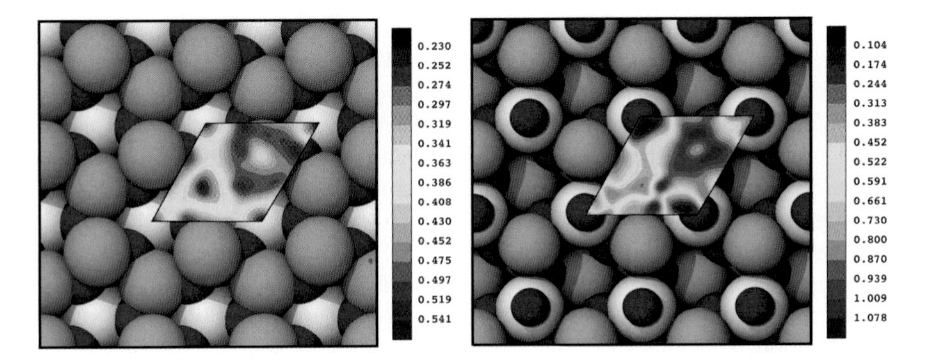

Fig. 1 PES for the adsorption of a single H_2O molecules on the positive (*left*) and negative (*right*) LN(0001) surface. Li atoms are *gray*, Nb *white* and oxygen *red*. Adsorption energies are in eV

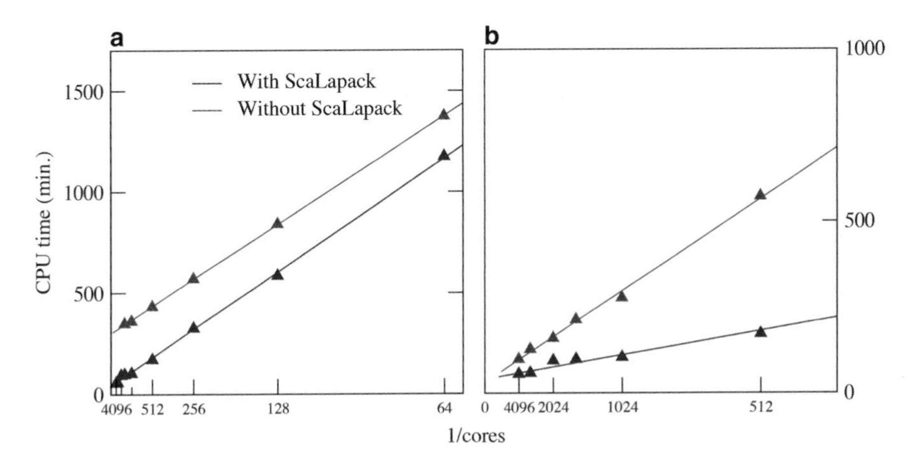

Fig. 2 CPU time required for a total energy calculation of stoichiometric $LiNbO_3$, modelled by a supercell containing 360 atoms and 2,160 electrons. (**a**) The effect of using the ScaLAPACK libraries in VASP. (**b**) Comparison between calculations performed by different simulation packages (VASP/*blue* and QuantumEspresso/*red*). The *blue* and the *red lines* are the linear interpolation of the calculated points and only serve to guide the eyes in order to appreciate the linear scaling at a first sight

for the positive face and 128 for the negative) and a vacuum region of ca. 16 Å. The lateral dimension of the unit cell largely reduces the unwanted interactions between the adsorbates and their periodic images. The dipole correction described in [21, 22] has been used to correct the artificial forces generated by the slab images. A Γ-centered $2 \times 2 \times 1$ k-point mesh was used to carry out the integration in the Brillouin zone. The adsorbate and the 6 uppermost surface layers were allowed to relax until the forces were lower than 20 meV/Å.

Turning to the numerics, the density functional implementation used in the present study makes highly efficient use of the HLRS CRAY XE6, which presents the main computational resource for the present calculations. As shown in Fig. 2, the scaling with the number of cores is roughly linear up to 4,096 cores. A substantial speedup is achieved by using the ScaLAPACK library. It is used for the LU decomposition and diagonalization of the sub space matrix, the dimension of which is given by the number of electronic states. These operations are very fast in the serial version, where they account for only about 2 % of the processor time, but become a bottleneck on massively parallel machines. While on slow networks and PC clusters, it is not recommended to use ScaLAPACK, it usage for the XE6 is obviously advantageous. In addition to the VASP simulation package [16], we also performed tests with the QuantumEspresso DFT implementation, parts of which were developed in our group [23]. Here we observe a superior scaling compared to VASP, although the total CPU time required is measurable larger, mainly due to the use of different pseudopotentials.

3 Results

We started our investigation with the determination of the favored adsorption site for single H_2O molecules on the considered model structure. We follow [18, 20, 24] concerning the convention for discriminating positive and negative surfaces. In a first step we have therefore calculated the potential energy surface (PES) for single adsorbates, which gives an approximate idea of the stable adsorption sites and a map of the different energy minima on the surface. Thereby we have evaluated the adsorption energy for 48 possible positions, and three different starting configurations, namely with the water dipole moment parallel, antiparallel and perpendicular to the spontaneous polarization of the substrate. The results are reported in Fig. 1 for the adsorption at the positive and negative side, respectively. The PESs are relatively corrugated, indicating a low surface mobility of the adsorbate. This holds in particular for the negative surface. Several minima and maxima of the adsorption energy are present at both sides. As a general feature, H_2O avoids a position right on top of the topmost Li atoms, and prefers an adsorption site between cations, above the lower lying oxygen atoms (second oxygen layer from the top).

In the energetically most favored configuration, the water molecule adsorbs tilted on the positive surface (see Fig. 3a) between one Li and one oxygen of the surface, with atomic distances $d(\text{O–Li}) = 2.06\,\text{Å}$ and $d(\text{O–H}) = 1.76\,\text{Å}$. Both the Li–O and O–H–O direction lie in the $(1\bar{1}00)$ plane. The water adsorption at the positive side does not substantially affect the substrate geometry. An analysis of the charge density reveals a polarization cloud between a water hydrogen and the surface oxygen it points to, as well as the negative charge accumulation at the oxygen side between molecule and surface Li (see Fig. 3a). The charge distribution, the interatomic distances and the adsorption geometry suggest that water molecules form both a Li–O bond of ionic character and an hydrogen-bond at the positive surface.

In the case of the negative surface the oxygen of the water molecule adsorbs close to a surface Li, at a distance $d(\text{O–Li}) = 1.83\,\text{Å}$. One of the two hydrogen atoms

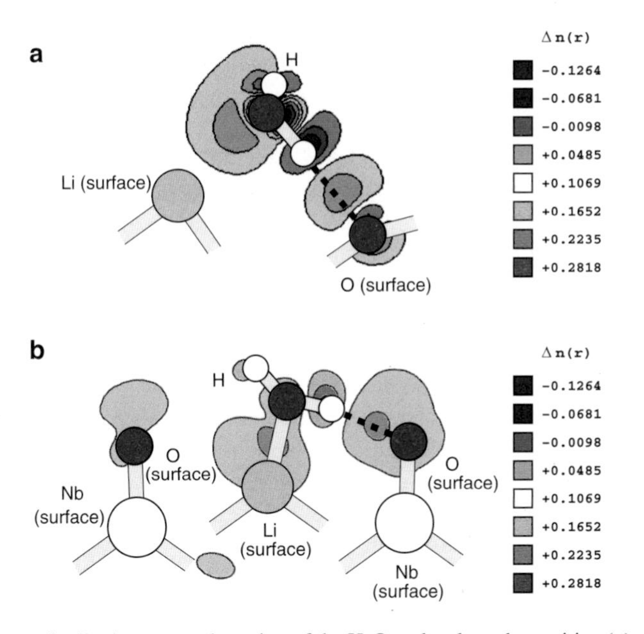

Fig. 3 Charge redistribution upon adsorption of the H_2O molecule at the positive (**a**) and negative (**b**) (0001) surface. Charge isolines in the $(1\bar{1}00)$ plane (which is y-plane as defined in [18]) are plotted. This plane contains the two bonds formed by the water molecule with the LN surface. *Red*/*blue* indicate regions with charge accumulation/depletion

points to a neighboring surface oxygen, with distance $d(\text{O–H}) = 1.50\,\text{Å}$, while the other points out from the surface, as represented in Fig. 3b. Charge distribution, interatomic distances and geometry are compatible with a Li–O bond and a O–H hydrogen bond. The presence of the adsorbate induces some relaxation of the surface atoms. The adsorption pulls the surface Li out of its relaxed surface position and elongates its three bonds to neighboring oxygen ions. This is very similar to the effect of water adsorption at the non-polar $Al_2O_3(0001)$ surface, which was recently found to significantly disrupt the clean Al_2O_3 surface geometry [25]. We notice that another configuration of very similar energy can be created. In this case the oxygen of the water molecule again adsorbs close to a surface Li, at a distance $d(\text{O–Li}) = 1.89\,\text{Å}$. However, both hydrogen atoms point roughly to a neighboring oxygen, with distances $d(\text{O}_1\text{–H}) = 1.88\,\text{Å}$, and $d(\text{O}_2\text{–H}) = 2.16\,\text{Å}$. In this second configuration all the water atoms are bound to the surface, with an ionic Li–O bond and two O–H hydrogen bonds (of different strength). In all the stable configurations, the z-component of the molecular dipole moment is directed against the spontaneous polarization of the substrate, thus reducing the total polarization.

The adsorption energy, as calculated from the difference

$$E_{ads} = E_{slab}(H_2O@LN) - E_{slab}(LN) - E_{gas}(H_2O) \tag{1}$$

amounts to 0.61 and 1.28 eV for the adsorption at the positive and negative surface respectively. The calculated energy difference is in qualitative agreement with the temperature programmed desorption (TPD) measurements of Garra et al. [9]. However, the measured adsorption energy difference (estimated to be between 2.8 and 4.0 kJ/mol, corresponding to 0.029–0.041 eV) is lower than the values predicted by the theory (0.67 eV) by one order of magnitude. This discrepancy may partially be explained by the relatively large error bars affecting both the measurements and the calculations. At one side, the experimental value is obtained modeling the TPD spectra within the Polany–Wigner relations, which contain pre-exponential factors to be determined, and for which values scattered over several orders of magnitudes have been reported. It is also not clear to what extent the experimental preparation conditions result in the (thermodynamically stable) surface atomic structures supposed in our study. From the theoretical point of view it has to be said that adsorption energies do strongly depend on the parameterization of the exchange-correlation functional, both directly and indirectly through their geometry dependence. We mention that a recent theoretical study [26] on the adsorption of methanol on lithium niobate Z-cut surfaces also found larger adsorption energy differences between positive and negative surface than concluded from the experimental data. The sizeable adsorption energy difference calculated in this work can be understood from an atomistic and an electrostatic perspective. The disparity of the values can be traced back to the different bonding scenarios at the two faces. The bond at the negative Z-cut is shorter, i.e. stronger than at the positive surface. This is due to the different stoichiometries and the different morphologies of the two (0001) faces. From an electrostatic perspective, it must be considered that the work function at the positive Z-cut is by about 2 eV larger than at the negative surface [4, 18]. This could contribute to the difference in adsorption energy by affecting the electron transfer between molecule and surface, thus explaining why the H_2O adsorption at the negative side is favored with respect to the adsorption at the positive side.

Depending on the experimental conditions, surface adsorbed water may form different low-dimensional structures, ranging from isolated monomers and clusters to one-dimensional (1D) chains and two-dimensional (2D) overlayers (see, e.g. [27]). With increasing coverage, water may form networks of hydrogen-bonded molecules, water multilayers and bulk ice-like structures. In order to study the water adsorption at higher coverage, we systematically increased the number of water molecules up to 4 per surface unit cell. Different adsorption configurations as well as (partially) dissociated adsorption models were probed. A number of different structures have been found to be (meta)stable, the most relevant among them are shown in Figs. 4 und 5. At the positive surface, with two water molecules per unit cell, both highly regular honeycomb structures (similar to the water hexagons formed on many metal oxide or metal surfaces [25,28]) or water chains (as observed on Rutile or ZnO [29]) have been found. They are shown in Fig. 4a, b, respectively. In both configurations the water hydrogen atoms point alternating to a surface oxygen and to the oxygen atom of the next water molecule. Three molecules per unit cell lead to the formation of a slightly distorted form of hexagons or chains plus an isolated water monomer. Four or more molecules per unit cell give rise to three-dimensional ice-like structures. The most stable of them are reminiscent of regular ice Ih.

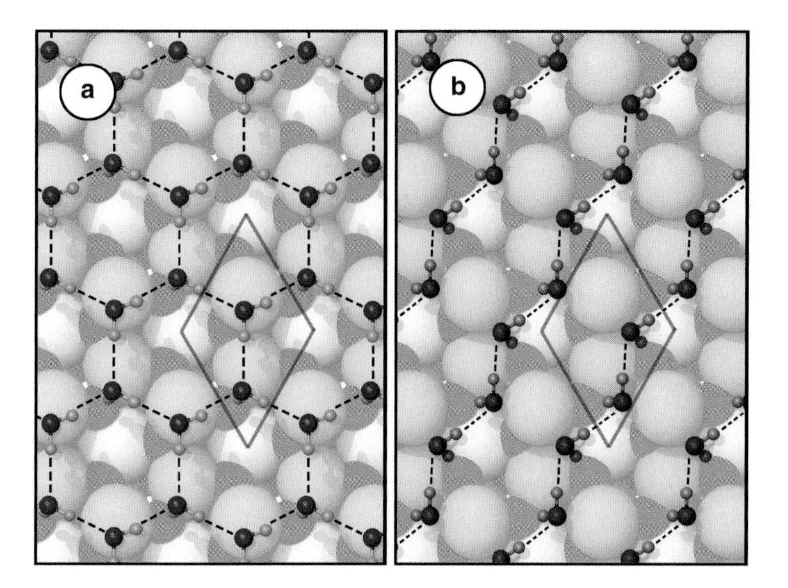

Fig. 4 Possible water configurations at the positive (0001) surface include (**a**) two dimensional regular hexagonal structures and chains (**b**). The rhombohedral unit-cell is highlighted

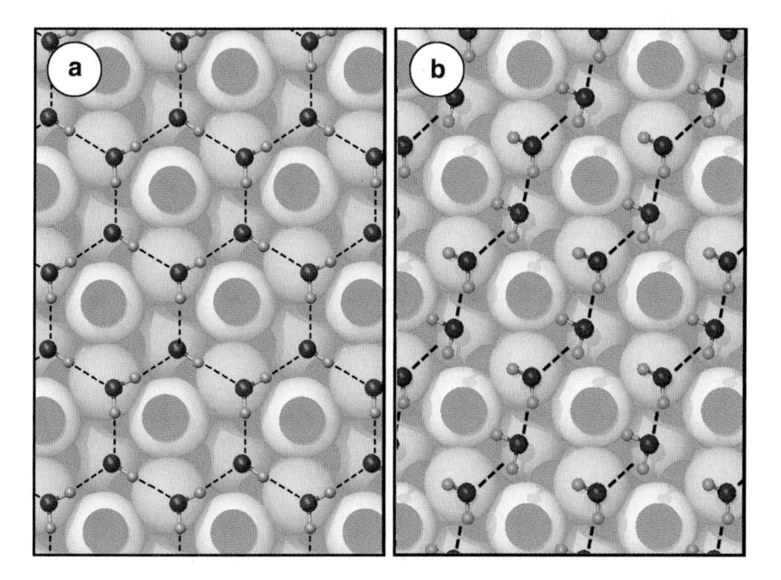

Fig. 5 Possible water reconstructions at the negative (0001) surface include two dimensional (**a**) hexamer structures and less regular chain structures

The negative (0001) surface is by far not as flat as the positive surface, it is rather characterized by a stronger surface corrugation. This hinders the uniform and regular adsorption of water films in regular patterns as in the case of the positive surface. Most of all, no more than two water molecules per unit cell can be adsorbed. Increasing the number of water molecules results in the formation of ice layers separated from the surface (negative adsorption energy). Among the stable structures, honeycomb films and different kinds of distorted hexamers can be formed, as shown in Fig. 5. The dotted lines joining the water molecules are arbitrarily drawn guides for the eyes and do not represent a chemical bond. We remark that our adsorption models are based on ideal surfaces created in vacuum. Real surfaces will be characterized by steps and other surface defects (primarily oxygen and Li vacancies), by the presence of other adsorbates and, in case of extremely O-rich growth conditions, even by a different termination.

The particular water structure occurring on the LN surface depends on the water availability. To compare different surface water structures energetically, we use the thermodynamic grand-canonical potential

$$\Omega(\mu^{H_2O}) = F(n) - n\mu^{\mathrm{H_2O}} \approx E(n) - n\mu^{\mathrm{H_2O}}. \tag{2}$$

Here n is the number of water molecules and $F(n)$ the free energy of a surface with n adsorbed water molecules. In our work the free energy is approximated by the DFT total energy $E(n)$, which is a reasonable approximation if we assume similar entropy contributions for different adsorption configurations. The water chemical potential $\mu_{\mathrm{H_2O}}$ can be directly related to the experimental conditions. In the following $\Delta\mu_{\mathrm{H_2O}}$ refers to the difference between the water chemical potential $\mu_{\mathrm{H_2O}}$ and its value in the ice phase $\mu_{\mathrm{H_2O}}^{[\mathrm{ice}]}$. The dependence of $\Delta\mu_{\mathrm{H_2O}}$ on temperature and pressure can be calculated in the approximation of a polyatomic ideal gas [30] as

$$\Delta\mu_{\mathrm{H_2O}}(p,T) = k_B T \left[\ln\left(\frac{pV_Q}{k_B T}\right) - \ln Z_{\mathrm{rot}} - \ln Z_{\mathrm{vib}} \right], \tag{3}$$

where k_B is the Boltzmann constant, T the temperature and p the pressure. V_Q is the quantum volume and is equivalent to λ^3, whereby λ is the de Broglie thermal wavelength of the water molecule with mass m

$$\lambda = \sqrt{\frac{2\pi\hbar^2}{mk_B T}}. \tag{4}$$

Here

$$Z_{\mathrm{rot}} = \frac{(2k_B T)^{\frac{3}{2}}(\pi I_1 I_2 I_3)^{\frac{1}{2}}}{\sigma\hbar^3} \tag{5}$$

and

$$Z_{\mathrm{vib}} = \prod_{\alpha} \left[1 - \exp\left(-\frac{\hbar\omega_\alpha}{k_B T}\right)\right]^{-1} \tag{6}$$

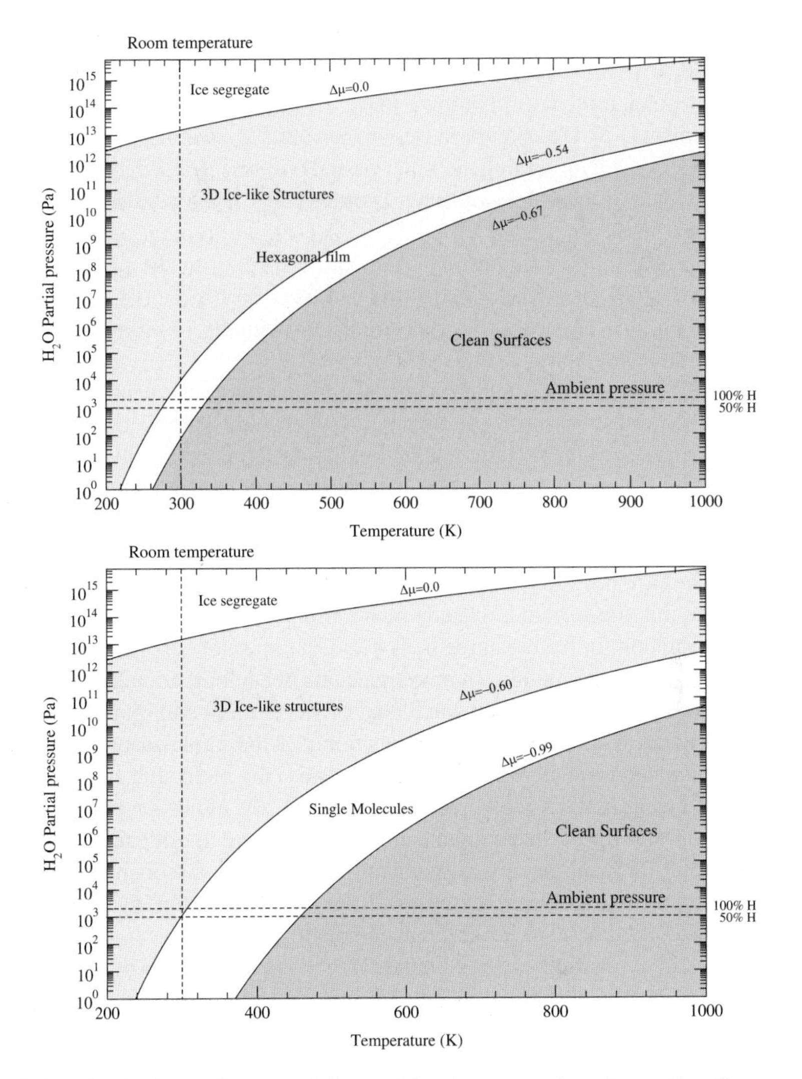

Fig. 6 Calculated phase diagram of the positive (*upper part*) and negative (*lower part*) $LiNbO_3(0001)$ surface as a function of the temperature and pressure. *Dotted lines* indicate the ambient conditions. The values of the chemical potential variations $\Delta\mu^{H_2O}$ (given with respect to ice Ih) at which a particular structure is formed, are indicated by *solid lines*

are the rotational and vibrational partition functions, respectively. We used the experimental values of the momenta of inertia I_i and of the vibrational frequencies ω_α of the water molecule [31]. The geometrical parameter σ takes the symmetry of the molecule into account. For H_2O (equal-sided triangle) it holds $\sigma = 2$. The calculated phase diagram of H_2O at the positive LN(0001) surface as a function of pressure and temperature is shown in Fig. 6 (top). According to our calculations, a water film with hexagonal symmetry is present at the LN surface at ambient condition with a relative humidity of about 50 %. This is not surprising, as the adsorption of polar molecules is known to be the major external charge compensation mechanism [32]. Heating the system above 100 °C, clean LN (i.e. water free) surfaces are recovered. Increasing the water partial pressure above the room pressure leads to the growth of water bilayers and 3D-structures. At the negative side (Fig. 6, lower part) water monomers or ice seeds are present at ambient conditions. Our calculations predict a very interesting behavior: according to the phase diagrams in Fig. 6, water shall freeze at different temperatures on differently polarized LN surfaces. Indeed, this phenomenon has been experimentally observed recently for the isostructural ferroelectric $LiTaO_3$ [33].

Garra et al. have proposed that water adsorbs on the LN surfaces through a bonding interaction between each molecule's oxygen and a surface cation (Li, Nb), as well as through the electrostatic interaction between molecule's hydrogen atoms and oxygen atoms on the surface [9]. Our microscopic calculations clearly confirm/validate this model. In qualitative agreement with the experiment, the relative adsorption energy per water molecule decreases with the water coverage. The energetic difference between adsorption geometries is, however, smaller than 15 %. This indicates that the surface-molecule interaction is dominant over the molecule-molecule interaction. In further agreement with [9], the water adsorption is found to occur mainly non-dissociatively. To investigate this behavior the total energy of dissociated water molecules has been calculated in 30 different configurations at each side. Thereby water fragments have been positioned at the stable adsorption sites for O and OH radicals as calculated in [34]. At both sides the molecular adsorption was favored over the dissociated adsorption by at least 0.02 eV. However, the presence of different adsorbates in the atmosphere could modify the thermodynamic stability of the surfaces and favor dissociative adsorption. Furthermore, dissociative adsorption might occur close to a surface defect or in the vicinity of a step, as in the case of MgO(100) and other metal oxide surfaces [27, 35].

Regarding the electrostatics, the calculation of the work function upon adsorption yields insights into the charge compensation mechanisms, which play a crucial role in the physics of ferroelectric surfaces. Indeed, it allows us to determine the direction of the charge transfer between surface and adsorbates. An estimation of the surface charge of ideal, relaxed surfaces, can be found in [20]. The authors present a simple electrostatic model, predicting a positive charge of $+e/4$ at the LN(000$\overline{1}$) (so called negative surface) and a negative charge of $-e/4$ at the LN(0001) (so called positive surface). This charge is expected to be compensated by adsorbates. We predict a decrease of the work function (of 0.34 eV) at the negative side and an increase of the work function (of 0.49 eV) at the positive side upon adsorption of a single molecule. This corresponds to an electron transfer directed from the molecule to the surface at the positively charged side and vice versa from the surface to the adsorbate at the negatively charged surface. Thus, the water adsorption has in both cases a stabilizing influence as it reduces the surface charge.

4 Conclusions

We have performed density functional calculations to model the water adsorption at the $LiNbO_3$ surface. Isolated H_2O molecules are characterized by an adsorption energy of 0.61 and 1.28 eV at the positive and negative side, respectively. The adsorption configuration and adsorbate mobility are strongly polarization dependent, in qualitative agreement with temperature programmed desorption measurements [9]. The adsorption energy differences are due to different molecular bonding geometries on the two structurally distinct surfaces as well as different electrostatics. Various structures are formed with increasing water coverage, the adsorption energy per molecule of which is in qualitative agreement with the TPD data. Also in agreement with [9], the water adsorption is found to occur mainly non-dissociatively, independently on the coverage. At ambient condition and assuming a relative humidity of 50 %, we expect water molecules adsorbed at the LN(0001) surface, either in form of thin films with honeycomb symmetry or in small clusters.

Acknowledgements The calculations were done using grants of computer time from the Höchstleistungs-Rechenzentrum Stuttgart (HLRS) and the Paderborn Center for Parallel Computing (PC^2). The Deutsche Forschungsgemeinschaft is acknowledged for financial support.

References

1. A. Räuber, in *Chemistry and Physics of Lithium Niobate*. Current Topics in Materials Science (North-Holland Publ. Company, Amsterdam, 1978)
2. G. Namkoong, K.K. Lee, S.M. Madison, W. Henderson, S.E. Ralph, W.A. Doolittle, Appl. Phys. Lett. **87**(17), 171107 (2005)
3. Y. Watanabe, M. Okano, A. Masuda, Phys. Rev. Lett. **86**(2), 332 (2001)
4. W.C. Yang, B.J. Rodriguez, A. Gruverman, R.J. Nemanich, Appl. Phys. Lett. **85**(12), 2316 (2004)
5. B. Rosenblum, P. Bräunlich, J.P. Carrico, Appl. Phys. Lett. **25**(1), 17 (1974)
6. K. Nassau, H.J. Levinstein, G.M. Loiacono, J. Phys. Chem. Solids **27**, 983 (1966)
7. C.L. Sones, S. Mailis, W.S. Brocklesby, R.W. Eason, J.R. Owen, J. Mater. Chem. **295**, 12 (2002)
8. Y. Yun, L. Kampschulte, M. Li, a D. Liao, E.I. Altman, J. Phys. Chem. C **111**, 13951 (2007)
9. J. Garra, J.M. Vohs, D.A. Bonnell, Surf. Sci. **603**, 1106 (2009)
10. Y. Inoue, I. Yoshioka, K. Sato, J. Phys. Chem. **88**, 1148 (1984)
11. S.V. Kalinin, D.A. Bonnell, T. Alvarez, X. Lei, Z. Hu, J.H. Ferris, Q. Zhang, S. Dunn, Nano Lett. **2**(6), 589 (2002)
12. D.D. Fong, A.M. Kolpak, J.A. Eastman, S.K. Streiffer, P.H. Fuoss, G.B. Stephenson, C. Thompson, D.M. Kim, K.J. Choi, C.B. Eom, I. Grinberg, A.M. Rappe, Phys. Rev. Lett. **96**(12), 127601 (2006)
13. S. Rode, R. Hölscher, S. Sanna, S. Klassen, K. Kobayashi, H. Yamada, W. Schmidt, A. Kühnle, Phys. Rev. B **86**, 075468 (2012)
14. S. Sanna, R. Hölscher, W.G. Schmidt, Phys. Rev. B **86**, 205407 (2012)
15. J.P. Perdew, W. Yue, Phys. Rev. B **33**(12), 8800 (1986)
16. G. Kresse, J. Furthmüller, Phys. Rev. B **54**(16), 11169 (1996)
17. L. Giordano, J. Goniakowski, J. Suzanne, Phys. Rev. Lett. **81**, 1271 (1998)

18. S. Sanna, W.G. Schmidt, Phys. Rev. B **81**(21), 214116 (2010)
19. P.E. Blöchl, Phys. Rev. B **50**(24), 17953 (1994)
20. S.V. Levchenko, A.M. Rappe, Phys. Rev. Lett. **100**(25) (2008)
21. J. Neugebauer, M. Scheffler, Phys. Rev. B **46**(24), 16067 (1992)
22. L. Bengtsson, Phys. Rev. B **59**(19), 12301 (1999)
23. P. Giannozzi, S. Baroni, N. Bonini, M. Calandra, R. Car, C. Cavazzoni, D. Ceresoli, G.L. Chiarotti, M. Cococcioni, I. Dabo, A. Dal Corso, S. de Gironcoli, S. Fabris, G. Fratesi, R. Gebauer, U. Gerstmann, C. Gougoussis, A. Kokalj, M. Lazzeri, L. Martin-Samos, N. Marzari, F. Mauri, R. Mazzarello, S. Paolini, A. Pasquarello, L. Paulatto, C. Sbraccia, S. Scandolo, G. Sclauzero, A.P. Seitsonen, A. Smogunov, P. Umari, R.M. Wentzcovitch, J. Phys. **CM 21**(39), 395502 (2009)
24. R.S. Weis, T.K. Gaylord, Appl. Phys. A Mater. **37**(4), 191 (1985)
25. P. Thissen, G. Grundmeier, S. Wippermann, W.G. Schmidt, Phys. Rev. B **80**, 245403 (2009)
26. A. Riefer, S. Sanna, W.G. Schmidt, Phys. Rev. B **86**, 125410 (2012)
27. S. Meng, E.G. Wang, S. Gao, Phys. Rev. B **69**, 195404 (2004)
28. A. Michaelides, K. Morgenstern, Nat. Mater. **6**(8), 597 (2007)
29. O. Dulub, B. Meyer, U. Diebold, Phys. Rev. Lett. **95**, 136101 (2005)
30. L.D. Landau, E.M. Lifshitz, *Statistical Physics, Part I*, 3rd edn. (Butterworth-Heinemann, Oxford, 1981)
31. V.W. Laurie, D.R. Herschbach, J. Chem. Phys. **37**, 1687 (1962)
32. F. Johann, E. Soergel, Appl. Phys. Lett. **95**, 232906 (2009)
33. D. Ehre, E. Lavert, M. Lahav, I. Lubomirsky, Science **327**, 672 (2010)
34. R. Hölscher, S. Sanna, W.G. Schmidt, Phys. Stat. Sol. (c) **9**, 1361 (2012)
35. M.J. Stirniman, C. Huang, R.S. Smith, S.A. Joyce, B.D. Kay, J. Chem. Phys. **105**(3), 1295 (1996)

Ab-Initio Calculations of the Vibrational Properties of Nanostructures

Gabriel Bester and Peng Han

Abstract The computational facility made available to us in the year 2012 at the High Performance Computing Center Stuttgart (HLRS), enabled us to calculate confinement and surface effects on the vibrational properties of colloidal semiconductor nanoclusters based on first-principles density functional theory (DFT). In this reporting period, we made four important contributions to the field through the use of high performance computing and the development of new algorithms. The four areas we have covered are (a) confinement effects on the vibrational properties of III–V and II–VI nanoclusters; (b) insights about the surface of colloidal nanoclusters from their vibrational and thermodynamic properties; (c) first-principles calculation of the electron–phonon interaction in semiconductor nanoclusters; and (d) vibron–vibron coupling from ab-initio molecular dynamics simulations.

1 Introduction

Colloidal semiconductor nanoclusters (NCs), or quantum dots (QDs), have a realistic potential to be used for bio-labeling, photovoltaic or as optoelectronic devices. Some successful applications in all these areas were already reported [1–7]. It is the intense research effort towards the fabrication (the growth) of structures with increasingly favorable properties that has helped to establish most of the knowledge base we rely on today. Theory has significantly contributed in the understanding of the electronic structure and the resulting optical properties [8, 9]. However, experiments have repeatedly shown that the surface can exert a dominating influence on many of the properties, [10–13] a fact that is difficult to take into account theoretically because of three reasons. One (i) being simply that the atomic structure

G. Bester (✉) · P. Han
Max-Planck-Institut für Festkörperforschung, Stuttgart, Germany
e-mail: g.bester@fkf.mpg.de; p.han@fkf.mpg.de

W.E. Nagel et al. (eds.), *High Performance Computing in Science and Engineering '13*, DOI 10.1007/978-3-319-02165-2_13,

of the surface still remains under debate today. The other (ii), being that surface reconstruction and strong ligand-to-dot bonding, probably of a coordination-type, requires the best available tools including self-consistent effects and correlations. Unfortunately, these tools, such as density functional theory (DFT), fail to address the experimental size regime if the ligands are fully taken into account. Using very small clusters with typically 66 core atoms allows for the treatment of truncated ligands (where only the active group is kept) at the DFT level [14–17]. However, confinement in the core is so strong, due to the reduced size, that the highest occupied, and lowest unoccupied molecular orbitals (HOMO and LUMO) tend to delocalize over the entire cluster, or even localize onto the ligands. This corresponds to a situation that qualitatively differs from the case with experimental size cores where HOMO and LUMO states do localize well inside the QD, leading to the known strong optical absorption. The last point (iii) applies to electronic and optical properties, where we have to expect poor results, even from DFT, due to the large errors in the effective masses. Moreover, if the spin–orbit interaction is neglected (as usually the case) the order and symmetry of the HOMO states (and hence the optical properties) are qualitatively defective [18, 19]. Our approach to remedy (iii) is to look at vibrational properties, these are knowingly very well described by DFT (neglecting spin–orbit coupling). From considerations (ii), we need to address the experimentally relevant core sizes, at DFT level. We therefore push the computational limit and use an approximation for the passivation (allowing us to treat 633 core atoms, at the lower bound of experimentally relevant structures) that we subject to variations within an experimentally reasonable range, thereby addressing issue (i).

The electron–phonon (e–ph) interaction has been the subject of theoretical investigation since the middle of last century [20]. Following the pioneering investigations of phenomenological deformation potentials, piezoelectric potentials, and Fröhlich interactions on the base of continuum models [18, 20], the atomistic methods such as the frozen–phonon approach [21–25], ab-initio density functional perturbation theory (DFPT) [26–31], and Born–Oppenheimer molecular dynamics (BOMD) [32, 33] have been proposed to study the e–ph interaction. Among these methods, ab-initio DFPT calculations represent the state-of-the-art and provides accurate e–ph coupling information. BOMD simulations, which monitor the change in the electronic levels under the influence of vibrations, can in principle describe the e–ph interaction beyond the harmonic approximation. However, both of these accurate first-principles methods are currently only possible for bulk and very small NCs due to the high computational demand.

In contrast to DFPT, where the change of potential is treated as the linear response of the charge density to the atomic displacement, the frozen–phonon approach is a direct method for the calculation of e–ph coupling. In this approach, the perturbation of specific phonon modes on the electronic states are handled through a frozen atomic displacement pattern [21–23]. The frozen–phonon approach was initially proposed as a method for the calculation of bulk phonon frequencies [34, 35]. In this approach, the total energy of a system is obtained for a specific atomic displacement pattern according to the phonon eigenmode. The phonon frequency is then calculated from energy difference between the system with and without the atomic displacement. In this approach, the computation of long wavelength bulk phonon frequencies is limited by the size of the cell. Since the DFPT linear-response approach provides accurate results of phonon eigenmode with low computational demand [36], the frozen–phonon method is rarely applied to calculate the phonon frequency now. With the development of first-principles calculation, the frozen–phonon approach was extended to self-consistent calculations of the e–ph interaction in bulk [21–23]. In these calculations, the change of potential caused by a phonon is computed self-consistently from two structures, the perfect crystal and the crystal with a frozen phonon [21]. Thereafter, the calculations were performed using empirical pseudopotentials [37,38], rigid-muffin-tin approximations [39], and ab-initio DFT [22, 25] along with phonon modes obtained from empirical valence force field models or first-principles linear response computations.

Despite the great success of the frozen–phonon approaches for the e–ph coupling in bulk, alloys and graphene [21–23, 25, 38] these methods have not been applied to study the e–ph coupling in colloidal semiconductor NCs. In this work, we extend the ab-initio frozen–phonon approach to the computation of e–ph interaction in colloidal NCs. In the calculations discussed here, the electronic states and the potential perturbed through the vibrations are self-consistently calculated using ab-initio DFT [40], and the vibrational eigenmodes are obtained from DFT computation implemented with a finite-difference approach [41].

Another class of important effects, that remains mainly untouched by atomistic theories, has to do with the anharmonic effects of lattice vibrations. These are important for real world applications of nanostructures, where the physical properties, such as thermal conductivity in nanowires, non-radiative transition via phonon in NCs, and Raman spectra broadening in nanostructures are dominated by the phonon lifetime. We have used ab-initio molecular dynamics (AIMD) to address these type of problems. Based on the accurately calculated forces from the electronic structure calculations, AIMD enables us to monitor the changes in the electronic and vibrational properties on-the-fly, with thermal effects included directly. Although AIMD simulations have been successfully applied to study a variety of problems, and the phonon lifetime in bulk has been studied using classical molecular dynamics simulations, a vibron–vibron interaction extracted from AIMD simulation is still lacking. During this reporting period, we performed AIMD simulations within the Born-Oppenheimer approximation of a $Si_{10}H_{16}$ cluster to study the geometry and the vibrational spectra from Fourier transformed velocity auto-correlation functions.

2 Computational Methods

2.1 *Vibrational Eigenmodes for Large Nanoclusters*

We construct spherical QDs made of material with zinc-blende type structure, centered on a cation with T_d point group symmetry. The dangling bonds are terminated by pseudo-hydrogen atoms H* with a fractional charge of 1/2, 3/4,

5/4, and 3/2 for group VI, V, III and II atoms, respectively. We use local density approximation (LDA) and Trouiller–Martin norm-conserving pseudopotentials with an energy cutoff of 30 Ry for InP and 40 Ry for CdSe clusters [41]. In order to calculate the vibrational eigenmodes of CdSe clusters with thousand atoms, we apply the non-linear core correction (NLCC) to Cd atoms instead of including Cd $4d$ states in the valence. The calculated transverse optical (TO) and longitudinal optical (LO) frequencies of bulk CdSe at the Γ point are 179.4 and 204.1 cm^{-1} with NLCC-LDA. Including d-states in the valence, we obtain 174.8 and 205.2 cm^{-1} in close agreement with NLCC results. The forces are minimized to less than 3×10^{-6} a.u. (5×10^{-4} a.u.) under constrained symmetry for the passivated (unpassivated) NCs. We calculate the bulk phonon states using density functional perturbation theory [36]. The vibrational properties are obtained from the eigenvalue equation [18]:

$$\sum_J \frac{1}{\sqrt{M_I M_J}} \frac{\partial^2 V(\boldsymbol{R})}{\partial \boldsymbol{R}_I \partial \boldsymbol{R}_J} \boldsymbol{U}_J = \omega^2 \boldsymbol{U}_I \quad (1)$$

where M are the atomic masses (and we will vary the mass of the passivant), V the potential, $\boldsymbol{R}$ the atomic positions, $\boldsymbol{U}$ the eigenvectors and ω the frequencies. We use ab-initio DFT implemented in the CPMD code [41] to optimize the geometry and to calculate the vibrational eigenmodes. The supercell is simple cubic with 3 Å space between the outmost atoms and the boundary. To analyze the eigenmodes in terms of bulk and surface contributions, we calculate the projection coefficients

$$\alpha^{\nu}_{c,s,p} = \frac{\sum_I^{(N_c,N_s,N_p)} |\boldsymbol{X}^{\nu}(I)|^2}{\sum_{I=1}^{N} |\boldsymbol{X}^{\nu}(I)|^2}, \quad (2)$$

where, N_c, N_s, N_p and N are the core, surface, passivant, and total number of atoms, $\boldsymbol{X}^{\nu}(I)$ represents the three components that belong to atom I from the $3N$-component eigenvector. We define the surface atoms as the atoms belonging to the outermost seven layers of the cluster (around 3 Å thick). From the phonon DOS $D(\omega)$ we obtain the specific heat according to:

$$C_v = N_A k_B \int_0^{\infty} \left(\frac{\hbar\omega}{k_B T}\right)^2 \frac{e^{\hbar\omega/k_B T}}{(e^{\hbar\omega/k_B T} - 1)^2} D(\omega) d\omega \quad . \quad (3)$$

2.2 *Calculation of Electron–Phonon Coupling Matrix Elements*

Within the Born–Oppenheimer approximation, the many-body Hamiltonian of the NC is decomposed into an electronic part, an ionic part and the coupling of the electron system with the lattice vibrations. The first two parts deal with the motions of electrons and ions separately while the third part describes the *electron–phonon interaction*. The e–ph interaction Hamiltonian can be expressed as a Taylor series expansion of the electronic potential [18]:

$$\Delta V^{\nu}(\boldsymbol{r}, \boldsymbol{R}) = \sum_{I} \frac{\partial V}{\partial \boldsymbol{R}_I} \cdot \boldsymbol{u}_I^{\nu} \tag{4}$$

where, V is the electronic potential and $\boldsymbol{R}_I$ denotes the nuclear position of atom I. The displacement vector $\boldsymbol{u}_I^{\nu}$, which belongs to atom I and the vibrational eigenmode ν, can be written in terms of the normal coordinates Q^{ν} [20]

$$\boldsymbol{u}_I^{\nu} = \frac{1}{\sqrt{M_I}} Q^{\nu} \boldsymbol{X}_I^{\nu} \tag{5}$$

with the occupation number representation,

$$Q^{\nu} = \sqrt{\frac{\hbar}{2\omega^{\nu}}} (a_{\nu}^{\dagger} + a_{\nu}) \tag{6}$$

where, M_I is the mass of atom I, $\hbar$ is the reduced Planck constant, ω^{ν} is the frequency of the vibrational mode ν, $\boldsymbol{X}_I^{\nu}$ are the three components of the vibrational eigenmode belonging to atom I, $a_{\nu}^{\dagger}$ and a_{ν} denote the creation and the annihilation operators of the ν-mode phonon.

According to (4) and (5), the e–ph coupling matrix elements for the transition from the initial state $|\psi_n, 0\rangle$ to the finial state $|\psi_m, 1^{\nu}\rangle$ with emission of a ν-mode phonon has the form

$$g^{\nu}(m, n) = \sum_{I} \sqrt{\frac{\hbar}{2M_I\omega^{\nu}}} \langle \psi_m, 1^{\nu} | \frac{\partial V}{\partial \boldsymbol{R}_I} \cdot \boldsymbol{X}_I^{\nu} (a_{\nu}^{\dagger} + a_{\nu}) | \psi_n, 0\rangle, \tag{7}$$

where, the polaron state $|\psi_m, i^{\nu}\rangle$ is composed of an electronic state $|\psi_m\rangle$ and a vibrational state $|i^{\nu}\rangle$.

Based on the frozen–phonon approach for the e–ph coupling, the change of the potential caused by a phonon distortion in (7) is replaced by [21–23]

$$\sum_{I} \frac{\partial V}{\partial \boldsymbol{R}_I} \cdot \boldsymbol{X}_I^{\nu} \approx \frac{V_{scf}^{\nu}(\boldsymbol{r}) - V_{scf}^{0}(\boldsymbol{r})}{u^{\nu}}, \tag{8}$$

where, $u^{\nu} = \sqrt{\sum_I \frac{1}{M_I} \boldsymbol{X}_I^{\nu 2}}$ is a scale describing effectively the frozen–phonon displacement caused by the lattice vibration, $V_{scf}^{\nu}(\boldsymbol{r})$ and $V_{scf}^{0}(\boldsymbol{r})$ are the self-consistent potentials with and without the phonon distortion [21–23].

In DFT, the self-consistent potential $V_{scf}(\boldsymbol{r})$ has the form [42],

$$V_{scf}(\boldsymbol{r}) = V_{ion}(\boldsymbol{r}) + e^2 \int \frac{\rho(\boldsymbol{r}')}{|\boldsymbol{r}-\boldsymbol{r}'|} d\boldsymbol{r}' + \frac{\delta E_{xc}[\rho(\boldsymbol{r})]}{\delta\rho(\boldsymbol{r})} \tag{9}$$

where, the ionic potential $V_{ion}(\boldsymbol{r})$ is typically decomposed into a local part $V_{ion}^{loc}(\boldsymbol{r})$ and a non-local part $V_{ion}^{NL}(\boldsymbol{r},\boldsymbol{r}')$, the Hartree potential $e^2 \int \frac{\rho(\boldsymbol{r}')}{|\boldsymbol{r}-\boldsymbol{r}'|} d\boldsymbol{r}'$, and the exchange-correlation potential $\frac{\delta E_{xc}[\rho(\boldsymbol{r})]}{\delta\rho(\boldsymbol{r})}$ are obtained self-consistently with $\rho(\boldsymbol{r}) = \sum_n^{occ} |\psi_n(\boldsymbol{r})|^2$. The local part of the self-consistent potential $V_{scf}^{loc}(\boldsymbol{r})$, which includes the local ionic potential $V_{ion}(\boldsymbol{r})$, the Hartree potential and the exchange-correlation potential, is computed self-consistently using standard DFT. The non-local potential is calculated using the real space representation of the Kleinman–Bylander form:

$$\langle \boldsymbol{r}|V_{ion}^{NL}|\boldsymbol{r}'\rangle = \sum_{I,l,m} \frac{\langle \boldsymbol{r}|\delta V_l^I \phi_{lm}^I\rangle\langle\phi_{lm}^I \delta V_l^I|\boldsymbol{r}'\rangle}{\langle\phi_{lm}^I|\delta V_l^I|\phi_{lm}^I\rangle} \tag{10}$$

where, l and m label the angular and magnetic moments, $\phi_{lm}^I(\boldsymbol{r}-\boldsymbol{R}_I)$ are the pseudo-wavefunctions centered on the atom position $\boldsymbol{R}_I$, and the potential $\delta V_l^I(|\boldsymbol{r}-\boldsymbol{R}_I|) = V_l^I(|\boldsymbol{r}-\boldsymbol{R}_I|) - V_{loc}^I(|\boldsymbol{r}-\boldsymbol{R}_I|)$. In contrast to the local potential, the Kleinman–Bylander non-local potential in (10) is a projector. The contribution of the non-local potential to (8) takes the form,

$$\sum_I \sqrt{\frac{\hbar}{2M_I\omega^\nu}} \sum_{r,r'} [\langle\psi_m|\boldsymbol{r}\rangle\langle\boldsymbol{r}|V_{ion}^{NL,\nu}|\boldsymbol{r}'\rangle\langle\boldsymbol{r}'|\psi_n\rangle - \langle\psi_m|\boldsymbol{r}\rangle\langle\boldsymbol{r}|V_{ion}^{NL,0}|\boldsymbol{r}'\rangle\langle\boldsymbol{r}'|\psi_n\rangle] \tag{11}$$

where, $V_{ion}^{NL,\nu}$ and $V_{ion}^{NL,0}$ represent the non-local potential with and without phonon distortion.

Finally, using (7)–(11) with the phonon distorted and un-distorted atom positions $\{\boldsymbol{R}_I + \boldsymbol{X}_I^\nu/\sqrt{M_I}\}$ and $\{\boldsymbol{R}_I\}$, the e–ph interaction matrix elements of the NCs can be calculated at the ab-initio level without additional parameters.

2.3 *Vibron–Vibron Coupling from Ab-Initio Molecular Dynamics*

To obtain vibron–vibron coupling matrix elements we performed AIMD simulations at a constant temperature (NVT-ensemble) using the Nosé-Hoover chain thermostat with time step of 20 a.u. (about 0.48 fs) in order to equilibrium the system. Following the 2 ps equilibration in the NVT-ensemble, we start a constant energy (NVE-ensemble) simulation and the trajectories are recorded at each time step. All the AIMD simulations are performed with the CPMD code [41]. In order to improve the statistics, we chop the NVE-ensemble simulation into several slices. Each slice starts from a different NVT-ensemble equilibrium time and ends with the same number of NVE-ensemble simulation time steps.

In order to obtain the temperature-dependent vibrational DOS from the AIMD simulation, we calculate the velocity auto-correlation function $C(t)$,

$$C(t) = \frac{\langle \boldsymbol{v}(t_0) \cdot \boldsymbol{v}(t_0 + t) \rangle}{\langle \boldsymbol{v}^2(t_0) \rangle} = \frac{\sum_{k=1}^{n_s} \sum_{j=1}^{n_t} \sum_{i=1}^{N} \boldsymbol{v}_i(t_{0j}^k) \cdot \boldsymbol{v}_i(t_{0j}^k + t)}{\sum_{k=1}^{n_s} \sum_{j=1}^{n_t} \sum_{i=1}^{N} \boldsymbol{v}_i^2(t_{0j}^k)}, \tag{12}$$

where, $\langle\ \rangle$ denotes the time averaged value calculated along the entire trajectory, $\boldsymbol{v}_i(t_{0j}^k)$ is the velocity vector of atom i in slice k at time origin point j. The number of atoms in the cluster, the number of time origin points, and the number of slices are labeled as N, n_t, and n_s, respectively. In the present work, the number of slices n_s is taken to be 15 with 6,400 time steps and 3,200 time origin points in each slice. The NVE-ensemble simulations are performed for a total of 96,000 time steps corresponding to about 46 ps simulation time. One slice corresponds to a simulation time of approximately 3 ps.

The vibrational DOS are calculated using the Fourier transform

$$VDOS(\omega) = \frac{1}{\sqrt{2\pi}} \int_{-\infty}^{+\infty} C(t) e^{-i\omega t} dt. \tag{13}$$

Since the limited statistic we obtained from AIMD at low temperature even for a small cluster, we use the DFPT results based on the harmonic approximation of lattice dynamics as a low temperature limit. In this case, the DFPT represents a very good classical (neglecting zero-point motion) approximation. The vibrational frequencies ω and the vibrational eigenvectors $\boldsymbol{U}_i$ are obtained from the eigenvalue equation given in (1).

In the quasi-harmonic approximation, where the changes in bond length due to thermal expansion are included but further anharmonic effects are excluded, the total energy of each vibrational mode $E^\nu(t)$—proportional to the occupation number of the vibrational mode—is expressed in terms of the time-dependent normal coordinates $Q_\nu(t)$

$$E^\nu(t) = \frac{1}{2}[\dot{Q}_\nu^*(t)\dot{Q}_\nu(t) + \omega_\nu^2 Q_\nu^*(t) Q_\nu(t)], \tag{14}$$

where

$$Q_\nu(t) = \sum_i^N \sqrt{\frac{M_i}{N}} \boldsymbol{U}_i^\nu \cdot [\boldsymbol{R}_i(t) - \boldsymbol{R}_i^0]. \tag{15}$$

The three-component vectors $\boldsymbol{U}_i^\nu$ in (15) represent the three components belonging to atom i of the vibrational eigenvectors obtained from (1). $\boldsymbol{R}_i^0$ is the equilibrium position of atom i obtained from the trajectory of the AIMD simulation as,

$$\boldsymbol{R}_i^0 = \frac{1}{n_t n_s} \sum_{k=1}^{n_s} \sum_{j=1}^{n_t} \boldsymbol{R}_i(t_j^k). \tag{16}$$

Based on the quasi-harmonic approximation, the vibrational vectors $\boldsymbol{U}_i^\nu$ used in (15) are calculated using DFPT with the atomic positions obtained from (16), i.e. from an AIMD simulation at a certain temperature.

The attenuation of the vibrational amplitude reflects the mode relaxation processes and can be described by the auto-correlation function of the energy fluctuation, written as

$$C_E^\nu(t) = \frac{\langle \delta E^\nu(t_0) \delta E^\nu(t_0 + t) \rangle}{\langle \delta E^\nu(t_0) \delta E^\nu(t_0) \rangle} \tag{17}$$

where, $\langle\ \rangle$ denotes the time averaged value calculated along the entire trajectory and $\delta E^\nu(t) = E^\nu(t) - \overline{E^\nu}$ is the deviation from the average vibrational energy $\overline{E^\nu}$.

2.4 Computational Details

The nanoclusters we studied are constructed by cutting a sphere, centered on a cation with T_d point group symmetry, from the zinc-blende bulk structure and removing the surface atoms having only one nearest-neighbor bond. The surface dangling bonds are terminated by pseudo-hydrogen atoms H* with a fractional charge of 1/2, 3/4, 5/4, and 3/2 for group VI, V, III, and II atoms, respectively. The calculations are performed using the LDA, Trouiller–Martin norm-conserving pseudopotentials with an energy cutoff of 30 Ry for III-Vs and 40 Ry for II-VIs.

The geometry relaxation is performed using the Broyden–Fletcher–Goldfarb–Shano procedure for the optimization of the ionic positions. The forces are minimized to less than 3×10^{-6} a.u. (5×10^{-4} a.u.) under constrained symmetry for the passivated (unpassivated) nanoclusters. In the calculation, the charge density $\rho_R(\mathbf{r})$ are obtained by solving the Kohn–Sham equation self-consistently and the values of $\frac{\partial \rho_R(\mathbf{r})}{\partial \mathbf{R}_J}$ are calculated using a finite difference approach. In principle we need $3N$ atomic displacements to obtain all the elements of the dynamical matrix (N being the number of atoms). In practice we calculate a significantly lower number of displacements ($3N/24$) and use the symmetry elements of the point group to deduce the missing elements. This is a key points to be able to treat these large structures.

All the calculations are performed with the CPMD code package developed at the Max Planck Institute in Stuttgart and at IBM [41]. The CPMD code is a high performance parallelized plane wave/pseudopotential implementation of DFT. It offers, at the moment, the best scaling among the DFT codes using a hybrid scheme of MPI and OpenMP. In this mixed parallelization, MPI is used across processors with distributed memory, while the shared memory architecture within one of these groups is parallelized through OpenMP. In this project, all the calculations were carried out on the NEC Nehalem Cluster at HLRS with 2.8 GHz and 12 GB memory per nodes, and infiniband interconnects. The optimum job sizes for the computations reported here are between 64 and 128 tasks with runtimes around 24 h. The total memory requirements are between 4 and 192 GB, depending on system type and size. We usually run using the maximum available memory on each note, i.e., 1.5 GB per task. The details of the scaling behavior and the performance per CPU for CPMD code have been given elsewhere [43].

3 Results

3.1 *Insights About the Surface of Colloidal Nanoclusters from Their Vibrational and Thermodynamic Properties*

We calculated the vibrational properties of colloidal quantum dots with up to 1,000 atoms from first principles. We studied the effect of the passivant and extract the following trends: The modes with optical character and significant surface contributions counter intuitively blue shift with increasing passivant mass, due to an under-coordination effect (see Fig. 1). The surface optic modes (for materials with acoustic-optic phonon gap) acquire passivant character and red shift with increasing passivant mass. Acoustic type modes are mostly unaffected by the passivants. We suggest that the low temperature specific heat should be a promising avenue to study the surface properties of NCs and offer a prediction for the onset temperatures and the temperature dependence in the Debye regime. A softening of the surface modes with decreasing cluster size and a negative Grüneisen parameter for TA-modes leads to a surprising larger specific heat for smaller clusters. We finally compare the vibrational properties of zinc-blende and wurtzite CdSe NCs with similar sizes and find significant differences in the optical-like branches and passivant modes. This fact reflects the distinct geometrical structure of the core and the surface and possibly of the spontaneous polarization [45, 46].

3.2 *First Principles Calculation of the Electron–Phonon Interaction in Semiconductor Nanoclusters*

To study the e–ph interaction in semiconductor nanoclusters, we have extend the ab-initio frozen–phonon approach to the computation of e–ph interaction in

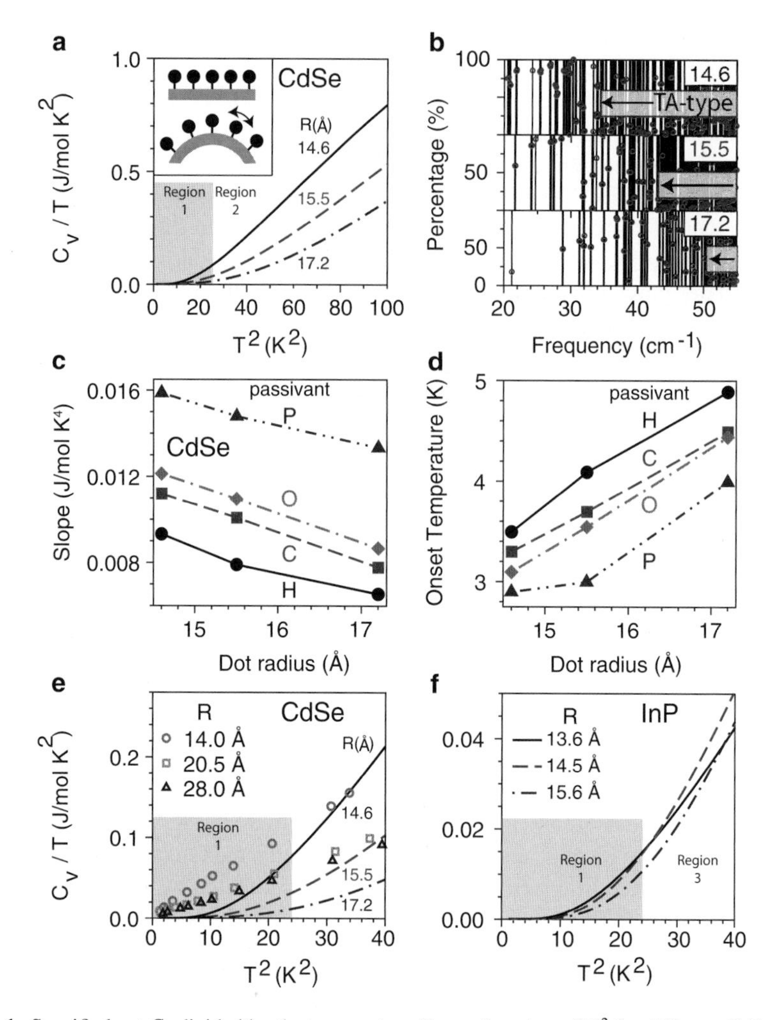

Fig. 1 Specific heat C_v divided by the temperature T as a function of T^2 for different CdSe NCs (**a**–**e**) and for InP clusters of different sizes (**f**). In (**e**) the experimental values [44] are given as symbols. In (**b**) the low frequency eigenmodes are drawn as *vertical lines* for three different CdSe NCs. The *solid circle* gives the percentage of surface character of each mode. In (**c**) we estimated the slopes in the (nearly) linear regime of panel (**a**) for CdSe NCs, changing the passivant mass. In (**d**) we report the corresponding onset temperature of C_v

colloidal NCs. In the calculations discussed here, the electronic states and the potential perturbed through the vibrations have been self-consistently calculated using ab-initio DFT [40], and the vibrational eigenmodes have been obtained from DFT computation implemented with a finite-difference approach [41]. A good agreement between the e–ph interaction matrix elements of small NC calculated using DFPT linear-response and our frozen–phonon method have confirmed the applicability of our approach. Moreover, our calculated e–ph transitions have been shown to obey selection rules strictly. Furthermore, we extended the calculations for the e–ph interaction to silicon NCs with up to 1,000 atoms and analyzed the phonon-induced carrier relaxation based on the Wigner–Weisskopf approach. We compared the intraband relaxation rate of silicon NC from our calculations to the experiments. We found that: (a) The frozen–phonon approach can provide the e–ph interaction matrix elements of semiconductor NCs with up to thousand atoms at the DFPT linear-response level. (b) The e–ph coupling strength was found to decrease with increasing cluster size. (c) We predicted a decaying Rabi-like oscillation between the LUMO and the second conduction states of NC with emission and absorption of a phonon. (d) The calculated intraband relaxation rate of silicon NCs were comparable to the experimental result, and this decay rate was shown to be determined by the phonon lifetime and the electron trapping time of NC surface states [47] (see Fig. 2).

3.3 Ab-Initio Molecular Dynamics Simulations of the Temperature Dependent Dynamical Processes in a Silicon Cluster

In order to investigate the temperature dependent dynamical processes, we performed AIMD simulations within the Born-Oppenheimer approximation of a $Si_{10}H_{16}$ cluster. We studied the geometry and the vibrational spectra using Fourier transformed velocity auto-correlation functions. The converged vibrational modes were obtained from a trajectory of about 46 ps (corresponding to 96,000 steps). We found that the high frequency Si–H stretching modes converge after around 1,800 vibrational cycles (around 30 ps) (see Fig.3). This comparably rapid convergence is attributed to the isolated frequency band, where interactions with other atoms is weak. Other modes require trajectories of more than 35 ps. Due to the limited statistic, the AIMD simulations cannot be applied to the low temperature region even for our small cluster. In this case we use DFPT that represents a very good classical (neglecting zero-point motion) approximation. We find a blue-shift of the Si–Si vibrational modes with transverse acoustic (TA) characters and a red-shift of the other vibrational modes with increasing temperature. We also see a broadening of all the vibrational modes with increasing temperature. The former can be linked to the negative (blue-shift) and positive (red-shift) Grüneisen parameters along with the extended bond lengths. The latter is attributed to the low symmetry (proximity to the surface) enhancing anharmonic effects at high temperatures. We also find that the Si–H bonds are more sensitive to temperature than the Si–Si bonds. This phenomenon is attributed to the free outward movement of the surface hydrogen atoms, which introduces a potential asymmetry towards the vacuum side and increases the anharmonicity. We also study the vibrational cooling process of selected modes, which corresponds to the phonon (or vibron) relaxation, by

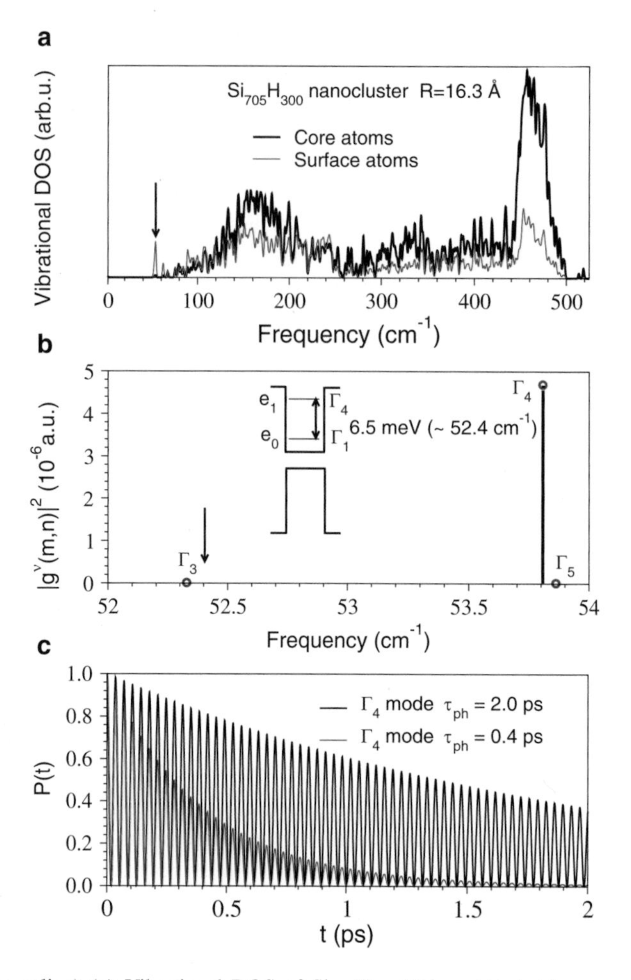

Fig. 2 (Color online) (**a**) Vibrational DOS of $Si_{705}H_{300}$ NC contributed by core (*thick black*) and surface atoms (*thin red/gray*) with a broadening of $0.8\,\text{cm}^{-1}$, (**b**) e–ph matrix elements for the interaction between e_0, e_1 states and the vibrations with the phonon energy $\hbar\omega^{\nu}$ around the electron energy level spacing $E_1 - E_0$, and (**c**) probability of e_1 state interacted by vibration with Γ_4 symmetry with phonon lifetime as 2.0 and 0.4 ps

calculating the energy auto-correlation functions. We extract the vibrational cooling time and find that the vibrational energy of the Si–H stretching mode decays rapidly, within 50 fs, while the cooling times of the other modes are around 1 ps. These relaxation times (see Fig. 4) are very comparable to phonon lifetimes in bulk silicon. The rapid cooling of the high frequency Si–H stretching mode is attributed to the existence of many low frequency relaxation channels. We also quantify the decrease of the vibrational cooling time with increasing temperature for different types of vibrational modes and find a stronger decrease for acoustic-like modes [48].

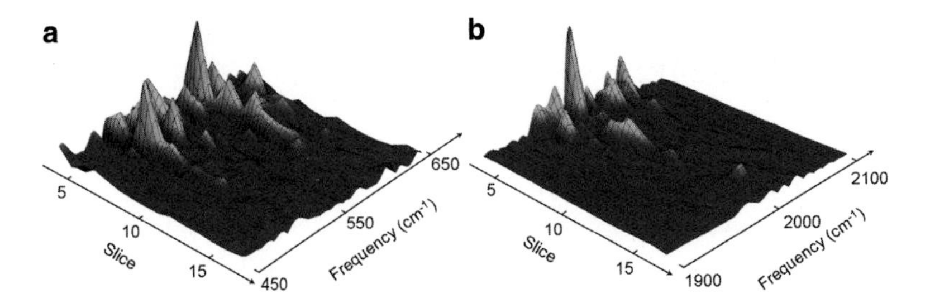

Fig. 3 Convergence of the vibrational modes $\Delta VDOS_k(\omega)$ with frequency between (**a**) 450 and 650 cm^{-1} (bending Si–H and rotation H–Si–H modes), and (**b**) 1,900 and 2,100 cm^{-1} (stretching Si–H modes) at $T = 800$ K as function of the simulation time given in unit of slices (3 ps). The *vertical axis* describes the deviation of the vibrational DOS between a simulation with k and a simulation with $k-1$ time slices. The *red color* (maximum deviations) reflects a deviation of 30 %

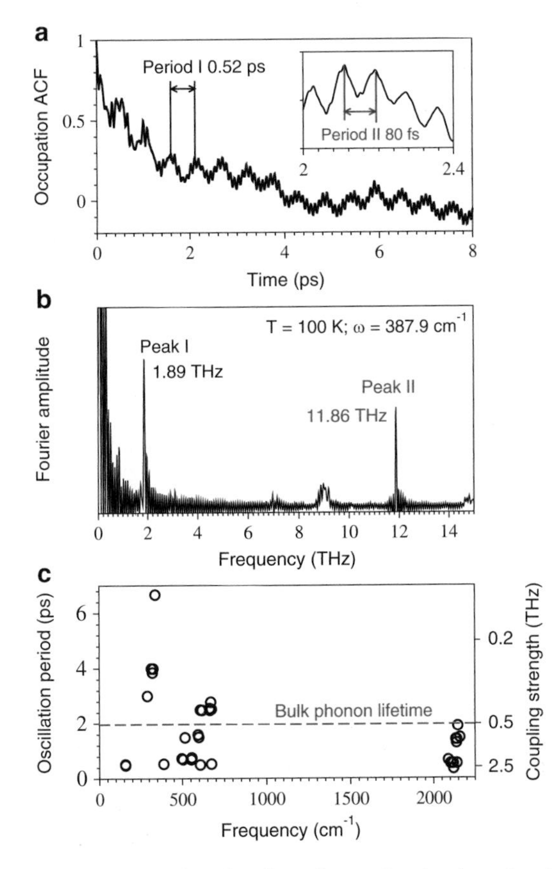

Fig. 4 (**a**) Occupation auto-correlation function of one vibrational mode with labeled vibron–vibron oscillation period (Period I) and vibrational oscillation period (Period II). (**b**) Fourier transform of the occupation auto-correlation function of (**a**) vibrational mode with labeled Peak I corresponding to the vibron–vibron coupling strength and peak II corresponding to the vibrational frequency. (**c**) The vibron–vibron oscillation period for all the vibrational modes (*circles*) at 100 K along with the bulk phonon lifetime (*dashed line*) from experiment

4 Publications Resulting from the HLRS Support for the Year 2012

[P1] P. Han and G. Bester. *Confinement effects on the vibrational properties of III–V and II–VI nanoclusters.* Phys. Rev. B **85**, 041306(R) (2012).
[P2] P. Han and G. Bester. *Insights About the Surface of Colloidal Nanoclusters from their Vibrational and Thermodynamic Properties.* J. Phys. Chem. C **116**, 10790 (2012).
[P3] P. Han and G. Bester. *First principles calculation of the electron–phonon interaction in semiconductor nanoclusters.* Phys. Rev. B **85**, 235422 (2012).
[P4] P. Han, L. Vilciauskas and G. Bester. *Vibron–vibron coupling from ab-initio molecular dynamics simulations of a silicon cluster.* New J. Phys. **15**, 043039 (2013).

Acknowledgements We gratefully acknowledge supercomputer time provided by the Höchst Leistungs Rechenzentrum Stuttgart (HLRS).

References

1. J.J.H. Pijpers et al., J. Phys. Chem. C **111**, 4146 (2007)
2. V.I. Klimove, S.A. Ivanov, J. Nanda, M. Achermann, I. Bezel, J.A. McGurie, A. Piryatinski, Nature **447**, 441 (2007)
3. R.M. Kraus, P.G. Lagoudakis, A.L. Rogach, D.V. Talapin, H. Weller, J.M. Lupton, J. Feldmann, Phys. Rev. Lett. **98**, 017401 (2007)
4. N. Gaponik, S.G. Hickey, D. Dorfs, A.L. Rogach, A. Eychmüller, Small **6**, 1364 (2010)
5. D.V. Talapin, J.-S. Lee, M.V. Kovalenko, E.V. Shevchenko, Chem. Rev. **110**, 389 (2010)
6. J.B. Sambur, T. Novet, B.A. Parkinson, Science **330**, 63 (2010)
7. D. Mocatta, G. Cohen, J. Schattner, O. Millo, E. Rabani, U. Banin, Science **332**, 77 (2011)
8. G. Allan, C. Delerue, M. Lannoo, Phys. Rev. Lett. **76**, 2961 (1996)
9. A. Franceschetti, H. Fu, L.-W. Wang, A. Zunger, Phys. Rev. B **60**, 1819 (1999)
10. P. Guyot-Sionnest, B. Wehrenberg, D. Yu, J. Chem. Phys. **123**, 074709 (2005)
11. A.L. Rogach, A. Eychmüller, S.G. Hickey, S.V. Kershaw, Small **3**, 536 (2007)
12. V.M. Huxter, G.D. Scholes, J. Chem. Phys. **132**, 104506 (2010)
13. A. Cros-Gagneux, F. Delpech, C. Nayral, A. Cornejo, Y. Coppel, B. Chaudret, J. Am. Chem. Soc. **132**, 18147 (2010)
14. A. Puzder, A.J. Williamson, N. Zaitseva, G. Galli, L. Manna, A.P. Alivisatos, Nano Lett. **4**, 2361 (2004)
15. S.V. Kilina, D.S. Kilin, O.V. Prezhdo, ACS Nano **3**, 93 (2009)
16. M.D. Ben, R.W.A. Havenith, R. Broer, M. Stener, J. Phys. Chem. C **115**, 16782 (2011)
17. V.V. Albert, S.A. Ivanov, S. Tretiak, S.V. Kilina, J. Phys. Chem. C **115**, 15793 (2011)
18. P.Y. Yu, M. Cardona, *Fundamentals of Semiconductors* (Springer, Berlin, 2010)
19. G. Bester, S. Nair, A. Zunger, Phys. Rev. B **67**, 161306 (2003)
20. O. Madelung, *Introduction to Solid-State Theory* (Springer, Berlin, 1996)
21. M.M. Dacorogna, M.L. Cohen, P.K. Lam, Phys. Rev. Lett. **55**, 837 (1985)
22. P.K. Lam, M.M. Dacorogna, M.L. Cohen, Phys. Rev. B **34**, 5065 (1986)
23. K.J. Chang, M.L. Cohen, Phys. Rev. B **34**, 4552 (1986)
24. F. Murphy-Armando, S. Fahy, Phys. Rev. B **78**, 035202 (2008)

25. J.-A. Yan, W.Y. Ruan, M.Y. Chou, Phys. Rev. B **79**, 115443 (2009)
26. S.Y. Savrasov, D.Y. Savrasov, O.K. Andersen, Phys. Rev. Lett. **72**, 372 (1994)
27. F. Mauri, O. Zakharov, D. de Gironcoli, S.G. Louie, M.L. Cohen, Phys. Rev. Lett. **77**, 1151 (1996)
28. A.Y. Liu, A.A. Quong, Phys. Rev. B **53**, R7575 (1996)
29. R. Bauer, A. Schmid, P. Pavone, D. Strauch, Phys. Rev. B **57**, 11276 (1998)
30. J. Sjakste, N. Vast, V. Tyuterev, Phys. Rev. Lett. **99**, 236405 (2007)
31. F. Giustino, M.L. Cohen, S.G. Louie, Phys. Rev. B **76**, 165108 (2007)
32. J. Gavartin, A. Shluger, Phys. Stat. Sol. (c) **3**, 3382 (2006)
33. D.M. Ramo, A.L. Shluger, J.L. Gavartin, G. Bersuker, Phys. Rev. Lett. **99**, 155504 (2007)
34. D.T. Devreese, P. van Camp, *Electronic Structure, Dynamics, and Quantum Structural Properties of Condensed Matter* (Springer, New York, 1985)
35. W. Frank, C. Elsässer, M. Fähnle, Phys. Rev. Lett. **74**, 1791 (1995)
36. S. Baroni, S. de Gironcoli, A. Dal Corso, P. Giannozzi, Rev. Mod. Phys. **73**, 515 (2001)
37. P.B. Allen, M.L. Cohen, Phys. Rev. **187**, 525 (1969)
38. S. Zollner, S. Gopalan, M. Cardona, J. Appl. Phys. **68**, 1682 (1990)
39. D.A. Papaconstantopoulos et al., Phys. Rev. B **15**, 4221 (1977)
40. P. Giannozzi et al., J. Phys. Condens. Matter. **21**, 395502 (2009). http://www.quantum-espresso.org
41. The CPMD consortium page, coordinated by M. Parrinello and W. Andreoni, Copyright IBM Corp 1990–2008, Copyright MPI für Festkörperforschung Stuttgart 1997–2001. http://www.cpmd.org
42. R.M. Martin, *Electronic Structure Basic Theory and Practical Methods* (Cambridge University Press, Cambridge, 2004)
43. G. Bester, P. Han, *High Performance Computing in Science and Engineering'11*, ed. by W. Nagel, D. Kröner, M. Resch (Springer, Berlin, 2012)
44. S. Neeleshwar, C.L. Chen, C.B. Tsai, Y.Y. Chen, C.C. Chen, S.G. Shyu, M.S. Seehra, Phys. Rev. B **71**, 201307 (2005)
45. P. Han, G. Bester, Phys. Rev. B **85**, 041306(R) (2012)
46. P. Han, G. Bester, J. Phys. Chem. C **116**, 10790 (2012)
47. P. Han, G. Bester, Phys. Rev. B **85**, 235422 (2012)
48. P. Han, L. Vilciauskas, G. Bester, New. J. Phys. **15**, 043039 (2013)

Part III
Chemistry

Christoph van Wüllen

In this section, there are three contributions from two different fields. The first paper, by Stegmüller and Tonner, is a semiconductor material science study, where initial stages in the metal-organic vapor phase epitaxy process are studied with high-level quantum chemical methods. In this process, GaP (a III/V semiconductor material) grows on a silicon substrate by adsorption of volatile gallium and phosphorus containing precursors (such as triethyl gallane and tri-tert-butyl phosphine) from the gas phase, followed by their decomposition on the surface and incorporation of the gallium and phosphorous atoms into the growing semiconductor material. The investigations presented here concentrate, in great detail, on simpler model gallanes and phosphines. A similar reaction of the silicon surface with hydrocarbons can be used to functionalize the surface through "decoration" with organic molecules captured from the gas phase. The calculations were performed both on a periodic system and on a cluster model which models adsorption at a cut-out of the silicon substrate whose dangling bonds are then saturated with hydrogen. The cluster model has the distinct advantage that one can do high-level wave function based calculations beyond density functional theory. This provides an independent test of the accuracy of the latter.

The following two contributions deal with the calculation of molecular vibrations. This is extremely easy within the harmonic approximation, because then the problem decomposes into independent harmonic oscillators which can be treated analytically. Beyond the harmonic approximation however, the problem becomes extremely difficult and explodes exponentially with the number of coordinates. The MCTDH ansatz pioneered by H.-D. Meyer reduces this complexity by expanding the vibrational wave function as a linear combination of simple products of single-particle functions. This is a prerequisite for handling molecules such as malon

C. van Wüllen
Fachbereich Chemie, Technische Universität Kaiserslautern, Erwin-Schrödinger-Straße, D-67663 Kaiserslautern, Germany
e-mail: vanWullen@Chemie.Uni-KL.de

aldehyde with 9 atoms and 21 internal coordinates. It has to be stressed that malon aldehyde is a "large molecule" if it comes to a highly accurate calculation of its vibrations, especially the tunneling splitting associated with the double-minimum proton transfer potential.

It should not be forgotten that the MCTDH calculations are based on a pre-computed potential energy surface which has been obtained as a fit to more than 11,000 single-point energies calculated with highly accurate quantum chemical methods. Usually, the large number of data points is not the only problem: one needs an analytical fit to these data points which requires much "hand-work" and adaptation to the case at hand. The contribution by Rauhut, Resch, and coworkers addresses the problem how potential energy surfaces can be computed automatically using grid computing. Even then, explicit consideration of each new case is necessary, because the potential energy surface is expressed by one-, two-, and three-body terms, and these are calculated with quantum chemical methods of different accuracy. Which accuracy is actually required for which term is evaluated beforehand by calibration studies on smaller systems. Still, it is remarkable that a vibrational spectrum at VMP2 level could be obtained for the 16-atom Li_8F_8 cluster.

At first sight, the systems investigated in this section seem to be rather small: adsorption of GaH_3 on silicon has been investigated instead of triethyl gallane, the molecule actually used in reality. Both malon aldehyde (9 atoms) and the Li_8F_8 cluster can be considered as small molecules, given that vibrational frequencies within the harmonic approximation can today routinely be calculated for molecules with some hundredth of atoms. But one must not forget that in all cases, the target is to reach a yet unprecedented level of accuracy. Realistic computational modeling beyond density functional theory and the harmonic approximation calls for gargantuan computational resources, but applications which seem esoteric at present may well become routine in the near future, but this is only possible if the methodological developments are done today. In this respect, the availability of high-performance computing facilities drives the development whose impact on science, economics and society can only be judged in hindsight.

From Molecules to Thin Films: GaP Nucleation on Si Substrates

Andreas Stegmüller and Ralf Tonner

1 Introduction

Silicon is prominently used as a substrate for a variety of functionalized materials. Those are tuned towards direct band gap semiconductor materials or for subsequent adsorption and growth of optically active organic compounds.

Bulk materials with direct optical gaps are heterogeneous mixtures of III/V materials which can be grown on Si [1,2]. Those find application as highly efficient solar cells or lasers for communication purposes. The metal-organic vapour phase epitaxy (MOVPE) is a frequently applied growth procedure for those materials [3]. The quality of the resulting films and bulk structures are highly dependent on the successful growth. Precursor decompositions and the hydrogenated substrate surface have to be clean to avoid carbon or oxygen incorporation [4], adsorption and surface migration processes need to be understood in order to predict the growth process and the characteristic steps determining the material's structural quality [5].

In order to evaluate certain structural features appearing in this early growth period, the thermodynamics and kinetics of decomposition and adsorption processes have to be known [6]. Chemical insight will be generated by calculating geometry parameters, molecular properties and reaction paths with quantum chemical methods. This will complement experimental findings and make the growth process controllable and predictable. The theoretical studies will as well inspire experimentalists to vary growth procedures, the applied precursors and help understand the measured properties [4,7–11].

A. Stegmüller · R. Tonner (✉)
Fachbereich Chemie, Philipps-Universität Marburg, Hans-Meerwein-Straße,
35032 Marburg, Germany
e-mail: tonner@chemie.uni-marburg.de

W.E. Nagel et al. (eds.), *High Performance Computing in Science and Engineering '13*, DOI 10.1007/978-3-319-02165-2_14,

As for the III/V semiconductor materials, the investigation of single molecular and sub-monolayer organic adsorbates is an experimentally demanding task [12]. Quantum-chemical methods can model the atomic and electronic structure with sufficient accuracy and will provide insight in to the nature of the surface-adsorbate interactions, chemical processes during the deposition and the properties of the Si substrate - organic adsorbate systems. The second subproject in this report describes a system of triple bond organic species adsorbed on non-hydrogenated Si(001), whose exceptional electronic structure is investigated prior to adsorption studies.

As many of the described calculations involve a significant number of atoms, this study is demanding in computational resources. A number of sub-projects are still in progress which include periodic GGA calculations of several elementary processes on the Si surface that have been carried out at the HLRS. The following chapters show some highlights of the research by our group using computational resources of the HLR Stuttgart in 2012.

2 Absorption Study Ga- and P-Precursors on Si(001)

2.1 Introduction

In order to understand the processes of adsorption during the early growth phase of GaP on silicon, the identification of thermodynamically stable adsorbates on Si(001) sites is of major importance. The initial adsorbate configurations might play a key role in the growth process of high quality III/V materials. Relevant adsorbates have been identified in a gas phase decomposition study and in the following subproject their adsorption behaviour on the Si(001)2x1 reconstruction is examined.

Due to the MOVPE reactor conditions, the silicon surface will be saturated with hydrogen. This well-known state of the Si surface [13–15] has two Si atoms to form dimers that are parallely aligned in rows along the surface plane. In the unit cell, one dimer is present with the Si–Si bond parallel to the surface plane and the hydrogen atoms standing about upright on-top of the Si atoms.

Possible adsorbates are not exclusively products from gas phase decompositions, molecules will also undergo surface-mediated dissociations [16]. At higher temperatures, hydrogen can desorb from the surface by recombinative desorption [17], which makes the surface more reactive with respect to Ga and P species. Multiple adsorbed atomic Ga and P form dimers on Si(001) very similar to the Si–Si-dimers [18–21].

This subproject focuses on the thermodynamic stabilities of single and double Ga- and P-atom adsorption on various (relative) adsorption positions. Secondly, the initial precursor of the described MOVPE process for growth of GaP on Si(001), tert-butyl phosphine, and the decomposition products PH_2 and GaH_2 will be examined in a sufficiently sized supercell of the Si(001)2x1 reconstruction [22, 23].

2.2 Methods

The Si(001) surface was modelled as a cluster saturated with H applying *Gaussian Type Orbital* bases and as a periodic slab with a plane wave basis set [24, 25].

For the cluster calculations, the def2-TZVPP basis sets [26] were used and adsorbate geometry optimizations were performed with fixed Si atoms to preserve the surface reconstruction. For the rotation profiles (see below) the tetrahedral angles were constraint and sampled for tert-butyl phosphine. The cluster approach was also used to estimate thermodynamic corrections to electronic energies for the stability of adsorption sites and relative adsorbate positions and orientations. The corrections were applied for realistic MOVPE growth conditions of 0.05 atm and 673, 773 and 948 K, respectively.

CCSD(T)/def2-TZVPP [27] calculations were used to confirm DFT-energies set as reference method. Bonding analyses was carried out with the *Natural Bond Orbital* (NBO) method as performed with NBO 5.9 [28] and the *Atoms In Molecules* (AIM) method.

Adsorption geometries and reaction energies were calculated with the more accurate slab model in a periodic plane wave basis with VASP 5.2.12 [29]. The PBE functional [30] was used as a standard method and the influence of dispersion effect on geometries and energies was modelled with the D2 correction [31]. A PAW-basis set [24, 25] with a cutoff at 350 eV of kinetic energy was chosen for the calculations performed. The slab unit cell was set up as a Monkhorst-Pack k-grid with 6 k-points in x- and y-direction.

2.3 Results

Adsorption Reactions and Geometries

For the relevant GaP adsorbates there are two possible adsorption mechanisms investigated in this work.

This is, firstly, the insertion of an adsorbate species into a Si–H bond at the surface and, secondly, the substitution of a H atom at the Si surface with an adsorbate molecule or atom.

The insertion mechanism involves the breaking of a Si–H bond at the fully hydrogenated Si(001) surface, the formation of the adsorbate–Si bond and, lastly, the attachment of the hydrogen atom to the adsorbed species. This mechanism is assumed to be a one-step process, which is represented by reactions Ads4, Ads5, Ads10 and Ads11 in Table 1. These reactions were found to be electronically favourable both for Ga and P species.

The substitution of H from a surface site results in the adsorbate covalently bonded to the surface Si atom and the elimination of a hydrogen molecule H_2 (or alkane species, Ads8). Energetically, those reactions are unfavourable, but as an additional volatile species is eliminated the steps will be entropically promoted.

Table 1 Single adsorption of Ga and P species onto a hydrogenated Si(001)6x3 supercell

	Reaction index	ΔE
Ads1	$[Si]{-}H + GaH_3 \longrightarrow [Si]{-}GaH_2 + H_2$	389.3
Ads2	$[Si]{-}H + GaH_2\cdot \longrightarrow [Si]{-}GaH\cdot + H_2$	365.0
Ads3	$[Si]{-}H + GaH \longrightarrow [Si]{-}Ga + H_2$	362.3
Ads4	$[Si]{-}H + GaH \longrightarrow [Si]{-}GaH_2$	−112.9
Ads5	$[Si]{-}H + Ga\cdot \longrightarrow [Si]{-}GaH\cdot$	−82.3
Ads6	$[Si]{-}H + H_2PBu \longrightarrow [Si]{-}HPBu + H_2$	371.1
Ads7	$[Si]{-}H + H_2PBu \longrightarrow [Si]{-}PH_2 + BuH$	83.7
Ads8	$[Si]{-}H + PH_2\cdot \longrightarrow [Si]{-}PH\cdot + H_2$	393.3
Ads9	$[Si]{-}H + PH \longrightarrow [Si]{-}P + H_2$	380.9
Ads10	$[Si]{-}H + PH \longrightarrow [Si]{-}PH_2$	−277.2
Ads11	$[Si]{-}H + P\cdot \longrightarrow [Si]{-}PH\cdot$	−254.6

Electronic reaction energies are given in kJ/mol on the PBE-D2/PAW level if theory

Table 2 Interaction energies of various species with the Si slab

Species	ΔE_{int}	d(Si–E)	d(Si–Si)
PBuH	−288.4	2.272	2.408
PBu	−277.5	2.251	2.424
PH_2	−275.4	2.277	2.406
PH	−272.2	2.231	2.404
P	−266.4	2.199	2.420
GaH_2	−275.4	3.396	2.397
GaH	−236.3	2.452	2.396
Ga	−248.1	2.582	2.409
H	−333.9	1.501	2.404

Energies are given in kJ/mol, the distances between the Si adsorption site and the adsorbate E and the Si–Si dimer distance are presented in Å

Moreover, at high temperatures this entropic effect will increase, however, the fraction of unsaturated Si sites will increase and yet another adsorption mechanism involving the reactive unsaturated site will be available. The selected substitution reactions are represented with Ads1–3 and Ads6–9 in Table 1, where electronic reaction energies of the two reaction mechanisms for adsorbates like gallane (GaH_3), $GaH_2\cdot$, GaH, atomic Ga·, tert-butyl phosphine (H_2PBu), $PH_2\cdot$, PH and atomic P· are presented.

In Table 2 the interaction energies of various species with a periodic surface slab are presented which show the influence of the adsorbates' geometries and energies on the surface reconstruction. Both, the examined Ga and P species are covalently bonded to a Si atom, substituting the hydrogen atom (which is the most stable adsorbate). However, the bonds of [Si]–GaH ($\Delta E = -236.6$ kJ/mol) and atomic [Si]–Ga ($\Delta E = -248.1$ kJ/mol) are somewhat lower than the average bond strength of P species. The Si–Si dimer bond length is not significantly affected by the other substitution reactions. The relevant adsorption geometries are presented in Fig. 1.

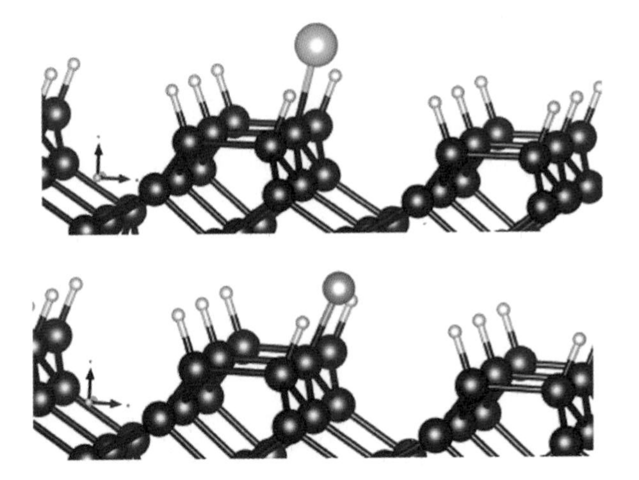

Fig. 1 Adsorbate geometries of single atoms on the hydrogenated Si(001)6x3 supercell representation calculated by PBE-D2/PAW. Colours are as above, *green*: Ga, *blue*: Si, *white*: H

At low coverages, lateral interactions between adsorbates are minimized. The chosen Si(001)6x3 supercell has an area of 175.2 Å^2 where adsorbates will almost exclusively interact with the substrate.

Multi-adsorption

The secondary adsorption was studied with the Si(001)6x3 supercell. While small adsorbate species such as GaH_x and PH_x ($x = 0, 1, 2$) were found to adsorb on neighbouring sites without significant hindrance, species with tert-butyl groups (HPtBu· and PtBu) block some positions by about 35–74 kJ/mol of energy discrimination. It is important to note that these species can not freely rotate around the surface normal.

Full Atomic Monolayer Coverage

A complete coverage with atomic Ga and P, respectively, was examined by the Si(001)2x1 slab and the Si bulk structure. Compared to the bulk structure with an atomic adlayer, the Si–P ($\Delta E = +97.6$ kJ/mol) and Si–Ga ($\Delta E = +27.5$ kJ/mol) bonds of the 2x1 reconstruction are less stable, respectively. The Si–P bond in one monolayer on the bulk structure is even stronger than the Si–H bond in the stable 2x1 reconstruction ($\Delta E = -71.1$ kJ/mol), which has the most stable substrate-adlayer interaction arrangement ($\Delta E = -165.1$ kJ/mol per H atom).

The surface reconstruction from Si bulk structure to Si(001)2x1 is about 60 kJ/mol per adatom. This is only compensated by an adlayer of P over the reconstructed Si surface due to the strong Si–P bond. The reconstruction with an atomic Ga monolayer is slightly endergonic, although lateral interactions are stabilizing the system. However, following this electronic argument the 2x1 surface reconstruction will disappear during growth as soon as an atomic monolayer of P is adsorbed. Thermodynamic and mechanistic arguments will be addressed in a follow-up study.

2.4 Summary

The available methodology (PBE-D) leads to a satisfying description of bulk and surface properties with the same density functional. Lattice constants and bulk moduli of the Si bulk lattice were calculated close to experimental values.

A cluster model of the surface needs to be sufficiently sized in order to produce realistic data, which intrinsically leads to high computational effort. However, periodic vibrational corrections for phonons and molecular vibrations can be obtained from plane-wave DFT methods, which as well requires extensive resources and will be performed for this system in the future.

The adsorption mechanism modelled in this study unveiled important chemical trends. The substitution of H at the fully hydrogenated Si surface is found to be energetically unfavourable for the examined Ga and P species. Reactions with the free vacancy at a Si atom might appear at high temperatures as recombinative desorption can expose highly reactive Si sites to the adsorbates. Those were not examined here.

In the future, we will focus on adsorption and decomposition mechanisms at the surface of relevant adsorbates, determine kinetic rates of those processes next to diffusion processes, multiple adsorption and layer-by-layer growth.

3 Adsorption Study of Acetylene and Cyclooctyne: Functionalization of Si(001)

3.1 Introduction

Cyclooctyne is the smallest cyclic hydrocarbon compound with a triple bond stable under lab conditions [32]. The most stable configuration is a linear geometry for the R–C≡C–R-chain along the triple bond, which can not be obtained for the (small) cyclic system of cyclooctyne. Ring strain leads to a deformation of the linear triple bond geometry. The distortion angle between the two ligands, R, and the center of the triple bond is proportional to the strain energy of the ring structure (83.3 kJ/mol [33]). Cyclo*octane* is more stable (strain energy of 51.5 kJ/mol). For that, cyclooctyne is more reactive which is utilized e.g. for the modification of biomolecules by [3 + 2]-cyclo additions [34, 35].

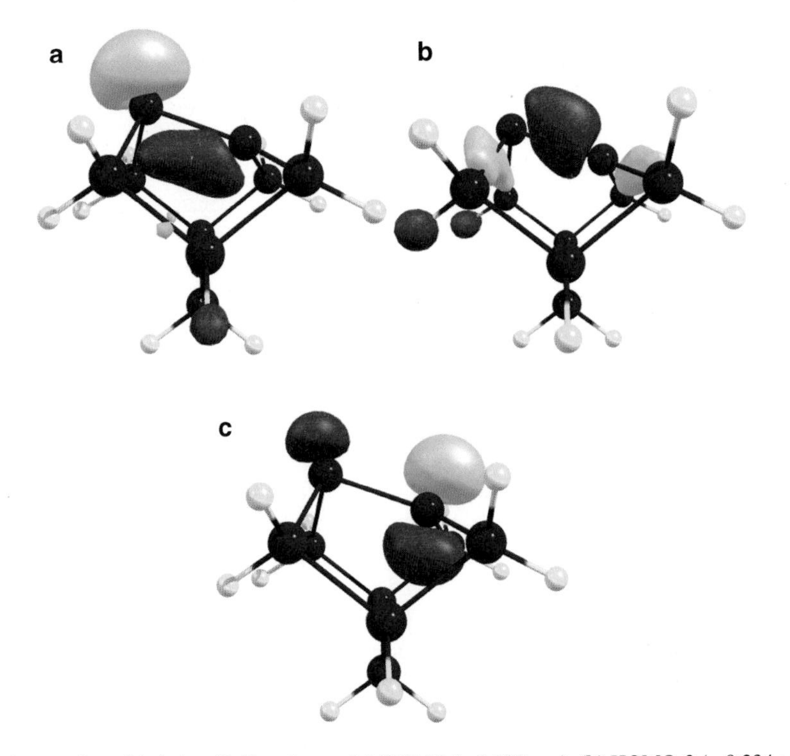

Fig. 2 Frontier orbitals in a Si_9H_{12}-cluster. (**a**) HOMO (−0.192 a.u.). (**b**) HOMO-2 (−0.234 a.u.). (**c**) LUMO (−0.140 a.u.)

Semiconductor materials can be functionalized by organic molecules. The Si(001)c4x2 reconstruction as one dominant non-hydrogenated surface modification is therefore an important object for adsorption studies [12]. This surface has been of major interest in surface science. In contrast to the previous subproject Si(001) is not hydrogenated for the adsorption of organic molecules. The surface reconstruction is therefore significantly different. On the surface Si(001)c4x2 the top Si atoms form dimers, that are tilted as shown in Fig. 2.

This study aims at answering the following questions related to adsorption of organic molecules on the surface reconstruction Si(001)2x1. If acetylene and cyclooctyne form stable adsorbates on the Si(001) surface, what are the adsorption geometries? How stable are systems with multiple molecules adsorbed (at neighbouring sites)? STM experiments revealed peculiar arrangements of multiple (neighbouring) cyclooctyne adsorbates. Can those results be understood with quantum chemical methods? [36–38].

Table 3 Selected convergence parameters of the Si bulk and the Si(001) surface slab unit cell chosen for periodic DFT calculations

	Value	E/atom (kJ/mol)	Comment
k-Points	4 4 4	0.14	Bulk
k-Points	2 4 1	0.06	Surface
Cutoff (eV)	350	0.39	
Vacuum (Å)	6.564	−0.07	
$\sum$ Si-layers	6	−0.02	

3.2 Methods

For the investigations performed in this study, periodic slab calculations were performed as well as molecular cluster models. For the periodic DFT calculations a plane wave PAW basis set was used and reciprocal k-points were generated in the Monkhorst-Pack procedure.

Next to the energy convergence of the basis size and the k-point density, the convergence with respect to the vacuum space between the silicon slabs was evaluated. Secondly, it was necessary to examine the convergence behaviour of the number of Si layers the slab consists of in order to gain correct surface geometries applying the slab model. The combination of these two parameters determines the absolute size of the model unit cells. Table 3 shows converged values for all four parameters, which were used in this study of cyclooctyne on non-hydrogenated Si(001)2x1. The determination of these converged parameters necessitated extensive use of computational resources.

In the periodic as well as the cluster calculations, the PBE functional was used. Molecular calculations and the AIM and NBO analyses were performed with the def2-SVP and def2-TZVPP basis sets.

3.3 Results

Electronic Structure of the Si–Si Dimer Within the Si(001)2x1 Surface Reconstruction

One dimer from the Si(001)2x1 surface reconstruction was modelled in a Si_9H_{12}-cluster to study the nature of the Si–Si dimer bond and the distribution of charge. The frontier orbitals of this cluster were calculated at the PBE/def2-TZVPP level of theory and are presented in Fig. 2. The Si atoms of the cluster were constraint to the geometry gained by a periodic PBE calculation.

The HOMO (Fig. 2a) indicates that electron density is accumulated at the higher lying dimer atom, Si_{up}, which has p-orbital character, and below the dimer bond plane. The HOMO-2 can be interpreted as a σ bond and the LUMO is mainly located at the lower lying dimer atom, Si_{down}. As reactivity studies of the Si(001) surface [39, 40] revealed, electron density seems to be asymmetrically distributed within the dimer.

Table 4 Bonding parameters of acetylene, Si(001) and two adsorption modes of a single acetylene on the surface

	Si–Si	C–C	Si_{up}–C	Si_{down}–C	∠HCC (°)
On-top	2.365 (2.365)	1.358 (1.358)	1.896 (1.897)	1.913 (1.916)	123.5
End-bridge	–	1.362 (1.363)	1.914 (1.918)	1.923 (1.925)	118.3
Acetylene	–	1.208 (1.208)	–	–	180
Ethylene	–	1.334	–	–	121.7
Ethane	–	1.538	–	–	111.8
Si(bulk)	2.371 (2.363)	–	–	–	–

Bond lengths are given in Å, values in brackets represent results from periodic PBE-calculations

The calculated bond length (2.37 Å) is comparable to other Si–Si single bond compounds like disilylenes in [41, 42].

The accuracy of the performed DFT calculations was compared to previous experimental and theoretical work. The Si–Si-dimer bond lengths of the Si(001)2x1 reconstruction derived by several methods [43–50]. This work provides a bond length in good agreement with recent theoretical work and SXD [45] measurements.

Acetylene on Si(001)

The first organic compound investigated in this study of adsorption on Si(001) is acetylene. For this study, the Si(001)c4x2 cell was chosen which contains four Si–Si-dimers and determines the coverage as 1/4 per adsorbate. In Table 4 geometry parameters are summarized as obtained from periodic PBE calculations of a single adsorbate. Acetylene's triple bond can interact with two major adsorption sites. Parallel to the Si-Si dimer of the surface is indicated as *on-top*, orthogonal to the dimer bond connecting two Si atoms of different dimers is indicated as *end-bridge*. The triple bond may as well be oriented between two Si–Si-dimers either parallely (*pedestal*) or orthogonal to the dimers connecting the Si–Si bond centers (*rotated pedestal*). The latter are less stable compared to the *on-top* and *end-bridge* adsorption modes.

Due to an adsorption of acetylene *on-top* of a Si–Si dimer, the respective dimer bond is lifted and no longer tilted but lies now parallel to the actual Si(001) plane. Any unoccupied, neighbouring Si–Si dimers remain tilted.

Cyclooctyne on Si(001)

The cyclooctyne molecule is subject to adsorption on the Si(001) surface as it can be chemically functionalized for a variety of potential applications. As already indicated, the molecule has a certain ring strain which constraints the geometry to only two stable conformations, chair and twist-boat. The relative Free Energy

Table 5 Adsorption energies for acetylene and cyclooctyne for the adsorption on various Si(001) sites, given in kJ/mol

E_{ads}	
Cyclooctyne	
On-top	−306.7
End-bridge	−279.4
Acetylene	
On-top	−268.6
End-bridge	−263.9

is $\Delta G = 12.5\,\text{kJ/mol}$, while the conformation transition Free Energy values to $\Delta G^{\#} = 32.7\,\text{kJ/mol}$. It is assumed that only the more stable conformation is relevant for adsorption.

Quantum chemical bond analyses (NBO, AIM) of the triple bonds in acetylene and cyclooctyne revealed that those bonds behave similarly in the gas phase, although the molecular structures are significantly different. Cyclooctyne's triple bond structure is influenced by a significant ring strain not present in acetylene. As a consequence, an open question was if the bonding situation with Si(001) will be rather similar for acetylene and cyclooctyne. The potential adsorption sites can be regarded identical with acetylene's as the triple bond of cyclooctyne is responsible for the bonding interaction with the Si–Si-dimer. The relevant adsorption modes are *on-top*, *end-bridge* and *pedestal*. In the Si(001)c4x2 cell the coverage of one adsorbate molecule is 1/4. In Table 5 the adsorption energies for several modes are presented. It was found that the adsorbate preparation energies are much higher for end-bridge adsorption modes than for on-top modes. This trend is also true for the substrate's preparation energy, which has, however, a smaller contribution to the overall interaction energy. The preparation energy of acetylene is mainly due to a bending of the H–C-C–H angle, which is changed by 56.6° upon adsorption (only 27.0° for cyclooctyne).

It is remarkable, that the interaction energies of both adsorbate compounds are very similar for the end-bridge mode. This indicates that the energy is mainly dependent on the interaction between the adsorbates' triple bond and the adsorption sites.

Multiple Adsorption of Cyclooctyne on Si(001)

One to four adsorbates can be regarded as multiples of 1/4 coverages within the Si(001)c4x2 cell. The higher the coverage, the more densely packed are the molecules and lateral interactions will have increasing contribution to the total interaction energy as shown in Fig. 3. As steric effects of the large hydrocarbon rings can be repulsive this will compete the energy gain of the system due to additional substrate–adsorbate bonds. For this study on multiple cyclooctyne adsorption stabilities, only the preferred *on-top* site was investigated.

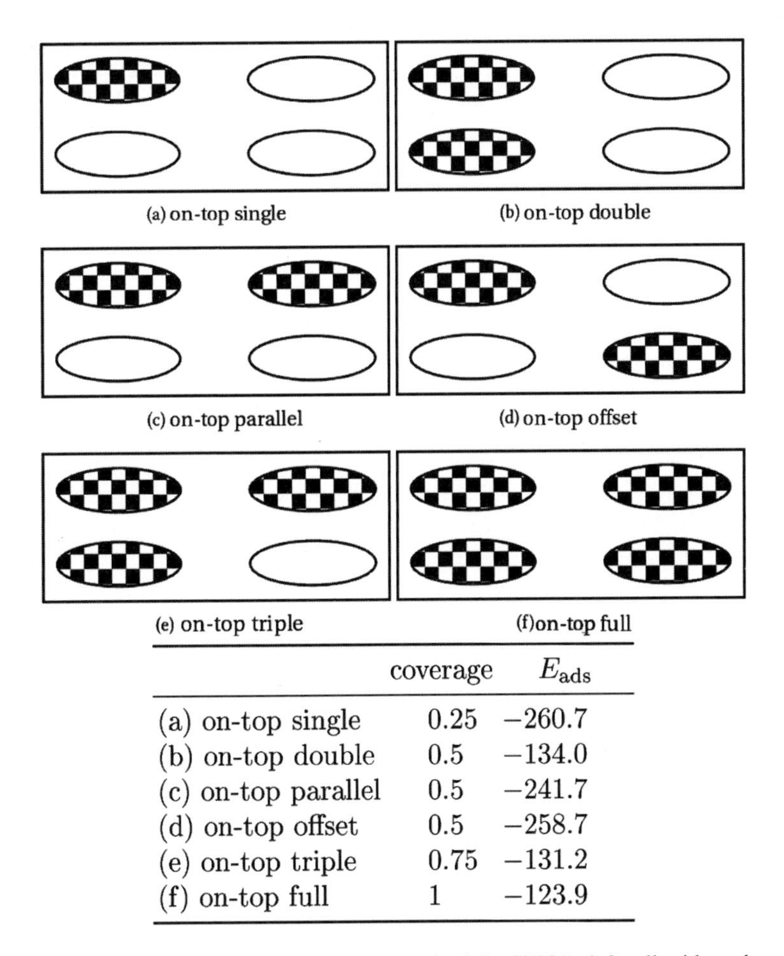

	coverage	E_{ads}
(a) on-top single	0.25	−260.7
(b) on-top double	0.5	−134.0
(c) on-top parallel	0.5	−241.7
(d) on-top offset	0.5	−258.7
(e) on-top triple	0.75	−131.2
(f) on-top full	1	−123.9

Fig. 3 Schematic representation of n∗1/4 coverages of the Si(001)c4x2 cell with cyclooctyne above the Si−Si-dimers. Possible on-top adsorption sites are indicated as *black-and-white grid*. Table: Adsorption energies in kJ/mol for various coverages of the Si(001)c4x2 cell. Energies read as energetic gain relative to the previous coverage state, respectively (triple coverage (*e*) relative to (*d*) on-top offset)

The system is stabilized on further adsorption, but the energy gain decreases with increasing coverage. This can be interpreted as tendency of neighbouring cyclooctyne rings to repel each other, which is best observed following the E_{ads} sequence of doubly occupied states Fig. 3b, c, d with identical coverage (0.5).

Regarding lower coverages (1/8 and 1/16 per adsorbate molecule, respectively) the supercells Si(001)c4x4 and Si(001)c8x4 were chosen. Smaller coverages lead to higher degrees of freedom (various adsorption arrangements, modifications) which, by itself, slightly stabilizes the system and might unveil more realistic adsorption patterns. Table 6 shows adsorption energies of multiple adsorbates at lower coverages of cyclooctyne on Si(001).

Table 6 Adsorption energies in kJ/mol of multiple on-top adsorbates on Si(001)c4x4 at lower coverages are presented

	Coverage	E_{ads}
Single	0.125	−263.1
Double	0.250	−262.5
Triple	0.375	−234.5
Triple[a]	0.1875	−250.9

Energies read as energetic gain relative to the previous coverage state, respectively. *Bottom*: Triple adsorption in Si(001)c8x4 cell
[a]Adsorption on Si(001)c8x4. Energies are given with respect to the doubly covered Si(001)c4x4 cell

For the triple adsorption at neighbouring on-top sites, (c) and (d), the central molecule has limited deformation potential. However, for increasing cell sizes (decreasing coverage) this is still stable. For low and medium coverages (large and medium cells) the secondary adsorption stabilizes the system. For low coverages, even the third adsorption is favourable.

It was concluded that acetylene as well as cyclooctyne are both chemisorbed on Si(001). For cyclooctyne, however, the on-top adsorption mode is far more stable than the end-bridge mode while the difference is small for acetylene.

3.4 Summary

The atomic and electronic structure of the Si(001) surface next to the adsorption of acetylene and cyclooctyne on this surface was examined with quantum chemical methods. The Si–Si bond length is elongated with respect to the Si bulk bond distance and in the order of gas phase single Si–Si bonds. Upon adsorption, the Si–Si bond length is not changed.

For the adsorption studies on silicon periodic slab models were created. A sufficient k-grid density and PAW-basis size are needed, but with the systematically refined parameters this study could successfully reproduce other theoretical and experimental data. Moreover, the choice of specific k-points is crucial, so that the correct adsorption energies can only be obtained if either the Γ-point or a large number of k-points centered at other (reciprocal) positions are included.

Two molecular conformations of the cyclooctyne molecule were calculated and the most stable one was used for the study of adsorption. In contrast to acetylene, this conformation of cyclooctyne adsorbs preferably at the *on-top* adsorption site as opposed to the *end-bridge* site. Secondary and tertiary adsorption is favourable for low coverages in the examined model cells, which is probably as well caused by the rings' steric effects and lateral interactions. Higher coverages, especially for more than two neighbouring adsorption sites (Si–Si-dimers) are less stable because ring repulsion becomes dominant.

This study provided microscopic understanding of the adsorbate structures based on a detailed analysis of electronic surface and adsorbate energies.

4 CPU Scaling: VASP Parallelization NPAR/LPLANE/NSIM

Some of the VASP specific parallelization parameters were tested specifically on the present HLRS system. CPU time was recorded for a standard job with respect to the NPAR, LPLANE, NSIM.

Memory

For larger systems memory problems can occur if none of the above options is used. The following are options for CRAY systems and recommended are

NPAR = $\sqrt{\text{number of CPUs}}$	# Care: CPUs divided by chosen NPAR (# CPUS/NPAR) has to be an integer number
LPLANE = .TRUE.	# VASP manual recommends .FALSE. for CRAY, but tests showed better performance with .TRUE.

Time Savings

The use of above options can lead to shorter wallclock times, which is presented in the following section for the HERMIT(Cray)-machine in Stuttgart. Note that the time can significantly be reduced with the right options. Test system was a Si(001)-surface cell with adsorbed cyclooctyne (96 Si, 56 H, 16 C) with a total number of 168 atoms in the periodic supercell. The wallclock times were recorded for a singlepoint calculation of the electronic structure by DFT methods in a sufficiently sized plane wave basis with standard k-point grid (see above subprojects).

NSIM = 1

NPAR	LPLANE	Tot. CPU / min
4	T	37
2	T	41
8	T	41
16	T	43
8	F	51
1	T	55
16	F	56
32	F	60
64	F	70
default	-	71

References

1. B. Kunert, K. Volz, I. Nemeth, W. Stolz, J. Lumin. **121**(2), 361–364 (2006)
2. B. Kunert, K. Volz, W. Stolz, Phys. Status Solidi (B) **244**(8), 2730–2739 (2007)
3. B. Kunert, K. Volz, J. Koch, W. Stolz, J. Cryst. Growth **298**, 121–125 (2007)
4. C. Wang, J. Cryst. Growth **272**(1–4), 664–681 (2004)
5. P. Gibart, Rep. Prog. Phys. **67**(5), 667–715 (2004)
6. N. Bahlawane, F. Reilmann, L.-C. Salameh, K. Kohse-Höinghaus, J. Am. Soc. Mass Spectrom. **19**(7), 947–954 (2008)
7. M.G. Jacko, S.J.W. Price, Can. J. Chem. **41**(6), 1560–1567 (1963)
8. A. Brauers, Prog. Cryst. Growth Charact. Mater. **22**, 1–18 (1991)
9. M. Trachtman, S. Beebe, C. Bock, J. Phys. Chem. **99**, 15028–15034 (1995)
10. D. Moscatelli, P. Caccioppoli, C. Cavallotti, Appl. Phys. Lett. **86**(9), 091106 (2005)
11. R. Schmid, D. Basting, J. Phys. Chem. A **109**(11), 2623–30 (2005)
12. R. Tromp, R. Hamers, J. Demuth, Phys. Rev. Lett. **557**(12), 1303–1307 (1985)
13. D. Allan, E. Mele, Phys. Rev. B **31**(8), 1–4 (1985)
14. A. Beyer, J. Ohlmann, S. Liebich, H. Heim, G. Witte, W. Stolz, K. Volz, J. Appl. Phys. **111**(8), 083534 (2012)
15. A. Beyer, I. Németh, S. Liebich, J. Ohlmann, W. Stolz, K. Volz, J. Appl. Phys. **109**(8), 083529 (2011)
16. S.R. Schofield, N.J. Curson, O. Warschkow, N.a. Marks, H.F. Wilson, M.Y. Simmons, P.V. Smith, M.W. Radny, D.R. McKenzie, R.G. Clark, J. Phys. Chem. B **110**(7), 3173–3179 (2006)
17. U. Höfer, L. Li, T. Heinz, Phys. Rev. B **45**(16), 9485–9488 (1992)
18. P. Sen, B. Gupta, I. Batra, Phys. Rev. B **73**(8), 085319 (2006)
19. Z. Aydugan, c. Kaderoglu, B. Alkan, M. Çakmak, Surf. Sci. **603**(15), 2271–2275 (2009)
20. S. Hara, K. Irokawa, H. Miki, A. Kawazu, H. Torii, H.I. Fujishiro, J. Appl. Phys. **98**(8), 083513 (2005)
21. S. Hara, H. Kobayashi, K. Irokawa, H.I. Fujishiro, K. Watanabe, H. Miki, A. Kawazu, Surf. Sci. **601**(12), 2415–2419 (2007)
22. A. Stegmüller, P. Rosenow, R. Tonner (in preparation)
23. P. Rosenow, Theoretische Untersuchungen zur Adsorption von GaP-Vorläufern auf der Si(001)2x1-Oberfläche, Master's Thesis, 2012
24. P. Blöchl, Phys. Rev. B **50**(24), 17953–17979 (1994)
25. G. Kresse, D. Joubert, Phys. Rev. B **59**(3), 11–19 (1999)
26. F. Weigend, R. Ahlrichs, Phys. Chem. Chem. Phys.: PCCP **7**(18), 3297–305 (2005)
27. A. Hellweg, C. Hättig, S. Höfener, W. Klopper, Theor. Chem. Acc. **117**(4), 587–597 (2007)
28. F. Weinhold, C.R. Landis, Chem. Educ. Res. Pract. **2**(2), 91 (2001)
29. G. Kresse, J. Furthmüller, Phys. Rev. B Condens. Matter **54**(16), 11169–11186 (1996)
30. J. Perdew, K. Burke, M. Ernzerhof, Phys. Rev. Lett. **77**(18), 3865–3868 (1996)
31. S. Grimme, J. Comput. Chem. **16**, 1787–1799 (2006)
32. I. Yavari, F. Nasiri, H. Djahaniani, A. Jabbari, Int. J. Quantum Chem. **106**(3), 697–703 (2006)
33. R.D. Bach, J. Am. Chem. Soc. **131**(14), 5233–5243 (2009)
34. N.J. Agard, J.A. Prescher, C.R. Bertozzi, J. Am. Chem. Soc. **126**(46), 15046–15047 (2004)
35. S.T. Laughlin, J.M. Baskin, S.L. Amacher, C.R. Bertozzi, Science **320**(5876), 664–667 (2008)
36. C. Schober, R. Tonner (in preparation)
37. G. Mette, M. Dürr, R. Batholomäus, U. Koert, U. Höfer, Chem. Phys. Lett. **556**, 70–76 (2013)
38. C. Schober, Theoretische Untersuchung der Adsorption von Ethin und Cyclooctin auf der Si(001)-Oberfläche, Master's Thesis, 2012
39. S.F. Bent, J. Phys. Chem. B **106**(11), 2830–2842 (2002)
40. J. Yoshinobu, Prog. Surf. Sci. **77**(1), 37–70 (2004)
41. M. Kira, T. Maruyama, C. Kabuto, K. Ebata, H. Sakurai, Angew. Chem. Int. Ed. Engl. **33**(14), 1489–1491 (1994)
42. W. Mönch, *Semiconductor Surfaces and Interfaces*, 3rd edn. (Springer, Berlin, 2001)

43. H. Over, J. Wasserfall, W. Ranke, C. Ambiatello, R. Sawitzki, D. Wolf, W. Moritz, Phys. Rev. B **55**(7), 4731–4736 (1997)
44. R. Gunnella, E.L. Bullock, L. Patthey, C.R. Natoli, T. Abukawa, S. Kono, L.S.O. Johansson, Phys. Rev. B **57**(23), 14739–14748 (1998)
45. M. Takahasi, S. Nakatani, Y. Ito, T. Takahashi, X.W. Zhang, M. Ando, Surf. Sci. **338**(1), L846–L850(1995)
46. P. Krüger, J. Pollmann, Phys. Rev. Lett. **74**(7), 1155–1158 (1995)
47. J. Dabrowski, M. Scheffler, Appl. Surf. Sci. **56**(0), 15–19 (1992)
48. J.E. Northrup, Phys. Rev. B **47**(15), 10032–10035 (1993)
49. N. Takeuchi, Surf. Sci. **601**(16), 3361–3365 (2007)
50. P.T. Czekala, H. Lin, W.A. Hofer, A. Gulans, Surf. Sci. **605**(15–16), 1341–1346 (2011)

Numerical Studies of the Tunneling Splitting of Malonaldehyde and the Eigenstates of Hydrated Hydroxide Anion Using MCTDH

Markus Schröder, Daniel Peláez, and Hans-Dieter Meyer

1 Introduction

A detailed and accurate description of vibrations of molecules and chemical reactions in the field of physical chemistry often requires a full quantum mechanical treatment of the system of interest. This usually implies that the time-dependent or the time-independent Schrödinger equation of the nuclear degrees of freedom (DOF) has to be solved explicitly. For small systems (up to six internal DOF) this can be done with standard methods, i.e., by directly sampling the quantum mechanical wavefunction on a (product-) grid and solving the Schrödinger equation on these grid points. Numerically, within the standard method the multi-dimensional quantum mechanical wavefunction is stored as an f-way tensor, where f is the number of DOF. Due to the linearity of the Schrödinger equation the resulting numerical tasks then usually reduce to standard problems such as the calculation of eigenvalues or solving first order differential equations.

With growing numbers f of internal DOF, however, the amount of data that has to be stored and processed quickly exceeds today's numerical capacities. The multi-configuration time-dependent Hartree algorithm therefore takes a different route. Here the wavefunction is expanded in terms of low-dimensional but optimal basis functions, each of which spanning typically one to four internal DOF. Numerically, this corresponds to combining a few (one to four) indices of the aforementioned f-way tensor into one single index and representing the original f-way tensor as a superposition of products of these basis functions. As the basis is chosen optimal for a given problem, only a few most important basis functions and their expansion coefficients have to be stored. This leads to an enormous reduction of data. In

M. Schröder · D. Peláez · H.-D. Meyer (✉)
Theoretische Chemie, Physikalisch-Chemisches Institut, Universität Heidelberg, INF 229, D-69120 Heidelberg, Germany
e-mail: Hans-Dieter.Meyer@pci.uni-heidelberg.de

W.E. Nagel et al. (eds.), *High Performance Computing in Science and Engineering '13*, DOI 10.1007/978-3-319-02165-2_15,

other fields, the resulting representation of the wavefunction (in tensor form) is also referred to as *Tucker-decomposition*. This representation of the wavefunction, however, also leads to much more involved equations that need to be solved.

Nevertheless, with this ansatz for the wavefunction the MCTDH algorithm is a general method to solve the time-dependent as well as the time-independent Schrödinger equation and has been successfully applied to a variety of problems. Shared as well as distributed memory parallelization has been implemented into the Heidelberg MCTDH package in recent years [1, 2] so that systems with more then 12 DOF become feasible to treat. One important example is the Zundel cation [2–7], $H_5O_2^+$, a flexible molecule with 15 internal DOF.

Currently, we are working on the calculation of eigenstates of malonaldehyde (21 DOF) and the hydrated hydroxide anion ($H_3O_2^-$, 9 DOF). Malonaldehyde exhibits an internal proton transfer reaction within a symmetric double well potential which gives rise to a characteristic tunneling splitting. In a recent publication [8] we reported results for the ground state energy as well as the ground state tunneling splitting. Preliminary results for excited state energies and tunneling splittings have been obtained. The purpose of the study of $H_3O_2^-$ is twofold. First, this system served as a benchmark for testing the capabilities of our recently published tensor-decomposition variational algorithm for large systems: Multigrid POTFIT (MGPF). And, secondly, using MGPF potential expansions we have achieved the most accurate description, up to the present date, of the gas-phase proton transfer of $H_3O_2^-$.

This report is structured as follows: in Sect. 2 the MCTDH algorithm is reviewed and parallelization strategies are outlined. In Sects. 3 and 4 we present recent results obtained for the tunneling splitting of malonaldehyde and vibrational eigenstates of hydrated hydroxide, respectively, and summarize in Sect. 5.

2 The Multi-configuration Time-Dependent Hartree Algorithm

2.1 MCTDH Equations of Motion and Improved Relaxation

Within the MCTDH algorithm f-dimensional wavefunction is approximated by the following sums-of-products *ansatz*:

$$
\begin{aligned}
\Psi(q_1,\cdots,q_f,t) &\equiv \Psi(Q_1,\cdots,Q_p,t) \\
&= \sum_{j_1}^{n_1}\cdots\sum_{j_p}^{n_p} A_{j_1,\cdots,j_p}(t)\prod_{\kappa=1}^{p}\varphi_{j_\kappa}^{(\kappa)}(Q_\kappa,t) \\
&= \sum_J A_J\,\Phi_J\ .
\end{aligned}
\tag{1}
$$

Here f denotes the number of physical degrees of freedom q_i and p is the number of composite coordinates or *particles* Q_i. n_κ is the number of basis functions, the so-called single-particle functions (SPF), $\varphi_{j_\kappa}^{(\kappa)}(Q_\kappa, t)$ for the κth (composite) coordinate and $A_J \equiv A_{j_1 \cdots j_p}$ also called the "A-vector" (or "core tensor") denote the MCTDH expansion coefficients of the total wavefunction.

Note, that the SPF as well as the A-vector are taken as being time-dependent quantities within this ansatz. As mentioned in the introduction, the SPF are chosen optimally to represent the wavefunction such that the A-vector is a small as possible. The SPFs themselves are represented by linear combinations of time-independent primitive basis functions $\chi_{l_j}^{(\kappa)}(q_{j,\kappa})$ with expansion coefficients $c_{j_\kappa l_1 \cdots l_d}^{(\kappa)}$:

$$\varphi_{j_\kappa}^{(\kappa)}(Q_\kappa, t) = \sum_{l_1=1}^{N_{1,\kappa}} \cdots \sum_{l_d=1}^{N_{d,\kappa}} c_{j_\kappa l_1 \cdots l_d}^{(\kappa)}(t)\, \chi_{l_1}^{(\kappa)}(q_{1,\kappa}) \cdots \chi_{l_d}^{(\kappa)}(q_{d,\kappa}). \tag{2}$$

In practice, $l_1 \cdots l_d$ are the aforementioned grid points on which the SPF are sampled.

Inserting the ansatz (1) into the Dirac-Frenkel variational principle [9, 10] for solving the time-dependent Schrödinger equation then leads to equations of motion (EOM) for the constituents of the wavefunction, in particular ($\hbar = 1$)

$$i \frac{\partial}{\partial t} A_J = \sum_L \langle \Phi_J | \hat{H} | \Phi_L \rangle A_L, \tag{3}$$

for the A-vector and

$$i \frac{\partial}{\partial t} \varphi_j^{(\kappa)} = \left(1 - P^{(\kappa)}\right) \sum_{k,l=1}^{n_\kappa} \left(\rho^{(\kappa)}\right)_{j,k}^{-1} \left\langle \hat{H} \right\rangle_{k,l}^{(\kappa)} \varphi_l^{(\kappa)}, \tag{4}$$

for the SPF. Here $\hat{H}$ is the Hamiltonian of the system, $P^{(\kappa)}$ is a projector on the subspace spanned by the SPF of the κ-th composite coordinate and $\rho^{(\kappa)}$ are reduced single-particle density matrices. The quantities $\left\langle \hat{H} \right\rangle_{k,l}^{(\kappa)}$ denote the so-called mean-field operators (cf. e.g. [11] or [12]). Note that the latter are obtained by numerically demanding multi-dimensional integrals over all but the κ-th coordinate and that the EOMs (3) and (4) are coupled differential equations. The differential equations (3) and (4), however, can be decoupled over an update time-step, the so-called constant mean-field approach.

The MCTDH ansatz also allows the calculation of eigenstates. In this case the expectation value of the Hamiltonian is minimized which leads to an eigenvalue problem for the A-vector

$$\sum_L \langle \Phi_J | \hat{H} | \Phi_L \rangle A_L = E\, A_J \tag{5}$$

and a propagation of the SPF in negative imaginary time $\tau = -it$, i.e., a relaxation, with the EOM

$$\frac{\partial}{\partial \tau}\varphi_j^{(\kappa)} := -\left(1 - P^{(\kappa)}\right) \sum_{k,l=1}^{n_\kappa} \left(\rho^{(\kappa)}\right)^{-1}_{jk} \left\langle \hat{H} \right\rangle^{(\kappa)}_{kl} \varphi_l^{(\kappa)} \rightarrow 0. \tag{6}$$

Since both equations have to be fulfilled simultaneously, one again uses the constant mean field approach to iteratively solve both equations until convergence is achieved.

Note, that (5) is an ordinary eigenvalue equation and can be solved using a Krylov subspace method. This also allows the selection of particular eigenstates and thus the calculation of excited states. Furthermore, the so-called block-improved relaxation [13, 14] can be used where set a few eigenstates is calculated simultaneously with the same set of SPF. The mean-fields (which themselves determine the SPF in (4)) are then state averaged mean-fields.

2.2 Parallelization

While the memory demands of the MCTDH wavefunction is rather moderate, the evaluation of the EOMs (3) and (4) is rather costly, even after applying the constant mean field approximation. As the calculation of the mean fields and matrix elements involves multi-dimensional integrals over all physical DOF, a significant speedup is achieved if the Hamiltonian operator can be expressed in terms of products of low-dimensional terms $\hat{h}_r^{(\kappa)}$ such that

$$\hat{H} = \sum_{r=1}^{s} c_r \prod_{\kappa=1}^{p} \hat{h}_r^{(\kappa)}, \tag{7}$$

where the $\hat{h}_r^{(\kappa)}$ operate only on the κth particle. With this ansatz the multi-dimensional integrals reduce to sums of products of low-dimensional integrals, for instance,

$$\langle \Phi_J | \hat{H} | \Phi_L \rangle = \sum_{r=1}^{s} c_r \prod_{\kappa=1}^{p} \langle \varphi_{j_\kappa}^{(\kappa)} | \hat{h}_r^{(\kappa)} | \varphi_{l_\kappa}^{(\kappa)} \rangle \ . \tag{8}$$

The evaluation of the single terms of the right hand side of (8) can be easily parallelized. To this end, within the MCTDH software, both, shared memory parallelization using POSIX threads and distributed memory parallelization using the message parsing interface (MPI), is used [2].

For the eigenvalue problem (5) the Davidson algorithm, which is a Krylov subspace method, is applied. Within the Davidson algorithm, temporary A-vectors are generated and stored in memory. The present implementation within the MCTDH algorithm allows distributing the temporary vectors among different MPI-processes—and hence compute nodes—which lowers the average memory requirements per compute node.

3 Tunneling Splitting of Malonaldehyde

As a prototype system for studying proton transfers, malonaldehyde has been subject of a number of theoretical investigations over the last few decades. Proton transfer processes can be categorized as large amplitude motions and play an important role in many chemical and biological reactions such as mutations in DNA, photo synthesis, hydrogen tunneling in enzymes or acid base reactions. One of the main difficulties when studying these transfer reactions is that the re-localization of the proton triggers reorganization processes in the surrounding material. This leads to highly correlated wavefunctions which are hard to treat numerically.

Malonaldehyde consists of a horseshoe shaped chain of three carbon atoms with conjugated double bonds, saturated with hydrogen and two oxygen atoms located on each end of the chain. Between the oxygen atoms resides a singe proton which can travel between two equivalent minima in close proximity to each of the oxygen atoms, respectively. The transfer process is outlined in Fig. 1. Note, that the aforementioned reorganization processes here manifest in form of an interchange of the double bonds as the proton travels from one minimum to the other. This interchange leads to large shifts of the inter-atomic distances and the aforementioned correlations. In the past, therefore, a number of approximative methods [15–17] and more general methods [18–22] have been applied to study this process.

Recently, also full-dimensional quantum mechanical calculations on the ground state energy and the tunneling splitting using different flavors of the MCTDH algorithm have been reported, including our own work [8]. Within these studies (and also the present work) used a recently published [22] potential energy surface (PES) which is following the authors, much more accurate then previously published PES.

3.1 Coordinates and Hamiltonian

The calculations of vibrational states in malonaldehyde was performed using of mass- and frequency-scaled normal-mode coordinate $\tilde{q}_i$ obtained at the transition state. To minimize the effects of the reorganization of the inter-atomic distances and to improve the quality of the description of the two minima, the coordinates have been modified following the approach Manthe et al. [23] according to

$$\begin{aligned} q_i &= \tilde{q}_i - F_i(\tilde{q}_{21}), \quad i = 1 \dots 20, \\ q_{21} &= \tilde{q}_{21}. \end{aligned} \tag{9}$$

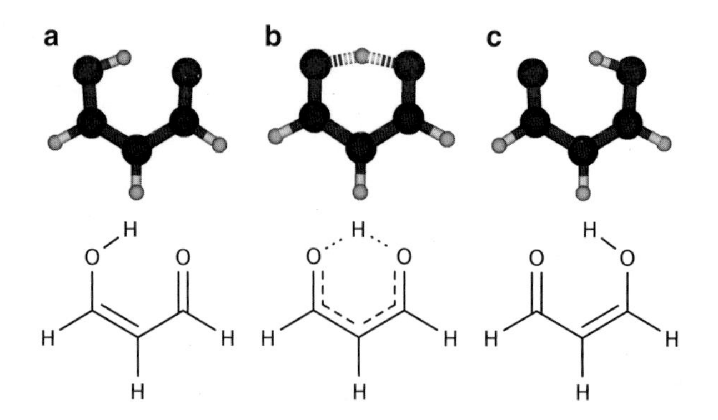

Fig. 1 Proton transfer in malonaldehyde: the proton located in one of the minimum positions (**a**) travels via a transition state (**b**) to the equivalent opposite minimum position (**c**)

Here q_{21} is the reaction coordinate describing the distance of the proton from the transition state and the F_i are analytic functions exclusively defined in this DOF. The F_i were obtained by minimizing the potential energy for a set of given configurations along the reaction path and subsequent fitting to polynomials of 7th order for asymmetric F_i and 8th other for symmetric F_i.

After applying the transformation (9) and separating global rotation and neglecting rotation-vibration coupling terms the KEO reads

$$\hat{T} = -\frac{1}{2}\left[\sum_{i=1}^{21} \omega_i \frac{\partial^2}{\partial q_i^2} + \omega_{21} \sum_{i,j=1}^{20} F_i'(q_{21}) F_j'(q_{21}) \frac{\partial^2}{\partial q_i \partial q_j} \right. \\ \left. -\omega_{21} \sum_{i=1}^{20} \left(F_i'(q_{21}) \frac{\partial}{\partial q_{21}} + \frac{\partial}{\partial q_{21}} F_i'(q_{21}) \right) \frac{\partial}{\partial q_i} \right], \tag{10}$$

where $F_i'(q_{21}) = \frac{\partial}{\partial q_{21}} F_i(q_{21})$ are the first derivatives of F_i with respect to the transfer coordinate and ω_i are the normal mode frequencies (ω_{21} is set to the absolute value of the imaginary normal mode frequency).

Additionally, following Hammer et al. [23], symmetry considerations have been used to further reduce the numerical complexity. As the PES is symmetric with respect to the transfer coordinate, it is (for the calculation of eigenstates) sufficient to describe only one of the two wells explicitly and reconstruct the other one via symmetry operations. To this end projection operators h and $h' = 1 - h$ that project onto one half or the other of the double well, respectively, and a reflection operator R that reflects the wavefunction under preserving the symmetry of the Hamiltonian are introduced. This leads to two effective Schrödinger equations for states which are symmetric ($|\Psi_+\rangle$) and anti-symmetric ($|\Psi_-\rangle$) with respect to the transfer coordinate (q_{21}) [23]:

$$H_+ |\Psi_+\rangle = \left(h\hat{H}h + h\hat{T}h'R \right) |\Psi_+\rangle = E_+ |\Psi_+\rangle, \\ H_- |\Psi_-\rangle = \left(h\hat{H}h - h\hat{T}h'R \right) |\Psi_-\rangle = E_- |\Psi_-\rangle. \tag{11}$$

With these two equations it is possible to treat symmetric and anti-symmetric wavefunctions with different sets of SPF.

Cluster Expansion of the Potential

As discussed with help of (7) the product form of the system Hamiltonian is of crucial importance for the numerical efficiency of the MCTDH equations. While the kinetic energy operator is usually given analytically and in product form this is in most cases not true for the PES. For smaller systems one can use the POTFIT algorithm [12, 24–27] or its multi-grid extension to transform the PES into product form. Due to the need of calculating multiple integrals over the complete grid, however, POTFIT can only be used up to a certain number of DOF. For larger systems such as malonaldehyde one needs to use alternative techniques.

In the following we use the so-called reaction path cluster expansion (RPCE), a variant of the n-mode representation, also called high-dimensional model representation [3, 28–33]. Within the RPCE the potential is approximated by n-particle interaction terms with one distinguished coordinate $\bar{Q}_{\mathrm{ref}}$, here, of course, the reaction coordinate. These terms are called clusters. Again the $\bar{Q}_\alpha$ refer to the composite coordinates as detailed in Sect. 2.2. The PES is approximated by the expansion

$$\begin{aligned} V(\bar{\mathbf{Q}}) = v_{\mathrm{ref}}^{(0)}(\bar{Q}_{\mathrm{ref}}) + \sum_{\alpha=1}^{p-1} v_{\alpha,\mathrm{ref}}^{(1)}(\bar{Q}_\alpha, \bar{Q}_{\mathrm{ref}}) + \sum_{\alpha<\beta}^{p-1} v_{\alpha,\beta,\mathrm{ref}}^{(3)}(\bar{Q}_\alpha, \bar{Q}_\beta, \bar{Q}_{\mathrm{ref}}) \\ + \sum_{\alpha<\beta<\gamma}^{p-1} v_{\alpha,\beta,\gamma,\mathrm{ref}}^{(3)}(\bar{Q}_\alpha, \bar{Q}_\beta, \bar{Q}_\gamma, \bar{Q}_{\mathrm{ref}}) + \cdots \end{aligned} \tag{12}$$

with

$$\begin{aligned} v_{\mathrm{ref}}^{(0)}(\bar{Q}_{\mathrm{ref}}) &= V(0, \ldots, \bar{Q}_{\mathrm{ref}}) \\ v_{\alpha,\mathrm{ref}}^{(1)}(\bar{Q}_\alpha, \bar{Q}_{\mathrm{ref}}) &= V(0, \ldots, \bar{Q}_\alpha, 0, \ldots, \bar{Q}_{\mathrm{ref}}) - v_{\mathrm{ref}}^{(0)}(\bar{Q}_{\mathrm{ref}}), \end{aligned} \tag{13}$$

etc.

The major advantage of this technique is that in many cases the series can be truncated at low orders n and that the n-particle interaction terms $v^{(n)}$ only depend on a few (composite) coordinates such that POTFIT can be used to transform these terms into product form. A serious disadvantage, however, is that the expansion (12) is not variational and that taking into account further terms does not lead to a guaranteed improvement of the representation of the PES. In particular the RPCE is only accurate in close proximity to $\bar{Q}_{\mathrm{ref}}$. The functions F_i have therefore been chosen such that the reaction coordinate q_{21} crosses the potential minima where the wavefunction mainly resides.

Furthermore, PES representations according to (12) often exhibit "holes", i.e., regions in which the potential is unphysically strong negative and further contributions need to be taken into account to compensate for these regions.

Table 1 Mode combinations for the reaction path cluster expansion

Logical coord.	Physical coord.	No. grid points
$\tilde{Q}_1$	q_1, q_2	143
$\tilde{Q}_2$	q_3, q_6	121
$\tilde{Q}_3$	q_4	25
$\tilde{Q}_4$	q_5, q_{20}	143
$\tilde{Q}_5$	q_7, q_8	121
$\tilde{Q}_6$	q_9, q_{10}	143
$\tilde{Q}_7$	q_{11}	11
$\tilde{Q}_8$	q_{12}, q_{15}	143
$\tilde{Q}_9$	q_{13}, q_{14}	143
$\tilde{Q}_{10}$	q_{16}	11
$\tilde{Q}_{11}$	q_{17}	13
$\tilde{Q}_{12}$	q_{18}, q_{19}	121
$\tilde{Q}_{13} = \tilde{Q}_{\text{ref}}$	q_{21}	24

Unfortunately, one does not know beforehand, which of the terms in (12) are actually important to represent the PES and which of the coordinates are advantageously combined into composite coordinates. To this end we used a diffusion Monte Carlo (DMC) algorithm to obtain the ground state density in terms of an ensemble of random walkers. This set of random walkers was subsequently used to identify candidates to form composite coordinates and estimate the contributions of the interaction terms.

In the present case we used a set of 13 composite particles as outlined in Table 1 have been used to describe the PES. Here we combined those coordinates that show strong interaction with each other but with only a few others. Those coordinates that interact strongly with many others were not combined. This ensures that a large number of terms $v^{(n)}$ with different combinations of coordinates can be computed while the $v^{(n)}$ do not become too large to be transformed with POTFIT. Note, however, that a third order term depends on 4–7 DOF because of mode combination (1) and the inclusion of the transfer coordinate (see (12)).

To represent the PES according to (12) we took into account all contributions up to second order, plus a selection of third to fifth order terms. These terms were selected if their absolute mean contribution, calculated using the DMC density, was larger then 1 cm^{-1} or if the RMS value was larger then 25 cm^{-1}. In addition, some clusters were included to correct for negative regions of the PES ("holes"). The expectation value of PES according to the DMC calculation has been obtained as 1.3 cm^{-1} too large. The RMS error has been obtained as 171.86 cm^{-1}. We believe, however, that the RMS value is too large, as some random walkers which are located remotely from the reaction path may contribute too much to the RMS error.

Nevertheless, the error of the expanded PES is rather uncontrollable and most likely the main source of error for the present contribution.

Table 2 Mode combinations for the actual MCTDH calculations

Logical coord. (MCTDH)	Logical coord. (clusters)	Physical coord.	Grid points
Q_1	$\tilde{Q}_1, \tilde{Q}_3$	q_1, q_2, q_4	3575
Q_2	$\tilde{Q}_2, \tilde{Q}_4$	q_3, q_5, q_6, q_{20}	17303
Q_3	$\tilde{Q}_5, \tilde{Q}_6$	q_7, q_8, q_9, q_{10}	17303
Q_4	$\tilde{Q}_8, \tilde{Q}_9$	$q_{12}, q_{13}, q_{14}, q_{15}$	20449
Q_5	$\tilde{Q}_7, \tilde{Q}_{12}$	q_{11}, q_{18}, q_{19}	1331
Q_6	$\tilde{Q}_{10}, \tilde{Q}_{11}, \tilde{Q}_{13}$	q_{16}, q_{17}, q_{21}	3432

3.2 *Zero Point Energy, Excited States and Tunneling Splitting*

For the actual MCTDH calculations the composite coordinates outlined in Table 1 have been further combined into six particles to reduce the size of the A-vector of the MCTDH wavefunction, (1), as outlined in Table 2.

Ground State: Convergence Checks and Upper Bounds

As in our previous work [8], which was done using a different mode-combination scheme, we used an extrapolation scheme to estimate the ground state energy. As outlined before, this is necessary because the main part of the computational effort is spent in calculating the mean fields and within the Davidson algorithm. Therefore the computation time critically depends on the length of the A-vector which has to be kept as small as possible. To estimate the true ground state energy one, therefore, performs a series of independent calculations using the variational properties of the MCTDH ansatz: taking into account more SPF will always lead to lower energies. Starting from an (in terms of SPF) almost converged wavefunction and assuming that the energy drop that is achieved by adding SPF in one mode is (almost) independent from all other modes leads an extrapolation scheme. Starting from a reference calculation we doubled the number of SPF in all modes independently and added the energy differences. This scheme is outlined in Table 3. We obtain a ground state energy of 14664.5 cm^{-1} and a tunneling splitting of 22.9 cm^{-1} which is in good agreement with our previous results [8] (14667.3 and 23.2 cm^{-1}, respectively). Note, that we consider the numerical representation of the potential to be better then in [8] as we have taken into account a larger number of clusters then previously and in particular used a better strategy to identify important terms based in a DMC ground state density.

Excited States: Preliminary Results

For the calculation of excited states we used a block Davidson algorithm to calculate the lowest vibrational eigenstates. In Table 4 preliminary results and assignments for

Table 3 Extrapolation of the ground state energy

No. of SPF for										
Q_1	Q_2	Q_3	Q_4	Q_5	Q_6	Energy (+)	ΔE	Energy (−)	ΔE	Splitting
16	10	11	10	10	18	14670.016	(ref)	14693.313	(ref)	23.297
32	10	11	10	10	18	14668.363	−1.653	14691.399	−1.914	23.036
16	20	11	10	10	18	14669.158	−0.858	14692.343	−0.970	23.185
16	10	22	10	10	18	14669.462	−0.554	14692.692	−0.621	23.230
16	10	11	20	10	18	14669.033	−0.983	14692.315	−0.998	23.281
16	10	11	10	20	18	14669.265	−0.751	14692.467	−0.846	23.202
16	10	11	10	10	36	14669.390	−0.626	14692.629	−0.684	23.239
						Sum	−5.424		−6.033	
						Extrapolated energy	14664.591		14687.280	22.689

ΔE denotes the difference to the reference state. All energies are in cm^{-1}

Table 4 First four eigenstates of the effective Hamiltonians H_+ and H_- (11)

Energy (+)	Energy (−)	Splitting	Assignment
14673.7	14697.4	23.7	GS
14932.5	14999.5	67.0	q_4
14964.6	14981.3	16.7	q_1
15083.2	15111.2	28.0	q_2

Numbers of SPF: Q_1:16; Q_2:10; Q_3:10; Q_4:9; Q_5:9; Q_6:16. Energies in cm^{-1}

the first eigenstates of the effective Hamiltonians H_+ and H_- (11) are outlined. All states were calculated using the same number of SPF: Q_1:16, Q_2:10, Q_3:10, Q_4:9, Q_5:9 and Q_6:16. The assignments indicate the main axis in terms of physical DOF along which a node is present.

In Table 4, the first row reproduces the ground state energy and the ground state tunneling splitting as in Table 3. Note, that the ground state energy is approximately 9 cm^{-1} above the one estimated in the previous section. This is mainly because the set of SPF is used to represent all four states simultaneously such that they are not optimal for representing a particular single state. It can be rated as an indication that the calculations are not fully converged in terms of numbers of SPF.

The second row in Table 4 states the first excited state. This state exhibits a node along the physical coordinate q_4 which mainly governs the O-O stretching. However, also q_{21}, the transfer coordinate, is involved as well, as the proton transfer is strongly coupled to the O-O distance. The following two rows 3 and 4 the fundamentals of q_1 and q_2 are stated. The two normal modes q_1 and q_2 are strongly coupled, indicated by the orientation of the node planes. This strong coupling was already found during the DMC calculations, hence the combination of q_1 and q_2 in one composite coordinate (cf. Table 1). q_1 and q_2 describe out-of-plane motions of the carbon backbone and the oxygen atoms.

The view that the wavefunctions are not fully converged in terms of numbers of SPF is supported when one reduces the number of states that are calculated simultaneously or if one increases the number of SPF. Doing this one can observe

Table 5 MCTDH setup and numerical demands for a selection of calculations

No. of SPF for						No. Packets	Memory in MB			CPU Time per iter. (h)	Percent CPU for		
Q_1	Q_2	Q_3	Q_4	Q_5	Q_6		A	SPF	WF		M-fields	DAV	RK8
16	10	11	10	10	18	1	24.1	5.3	29.4	742	28.1	65.6	0.6
32	10	11	10	10	18	1	48.3	5.8	54.1	2304	34.4	60.2	0.2
16	10	10	9	9	16	4	63.3	5.0	68.3	1882	29.1	64.4	0.3
14	11	10	11	10	14	6	108.6	5.3	113.9	4864	73.2	21.4	0.1

The first six columns outline the number of SPF used per combined mode and the number of packets (eigenstates calculated simultaneously), followed by the memory demands for the A-vector (A), the SPF and the complete wavefunction (WF). Column 11 shows the total CPU time in hours needed to complete one iteration step of the relaxation, i.e., calculation of the mean-fields plus solving the eigenvalue problem of the A-Vector (DAV) plus propagating the SPF using a 8th order Runge-Kutta integrator (RK8). The following columns outline the fraction of CPU spent for these three sub-steps. Note, that the timing data represents typical averaged values and may vary a few percent for particular setups

large energy drops in the order of 10^1 cm^{-1} and also changes of the splittings. In a further step we will use the state vectors calculated above as stating vectors for single relaxations such that the SPF basis is optimized for each state individually. From this we expect much more accurate data.

Computational Effort

From the numerical point of view the vast majority of CPU time is spent for the calculation of the mean fields and within the Davidson routine while the Runge-Kutta integration, which has been recently parallelized using MPI, needs a negligible amount of time.

Table 5 outlines the computational demands for a selection of calculations performed. The absolute CPU times given are calculated from the real-time used to complete one iteration consisting of the calculation of the mean fields, one diagonalization of the eigenvalue problem (5) and the 8th order Runge-Kutta integration of the SPF. This time is multiplied by the number of CPU cores used for the particular setup.

Depending on the particular setup, approximately 10–20 iterations are needed until convergence, where we observed that the block relaxations (more then one state) tend to converge slower than the singe relaxations.

Note, that the memory demands outlined in Table 5 only accounts for the wavefunction. In addition to this, approximately 300 MB RAM per MPI process had to be allocated to store the Hamiltonian operator. Also temporary workspace (amongst others for the distributed Davidson vectors) in the order of 2–10 GB per MPI-process had to be provided where typically 8–32 MPI-processes had been invoked, each of which using 16 POSIX threads.

4 Proton Transfer in Hydrated Hydroxide

The $H_3O_2^-$ anion, which has been observed and characterized both in the gas phase [34, 35] and in solution [36], it is believed to be responsible for the anomalous mobility of hydroxide anion in solution [37]. With respect to its proton transfer, Diken et al. [35] registered the argon predissociation spectrum of $H_3O_2^-$ and assigned a very intense and narrow feature at 697 cm^{-1} to this motion by comparison with DMC calculations [38]. From a theoretical perspective, the ground and vibrationally excited states of $H_3O_2^-$ have been studied by several groups [39–41] at different levels of calculation. However, the agreement exhibited among their results themselves and with experiment is not satisfactory.

For our MCTDH calculations it is necessary to express the PES in product form. In the following section, we introduce our new tensor-decomposition scheme, Multigrid POTFIT (MGPF), which allows us to obtain the desired form for large systems.

4.1 *MGPF Expansion of the Potential*

The problems associated with an RPCE of the PES, Sect. 3.1, have led us to devise a new variational algorithm to re-fit PES-tensors in MCTDH (product) form as in (1): Multigrid POTFIT [42]. This method is based on a multigrid formalism and relies on a series of underlying POTFIT calculations on subsets of the full (primitive) grid. The resulting expansion is formally identical to that of POTFIT and hence it is fully compatible with the MCTDH software package. Moreover, MGPF inherits the variational character of POTFIT and MCTDH and, hence, provides us us with an efficient error control. In its present implementation, MGPF allows the treatment of systems up to $\sim$12 physical DOF.

4.2 *Numerical and Computational Details*

System Definition: Sets of Coordinates

To describe the system, we have considered two different coordinate frames: (1) a Jacobi set (Fig. 2a), and (2) a valence one (Fig. 2b). Moreover, in order to avoid configurations leading with very high energy values of the potential (irrelevant for our purposes), we have redefined the z-bridging proton transfer coordinate as previously suggested [5]:

$$z_{\mathrm{red}} = \frac{z}{(R - 2d_0)} \tag{14}$$

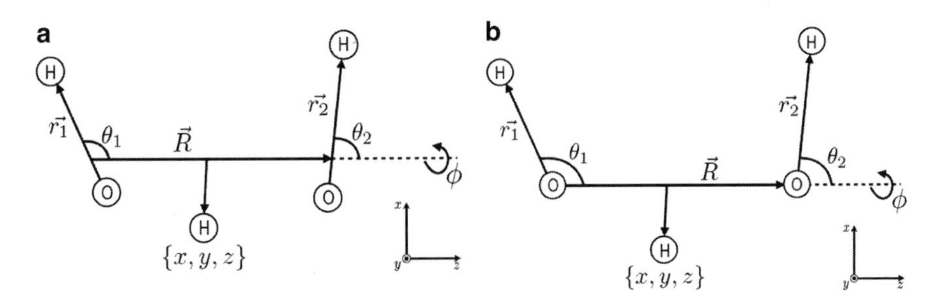

Fig. 2 Set of: (**a**) Jacobi and (**b**) valence coordinates for the $H_3O_2^-$ system

Table 6 Definition of the fine (primitive) grids for the Jacobi and valence sets of coordinates

	Jacobi			Valence		
DOF	DVR	N	Range	DVR	N	Range
r_1	HO	13	[1.400, 2.400]	HO	13	[1.400, 2.400]
r_2	HO	13	[1.400, 2.400]	HO	13	[1.400, 2.400]
R	HO	13	[4.200, 5.500]	HO	11	[4.150, 5.500]
x	HO	11	[−0.800, 0.800]	HO	10	[−0.800, 0.800]
y	HO	11	[−0.800, 0.800]	HO	10	[−0.800, 0.800]
z_{red}	HO	19	[−0.500, 0.500]	HO	20	[−0.500, 0.500]
u_1	sin	13	[−0.800, 0.350]	sin	12	[−0.800, 0.350]
u_2	sin	13	[−0.350, 0.800]	sin	12	[−0.350, 0.800]
ϕ	exp	21	$[0, 2\pi]$	exp	21	$[0, 2\pi]$

HO denotes harmonic oscillator (Hermite) DVR and exp an exponential one. N is the number of grid points. The range is the first and last grid point in atomic units, except for ϕ which is given in radians. For the definition of the coordinates, see Fig. 2. Cosines of the valence angles have been used: $u_i = \cos\theta_i$. The reduced variable z_{red} defined as $z_{red} = z/(R-2d_0)$. The mode combination scheme is $[r_1, r_2]$, $[x, y, \phi]$, $[u_1, u_2]$, $[R, z_{red}]$

System Definition: Grids

In this work, the PES values are provided by the PES3C routine [39]. The full primitive grids (Discrete Variable Representations, DVR) for both set of coordinates are displayed in Table 6. Note that both share the coordinate labelling and ordering: $\{r_1, r_2, R, x, y, z_{red}, u_1, u_2, \phi\}$. They consist in $1.79{\cdot}10^{10}$ and $1.12{\cdot}10^{10}$ grid points, respectively. By considering that the energies must be given in an 8 bytes real representation for the required accuracy of our calculations, if we were to store the full-tensor potentials 133.6 GB and 83.8 GB, respectively, would be necessary. It is evident that this aspect severely limits our computational efficiency and that tensor-decomposition schemes become demanding.

System Definition: Computational Demand

The Improved Relaxation [43] algorithm has been used to compute the GS energies and, its state-average variant, the Block Improved Relaxation method, was used to compute vibrationally excited eigenstates. In the former [11/55/25/18] SPF were needed, whereas in the latter a larger set was needed: [11/80/35/25]. The reason for this difference lies in the fact that in the block formalism the same set of SPF accounts for several states. It should be indicated that the full MGPF PES-tensor occupies 1.4 GB and therefore working with it is impractical. For our different purposes, we trimmed the MGPF tensor-expansion down to smaller ones which ranged from 1.2 to 62 MB. In the case of Improved Relaxations, we could afford the use of the largest size expansions, unfortunately, this is not possible in Block Relaxations.

Our highest accuracy, as well as largest affordable, GS energy calculation consisted in an Improved Relaxation calculation starting from a very high-quality reference wavefunction. The final converged value of 6602.14 cm^{-1} was obtained after 17 iterations, each of them requiring a bit more than 1 h on 64 processors. On the other hand, for the Block Relaxations, we could not afford the previous potential and a 6.8 MB tensor-expansion was used. The energetically highest 6-state block relaxation for this potential was also started from a good-quality set of wavefunctions from a previous run. The converged results were obtained after 18 iterations in 112 h of computation on 64 processors.

4.3 *Results: Ground and Vibrationally Excited States of $H_3O_2^-$*

In order to illustrate the stability of the MGPF decompositions, GS energies have been computed for 20 td-MGPF PES. The results are collected in Table 7. As expected from the variational character of the MGPF approach, the GS values for higher accuracies become very stable, that is, the difference between the corresponding average value and the individual one becomes negligible and independent of the set of coordinates or the grid. Our best GS energy value, which we take as the average in the most accurate series (6,602 cm^{-1}), is in excellent agreement with the DMC value of $6605 \pm 5\ \mathrm{cm}^{-1}$ [39]. It should be indicated that all previous calculations [39, 40] have yielded consistently values about 20 cm^{-1} too high with respect to the reference DMC.

In order to investigate the proton transfer feature experimentally observed around 697 cm^{-1} [35], we have computed, by means of Block Improved Relaxation, the lowest lying vibrational states of $H_3O_2^-$ up to $\sim$800 cm^{-1} using selected td-MGPF potentials. The following blocks of states were used: [0–10], [10–15], [15–19], and [19–24]. It should be noted that an overlapping state in each set is necessary in order to provide a sensible initial guess. Furthermore, for all these eigenstates, assignments have been also proposed following [7]. The results are collected in Table 8.

Table 7 Comparison of the ground state energies for the $H_3O_2^-$ complex in the valence and Jacobi pictures for two different values of the d_0 parameter (see Fig. 2)

		Ground state energies (cm^{-1}) Maximum natural population				
Coord.	d_0	10^{-1}	10^{-2}	10^{-3}	10^{-4}	10^{-5}
Jacobi	1.5	6572.97	6600.15	6601.60	6602.26	6602.29
Valence	1.5	6606.33	6597.32	6600.96	6600.58	6601.30
Jacobi	1.6	6594.53	6604.07	6600.05	6601.60	6602.48
Valence	1.6	6597.46	6600.95	6600.33	6602.14	6602.50
Jacobi	1.5	1.3	4.1	15.0	42.0	127.0
Valence	1.5	0.6	2.7	9.3	34.0	101.0
Jacobi	1.6	0.8	3.0	10.0	28.0	79.0
Valence	1.6	0.5	1.7	6.8	22.0	62.0

The energies (in cm^{-1}) are grouped according to the natural population cutoff in the td-MGPF expansion. In each series, the full MGPF expansion is trimmed down to a maximum value of the natural population. The actual value is obtained by multiplying the corresponding power of ten at the column header times 85230 $(cm^{-1})^2$ for the Jacobi set, and 121500 $(cm^{-1})^2$ for the valence one. The last four rows display the size of the MGPF potential expansions in MB

The results, collected in Table 8, exhibit a very good agreement irrespective of the value of d_0 or the trimming threshold. This makes us confident about the reliability of the present calculations. Moreover, in agreement with previous works [39,40], we have found the following fundamentals (in order of increasing energy): OH torsion, OO stretching, OH wagging, OH rocking, and BH stretching. Compare to previous studies [39, 40], a nice agreement is observed, however our results are consistently lower in energy. This is particularly severe in the case of the bridging hydrogen stretching (z) mode (proton transfer) for which these previous studies yield too different values, either too high or too low. On the contrary, our MCTDH/MGPF value of 691.69 cm^{-1} perfectly agrees with the experimentally observed value.

5 Summary

Full dimensional calculations of the zero point energy and the ground state tunneling splitting of malonaldehyde and lowest vibrational eigenstates of hydrated hydroxide have been performed with the MCTDH algorithm, using both, shared and distributed memory parallelization.

Due to the numerical demands for (in terms of numbers of SPF) fully converged calculations, in malonaldehyde an extrapolation scheme has been used to estimate upper bounds for the zero point energy and the splitting. The zero point energy has been obtained as 14664.6 cm^{-1} which is approximately 15 cm^{-1} lower then obtained by Wang et al. in [22]. The inversion splitting has been obtained as 22.7 cm^{-1} which is in excellent agreement with their results and those published by

Table 8 Excitation energies of the lowest lying vibrational eigenstates of $H_3O_2^-$

		Excitation energy (cm^{-1})			
		$d_0 = 1.5$		$d_0 = 1.6$	
		Maximum natural population in $(cm^{-1})^2$			
State	Assignment	1215	121.5	1215	121.5
Reference GS energy (cm^{-1})		6597.36	6600.99	6600.98	6600.36
1	$(0000000)^-$	21.64	18.13	20.79	18.43
2	$(0000001)^+$	130.33	131.71	130.13	131.20
3	$(0000001)^-$	219.86	217.88	218.89	217.79
4	$(0000002)^+$	275.26	273.39	273.78	273.20
5	$(0000100)^+$	440.69	439.90	441.27	439.66
6	$(0000002)^-$	442.22	442.50	443.40	443.07
7	$(0000003)^+$	449.30	447.54	448.30	447.31
8	$(0000100)^-$	482.52	480.45	484.02	481.74
9	$(1000000)^+$	485.18	485.07	485.56	485.28
10	$(1000000)^-$	506.16	503.34	505.87	503.80
11	$(0000010)^+$	572.20	573.07	569.65	571.87
12	$(0000010)^-$	583.13	583.82	580.29	582.55
13	$(0000101)^+$	595.94	595.33	596.88	596.03
14	$(1000001)^+$	615.02	616.29	615.08	616.19
15	$(0001000)^+$	691.91	691.69	694.23	692.69
16	$(0000101)^-$	697.38	696.11	697.44	697.21
17	$(1000001)^-$	701.25	698.98	700.48	699.12
18	$(0000011)^+$	713.12	716.41	710.93	716.58
19	$(0001000)^-$	724.06	721.40	725.82	722.47
20	$(0000003)^-$	728.39	725.95	727.53	725.67
21	$(0000004)^+$	728.92	726.81	728.44	726.44
22	$(1000002)^+$	757.40	756.11	755.93	756.17
23	$(0000102)^+$	783.38	783.62	781.09	784.41
24	$(0000011)^-$	790.34	793.21	787.74	792.95

Columns one and two display the number of the state and its assignment. Parity: + (even) or − (odd). Ordering of modes: $[R, x, y, z, u_1 + u_2, u_1 - u_2, \phi]$. The MGPF expansions trimmed down to the indicated natural population

Manthe et al. in [23]. Yet, the tunneling splitting reported in the present contribution differs from the experimental value of 21.6 cm^{-1}.

We also report preliminary results for excited states and their tunneling splitting as well as assignments to normal modes. Those results were obtained using the block relaxation scheme of the Heidelberg MCTDH software package. We observed that some of the state energies exhibit large energy drops as SPF are added or packets are removed. This indicates that the wavefunctions are not yet converged in terms of the number of SPF and that further calculations are needed to obtain converged results. Nevertheless, the results obtained so far are in reasonable agreement with the results of Hammer and Manthe [44].

For bihydroxide complex, the lowest vibrational eigenstates have been studied. To this end, we have made use of the newly developed Multigrid POTFIT (MGPF) algorithm [42], for tensor decomposition of high-dimensional PES in product form. The quality of the MGPF approach has been carefully checked by rms-error analysis and comparison of the GS energies obtained on each of the MGPF PES.

Additionally, Improved Relaxations and Block Improved Relaxations have been carried out. The obtained energies are in nice agreement with previous calculations. In particular, our MCTDH/MGPF are, up to the present day, the only which have reproduced the experimental value of the proton transfer frequency.

References

1. M. Brill, O. Vendrell, F. Gatti, H.-D. Meyer, in *High Performance Computing in Science and Engineering 07*, ed. by W.E. Nagel, D.B. Kröner, M. Resch (Springer, Heidelberg, 2008), pp. 141–156
2. M. Brill, O. Vendrell, H.-D. Meyer, in *High Performance Computing in Science and Engineering 09*, ed. by W.E. Nagel, D.B. Kröner, M. Resch (Springer, Heidelberg, 2010), pp. 147–163
3. O. Vendrell, F. Gatti, D. Lauvergnat, H.-D. Meyer, J. Chem. Phys. **127**, 184302 (2007)
4. O. Vendrell, F. Gatti, H.-D. Meyer, J. Chem. Phys. **127**, 184303 (2007)
5. O. Vendrell et al., J. Chem. Phys. **130**, 234305 (2009), see supplementary material, EPAPS document E-JCPSA6-130-023924, which can be downloaded from: ftp://ftp.aip.org/epaps/journ_chem_phys/E-JCPSA6-130-023924/
6. O. Vendrell, F. Gatti, H.-D. Meyer, Angew. Chem. Int. Ed. **48**, 352 (2009)
7. O. Vendrell, F. Gatti, H.-D. Meyer, J. Chem. Phys. **131**, 034308 (2009)
8. M. Schröder, F. Gatti, H.-D. Meyer, J. Chem. Phys. **134**, 234307 (2011)
9. P.A.M. Dirac, Proc. Camb. Philos. Soc. **26**, 376 (1930)
10. J. Frenkel, *Wave Mechanics* (Clarendon Press, Oxford, 1934)
11. U. Manthe, H.-D. Meyer, L.S. Cederbaum, J. Chem. Phys. **97**, 3199 (1992)
12. M.H. Beck, A. Jäckle, G.A. Worth, H.-D. Meyer, Phys. Rep. **324**, 1 (2000)
13. L.J. Doriol, F. Gatti, C. Iung, H.-D. Meyer, J. Chem. Phys. **129**, 224109 (2008)
14. H.-D. Meyer, Wiley Interdiscip. Rev. Comput. Mol. Sci. **2**, 351 (2012)
15. T. Carrington Jr., W.H. Miller, J. Chem. Phys. **84**, 4364 (1986)
16. N. Shida, P.F. Barbara, J.E. Almöf, J. Chem. Phys. **91**, 4061 (1989)
17. D. Tew, N. Handy, S. Carter, J. Chem. Phys. **125**, 084313 (2003)
18. M.D. Coutinho-Neto, A. Viel, U. Manthe, J. Chem. Phys. **121**, 9207 (2004)
19. A. Viel, M.D. Coutinho-Neto, U. Manthe, J. Chem. Phys. **126**, 024308 (2007)
20. A. Hazra, J.H. Skone, S. Hammes-Schiffer, J. Chem. Phys. **130**, 054108 (2009)
21. Y. Wang, J.M. Bowman, J. Chem. Phys. **129**, 121103 (2008)
22. Y. Wang et al., J. Chem. Phys. **128**, 224314 (2008)
23. T. Hammer, M.D. Coutinho-Neto, A. Viel, U. Manthe, J. Chem. Phys. **131**, 224109 (2009)
24. A. Jäckle, H.-D. Meyer, J. Chem. Phys. **104**, 7974 (1996)
25. A. Jäckle, H.-D. Meyer, J. Chem. Phys. **109**, 3772 (1998)
26. H.-D. Meyer, F. Gatti, G.A. Worth (eds.), *Multidimensional Quantum Dynamics: MCTDH Theory and Applications* (Wiley-VCH, Weinheim, 2009)
27. F. Gatti, H.-D. Meyer, Chem. Phys. **304**, 3 (2004)
28. S. Carter, S.J. Culik, J.M. Bowman, J. Chem. Phys. **107**, 10458 (1997)
29. H. Rabitz, O.F. Alis, J. Math. Chem. **25**, 197 (1999)
30. J.M. Bowman, S. Carter, X. Huang, Int. Rev. Phys. Chem. **22**, 533 (2003)
31. O.F. Alis, H. Rabitz, J. Math. Chem. **29**, 127 (2001)

32. G.Y. Li et al., J. Phys. Chem. A **110**, 2474 (2006)
33. S. Manzhos, T. Carrington, J. Chem. Phys. **125**, 084109 (2006)
34. E.A. Price et al., Chem. Phys. Lett. **366**, 412 (2002)
35. E.G. Diken et al., J. Phys. Chem. A **109**, 1487 (2005)
36. S.T. Roberts et al., Proc. Natl. Acad. Sci. **106**, 15154 (2009)
37. M.E. Tuckerman, D. Marx, M. Parrinello, Nature **417**, 925 (2002)
38. A.B. McCoy et al., J. Chem. Phys. **122**, 061101 (2005)
39. A.B. McCoy, X. Huang, S. Carter, J.M. Bowman, J. Chem. Phys. **123**, 064317 (2005)
40. H.-G. Yu, J. Chem. Phys. **125**, 204306 (2006)
41. A.B. McCoy, E.G. Diken, M.A. Johnson, J. Chem. Phys. **113**, 7346 (2009)
42. D. Peláez, H.-D. Meyer, J. Chem. Phys. **138**, 014108 (2013)
43. H.-D. Meyer, F. Le Quéré, C. Léonard, F. Gatti, Chem. Phys. **329**, 179 (2006)
44. T. Hammer, U. Manthe, J. Chem. Phys. **134**, 224305 (2011)

Efficient Calculation of Multi-dimensional Potential Energy Surfaces of Molecules and Molecular Clusters

Michael Neff, Dominik Oschetzki, Yuriy Yudin, Yevgen Dorozhko, Natalia Currle-Linde, Michael Resch, and Guntram Rauhut

Abstract Highly accurate multi-dimensional potential energy surfaces have been computed in a fully automated fashion using newly implemented grid computing capabilities, which allow for the use of an unlimited number of cores. This new feature, which has been interfaced to our potential energy surface generator, allows for the accurate investigation of molecular systems, which are significantly larger than reported in the recent literature. Multi-dimensional potential energy surfaces at the coupled-cluster level were generated for systems of up to 16 atoms, which were used to compute accurate anharmonic vibrational spectra, which can directly be compared with experimental data.

1 Introduction

The calculation of multidimensional Born-Oppenheimer potential energy surfaces (PES) of molecules or molecular clusters by means of quantum mechanical methods is a computationally highly demanding task for several reasons. (a) An analytical form of the PES is not known a priori and thus the PES needs to be determined at predefined grid points. (b) The number of grid points of a 3N-6 dimensional molecular system (N being the number of atoms) grows exponentially and thus a complete surface can only be evaluated for the smallest systems with not more than 4 or 5 atoms. Even if some kind of truncated expansion, e.g. a Taylor or multimode expansion, is used, the number of grid points is vast for systems with

M. Neff · D. Oschetzki · G. Rauhut (✉)
Institute of Theoretical Chemistry, University of Stuttgart, Pfaffenwaldring 55, 70569 Stuttgart, Germany
e-mail: rauhut@theochem.uni-stuttgart.de

Y. Yudin · Y. Dorozhko · N. Currle-Linde · M. Resch
High Performance Computing Center Stuttgart (HLRS), Nobelstr. 19, 70569 Stuttgart, Germany
e-mail: linde@hlrs.de

W.E. Nagel et al. (eds.), *High Performance Computing in Science and Engineering '13*, DOI 10.1007/978-3-319-02165-2_16,

more than 10 atoms. (c) The accurate determination of an individual grid point by high-level ab initio methods, e.g. coupled-cluster methods, is a demanding task in itself and may require several hours of CPU time. (d) Low level methods, like density functional theory, cannot be used for all purposes as the accuracy offered by these methods is not sufficient in order to achieve spectroscopic accuracy. As we focus here on the prediction of accurate vibrational (infrared) spectra, wave function based approaches appear mandatory.

In the recent past, programs for the automated generation of PESs have been developed by several research groups [1–3]. All these programs use several tricks at the same time to fight the number of ab initio calculations and to reduce the computational effort within the calculations of the individual grid points. These tricks embrace iterative interpolation schemes, prescreening techniques, multi-level approaches, the exploitation of molecular symmetry and so forth to name just a few of them. The interested reader is directed to the original literature [1–3]. An important aspect within this context is, of course, the parallelization and/or the grid computing capabilities of these PES generators. As the evaluation of total electronic energies at different grid points is a completely decoupled task, embarrassingly parallel implementations can be realized without major problems, which allow for the use of many thousands of cores [4]. Nevertheless, such an implementation is hampered by several aspects. (1) As the total number of grid points needed for representing the PES is not known from the very beginning within the iterative interpolation scheme, intermediate synchronization steps need to be introduced, which prohibit a consecutive evaluation of the individual terms within the expansion of the PES. However, this problem can be circumvented by a simultaneous determination of all terms at the same time. (2) The I/O demands of the individual electronic structure calculations can easily be in the range of 10–100 GiB. Once thousands of cores are used, this may lead to a substantial I/O bottleneck if local disks are not available. As a consequence a parallelization of the *individual electronic structure calculations* is highly desirable, which would diminish the latter problem. However, a combination of both parallelization schemes within a single calculation is not trivial, but can be realized quite easily within the framework of grid computing. Grid computing in the context of PES generation means, that batches of input files for the individual electronic structure calculations are generated by the master process of the PES generator, which will be passed independently to the electronic structure code. In a subsequent step, the results of these calculations will be assembled and passed back to the master process. In addition to the twofold parallelization this approach allows for the implementation of error correction schemes due to hardware failures and/or failures arising from the electronic structure calculations, e.g. not converging Hartree-Fock calculations. These techniques have been described in detail elsewhere [5]. As a consequence, the generation of potential energy surfaces by means of grid computing is more robust than a simple embarassingly parallel implementation of the PES generator, but suffers from two aspects. In order to generate smooth PESs many electronic structure methods, e.g. local correlation methods, require passing information, obtained at the reference structure, to electronic structure calculations for distorted geometries. This problem has been solved by distributing an additional (binary) file along with the input file for the different electronic structure calculations. More importantly, the file and process handling of several thousands of individual calculations becomes a tedious task once load balancing is supposed to be optimized. For that reason we have interfaced the MOLPRO package of ab initio programs [6], which comprises also the PES generator, to the SEGL grid computing client [7]. As a consequence the entire determination of the PES generation can be controlled by the graphical control panel of SEGL.

Within this project high-level PESs for systems with up to 16 atoms have been generated, which allow for the accurate calculation of their vibrational spectra. Two examples will be provided, the first one (OD_3^+) demonstrating the accuracy of our calculations and the second (LiF clusters) showing the range of applicability of these state-of-the-art calculations. The first example is a particular challenging one, as the system constitutes a double-well potential, which gives rise to potential-sensitive tunneling splittings. To the best of our knowledge the 2nd system is the largest system, which has been computed by coupled-cluster approaches. As PESs cannot directly be compared with experimental data, we have based our comparison on vibrational frequencies, which are very sensitive with respect to changes of the PES. All vibrational structure calculations presented here employed vibrational self-consistent field theory (VSCF) [8,9] and subsequent vibration correlation methods and made use of the Watson-Hamiltonian [10]

$$\hat{H} = \frac{1}{2}\sum_{\alpha\beta} \hat{\pi}_\alpha \mu_{\alpha\beta} \hat{\pi}_\beta - \frac{1}{8}\sum_{\alpha} \mu_{\alpha\alpha} - \frac{1}{2}\sum_{i} \frac{\partial^2}{\partial q_i^2} + V(q_1, \ldots, q_{3N-6}) \tag{1}$$

for solving the Schrödinger equation.

2 Computational Details

All potential energy surfaces, $V(\mathbf{q})$, of this study were expanded in terms of normal coordinates, $\mathbf{q}$, being determined in preceding harmonic frequency calculations. Multimode expansions

$$V(\mathbf{q}) = \underbrace{\sum_{i} V_i(q_i)}_{1D} + \underbrace{\sum_{i<j} V_{ij}(q_i, q_j)}_{2D} + \underbrace{\sum_{i<j<k} V_{ijk}(q_i, q_j, q_k)}_{3D} + \ldots \tag{2}$$

truncated after the 3D or 4D contributions, i.e. V_{ijk} or V_{ijkl}, respectively, have been used throughout. In this notation V_i, V_{ij}, etc. denote difference surfaces, i.e.

$$V_i(q_i) = V_i^0(q_i) - V_0, \tag{3}$$

$$V_{ij}(q_i, q_j) = V_{ij}^0(q_i, q_j) - \sum_{r \in i,j} V_r(q_r) - V_0, \tag{4}$$

$$V_{ijk}(q_i, q_j, q_k) = V_{ijk}^0(q_i, q_j, q_k) - \sum_{\substack{r,s \in i,j,k \\ r>s}} V_{rs}(q_r, q_s) - \sum_{r \in i,j} V_r(q_r) - V_0 \tag{5}$$

and thus high-order coupling terms are corrections to the low-order contributions. V_0 is the total energy of the equilibrium structure and the superscript 0 denotes the surfaces, which are obtained from electronic structure calculations. Using this expansion and a truncation after the 3D terms, one would formally have to perform about $47 \cdot 10^6$ electronic structure calculations for the lithium fluoride cluster (Li_8F_8) presented below, once 16 grid points are used along each coordinate. As the importance of the difference surface decreases with increasing order, we have used multi-level schemes in order to reduce the computational effort. In these schemes different electronic structure methods are used for the different orders of the expansion.

2.1 Computational Details for OD_3^+

In our calculations for OD_3^+ the 1D and 2D terms were evaluated from explicitly correlated coupled-cluster theory in combination with a quadruple-ζ basis set, i.e. CCSD(T)-F12a/cc-pVQZ-F12. This scheme corresponds to conventional coupled-cluster calculations at the basis set limit [11]. Core correlation and scalar relativistic effects were obtained at the CCSD(T)/cc-pCVQZ-dk level using the Douglas-Kroll-Hess Hamiltonian. High-order contributions to the coupled-cluster expansion were determined by Kallay's MRCC program [12, 13] interfaced to MOLPRO for the valence electrons at the CCSDT/cc-pVTZ and CCSDT(Q)/cc-pVTZ(d/p) levels. The latter basis set refers to Dunning's cc-pVTZ basis [14], but without f functions for oxygen and without d functions for hydrogen atoms. Due to the closed-shell character of the molecules, spin-orbit effects (SO) were considered to be very small and have thus been neglected. Likewise diagonal Born-Oppenheimer corrections (DBOC) were not included as it is known from highly accurate calculations on OH_3^+, that these corrections have very little impact on the vibrational levels. For the calculation of the 3D and 4D terms core-correlation effects, relativistic corrections and high-order terms have been neglected and the orbital basis within the explicitly correlated coupled cluster calculations has been reduced to the triple-ζ level. The MP2-F12 calculations, which are the first step in CCSD(T)-F12a calculations, were performed using the MP2-F12/**3**C(FIX) method described in detail in [15, 16]. In this method the numerous two-electron integrals are computed using robust density fitting approximations [17, 18]; the aug-cc-pVQZ/MP2FIT basis sets of Weigend et al. [19] were used as auxiliary base. For the resolution of the identity and Fock matrix density fitting, the cc-pVQZ/JKFIT basis sets [20] were used; previous work has shown that these are well suited for this purpose [15,21]. Density fitting was not used in the Hartree-Fock and CCSD(T)-F12a calculations apart from the MP2-F12 part. The perturbative CABS (complementary auxiliary basis set) singles correction for the Hartree-Fock basis set error as described in [16, 22] was applied in all F12 calculations. Tight CABS thresholds were used so that no functions were deleted. This guarantees smooth potential energy surfaces.

The D_{3h} transition state of the inversion reaction was taken as the reference point within the expansion of the surfaces. Displacement vectors belonging to the degenerate eigenvalues of the Hessian were rotated against each other by 45° in order to retain the degeneracy for not converged potentials. Thirty six grid points were generated along each normal coordinate, i.e. along the inversion mode 18 grid points were used for describing each minimum, which is in the typical range. On average not more than 10 ab initio single points along each coordinate have been used in the coarse grid prior to fitting. For representing the PES at most 4-mode coupling terms were used. In a first step vibrational wave functions and state energies were determined by vibrational self-consistent field theory (VSCF). In these calculations a non-orthogonal basis of 36 distributed Gaussians—centered at the grid points—has been used [23]. The Watson correction term was added as a mass-dependent pseudopotential like contribution to the PES. Vibrational angular momentum terms did not enter into the vibrational wave functions but were added perturbationally to the state energies a posteri. All calculations were computed state-specifically, i.e. the modals were individually optimized for each state of interest. In subsequent vibrational configuration interaction calculations 46,493 configurations (up to quintuple excitations, i.e. VCISDTQ5) were used. We used a configuration selective VCI approach, which has been described elsewhere [24]. In these calculations excitations for each mode up to the 9th root have been considered. This is a rather high value, but was found to be necessary as (−)-states constitute excited states up to the 4th root in the VCI calculations. A 2nd order multimode expansion has been used for representing the vibrational angular momentum terms within the Watson Hamiltonian, i.e. the μ-tensor has been expanded accordingly (for details see [25]).

2.2 Computational Details for the Lithium Fluoride Clusters

Due to the size of the LiF clusters, Li_7F_7 and Li_8F_8, a different strategy has been chosen to compute these systems. 1D terms of the potential were evaluated from density fitting explicitly correlated local coupled-cluster calculations in combination with a cc-pVTZ-F12 basis, i.e. df-LCCSD(T)-F12/cc-pVTZ-F12 (848 basis functions for Li_8F_8). Each single-point calculation required about 6 h of CPU time and requested 29 GB of scratch space. Without the use of density fitting this calculation would require about 1 TB of scratch space and thus prohibit the use of large parallel architectures. The 2D terms were determined at the same level of electronic structure theory but in combination with a small orbital basis set, i.e. cc-pVDZ-F12. These calculations made use of the **3**A(`FIX`,`NOX`) ansatz with fixed amplitudes and neglecting the X terms. In contrast to the calculations for OD_3^+ density fitting was also used in the preceding Hartree-Fock calculations, which employed a def2-QZVPP/JKFIT basis. 1D and 2D terms were computed using the grid computing feature as outlined above. Due to the vast number of 3D terms, a modeling scheme based on semiempirical MO theory has been used for determining these contributions [26]. In this scheme, semiempirical parameters are reoptimized with respect to the 1D and 2D surfaces. Details of this procedure are presented elsewhere. Benchmark calculations for small lithium fluoride clusters [5], i.e. Li_2F_2, showed that the combination of density fitting explicitly correlated local coupled-cluster calculations for the 1D and 2D terms and the modeling scheme for the 3D terms results in deviations of about one wavenumber in comparison to conventional coupled-cluster calculations for the 1D, 2D and 3D terms and are thus negligible. Dipole surfaces as needed for the evaluation of infrared intensities were truncated after the 2D terms and were determined at the density fitting Hartree-Fock level. We have shown quite recently that this rather crude approximation yields results, which are not only qualitatively correct, but which are in good agreement with accurate results [5]. In a subsequent step, the potential being represented by grid points was transformed to polynomials up to 8th order. χ^2 values within the fitting were as low as 10^{-13} E_H for the 3D terms and as low as 10^{-19} E_H for the 1D terms. Vibrational angular momentum terms were evaluated with a constant μ-tensor. Vibration correlation effects were accounted for by 2nd order vibrational Møller-Plesset perturbation theory, VMP2.

3 Results

Both systems studied here must be considered rather challenging systems, although for completely different reasons. As mentioned above, the first one, i.e. OD_3^+, shows a double-minimum potential and thus cannot be studied by standard methods like vibrational perturbation theory, VPT2, based on a quartic force field. Each individual electronic structure calculation is very demanding and requests several hours of CPU time. In order to represent the potential as accurately as possible many more grid points have been computed than used in standard calculations for 4-atomic systems. The 2nd system, i.e. LiF clusters, became only feasible due to the use of local correlation methods as the computational demand of these approaches scales almost linearly with respect to the system size. Nevertheless, the number of 3D grid points to be determined was so excessive, that even faster methods had to be used for this part of the calculation (see above).

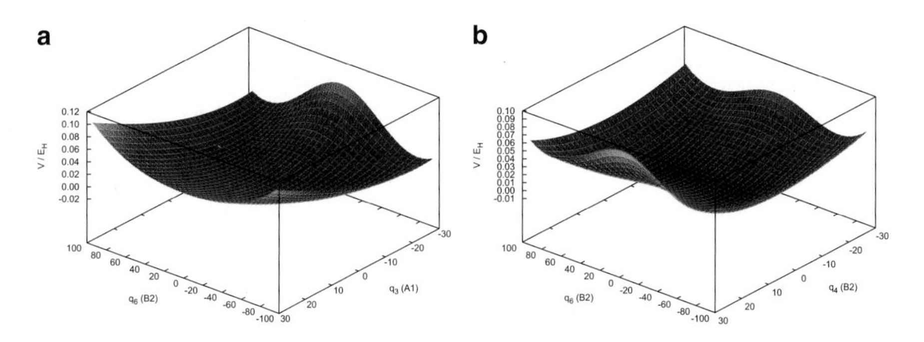

Fig. 1 Two-dimensional cuts of the potential energy surface of OD_3^+

3.1 *Benchmark Calculations on OD_3^+*

Quite recently we have presented results for NH_3 and OH_3^+ using the same recipe as employed here [27]. We were able to show, that normal coordinates can be used without problems for expanding the PES. However, due to the mass-dependence the use of normal coordinates requires the recalculation of the PES for isotopologues and thus our previously determined surface for OH_3^+ could not be used for the calculations presented here. Figure 1 shows the two most important cuts of the six-dimensional PES. The lhs of the figure shows the 2D surface along the inversion mode and the symmetric stretching vibration, while the rhs shows the surface spanned by the inversion mode and the asymmetric stretching. Note, that 1D contributions to the potential are included in both plots. In contrast to all other benchmark calculations on this system, the PES contributions shown in Fig. 1 were generated in a fully automated manner, i.e. the calculation was controlled by just a single input file and has not been corrected for deficiencies upon inspection of the PES. This capability opens the way for the automated and reliable calculation of PESs for extended and complex systems, which prohibit visual inspection due to the dimensionality of the system or simply by the amount of data, as for example in the case of LiF clusters. The results, obtained from subsequent VCI calculations, are summarized in Tables 1 and 2. While Table 1 shows the dependence of the vibrational state energies and tunneling splittings on the expansion of the potential, Table 2 shows the results obtained from a 4D potential, but in dependence on the expansion of the μ-tensor as arising in the Watson-Hamiltonian. First of all, our final 4D results are in excellent agreement with experimental data or the most reliable computed data of Halonen and co-workers [28]. The only larger deviation ($5\,cm^{-1}$) with respect to the tunneling splittings can be seen for the inversion mode, while for all other states the deviation in the splittings was less than one wavenumber. Most likely the deviation observed for the inversion mode arises either from fitting problems in the region of the transition state or the neglect of diagonal Born-Oppenheimer corrections in the generation of the PES, which for example

Table 1 VCI State energies of OD_3^+ relative to the vibrational ground state in dependence on the expansion of the potential energy surface

		2D		3D		4D		[28]		Exp[a]	
State	Sym.	$\Delta\tilde{\nu}$	$\Delta\Delta\tilde{\nu}$	$\Delta\tilde{\nu}$	$\Delta\Delta\tilde{\nu}$	$\Delta\tilde{\nu}$	$\Delta\Delta\tilde{\nu}$	$\Delta\tilde{\nu}$	$\Delta\Delta\tilde{\nu}$	$\Delta\tilde{\nu}$	$\Delta\Delta\tilde{\nu}$
$00^+0^00^0$	A_1	0.0		0.0		0.0		0.0		0.0	
$00^-0^00^0$	A_1	25.0	25.0	14.8	14.8	15.4	15.4	15.8	15.8	15.4	15.4
$01^+0^00^0$	A_1	425.6		452.1		451.7		453.5		453.7	
$01^-0^00^0$	A_1	690.7	240.1	645.2	193.1	648.1	196.5	646.6	193.1	645.1	191.4
$00^+1^10^0$	E′	2634.6		2623.0		2629.7		2628.7		2629.7	
$00^-1^10^0$	E′	2682.0	22.4	2638.9	15.9	2639.7	10.0	2638.9	10.2	2639.6	9.9
$10^+0^00^0$	A_1	2388.4		2484.2		2483.7		2485.1			
$10^-0^00^0$	A_1	2427.7	14.3	2497.8	13.7	2497.8	14.1	2499.4	14.3		
$00^+0^01^1$	E′	1144.8		1190.0		1195.8		1196.5			
$00^-0^01^1$	E′	1195.0	25.2	1207.2	17.2	1214.4	18.6	1215.7	19.1		

[a] Data taken from [31, 32]

Table 2 VCI State energies of OD_3^+ relative to the vibrational ground state in dependence on the expansion of the μ-tensor

		μ=0D		μ=1D		μ=2D		[28]		Exp[a]	
State	Sym.	$\Delta\tilde{\nu}$	$\Delta\Delta\tilde{\nu}$	$\Delta\tilde{\nu}$	$\Delta\Delta\tilde{\nu}$	$\Delta\tilde{\nu}$	$\Delta\Delta\tilde{\nu}$	$\Delta\tilde{\nu}$	$\Delta\Delta\tilde{\nu}$	$\Delta\tilde{\nu}$	$\Delta\Delta\tilde{\nu}$
$00^+0^00^0$	A_1	0.0		0.0		0.0		0.0		0.0	
$00^-0^00^0$	A_1	14.5	14.5	15.3	15.3	15.4	15.4	15.8	15.8	15.4	15.4
$01^+0^00^0$	A_1	452.9		451.9		451.7		453.5		453.7	
$01^-0^00^0$	A_1	645.0	192.0	648.1	196.2	648.1	196.5	646.6	193.1	645.1	191.4
$00^+1^10^0$	E′	2625.8		2629.7		2629.7		2628.7		2629.7	
$00^-1^10^0$	E′	2635.3	9.5	2639.6	10.0	2639.7	10.0	2638.9	10.2	2639.6	9.9
$10^+0^00^0$	A_1	2483.3		2483.6		2483.7		2485.1			
$10^-0^00^0$	A_1	2496.5	13.2	2497.7	14.1	2497.8	14.1	2499.4	14.3		
$00^+0^01^1$	E′	1193.8		1195.7		1195.8		1196.5			
$00^-0^01^1$	E′	1211.0	17.2	1214.2	18.5	1214.4	18.6	1215.7	19.1		

[a] Data taken from [31, 32]

lower the reaction barrier by $-11\,\mathrm{cm}^{-1}$ [29]. With respect to the dependence of the tunneling splittings on the expansion of the μ-tensor, Table 2 shows, that it is fully sufficient to truncated the expansion after the 1D terms. As the computational demands rise steeply with increasing orders of the expansion, this is an important aspect. Moreover, these calculations show also, that the potential can be described sufficiently by coupling terms up to 4th order, terms of higher orders appear not to necessary. However, this picture may alter for different systems, in which the transition from one minimum to the other involves more coordinates than just two (i.e. the inversion mode and the symmetric stretching mode) as in this example here.

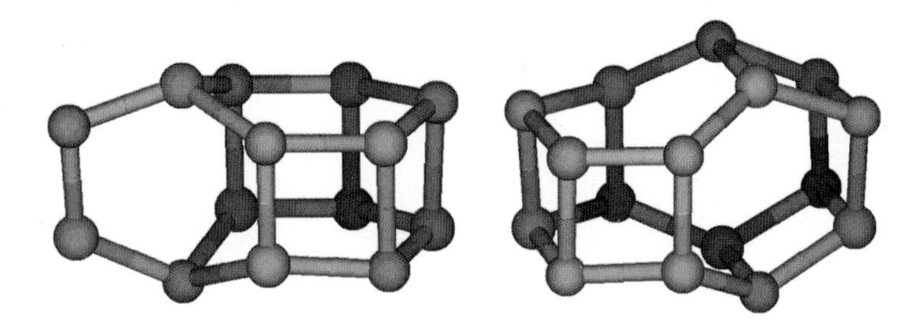

Fig. 2 Structures of the Li_7F_7 and Li_8F_8 clusters

3.2 Vibrational Spectra of Lithium Fluoride Clusters

The challenges arising within the calculation of the LiF clusters are of completely different nature. LiF clusters have been studied recently by Doll, Schoen and Jansen at the Hartree-Fock and DFT levels [30]. Their main focus was on an energetical survey of the isomeric space. The clusters may differ considerably in structure, symmetry and energetics. In contrast to the work of Doll et al., in this study here we focused mainly on the vibrational structure of energetically low lying clusters being built up by 7 and 8 LiF entities. Figure 2 shows the corresponding structures. While the Li_7F_7 cluster shows C_s symmetry, the Li_8F_8 cluster belongs to the non-Abelian S_4 group. As will be discussed below this has strong impact on the vibrational spectra. The PESs of both systems have been computed using the grid computing capabilities of MOLPRO on the Cray XE6. For the calculation of the 1D terms of the LiF clusters it was necessary to use compute nodes with 64 GB memory. In contrast, for the 2D terms it was sufficient to use nodes with 32 GB memory. Tests on larger LiF clusters i.e. $Li_{10}F_{10}$, showed that the compute nodes with 32 GB memory cannot efficiently be used anymore, because each core has only fast access to 1 GB of memory once all cores of a node are used. This, of course, is due to the use of all 32 cores per node. The same holds true for the 1D calculations and the compute nodes with 64 GB memory. Up to 10.000 cores have been used at the same time. When using that many cores, the request for scratch space increases dramatically although a single electronic structure calculation needs only a few GB as discussed above. As the number of grid points is vast only for the 2D terms and higher coupling contributions, the needed scratch space is significantly reduced due to the use of the multi-level ansatz. For example, for Li_8F_8 one 2D grid point requests just 13 GB of scratch space in comparison to 29 GB for a 1D grid point (see above). Using this approach the calculation of the PES for Li_7F_7 took less than 24 h. Currently, these calculations are at the cutting edge of feasibility, but we believe, that in a few years distance they will become routine applications, once standard personal computers provide a few hundreds of cores. For that reason, the computational bottlenecks of these calculations pinpoint those parts of the algorithm, which still need to be improved. In that respect these calculations allowed for an important conclusion: while the PES generator appears to be well adapted

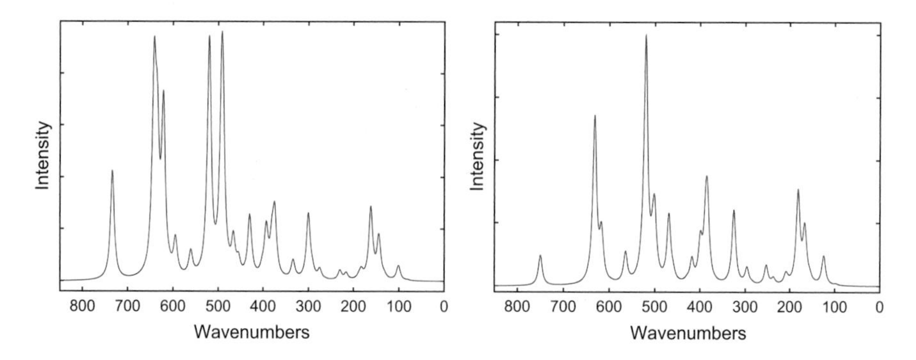

Fig. 3 VSCF (*lhs*) and VMP2 (*rhs*) vibrational spectra of Li_7F_7

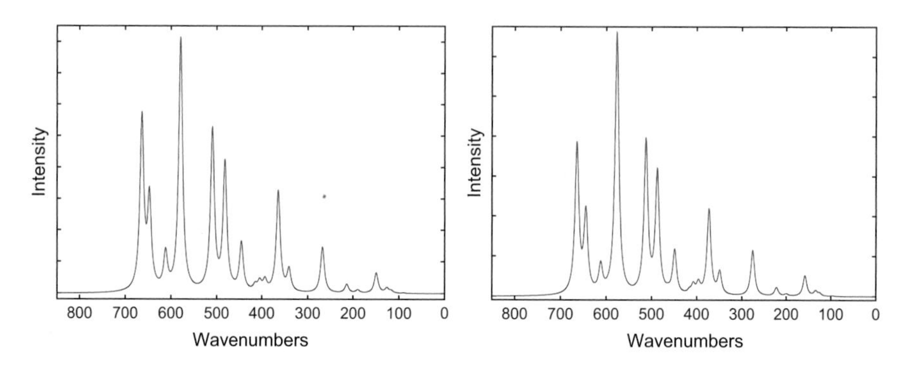

Fig. 4 VSCF (*lhs*) and VMP2 (*rhs*) vibrational spectra of Li_8F_8

for the situation of treating large systems, the vibration correlation calculations need further improvement with respect to computational speed. Consequently, new approximations and parallelization schemes need to be introduced in order to make these algorithms competitive to other approaches. Figure 3 shows the vibrational spectra obtained for Li_7F_7 at the VSCF and VMP2 levels. In contrast to the spectra for Li_8F_8 distinct differences can be seen. This shows, that vibration correlation effects are quite important for this system. Obviously strong couplings for the states above $500\,cm^{-1}$ lead to strong changes in the intensities, while for the lower lying states this effect cannot be observed. Most likely, the increasing state density is one aspect, which has certain impact. This phenomenon cannot be seen for Li_8F_8 (cf. Fig. 4). In contrast to Li_7F_7 not all vibrations of Li_8F_8 are IR active. Moreover, as a consequence of the S_4 point group many states are pairwise degenerate and thus the spectrum is much clearer. The VMP2 calculation for Li_8F_8 included up to quadruple excitations and thus more than 14.5 million configurations. The fundamentals of Li_8F_8 are shown in Table 3. Vibration correlation effects were found to have a significantly smaller impact than for Li_7F_7. Most importantly, the VMP2 spectrum of Li_8F_8 looks completely different from the spectrum of Li_7F_7 and

Table 3 Fundamentals of Li_8F_8 at the VMP2 level

Mode	Sym.	$\Delta\tilde{\nu}$	Mode	Sym.	$\Delta\tilde{\nu}$	Mode	Sym.	$\Delta\tilde{\nu}$	Mode	Sym.	$\Delta\tilde{\nu}$
32	B	657.1	24	B	479.9	16	B	388.8	8	E	214.1
31	E	638.0	23	B	457.2	15	E	364.9	7	B	191.8
30	B	608.8	22	A	435.4	14	B	340.8	6	B	174.7
29	B	605.1	21	E	441.7	13	A	309.8	5	E	149.6
28	A	592.5	20	B	409.1	12	E	267.8	4	B	126.0
27	E	570.3	19	E	400.5	11	B	266.2	3	E	116.8
26	E	505.4	18	A	404.5	10	B	265.1	2	A	94.8
25	A	507.5	17	A	385.8	9	A	248.5	1	B	89.1

thus FTIR spectroscopy can be used to distinguish the different clusters from each other. Guided by accurate quantum chemical calculations this should allow for the identification for structural motifs of these complex systems.

3.3 *Summary and Conclusions*

Grid computing capabilities have been implemented in the MOLPRO suite of ab initio programs, which allow for the fully automated generation of multi-dimensional high-level potential energy surfaces using several thousands of cores on massively parallel architectures or even distributed systems. These PESs have been used to compute the vibrational spectra for two challenging systems. The first one is a benchmark system, which is known to be extremely sensitive with respect to the accuracy of the underlying PES. We were able to show, that our PES generator is capable of treating even such troublesome cases quite accurately. The 2nd study is a state-of-the-art application comprising the investigation of the vibrational spectra of extended systems without distinct structural motifs or functional groups. The latter calculations became only feasible due to a combination of several new developments in electronic and vibrational structure calculations and the use of the newest massively parallel hardware. As a result, the calculations presented here are the largest currently performed at this very high level of theory.

Acknowledgements The authors like to thank the German Research Foundation (DFG) for financial support of the project within the Cluster of Excellence in Simulation Technology (EXC 310/1).

References

1. M. Sparta, D. Toffoli, O. Christiansen, Theor. Chem. Acc. **123**, 413–429 (2009)
2. F. Richter, P. Carbonniere, A. Dargelos, C. Pouchan, J. Chem. Phys. **136**, 224105 (2012)
3. G. Rauhut, J. Chem. Phys. **121**, 9313 (2004)

4. T. Hrenar, H.-J. Werner, G. Rauhut, J. Chem. Phys. **126**, 134108 (2007)
5. D. Oschetzki, M. Neff, P. Meier, F. Pfeiffer, G. Rauhut, Croat. Chem. Acta **85**, 379–390 (2012)
6. H.-J. Werner, P.J. Knowles, G. Knizia, F.R. Manby, M. Schütz, WIREs Comput. Mol. Sci. **2**, 242–253 (2012)
7. N. Currle-Linde, U. Küster, M. Resch, B. Risio, in *Parallel Computing: Current & Future Issues of High-End Computing*, 2005, p. 49
8. J.M. Bowman, Acc. Chem. Res. **19**, 202 (1986)
9. R. Gerber, M. Ratner, Adv. Chem. Phys. **70**, 97 (1988)
10. J.K.G. Watson, Mol. Phys. **15**, 479 (1968)
11. G. Rauhut, G. Knizia, H.-J. Werner, J. Chem. Phys. **130**, 054105 (2009)
12. M. Kallay, Ph.D. Thesis, Budapest University of Technology and Economics, 2001
13. M. Kallay, P. Surjan, J. Chem. Phys. **115**, 2945 (2001)
14. D.E. Woon, T.H. Dunning, J. Chem. Phys. **98**, 1358 (1993)
15. H.-J. Werner, T.B. Adler, F.R. Manby, J. Chem. Phys. **126**, 164102 (2007)
16. G. Knizia, H.-J. Werner, J. Chem. Phys. **128**, 154103 (2008)
17. F.R. Manby, J. Chem. Phys. **119**, 4607–4613 (2003)
18. A.J. May, F.R. Manby, J. Chem. Phys. **121**, 4479–4485 (2004)
19. F. Weigend, A. Köhn, C. Hättig, J. Chem. Phys. **116**, 3175 (2002)
20. F. Weigend, Phys. Chem. Chem. Phys. **4**, 4285 (2002)
21. F.R. Manby, H.-J. Werner, T.B. Adler, A.J. May, J. Chem. Phys. **124**, 094103 (2006)
22. T.B. Adler, G. Knizia, H.-J. Werner, J. Chem. Phys. **127**, 221106 (2007)
23. G. Rauhut, T. Hrenar, Chem. Phys. **346**, 160 (2008)
24. M. Neff, G. Rauhut, J. Chem. Phys. **131**, 124129 (2009)
25. M. Neff, T. Hrenar, D. Oschetzki, G. Rauhut, J. Chem. Phys. **134**, 064105 (2011)
26. G. Rauhut, B. Hartke, J. Chem. Phys. **131**, 014108 (2009)
27. M. Neff, G. Rauhut, Spectrochim. Acta A, http://dx.doi.org/10.1016/j.saa.2013.02.033
28. T. Rajamäki, A. Miani, L. Halonen, J. Chem. Phys. **118**, 6358 (2003)
29. T. Rajamäki, M. Kallay, J. Noga, P. Valiron, L. Halonen, Mol. Phys. **123**, 134308 (2005)
30. K. Doll, J.C. Schoen, M. Jansen, J. Chem. Phys. **133**, 024107 (2010)
31. M. Araki, H. Ozeki, S. Saito, J. Chem. Phys. **109**, 5707 (1998)
32. H. Petek, D.J. Nesbitt, J.C. Owrutsky, C.S. Gudeman, X. Yang, D.O. Harris, C.B. Moore, R.J. Saykally, J. Chem. Phys. **92** 3257 (1990)

Part IV
Reactive Flows

Dietmar Kröner

Turbulence for reactive flows are the main subject of the following three papers. The combination of RANS and LES, the influence of curvature and strain on the flame velocity and the influence of grid refinement on turbulent combustion and combustion noise is studied. All computations are based on the compressible Navier-Stokes-equations and transport equations for the species.

The simulation of a rotationally symmetric free jet in 3d is the main result of the contribution about "Delayed detached eddy simulations of a rotationally symmetric supersonic jet" by Kindler, Lempke, Gerlinger and Aigner. The underlying numerical method is a hybrid RANS/LES-method that uses the RANS-approach in the near wall region and the LES-modeling otherwise. This method is used for the simulation of a supersonic Mach 1.8 jet in a nozzle. The computed results are compared to measured values. The agreement between experiment and simulation is quite satisfying with some problems in the transition region. Furthermore the influence of the variation of a certain model constant and the grid refinement to the simulation is investigated. It turns out that the effect is not significant. The computation has been performed on the CRAY XE6-system and 10 % peak performance has been obtained. The scaling behavior was nearly ideal.

In the paper "DNS of lean premixed flames" Denev and Bockhorn describe their investigations of estimating the burning velocity of flames and the mechanism by which turbulence can influence it. It is known that turbulence can change the flame area by the effects of curvature and strain. It remains the question whether both curvature and strain effects the burning velocity to the same extent. These questions are considered for reactive flows with chemical mechanisms based on nine species and 38 reaction steps for hydrogen flames. The numerical code uses six order central differencing in space and fourth order Runge-Kutta-methods with respect

D. Kröner
Abteilung für Angewandte Mathematik, Universität Freiburg, Hermann-Herder-Str. 10, 79104, Freiburg, Germany
e-mail: dietmar@mathematik.uni-freiburg.de

to time. The computation has been performed on the CRAY XE 6 super computer in Stuttgart. The scaling which is documented in the paper seems to be o.k.

The main focus of the project about "Impact of grid refinement on turbulent combustion and combustion noise modeling with large eddy simulations" by Zhang, Habisreuther and Bockhorn lies on the investigation of the influence of grid resolution on the noise radiation in LES for turbulent combustion. The computation has been done on the HP XC 4000 cluster of the Steinbuch Center for Computing at the KIT with the resolution up to 10 million cells. It turned out that the noise level derived from a finer resolved LES is higher than that provided by a coarse LES. The solutions obtained by LES on the finest mesh show the best agreement with experimental data.

Two of these projects have been supported by the DFG.

Delayed Detached Eddy Simulation of a Rotationally Symmetric Supersonic Jet

Markus Kindler, Markus Lempke, Peter Gerlinger, and Manfred Aigner

Abstract A Delayed Detached Eddy Simulation (DDES) of a rotationally symmetric free jet is performed with the scientific in-house code TASCOM3D (Turbulent All Speed Combustion Multigrid 3D). The DDES is a hybrid RANS/LES method that employs the RANS approach in the near-wall region and LES modeling otherwise. The transition is achieved by filtering depending on the grid spacing. To preserve the RANS mode in the boundary layer an appropriate blending function is used. The investigated experiment is a supersonic Ma 1.8 jet exiting a nozzle into ambient conditions. The simulation results for the axial velocity, temperature, density and density fluctuations are compared to measured values. Several simulations have been performed to analyze the effect of the modeling constant C_{DES}, grid refinement, and the addition of artificial fluctuations in the boundary layer that are not created inherently in the used approach. The overall agreement between experiment and simulation is quite satisfying although the transition is apparently predicted too far upstream in all simulations.

1 Introduction

Despite the advances in high performance computing Large Eddy Simulations (LES) of practical configurations are very expensive in terms of computing time. Spalart estimates the requirements for the flow around an airfoil to be 10^{11} cell volumes and 10^7 time steps [1]. Due to these orders of magnitude the LES seems infeasible for several more decades. Especially in the near-wall region the LES requires very high resolutions depending on the Reynolds number ($N_{LES} \propto Re_l^{13/7}$)

M. Kindler · M. Lempke · P. Gerlinger (✉) · M. Aigner
Institut für Verbrennungstechnik der Luft- und Raumfahrt, Universität Stuttgart, Pfaffenwaldring 38-40, 70569 Stuttgart, Germany
e-mail: peter.gerlinger@dlr.de

W.E. Nagel et al. (eds.), *High Performance Computing in Science and Engineering '13*, DOI 10.1007/978-3-319-02165-2_17,

due to the small Kolmogorov-scales [2]. The traditional RANS-methods, however, are well suited to accurately simulate wall-bound flows. Therefore it is desirable to combine the positive attributes of the two approaches by employing hybrid RANS/LES approaches.

Sagaut et al. [3] classify such approaches in three categories: the instationary statistical model, the zonal hybrid approaches, and the global hybrid approaches. The first group includes the classical URANS (Unsteady RANS) as well as more advanced methods such as the Scale Adaptive Simulation (SAS) [4]. In zonal hybrid RANS/LES approaches the spatial interfaces between the two methods are predefined and the bilateral exchange at this interface is required. The major drawback of the zonal approach is the spatially fixed interface that might result in discontinuities between the two different approaches [3]. The global hybrid RANS/LES approaches on the other hand are based on a single set of equations. The probably most widely used global techniques are summarized as Detached Eddy Simulation (DES) and originate in the work of Spalart [2]. The basic principle of the DES is the modification of the turbulent length scale l_t in the RANS turbulence model to achieve a transition to a LES model. In the original formulation the transition only depends on the spatial discretization so that under certain circumstances the transition to LES can occur within the boundary layer. The Delayed Detached Eddy Simulation (DDES) used in this work prevents this by the use of an appropriate blending function.

2 Numerical Scheme

TASCOM3D has been employed at the Institute for Combustion Technology in Aerospace Engineering (IVLR) for over two decades for investigations related to scramjet and rocket combustion. The code solves transport equations for mass, momentum, total energy, turbulence variables and species mass fractions. In this work turbulence closure has been achieved by the DDES model which employs the k-ω-model of Wilcox [5] in RANS mode. TASCOM3D uses an implicit up to third order in time BDF (Backward Differentiation Formula) LU-SGS (Lower-Upper Symmetric Gauss-Seidel [6, 7] finite-volume algorithm. The code employs finite-rate chemistry fully coupled with the fluid motion. Turbulence-chemistry interaction is incorporated via an assumed PDF approach. The code has been parallelized though MPI (Message Passing Interface) and shows good performance on both vector processor and massively parallel scalar architectures [8].

2.1 *MLP^{ld}*

In case of LES the use of higher order methods rather than second order schemes is very beneficial because the requirements for the computational grid are reduced. On the other hand supersonic flows require discretization techniques that are able to resolve shock waves without significant oscillations but also without smearing. In classical high order MUSCL (Monotonic Upstream-centered Scheme for Conservation Laws) [9] approaches, TVD (Total Variation Diminishing) limiters [10] are used to avoid oscillations at discontinuities. Because high order numerical schemes usually suffer from stability problems and TVD approaches often prevent convergence to machine accuracy, the multi-dimensional limiting process (MLP) [11] is employed. It interacts with the TVD limiter in such a way that local extrema at the corner points of the volume are avoided. This stabilizes the numerical scheme and enables convergence in cases where standard limiters fail to converge.

An improved MLP version (MLP^{ld}, low diffusion) [12] has been developed and implemented into TASCOM3D which offers high accuracy and robustness while keeping the computational costs low. In case of DDES the code uses a blend between the fifth order MLP^{ld} discretization and a sixth order central stencil. Especially at shock waves the fifth order upwind TVD scheme is activated, whereas away from shock waves the sixth order scheme is employed. This blend is performed by a sensor based on local values of vorticity. The fluxes through the cell interfaces are calculated using the $AUSM^{+}$-up flux vector splitting of Liou [13]. A more detailed explanation of the MLP^{ld} approach is given in [12].

2.2 *Delayed Detached Eddy Simulation*

As mentioned before in the original DES approach the distinction between the two modes (RANS and LES) only depends on the comparison of the local grid spacing Δ to the wall distance d. This causes problems in the boundary layer where the grid spacing becomes often much smaller than the wall distance. To maintain the RANS mode in the boundary layer Spalart introduced a blending function

$$f_d = 1 - tanh\left([8r_d]^3\right), \tag{1}$$

$$r_d = \frac{\nu_t + \nu}{\sqrt{\frac{\partial u_i}{\partial x_j}\frac{\partial u_i}{\partial x_j}\kappa^2 d^2}}, \tag{2}$$

that depends on local flow conditions. Hereby ν_t and ν denote the kinematic eddy viscosity and molecular viscosity, respectively. The van Kármán constant κ has the value 0.41 and d is the wall distance. Please note that the term under the square root is written in the Einstein notation. Another improvement is that the blending in the DDES is time-resolving due to the time dependence of the variables in (2). The resulting length scale is

$$l_{DDES} = d - f_d max(0, d - C_{DES}\Delta) \tag{3}$$

with the grid spacing Δ and the modeling constant C_{DES}.

The original formulation of the DDES is based on the Spalart-Allmaras turbulence model. However, Travin et al. [14], employed the approach by modifying the dissipative term of the k-equation in Wilcox' k-ω turbulence model. This approach is used in the present work. More details concerning the employed DDES approach are given in [15].

3 Supersonic Ma 1.8 Jet

3.1 Experiment

The simulation is based on the experimental work of Panda and Seasholtz [16, 17]. From the variety of conducted experiments the Ma 1.8 free jet was chosen for numerical investigation. The free jet is created by the nozzle depicted in Fig. 1. The nozzle exit diameter that will later be used for normalization is $d_i = 50.9\,\text{mm}$. More information on the dimensions of the nozzle is also given in Fig. 1. The experimental boundary conditions are summarized in Table 1.

Experimental data is available for the mean values of the density, the axial velocity, and the temperature. Furthermore, the rms-values of the density and its spectra have been measured. According to the authors the uncertainty of the measurements are $\pm 10\,\text{m/s}$ and $\pm 1\,\%$ for the velocity and the density measurements, respectively. Although in the original work the uncertainty in the temperature measurement was estimated to be ± 8 K the authors later stated that due to problems with condensation the uncertainty was indeed much higher.

3.2 Numerical Setup

In order to reduce the computational cost the simulation is performed in two steps. First, the nozzle flow is calculated in a RANS simulation. Thus the inflow conditions for the second step, the DDES of the free jet, are determined. The grids for the two successive calculations are depicted in Fig. 2. For the sake of clarity only a strongly reduced number of cell volumes are shown. Both grids are clustered towards the wall in order to adequately resolve the turbulent quantities and velocity profiles.

The nozzle grid comprises 1.14 million cell volumes. The inflow conditions (subsonic inflow) are chosen according to the conditions in the plenum in the experiment in Table 1. Additionally the initial axial velocity was set to $u = 10\,\text{m/s}$ to avoid instabilities in the simulation. The computational domain ends at the nozzle exit that is numerically represented by a supersonic outflow condition. The nozzle walls are simulated as adiabatic no-slip walls.

The computational domain for the hybrid RANS/LES overlaps the nozzle over a distance of 12.83 mm and extends downstream for 458.1 mm. The radial extension is 500 mm. At the inflow boundary inside the nozzle the complete variable vector is set to the values extracted from the RANS simulation. At the ambient inflow

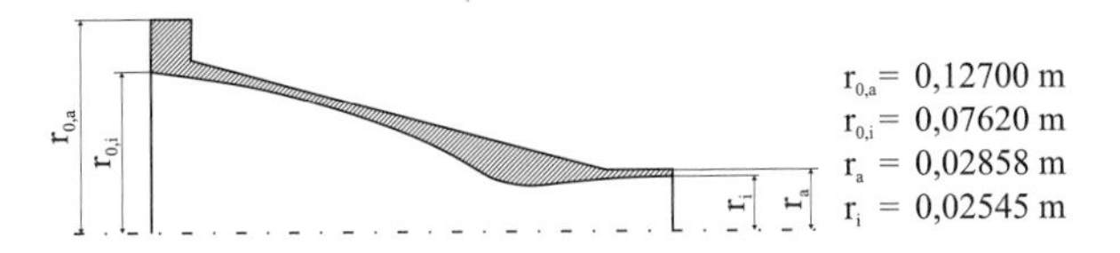

Fig. 1 Sketch of the nozzle creating the supersonic Ma 1.8 rotationally symmetric jet

Table 1 Boundary conditions in the experiments conducted by Panda and Seasholtz [16, 17]

Ambient temperature T_a	297 K
Ambient pressure p_a	98,950 Pa
Total chamber temperature T_0	300 K
Total chamber pressure p_0	564,200 Pa
Static jet temperature T_j	182.4 K
Jet density ρ_j	1.89 kg/m^3
Jet velocity u_j	486 m/s

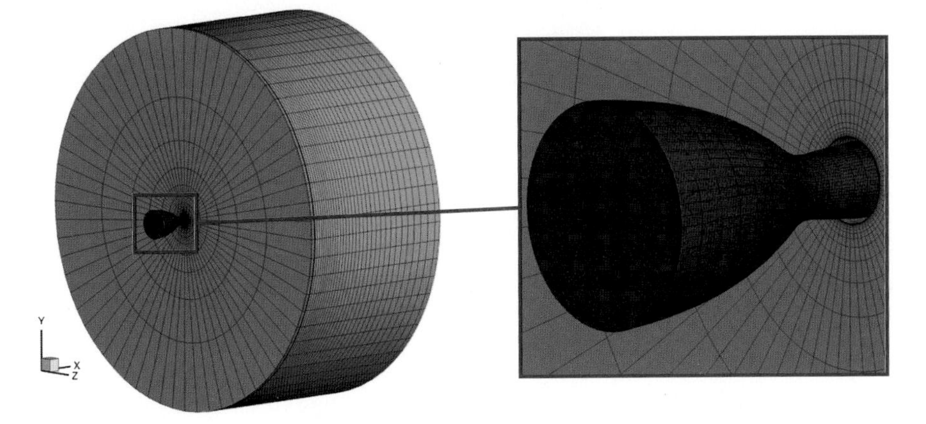

Fig. 2 Representation of the computational domain and the meshes employed for the simulation of the supersonic jet. Please note that for the sake of clarity the number of shown cell volumes is strongly reduced. The meshes for the RANS and the hybrid RANS/LES approach are depicted in *red* and *green*, respectively

boundary the corresponding values from Table 1 are prescribed. Again a small value of 10 m/s is set for the axial velocity in order to stabilize the numerical simulation. Depending on the local Mach number super- and subsonic regions are distinguished at the outflow boundary. In order to minimize the effect of the boundary condition on the jet the radial extent of the domain is quite large. A symmetry boundary condition is set at the radial boundary of the computational domain. Overall the computational domain comprises 7.1 million cell volumes with moderate coarsening in axial direction. For verification purposes another mesh that is strongly refined in axial direction with a total of 12.15 million cell volumes has also been employed. The temporal resolution is chosen in such a way that the CFL number is less than 1 throughout the domain. Converged RANS solutions have been used as initial condition for all DDES computations.

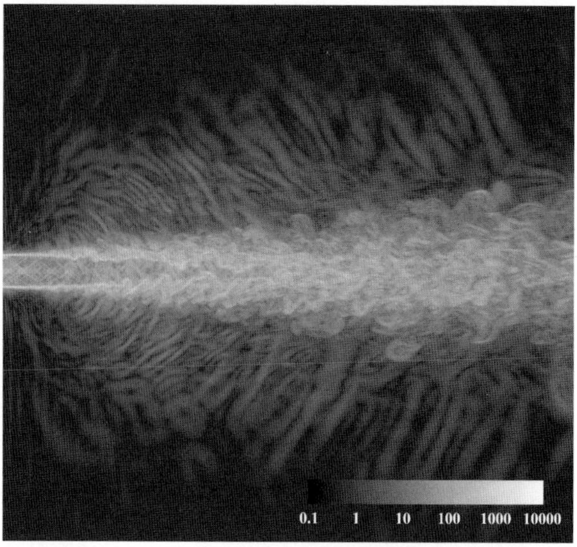

Fig. 3 Numerical Schlieren picture (absolute value of the density gradient) of the supersonic jet in the central (z = 0) cross section

4 Results

Figure 3 depicts the absolute value of the pressure gradient of the jet in the central cross section (z = 0). Directly downstream the nozzle exit the shock train system that results from the adjustment of the jet to ambient conditions is clearly visible. Moreover, the propagation of sonic waves induced by the free jet can be observed.

In order to assure a meaningful comparison between the statistical data of the simulation and the experiment a temporal averaging of the simulation results over at least 10 residence times is performed. Figure 4 shows the comparison of experimental and numerical data on the centerline of the jet. All coordinates are normalized with the inner diameter ($d = 50.9\,\text{mm}$) of the nozzle. In addition to the regular DDES a simulation with added artificial fluctuations within the boundary layer has been investigated. This is to mimic natural fluctuations originating in the boundary layer that can not be reproduced by the employed hybrid RANS/LES approach. The magnitude of the artificial fluctuations is set to approximately 20 % of the axial velocity. It has to be mentioned that due to its simplicity the chosen approach is only able to show qualitative tendencies of how natural fluctuations would affect the simulation result.

Close to the nozzle ($x/D \leq 5$) the simulations show oscillations in the mean values of the axial velocity, temperature and density. These are due to the pressure adaption to ambient conditions (shock train) and are also visible in the experimental data. In this region the agreement between the simulation and the experiment is very satisfying as far as the density and the velocity are concerned. The temperature, however, is overestimated in both simulations. The reduction in density and velocity at around 10 nozzle diameters downstream the nozzle in the experimental data takes

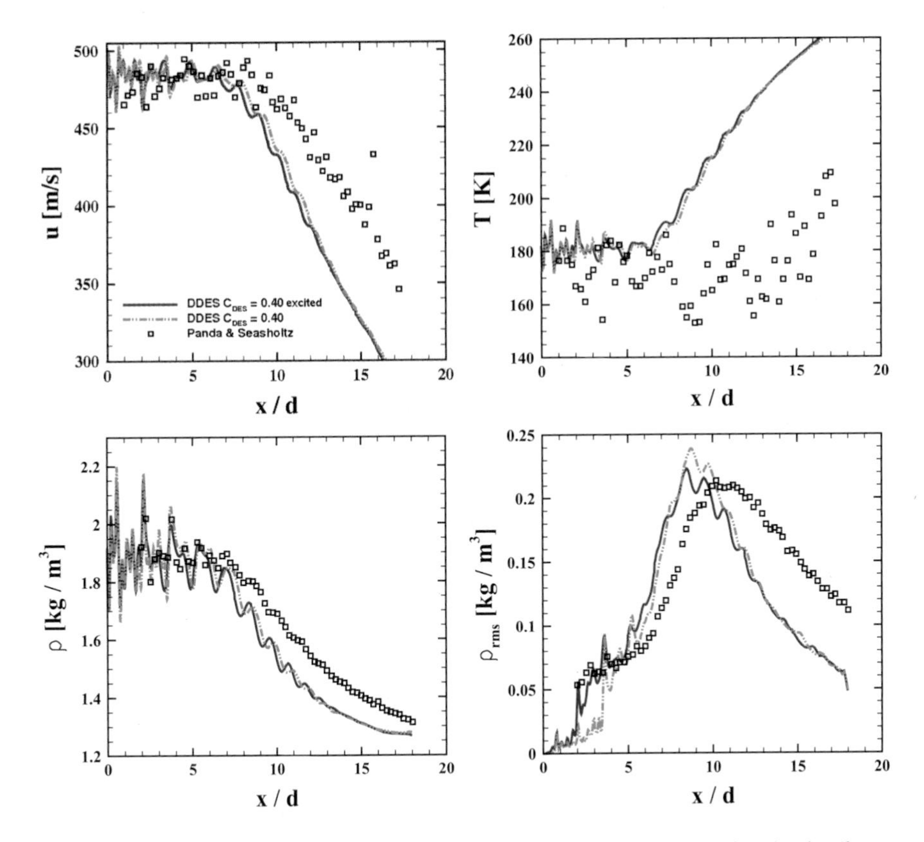

Fig. 4 Axial distribution of axial velocity (*upper left*), temperature (*upper right*), density (*lower left*) and density fluctuations (*lower right*) along the symmetry axis. The regular simulation is depicted in *green* whereas the simulation with added artificial fluctuations in the boundary layer is shown in *red*

place prematurely (approximately 1–2 nozzle diameters) in the simulations. This might be attributed to the more pronounced density fluctuations and therefore elevated turbulent kinetic energy. Consequently the transition region in the simulation is reduced and the simulations show pronounced deviations from the experimental data further downstream. The artificial excitation in the boundary layer leads to a better agreement of the density fluctuations close to the nozzle exit. The influence on the mean quantities, however, is minor.

The comparison of experiment and simulations along radial profiles in Fig. 5 confirms the previous observations. Although the difference between the artificially stimulated and regular simulation is more pronounced in the density fluctuations for $x/d \leq 8$ these differences vanish further downstream and the simulations become virtually identical. However, close to the nozzle exit the agreement of the density fluctuations in the experiment and the artificially excited simulation is much better than with the regular simulation.

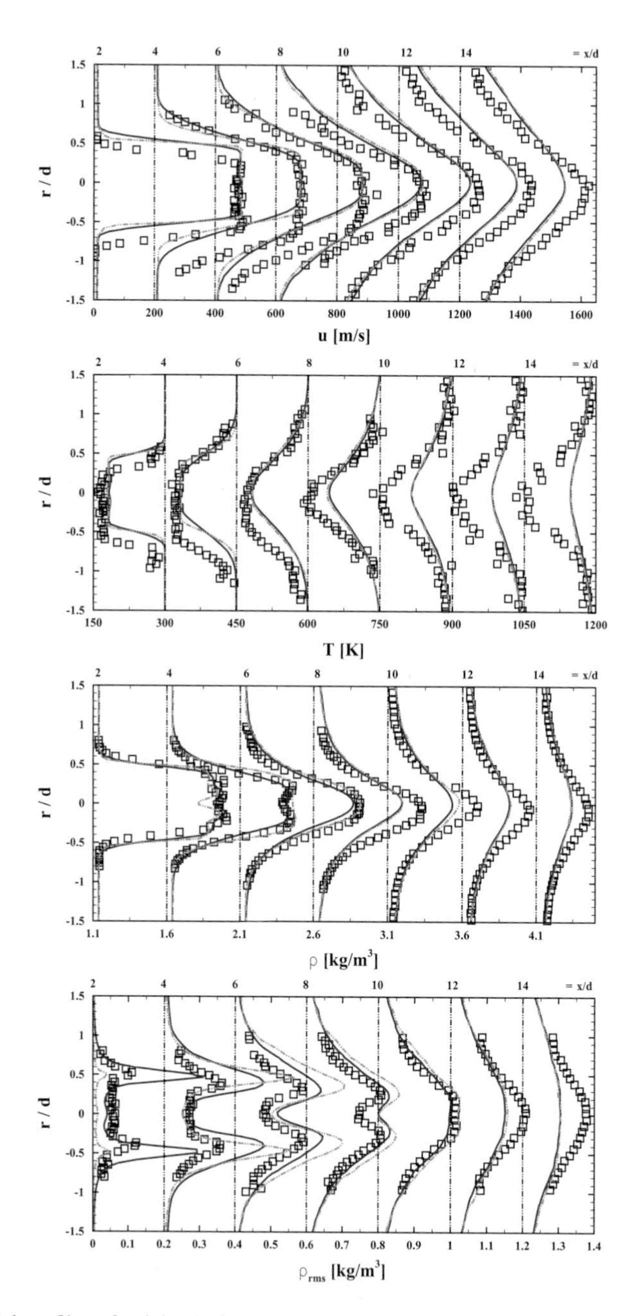

Fig. 5 Radial profiles of axial velocity, temperature, density and density fluctuations at several axial locations. The regular simulation and the simulation with artificial stimulation are depicted in *green* and *red*, respectively

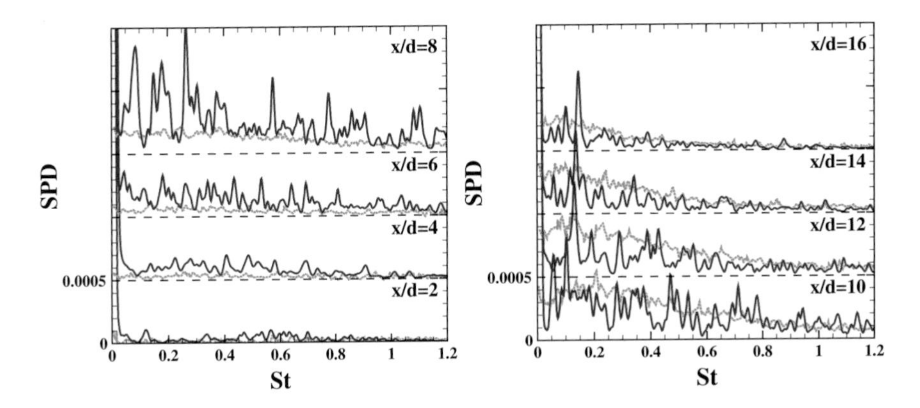

Fig. 6 Experimental and numerical spectra of the density fluctuations over the Strouhal-number St at several axial locations x/d. *Green color* denotes the measured values whereas the simulation is depicted in *red*

The spectra of the density fluctuations at different axial locations were obtained by a Fourier-analysis of the recorded density data. Figure 6 shows the comparison of the numerical (baseline case) with the experimental data in dependence of the Strouhal number at various axial locations. The Strouhal number

$$St = \frac{fd}{u_j} \tag{4}$$

is calculated from the frequency, the nozzle exit diameter d and the jet velocity u_j. In compliance with the previous findings it can be observed that density fluctuations are overestimated for x/d $\leq$ 8 and underestimated further downstream. The qualitative agreement between experiment and simulation, however, is satisfying.

For further verification two more simulations have been carried out. The simulations with $C_{DES} = 0.2$ and the refined mesh (12.15 million cell volumes, $C_{DES} = 0.4$), respectively, are depicted in Fig. 7 together with the regular simulation. However, both variations have a limited effect on the results. Therefore it can be assumed that for the present case the influence of the modeling constant C_{DES} is low. Furthermore it seems evident that a moderate grid refinement is not able to improve the result with the current hybrid RANS/LES approach.

5 Computation Details

All calculations have been performed on the Cray XE6-System (Hermit1) at the HLRS. A very detailed performance analysis of TASCOM3D on this system was performed in 2011 [8]. The major result of this investigation was the importance of the maximum block size because the number of cache misses increases with the

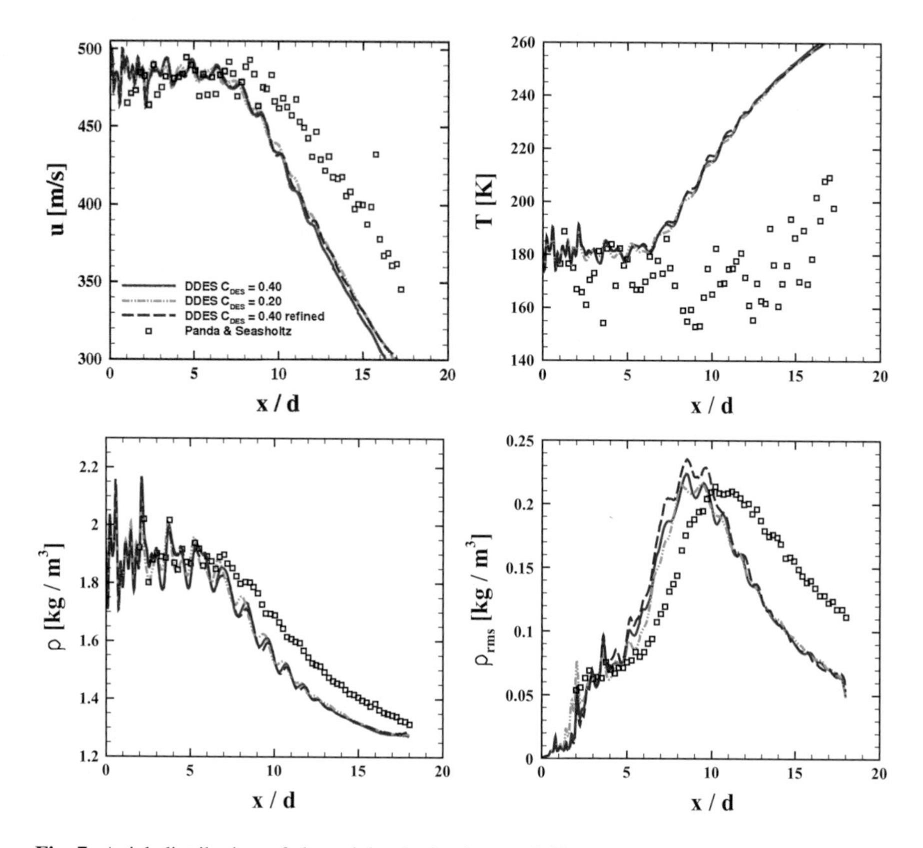

Fig. 7 Axial distribution of the axial velocity (*upper left*), temperature (*upper right*), density (*lower left*) and density fluctuations (*lower right*) along the center line. The measured values are compared with three different simulations: the regular case with $C_{DES} = 0.4$ (*red*), a simulation with $C_{DES} = 0.2$ (*green*) and the simulation with refined mesh (*blue*)

number of cell volumes per CPU. If attention is paid to this aspect good performance values of around 10 % peak performance can be attained. Furthermore it was shown that the strong scaling behavior is nearly ideal (deviation of less than 3 % for 512 CPUs compared to a single CPU). Although no detailed and systematic investigation was performed for the presented test case the observations are in accordance with the previous findings.

Despite the fact that the presented test case deals with a non-reactive flow the computational demand is already very high due to the employed hybrid RANS/LES approach. A total of approximately 45,000 and 75,000 Core-hours was needed for the simulation of the supersonic jet on the baseline and refined mesh, respectively. Compared to this the requirements of the RANS simulation of the nozzle are quite low with only a few CPU hours. In order to reduce the number of cell volumes per processor (to reduce cache misses and due to RAM constraints) the DDES was distributed over 583 (regular mesh) and 989 (refined mesh) CPUs.

6 Conclusion and Outlook

A hybrid RANS/LES simulation of a rotationally symmetric supersonic Ma 1.8 jet has been performed with the DDES approach in the framework of Wilcox' k-ω turbulence model. The results are qualitatively in good agreement with measured values. However, the transition to a fully turbulent flow is predicted too far upstream so that the quantitative agreement becomes worse good downstream of $x/d \approx 8$. The deviation is especially profound for the temperature distribution. However, it has to be kept in mind that the high uncertainty in the temperature measurement was also an issue addressed by the authors [16]. The addition of artificial fluctuations to the boundary layer is able to improve the agreement of the density fluctuations between simulation and experiment close to the nozzle exit but shows no significant effect otherwise. Also a variation of the modeling constant C_{DES} and grid refinement only affect the simulation result insignificantly for the present case.

In the literature some numerical investigations of similar free jet experiments with LES-based methods can be found (e.g. Lo et al. [18], Bodony and Lele [19], Boersma [20]). Interestingly similar phenomena are reported. To the knowledge of the authors only Mendez et al. [21] achieved better agreement with a similar free jet configuration employing a LES of the whole domain including the nozzle on a mesh with 28 million volumes.

For more information about the implemented DDES approach in TASCOM3D and further validation test cases the interested reader is referred to Kindler [15]. Based on these findings the authors are confident in the implemented DDES approach and strive to employ the methodology for ongoing simulations with combustion and more practical configurations.

References

1. P.R. Spalart, Detached Eddy simulation. Annu. Rev. Fluid Mech. **40**, 181–202 (2009)
2. P.R. Spalart, W.H. Jou, M. Strelets, S.R. Allmaras, Comments on the feasibility of LES for wings, and on a hybrid RANS/LES approach, in *Advances in DNS/LES, Proceedings of the First AFOSR International Conference on DNS/LES*, 1997
3. P. Sagaut, S. Deck, M. Terracol, *Multiscale and Multiresolution Approaches in Turbulence* (Imperial College Press, London, 2006)
4. F.R. Menter, Y. Egorov, Scale-adaptive simulation method for unsteady turbulent flow predictions. Part 1: theory and model description. Flow, Turbul. Combust. **85**(1), 113–138 (2010)
5. D.C. Wilcox, Formulation of the k-ω turbulence model revisited. AIAA J. **46**, 2823–2838 (2008)
6. A. Jameson, S. Yoon, Lower-upper implicit scheme with multiple grids for the euler equations. AIAA J. **25**, 929–937 (1987)
7. J.S. Shuen, Upwind differencing and LU factorization for chemical non-equilibrium Navier-Stokes equations. J. Comput. Phys. **99**, 233–250 (1992)
8. M. Kindler, P. Gerlinger, M. Aigner, Delayed detached Eddy simulations of compressible turbulent mixing layer and detailed performance analysis of scientific in-house code TASCOM3D, in *High Performance Computing in Science and Engineering* (Springer, Berlin, 2011)

9. B. Van Leer, Towards the ultimate conservative difference scheme II: monotonicity and conservation combined in a second order scheme. J. Comput. Phys. **14**, 361–370 (1974)
10. A. Harten, High resolution schemes for hyperbolic conservation laws. J. Comput. Phys. **49**, 357–393 (1983)
11. S.-H. Yoon, C. Kim, K.-H. Kim, Multi-dimensional limiting process for three-dimensional flow physics analyses. J. Comput. Phys. **227**, 6001–6043 (2008)
12. P. Gerlinger, High-order multi-dimensional limiting for turbulent flows and combustion. J. Comput. Phys. **231**(5), 2199–2228 (2012)
13. M.-S. Liou, A sequel to AUSM, Part II: AUSM$^+$-up for all speeds. J. Comput. Phys. **214**, 137–170 (2006)
14. A. Travin, M. Shur, M. Strelets, P. Spalart, Physical and numerical upgrades in the detached-Eddy simulation of complex turbulent flows. Adv. LES Complex Flows **65**(5), 239–254 (2004)
15. M. Kindler, Instationäre Verbrennungssimulation in kompressiblen Strömungen, Ph.D. Thesis, Universität Stuttgart, to be published
16. J. Panda, R.G. Seasholtz, Velocity and temperature measurement in supersonic free jets using spectrally resolved Rayleigh scattering. AIAA Paper 99-0296, 1999
17. J. Panda, R.G. Seasholtz, Density fluctuation measurement in supersonic fully expanded jets using Rayleigh scattering. AIAA Paper 99-1870, 1999
18. S. Lo, G. Blaisdell, A. Lyrintzis, Numerical simulation of supersonic jet flows and their noise. AIAA Paper 2008-2970, 2008
19. D.J. Bodony, S.K. Lele, On using large-Eddy simulation for the prediction of noise from cold and heated turbulent jets. Phys. Fluids **17**(8) (2005)
20. B.J. Boersma, Large Eddy simulation of the sound field of a round turbulent jet. Theor. Comput. Fluid Dyn. **19**, 161–170 (2005)
21. S. Mendez, M. Shoeybi, A. Sharma, F.E. Ham, S.K. Lele, P. Moin, Large-Eddy simulations of perfectly expanded supersonic jets using an unstructured solver. AIAA J. **50**, 1103–1118 (2012)

DNS of Lean Premixed Flames

Jordan A. Denev and Henning Bockhorn

Abstract The paper presents results from the Direct Numerical Simulation of lean premixed hydrogen and methane flames. A new turbulence-generating technique, based on step-wise body forcing in physical space is introduced. It is stable, easy to implement and straight forward to use in the case of parallelization strategy based on domain decomposition.

Flame-vortex interactions are also studied, both in two and three dimensions. A non-stationary method for straining the flame front, based on artificially generated Couette-like flow fields, is applied. It results pure strain without curvature. For the cases studied, much larger strain rate is achieved than that for the extinction limit in the case of twin opposed stationary flames. Despite the quite high strain rates achieved, no local extinction of the flame front has been achieved for the present simulations.

All computations are carried out on the new CRAY-XE6 supercomputer at the High Performance Computing Center Stuttgart (HLRS). Numerical issues and results from a three-dimensional scaling test are presented and discussed.

1 Introduction

In the present work results from Direct Numerical Simulation (DNS) of lean premixed flames are presented. In premixed combustion it is of general interest to estimate the burning velocity and the mechanisms by which turbulence can influence it. Therefore, it is important to first generate turbulence flow fields with well-controlled properties. As discussed later, while a variety of methods exist in the

---

J.A. Denev (✉) · H. Bockhorn
Engler-Bunte Institute, Combustion Division, Karlsruhe Institute of Technology Engler-Bunte Ring 1, D-76131 Karlsruhe, Germany,
e-mail: jordan.denev@kit.edu; henning.bockhorn@kit.edu

W.E. Nagel et al. (eds.), *High Performance Computing in Science and Engineering '13*, DOI 10.1007/978-3-319-02165-2_18,

literature, there is still a potential for further improvements of the way turbulence is generated numerically. Here a new method for non-decaying turbulence is proposed which is suitable for the study of premixed combustion in DNS. With the proposed method, the turbulent eddies develop within short physical times, they are physically correct (obey all transport equations and continuity) and their location and size are easy to control algorithmically. There is no decay of the eddies between a domain boundary and the flame front, because of the fact that the eddies are generated directly at the location where they are required, i.e. near the flame front. The application of the method for supercomputing is straightforward because there is no need in special efforts for its parallelization. The paper presents and comments results obtained with this method in Sect. 3.

Further, effects of pure, non-stationary strain on the flame behaviour are studied. It is known from the literature [16] that turbulence changes the flame area through the effects of curvature and strain. Both curvature and strain are parts of stretch, but the question that emerges is whether both curvature and strain affect the burning velocity to the same extend. Law et al. [12] analyzed results from twin methane flames in a symmetrical counterflow configuration—a stationary flame configuration with no curvature of the flame front. They argue that the "scalar structure of the flame, and thereby the flame thickness, are insensitive to strain rate variations for these purely strained flames, and these flames cannot be extinguished by straining alone". For the same configuration of hydrogen flames the conclusion is that "the sensibility is noticeable, but still small as compared to the extent of the strain rate variation", [20]. On the other hand, studying the correlation between flame speed and stretch in premixed methane/air flames, the authors of [6] point out that curvature has quite a strong influence. They conclude that "exceedingly large negative values of stretch can be obtained solely through curvature effects, which give rise to an overall nonlinear correlation of the flame speed with stretch". First own results [13] show also that, although both strain and curvature contribute to stretch, their effect on the flame extinction is quite different. In order to study this in a consistent way, a new setup is introduced in Sect. 4. In this setup the velocity field is similar to the one in the Couette flow. The effect on the flame front is non-stationary and allows to achieve very high strain rates which are several times higher than those, at which extinction appears in the setup of twin opposed stationary flames.

The next part of the present work reports simulation results with large chemical mechanisms. The target of this part of the study is the more precise modelling of fuels utilizing chemical mechanisms with a large number of species and chemical reactions. Results from the corresponding simulations are presented and discussed.

Finally, a scaling test and numerical issues of the computations carried out on the new CRAY-XE6 supercomputer at the High Performance Computing Center Stuttgart (HLRS) are presented and discussed.

2 Numerical Method

The research code PARCOMB-3D [21] for fully compressible three-dimensional flows is applied. The code is specialized for the detailed solution of non-stationary combustion phenomena. It solves the coupled system of the Navier-Stokes, species- and energy conservation equations. The simulation code uses 6th order central differencing in space, along with 3rd order differencing on the boundaries on carthesian numerical grids. The time advancement is explicit utilizing a 4th order Runge-Kutta scheme. The time-step is controlled through the Courant-Friedrichs-Lewy (CFL) number, supplied by the user. Usually CFL-numbers around unity are feasible, however, when the mechanism for the chemical reactions utilized is stiff, the CFL-value needs to be reduced considerably, in some cases even more than 100 times.

In the present work the chemical mechanism of Miller et al. [14] with 9 species and 38 reaction steps is used for the hydrogen flames. For the methane combustion the reduced chemical kinetics consists of 16 species and 50 reaction steps [4, 19]. Both mechanisms are numerically very stable and allow CFL-numbers around unity for guiding the explicit time-stepping in the simulations. The species transport data are simulated in detail (the so-called "complete multicomponent diffusion treatment" is applied) and the Soret effect (diffusion due to the presence of temperature gradients) is also taken into account for all hydrogen flames. Radiative heat transfer is not accounted for in the simulations.

The Navier-Stokes characteristic boundary conditions [3, 15] are used at the inlet and outlet boundaries which take into account also detailed chemistry equations. Other types of boundary conditions used in the present work are symmetrical or periodical. Single votexes are generated as initial conditions by means of two-dimensional streamline functions which ensure that the resulting flow fields are divergence-free.

PARCOMB-3D is a code written in FORTRAN90. It uses domain decomposition which is carried out automatically from specialized routines. For the parallelization MPI 2.0 is used. All simulations were carried out on the CRAY-XE6 at the High Performance Computing Center Stuttgart (HLRS).

3 A New Method for the Generation of Non-decaying Turbulent Flow Fields

In the present work, the interaction between the flame front and turbulent eddies is studied. Generally, in combustion devices, highly turbulent conditions exist and hence, the turbulent vortexes span a wide range of length scales—from very large, geometrically defined, to very small, in the range of microns [9]. In DNS, even with high-performance computing, only a very small portion of the physical domain is simulated—typically of the order of 1 cm^3 or less. This length is often comparable with the integral length scale reported in the experiments [23]. In the simulations, from one side, it is difficult to create turbulent eddies which are larger than the computational domain and, from the other side, the appropriate experimental data are generally missing, therefore special methods for generation of turbulent eddies need to be utilized. One common approach is to apply decaying turbulence conditions, which means that the turbulence is prescribed only as an initial condition [7, 18]. Another method consists of prescribing well-correlated turbulent perturbations with controlled spectral features at the inflow boundary of the computational domain. This is a common approach and therefore a great variety of its variants exists in the literature [10, 11], including some in-house applied [8] or even in-house developed variants of it [5]. However, one drawback of the above methods is that the turbulence generated this way decays (though to a limited extend) between the inflow boundary and the flame front.

In the present work, a new method for generating non-decaying turbulent fields is presented. The method is generally applicable and can be used to specify turbulent fields with any desired length-scale and spectra. However, for the specific purposes of the present work, the new method was used to specify a row of turbulent vortexes with an in-advance prescribed length-scale.

The main idea of the method is that vortexes are generated through the abrupt change of a force. The force, which creates the vortexes, is a body force and the term that corresponds to it is present in the momentum equations (e.g. the term for the gravity force). The force could be constant, or, it can change its intensity throughout the computational domain. In the present method this body force is set to zero in certain areas (lets call them "islands") of this domain. Through this, the vortexes appear near the "force-islands" which is similar to the way they appear near rigid walls in the flow field. The length scale, time of emergence and location of the vortexes can be controlled simply by controlling the location, size and duration of the "force-islands" and of the strength and direction of the body force in the computational domain. The "islands" are just in terms of the imposed body force, there is no direct affect of the islands on the flow (on velocity, pressure, energy and species) in any other way.

The proposed method has several benefits. First, it does not "disturb" any of the partial differential equations of the system solved, including continuity and momentum equations. Therefore, all the eddies generated, are "divergence-free". Another advantage is that no special treatment at the flow boundaries is required. The vortexes are generated in a controlled way at the location where they are needed (usually close to the flame front) and hence they do not decay prior to interacting with the flame front. The method is also very economical computationally—it requires few additional lines of code only; this code is compact and is programmed in the routine where the body forces are treated. And there is no need in additional efforts for its parallelization, neither there is a need for special exchange of variables between processors—the algorithm for body forces is already implemented for parallel computations in the code.

Results from this new approach are shown in Fig. 1. In this figure, all vortexes are generated by "islands" of a square shape. The top part of the figure shows that

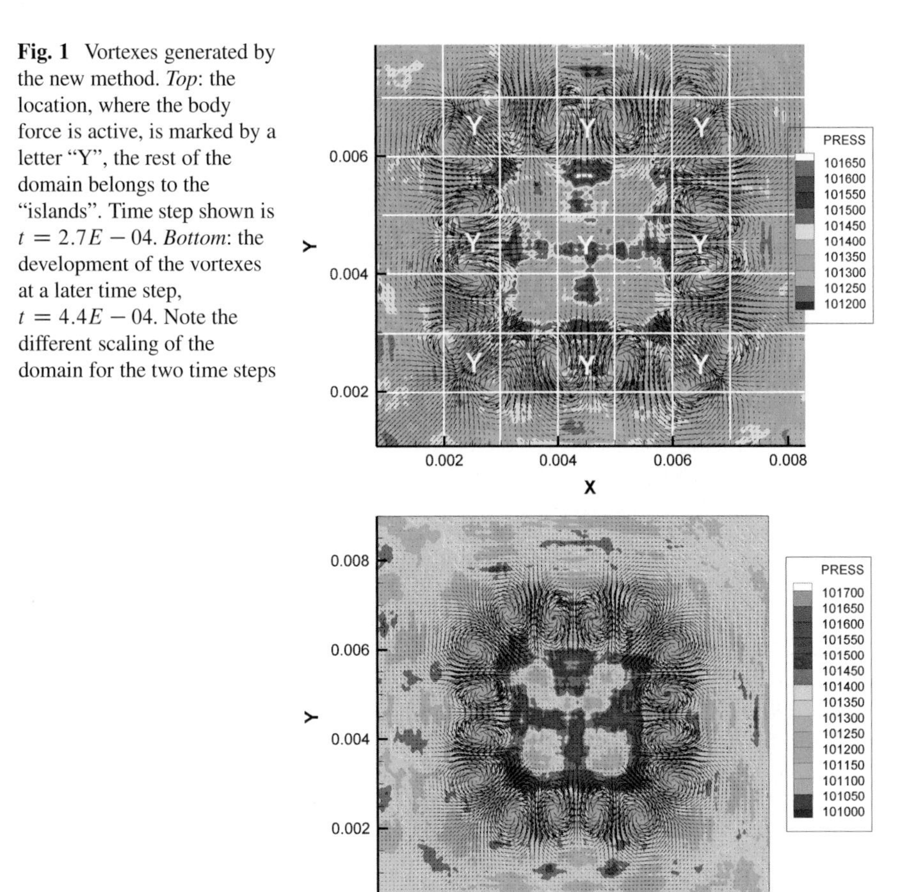

Fig. 1 Vortexes generated by the new method. *Top*: the location, where the body force is active, is marked by a letter "Y", the rest of the domain belongs to the "islands". Time step shown is $t = 2.7E - 04$. *Bottom*: the development of the vortexes at a later time step, $t = 4.4E - 04$. Note the different scaling of the domain for the two time steps

centers of the vortexes are located at the side lines of the "islands"—either in the middle of these lines, or at their corners. When the time advances, which is shown in the bottom part of the figure, the vortexes get more stronger, and the pressure range in the computational domain increases (cf. range of pressure fields in the legends). The vortexes are arranged in pairs, i.e. each "island" produces one vortex pair. The pressure minima in the figure indicate the location of the vortexes. The vortex size equals the size of the "islands" which was set to 1 [mm]. This example shows the vortex generation in a cold flow.

In the second example shown in Fig. 2, the interaction of the vortexes, generated with the new method, and a flame front is presented. In the figure, two different flow examples of lean hydrogen flames are given. The "islands" in both examples have a square shape and are oriented like in the previous example. The size of these islands is the same as the distance between them and the resulting vortexes also have the same size. Depending on the strength of the force, vortices can build up

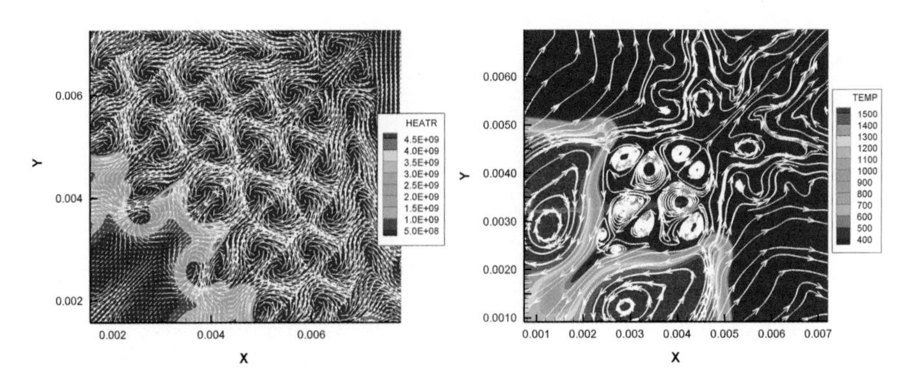

Fig. 2 Vortexes generated by the new method. *Left*: vectors showing regular vortexes in a row with a length scale of 0.5 mm. *Colours* present the heat release in the flame front, [J/(m^3 s)]. *Right*: streamlines showing vortexes with larger size in the region of burned gases. *Colours* present the temperature [K]

quite quickly. In the example shown in the left part of the figure, the strength of the force changes between 50 and 90 [kN/m^3] throughout the computational domain. As a consequence, for a time of 2.2E−3 [s], the flame front moved a bit less than its thickness until the vortexes were completely build up.

It is a known property of the flame that it can "destroy" the small-scale turbulent vortexes due to the increased viscosity of the hot gases. This property is clearly seen in both time-steps presented in Fig. 2: the region with the burned gases, which is at the lower left corner in both pictures, contains only larger vortexes, despite that the vortex generation method applied was the same for the vortexes in the burnt and in the fresh gases. Therefore, it may be concluded that the method proposed works physically correct (at least qualitatively) also for the case of vortexes interacting with the flame front.

4 Separating the Effects of Strain and Curvature

In theory of turbulent flames, the effects of stretch on the flame speed are important and investigated for a long time, see e.g. [2]. The stretch κ, [s^{-1}] is defined with the rate of change of the area A of the flame front:

$$\kappa = 1/A * dA/dt \tag{1}$$

The stretch defined with the above equation is treated on a global scale, e.g. in the whole computational domain or for the whole experimental image. However, flame stretch can also be calculated on a local basis, for each single point of the flame front. The mathematical derivation of this is given in [16] and has two additive terms which separate the stretch in contribution of curvature and of strain:

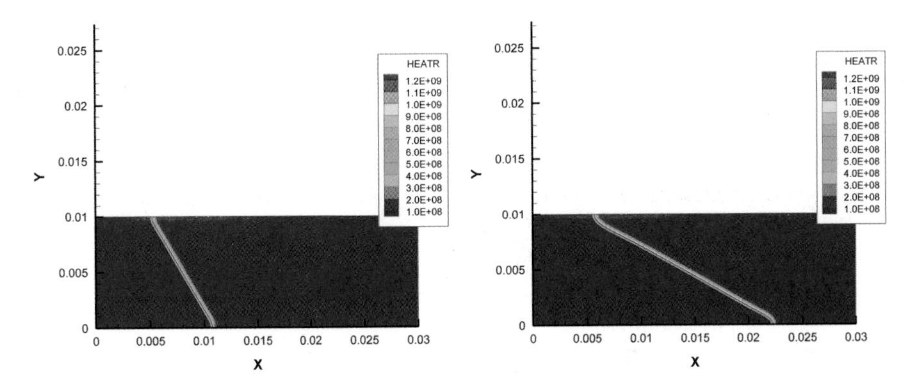

Fig. 3 The shape of the flame front at two time instances for the case with a maximum velocity of 5 [m/s]. The velocity profile is linearly increasing from the top to the bottom. At the beginning of the simulation the flame is vertical. *Left*: time step $t = 0.0012$ s. *Right*: time step $t = 0.0036$ s

$$\kappa = a_T + S_d \nabla . n \tag{2}$$

In the above equation, a_T is the tangential strain rate (in the flame tangent plane), S_d is the speed with which the flame propagates, called displacement speed, and $\nabla .n$ is the curvature with "n" being the normal vector of the flame front pointing towards the fresh gases.

The present investigation is part of a more general research attempting to clarify the separate influence of strain and curvature on the flame speed during transient combustion processes. Here, only the part with the effect of pure strain is considered. For this purpose, a special case of flame distribution was considered, namely, the straight flame distribution in a Couette-like flow field. In this configuration, one end of the flame was not moving, while the other one was strained with a constant speed—the maximum speed in the Couette-like flow. The fluid flow changes linearly as in a real Couette flow and the flame front has the shape of a straight line as shown in Fig. 3. The difference between the present flow field (called Couette-like field) and the real Couette flow is that no rigid walls exist here, instead, symmetry boundary conditions are imposed. The result is that all the points along the flame front experience the same strain rate a_T without the presence of curvature (except few points close to the symmetry boundaries which experience also the effect of curvature).

Depending on the maximum speed of the Couette-like flow, imposed as an initial condition in the computational domain, different strain rates could be achieved. This was varied for both the methane and the hydrogen flames at lean conditions. Several simulations have been carried out with maximum flow velocities equal to 5, 10, 100 or 150 [m/s]. The equivalence ratio for the hydrogen flames was $\Phi = 0.33$ and $\Phi = 0.5$ and for the methane flames—$\Phi = 0.55$ and $\Phi = 0.67$. Typical domain dimensions were 6×1 cm and the number of grid points was typically 1440×240.

In all cases the simulation was carried out until the flame reached the outflow boundary which is located at the r.h.s. of the computational domains shown in Fig. 3.

Results show that with this setup of the simulations, where the velocity field at the inflow remains constant, the strain rate (which is equal to the stretch in the considered problem) increases very quickly at the beginning. After this, at later times, the strain rate decreases with time but it still remains quite large. For the simulations with the hydrogen flame (150 [m/s], $\Phi = 0.5$) the maximum value reached was $\kappa = a_T = 15{,}000$ [s^{-1}]. Until the final time simulated $t = 0.00035$ [s], the stretch rate was continuously higher than 3,000 [s^{-1}], but the flame did not extinguish. Very similar stretch rates were obtained also for the methane flame, which also did not extinguished during the simulation. For comparison, [17] reports that the quenching limit for a stationary twin-flame counterflow configuration of methane at stoichiometric conditions occurs at stretch rates $\kappa = 2{,}280$ [s^{-1}]. Thus, the present flames—for the speed of 150 [m/s]—have strain (stretch) values which are higher than the reported quenching limit during the whole simulation time. Because the simulated physical time might be not long enough to reach the state of extinction, future simulations are going to clarify the question whether the flames, if exposed to high instationary stretch for a longer time interval, will extinguish or not.

As it can be seen also from the comparison of the two time steps in Fig. 3, the heat release is not affected by the presence of strain. This corresponds well with the conclusions from the work of Law et al. [12], cited in the introduction.

5 DNS and Chemical Mechanisms with a Large Number of Species: Resources

Modern combustion devices, e.g. gas turbines, support the use of different kinds of fuels. At the same time new biofuels from different sources arise on the market, the properties of which need to be investigated. For this purpose a large variety of chemical mechanisms are developed, containing a large number of chemical species and reactions. Even when appropriately reduced, from the viewpoint of DNS, such mechanisms remain quite resource-intensive and this can hinder the use of the mechanisms. Some tests for utilizing different fuels with a large number of species, were investigated here.

The first mechanism tested was that of [22] with 50 species and 488 elementary reactions. It is designed to focus on conditions relevant to flames, high temperature ignition and detonations for fuels with $H2/O2/C1 - C3$-chemistry. The mechanism was used for the combustion of propane at lean conditions, $\Phi = 0.67$. The spherically expanding flame was simulated on a two-dimensional numerical grid with 1024×1024 points. The domain was 2×2 cm. The computations consisted of 7 runs (each one—for 24 h and with 3,072 processors).

The main conclusions from the simulation are as follows:

- in combination with the explicit time-stepping of the code PARCOMB-3D, the mechanism turned out to be stiff, which required very small time steps in order to prevent divergence.
- the typical CFL-number, which governs the time-step, was $CFL = 0.007$. Such a low value (compared to similar simulations with hydrogen, where $CFL = 1.0$) increases unacceptably the resources for the computations; however, the decrease was necessary for the stability of the algorithm. The resulting time-step was $\Delta t = 1.0E - 10$.
- this small time-step made the whole simulations to advance quite slow and practically prevented this part of the work, eventhough it was conducted in two dimensions only. The simulations were stopped after $3.8E + 6$ iterations and the physical time reached was only 0.0006 [s].

The second chemical mechanism tested was the mechanism of [1], containing 101 species and 1,021 elementary reactions. It was tested on a stoichiometric methane flame, on the problem of a freely propagating planar laminar flame (one dimension). The CFL-number had a value of $CFL = 0.03$ and the typical time-step was $\Delta t = 1.1E - 9$. Although this mechanism performed better in a numerical sense than the previous one, the conclusions here are similar to those for the first chemical mechanism. Obviously, some more research is needed to make the chemical mechanisms with a large number of species more robust for DNS-codes with explicit time stepping.

6 Supercomputing Issues and Code Scaling for a Three-Dimensional Test Problem

A sequence of tests for the weak scaling of the parallel algorithm in the code PARCOMB-3D was carried out. The tests correspond to a hydrogen flame which interacts with a sinusoidal three-dimensional vortex. The fuel-equivalence ratio was $\Phi = 0.50$. The domain size was $1\times0.5\times0.5$ cm and the grid points were $301\times151\times151$. The flame displacement speed was compensated by an inflow boundary condition with a constant velocity. The opposite boundary had outflow conditions while all other boundary conditions were periodic.

The simulations have been made with a minimum of 16 processors (cores) on the CRAY-XE6 and the number of processors was increased accordingly, see Fig. 4. Each computation had a duration of 2 h wall-clock time. There was no data writing during the tests—the purpose was to count the number of time-steps advanced under the assumption that each time-step requires approximately the same CPU-time. For the job-submission the options *#PBS−l mppnppn* = 32 were applied. The compiler environment was set to *PrgEnv-pgi* and the compiler options used were $-c - r8 - O4 - Mipa = fast, inline$.

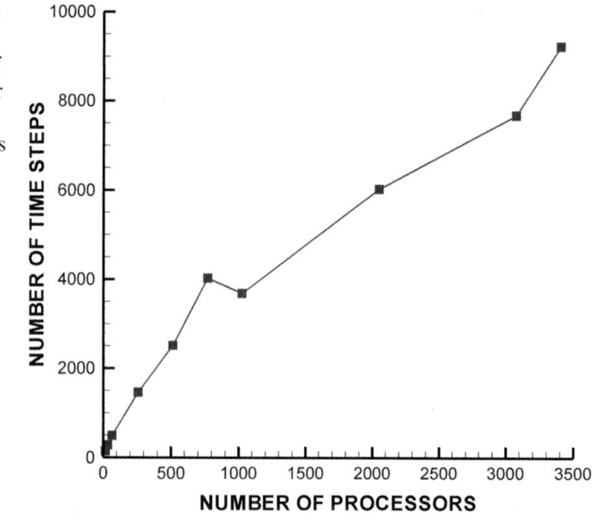

Fig. 4 Number of time steps computed for 2 h wall-clock time by a different number of processors. For this particular test case only, the numerical efficiency of the computations is greatly influenced by the increased number of dummy cells at the block boundaries for all cases with more than 1,000 processors

The results show some irregularity in the scaling, therefore the runs for 768 and 1,024 processors were submitted twice, but the result did not change. The reason was found in the number of dummy cells which are additionally set around each numerical block (required by the domain decomposition algorithm). The numerical values of all variables in these cells are exchanged during each time step, which leads to a corresponding increase of the number of sent and received MPI-messages. In the given example, the number of dummy cells increases by 60 % between 768 and 1,024 processors and it remains very high in the consequent runs with a large number of processors. The unfavourable number of dummy cells appears when only a small number of grid nodes per processor remain (in the particular case—less than 6,800 grid nodes per processor). With a larger number of grid nodes per processor the relative weight of the dummy cells decreases. There is obviously an optimum of processor numbers for each computational grid which needs to be determined individually for each case simulated. The knowledge gained from the above scaling-test simulations was then used to increase the efficiency of all consequent computations.

Other experiences from the use of the CRAY-XE6 are itemized here:

- When in the header of the makefile the option *FF_FLAGS* = *−mcmodel* = *medium* is included, the computations run much faster, and hence this feature was permanently included for code compilations with the PGI compiler environment.
- Saving of intermediate results was made utilizing special libraries for the TECPLOT-software which were precompiled on an external unix environment for a sequential (non-parallel) use. The use of these libraries is acceptable for small two-dimensional problems, but the sequential writing turns out to be very inefficient for large files in the case of three-dimensional computations. The experience shows that the time for writing a 16 Gb result-file in TECPLOT format

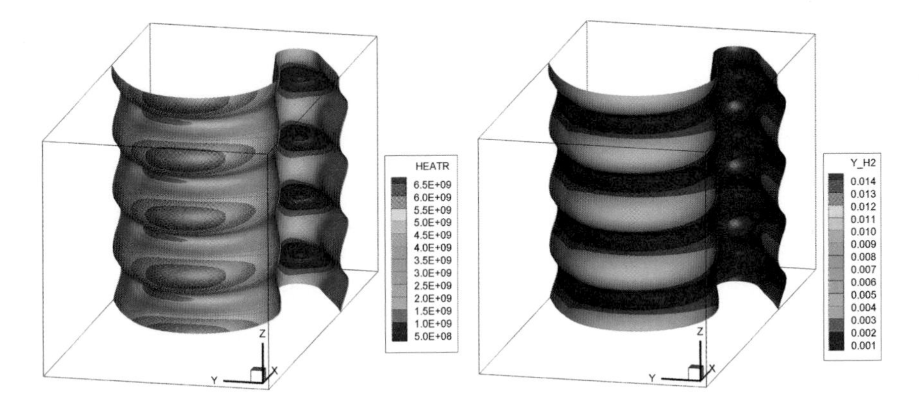

Fig. 5 Three-dimensional isosurface of the O_2 mass fraction of the sinusoidal three-dimensional flame front. The isosurface is coloured by the heat release, [J/(m^3 s)] (*left*) and by the mass fraction of the fuel, H_2 (*right*)

takes more than 20 min, which completely deteriorates the parallel computations. At the same time, even in sequential writing mode, the standard binary-writing routine of the code PARCOMB-3D is sufficiently efficient and completes the task for saving the results within one and a half minutes. Therefore, it was decided that the creation of TECPLOT-files is left for an additional postprocessing step and that all three-dimensional results are saved only in the standard binary format.

- For the particular test case in this section, the size of the TECPLOT-file for the postprocessing is 1.6 Gb. The file contains 30 three-dimensional variables. The postprocessing is made on linux computers with 8 Gb RAM.

Three-dimensional results for the simulation, for which the above scaling test is reported, are given in Fig. 5. The figure presents the interaction of a plane flame front with a three-dimensional sinusoidal vortex. The isosurface of O_2 with a mass fraction value equal to 0.142 is shown which corresponds to our definition of the location of the flame front. The viewpoint in Fig. 5 is from the side of the fresh gases and the burnt gases are behind the surface. The surfaces which have a positive curvature towards the fresh gases (i.e. surfaces that are convex toward the fresh gases) have a high value of heat release, while concave (with respect to the fresh gases) surfaces have a heat release which is approximately an order of magnitude smaller. The reason for this lies in the local mass fraction of the fuel shown at the right hand side of the figure. It is seen that the regions of small heat release are quite lean and that the lean locations are those which are concave towards the fresh gases. In these locations the diffusion of fuel towards the burnt gases is quite quick and the flame locations remain with a relatively small amount of fuel. This in turn slows down the chemical reaction in such flame regions.

7 Conclusions

The present work reports results from the Direct Numerical Simulation of lean premixed hydrogen and methane flames in two and three dimensions. The main issues can be summarized as follows:

- A new turbulence-generating technique, based on the abrupt change of a body force in physical space is introduced. It is stable, easy to implement and straight forward to use in the case of parallelization strategy based on domain decomposition. It is suitable for both two-dimensional and three-dimensional simulations.
- Flame-vortex interactions are also studied with the aim to clarify the separate effects of strain and curvature. A non-stationary method for straining the flame front, based on artificially generated Couette-like flow fields, is applied. It results pure strain of the flame front without the presence of a curvature. The strain rate achieved in this flow is much larger than the strain rate at the extinction limit in the case of twin opposed stationary flames. Despite this, no extinction was yet achieved for flows with a maximum speed of up to 150 [m/s].
- Large chemical mechanisms of up to 1,021 elementary reactions have been simulated. While the code is capable of handling this increased complexity, a main obstacle appears the stiffness of the mechanisms when related to the explicit time-stepping algorithm. This produced unacceptably low time-steps which in turn were necessary to guarantee the stability of the numerical algorithm.
- Numerical issues and results from a scaling test on the CRAY-XE6 super-computer at the High Performance Computing Center Stuttgart (HLRS) are presented and discussed. In connection with the parallelization strategy of the code PARCOMB-3D it turned out that there is an optimum number of processors which needs to be determined individually for each case simulated. The reason is in the sudden increase of the additional (dummy) cells, required for the domain decomposition in the case when the number of grid nodes per processor are below approx. 7,000.

Acknowledgements The present work was enabled through the funding of the DFG collaborative research center SFB606 "Non-stationary combustion: transport phenomena, chemical reactions and technical systems", subproject $B8$. The simulations were performed on the national super computer CRAY-XE6 at the High Performance Computing Center Stuttgart (HLRS) under the grant with acronym "DNSPREM" which is highly appreciated.

References

1. J. Appel, H. Bockhorn, M. Frenklach, Kinetic modeling of soot formation with detailed chemistry and physics: laminar premixed flames of c2 hydrocarbons. Combust. Flame **121**, 122–136 (2000)

2. K.T. Aung, M.I. Hassan, G.M. Faeth, Flame stretch interactions of laminar premixed hydrogen/air flames at normal temperature and pressure. Combust. Flame **109**, 1–24 (1997)
3. M. Baum, T. Poinsot, D. Thevenin, Accurate boundary conditions for multicomponent reactive flows. J. Comput. Phys **116**, 247–261 (1994)
4. R. Bilger, M. Esler, S. Starner, On reduced mechanisms for methane-air combustion, in *Lecture Notes in Physics*, vol. 384 (Springer, Berlin, 1991), pp. 86–110
5. H. Bockhorn, N. Zarzalis, M. Lecanu, *DFG Collaborative Research Centre 606, Non-stationary Combustion: Transport Phenomena, Chemical Reactions, Technical Systems.* DFG Funding Proposal 1.1.2009–31.12.2012, Subproject B8, 2008
6. J.H. Chen, H.G. Im, Correlations of flame speed with stretch in turbulent premixed methane/air flames, in *Twenty-Seventh Symposium (International) on Combustion*, The Combustion Institute, 1998, pp. 819–826
7. J.H. Chen, H.G. Im, Stretch effects on the burning velocity of turbulent premixed hydrogen/air flames. Proc. Combust. Inst. **28**, 211–218 (2000)
8. J.A. Denev, J. Fröhlich, H. Bockhorn, Large eddy simulation of a swirling transverse jet into a crossflow with investigation of scalar transport. Phys. Fluids **21**, 015101 (2009)
9. E. Hassel, N. Kornev, V. Zhdanov, A. Chorny, M. Walter, Analysis of mixing processes in jet mixers using les under consideration of heat transfer and chemical reaction, in *Springer Series on Heat and Mass Transfer (Volume Title: Micro and Macro Mixing)*, ed. by H. Bockhorn et al. (Springer, Berlin, 2010), pp. 165–184
10. M. Klein, A. Sadiki, J. Janicka, A digital filter based generation of inflow data for spatially developping direct numerical of large eddy simulation. J. Comput. Phys. **186**, 652–665 (2003)
11. N. Kornev, E Hassel, Synthesis of homogeneous anisotropic divergence-free turbulent fields with prescribed second-order statistics by vortex dipoles. Phys. Fluids **19**(6), 068101 (1)–068101(4) (2007)
12. C.K. Law, C.J. Sung, G. Yu, R.L. Axelbaum, On the structural sensitivity of purely strained planar premixed flames to strain rate variations. Combust. Flame **98**, 139–154 (1994)
13. R. Lepper, Analyse der interaktion von turbulenzwirbeln und vorgemischten flammenfronten mittels direkter numerischer simulationen. *Diplomarbeit KIT, Engler-Bunte-Institut Bereich Verbrennungstechnik, Prof. Dr.-Ing. H. Bockhorn, Ord. und Institut fr Thermische Strmungsmaschinen Prof. Dr.-Ing. H.-J. Bauer, Ord.*, p. 120, 2012
14. J.A. Miller, R.E. Mitchell, M.D. Smooke, R.J. Kee, Toward a comprehensive chemical kinetic for the oxidation of acetylene: comparison of model predictions with results from flame and shock tube experiments. Proc. Combust. Inst. **19**, 181–196 (1982)
15. T. Poinsot, S. Lele, Boundary conditions for direct simulations of compressible viscous flows. J. Comput. Phys **101**, 104–129 (1992)
16. T. Poinsot, D. Veynante, *Theoretical and Numerical Combustion*, 2nd edn. (Edwards, Philadelphia, 2005)
17. B. Rogg, Response and flamelet structure of stretched premixed methane-air flames. Combust. Flame **73**, 45–65 (1988)
18. J. Savre, H. Carlsson, X.S. Bai, Tubulent methane/air premixed flame structure at high karlovitz numbers. Flow Turbul. Combust. **90**, 325–341 (2013)
19. M.D. Smooke, V. Giovangigli. Premixed and nonpremixed test problem results, in *Lecture Notes in Physics*, vol. 384 (Springer, Berlin, 1991), pp. 29–47
20. C.J. Sung, J.B. Lliu, C.K. Law, On the scalar structure of nonequidiffusive premixed flames in counterflow. Combust. Flame **106**, 168–183 (1996)
21. D. Thévenin, F. Behrendt, U. Maas, B. Przywara, J. Warnatz, Development of a parallel direct simulation code to investigate reactive flows. Comput. Fluids **25**, 485–496 (1996)
22. UCSD Chemical-Kinetic Mechanisms for Combustion Applications, http://web.eng.ucsd.edu/mae/groups/combustion/mechanism.html. Mechanical and Aerospace Engineering (Combustion Research), 2012
23. M. Weiss, N. Zarzalis, R. Suntz, Experimental study of markstein number effects on laminar flamelet velocity in turbulent premixed flames. Combust. Flame **154**, 671–691 (2008)

Impact of Grid Refinement on Turbulent Combustion and Combustion Noise Modeling with Large Eddy Simulation

Feichi Zhang, Henning Bonart, Peter Habisreuther, and Henning Bockhorn

Abstract For Large Eddy Simulation (LES) of turbulent combustion, as the turbulent flow as well as the thin flame front are directly filtered by the cut-off scale, resolution of the computational grid plays a very important role in this case and represents always a quality determining factor. In addition, the fluctuation of heat release is found to be the main source for noise generation from turbulent combustion, which is attributed to the interaction of the turbulent flow and the combustion reaction. As the flame is thickened or filtered respectively by the computational grid, it becomes less sensitive to fluctuations of the flow as well, so that the emitted noise from the flame due to unsteady heat release is affected by the grid resolution in LES combustion modeling as well. The current work represents an investigation with respect to these aspects. For this purpose, LES and DNS simulations for a realistic jet flame at moderately turbulent condition have been carried out. The LES calculations have been performed on computational grids with different resolutions (0.4/3.2/10.7 million cells) on the HP XC4000 cluster of the Steinbuch Centre for Computing (SCC) at the KIT. In order to assess predictability of the LES methodology, a three-dimensional DNS simulation on a grid with 54 million cells has been carried out on the Cray XE6 (HERMIT) of the High Performance Computing Center Stuttgart (HLRS). The comparison of the LES solution with experimental and DNS data allows an evaluation of the influence of the grid refinement to a great extent.

F. Zhang (✉) · H. Bonart · P. Habisreuther · H. Bockhorn
Division of Combustion Technology, Engler-Bunte-Institute, Karlsruhe Institute of Technology, Engler-Bunte-Ring 1, 76131 Karlsruhe, Germany
e-mail: feichi.zhang@kit.edu

W.E. Nagel et al. (eds.), *High Performance Computing in Science and Engineering '13*, DOI 10.1007/978-3-319-02165-2_19,

1 Introduction

Large eddy simulation (LES) has become a popular method for numerical modeling of turbulent combustion flows in recent years. This is attributed to the need for more predictive simulation methods and the tremendous progress in computational powers. Combustion in practical devices usually occurs with a fast mixing progress and a short reaction time, so that stabilization of the flame generally takes place in very complicated flow patterns, such as swirling flows, breakdowns of large-scale vortical structures, and recirculation regions. For this case, the traditional Reynolds-averaged Navier-Stokes (RANS) method often cannot satisfy the required accuracy [1]. In LES, the turbulent flow field is divided into a large-scale resolved and a small-scale unresolved contribution. This is done by spatially filtering the field variables, hence, removing turbulent motions of length scales smaller than the filter size. Compared to the conventional RANS models, LES is more expensive in computational cost. The advantage of LES is however that the large-scale energy containing motion of the turbulence is resolved directly, whereas the small-scale vortices exhibit more universal features and modeling of these fine structures is more suited [2].

For numerical computation of combustion noise, the hybrid CFD (Computational Fluid Dynamics)/CAA (Computational Aero-Acoustics) method remains as the most commonly used tool until now [3]. In this technique, the noise-generating acoustic sources (e.g. the turbulent flame) are calculated with help of CFD combustion modeling, which serve as input data for the subsequent CAA computation to describe propagations of sound waves emitted from these noise sources into the far field [4]. On the other side, the acoustic radiation from a turbulent flame can be obtained by solving the compressible Navier-Stokes (NS) equations directly, for example, in framework of a DNS (Direct Numerical Simulation) or a LES. The major shortcoming of these concepts is that the computational domain is often restricted to a relatively small region close to the flame to reduce computational cost which would be very high in case of using a large domain due to the disparity between the length scales of the reactive flow (thickness of a flame $O(\delta_f) \cong 0.1$ mm) and the propagating acoustic waves (wave length in the far field $O(\lambda) \cong 1$ m).

In comparison with the conventional modeling methodologies like RANS or LES, the application of DNS for combustion related problems is mainly limited to research purpose, for example, for flows with a very low Reynolds number (Re) or restricted to a small and simple computational domain, due to a very high computational effort. On the other hand, a DNS solves the governing equations exactly without any empirical models and provides a detailed insight into the interaction of the turbulent flow with the chemical kinetics, so that it can be used to validate and to improve subordinate modeling methods.

2 Combustion Modeling

The current work applies an extended version of the reaction model of Schmid et al. [5] to describe the combustion reaction and its interaction with the turbulent flow. The concept makes use of the progress variable θ to follow the combustion reaction progress, which is defined in our work by means of the chemically bounded oxygen. It quantifies the transition from unburnt ($\theta = 0$) to burnt state ($\theta = 1$) by:

$$\theta = \frac{y_{O_2} - y_{O_2,ub}}{y_{O_2,br} - y_{O_2,ub}} \tag{1}$$

In (1), $y_{O_2,ub}$ and $y_{O_2,br}$ indicate mass fractions of O_2 in the unburnt and the completely burnt state (of equilibrium). The LES averaged formulation of the transport equation for θ is given as [6]:

$$\frac{\partial \bar{\rho}\tilde{\theta}}{\partial t} + \frac{\partial \bar{\rho}\widetilde{u_i}\tilde{\theta}}{\partial x_i} = \frac{\partial}{\partial x_i}\left(\frac{\mu_l + \mu_t}{Sc_t}\frac{\partial \tilde{\theta}}{\partial x_i}\right) + \bar{\dot{\omega}}_\theta \tag{2}$$

where $\bar{\cdot}$ denotes a spatially filtered value and $\tilde{\cdot}$ a Favre-filtered value. The turbulent flame speed S_t in the source term, $\bar{\dot{\omega}}_\theta$ has to be modeled to close the equation mathematically. For this reason, it belongs to the class of turbulent flame-speed closure (TFC) combustion models [6]. The rate term in (2) is given as follows [5]:

$$\bar{\dot{\omega}}_\theta = \rho_u \frac{S_t^2}{D_l + D_t}\tilde{\theta}(1 - \tilde{\theta}), \quad \frac{S_t}{S_l} = 1 + \frac{u'}{S_l}\left(1 + Da^{-2}\right)^{-1/4} \tag{3}$$

with the laminar burning velocity S_l, the density of the unburnt mixture ρ_u, the turbulence intensity u' and the turbulent length scale L_t. D_l and D_t are the laminar and turbulent diffusion coefficient which are related to the laminar and turbulent viscosities through the Schmidt number; the Dahmköhler number Da is evaluated by the ratio of the turbulent time scale $\tau_t = L_t/u'$ to the chemical time scale $\tau_c \propto a/S_l^2$ with the thermal diffusivity a.

To account for mixing effects, a transport equation for the mixture fraction ξ is solved additionally. Assuming that the mixing of fuel and oxidizer takes place before the chemical reaction, the entire turbulent flame can be considered as an ensemble of distinct reaction zones with different equivalence ratios. Thereby, the mixing process is controlled by the mixture fraction and the subsequent chemical reaction by the progress variable. Structures of the individual premixed reaction layers can then be pre-computed by solving 1D flame equations, for example, using the CHEMKIN program package [7]. To take the effect of turbulence on mixing into account, the pre-defined profiles of these flame sheets are averaged by a probability density function (PDF) $P(\xi)$ determined by the mean value of $\tilde{\xi}$ and its variance $\widetilde{\xi''^2}$ with a presumed principal shape of a β-function [8]. Finally, all averaged (i.e. over the PDF integrated) quantities are picked up in a look-up table depending on the three control parameters: $\tilde{\xi}$, $\widetilde{\xi''^2}$ and $\tilde{\theta}$.

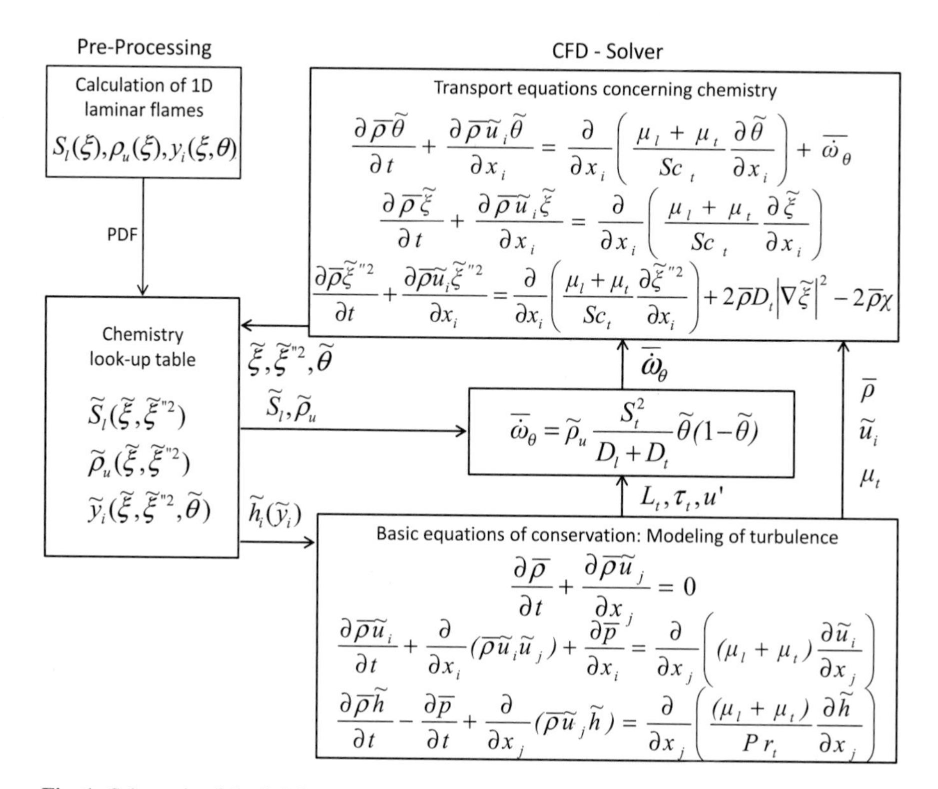

Fig. 1 Schematic of the link between the look-up chemistry table with the CFD solver

As illustrated in Fig. 1, internal structures of the reaction zones at different equivalence ratios are calculated in a pre-processing step with a detailed reaction mechanism. The flow solver and the pre-computed flame profiles, i.e. the look-up chemistry table, are linked through exchange of $\tilde{\xi}$, $\widetilde{\xi''^2}$ and $\tilde{\theta}$, where multi-dimensional interpolations have been applied in order to obtain intermediate values. The source term of the variance transport equation $\bar{\dot{\omega}}_{\xi''^2}$ is given by [8]:

$$\bar{\dot{\omega}}_{\xi''^2} = 2\bar{\rho} D_t |\nabla \tilde{\xi}|^2 - 2\bar{\rho}\tilde{\chi} \tag{4}$$

with the scalar dissipation rate $\tilde{\chi} \propto 1/\tau_t$.

To use the concept for LES combustion modeling, the turbulence parameters in S_t (i.e. u', τ_t and L_t) have been evaluated at the cut-off scale [9]. In doing so, the unresolved scalar flux increases the flame thickness, either through the turbulent or numerical diffusion, so that the thickened (implicitly filtered) flame front can be resolved on the LES mesh. Applications of the concept in connection with LES and RANS turbulence models have already been successfully performed for different flame regimes and validated by experiments [10, 11] so that it may be adequately used for the current study.

For DNS simulation, the three-dimensional conservation equations for total mass, momentum, energy and species masses, have been directly solved without using any simplifications. These constitute a large number of non-linear, partial differential equations which have to be solved numerically together with the thermodynamic equation of state. The reaction rates for the species equations are thereby calculated by means of the kinetic law of chemical reactions together with the extended Arrhenius law, whereas the ordinary species flux is given either by an exact but very time-consuming multi-component or a simplified mixture-averaged diffusion approach [12]. The molecular viscous and heat fluxes are evaluated with the exact transport coefficients.

3 Experimental and Numerical Setups

3.1 Case Description

The burner system shown in Fig. 2 has been designed and built at the Technische Universität Berlin in order to experimentally investigate the combustion-induced noise [13]. Therefore, its setup is deliberately kept simple by having an annular injector at the very bottom, a mixing plenum, a converging nozzle, and a cooled plate at the outlet of the nozzle. In order to reduce computational efforts even further, the Bunsen-type flame is stabilized by flame propagation itself. After fuel and air are well mixed in the plenum, the mixture flows through the nozzle and leaves the burner at the diameter D. The flame is then operated at atmospheric pressure and temperature, employing pure methane as fuel and the equivalence ratio $\phi = 1.3$ (fuel-rich). By passing through the nozzle, the flow accelerates yielding a Reynolds number of $Re = 7,500$ at the exit plane of the burner. Particle Image Velocimetry (PIV) and microphone measurements in an anechoic environment have been carried out for investigations of the flow field and the acoustic field, respectively.

3.2 Numerical Setups

As the considered case is represented by a round jet configuration, the computational domain is chosen to be constructed with the convergent nozzle part to allow development of the turbulence within the burner and the velocity changes directly at the nozzle exit, and a large cylindrical domain ($length \times diameter = 60D \times 90D$) downstream of the burner where mixing and combustion take place. Due to the simple geometries of the computational domain and in order to achieve best accuracy, the computational grids are built up in a block-structured way employing hexahedral shaped elements, as illustrated in Fig. 3. To study the effect of grid resolution on the LES combustion modeling, meshes with different refinements (but the same topology) have been conducted with 0.4/3.2/10.7 million cells. These are

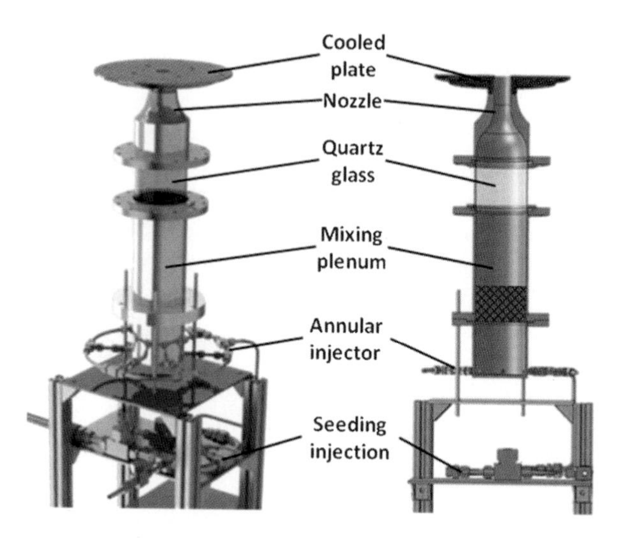

Fig. 2 Schematic of the experimental setup of the burner system [13]

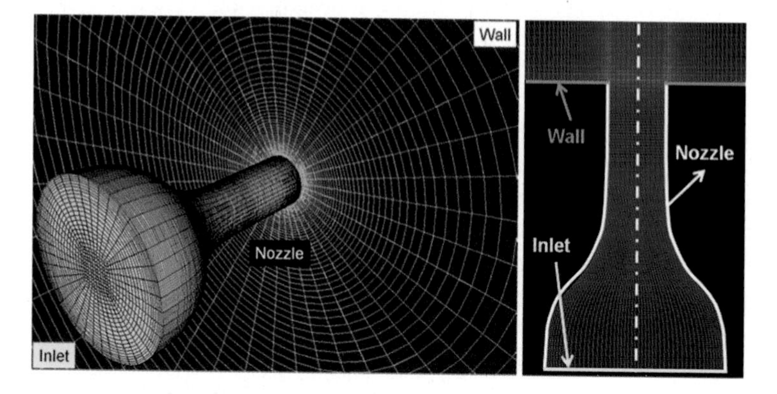

Fig. 3 Computational grid used for the LES

systematically refined along the burner wall and the shear layer continuously to a higher ordered refinement level. The smallest mesh sizes used for these critical regions are (from the coarsest to the finest grid) 0.7, 0.35, 0.175 mm. The x- and r-coordinates indicate always the streamwise and radial axis of the cylindrical coordinate system.

The DNS simulation has been restricted to a compact domain close to the flame ($length \times diameter = 23D \times 23D$) in order to save computational time. The computational grid is built up by approx. 54 million finite volumes. The smallest mesh size of 0.1 mm is used to resolve the reaction zone and the shear layer. The ordinary diffusion flux is described by the mixture-averaged approach and the combustion reaction covers a complex chemical mechanism with 18 species and 69 elementary reactions including the reaction chain of the short-lived OH^* free radical [14].

The open source CFD code OpenFOAM [15] has been used to solve the filtered and the exact conservation equations in LES and DNS, employing the finite volume method with a cell-centered storage arrangement. A fully implicit compressible formulation was applied together with the pressure-implicit split-operator (PISO) technique for the pressure correction [16]. Discretisation of convective flux terms is based on a second order interpolation scheme. The boundary conditions have been set according to the experimental setup, i.e. methane/air mixture of the equivalence ration of $\phi = 1.3$ flows from the inlet into the computational domain and becomes burnt after the convergent nozzle by a Re of 7,500 under atmospheric condition.

A turbulence inflow generator [17] was implemented in OpenFOAM and used for the LES and DNS simulations, which is based on digital filtering of random data series with pre-defined correlation functions. By doing so, it provides transiently correlated velocity components at the inlet boundary for each time step. The turbulence properties thereby was set according to LDA measurement of the velocity field within the quartz glass (see Fig. 2). The partially non-reflecting boundary condition (NRBC) proposed by Poinsot and Lele [18] has been applied to all inlet, outlet and entrainment boundaries in order to avoid unphysical reflection of pressure waves. One additional issue of the simulations is to include the buoyant force caused by the large density change through the flame front. As density of the unburnt mixture upstream of the flame front is much higher than that of the burnt mixture, the fresh gas upstream of the flame front is heavier than the hot product gas, leading to an upwardly buoyant force. The effect of buoyancy can be neglected for combustion at high Re as it causes only minor difference in that case. As the Re is relatively low in the current case, an acceleration of the flow through buoyant force is discernible in addition to the thermal expansion.

The combustion model proposed in Sect. 2 has been implemented into the OpenFOAM code, which was used together with the Smagorinsky subgrid scale (sgs) model [2] for LES combustion modeling, assuming a constant turbulent Schmidt/Prandtl number $Sc_t = Pr_t = 0.7$. For the LES modeling the CHEMKIN program package [7] has been used to calculate the internal flame structure applying the GRI 3.0 (Gas Research Institute) mechanism containing 325 reactions and 53 species [24] for methane/air combustion. The Implementation of the DNS solver is based on coupling the standard OpenFOAM applications with the thermo-chemical libraries of Cantera [19]. The OpenFOAM based program is thereby used for solving the governing equations together with the state equation in parallel, whereas the thermodynamic state in terms of pressure, temperature and species compositions, serves as input for the Cantera algorithms, which evaluate the necessary transport properties (e.g. viscosity, diffusion coefficients and thermal conductivity) and the reaction rates needed for solution of the conservation equations in OpenFOAM [20, 21].

3.3 *Computational Effort and Performance*

The LES on the coarsest grid with 0.4 mil. nodes has been made parallel with eight CPU cores (i7-860) on the in-house Beowulf Linux cluster at the Division

Table 1 Mesh parameters and simulation setups used for LES and DNS simulations

CFD	No. cells	Δ_{min} (mm)	N_{inlet}	Δt (μs)	t_{total} (s)	$N_{\Delta t}$	No. cores	$t_{clock}/\Delta t$ (s)	Core hours
LES	0.4 mil.	0.700	513	50	8	160.000	8 (BEOWULF)	2.5	900
LES	3.2 mil.	0.350	2,052	25	4	160.000	256 (XC4000)	2.0	23.000
LES	10.7 mil.	0.175	5,700	10	2.4	240.000	512 (XC4000)	2.5	85.000
DNS	53.9 mil.	0.1	19,305	2	0.2	100.000	8192 (CRAY)	10	2.3 mil.

of Combustion Technology in KIT. As the own cluster applies a Gigabit Ethernet Interconnect between the nodes, it is not able to speed up efficiently by increasing used number of processors. For this reason, the LES simulations on the finer grids (with 3.2 mil. and 10.7 mil. grid points) have been conducted on the HP XC4000 cluster of the Steinbuch Centre for Computing (SCC) at the KIT [22]. As the computational cost for DNS is much higher than that needed for LES, the DNS simulation has been carried out on the Cray XE6 (HERMIT) cluster in HLRS [23] with 8,192 cores. A total computing time of approx. 12 days has been conducted for 100.000 time steps.

Table 1 gives an overview on simulation setups used for the different cases, thereby Δ_{min} denotes the minimum mesh size applied for each computational grids and N_{inlet} indicates the used number of elements to resolve the inlet boundary patch. In order to obtain a temporally well resolved progress of the field variables and to fulfill the criterion of numerical stability, time steps Δt for the LES simulations were set to allow the overall CFL no. (Courant-Friedrichs-Lewy condition) to be below 0.2. After running the LES for some volumetric residence (or through flow) times, statistical averaging of the flow variables and sampling of the sound signals have been conducted over a certain simulation time of t_{total}. The number of processors used for LES on the HP XC4000 Cluster in SCC at KIT are typically chosen to run a whole simulation with less than two jobs using the maximal running time of 4,320 min. For LES simulations considered for the current work, where at least 100.000 time steps are necessary to get well averaged statistics, up to 512 processors are needed depending on the mesh size to achieve a computing speed of approximately 2.5 s per time step. Comparing performances of the LES calculations with 3.2 mil. and 10.7 mil. cells on the HP XC4000 cluster, the speedup is higher by using the finer mesh because the number of cells per CPU core is larger in this case.

In addition, DNS calculations have been carried out to obtain information about parallel efficiency and scalability on the Cray XE6 platform on HLRS. The method for parallel computing used by OpenFOAM is known as domain decomposition, where the computational grid associated with internal cell volumes and boundary patch elements are broken into pieces and allocated to a number of processors for separate solutions. Therefore, the scale-up factor depends strongly on the number of allocated cells to each processor. Figure 4 shows the intra- (left) and inter-node (right) performance by running the DNS solver on a computational grid with two million cells. It is clear, that the intra-node speedup is far from the ideal linear speedup. This indicates that some bottleneck due to the communication between the single cores is reached. Thus, it is not worthwhile using the Cray XE6 cluster with

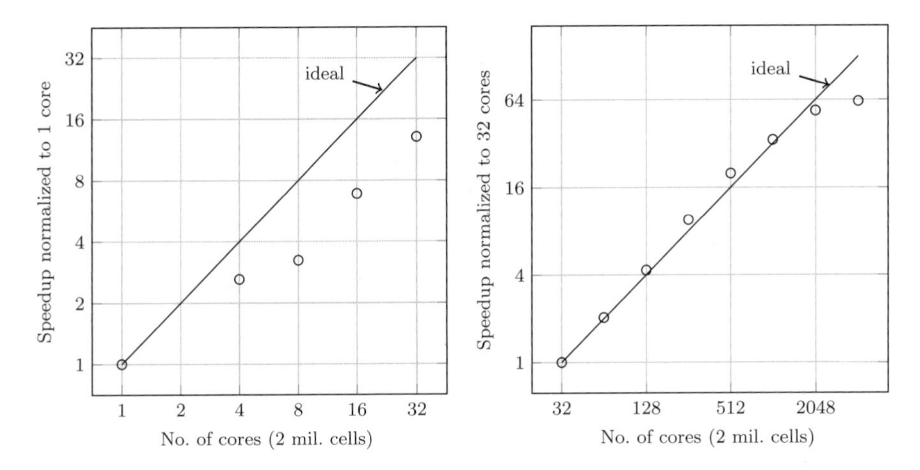

Fig. 4 Relative speedup factor by running DNS on a computational grid with two mil. cells: intra-node (*left*) and inter-node performance

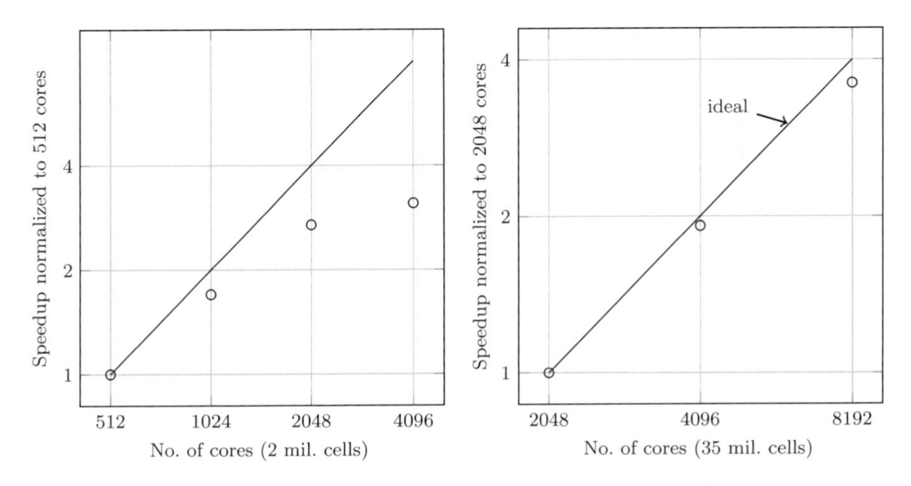

Fig. 5 Relative speedup factors by running DNS on two grids with 2 (*left*) and 35 mil. (*right*) cells

less than 32 processors on one node. On the contrary, the inter-node performance shows a very good scalability up to 2,048 processors. From 64 to 1,024 processors the scale-up is even superlinear.

Figure 5 compares the scaling performance of the DNS simulation on two different computational grids with approx. 2 million and 35 million cells. It is clear from Fig. 5 on the left that the solver does not scale well for a low number of cells per core in case of using the coarse grid. On the contrary, the scaling factor is excellent by use of the fine grid with a high number of cells per core, as one can see on the right of Fig. 5. In this case, different threads of the solver take the most time for calculation than for communication and synchronization between the CPU cores.

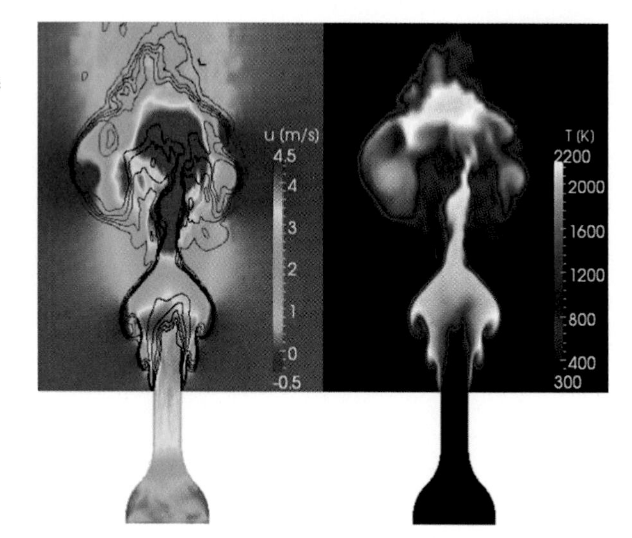

Fig. 6 Instantaneous contours of the streamwise velocity and the temperature

4 Simulation Results

4.1 Turbulent Reactive Flow

Figure 6 shows meridian planes passing through the centerline axis and being derived from the LES on the finest grid with 10.7 million nodes. The flame front is indicated by isolines of the progress variable $\tilde{\theta} = 0.1/0.3/0.5/0.7/0.9$ in contours of the streamwise velocity on the left of Fig. 6, which is corrugated and stretched by the turbulent flow. Both an inner and an outer reaction zone can be identified, which are sustained by a premixed and a non-premixed combustion. The diluted mixture is pre-burned by a higher reaction rate from the premixed combustion, the remaining fuel or combustible intermediate species (e.g. CO) mixes thereafter with ambient air leading to a weaker diffusion flame. In this case, the flame is a so-called partially premixed flame, where both premixed and non-premixed combustion regimes exist at the same time.

By passing through the convergent nozzle, the unburnt mixture is accelerated and homogenized according to continuity, so that the flow is rather undisturbed there. As a result, the premixed flame front is not strongly corrugated. At the reaction zones of the primary premixed flame, the temperature increases from 300 to 2,200 K from unburned to burned state. The resulting thermal expansion and buoyancy lead again to an acceleration of the gas. The flow is however even less turbulent due to the strongly increased fluid viscosity at high temperature. As the gradient of flow velocity and the turbulence level are relatively low, large ring vortices or coherent structures are generated in the shear layer between the main jet and the surrounding air, which contribute to the transition process from laminar to turbulent flow regime. These flow instabilities result in a strong entrainment of ambient air into the primary

Fig. 7 Three-dimensional streamlines colored by contours of the streamwise velocity

exhaust gas, which enhance the mixing and combustion process in the secondary diffusion flame. Further downstream, the coherent vortices become unstable and break down leading to the fully turbulent flow regime.

The instantaneous three-dimensional trajectories of the flow field shown in Fig. 7 illustrate behaviors of these flow patterns clearly, where the streamlines are colored by values of the streamwise velocity and the flame surface of the external diffusion flame is indicated by isosurface of $T = 1,800\,\mathrm{K}$. On the left of Fig. 7, the surrounding air is sucked into the flame region following large-scale helical paths. Looking more closely at the near field region on the right of Fig. 7, sinks of the streamlines can be identified as the coherent ring vortices, which drive the ambient air into the secondary diffusion reaction zone. Moreover, shape of the flame surface evolves mostly with these large vortical motions as well, which is caused by interactions of the flame with the turbulent flow.

Figure 8 on the left compares intensity signals from the OH-chemiluminescence measurement which represents line-of-sight integrated values of the excited OH or OH^* concentration, with the calculated mass fraction of OH^* from DNS, which has been appropriately summed up along the lines-of-sight for each position of the diagram. Obviously, there is a qualitatively good agreement for OH^* intensity or concentrations given by both methods. The length of the flame is predicted accurately by the DNS simulation. In the following, time-mean velocities derived from the LES simulations with different grid resolutions will be compared. As the DNS calculation has only be run for 0.2 s (see Table 1) which is too short to resolve long-term coherent vortical flows, statistically converged mean fields cannot be accomplished yet for direct comparison with LES results.

4.2 *Influence of Grid Resolution*

Figure 9 demonstrates influence of the mesh size on the resolved time mean flow field for the LES simulation. It is obvious that the grid resolution imposes a major impact in this case where the flame length increases significantly by using a more

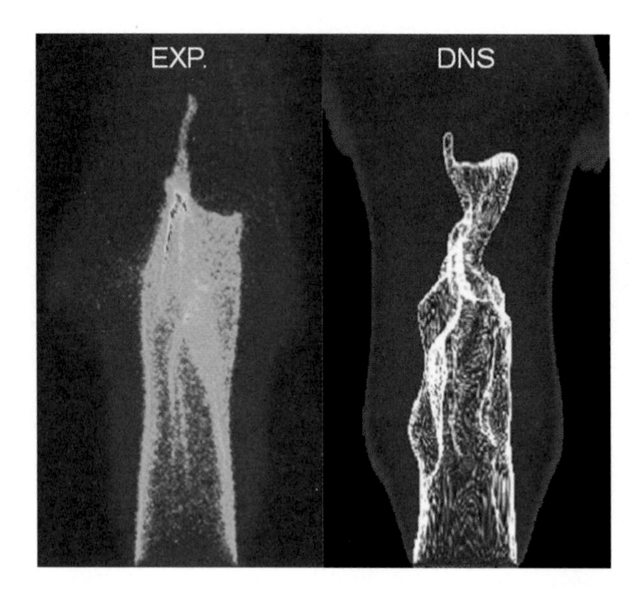

Fig. 8 Comparison of OH-chemiluminescence measurement with OH^* calculated by DNS

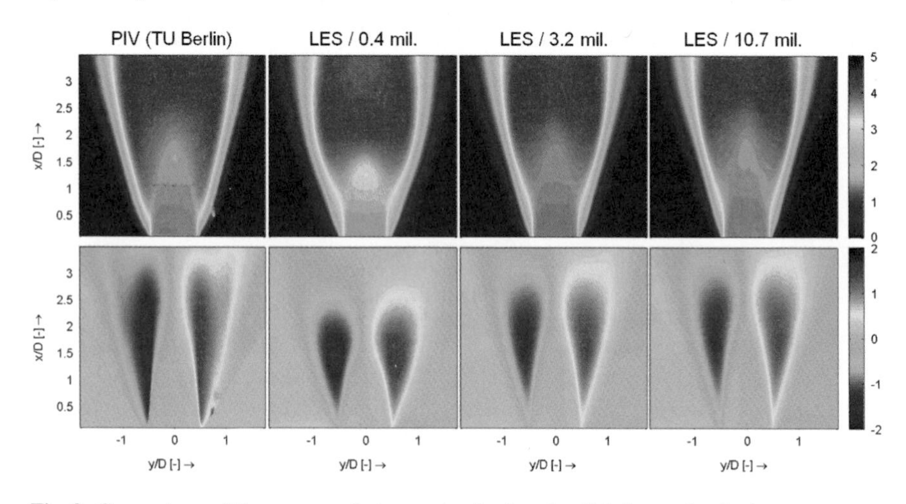

Fig. 9 Comparison of time averaged streamwise (*top*) and radial (*bottom*) velocity components derived from experiment and LES computations with different grid sizes

refined computational grid in LES. This is attributed to the fact that a smaller grid length leads to an overall attenuated turbulent diffusion ($\nu_{sgs} \propto \Delta^2$ according to the standard Smagorinsky subgrid scale model [2]) and numerical diffusion ($O(\nu_\varepsilon) \propto \Delta^2$). As shown in Fig. 9, the solution provided by LES on the finest mesh shows the best agreement with the experimental data [13], which however exhibits still a slightly shorter flame than the measured one. This difference is caused by the quasi-laminar flow nature of the considered case, where the LES fails to predict

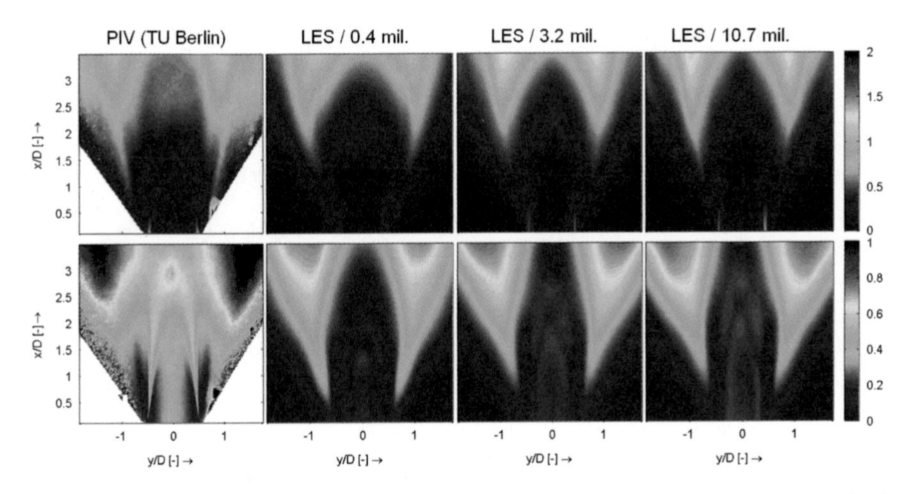

Fig. 10 Comparison of RMS streamwise (*top*) and radial (*bottom*) velocity fluctuations provided by experiment and LES computations with different grid sizes

the correct molecular transport coefficients (viscosity ν_l, thermal diffusivity a_l and diffusion coefficient D_l) used in the governing equation. Instead, the LES method applies the effective, superimposed values together with the turbulent or sgs part for these parameters, i.e. $\nu_{eff} = \nu_l + \nu_{sgs}|a_{eff} = a_l + a_{sgs}|D_{eff} = D_l + D_{sgs}$ (see Fig. 1), so that the transport fluxes are always overpredicted by the LES in regions where the flow is characterized by a laminar nature. This effect becomes weaker with a refined grid resolution, as the turbulent viscosity decreases strongly with the grid length.

A finer computational grid leads to a better resolved, thinner shear layer between the main jet and the free domain, which results in a higher gradient of the velocity and therefore a higher turbulence level. Consequently, the root mean square values (RMS) of velocity fluctuations or the roots of Reynolds stresses are generally larger in case of using a refined mesh, as shown in Fig. 10. The influence of the laminar natured coherent flow on the LES modeling applies to the calculated RMS velocity fluctuations as well, which are not as high as those from the measurement, because the turbulent transport fluxes are predicted too high. In this case, the flame front does not fluctuate as intensive as it is in the reality, so that a large difference can be identified directly in the vicinity of the flame surface (see lower part of Fig. 10).

As the compressible formulation for LES and DNS is used in the current work, the noise radiations or respectively the pressure fluctuations inside the computational domain can be directly calculated. In Fig. 11, sound pressure levels (SPL) derived from different simulation methods and the experiment show a good agreement. The DNS solution follows the measured profile very well in the whole considered frequency range. The SPLs from LES are predicted smaller than those of the DNS and the experiment in the low frequency range (up to 200 Hz). This is again attributed to the overestimated turbulent transports in the laminar, coherent flow

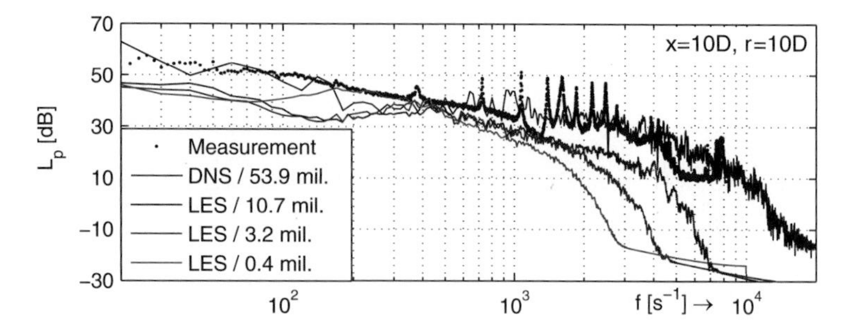

Fig. 11 SPL calculated from experiment, DNS and LES computations with different grid sizes

regions, which result in a stronger dissipation in the flow. The difference between LES and DNS is large particularly in the high frequency region. In this case, as the small scale turbulent motions can be resolved better by using a finer LES mesh, the intrinsic effect of grid refinement on the emitted sound is mostly characterized by a higher pressure level in the high frequency regions. As shown in Fig. 11, the pressure levels given by LES on different grids exhibit similar distribution in the low frequency range. However, progress of the SPLs diminishes with increasing frequencies much faster in case of using coarser computational grids.

5 Conclusions

The work presents a numerical study on LES and DNS simulations of turbulent combustion by means of different grid resolutions. The objective is to assess the influence of grid refinement on the resolved reactive flow field and the emitted noise level by means of LES combustion modeling. A realistic, moderately turbulent jet flame has been considered for this study, which was also experimentally investigated. It has been shown that the refinement of the computational grid plays an important role for predicting the combustion properties with LES, thereby the calculated length of the flame increases clearly by using a finer resolved grid due to attenuated numerical diffusion. In addition, the noise level derived from LES on a more refined grid is higher than that provided by LES on a coarse grid, this is discernible particularly in the high frequency domain. In fact, the filtering procedure implicitly employed by the LES method results in an overall enhanced numerical diffusion whose effect depends on the grid resolution, so that the predicted flame length as well as the sound level by LES is generally lower than those given by the DNS simulation. Although more computational effort is required for LES on a finer mesh, a better accuracy have been evidenced. Nevertheless, LES simulations on the coarse grids are able to reproduce the main dynamics of the flow appropriately so that they are also useful, for example, to gain a first insight of the flow patterns. With help of the current study, a more detailed estimation for LES of turbulent combustion and its noise propagation on a further refined grid level may be accomplished.

Acknowledgements The authors gratefully acknowledge the financial support by the German Research Council (DFG) through the Research Unit DFG-BO693 "Combustion Noise". Major part of the computation time was kindly provided by the High Performance Computing Center Stuttgart (HLRS) of the University of Stuttgart and the Steinbuch Centre for Computing (SCC) of the Karlsruhe Institute of Technology.

References

1. H. Pitsch, Large-Eddy simulation of turbulent combustion. Annu. Rev. Fluid Mech. **38**, 453–482 (2006)
2. J. Fröhlich, *Large Eddy Simulation Turbulenter Strömungen* (Teubner Verlag, Wiesbaden, 2006)
3. A. Schwarz, J. Janicka (eds.), *Combustion Noise* (Springer, Berlin, 2009). ISBN-10: 3642020372
4. S. Marburg, B. Nolte (eds.), *Computational Acoustics of Noise Propagation in Fluids* (Springer, Berlin, 2008). ISBN 978-3-540-77447-1
5. H. Schmid, P. Habisreuther, W. Leuckel, A model for calculating heat release in premixed turbulent flames. Combust. Flame **113**, 79–91 (1998)
6. T. Poinsot, D. Veynante, *Theoretical and Numerical Combustion*, 2nd edn. (Edwards, Philadelphia, 2005). ISBN 978-1-930217-10-2
7. J. Kee, F. Rupley, J. Miller, Chemkin-II: a Fortran chemical kinetics package for the analysis of gas-phase chemical kinetics, Report No. SAND89-8009B, Sandia National Laboratories (1989)
8. N. Peters, *Turbulent Combustion* (Cambridge University Press, Cambridge, 2000)
9. F. Zhang, P. Habisreuther, M. Hettel, H. Bockhorn, Modelling of a premixed swirl-stabilized flame using a turbulent flame speed closure model in LES. Flow Turbul. Combust. **82**, 537–551 (2009)
10. F. Zhang, P. Habisreuther, H. Bockhorn, A unified TFC combustion model for numerical computation of turbulent gas flames, in *High Performance Computing in Science and Engineering '12*, ed. by W.E. Nagel, D.B. Kröner, M.M. Resch (Springer, Berlin, 2013). ISBN 978-3-642-33374-3
11. F. Zhang, P. Habisreuther, M. Hettel, H. Bockhorn, A newly developed unified turbulent flame speed closure (UTFC) combustion model for numerical simulation of turbulent gas flames, 25. Deutscher Flammentag, Karlsruhe, Germany (2011)
12. R.J. Kee et al., *Chemically Reacting Flow: Theory and Practice* (Wiley, New York, 2003)
13. H. Nawroth, C.O. Paschereit, F. Zhang, P. Habisreuther, H. Bockhorn, Flow investigation and acoustic measurements of an unconfined turbulent premixed jet flame, in *43rd AIAA Fluid Dynamics Conference and Exhibit*, San Diego, 2013
14. T. Kathrotia et al., Experimental and numerical study of chemiluminescent species in low-pressure flames. Appl. Phys. B **107**, 571–584 (2012)
15. OpenCFD Ltd, *OpenFOAM User Guide, Version 2.0.1* (OpenFOAM® Documentation - The OpenFOAM® Foundation, 2011). www.openfoam.org
16. J. Ferziger, M. Periĉ, *Computational Methods for Fluid Dynamics* (Spinger, Berlin, 2002)
17. M. Klein, A. Sadiki, J. Janicka, A digital filter based generation of inflow data for spatially developing direct numerical or large eddy simulations. J. Comput. Phys. **286**, 652–665 (2003)
18. T. Poinsot, S. Lele, Boundary conditions for direct simulation of compressible viscous flows. J. Comput. Phys. **101**, 104–129 (1992)

19. D.G. Goodwin, *Cantera C++ Users Guide* (California Institute of Technology, Pasadena, 2002)
20. F. Zhang, H. Bonart, P. Habisreuther, H. Bockhorn, Prediction of combustion generated noise using direct numerical simulation. Euromech Colloquium 546, Villa Vigoni, Menaggio, Italy, 2013
21. F. Zhang, H. Bonart, P. Habisreuther, H. Bockhorn, Direct Numerical Simulations of Turbulent Combustion with OpenFOAM. 26. Deutscher Flammentag, Duisburg (2013)
22. Karlsruhe Institute of Technology (KIT), Karlsruhe, Germany, Rechenzentrum. HP XC4000 User Guide (2009)
23. High Performance Computing Center Stuttgart (HLRS) (2011): CRAY XE6 (HERMIT), http://www.hlrs.de/systems/platforms/cray-xe6-hermit/
24. G.P. Smith, D.M. Golden, M. Frenklach, N.W. Moriarty, B. Eiteneer, M. Goldenberg, C.T. Bowman, R.K. Hanson, S. Song, W.C. Gardiner Jr., V.V. Lissianski, Z. Qin, http://www.me.berkeley.edu/gri_mech/

Part V
Computational Fluid Dynamics

Ewald Krämer

As in the previous years, the highest demand for High Performance Computing (HPC) resources came from the CFD community. The spectrum of projects ranges from Direct Numerical Simulations (DNS) over Large Eddy Simulations (LES) to Reynolds Averaged Navier Stokes (RANS) simulations on the one hand, and from code development over fundamental research up to application oriented research on the other hand. In-house codes were used as well as commercial and even open source codes. All projects of the recent funding period were run on the supercomputers of the High Performance Computing Center Stuttgart (HLRS). Most of them used the new Cray XE-6 HERMIT, which offers a peak speed of more than 1 PFLOP/s. This significant step in computer performance allowed dealing with problems that were more complex than before in geometry or physics or both. Extremely fine mesh resolutions were realizable with numbers of grid cells beyond 50 million, and very small time steps in case of unsteady flows.

All proposed papers have undergone a peer review process, and, finally, 15 (which is about 2/3) have been selected for inclusion into this book.

The first two contributions deal with Discontinuous Galerkin (DG) methods, which may be considered as a combination of finite-element schemes (with a continuous higher order polynomial in each grid cell) and a finite-volume scheme (allowing for discontinuities at the cell faces), and which are supposed to have the potential to some day replace, at least for certain applications, the well-established finite volume production codes. The working group of Munz at the Institute of Aerodynamics and Gas Dynamics (IAG), University of Stuttgart, has built-up a DG-based high order simulation framework, which consists of four individual codes, each of them with a distinct focus. In the present paper, Andrea Beck et al. show results from their latest code, FLEXI, which uses a very efficient variant of a

E. Krämer
Institut für Aerodynamik und Gasdynamik, Universität Stuttgart, Pfaffenwaldring 21,
70550 Stuttgart, Germany
e-mail: kraemer@iag.uni-stuttgart.de

DG formulation, the so-called DG spectral element method. It has been built from ground up for high performance computing and solves three-dimensional advection dominated problems on unstructured meshes. Results are presented for flows past a cylinder and past a wing segment, both using more than 2,000 cores on the Cray XE-6. In case of the cylinder with high polynomial degrees this means that one grid cell was located on one processor. Additionally, a case with fluid-structure interaction was treated, showing the capability of the code to handle moving meshes. As one of the next steps, it is planned to extend the DG framework to be able to use numbers of cores of O(10,000) or even O(100,000).

The structured code HALO, which is another code of the above mentioned DG simulation framework of the IAG, and the zonal RANS-LES coupling finite-volume code of the RWTH Aachen were used in the Project STEDG2, which has aimed at assessing the feasibility and applicability of HPC systems to obtain highly accurate simulation results within very short turn around times in an industrial design process. Consequently, contributions to the present paper came from academia (Univ. Siegen, Stuttgart, RWTH Aachen), industry (Trumpf, Bosch), and from the HLRS itself. Using three different industrial test cases, the efficiency of a high-order DG scheme on a coarser mesh as well as of the zonal RANS-LES coupling code using solution adapted models is compared to lower order finite-volume schemes as employed e.g. in state-of-the-art commercial CFD-codes. The latter are known to suffer from an only limited scalability, which prohibits large scale deployment on massively parallel supercomputers. In this paper by Harlacher et al., a comprehensive chapter is dedicated to strategies to optimize the code performance of HALO on the former NEC Nehalem Cluster Laki. They also report about their porting experience from Laki to HERMIT and the scaling behaviour on this system, and about the dynamic load balancing, which is crucial to keep the high parallel efficiency especially for a large number of processes.

In the paper of Zhu et al., the primary breakup of an evolving inelastic non-Newtonian jet is analyzed by means of a DNS based on the Volume-of-Fluid (VOF) method. The in-house code FS3D of the Institute of Aerospace Thermodynamics (ITLR) of the University of Stuttgart was applied for this purpose. Beside the physical results in terms of the three-dimensional jet structure, the generation of ligaments and droplets and the non-Newtonian characteristics, also a comprehensive performance analysis on the Cray XE-6 is described in the paper. While the code has been run on the NEC SX-9 vector computer so far, the process of porting it to the massive parallel computer is ongoing. The current status of MPI optimizations already shows fairly good speed-ups.

DNS was also applied to study the boundary layer transition in a streamwise 90° corner flow. The work was done by Schmidt et al. in the working group of Rist at the IAG. The code solves the compressible Navier-Stokes equations by using 6th-order accurate biased compact differences for the convective terms. A perturbation formulation is employed that allows calculating the perturbation flow field solely upon a fixed steady base state. It was found that the transition takes place in form of turbulent wedge situated in the near-corner region. This is remarkable as it is in accordance with measurements but opposite to the linear theory, which does not predict a dominant instability mechanism due to the presence of a corner. Like the above mentioned FS3D code, the NS3D code used here was originally designed and optimized for a vector computer architecture. Hence, the performance on the Cray XE-6 was significantly worse. However, a recently developed and implemented sub-domain scheme, which decouples the discretization in x- and y-direction, shows an enormous gain in computational performance and will be used for future simulations.

Several contributions deal with the application of LES to unsteady flow problems, two of them to cavitating flows. In the first paper by Egerer et al. from the working group of Adams at the Technical University Munich (TUM), preliminary results of their ongoing work on LES of turbulent cavitating shear layers are presented. The focus is laid on the mutual interaction of turbulence and cavitation, the resulting modulation of turbulence due to phase change, and its influence on properties of the shear layer. The major findings up to now are differences in the evolution of the Reynolds stress tensor and the Reynolds stress anisotropy tensor when the cavitation number is reduced, although the overall level of turbulence is approximately independent of the investigated cavitation numbers. This requires further investigations.

At the Karlsruhe Institute of Technology (KIT), Magagnato and Dumont also look into cavitating flows. However, different from the work at the TUM, they apply a novel numerical approach using a stochastic field formulation. A probability density function is used for the vapour mass fraction, in which time evolution is described by pure Eulerian transport equations with stochastic source terms. This Eulerian description allows an efficient parallel numerical simulation on about 1,000 processors with a load-balancing of 99 % using an in-house finite-volume solver.

The third LES paper is concerned with the simulation of turbulent particle-laden two-phase flows based on the Euler-Lagrange approach. In the current work, the methodology developed by Alletto and Breuer from the Helmut-Schmidt-University in Hamburg, has now been applied to a practically relevant problem, i.e. the flow in a gas cyclone separator, which is very complex, even without particles. For the time being, the simulation is limited to a one-way coupled fluid-particle prediction, however, the full methodology is described including a sandgrain roughness model and a deterministic collision model. The predicted results of the LES combined with a point-particle approach are compared with measurements. Small discrepancies in the flow field were found that are attributed to the strong influence of the wall shear stress, which had been obtained by a wall model instead of a no-slip boundary condition. In order to analyze the performance of the collision-handling algorithm and to investigate the scaling of the computational costs with the number of particles inside the domain, a particle-laden pipe-flow was additionally computed, taking only the one-way coupling into account, but considering particle-particle collisions.

Two further papers are dedicated to particle-laden flow. Within the SimTurb project, Siewert et al. from Schröder's working group at the Institute of Aerodynamics, RWTH Aachen, have coupled their in-house "general purpose" flow solver, running in DNS mode, with a Lagrangian particle solver. The aim of this study has been to investigate the particle clustering in a turbulent flow. They present a new numerical setup to look into the combined effects of gravity and turbulence on the motion of small and heavy particles, where the turbulence is only forced at the inflow and advected through the domain. About 43 million particles were advanced in time, and the size of the Cartesian mesh amounted to 53 million cells. Since the influence of the particles on the flow is assumed to be low, a one-way coupling was applied. An excellent parallel performance could be demonstrated, with a speed-up nearly following the ideal linear scaling within a range of 1,024 to 4,096 cores.

Another approach is followed by the working group of Eberhard at the Institute of Engineering and Computational Mechanics (ITM) in Stuttgart. Instead of an Eulerian-Lagrangian formulations as in the before mentioned works, they use a Lagrangian-Lagrangian coupling, so two meshless methods are combined. The solid particles are modelled by the Discrete Element Method (DEM), and the Smoothed Particle Hydrodynamics (SPH) method is used for the simulation of the fluid. Some numerical experiments are presented in the paper of Beck et al. mainly to show that the parallel implementation works well on the used supercomputer cluster NEC Nehalem. For the pure particle code, the maximum resources of the cluster provided to the user could be exploited in a free-falling test case, resulting in more than 52 million particles. A coupled computation was also performed employing 36 nodes, in which the impact of a liquid free jet with solid particle loading on a turbine blade was simulated.

For an accurate prediction of unsteady aerodynamic, aero-structural, and/or aero-acoustical phenomena in helicopter simulations, a low numerical dissipation is mandatory in order to conserve the rotor wake over a long distance in space and time. The well-established second order schemes, as e.g. implemented in the widely used FLOWer code developed by the German Aerospace Center (DLR), usually do not meet this requirement, at least for a reasonable number of grid cells. Kowarsch et al. from the IAG in Stuttgart present a 5th order accurate WENO scheme for the reconstruction of the cell fluxes in conjunction with two different Riemann solvers. The new method has been tested for two cases, where interactions of the rotor wake play an important role, and which are, therefore, sensitive to correct wake preservation. Case 1 is a complete helicopter in fast forward flight, where the wake interacts with the aft fuselage (and eventually leads to the so-called tail shake), and case 2 is an isolated rotor in descent, where the blade-vortex-interaction causes high noise emission. In the latter case, a structured grid with about 52 million cells was used in order to resolve frequencies up to 1,200 Hz. All in all, significant gains in accuracy could be achieved, most notably when computing the acoustics.

Within the European aeronautical community, the hybrid TAU code developed by DLR is the standard CFD tool in industry as well as in academia for fixed wing applications. Gansel et al. form the IAG use this code to study the unsteady behaviour of the wake of a wing with flow separation and its impact on the horizontal tail plane. The computations were performed on a 20 million cells grid in URANS mode. Spectral wake analyses are provided and the presence of the spectral gap in the simulations is discussed. In a next step, it is planned to increase the accuracy by performing DES in conjunction with Reynolds Stress Models. The scaling behaviour of the TAU code on the Cray XE-6 was excellent up to 1,024 core. For 4,096 processors still an efficiency of 70 % was achieved.

The instantaneous flow inside a Francis turbine operating in part load conditions has been investigated by Krappel at al. at the Institute of Fluid Mechanics and Hydraulic Machinery (IHS) in Stuttgart. In order to better resolve the turbulent structures, the Scale Adaptive Simulation (SAS) by Menter was applied rather than a "conventional" RANS turbulence model. Two flow solvers were deployed, the commercial Ansys CFX code and the open source code OpenFOAM. Two meshes with 16 and 40 million elements, respectively, were used. As the flow phenomena in a Francis turbine require a long physical time to convect through the whole machine, high computational efforts were necessary. To obtain turbulent statistics, several months of computational time were needed on the Cray XE-6. However, it has to be mentioned that the scaling behaviour of both codes on the Cray is not ideal. Hence, only 192 cores were used for the CFX simulations and 768 for the OpenFOAM simulations, respectively.

The commercial Ansys CFX code has also been used in the work of Boose et al. at the Institute for Internal Combustion Engines and Automotive Engineering (IVK) in Stuttgart. They performed a 3D CFD simulation of a complete twin entry turbocharger under transient conditions generated by the working cycle of a four cylinder Otto engine. The simulation contained the turbine side and the compressor side. The mesh for the flow model consisted of just about 1.75 million nodes. The simulations were performed on the NEC Nehalem cluster. A comparison of the time per iteration shows that the time savings between using 32 and 256 CPUs is only marginal.

Zhang and Laurien from the Institute of Nuclear Technology and Energy Systems (IKE) in Stuttgart have performed numerical simulations of flow with volume condensation in a model containment. In their paper, they present a newly developed volume condensation model with a two-phase flow, which they apply together with the Ansys CFX code. The model is based on the two-fluid model, which simulates a continuous gas mixture and a dispersed droplet. Thermal and mechanical non-equilibrium are considered for the temperatures and velocities between both phases. As two-dimensional simulations were deemed sufficient for the validation of the new model, the computations for an industrial relevant test case could be performed on one node of the NEC Nehalem cluster.

The paper of Masilamani et al. from the University of Siegen is strongly focussed on HPC issues, that is to say they present a very detailed investigation into the performance and scalability of their code on the Cray XE-6. The scientific aim of their project is to better understand the physical phenomena occurring in an electrodialysis process, which uses ion exchange membranes to separate ionic species from sea water under the influence of an electric field. For this purpose, they use a Lattice Boltzmann Model, which they extended to a multi species variant. The performance analysis employing up to 32,768 cores were performed on periodic cubic domains with different compilers and communication patterns. For their actual production run, they used a mesh with 520 million elements. Two lakh twenty five thousand time steps took about 3 h computational time on 4,096 cores. The well-suited behaviour of the code for large scale runs with a high sustained performance is impressively demonstrated.

In summary, a large variety of ambitious project have been performed at the HLRS in the field of Computational Fluid Dynamics. Most of them were of very high quality and had not been realizable without an access to supercomputer facilities. This demonstrates the high value as well as the indispensability of supercomputing in this area.

Discontinuous Galerkin for High Performance Computational Fluid Dynamics

Andrea Beck, Thomas Bolemann, Hannes Frank, Florian Hindenlang, Marc Staudenmaier, Gregor Gassner, and Claus-Dieter Munz

Abstract In this report we present selected simulations performed on the HLRS clusters. Our simulation framework is based on the Discontinuous Galerkin method and consists of four different codes, each of which is developed with a distinct focus. All of those codes are written with a special emphasis on (MPI) based high performance computing. In this report, we show results of our newest code, FLEXI, which is tailored solely towards high performance computing of three dimensional advection dominated problems. Currently, we are interested in extending our direct numerical simulation framework to large eddy simulations and present computations and comparisons of our framework for the flow past a cylinder and the flow past a wing. Furthermore, first results on a MPI based multi-physics extension of our framework is presented. All simulations are typically performed on hundreds and thousands of CPU cores.

1 Introduction

The central goal of our research is the development of high order discretization schemes for a wide range of continuum mechanic problems with a special emphasis on fluid dynamics. Therein, the main research focus lies on the class of Discontinuous Galerkin (DG) schemes. The in-house simulation framework consists of four different Discontinuous Galerkin based codes with different features, such as structured/unstructured grids, non-conforming grids (h-adaptation),

A. Beck (✉) · T. Bolemann · H. Frank · F. Hindenlang · M. Staudenmaier · G. Gassner · Claus-Dieter Munz
Institute of Aerodynamics and Gasdynamics, Universität Stuttgart, Pfaffenwaldring 21, 70569 Stuttgart, Germany
e-mail: beck@iag.uni-stuttgart.de; bolemann@iag.uni-stuttgart.de; frank@iag.uni-stuttgart.de; hindenlang@iag.uni-stuttgart.de; staudenmaier@iag.uni-stuttgart.de; gassner@iag.uni-stuttgart.de; munz@iag.uni-stuttgart.de

W.E. Nagel et al. (eds.), *High Performance Computing in Science and Engineering '13*, DOI 10.1007/978-3-319-02165-2_20,

non-conforming approximations spaces (p-adaptation), high order grids (curved) for approximation of complex geometries, modal and nodal hybrid finite elements and spectral elements with either Legendre-Gauss or Legendre-Gauss-Lobatto nodes. The time discretization is an important aspect in our research and plays a major role in the computing performance of the resulting method. The simulation framework includes standard explicit integrators such as Runge-Kutta, a time accurate local time stepping scheme developed in-house and an implicit time discretization based on implicit Runge-Kutta methods. The general layout of the framework outsources all aspects of a specific physical problem to be solved (e.g. fluid dynamics) in an encapsulated module separated from the main code by clearly defined interfaces. Thus by exchanging this physical problem definition module, the framework is able to solve various partial differential equations such as the compressible Navier-Stokes equations (fluid dynamics), linearized Euler equations (aeroacoustics), Maxwells equations (electrodynamics) and Magnetohydrodynamics equations (Plasma simulation). One of the major foci in the group is the simulation of unsteady compressible turbulence in the context of Large Eddy Simulation (LES) and Direct Numerical Simulation (DNS). Due to the occurrence of multiple spatial and temporal scales in such problems and the resulting high demand in resolution for both, space and time, a high performance computing framework is mandatory. Each new code iteration is based and improved on findings and insights learned during the development of its predecessor. Our newest code iteration called FLEXI (unstructured successor of the structured code STRUKTI) is build from ground up with high performance computing in mind and is currently our main tool to simulate three dimensional PDE based problems.

2 Description of Methods and Algorithms

Discontinuous Galerkin (DG) schemes may be considered a combination of finite volume (FV) and finite element (FE) schemes. While the approximate solution is a continuous polynomial in every grid cell, discontinuities at the grid cell interfaces are allowed which enables the resolution of strong gradients. The jumps on the cell interfaces are resolved by Riemann solver techniques, already well-known from the finite volume community. Due to their interior grid cell resolution with high order polynomials, the DG schemes can use coarser grids. The main advantage of DG schemes compared to other high order schemes (Finite Differences, Reconstructed FV) is that the high order accuracy is preserved even on distorted and irregular grids. The formal order of the scheme can be set by a parameter, namely the polynomial degree of freedom N. Additionally to the formal order of convergence, the polynomial degree N has a strong influence on the dispersion and dissipation behavior. The artificial dissipation of the method is generated via the numerical flux function which is typically based on Riemann solver technology borrowed from the FV community. It can be shown that the overall dissipation in the resolved spectrum decreases with increasing polynomial degree N. Furthermore,

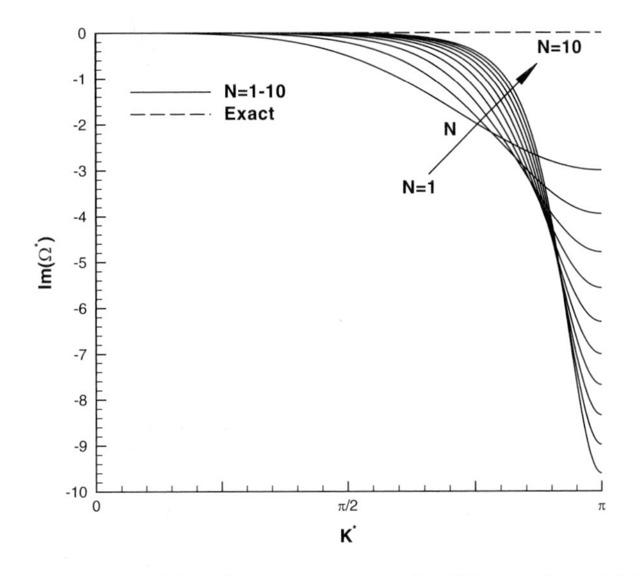

Fig. 1 Dissipation behavior of the DG-derivative operator for different polynomial degrees N

the dissipation behavior for large N acts like a cut-off filter at a certain wavelength, as can be seen in Fig. 1. We have low dissipation in the resolved spectrum and high dissipation in the underresolved part. This is the ideal situation for implicit large eddy simulation (iLES) of turbulent flow.

The low dissipation of high order DG methods also comes with a price: we have a higher sensitivity with respect to aliasing errors which may even lead to aliasing instabilities. A cure for those instabilities lies in the proper evaluation of all spatial integrals [6]. By choosing a sufficient accurate quadrature rule, which also accounts for the non-linearity of the flux function, stability is re-established, also in case of severe underresolution which is typical for large eddy simulations. In conclusion, we use a high order accurate Discontinuous Galerkin discretization with sufficient number of integration points to get a stable and accurate iLES type approach. We use the computing power provided by HLRS to investigate the performance and accuracy of this approach.

In the following subsections, our simulation codes are detailed. All of our codes are developed in FOTRAN 90 with MPI based parallelization.

2.1 High Order DGSEM Solver STRUKTI

A very efficient variant of a Discontinuous Galerkin formulation is the Discontinuous Galerkin spectral element method (DGSEM). This special variation of the DG-method is based on a nodal tensor-product basis with collocated integration and interpolation points on hexahedral elements, allowing for very efficient dimension-by-dimension element-wise operations.

An easy-to-use structured code (STRUKTI) was set up to test the performance of this method, especially for large scale calculations. As a time integration method, a five stage fourth order accurate explicit Runge-Kutta method is implemented and used for all test cases.

2.2 High Order DGSEM Solver FLEXI

To enable the efficient simulation of complex geometries, a second DGSEM based solver was developed. Sharing the same numerical discretization as STRUKTI, FLEXI is tailored to handle unstructured and even non-conforming hexahedra meshes. A base tool for grid pre-processing was developed. This program allows us to process grid files from different commercial grid generators and translate them into readable FLEXI meshes. Furthermore, a module for curved grid generation and for non-conforming grid connection is included in this tool [5]. As FLEXI shares the same efficient discretization as STRUKTI, the performance of both codes is comparable as in a high order method, the effort of managing the grid is negligible. The difference lies in the parallelization of both codes. FLEXI uses domain partition based on space filling curves, whereas STRUKTI is optimized for structured meshes. Benchmarking of STRUKTI and FLEXI and improvements of FLEXI's parallelization is an ongoing important task in the group.

However, the performance results we already have for FLEXI on the HLRS CRAY-XE6 cluster are at least promising and show great strong scalability of the framework. Figure 2 shows the strong parallel scaling of FLEXI up to 1,024 processors for different polynomial degrees N and a fixed mesh of 1,024 grid cells. This means that we are able to perform strong scaling up to the limit of domain decomposition based MPI, namely with only one grid cell on a processor left.

In DG, the load on the processor is not only determined by the number of grid cells, but also likewise by the polynomial degree N. In the left part of Fig. 2 the influence of the polynomial degree on the strong scaling capabilities is investigated. By choosing a polynomial degree of $N = 8$, we get a super-linear strong scaling of over 100 %. With only one or two grid cells on a processor left, the load but also the memory resources are so small, that strong caching effects can be observed. The right part of Fig. 2 shows investigations of the influence of the load per processor for a fixed number of processors (64 cores in this case) on the PID. The performance index (PID) is a very good measure to compare different simulation setups: it is the computational time needed to update one degree of freedom for one time step, and is computed from the total core-h, the total number of timesteps and degrees of freedom

$$\mathrm{PID} = \frac{\text{Wall-clock-time \#cores}}{\text{\#DOF \#Timesteps}} .$$

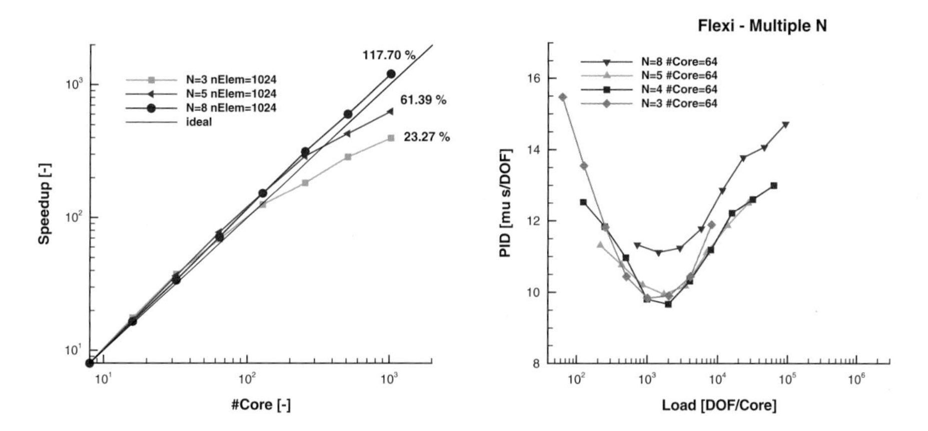

Fig. 2 Strong scaling results for discretizations with constant mesh size and different polynomial degree N. *Left figure* shows the total runtime speedup and right shows the behavior of the specific CPU time index PID

It helps to compare the performance of the different runs. For example, a perfect strong scaling would result in a constant performance index. However, the right part of Fig. 2 shows that there is a strong influence of the local processor load on the PID and that the PID drastically decreases when reducing the load on a processor although the number of processors is constant. This holds until a certain minimum load per processor (2000 DOF/processor) is reached. For even lower values communication overhead causes an increase of the PID. Our experience shows that the minimum value of load/processor depends on the HPC architecture, with 2000 DOF/processor (the lower this value, the better the architecture) being a very good value. When performing a strong scaling investigation where we reach the limits of MPI (one grid cell on a processor), the load on a processor is in the range of hundreds of DOF, thus the caching effects augment the increasing communication overhead resulting in a very good strong scaling capability. If choosing high order polynomials, i.e. $N = 7$, the load on a processor is in the range of the limit of positive caching effects resulting in a real super-linear speed up of the computation as can be seen in Fig. 2.

3 Flow Past a Cylinder at Reynolds number $Re = 3{,}900$

In this section we use a standard LES benchmark example, namely the flow past a cylinder with Reynolds number $Re = 3{,}900$, as a test case to validate our framework. There are many results in literature available, e.g. Kravchenko and Moin used a B-spline based spectral element method with an explicit Smagorinsky model [7] with a resolution varying in-between 1–2 mill DOF, Blackburn and Schmidt

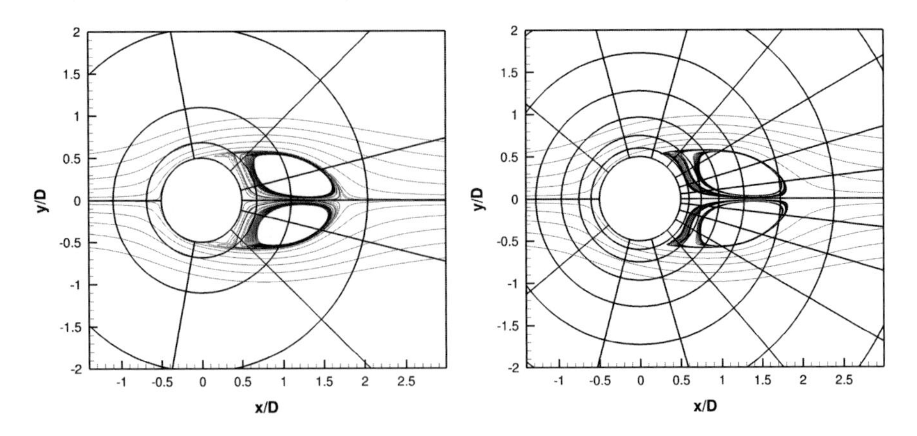

Fig. 3 Zoomed view of the computational grid with mean streamlines of the resulting simulation

used a continuous spectral element method with a dynamic Smagorinsky model [1] with about 1.5 mill DOF; Meyer, Hickel and Adams a finite volume based implicit LES method called ALDM with 6 mill DOF resolution [8]. To compare our results achieved with a compressible solver to the literature results which are all based on simulating the incompressible Navier-Stokes equations, we choose a low Mach number of $M = 0.1$.

For our investigations, we choose two different setups: a setup with a coarse grid of $8 \times 6 \times 6$ grid cells and polynomial degree $N = 11$ resulting in about 500k DOF and a setup with a medium grid of $16 \times 12 \times 12$ grid cells and polynomial degree $N = 7$ resulting in about 1.2 mill DOF. A zoomed view on the computational grid is plotted in Fig. 3. All computations are dealiased with approximately $2\,N$ quadrature points.

After the initial transient phase from uniform flow conditions to the development of the steady vortex shedding, turbulent statistics where gathered over 144 shedding cycles. The $N = 7$ computations were run on 2,304 processors on the Cray XE6, which means that only one element was located on a processors (limit of MPI scaling). The wall-clock time for this run was about 3 h. For the coarse grid calculation with $N = 11$, only 288 processors could be used due to the MPI limit of one grid cell on a processor. The wall-clock time for this run was about 17 h.

3.1 Results

In the first part of the results section, integral quantities are compared to results available in literature. As listed in Table 1, there is a certain spread of the values and no clear reference value can be determined. Both our computations, the computation

Table 1 Comparison of current results with simulations from literature

Author	$C_{p_{Base}}$	Str	C_D	L_r/D
Kravchenko and Moin	−0.94	0.21	1.04	1.35
Blackburn and Schmidt	−0.93	0.218	1.01	1.63
Meyer and Hickel	−0.92	0.21	1.05	1.38
Current: $N = 11$	−1.0	0.212	1.09	1.26
Current: $N = 7$	−0.93	0.208	1.04	1.37

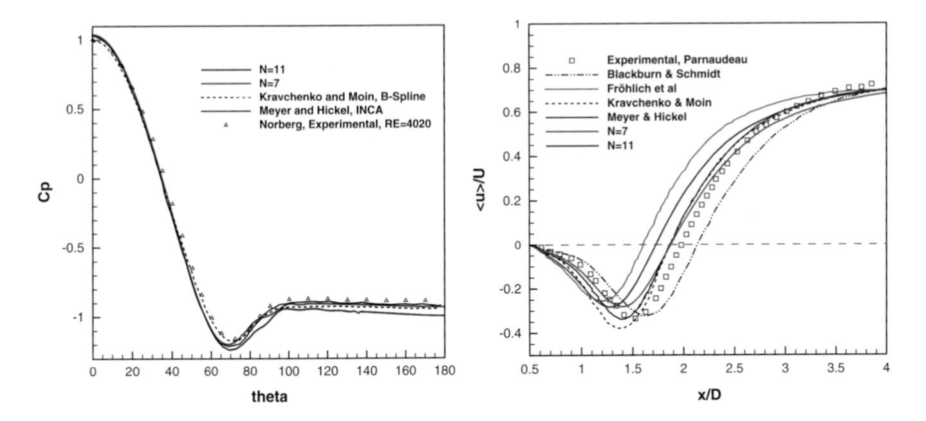

Fig. 4 Distribution of pressure and mean centerline velocity and comparison to results from literature

with coarse mesh and $N = 11$ and the medium mesh computation with $N = 7$, fit well to the range of literature results while using less degrees of freedom than the other approaches. This is also confirmed by the distribution of the pressure and the mean centerline velocity, which compares again well to the available literature results, see Fig. 4.

In a second step, the influence of our artificial dissipation mechanism on the numerical result is investigated. As stated above, artificial dissipation is introduced via the numerical flux functions which are typically based on approximate Riemann solvers. Two very common variants used in the DG community is either the very simple local Lax-Friedrichs flux (LLF) function which is generally known to be highly dissipative and the approximate Riemann solver of Roe which is used because of its low inherent dissipation. Averaged velocity profiles at different distances x/d are compared in Fig. 5 and reveal that the simulation based on the Roe dissipation mechanism is more accurate than the simulation based on the local Lax-Friedrichs flux.

This is an important result for us and will be the base for further investigations in future projects.

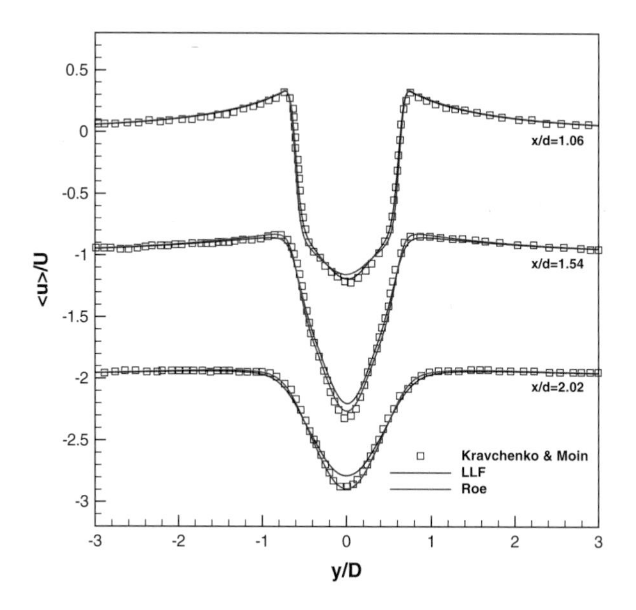

Fig. 5 Comparison of averaged velocity profiles at fixed distances x/d for different numerical flux functions

4 Flow Past an SD7003 Wing at Reynolds number *Re* = 60,000

Following the test case C3.3 from the High Order CFD Workshop 2013 (Köln), we perform simulations of the low Reynolds number flow around an SD7003 airfoil at high angle of attack ($Re = 60{,}000, \alpha = 8$). The Mach number is fixed at $Ma = 0.1$. The aim of this test case is to investigate the capability of implicit LES approaches for mixed laminar, transitional and fully turbulent flows. The described setup results in a short separation bubble at the leading edge of the suction side and subsequent turbulent reattachment.

4.1 Simulation Setup

For the discretization, we choose the polynomial degree $N = 3$ and use a fourth order accurate explicit Runge-Kutta time integration method. In this case, only $1.5\,N$ quadrature points are chosen for dealiasing, as the rather low polynomial degree $N = 3$ offers already a certain amount of artificial dissipation.

For the presented case, we use an unstructured mesh with $21,660$ cells, corresponding to $1.39\,10^6$ degrees of freedom at a polynomial degree of 3. Isothermal walls are applied at the wing surface, the geometry curvature is maintained through

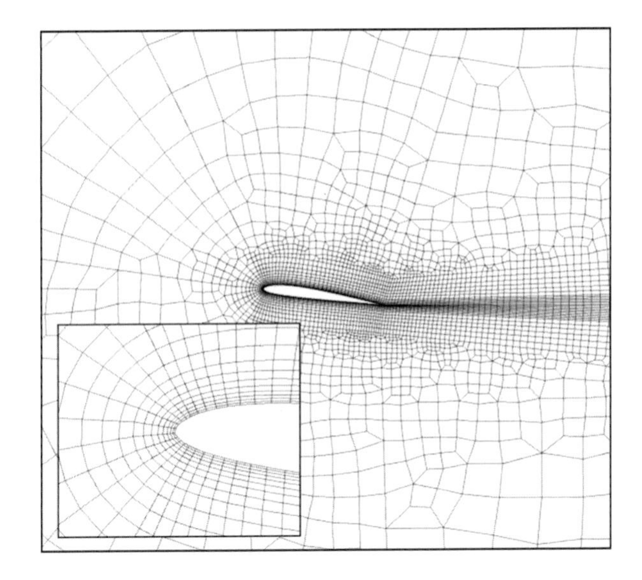

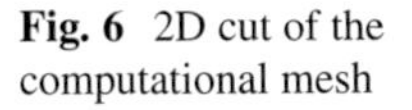

Fig. 6 2D cut of the computational mesh

curved boundaries. In the spanwise direction, we resolve 0.2 chords and employ periodic boundary conditions. The farfield boundary with free-stream conditions is located five to seven chords away from the wing. Since the influence of the farfield boundaries cannot be guaranteed to vanish with this domain size, further work will use larger domains and address this issue. The grid used in this simulation is shown in Fig. 6.

4.2 Results

In order to collect a sufficient amount of statistical data, the simulation is run for 20 convective times (C/U_∞ with the chord C and the free-stream velocity U_∞). The simulation was performed on the Cray XE6 Hermit cluster at HLRS with 2,272 processors and took 8.35 h.

As can be seen in the lift and drag coefficients evolution in Fig. 7, an unsteady transition phase is followed by a statistical steady state.

We plot iso-surfaces of the Q criterion at $t = 10$ in Fig. 8 as a qualitative visualization of the instantaneous near-wall shear flow. Vortical structures emerge at the leading edge of the suction side due to the high angle of attack, the remainder of the suction side's boundary layer remains turbulent.

More quantitative features of this flow are discussed using time averaged data. We extract the data in the time interval $t \in [8, 20]$. In Table 2, the mean aerodynamic loads (lift and drag coefficient) are compared to results from literature. Our results compare favorably to the published data.

Figure 9 (left) shows the computed negative pressure coefficient in comparison with LES results from literature. While the distribution on the pressure coefficients

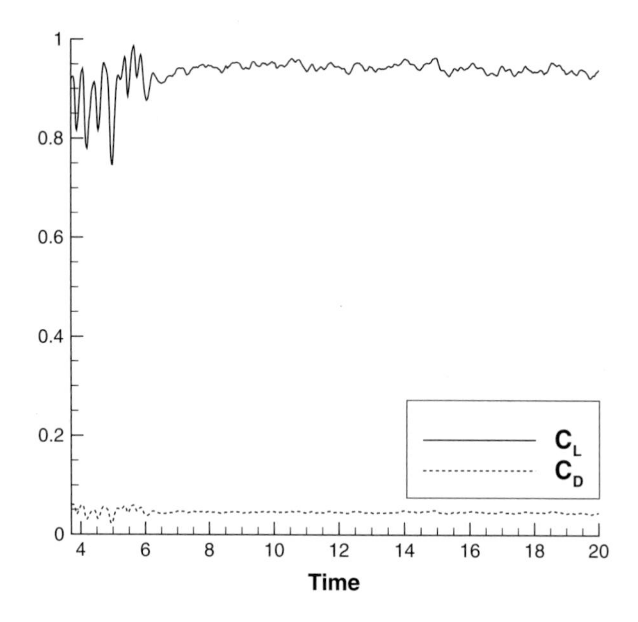

Fig. 7 Evolution of the drag and lift coefficient over time

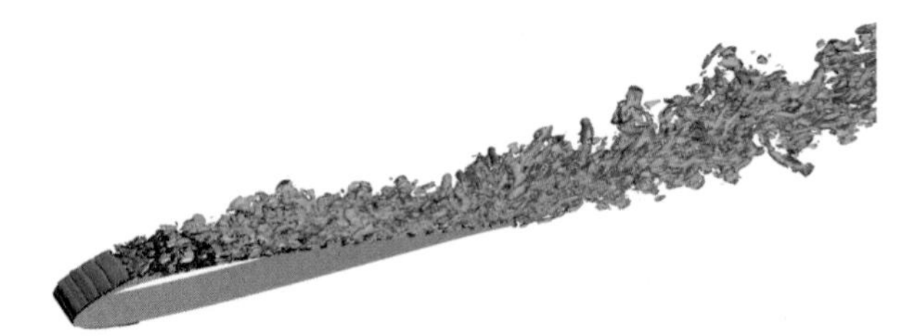

Fig. 8 Iso-surfaces of the Q criterion ($Q = 50$), colored with vorticity magnitude, after 10 convection times

Table 2 Mean aerodynamic loads and comparison to literature

	C_L	C_D
DGSEM, $N = 3$	0.943	0.0455
Garmann and Visbal [4]	0.9696	0.0391
Galbraith and Visbal [3]	≈0.91	≈0.043
Catalano and Tognaccini [2]	≈0.94	≈0.044

on the pressure side is in good agreement between all references and our solution, some differences are visible in the separation region on the first half of the suction side.

The friction coefficient shown in the left part of Fig. 9 reveals an even wider spread between the different results for this case. The negative peak in C_f is in good quantitative agreement with the results of Garmann and Visbal as well as Catalano

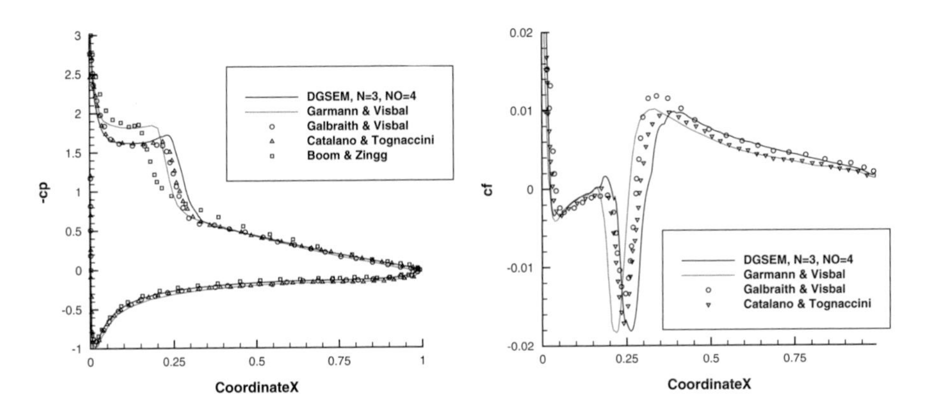

Fig. 9 Time-averaged pressure coefficient (*left*) and friction coefficient (*right*)

Table 3 Separation (x_1) and reattachment (x_2) locations and comparison to literature

	x_1	x_2
DGSEM, $N = 3, NO = 4$	0.025	0.312
Garmann and Visbal [4]	0.023	0.259
Galbraith and Visbal [3]	0.040	0.280
Catalano and Tognaccini [2]	0.030	0.290

and Tognaccini, while the location of this peak is slightly shifted. The computed separation and reattachment points on the suction side are listed in Table 3.

Evaluation of this test case is very difficult, since no DNS data is available for reference. We will continue our work on this case with mesh convergence investigations and higher polynomial degrees as well as larger computational domains.

5 Fluid-Structure Coupling

In the last example, we are presenting the first result obtained with an extension of our framework to a complex multi-physics application. We are considering a fluid-structure interaction, where the flow solver FLEXI is coupled using MPI to a structural solver. The problem is a membrane water pump. The membrane moves up and down and induces the flow to move through the pump, without the need of valves. A sketch of the pump is shown in Fig. 10. In Fig. 11, the unstructured hexahedral mesh, consisting of 1,162 elements, and the flow solution with a deformed membrane is shown. A polynomial degree of $N = 2$ was used, which leads to 31,374 degrees of freedom. The simulation was run with up to 1,162 cores, using only one element per core.

With this test case, the interaction of fluid and structure is investigated. Because of the unsteady nature of the problem and large deformations, a strong coupling

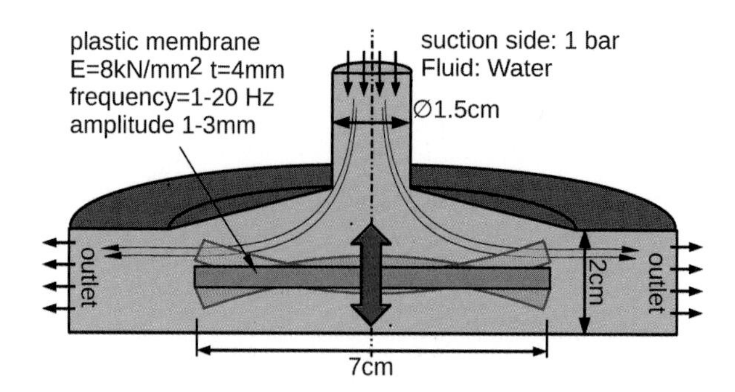

Fig. 10 Setup of the membrane pump fluid-structure interaction

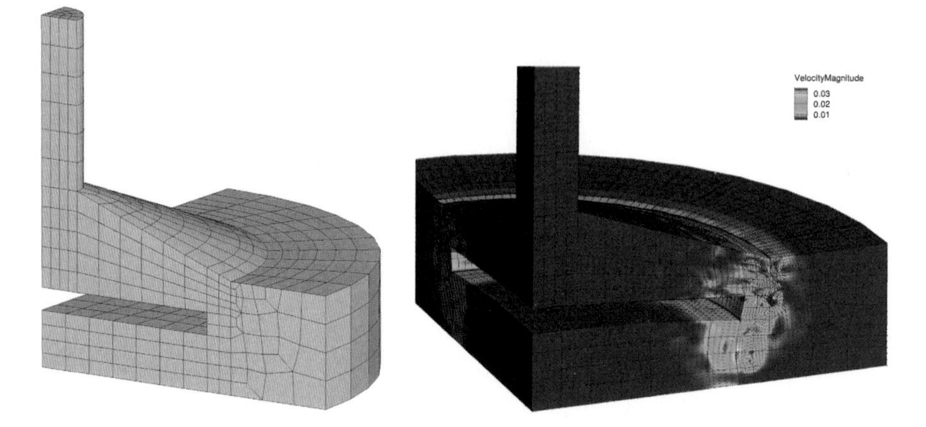

Fig. 11 Mesh of the quarter part of the pump, and flow velocity with a deformed membrane

is envisaged, explicitly in time. We need to solve not only for the flow dynamics, but also the structure dynamics, and as well realize a coupling between the two physics. We will follow a separated approach, where the flow solver works on a moving mesh, and is coupled via MPI to a structural solver. The flow solver FLEXI introduced in Sect. 2.2 was modified, to be able to treat moving meshes. Here, we implemented the approach presented in [9], which leads to a time-consistent DG method on moving meshes for the simulation of unsteady flows.

Up to now, the structure dynamics of the circular membrane is modeled via a one-dimensional fourth order equation, making use of the axial symmetry of the membrane. It is solved on one core with a fast Chebyshev-Collocation method and integrated implicitly in time. The pressure forces are integrated over the membrane surface and communicated from the flow solver to the structure solver in each time step. Then the one-dimensional deformation is passed via MPI messages to the flow solver, which is transformed to a three-dimensional mesh movement. A sketch of the coupling methodology is shown in Fig. 12. Both codes run separately, but inside the

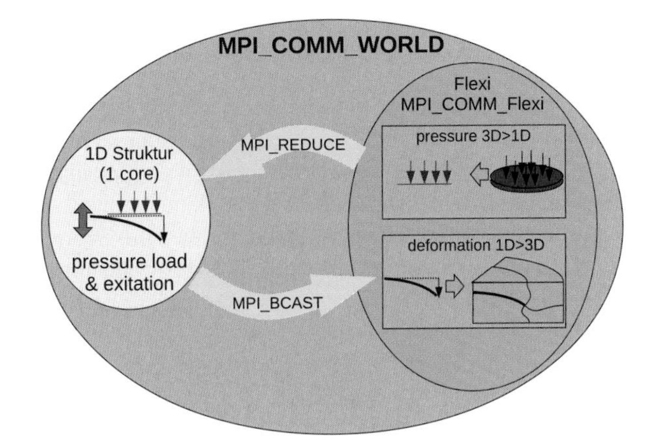

Fig. 12 MPI methodology for the coupling between flow and structure solver

Table 4 Pump simulations (FLEXI $N = 2$, coupled and not coupled) on the HLRS CRAY XE6 system

	#cores	DOF per core	Sim. time	Wall-clock time (h)	Core-h	PID (μs)
Not coupled	581	54	0.4	3.38	1,940	27
Not coupled	1,162	27	0.15	0.78	914	32
Coupled	581	54	0.3	0.13	76.2	28.5

global communicator `MPI_COMM_WORLD`. This is a multiple program multiple data approach (MPMD), and both executables are executed in the same MPI universe using the command

```
mpiexec -n 1  ./struct structure.ini : -n 512 ./flexi flow.ini
```

where the flow solver runs on 512 cores and the structural solver on a single core. This gives a lot of flexibility in both codes, since only the coupling data is interchanged via MPI communication, and the data structures of each program are not affected. In addition, since both codes run parallel to each other, and a timestep for the flow solution takes more time than the structure solution, the coupling data is always ready for communication, leading to a very low overhead, see the difference in the performance index (PID) in Table 4. The uncoupled reference PID is 27 μs whereas the coupled computation has a PID of 28 μs.

Table 4 summarizes the computational time spend on different runs of the problem. Factoring in the different physical runtimes of the simulation (sim. time in the table), a strong scaling from two (PID = 27 μs) to one element (PID = 32 μs) per core has still a scaling efficiency of 84 %.

6 Summary and Outlook

In this project, we have extended our framework to simulate compressible high Reynolds number flows by means of a large eddy type approach and successfully computed the turbulent flow past a cylinder and a wing with typical runs on $> 1{,}000$ processors. The available CPU resources are used to demonstrate the high accuracy of this novel approach in comparison to standard methods from the literature. Furthermore, in the last test case, an extension of this code to a multi-physics application where the MPI interface is used as a connector of the different field models is shown.

Our typical reliable "production" runs use $O(1{,}000)$ processors. In the future, we plan to extend and improve our framework to support reliable simulation runs on $O(10{,}000)$ and even on $O(100{,}000)$ processors to fully unleash the available (and projected) processing power of the HLRS CRAY-XE6 cluster.

Acknowledgements The research presented in this paper was supported in parts by Deutsche Forschungsgemeinschaft (DFG), amongst others within the Schwerpunktprogramm 1276: Met-Stroem and the Graduiertenkolleg 1095: Aerothermodynamische Auslegung eines Scramjet-Antriebssystems für zukünftige Raumtransportsysteme and the research projects IDIHOM within the European Research Framework Programme.

References

1. H.M. Blackburn, S. Schmidt, Large eddy simulation of flow past a circular cylinder, in *Proceedings of 14th Australasian Fluid Mechanics Conference*, Adelaide (2001)
2. P. Catalano, R. Tognaccini, Large eddy simulations of the flow around the SD7003 airfoil, in *Proceedings of AIMETA Conference*, Bologna (2011)
3. M.C. Galbraith, M.R. Visbal, Implicit large-eddy simulation of low Reynolds number flow past the SD 7003 airfoil, in *Proceedings of 46th AIAA Aerospace Sciences Meeting and Exhibit*, Reno (2008)
4. D.J. Garmann, M.R. Visbal, High-order solutions of transitional flow over the SD7003 airfoil using compact finite-differencing and filtering. Presented at the 1st International Workshop on High-Order CFD Methods, 2012, 50th AIAA Aerospace Sciences Meeting, Nashville
5. F. Hindenlang, G. Gassner, T. Bolemann, C.-D. Munz, Unstructured high order grids and their application in Discontinuous Galerkin methods, in *Conference Proceedings, V European Conference on Computational Fluid Dynamics (ECCOMAS CFD 2010)*, Lisbon, 2010
6. R.M. Kirby, G.E. Karniadakis, De-aliasing on non-uniform grids: algorithms and applications. J. Comput. Phys. **191** (2003)
7. A.G. Kravchenko, P. Moin, Numerical studies of flow over a circular cylinder at ReD = 3900. Phys. Fluids **12**, 403 (2000)
8. M. Meyer, S. Hickel, N.A. Adams, Assessment of implicit large-eddy simulation with a conservative immersed interface method for turbulent cylinder flow. Int. J. Heat Fluid Flow **31**(3) (2010)
9. C.A.A. Minoli, D.A. Kopriva, Discontinuous Galerkin spectral element approximations on moving meshes. J. Comput. Phys. **230**(5), 1876 (2011)

Industrial Turbulence Simulations at Large Scale

Daniel F. Harlacher, Sabine Roller, Florian Hindenlang, Claus-Dieter Munz, Tim Kraus, Martin Fischer, Koen Geurts, Matthias Meinke, Tobias Klühspies, Yevgeniya Kovalenko, and Uwe Küster

Abstract The most important aspect for simulations in industrial design processes is the time to solution. To obtain highly detailed results nevertheless, massive computational resources have to be deployed. Feasibility and applicability of HPC systems to this purpose is the main focus of this paper. Two different numerical approaches, implemented with parallelism in mind, are investigated with respect to quality as well as turn around times on large super computing systems. The one approach compares the efficiency of high order schemes on coarser meshes to lower order schemes on finer meshes. The second approach employs a zonal coupling of LES and RANS to limit the computational effort by using solution adapted models.

D.F. Harlacher · S. Roller (✉)
Simulationstechnik und Wissenschaftliches Rechnen, Universität Siegen, 57078 Siegen, Germany
e-mail: daniel.harlacher@uni-siegen.de; sabine.roller@uni-siegen.de

F. Hindenlang · C.-D. Munz
Institut für Aerodynamik und Gasdynamik, Universität Stuttgart, 70569 Stuttgart, Germany
e-mail: hindenlang@iag.uni-stuttgart.de; munz@iag.uni-stuttgart.de

T. Kraus · M. Fischer
Robert Bosch GmbH, 70049 Stuttgart, Germany
e-mail: Tim.Kraus2@de.bosch.com; Martin.Fischer7@de.bosch.com

K. Geurts · M. Meinke
Chair of Fluid Mechanics and Institute of Aerodynamics, RWTH Aachen University, 52056 Aachen, Germany
e-mail: k.geurts@aia.rwth-aachen.de; m.meinke@aia.rwth-aachen.de

T. Klühspies
Trumpf Werkzeugmaschinen GmbH + Co. KG, 71254 Ditzingen, Germany
e-mail: tobias.kluehspies@de.trumpf.com

Y. Kovalenko · U. Küster
Höchstleistungsrechenzentrum Stuttgart (HLRS), Nobelstr. 19, 70569 Stuttgart, Germany
e-mail: kovalenko@hlrs.de; kuester@hlrs.de

W.E. Nagel et al. (eds.), *High Performance Computing in Science and Engineering '13*, DOI 10.1007/978-3-319-02165-2_21,

Three industrial use-cases evaluate the performance and quality of these approaches. General optimizations are presented as well as solutions for load-balancing.

1 Introduction

In the context of the STEDG project the applicability of academic fluid dynamic solvers executed on distributed systems to industrial test-cases is investigated. It is shown that commercial solvers cannot fulfill the given requirements, both in simulation time and solution quality. A strong limitation here is the scalability of these commercial software solutions, which prohibit large scale deployments on massively parallel super computing systems. The focus of this paper therefore is on performance and scalability of two different numerical methods which are designed for efficient simulation of turbulent flows.

The outline of the paper is as follows: after a short introduction to the codes in Sect. 2, the strategies to enable efficiency on large scale distributed systems are presented in Sect. 3. First, we present general performance optimizations. We present how routines that prohibit optimal serial performance are identified and show the improvements in the adapted over the original implementation. Then the porting experience onto the new Hermit System is presented. The available compilers on this system are compared with respect to the speed of the resulting executable and how much time was spend to get the code compiling. Afterwards, intra-node scaling of the code is presented and compared to the behavior on different systems. Thereafter inter-node scaling is presented and compared. Sections 4–6 describe the industrial test cases with their results and performance evaluation before Sect. 7 briefly summarizes the experiences.

2 Description of Methods and Algorithms

We shortly describe the key features of the two numerical schemes. The one is a high order discontinuous Galerkin (DG), the other a second order finite volume (FV) scheme with special implementations regarding LES computations. Actually, finite volume schemes can be found in most commercial flow solvers, i.e. CFX or FLUENT, thus being state of the art in industrial development processes.

High Order Discontinuous Galerkin Scheme may be considered as a combination of finite volume (FV) and finite element (FE) schemes. While the approximate solution is a continuous polynomial in every grid cell as in FE, discontinuities at the grid cell interfaces are allowed which enables the resolution of strong gradients as in FV. The jumps on the cell interfaces are resolved by Riemann solver techniques, already well-known from the finite volume community. Due to their interior grid cell resolution with high order polynomials, the DG schemes may use coarser grids. The discontinuous Galerkin scheme developed in the group of Prof. Munz is implemented in the code HALO (Highly Adaptive Local Operator). The code runs on unstructured meshes consisting of hexahedra, prisms, pyramids and tetrahedra. The code is designed for the computation of unsteady flow problems using explicit time stepping. Each grid cell only needs direct neighbor information which allows a very efficient parallelization. The computation domain is decomposed by either ParMetis or recently also by the use of space filling curves and fully parallelized with MPI . A major disadvantage of an explicit DG scheme may be the global time step restriction to establish stability. This restriction depends on the grid cell size, on the degree of the polynomial approximation, on wave speeds for advection terms, and on diffusion coefficients for diffusion terms. In HALO, this drawback is overcome by a special time discretization, so called time-consistent local time stepping [2, 6]. The stability criterion is only locally applied to each grid cell, thus each cell runs with its optimal time step. Thus, the computational effort is concentrated on the grid cells with small time steps. On meshes with strong heterogeneity in grid cells or flow velocities, the number of operations is greatly reduced compared to an global time stepping approach.

Zonal RANS-LES Coupling on 2nd Order Finite Volume Scheme: TFS, the flow solver of the Institute of Aerodynamics Aachen solves the Navier-Stokes equations for compressible flows on a block-structured grid. A modified AUSM method as introduced in [5] is used for the Euler terms which are discretized to second-order accuracy by an upwind-based approximation. For the non-Euler terms, a centered approximation of second-order accuracy is used. The temporal integration is done by a second-order accurate explicit 5-stage Runge-Kutta method with coefficients optimized for maximum stability, [7]. The sub-grid scale modeling for the large-eddy simulations is based on an implicit ansatz, i.e., the MILES (monotone integrated LES) approach of Boris et al. [1]. For the RANS zones the Spalart-Allmaras turbulence model was chosen to close the Reynolds-averaged Navier-Stokes equations. To reduce computation time, the solution methods based on RANS and LES can be combined into a zonal method. The LES regions are used to resolve the leading and trailing edge region where flow separation occurs, while the RANS zone is used for the attached flow regions. The schematics of the overlapping zones is shown in Figs. 1 and 2, where the values that have to be communicated back and forth between the two different approaches are indicated. In the overlapping region, where the flow is directed from a RANS to LES zone, synthetic eddies are introduced to accelerate the generation of coherent turbulent structures using the method of Jarrin [4]. Control planes are used to drive the solution towards the correct turbulence level in the LES domain according to Spille and Kaltenbach [8]. When the flow is coming from the LES into the RANS domain, the RANS requires a definition of the eddy viscosity ν_t, where the value is reconstructed from time and spatial averaging of the LES data.

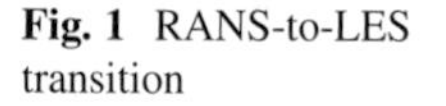

Fig. 1 RANS-to-LES transition

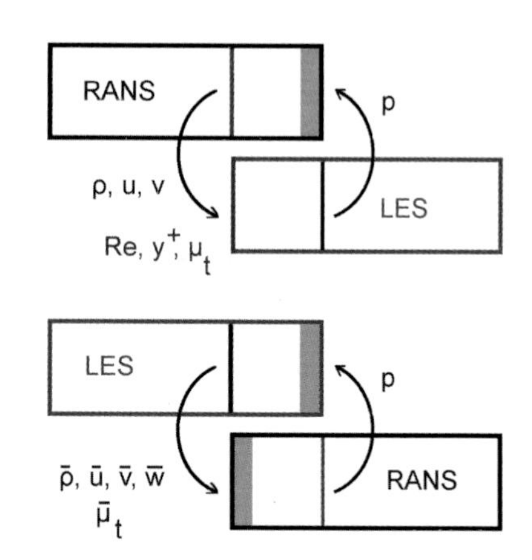

Fig. 2 LES-to-RANS transition

3 HPC: Issues, Techniques, and Optimization

3.1 General Performance Optimization

The efficiency of a scheme is influenced by its order of accuracy since this has a direct influence on the locality of the data and thus the memory access patterns. Therefore, serial performance was analyzed for three different orders of accuracy (2nd, 4th, and 6th) for the DG code HALO, executed on the NEC Nehalem Cluster Laki. With the help of Intel Vtune Amplifier XE 2011 the bottlenecks of the application could be identified. The code uses complicated nested data structures including pointers, which are not transparent to the compiler. Therefore compilers are not able to optimize the code to an acceptable degree. Global changes in the data model would result in considerable changes to the overall program itself. Thus, the data model was modified only in the most time-consuming places, which were consuming about 80 % of the run time . In those functions the intrinsic MATMUL-function was replaced by unrolled loops, data types were changed from Fortran pointers to Fortran allocatables. Those changes allow for both better hand and more aggressive compiler optimizations. Improvements in the run time of these functions of up to 70 % have been achieved. The reduction of the total execution time was 54 % when using the Intel 12.0 compiler with GNU 4.6 compiler.

Further optimization concerned the replacement of short arrays by variables and inlining of short procedures by hand. Some essential loops were switched between different sizes, hand-unrolled where possible, merged or simplified. Those code modifications enabled better compiler optimization and allowed to reduce the run time by 25 %. Further improvements would affect the data model and would imply large changes of the entire application.

The memory usage of the HALO application was analyzed with Valgrinds Massif tool. It measures how much heap memory the program uses over time. The peak memory consumption was measured, which includes not only the computational part but also the setup part of execution. Table 1 shows the relationship of chosen

Table 1 Memory usage

Example	Order	DOFs	No. elements	Memory for computations (MB)	Memory peak (MB)	Memory/element (MB)
	6	1,512	27	13.5	14.4	0.5
LES model	5	2,240	64	27.25	48.97	0.43
disabled	2	168,128	42,032	1146.9	1249.28	0.027
LES model	6	1,512	27	34.1	34.9	1.26
enabled	5	945	27	16.4	16.4	0.59
	4	540	27	7.8	7.8	0.26

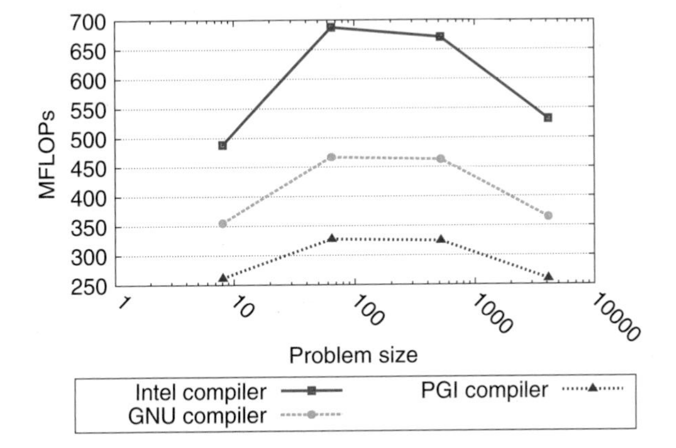

Fig. 3 Serial performance of HALO, compiled with different compilers (Intel, GNU, PGI)

scheme order and the element size. It is also shown that the usage of the LES model increases the memory-demand per element due to additional data which need to be stored in the elements.

3.2 *Hermit System at HLRS*

3.2.1 Porting

With the availability of the new Cray XE6 system at the HLRS the HALO code was ported and tested on the new environment. Unfortunately, we couldn't compile the HALO code with the Cray programming environment due to internal compiler errors. Therefore investigating the other available environments (Intel, GNU and PGI), we found large discrepancies regarding the speed of the different executables produced by the different compilers. Figure 3 shows the comparison of the serial performance for the different compilers. It can be seen that for the HALO the Intel compiler produces the fastest executable. GNU and PGI compiled executables only reach 68.5 % and 48.7 % respectively of the performance of the Intel executable.

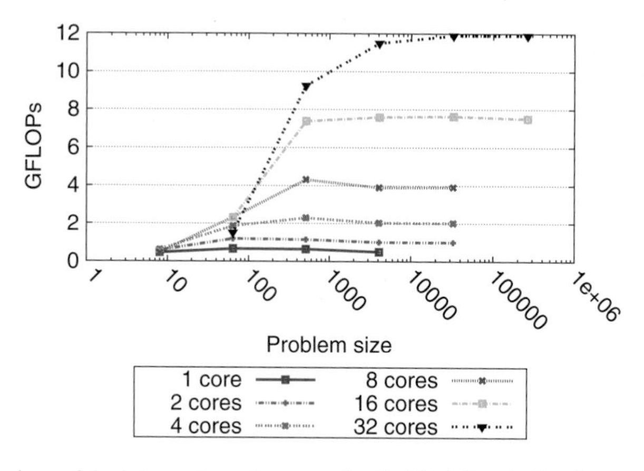

Fig. 4 Comparison of the intra-node performance for the HALO code (Intel)

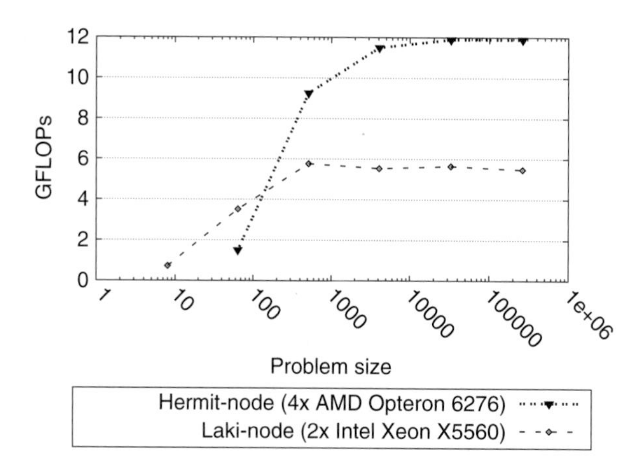

Fig. 5 Comparison of node performance of the HALO on Hermit and Laki

Being one of the first machines with the new AMD Interlagos architecture, the behavior of the code within a node was unclear. Especially due to the high number of cores per node it was not a given that the usage of every core is beneficial for the overall code performance. Therefore an intra-node scaling was performed using the Intel environment. Figure 4 shows that the code scales very well within the node. The usage of all cores on a node gives the best performance. With 12 GigaFlops the code achieves 4 % of the theoretical peak performance of 294.4 GigaFlops. To compare the node performance to the NEC Nehalem cluster (Laki) we performed the same intra-node test on the Laki system as well, using OpenMPI 1.5.4 and the Intel compiler 12.0.4. Figure 5 shows the comparison between a Hermit node using all available 32 cores and a Laki node using all available 8 cores. It can be seen, that a Hermit node achieves more than twice the performance of a Laki node—however the relation of the performance per core is exactly the reverse. Execution times of

Table 2 Single core runtime on Cray XE6 and NEC Nehalem Cluster

	Cray XE6		NEC Nehalem		Runtime reduction	
Examples	Intel -O2-g(s)	PGI -fast (s)	Intel -O2 -g (s)	PGI -fast (s)	Intel -O2 -g (%)	PGI -fast (%)
LES disabled	104.35	256.74	68.17	160.41	34.7	37.5
LES enabled	274.99	749.60	191.26	429.89	30.4	42.7

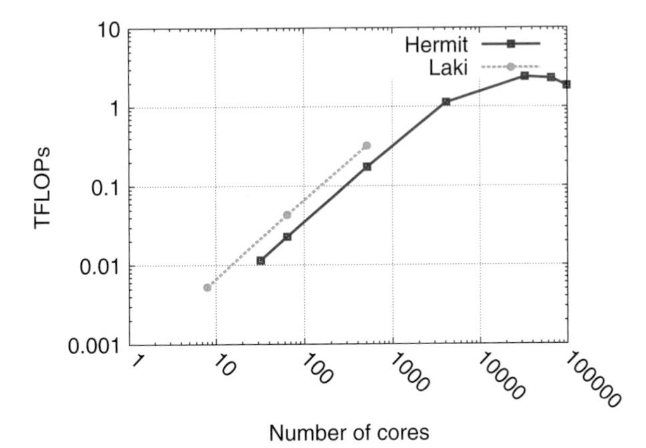

Fig. 6 Performance HALO for small problem sizes under strong scaling on Hermit and Laki

HALO on Laki and Hermit for Intel and PGI compilers are presented in Table 2. On Laki the single core run time is up to 42 % less than on Hermit.

3.2.2 Scaling

In this section we investigate the inter-node behavior of the HALO code on the Hermit system. Figure 6 shows the scaling behaviour for the largest mesh fitting onto a single Hermit node (262,144 elements) on Hermit and Laki. The results show that the scaling on Laki for up to 512 cores is identical to the behaviour on Hermit, but resulting in higher overall performance due to the stronger per-core performance on Laki. To achieve the same performance and therefore similar run time for this test case on Hermit around twice as many cores are required. This corresponds well with the data from the serial investigations. Additionally the figure contains the information for the maximum speedup that is effectively possible. A total speedup of 100 can be achieved for a problem size, that just fits into main memory, before the scaling degrades. To demonstrate the scaling capabilities of the HALO code a larger mesh (two million elements) for the test case was used to show scalability on up to 3,072 nodes on the Hermit system. Figure 7 shows the results, while indicating the minimum number of cores the problem needed to run. The code scales well up to 2,048 nodes and shows a drop in performance above due to the low number

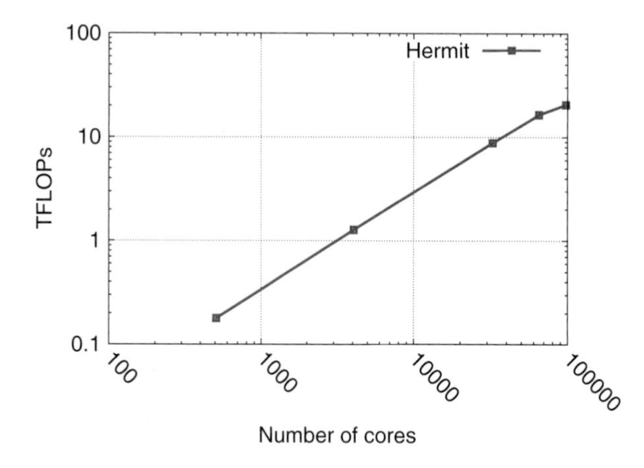

Fig. 7 Performance of the HALO code for large problems under strong scaling on Hermit

of elements per core/node left at this high numbers of processes. Again the total speedup that can be achieved for the problem is a factor of 100 which shows that the scaling behaviour of the code is not influenced by scaling to higher number of cores and shows a predictable performance within the complete Hermit system.

3.3 Dynamic Load-Balancing

The findings in Sect. 3.2 were achieved using examples that were designed to behave completely the same on all nodes. Therefore the results can be considered a upper bound of the performance of the code. For highly instationary flows in which the computational load shifts on the computing cores over time due to physical effects propagating within the domain this balance is not naturally given. Dynamic load-balancing becomes crucial to keep the high parallel efficiency, especially on large number of processes as e.g. a spatial concentrated load-imbalance gets worse when increasing the number of processes.

In the context of the project a partitioning algorithm based on space-filling curves (SFCs) was developed. This algorithm (SPartA) relies on the inherent given locality of SFCs and exploits the fact that the shifting of elements along the SFC only has minor influence on the locality of the partition. SPartA calculates the new partitions completely in parallel which guarantees good scalability. Scaling and memory investigations of the algorithm were done and presented in a separate paper [3]. Using SPartA within the HALO application provides high parallel efficiency even for highly dynamic systems. A supersonic free-stream was chosen to evaluate the possible gain in parallel efficiency. The used mesh consists of two million hexahedral elements. Fourth order in time and space were used to solve the problem with a LES model. The jet starts with a high imbalance at $t = 0$. Figure 8 shows the

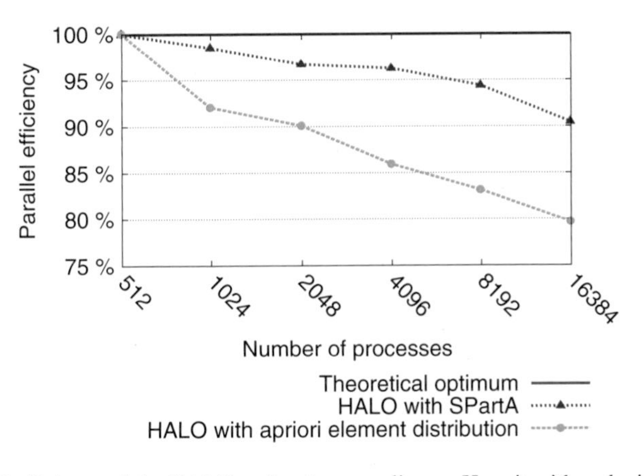

Fig. 8 Parallel efficiency of the HALO under strong scaling on Hermit with and without dynamic load-balancing

scalability of the HALO with and without dynamic load-balancing during the start-up phase of the simulation. It is shown that the parallel efficiency without dynamic load-balancing drops significantly with increasing number of cores. This is to be expected due to the concentrated imbalance introduced in the simulation. Using dynamic load-balancing however the code maintains high parallel efficiency over the start-up phase of the simulation.

4 Industrial Test-Cases I: Gas Injection Nozzle

In this test case setup, air is expanded through a gas injection valve from its operating pressure to ambient conditions ($\rho_\infty = 1.2\,\mathrm{kg/m^3}$ $u_\infty = v_\infty = w_\infty = 0.0\,\mathrm{m/s}$ and $p_\infty = 100{,}000\,\mathrm{Pa}$). The nozzle exit geometry of the valve, as depicted in Fig. 9, represents a silencer duct with a main-diameter of 6.11×10^{-3} m. In total, the length of the geometry is 3.74×10^{-3} m. At the outlet of the silencer, a notch reduces the inner diameter to 5.95×10^{-3} m. At the base of the duct a step narrows the flow domain to a diameter of 2.26×10^{-3} m. In the bottom of the duct, four kidney shaped orifices are symmetrically positioned around a flow separating truncated cone that arises in the middle of the silencer. These inlets are located at a radius of $r_0 = 1.71 \times 10^{-3}$ m (measured from middle-axis of the duct to the averaged free-stream center). The Reynolds number based on the width of an orifice is $Re = 52{,}000$ ($\mu = 3.7 \times 10^{-5}\,\mathrm{m^2/s}$, $Pr = 0.72$).

To capture the acoustic wave generation of the jets separately as well as the flow and wave interaction a simulation of the entire three-dimensional geometry is performed. Due to high flow speed, sophisticated device geometry and the small size of the shock-cells, a very high resolution of physics is required within the silencer geometry. Therefore this domain demands the largest part of computational

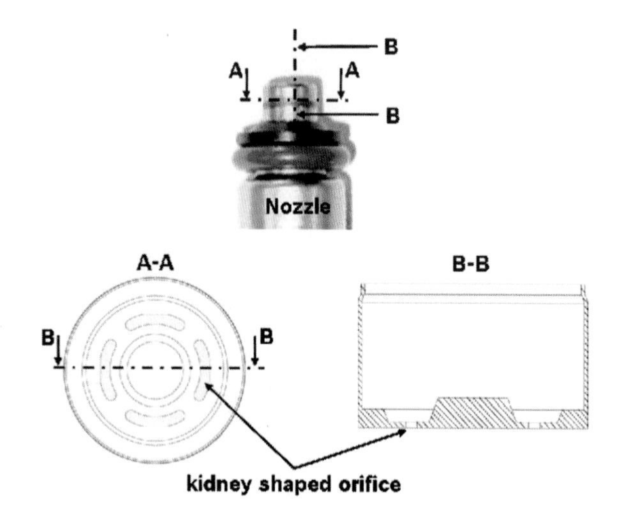

Fig. 9 Nozzle exit geometry

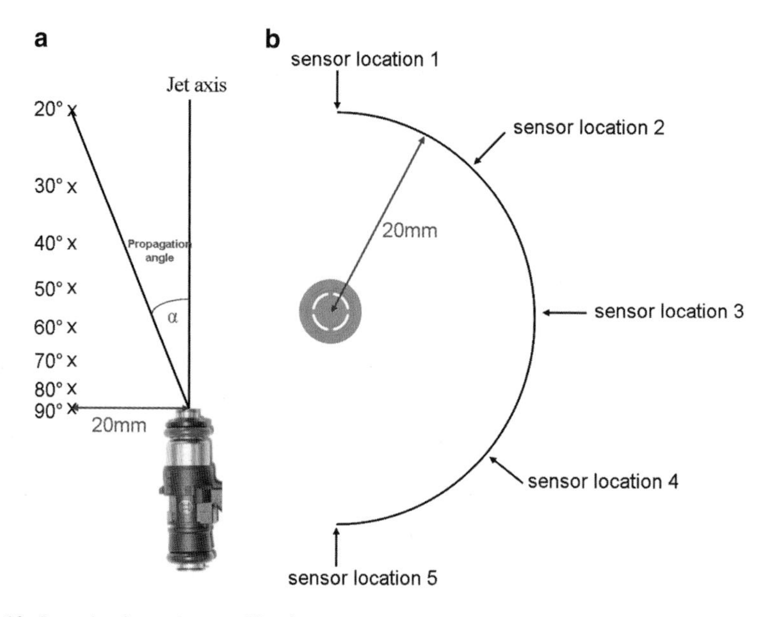

Fig. 10 Investigation point specification

efforts in the simulation. With a calculated Kolmogorov length of $\sim 10^{-7}$ m a full direct numerical simulation of all turbulence scales within the silencer is out of reach. Thus, all simulations are under-resolved, whereas the resolution significantly influences the quality of results which will be shown in Sect. 4.1.

Experimental data of pressure fluctuations at different propagation angles between 20° and 90° at a distance of 20 mm (see Fig. 10a) is available over

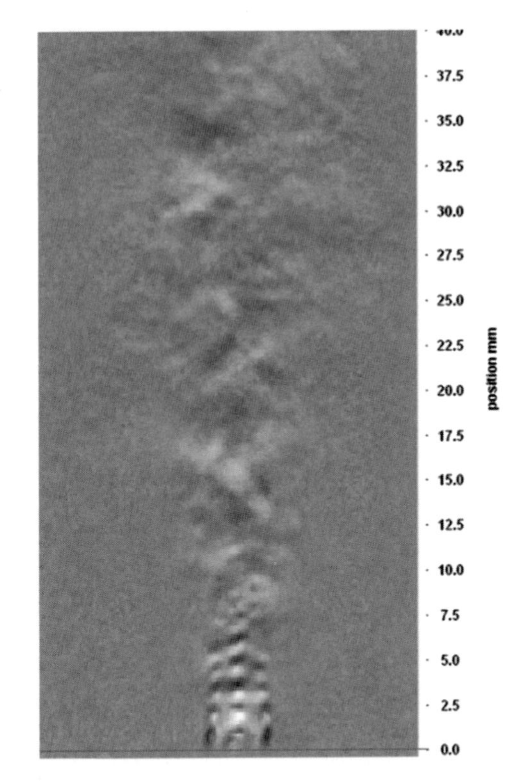

Fig. 11 Schlieren image

a time period of 5 ms. In order to accelerate statistical convergence, the data is averaged over five positions distributed in azimuthal direction around the jet axis (see Fig. 10b). The problem is computed by solving the compressible three dimensional Navier-Stokes equations with the commercial Solver Ansys CFX with a Smagorinsky subgrid scale model and with a stabilized discontinuous Galerkin scheme 2 of 4th order in space and a 3rd order time integration.

The grid for the DG Scheme consists 323.125 unstructured hexahedra, yielding a total of 6.462.500 degrees of freedom. Concerning the Ansys CFX simulation this grid was refined by a factor of three in all dimensions resulting in a total of 8.724.375 Cells. The simulations run for $T_{sim} = 5.5$ ms where the last 5 ms are used for averaging and data monitoring, allowing a reliable evaluation of the overall sound pressure level at the points specified in Fig. 10.

4.1 Results: Flow Field, Acoustics and Solver Performance

The supersonic nozzle exit conditions produce four separated under-expanded jets that are each dominated by a system of shock-cells. The collapse of these structures is observed at $z = 4r_o$ (see Fig. 11) where the four jets combine to a highly turbulent jet flow with an opening angle of 9.5° with respect to the jet axis.

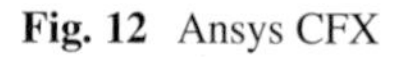

Fig. 12 Ansys CFX

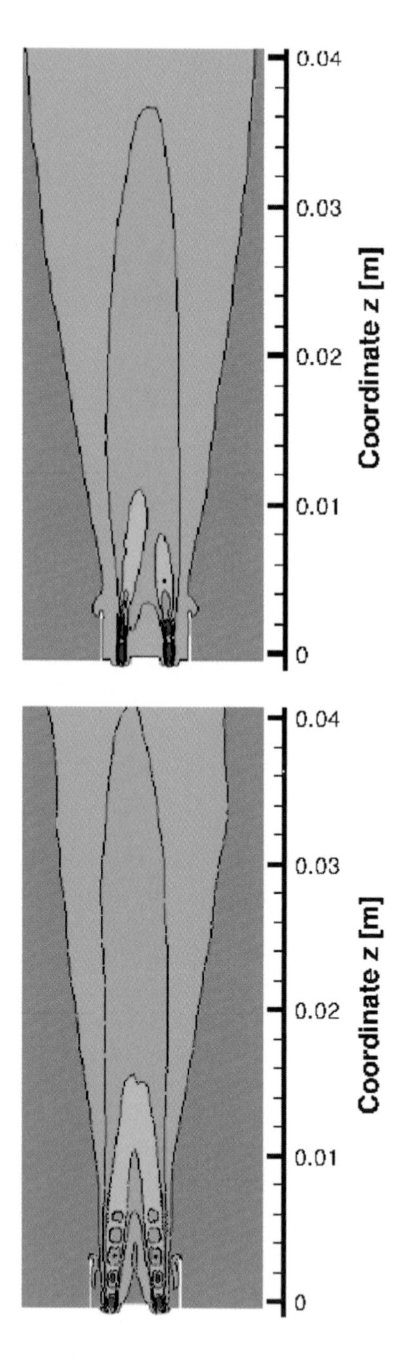

Fig. 13 HALO

In Figs. 12 and 13 an A-A cut plane shows the time-averaged Mach contours M = 0,0; 0,1; 0,5; 1,0; 1,5 and 2,0 in the jet region obtained by Ansys CFX and the HALO solver. The shock structures collapse at $z = 2.34r_0$ and $z = 3.8r_0$,

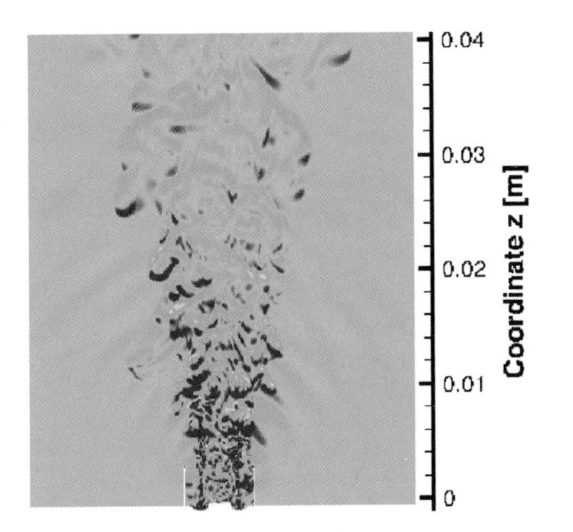

Fig. 14 Ansys CFX Results: stream-wise Density Gradient −200; 500 kg/m^4

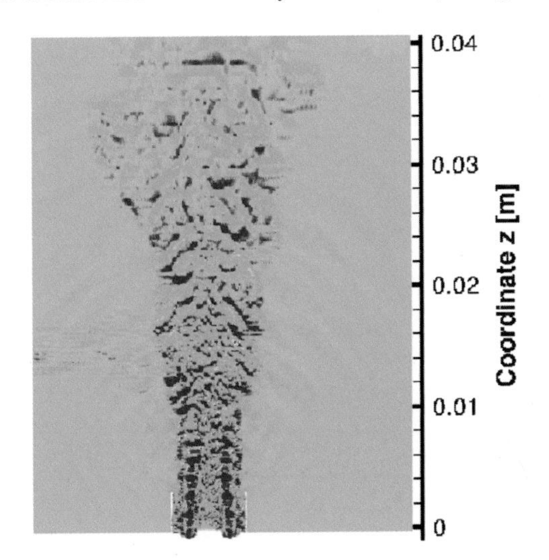

Fig. 15 HALO Results: stream-wise Density Gradient −200; 500 kg/m^4

respectively, showing that the HALO code is able to resolve the shocks properly in contrast to Ansys CFX. Due to the earlier break up of the shock structures the opening angle of the jet is slightly overpredicted by Ansys CFX with 10.6° while the HALO Solver result matches well with experimental data.

In Figs. 14 and 15 one can see a snapshot of the density gradient in stream-wise direction showing that the DG scheme of 4th order is able to resolve smaller turbulent structures on a coarse grid then Ansys CFX on a refined grid.

Figure 16 shows the directivity pattern of the overall sound pressure level. The levels detected from experimental data decrease nearly linear with larger

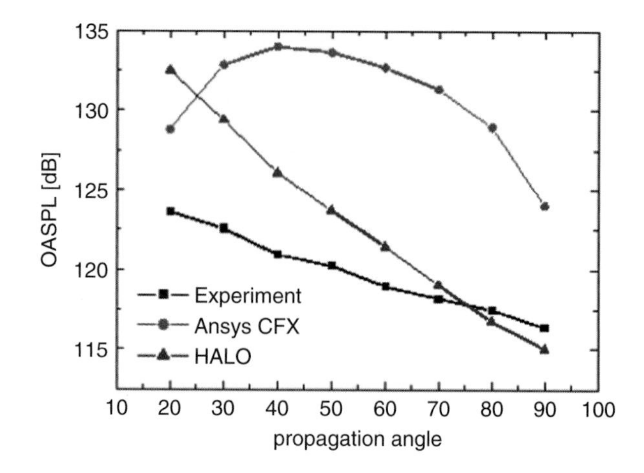

Fig. 16 Directivity for present simulations

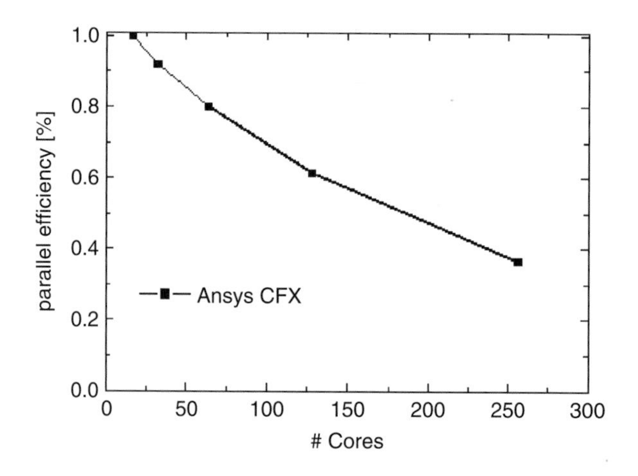

Fig. 17 CFX scaling results

observation angles from 123.5 to 116 dB. The acoustics predicted with Ansys CFX show a different pattern with an increasing level from 128 dB at 20° observation angle to 134 dB at 40° and a following reduction down to 125 dB at 90°. Thus, the overall sound pressure level is overpredicted at all observation points with a maximum of 15 dB at 50°. These discrepancies could be due to the under-resolution of the provided grid and the poor prediction of shock structures at the jet basis which strongly influence the correct prediction of relevant sound sources. Due to the better resolution of shocks and turbulences, the prediction of aeroacoustics with the HALO solver can reproduce the observed directivity pattern quite well. While the deviation of overall sound pressure level towards experiments at observation angles from 70° to 90° are already smaller than 2 dB, the levels at smaller angles are still overpredicted with a maximum of 7 dB at 20°. These results show that this simulation approach is still under-resolved and an improvement is expected with an approach of 5th order in space.

The computational costs per 1 ms simulation time were 113 CPU days with Ansys CFX and 186 CPU days with the HALO solver. In Fig. 17 the Strong scaling results of both solvers, concerning this testcase, are presented. While Ansys CFX shows a good parallel efficiency only up to 64 Cores (83 %). The usage of more than 128 Cores, with an acceptable efficiency of 67 %, is not reasonable.

5 Industrial Test-Cases II: Laser Cutting Device

In this study, the simulation of a laser cutting device is conducted with the discontinuous Galerkin code HALO, especially regarding to the unsteady behavior of the flow. Until now, only stationary simulations produced by FLUENT were known for this setup. The model consists of a nozzle ($d = 2.3$ mm) positioned over a plate ($t = 15$ mm) with an existing kerf, see Fig. 18a. The simulation is done for the half model with a symmetry boundary condition in the symmetry plane, the mesh is displayed in Fig. 18b. The nozzle has a fixed position and all wall boundary conditions are isothermal at ambient conditions, in order to reproduce the experimental setup. A high pressure configuration is investigated, in which the gauge pressure at the nozzle inlet is at 21 bar, yielding high Mach numbers and strong shocks inside the kerf. The boundary conditions are listed in Table 3, together with the fluid properties. The hybrid mesh consists of 755,908 cells (747,116 tetrahedra) resulting in 3.02×10^6 DOF using a 2nd order discretization. The simulations were performed on the HLRS CRAY XE6 system with 512 and 2,048 cores.

5.1 *Initialization*

The initialization of the unsteady simulation has to be done carefully. First, we simply initialized the domain with a constant pressure. Due to the high pressure ratio, this strongly increased the simulation time, since the transient start of the flow was be reproduced, making the simulation nearly unfeasible. Instead, a steady state solution produced by FLUENT, which is already available for this test case, was used to initialize the unsteady simulation. Now, the unsteady flow is fully developed after 0.8 ms simulation time, or accordingly $\approx$1,000 core-h. The initial flow field and the fully developed unsteady flow through the nozzle and the kerf is shown Fig. 19.

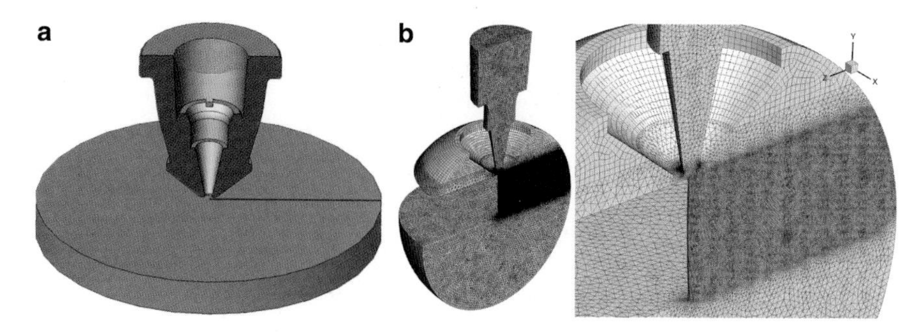

Fig. 18 (**a**) CAD geometry and (**b**) volume mesh of the half model

Table 3 Fluid properties and boundary conditions

Nitrogen, gas constant	$R = 296.8$ J/(kg K)	Viscosity	$\mu = 1.663 \times 10^{-5}$ m^2/s
Inlet pressure	$p_0 = 2,201,325$ Pa	Ambient pressure	$p_\infty = 101,325$ Pa
Inlet temperature	$T_0 = 293.15$ K	Ambient temperature	$T_\infty = T_{wall} = 293.15$ K
Inlet density	$\rho_0 = 25.3$ kg/m^3	Ambient density	$\rho_\infty = 1.164$ kg/m^3

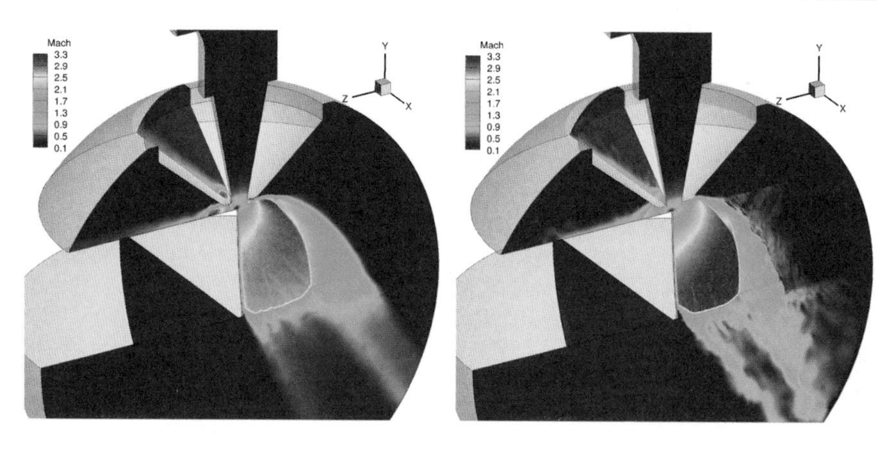

Fig. 19 Mach number distribution in the yz and xy plane. *Left*: steady state FLUENT solution. *Right*: instantaneous HALO solution after $t = 0.85$ ms

5.2 Results

In this section, the HALO simulation results are compared to the steady state FLUENT solution. Furthermore, the unsteady flow data inside the kerf is analyzed.

In areas of strong shocks, artificial viscosity is applied for stabilization of the HALO code. The location and the strength, detected by a density indicator is shown in Fig. 20. The shocks are well detected. In addition, high values at the nozzle exit, where no strong shocks occur, may indicate that the mesh in this area is too coarse.

For the direct comparison of HALO and FLUENT, the unsteady flow data is averaged in time, for $\Delta t_{avg} = [0.9\,\text{ms}, 3.0\,\text{ms}]$. In Fig. 21, the nozzle and the inside

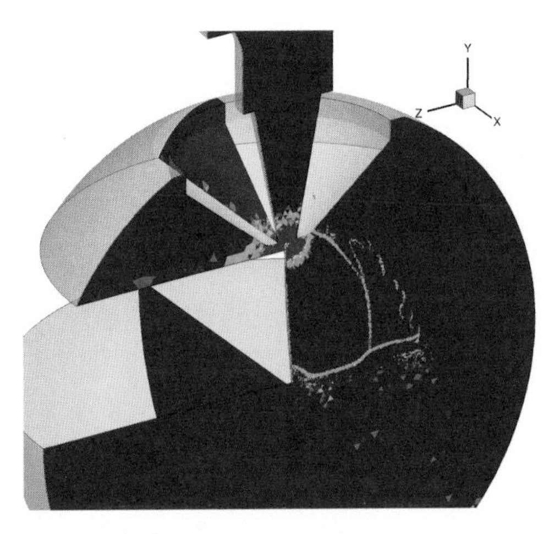

Fig. 20 Shock indicator showing strong variations in density

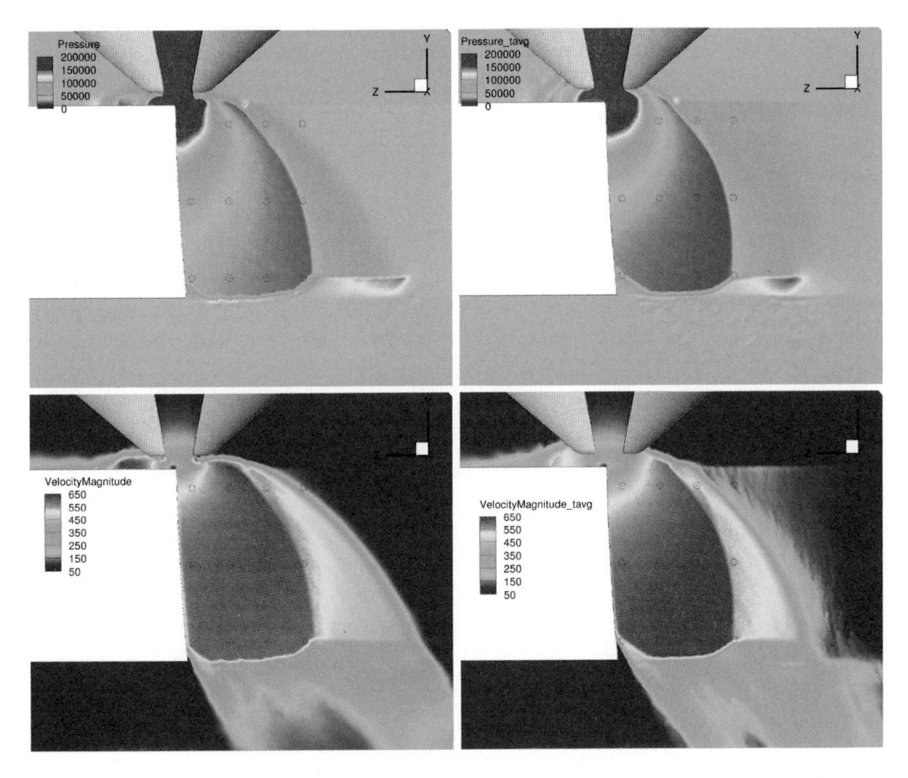

Fig. 21 Pressure (Pa) and velocity magnitude (m/s) in the symmetry plane. *Left*: steady state FLUENT solution. *Right*: time averaged HALO solution $\Delta t_{\text{avg}} = [0.9\,\text{ms}, 3.0\,\text{ms}]$

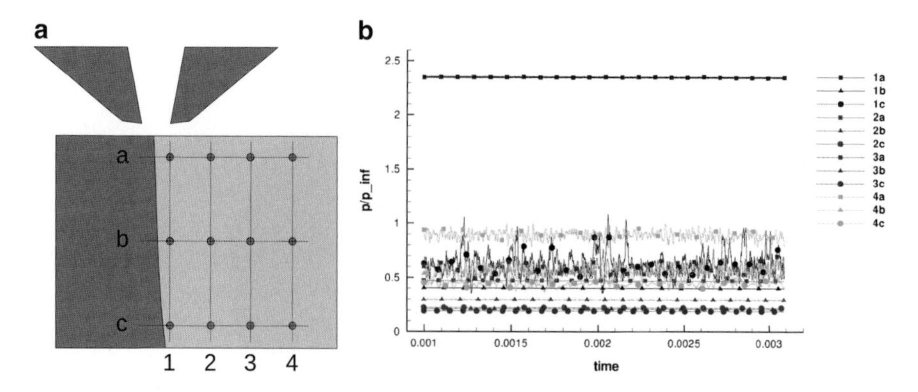

Fig. 22 (**a**) Nomenclature of the pressure tappings and (**b**) temporal evolution of the pressure at these positions

of the kerf is shown for both simulations. The solutions differ mainly at the nozzle exit, where the high flow velocity and the coarse mesh do not permit to resolve the boundary layers. The position of the shocks and the shear layer differ between FLUENT and HALO, though the magnitudes of the pressure and velocities are equal. At the cutting edge, the HALO solution separates earlier and the length of the outflow under the kerf is smaller.

At specific record points, pressure data is tracked in time during the simulation. The dots in Fig. 21 indicate the positions of the record points. They are chosen to match the pressure tappings of the experiment, with a diameter of $d = 0.5$ mm, see Fig. 22a. The pressure data is first averaged in time and then averaged spatially over the diameter on each tapping. The comparison between both codes and the experiment is displayed in Table 4. For simulation data, the mean as well as the maximum and minimum pressure at the tapping are given. If a shock is near the tapping, large pressure variation over the diameter occurs ($1c, 3a, 4c$). No agreement is found with the experiment, with differences up to 66 %, and differences between the codes up to 52 %. Notice that due to high pressure jumps over the shocks, a small difference in shock position can cause these deviations. However, further investigations using higher resolutions, especially at the nozzle exit, are necessary to confirm this assumption.

In the HALO simulation, it was found that the flow is inherently unsteady inside the kerf, with a turbulent shear layer and oscillating shocks. In Fig. 22b, the temporal evolution of the pressure is plotted. In Table 5, the root mean square of the pressure fluctuations is shown. The largest fluctuations are found at $1c, 3a, 4a, 4b, 4c$, which are all tappings lying in proximity to shocks or shear layers. At position $1c$, the strongest fluctuations are found. Here, we transformed the pressure fluctuations in the time interval $t = 0.1$–0.3 ms by a Fourier transform, see Fig. 23. The gray shaded area indicates the limit of the frequency resolution. The lowest resolved

Table 4 Comparison of mean pressure (in bar) at the pressure tappings

	Experiment	FLUENT $\binom{\max}{\min}$	Diff. to Exp.	HALO $\binom{\max}{\min}$	Diff. to Exp.	HALO/FLUENT
1a	2.210	2.267 $\binom{+19\,\%}{-15\,\%}$	+3 %	2.359 $\binom{+18\,\%}{-16\,\%}$	+7 %	+4 %
1b	0.618	0.517 $\binom{+4\,\%}{-4\,\%}$	−16 %	0.409 $\binom{+4\,\%}{-4\,\%}$	−34 %	−26 %
1c	0.741	0.339 $\binom{+3\,\%}{-3\,\%}$	−54 %	0.627 $\binom{+41\,\%}{-49\,\%}$	−15 %	+46 %
2a	0.501	0.475 $\binom{+12\,\%}{-12\,\%}$	−5 %	0.482 $\binom{+13\,\%}{-11\,\%}$	−4 %	1 %
2b	0.534	0.387 $\binom{+3\,\%}{-3\,\%}$	−28 %	0.304 $\binom{+4\,\%}{-4\,\%}$	−43 %	−27 %
2c	0.612	0.291 $\binom{+3\,\%}{-2\,\%}$	−53 %	0.228 $\binom{+2\,\%}{-2\,\%}$	−63 %	−27 %
3a	0.708	0.578 $\binom{+12\,\%}{-8\,\%}$	−18 %	0.607 $\binom{+17\,\%}{-10\,\%}$	−14 %	+5 %
3b	0.344	0.274 $\binom{+5\,\%}{-5\,\%}$	−20 %	0.217 $\binom{+3\,\%}{-2\,\%}$	−37 %	−26 %
3c	0.482	0.253 $\binom{+3\,\%}{-2\,\%}$	−47 %	0.198 $\binom{+2\,\%}{-2\,\%}$	−59 %	−28 %
4a	0.920	0.987 $\binom{+1\,\%}{-1\,\%}$	+7 %	0.911 $\binom{+2\,\%}{-1\,\%}$	−1 %	−8 %
4b	0.496	0.422 $\binom{+14\,\%}{-37\,\%}$	−15 %	0.539 $\binom{+6\,\%}{-6\,\%}$	+9 %	+22 %
4c	0.684	0.233 $\binom{+36\,\%}{-8\,\%}$	−66 %	0.480 $\binom{+68\,\%}{-44\,\%}$	−30 %	+52 %

Table 5 Pressure fluctuations at the pressure tappings

	1a	1b	1c	2a	2b	2c	*3a*	3b	3c	*4a*	*4b*	*4c*
mean (bar)	2.36	0.41	0.63	0.48	0.30	0.23	0.61	0.22	0.20	0.91	0.54	0.48
RMS (mbar)	7.5	0.1	**138**	0.3	0.09	0.05	**70.7**	0.06	0.05	**45.6**	**44.4**	**84.9**

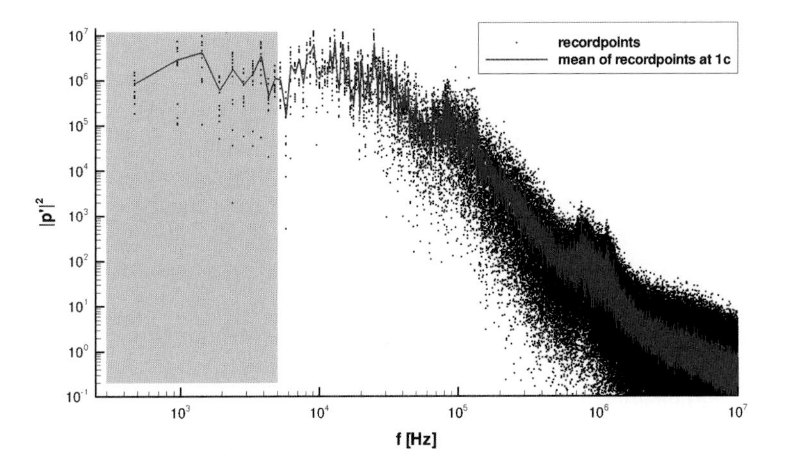

Fig. 23 Fourier transform of pressure fluctuations at tapping *1c* (*grey*: unresolved frequencies)

frequency for a time interval of 2 ms is 10/2 ms = 5,000 Hz (a minimum of 10 modes). For the tapping 1c, there is no unique peak frequency, but the maximum peak is around 10,000 Hz.

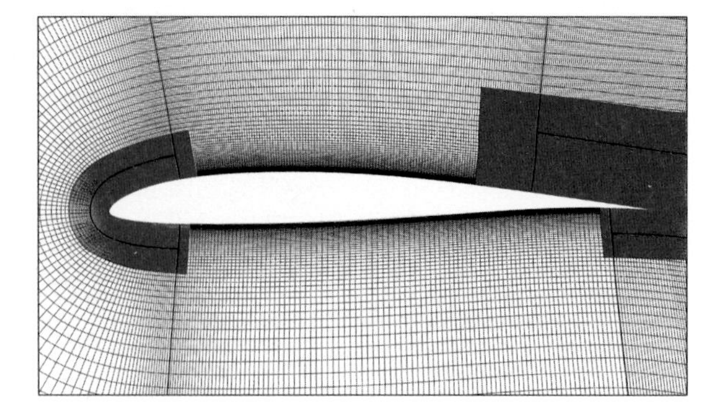

Fig. 24 Zonal grid (*red* = LES, *black* = RANS)

6 Industrial Test-Cases III: Airfoil Aerodynamics

6.1 HGR-01 Profile at High Angle of Attack

In this study, the configuration at an angle of attack of 12° with laminar separation bubble and trailing-edge separation is simulated. The Reynolds number based on the chord length is $Re_c = 0.65 \cdot 10^6$. Results of a pure LES computation are used as reference data for a fully coupled zonal RANS-LES solution. The transition at the upper surface is predicted by the LES while the lower surface is entirely laminar. The grid resolution for the pure LES computation and the LES domain of the zonal RANS-LES simulation is chosen according to Zhang et al. [10]. The resolution of the pure LES grid in the stream-wise, wall normal, and span-wise direction of $\Delta x^+ \approx 100$, $\Delta y^+_{min} \approx 1$ and $\Delta z^+ \approx 20$, respectively, results in a mesh with $51.4 \cdot 10^6$ grid points. The span-wise extension of the grid is $0.02\,c$. Using the same grid resolution and span-wise extension for the LES domains in the zonal RANS-LES grid, the total number of grid points was reduced by a factor of 4 to $13.2 \cdot 10^6$ grid points.

The complexity of this test case for a zonal RANS-LES approach not only lies in the simulation of the different flow phenomena but also in positioning the LES domains and the transition from the RANS into LES domains and vice versa. These transition regions are located at positions in the flow where different conditions exist, such as laminar and turbulent flow and both positive and negative pressure gradients. One LES domain surrounds the leading edge to capture the LSB and the laminar-to-turbulent transition. A second LES region is located at the trailing edge to accurately predict the highly unsteady behavior of the trailing edge separation. The rest of the computation domain is meshed with a RANS grid. An overview of the grid lay-out around the HGR-01 airfoil can be found in Fig. 24.

The flow dynamics simulated in the LES domains around the leading edge and the trailing edge are visualized in Fig. 25. The LSB and the laminar-to-turbulent transition are visualized by λ_2 structures in Fig. 25a. The vortex shedding of the LSB is clearly visible as well as the three-dimensionality of the flow after transition.

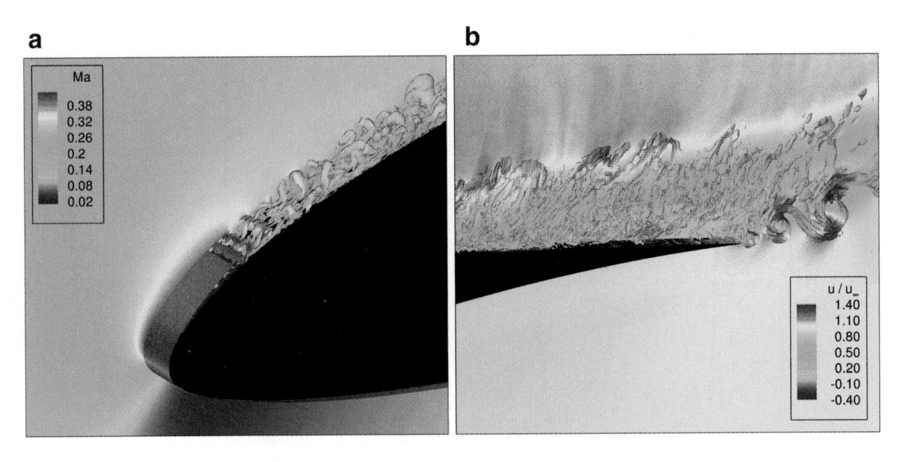

Fig. 25 λ_2 structures from the zonal computation showing the LSB and laminar-to-turbulent transition at the leading edge and the turbulent separation at the trailing edge, mapped on Mach number and stream wise velocity respectively. (**a**) Close-up of leading edge. (**b**) Close-up trailing edge

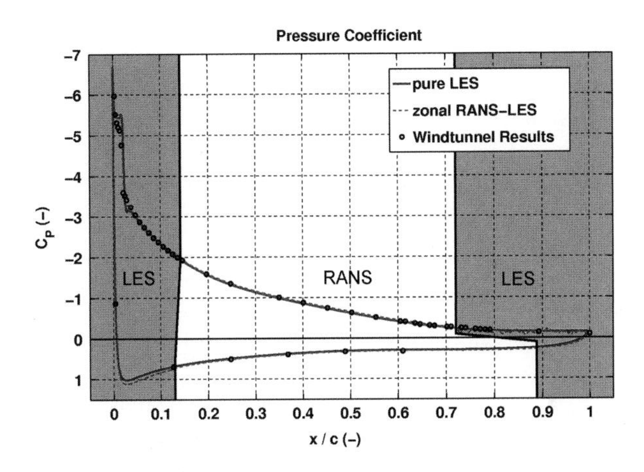

Fig. 26 Pressure coefficient c_p at upper and lower side of the HGR-01 airfoil for the zonal RANS-LES, pure LES computations, and experiments

Figure 25b visualizes the flow at the trailing-edge separation and the large vortex structures in the wake.

Figure 26 shows the averaged pressure coefficient c_p. The grey shaded areas represent the LES domains around the leading and the trailing edge. A smooth transition from the RANS- to the LES zone can be observed. Precise simulation of the LSB at the leading edge is essential for the flow dynamics of the entire airfoil. The difficulty herein comes from the position of the LES inflow boundary upstream of the leading edge, with a large negative pressure gradient from the incoming flow. The close-up clearly visualizes the ability of the zonal method to capture the position

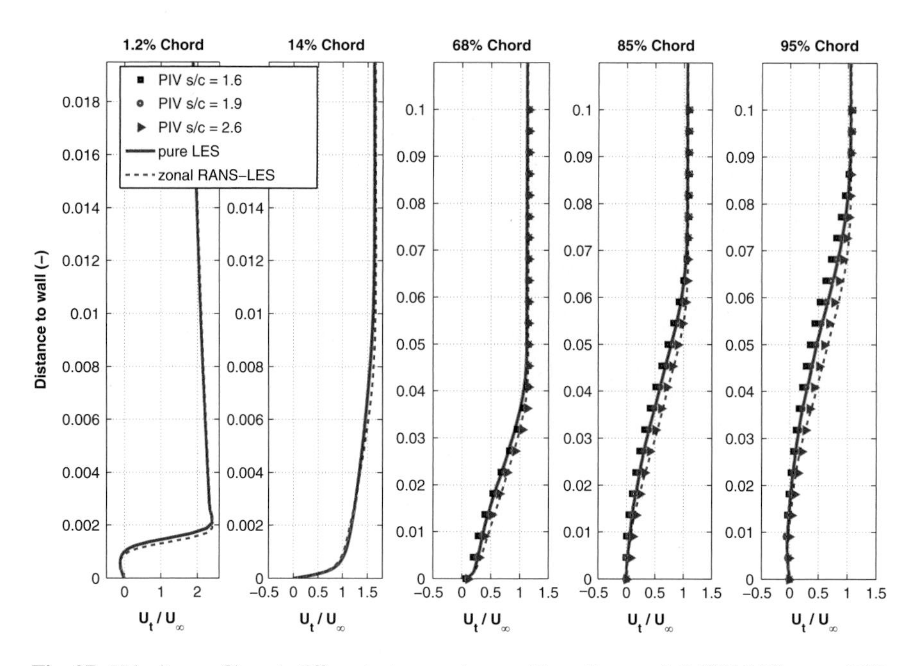

Fig. 27 Velocity profiles at different stream-wise positions for zonal RANS-LES, pure LES computations and experiments

and length of the LSB. The small deviation with respect to the experiments depends on the absence of free-stream turbulence in the pure LES and the zonal computation.

The velocity profiles in Fig. 27 are positioned at several stream wise positions on the upper surface of the HGR-01 airfoil. The profiles from left to right represent the LSB at $0.012\ c$, the LES-to-RANS transition at $0.14\ c$ and three profiles in the trailing edge LES region, i.e. at $0.68\ c$, $0.85\ c$ and $0.95\ c$. The results of the averaged zonal RANS-LES and the pure LES are compared, at the trailing edge separation region particle-image velocimetry (PIV) data are used to validate the numerical results. The PIV results depend on the span-wise position and show a small three-dimensional effect in the trailing-edge separation. The maximum span s of the experimentally investigated airfoil is $3.25\ s/c$ and the visualized PIV results represent the velocity profiles at 1.6, 1.9 and $2.6 s/c$ respectively. Both the reference pure LES computation as well as the zonal RANS-LES simulation show good agreement with the PIV measurements. The importance of this test case can be found in the correct simulation of the LSB together with the laminar-to-turbulent transition, as these phenomena influence the entire flow field around the airfoil. The position of the trailing-edge separation and the velocity profiles in the re-circulation zone show that flow characteristics are well transferred from the LES region around the leading edge to the RANS domain at the upper surface and again into the LES region at the trailing edge. The fact that the zonal RANS-LES method is capable of reproducing these phenomena with high precision demonstrates the capabilities

Table 6 Lift and drag comparison

	LES	ZONAL	Experiments	RANS
Lift C_l	1.366	1.426	1.370	1.53
Drag C_d	0.0403	0.0414	0.032	0.028

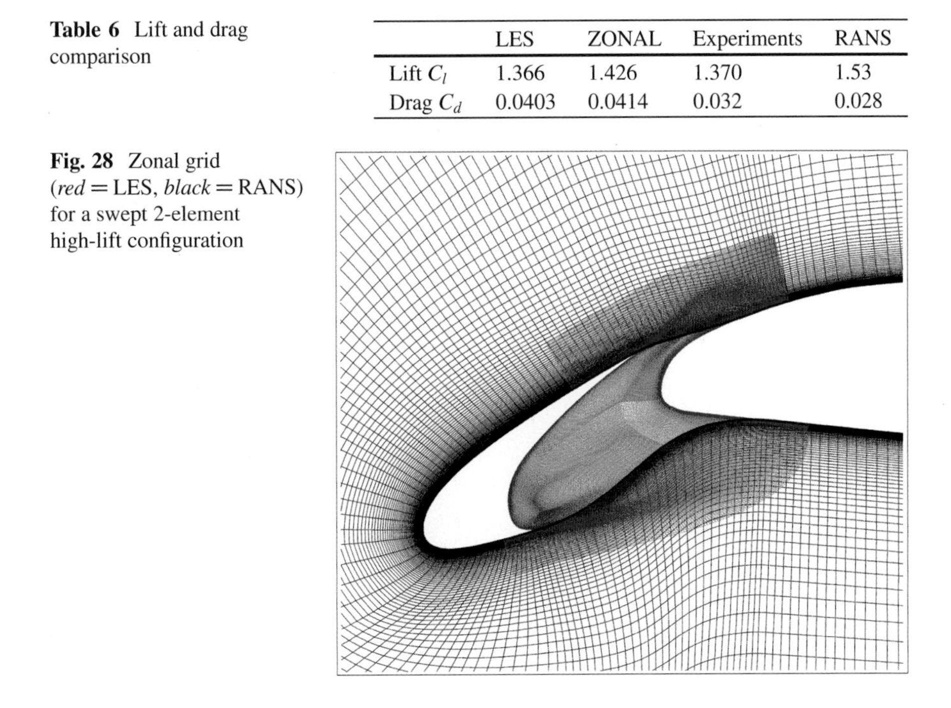

Fig. 28 Zonal grid (*red* = LES, *black* = RANS) for a swept 2-element high-lift configuration

of the zonal RANS-LES method to determine the airfoil aerodynamics such as the lift and drag coefficients at high angles of attack. Table 6 shows a comparison of the characteristic values with respect to the pure LES computation, the experiments, and RANS data [9]. The lower drag coefficient for both the RANS and the experiments can be explained by the smaller trailing-edge separation. The RANS overestimates the lift and underestimates the drag due to the fact that the RANS model predicts the turbulent separation point too far backwards to the trailing edge.

6.2 2-Element High-Lift Configuration

A second validation for the industrial application of airfoil aerodynamics, is the simulation of the flow around a 2-element high-lift configuration of an airfoil with extended flap at the leading edge. Using the zonal RANS-LES method here gives a reduction in grid size with a factor 10 with respect to a pure LES grid. Figure 28 shows the grid topology of the high-lift test-case. In Fig. 29, the time-averaged Mach number contours and streamlines around the profile leading edge and slat are shown. Smooth transitions between RANS and LES domains are attained.

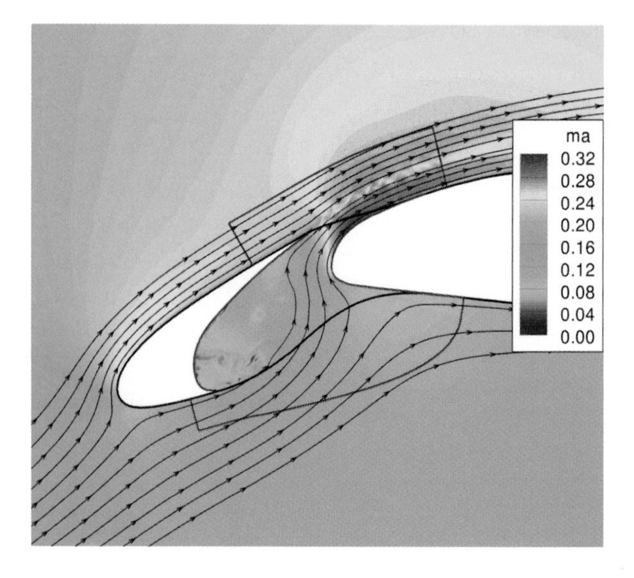

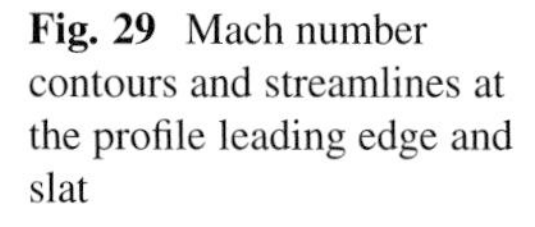
Fig. 29 Mach number contours and streamlines at the profile leading edge and slat

7 Summary and Outlook

Three industrial showcases highlighted the efficient utilization of modern highly distributed systems for large scale flow simulations of technical devices. A comparison of the older cluster system Laki was compared to the new Hermit system at HLRS showed the decreased performance per core, and emphasizes the need for highly scalable algorithms to exploit future super computing systems. With the help of highly sophisticated mechanisms like the dynamic load balancing it became possible to simulate complex and demanding phenomena on such systems with a great degree of parallelism.

Acknowledgements This work has been funded by Bundesministerium für Bildung und Forschung (Federal Ministry for Education and Research, BMBF) in the framework of the HPC software initiative in the project STEDG—"Hocheffiziente und skalierbare Software für die Simulation turbulenter Strömungen in komplexen Geometrien." We thank the Gauss Alliance of German Supercomputing Centers for the provided computing time. We are grateful for the ongoing kind support by HLRS and Cray in Stuttgart.

References

1. P. Boris, F.F. Grinstein, E.S. Oran, R.L. Kolbe, New insights into large eddy simulation. Fluid Dyn. Res. **10**, 199–228 (1992)
2. G. Gassner, F. Lörcher, C.-D. Munz, A discontinuous Galerkin scheme based on a space-time expansion II. Viscous flow equations in multi dimensions. J. Sci. Comput. **34**, 260–286 (2007)
3. D.F. Harlacher, C. Siebert, H. Klimach, S. Roller, F. Wolf, Dynamic load balancing for unstructured meshes on space-filling curves, in *Proceedings of the Workshop on Large-Scale Parallel Processing*, IPDPS 2012 (accepted for publication)

4. N. Jarrin, N. Benhamadouche, S. Laurence, D. Prosser, A synthetic-eddy-method for generating inflow conditions for large-eddy simulations. J. Heat Fluid Flow **27**, 585–593 (2006)
5. M.-S. Liou, C.J. Steffen, A new flux splitting scheme. J. Comput. Phys. **107**, 23–39 (1993)
6. F. Lörcher, G. Gassner, C.-D. Munz, A discontinuous Galerkin scheme based on a space-time expansion I. Inviscid compressible flow in one space dimension. J. Sci. Comput. **32**, 175–199 (2007)
7. M. Meinke, W. Schröder, E. Krause, Th. Rister, A comparison of second- and sixth-order methods for large-Eddy simulations. Comput. Fluids **31**, 695–718 (2002)
8. A. Spille, H.-J. Kaltenbach, Generation of turbulent inflow data with a prescribed shear-stress profile, in *Third AFSOR Conference on DNS and LES* (2001)
9. R. Wokoeck, N. Krimmelbein, J. Ortmanns, V. Ciobaca, R. Radespiel, A. Krumbein, RANS simulation and experiments on the stall behaviour of an airfoil with laminar separation bubbles. AIAA Paper AIAA-2006-0244 (2006)
10. Q. Zhang, W. Schröder, M. Meinke, A zonal RANS/LES method to determine the flow over a high-lift configuration. J. Comput. Fluids **39**(7) (2010)

Direct Numerical Simulation of Inelastic Non-Newtonian Jet Breakup

Chengxiang Zhu, Moritz Ertl, Christian Meister, Philipp Rauschenberger, Andreas Birkefeld, and Bernhard Weigand

Abstract Direct Numerical Simulations (DNS) based on the Volume of Fluid (VOF) method are carried out in this work, with the aim of analyzing the primary breakup of an inelastic non-Newtonian jet. Detailed description of the jet phenomena is provided firstly, followed by analysis on the generation of ligaments and droplets. The non-Newtonian characteristics of the jet are discussed as well, with a modified Ostwald de Waele model applied for the computation of the shear thinning liquid viscosity. Further, information about the current status of the MPI optimization regarding the Cray XE6 platform is revealed and a first performance analysis is presented.

1 Introduction

Sprays of fluids with non-Newtonian properties are found in chemical engineering, food processing, pharmaceutical production and other applications. For example, the production of well defined powders can be achieved through spray drying of non-Newtonian fluids. Even though these processes are in use today, many of the underlying processes are still not fully understood. In literature, a number of experimental [14, 17, 25] and numerical [3, 6, 15] investigations can be referred to when regarding the jet breakup of Newtonian fluids. However, studies of non-Newtonian fluids, especially numerical investigations of the jet breakup, are very sparse.

In this paper, direct numerical simulations of an evolving non-Newtonian jet are carried out, using the ITLR in-house code Free Surface 3D (FS3D). As the

C. Zhu (✉) · M. Ertl · C. Meister · P. Rauschenberger · A. Birkefeld · B. Weigand
Institut für Thermodynamik der Luft- und Raumfahrt, Universität Stuttgart, Pfaffenwaldring 31, 70569 Stuttgart, Germany
e-mail: chengxiang.zhu@itlr.uni-stuttgart.de

W.E. Nagel et al. (eds.), *High Performance Computing in Science and Engineering '13*, DOI 10.1007/978-3-319-02165-2_22,

simulation requires a lot of resources, it was carried out at the High Performance Computing Center Stuttgart (HLRS).

Non-Newtonian fluids are usually classified into two dominant types: power law fluids and viscoelastic fluids. The former type has only very weak elastic effects, which can be neglected [2]. The power law fluids can again be divided into two classes based on their viscosity behavior: shear thinning or shear thickening. The viscosity of shear thinning fluids decreases with increasing shear rate, while the viscosity of shear thickening fluids increases. In the current work, a power law shear thinning fluid is used as the liquid phase for the investigation.

2 Numerical Method

FS3D is a scientific numerical tool for incompressible multiphase flows based on the volume of fluid (VOF) method. It has been developed at the Institute of Aerospace Thermodynamics (ITLR), University of Stuttgart for about 20 years. Since then it has been used for investigations in several research areas including droplet deformation [10] and collision [5], splashing [16], Newtonian jet breakup [11], bubbles rising [23], etc. The code has also been expanded to resolve heat and mass transfer for the computation of evaporation effects [9, 19] and has been used by the Mathematical Modeling and Analysis group at the Center for Smart Interface, Technical University Darmstadt, which had added the capability to simulate chemical reactions [1].

FS3D is based on a finite volume discretization with velocities being in a staggered arrangement (according to the Marker and Cell (MAC) method [8]). Thereby the incompressible Navier-Stokes equations for the mass and the momentum conservation are solved:

$$\nabla \cdot \mathbf{u} = 0, \tag{1}$$

$$\frac{\partial(\rho\mathbf{u})}{\partial t} + \nabla \cdot [(\rho\mathbf{u}) \otimes \mathbf{u}] = -\nabla p + \nabla \cdot \left[\mu\left(\nabla\mathbf{u} + (\nabla\mathbf{u})^T\right)\right] + \rho\mathbf{g} + \mathbf{f}_\gamma. \tag{2}$$

Here, $\mathbf{u}$ denotes the velocity vector, t the time, ρ the density, μ the viscosity, p the pressure, $\mathbf{g}$ the external body forces and $\mathbf{f}_\gamma$ the volume forces due to surface tension. Two time integration schemes are implemented: The Euler explicit method and a second order Runge-Kutta method. Both methods are second order accurate in space. The Euler explicit method is used in the jet breakup calculations. The whole numerical scheme is described extensively e.g. in [18, 20]. Hence, only the key aspects necessary for the presented simulation will be explained in the following.

To compute the non-Newtonian viscosity, a modified Ostwald/de Waele power law model is applied, with

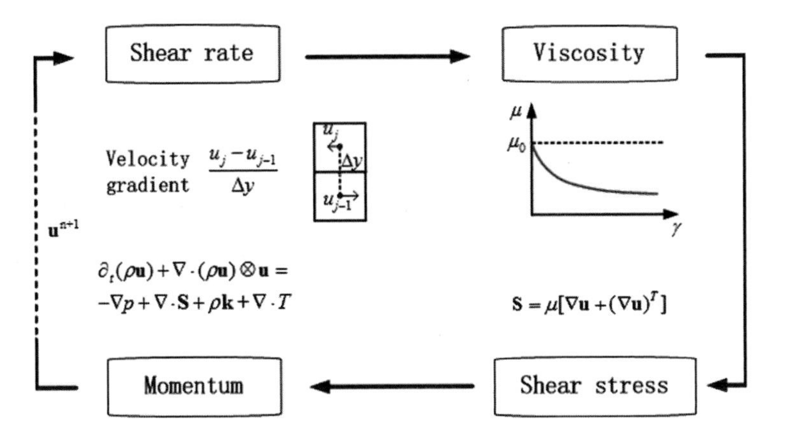

Fig. 1 Schematic for solving the momentum equation in consideration of a non-Newtonian viscosity

$$\mu(\dot{\gamma}) = \frac{\mu_0}{1 + \mu_0 k^{-1} \dot{\gamma}^{1-n}}, \tag{3}$$

where μ_0 is the zero shear viscosity, while k and n are model constants which correspond to different fluids and environmental conditions. Figure 1 provides a schematic illustration of the computation of this viscosity in FS3D. At first, the shear rate $\dot{\gamma} = \left\| \frac{1}{2}\left(\nabla \mathbf{u} + (\nabla \mathbf{u})^T\right) \right\|_2$ is calculated from the velocity field of the current time step. With the evaluated shear rate and the modified power law model, the local viscosity $\mu(\dot{\gamma})$ can be computed. Furthermore, the shear stress $\mathbf{S}$ is obtained, contributing to the momentum change which results in a new velocity field $\mathbf{u}^{n+1}$ for the next time step.

To be able to additionally study the destabilizing influence of inlet disturbances on the jet breakup, a turbulent inflow synthetization method implemented by Huber et al. [11] is applied for the inflow boundary condition. It is based on a method developed by Kornev and Hassel [12, 13] which superimposes the mean inflow velocity profile $\bar{u_i}(t, y, z)$ by velocity fluctuations $u_i'(t, y, z)$. Here, this method is applied to the energy spectrum for decaying turbulence [11].

3 Numerical Setup

In the current work, an inelastic non-Newtonian liquid is injected into quiescent ambient air with a mean inflow velocity of $U_0 = 50\,\text{m/s}$. The fluid properties are $\mu_0 = 2.559 \cdot 10^{-2}\,\text{kg/(ms)}$, $\rho_l = 1{,}049\,\text{kg/m}^3$, $\sigma = 6.437 \cdot 10^{-2}\,\text{kg/s}^2$, where ρ_l and σ denote the liquid density and the surface tension, respectively. The

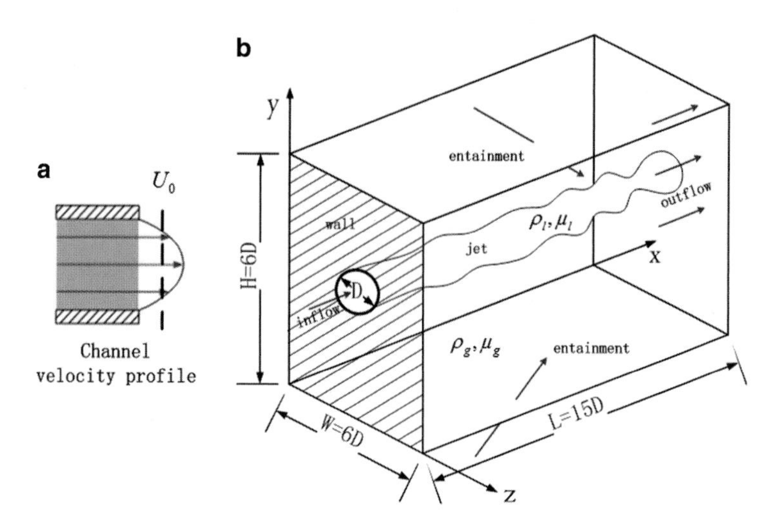

Fig. 2 The velocity profile for the inflow (**a**). The computational domain for the simulation (**b**)

constants k and n for the non-Newtonian model are selected to $n = 0.126$ and $k = 890.8\,\text{kg/(ms}^{2-n})$.

A rectangular computational domain is applied in this study. The nozzle inflow boundary condition is placed at the center of the left side while the remaining parts of the left side are set as a no-slip wall. A continuous boundary condition is chosen for the other sides of the domain. The dimensions are as follows: $6D$ in both directions of the nozzle plane (y and z) and $15D$ in the main flow direction of the jet (x), with a nozzle diameter $D = 0.4\,\text{mm}$. A channel velocity profile is used for the nozzle boundary condition and a turbulent intensity of $Tu = 10\,\%$ is applied to the inflow. Sixty four cells are used to resolve the nozzle diameter which results in a grid spacing of $6.25\,\mu\text{m}$, and a total number of cells adding up to $33.5 \cdot 10^6$ in this setup. The velocity profile and the computational domain are shown in Fig. 2.

The grid resolution is of major importance for DNS calculations. An estimation can be obtained from calculating the smallest dissipative length scale λ_k according to Kolmogorov's theory of the universal equilibrium:

$$\lambda_k = \frac{L_t}{Re_t^{0.75}}, \qquad \text{with} \quad Re_t = \frac{\rho_l u' L_t}{\mu_l}, \tag{4}$$

where the turbulent length scale $L_t \approx TuD$ and the turbulent velocity $u' \approx TuU_0$ can be estimated from the turbulent intensity Tu. Based on these, the Kolmogorov length scale can be calculated as $\lambda_k = 8.3\,\mu\text{m}$ for the simulation. Therefore, the current setup satisfies the spatial resolution necessary for DNS calculations.

At the inflow conditions given above, the corresponding Weber (We) and Ohnesorge (Oh) numbers ($We = \rho_l U_0^2 D/\sigma$, $Oh = \mu_0/(\rho_l \sigma D)^{0.5}$) can be calculated to be 16,296 and 0.1557 (at zero shear rate) for the current work. The simulation is carried out for a physical duration of 75 μs, which took 192 CPU hours on the NEC SX-9 at the HLRS.

4 Results

In this section the three dimensional (3D) jet structure, the generation of ligaments and droplets, as well as the non-Newtonian characteristics of the jet will be discussed to provide a first impression of the non-Newtonian jet behavior.

4.1 Surface Structure

Since high inflow turbulence is superimposed on this jet, interfacial fluctuations can be observed right at the nozzle exit, as shown in Fig. 3. In the vicinity of the nozzle exit oscillations are of a large number but with small amplitudes. For instance, the highest surface amplitude at $t^* = 6.3$ (dimensionless time $t^* = tU_0/D$) is only $D/20$ at $x/D = 0.2$. These oscillations are mainly generated due to inflow turbulence. Therefore, their amplitudes remain similar at all three time steps plotted. However, they are enhanced with increasing axial positions. At $t^* = 6.3$ for instance, the highest amplitude of the surface oscillations at $x/D = 3$ is nearly three times higher than that at $x/D = 0.2$ due to air friction. The Kelvin-Helmholtz instabilities caused by the velocity shear between the liquid phase and the gas phase contribute a lot to the deformation of the liquid surface. Gas is accelerated and rolled up near the liquid phase, enhancing the deformation of interfacial waves as well. For non-Newtonian fluids it is also noticeable that the liquid viscosity is actually reduced due to the existing local high shear rates, indicating weaker resistance of the liquid phase. Details about this non-Newtonian behavior will be explained in the next section.

Further downstream of the nozzle exit, ligaments and droplets are generated from the liquid core. At $t^* = 6.3$ the first separated liquid volume appears at $x/D = 4.8$, while at $t^* = 7.8$ it appears at $x/D = 3.8$. This position even decreases to $x/D = 3.3$ at $t^* = 9.3$. Therefore, it can be concluded that with increasing time disintegration is increasing correspondingly. Referring to Fig. 3, it can also be found that the number of separated volumes increases with time. For instance, the number of the generated droplets at $t^* = 9.3$ is 89, which is nearly twice the number of droplets (42) at $t^* = 7.8$.

Regarding the contours of the axial velocity in Fig. 3, the jet tip has a higher velocity than other positions. According to the channel velocity profile we applied in the current work, the liquid axial velocity at the jet center is two times larger than the average velocity U_0. But due to air resistance, the highest axial velocity of the jet tip reduces to $u/U_0 = 1.92$ at $t^* = 6.3$ and similarly at $t^* = 7.8$. At $t^* = 9.3$,

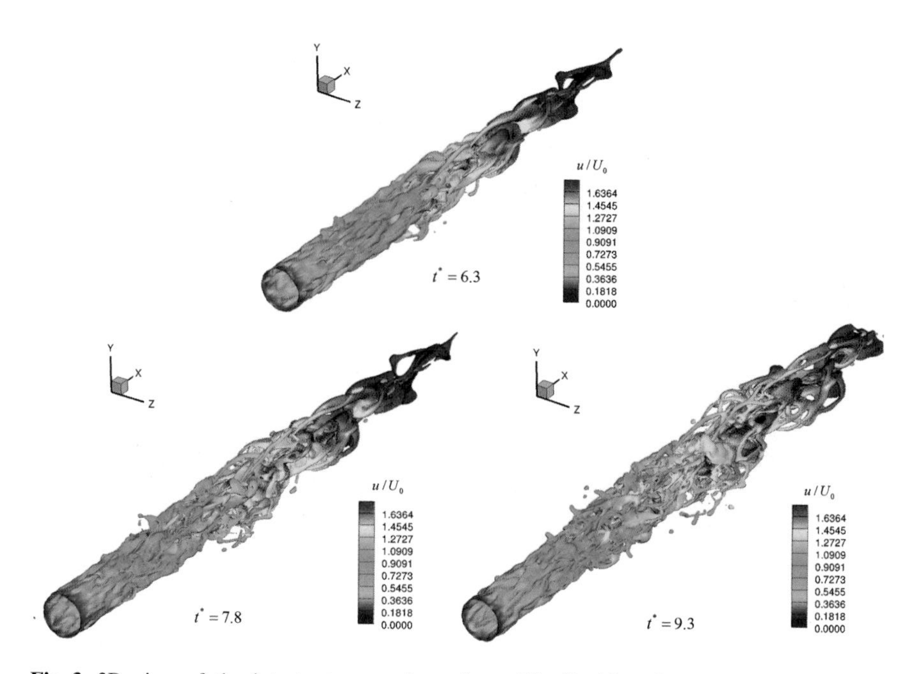

Fig. 3 3D view of the jet structure at three times. The liquid surface is contoured by the dimensionless axial velocity u/U_0

this velocity is not trackable any more since the jet tip has left the computational domain.

Looking inside the disintegration of this non-Newtonian jet, some intriguing phenomena can be detected. From Fig. 4b, it can be deduced that the intact length of the jet (measured from the nozzle exit to the disconnecting point in x-direction) at the depicted time is around $7.7D$, which differs greatly from conventional empirical formulas in literature [4]. For instance, the intact length can normally be calculated as $L \sim 0.5D(\rho_l/\rho_g)^{0.5}$ under large Weber (*We*) numbers, which results in a length of $14.75D$ under current conditions. This early jet breakup is caused by two main reasons. On one hand, strong inflow turbulence is employed in the current work. On the other hand, it is worth to mention again that the shear thinning viscosity contributes to this earlier breakup as well, since the resistance of the liquid phase is decreased.

The size of the generated ligaments and droplets can differ greatly. For example, the ligament A in Fig. 4 has a radius of $D/13$, while the largest radius of the ligament B is $D/10$. Droplets have the same regime: the diameter of droplet C is $D/12$, while that of droplet D is $D/10$, which varies by a factor of 1.3.

In order to show this more clearly, the distributions of the number of droplets and their sizes are depicted in Fig. 5. The vertical axis refers to the percentage of the droplet number within a certain range compared to the total number of droplets, while the horizontal axis represents the dimensionless droplet diameter D^*, which

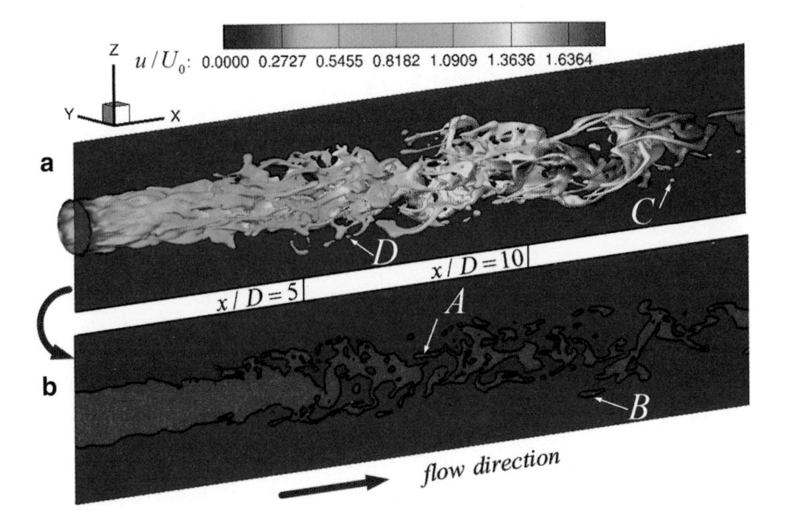

Fig. 4 (**a**) A detailed insight into the disintegration of the non-Newtonian jet at $t^* = 9.3$. The slice (**b**) is cut through the symmetric XZ-plane and reveals the 2D jet structure

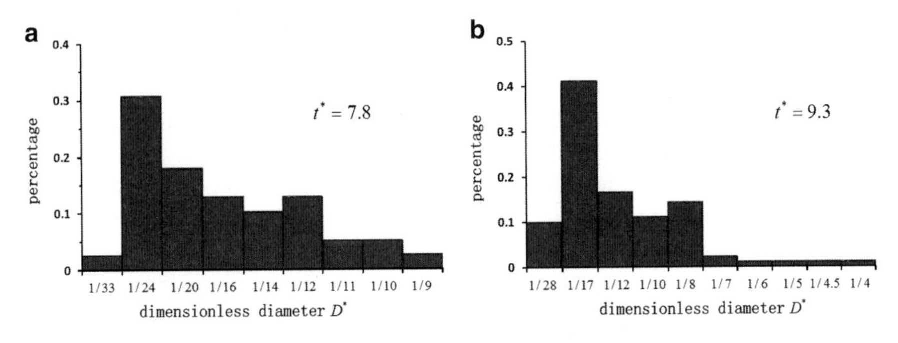

Fig. 5 Distributions of the number and the diameter of droplets at two times

denotes the droplet diameter normalized by the nozzle diameter. At $t^* = 7.8$, 30 % of the droplets generated from the jet have a diameter around $D/25$, while more than 80 % of the droplets are smaller than $D/12$. At $t^* = 9.3$, 40 % of the droplets have a diameter around $D/17$, while more than 90 % of them are smaller than $D/8$. Compared to the research by Shinjo and Umemura [21], a similar distribution of the droplet size is obtained. For droplet analysis, the Sauter mean diameter (SMD), which is defined as $D_{32} = \sum D_i^3 / \sum D_i^2$ (D_i denotes the diameter of each droplet), is used to evaluate the mean size of the generated droplets. The calculated D_{32} at $t^* = 7.8$ is $D/10.8$, while that at $t^* = 9.3$ is $D/8.3$. Combined with previous observations, it can be concluded that, not only more droplets are generated at $t^* = 9.3$, but also the mean droplet size is larger compared to $t^* = 7.8$.

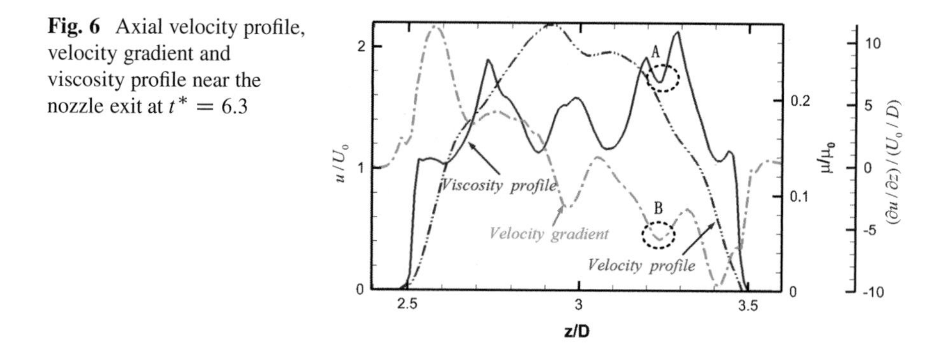

Fig. 6 Axial velocity profile, velocity gradient and viscosity profile near the nozzle exit at $t^* = 6.3$

4.2 Non-Newtonian Characteristics

As mentioned above, the viscosity of the liquid phase shows a shear thinning behavior. With increasing shear rate, which relates greatly to the velocity field, the liquid viscosity decreases. Therefore, a close connection between the velocity field and the viscosity field should be noticed. The relationship between the flow property and the physical property is also one of the most significant features of non-Newtonian fluids. Figure 6 depicts the axial velocity profile u, the velocity gradient $\partial u/\partial x$ and the viscosity profile μ at a plane near the nozzle exit at $t^* = 6.3$. The quantities are normalized by the mean jet velocity U_0, U_0/D and the zero shear viscosity μ_0, respectively.

Due to the inflow turbulence with a turbulence intensity of 10 %, the highest axial velocity is nearly 2.2 at the radial position of $z/D = 2.9$. This sudden acceleration of the liquid phase generates a strong velocity gradient here, resulting locally in a high shear rate. As a consequence, the liquid viscosity decreases correspondingly at position A. The rapid increase of the viscosity at the radial positions of $z/D = 2.5$ and 3.5 is due to the existence of the interface. Regarding the right vertical label of the viscosity, it can also be found that the highest liquid viscosity is only 0.27 times μ_0, indicating a strong shear thinning behavior of the fluid. According to the simulation data, the local shear rate is even higher than $2 \cdot 10^6$ 1/s around here.

However, it should be pointed out that only the axial velocity u and its gradient $\partial u/\partial x$ in x-direction are shown in Fig. 6. The shear rate, on which the liquid viscosity relies, actually relates to all nine components of the velocity gradient. Some variations of the liquid viscosity here might be caused by the three dimensional effect, which cannot be depicted in this figure.

Despite the velocity and the viscosity properties mentioned above, dimensionless groups will be analyzed as well in the following. For non-Newtonian fluids, the Deborah (*De*) number and the Elasto-Capillary (*Ec*) number, which both consider the elastic effect, are normally of interest. But for the inelastic fluid used in the current work, they are meaningless. Therefore, only the Ohnesorge (*Oh*) number and the Reynolds (*Re*) number will be discussed. Attention should be paid to the fact that both parameters are mean values (arithmetic mean over the whole computational

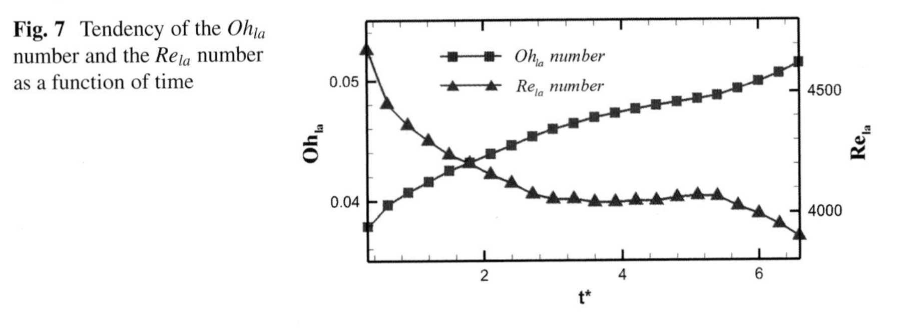

Fig. 7 Tendency of the Oh_{la} number and the Re_{la} number as a function of time

domain) whose definition is based on local properties: $Oh_{la} = \overline{\mu_{loc}/(\rho_l \sigma D)^{0.5}}$ and $Re_{la} = \overline{\rho_l u D/\mu_{loc}}$, where the subscript la refers to "local average". In addition, only the liquid phase is encountered for the parameters' calculation.

In general, the Oh_{la} number increases with time, while the Re_{la} number decreases. Regarding the jet structure this can be easily explained. At later time steps, the liquid axial velocity tends to be more uniform due to viscous forces, indicating a small local shear rate. Therefore, the local viscosity of the liquid phase increases correspondingly. From the definition of both parameters, the Oh_{la} number is proportional to the local viscosity, while the Re_{la} number is inverse proportional to it. Thus, the tendency as shown in Fig. 7 is obtained. From this figure, it can be found that the Re_{la} number decreases nearly 17 %, from 4,700 at $t^* = 0.3$ to 3,900 at $t^* = 6.3$, while the Oh_{la} number increases from 0.038 at $t^* = 0.3$ to 0.051 at $t^* = 6.3$. Both parameters reflect the impact of viscous forces, indicating a great deviation of the liquid local viscosity. Hence, for non-Newtonian fluids the actual flow properties could be totally different in the jet compared to Newtonian fluids, although the inflow conditions remain the same. In certain cases, the corresponding jet breakup regime can even change in time.

Figure 8 shows the local average parameters along the streamwise axis x at $t^* = 6.3$. The Oh_{la} number and the Re_{la} number are defined in the same way as before, but the averaging domain is restricted to the liquid phase in a certain 2D axial slice with $x/D = constant$, rather than the whole 3D computational field. From this figure, the Oh_{la} number increases with axial position, from 0.03 at the nozzle exit to 0.12 at $x/D = 12$. Recalling the Oh_{la} number (0.051) in the whole 3D computational domain at this time, the highest Oh_{la} number in a certain cross section deviates even over 100 %. It is hence confirmed again that the fluid analyzed in this work exhibits a strong shear thinning behavior.

In general, the Re_{la} number decreases with axial position, from 5,200 at the nozzle exit to only 1,980 at $x/D = 12$, while it is 3,900 in the 3D computational domain at this time as previously observed. However, it is noticeable that between $x/D = 1$ and $x/D = 6$, the Re_{la} number increases slightly, on contrary to the general tendency. Referring to the Oh_{la} number, it is clear that the average liquid viscosity $\overline{\mu_{loc}}$ still increases with the axial position within this range. The reason for

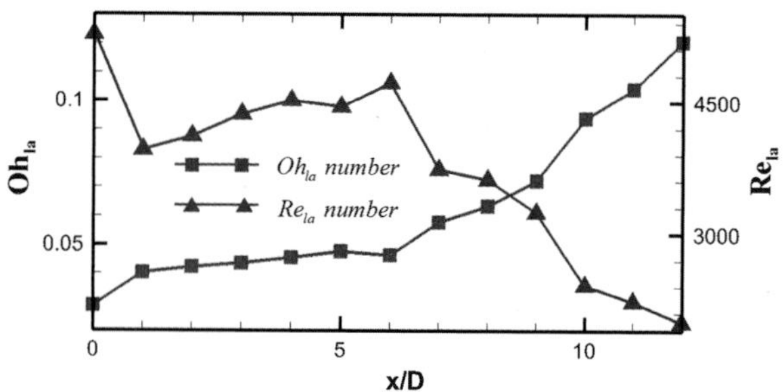

Fig. 8 Tendency of the Oh_{la} number and the Re_{la} number as a function of axial position

an increase of the Re_{la} number is the average axial velocity $\overline{u}$, which has a stronger influence than the viscosity here.

4.3 Comparison with a Newtonian Simulation

To show the influence of the Non-Newtonian characteristics, a Newtonian simulation with the same inflow conditions is compared to the Non-Newtonian simulation. In order to produce comparable results, the liquid viscosity is set to the value ($\mu_l = 0.01275\,\text{Pa s}$), obtained from spatially averaging the viscosity in the Non-Newtonian simulation. Figure 9a shows a 3D view of the jet at $t^* = 6.3$, contoured by the dimensionless velocity. A comparison with the Non-Newtonian jet from Fig. 3 at this time displays that less interfacial undulations are formed on the interfaces and liquid sheets are the dominant flow structure of this jet. In general terms, it can be concluded that the Newtonian jet performs more stable than the Non-Newtonian jet.

Since less disintegrations appear in the Newtonian jet, its surface area increases much slower than that of the Non-Newtonian jet. A quantitative sketch of this surface behavior is shown in Fig. 9b, in which the dimensionless quantity $S^* = S/(\pi U_0 Dt + \pi(D/2)^2)$ is shown, representing the dimensionless surface area of the jet. The surface area of the jet S is normalized by the ideal surface area of a column jet with a constant inflow velocity U_0. The term $\pi U_0 Dt$ denotes the lateral area of the column jet increasing with time, while the term $\pi(D/2)^2$ represents the tip area of the jet. It can be found that the surface area of the Non-Newtonian jet is about 2.5 times larger than that of the Newtonian jet at $t^* = 8$.

Recalling that the liquid viscosity is the only difference between two simulations, it is therefore also the only reason responsible for their specific flow features. Due to the shear thinning behavior of the liquid viscosity, the resistance in the Non-Newtonian jet is weaker than that in the Newtonian case. This actually has two impacts on this Non-Newtonian jet. First, a larger number of interfacial undulations can be formed consequently, as can be observed from Fig. 3. Second, the liquid phase is more distorted within the jet, which prepares a better condition for further downstream breakup.

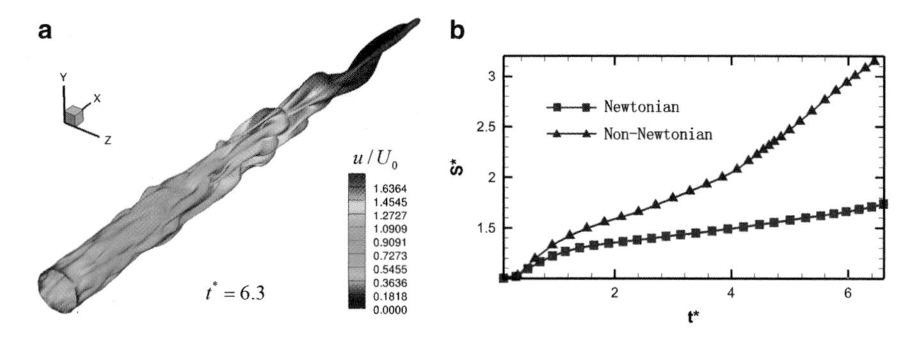

Fig. 9 Jet structure of the Newtonian fluid at $t^* = 6.3$ (**a**). Time dependent evolution of the jet surface area for both the Newtonian and the Non-Newtonian fluids (**b**)

5 Performance Analysis on Cray XE6

One major issue of incompressible fluid dynamics is the efficient solution of the arising pressure Poisson equation, which is an implication of the necessity to obtain a divergence-free velocity field at the end of a time step. The system of linear equations has dimensions of $N \times N$ with N being the number of grid cells. We employ a multigrid solver in V- or W-cycle for fast convergence [24], but solving the Poisson equation remains the most time consuming operation. A second source of computational costs in FS3D is the interface tracking. In interface cells, we reconstruct a plane separating the two phases by using the piecewise linear interface calculation (PLIC) algorithm. This has several advantages like a sharp interface (i.e. non-diffusive), precise phase flux calculation and knowledge of the exact interface position, which is crucial in phase change problems. However, the computational costs rise linearly with the number of interface cells. Hence, for problems involving splashing or jet breakup, where large numbers of small droplets emerge, interface reconstruction can become very costly (up to 30 % of total computation time).

As FS3D works on a structured Cartesian grid the computational domain can be easily subdivided into cubic blocks which allows to hide some of them if necessary (e.g. flow over a step with wake space). Hence, the parallelization of the code is two-fold. MPI is used on the superior level for communication between blocks and OpenMP works within blocks parallelizing loops and Fortan array assignments.

Hereafter, we give a detailed performance analysis on the high performance computer Cray XE6. This represents the current state in the ongoing process of porting the FS3D from the NEC SX-9 vector computer to the Cray XE6 massive parallel computing cluster.

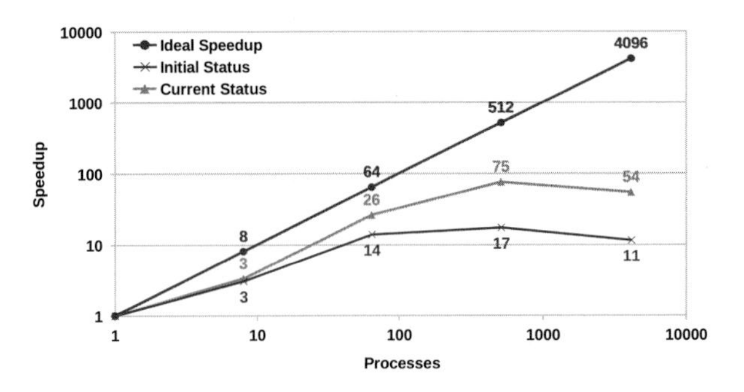

Fig. 10 Speedup comparison of the initial and the current status of MPI optimizations

5.1 Performance of Previous and Current MPI Implementation

We first compare the performance of the current MPI implementation with the initial status of the code before optimization to demonstrate our efforts until now. The reference test case consists of a cubic computational domain with 128^3 grid cells and represents a basic fluid dynamic process, namely a droplet in a free-stream. For every computation there was an equal amount of MPI processes in the three spatial directions, where each MPI process was assigned to one block of cubic shape. For this investigation only MPI parallelization is used. The resulting speedups as well as the ideal speedup are depicted in Fig. 10. The major modification between the two states of the code are:

- removal of MPI imbalance in user routines,
- removal of redundant MPI communication,
- removal of excessive file system access and
- optimizing the multigrid algorithm with respect to MPI.

The speedup curves reveal a maximum in the MPI performance when the number of grid cells per MPI processes is in the range of 32,768 (i.e. 64 MPI processes) down to 4,096 (i.e. 512 MPI processes). For a smaller amount, the cost of MPI communication exceeds the reduction of computation time and therefore speedup decreases substantially.

When evaluating these results, it has to be taken into account that FS3D has to treat the different parts of the governing equation system with different numerical schemes one after another due to their physical and mathematical properties. In consequence, a larger number of communications of smaller size is necessary, compared to a standard explicit solver, which can usually exchange all state variables for the dummy cells at once. Furthermore, the Poisson equation is solved with an iterative solver, which requires a dummy value exchange after each cycle. Finally, it has to be taken into account that the optimizations are still in progress.

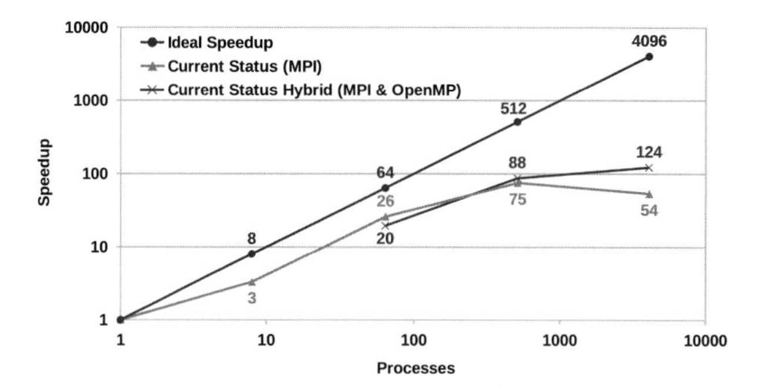

Fig. 11 Speedup comparison of pure MPI and Hybrid (MPI & OpenMP) implementation

5.2 *Performance Including OpenMP*

In addition, we did some performance analysis of the code using MPI in conjunction with OpenMP. In this setup, we split the total number of MPI processes into 4 MPI processes per node and assigned each of them 8 OpenMP threads. In Fig. 11 we compare the results with regard to the used hardware, namely the total number of cores. Therefore, a pure MPI run with 512 processes is compared to a hybrid MPI/OpenMP run consisting of 64 MPI processes and 8 OpenMP threads per MPI process. Due to the usage of OpenMP, large numbers of processes can be utilized in a more efficient way. The speedup curves reveal an optimal hybrid performance when the number of grid cells per MPI process is 4,096, resulting in 512 grid cells per OpenMP thread.

5.3 *Current and Future Work*

When parallelizing the multigrid algorithm, one faces the problem of extensive communication in between smoothing operations on the successively coarser grids. Before each coarsening or prolongation step, one row of dummy cells at the block boundaries must be exchanged in all three spatial directions. Hence, we currently suffer from a certain communication overhead. Nevertheless, the multigrid solver is very efficient even with an MPI implementation, since it converges to the desired residuum within very few iterations. Due to the increasing number of CPUs available, communication overhead is getting the relevant bottleneck instead of pure computational effort. Hence, a Poisson solver with less iterations, each being more expensive, will be the better choice for massive parallel computations. Following this way we shall investigate the performance of other methods such as e.g. the multigrid preconditioned conjugate gradient method (MGCG) [7, 22] in the near future.

For further improvement of the overall performance, we currently try to make extensive use of non-blocking MPI communication combined with latency-hiding measures allowing to reduce actual waiting time. Also the implementation of MPI-IO routines is in progress which allow an efficient output of the computational results avoiding the bottleneck of accessing multiple files at a time.

6 Conclusion

A direct numerical simulation of the primary jet breakup for a non-Newtonian fluid is carried out in this work on the NEC SX-9 at the HLRS. The 3D jet structures and the generation of ligaments and droplets as well as the non-Newtonian characteristics of the jet are discussed. With increasing time, disintegrations are increasing in both axial position and number. Due to the inflow turbulences employed, the undisturbed length of the jet is shortened as well. Most of the separated droplets tend to be of smaller sizes. The shear thinning viscosity of the non-Newtonian fluid is dominantly reverse proportional to the gradient in x-direction of the axial velocity in the current case. With regard to the local average properties (the velocity and the viscosity), the Oh_{la} number increases with time and axial position in general, while the Re_{la} number decreases. Also the ongoing MPI optimizations with regard to the Cray XE6 system already shows fairly good speedups.

Acknowledgements The authors greatly thanks the *High Performance Computing Center Stuttgart* (HLRS) for support and supply of computational time on the NEC SX-9 and Cray XE6 platforms under the Grant No. FS3D/11142. The authors also greatly appreciate financial support of this project from DFG within the priority program SPP 1423 "Prozess Spray", as well as the financial support from the China Scholarship Council (CSC), as well as from the SFB Transregio 75 and the DFG Cluster of Excellence Simulation Technologies at the University of Stuttgart.

References

1. D. Bothe, M. Kröger, H.J. Warnecke, A VOF-based conservative method for the simulation of reactive mass transfer from rising bubbles. Fluid Dyn. Mater. Process. **7**(3), 303–316 (2011)
2. C. Clasen, P.M. Phillips, L. Palangetic, J. Vermant, Dispensing of rheologically complex fluids: the map of misery. AIChE J. **58**, 3242–3255 (2012)
3. O. Desjardins, V. Moureau, H. Pitsch, An accurate conservative LES ghost fluid method for simulating turbulent atomization. J. Comput. Phys. **227**, 8395–8416 (2008)
4. J. Eggers, E. Villermaux, Physics of liquid jets. Rep. Prog. Phys. **71**, 036601 (2008)
5. H. Gomaa, N. Roth, J. Schlottke, B. Weigand, DNS calculations for the modeling of real industrial applications. Atomization Sprays **20**(4), 281–296 (2010)
6. M. Gorokhovski, M. Herrmann, Modeling primary atomization. Annu. Rev. Fluid Mech. **40**, 343–366 (2008)
7. A. Götz, I. Lenhardt, H. Obermaier, *Parallele Numerische Verfahren* (Springer, Berlin, 2002)

8. F.H. Harlow, J.E. Welch, Numerical calculation of time-dependent viscous incompressible flow of fluid with free surface. Phys. Fluids **8**(12), 2182–2189 (1965)
9. M. Hase, B. Weigand, A numerical model for 3D transient evaporation processes based on the Volume-of- Fluid method, in *ICHMT International Symposium on Advances in Computational Heat Transfer*, Norway (2004), pp. 1–23
10. M. Hase, M. Rieber, F. Graf, N. Roth, B. Weigand, Parallel computation of the time dependent velocity evolution for strongly deformed droplets, in *High-Performance Computing in Science and Engineering* (Springer, Berlin, 2001), pp. 1–10
11. C. Huber, H. Gomaa, B. Weigand, Application of a novel turbulence generator to multiphase flow computations, in *HLRS, High Performance Computing in Science & Engineering* (Springer, Berlin, 2010), pp. 1–15
12. N. Kornev, E. Hassel, Method of random spots for generation of synthetic inhomogeneous turbulent fields with prescribed autocorrelation functions. Commun. Numer. Methods Eng. **23**(1), 35–43 (2007)
13. N. Kornev, E. Hassel, Synthesis of homogeneous anisotropic divergence-free turbulent fields with prescribed second-order statistics by vortex dipoles. Phys. Fluids **19**(6), 068101 (2007)
14. P. Marmottant, E. Villermaux, On spray formation. J. Fluid Mech. **498**, 73–111 (2004)
15. Y. Pan, K. Suga, A numerical study on the breakup on the breakup process of laminar liquid jets into a gas. Phys. Fluids **18**, 052101 (2006)
16. M. Rieber, A. Frohn, A numerical study on the mechanism of splashing. Int. J. Heat Fluid Flow **20**(5), 455–461 (1999)
17. K.A. Sallam, Z. Dai, G.M. Faeth, Liquid breakup at the surface of turbulent round liquid jets in still gases. Int. J. Multiphase Flow **28**, 427–449 (2002)
18. J. Schlottke, Direkte Numersiche Simulation von Mehrphasenströmumgen mit Phasenübergang, Ph.D. Thesis, Universität Stuttgart, 2010
19. J. Schlottke, P. Rauschenberger, B. Weigand, C. Ma, D. Bothe, Volume of fluid direct numerical simulation of heat and mass transfer using sharp temperature and concentration fields, in *ILASS—Europe 2011, 24th European Conference on Liquid Atomization and Spray Systems*, Estoril, 2011
20. J. Schlottke, B. Weigand, Direct numerical simulation of evaporating droplets. J. Comput. Phys. **227**(10), 5215–5237 (2008)
21. J. Shinjo, A. Umemura, Simulation of liquid jet primary breakup: dynamics of ligament and droplet formation. Int. J. Multiphase Flow **36**, 513–532 (2010)
22. O. Tatebe, The multigrid preconditioned conjugate gradient method, in *6th Copper Mountain Conference on Multigrid Methods*, Copper Mountain, April 1993
23. H. Weking, J. Schlottke, M. Boger, C.D. Munz, B. Weigand, DNS of rising bubbles using VOF and balanced force surface tension, in *High Performance Computing on Vector Systems* (Springer, Berlin, 2010)
24. P. Wesseling, *An Introduction to Multigrid Methods* (Wiley, Chichester, 1991)
25. P. Wu, R. Miranda, G. Faeth, Effect of initial inflow conditions on primary break-up of nonturbulent and turbulent jets. Atomization Sprays **5**, 175–196 (1995)

Direct Numerical Simulation of Boundary Layer Transition in Streamwise Corner-Flow

Oliver Schmidt, Björn Selent, and Ulrich Rist

Abstract The process of laminar-turbulent transition in streamwise corner-flow is considered my means of direct numerical simulation (DNS). It is shown that transition triggered by harmonic forcing originates from the near-corner region in the shape of a turbulent wedge. The resulting mean flow deformation takes the shape of an outward bulge, and can be linked to experimental observations. A spectral analysis of the transient flow data is undertaken to elaborate on non-linear interactions, modal structures and spectral energy distribution. Additionally, the massive parallel performance of the DNS code on the Cray XE6 supercomputer is discussed and compared with the performance on previous vector architectures.

1 Introduction

The basic corner-flow setup consists of two semi-infinite perpendicular flat plates with the freestream parallel to the intersection axis as sketched in Fig. 1. It serves as a generic model for a variety of technical flows. Because of its practical relevance the corner-flow problem has been studied both, numerically and experimentally for more than 60 years starting with the early theoretical considerations by Carrier [3]. The reader is referred to the work of Zamir [17] for a comprehensive review of experimental work that was mainly conducted during that period. Until now, numerical studies were almost exclusively restricted to eigenvalue-based linear stability studies of the two-dimensional self-similar corner-flow solutions first obtained by Rubin and Grossman [14] and [8] for incompressible and general compressible flows, respectively. Overviews of the numerous numerical studies can be found, e.g. in the work of Galionis and Hall [5]. There exists a remarkable

O. Schmidt (✉) · B. Selent · U. Rist
Institute of Aerodynamics and Gas Dynamics, University of Stuttgart, Stuttgart, Germany
e-mail: o.schmidt@iag.uni-stuttgart.de

W.E. Nagel et al. (eds.), *High Performance Computing in Science and Engineering '13*, DOI 10.1007/978-3-319-02165-2_23,

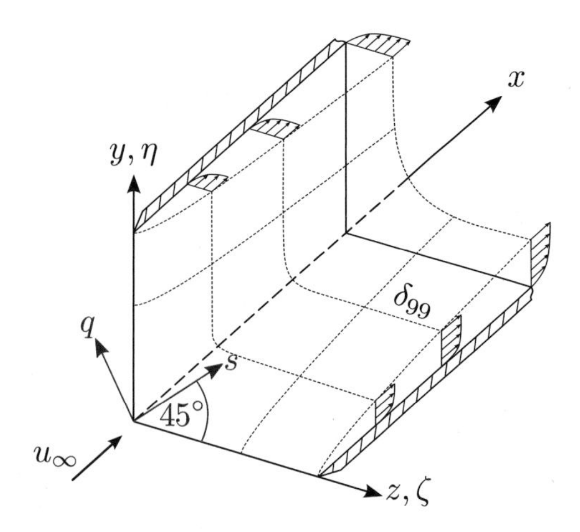

Fig. 1 Sketch of the flow in a streamwise corner: (x, y, z) denote the Cartesian coordinates with the x-coordinate in the streamwise direction, (η, ζ) the self-similar spanwise coordinates and (s, q) an auxiliary coordinate frame with the ordinate along the corner bisector [16]

discrepancy between experiment and theory that has not been resolved. In linear stability studies, a critical Reynolds number comparable to the one found for the closely related flat-plate scenario is found while rapid transition to turbulence at short distances after the leading edge is consistently observed in experiment. In the work at hand, we present the first direct numerical simulation of the transition process in axial corner-flow. The study is organized as follows: an introduction to the corner-flow problem is given in Sect. 1, the numerical setup an the laminar self-similar base state are discussed in Sect. 2, and the results are summarized in Sect. 3. The parallel performance of the direct numerical simulation code is addressed in Sect. 4. Final conclusions are given in Sect. 5.

2 Numerical Setup and Base State

Governing Equations The flow of a compressible fluid is governed by the full Navier-Stokes equations

$$\frac{\partial \rho}{\partial t} = -\nabla \cdot \rho \mathbf{u}, \tag{1}$$

$$\frac{\partial \rho \mathbf{u}}{\partial t} = -\frac{1}{2}\nabla \cdot (\mathbf{u} \otimes \rho \mathbf{u} + \rho \mathbf{u} \otimes \mathbf{u}) - \nabla p + \frac{1}{\mathrm{Re}}\nabla \cdot \tau, \tag{2}$$

$$\frac{\partial \rho e}{\partial t} = -\nabla \cdot \rho e \mathbf{u} + \frac{1}{(\gamma - 1)\mathrm{RePrMa}_\infty^2}\nabla \cdot k \nabla T - \nabla \cdot p\mathbf{u} + \frac{1}{\mathrm{Re}}\nabla \cdot \tau \mathbf{u}, \tag{3}$$

where ρ is the density, $\mathbf{u} = (u, v, w)^T$ the vector of Cartesian velocity components, p the pressure, e the total energy, T the temperature and t the time. The dynamic viscosity μ and the thermal conductivity k are fluid properties and $\tau = \mu\left(\nabla \mathbf{u} + \nabla \mathbf{u}^T\right) - \frac{2}{3}\mu(\nabla \cdot \mathbf{u})\mathrm{I}$ the viscous stress tensor. All flow quantities

Table 1 Dimensionless quantities and freestream properties

Pr (–)	γ (–)	p^*_∞ (hPa)	T^*_∞ (K)	c^*_p (J/kg K)	R^* (J/kg K)
0.714	1.4	1013.25	293.15	1,005	287

are non-dimensionalized by their respective dimensional free-stream values denoted by $(.)^*_\infty$ and the coordinates by the boundary layer displacement thickness δ^*_1. The dimensionless Reynolds number $\mathrm{Re} = \rho^*_\infty u^*_\infty \delta^*_1/\mu^*_\infty$, Prandtl number $\mathrm{Pr} = \mathrm{c}^*_\mathrm{p}\mu^*_\infty/k^*_\infty$, and Mach number $\mathrm{Ma} = u^*_\infty/a^*_\infty$ hence fully describe the flow. The set of governing equations (1) is closed by the ideal gas law $p = \rho T/(\gamma \mathrm{Ma}^2_\infty)$ and Sutherland's law $\mu^*(T) = \mu^*_{ref} T^{3/2}(1 + T_s)/(T + T_s)$ that empirically relates viscosity and temperature. The empirical constants are given as $\mu^*_{ref}(T^*_{ref} = 280\,\mathrm{K}) = 1.735 \times 10^{-5}\,\mathrm{kg/ms}$ and $T_s = 110.4\,\mathrm{K}/T^*_\infty$. A technically relevant flow case with $\mathrm{Ma} = 0.8$ and dry air at standard conditions is chosen for the study at hand. The corresponding Prandtl number, heat capacity ratio, and dimensional freestream properties are listed in Table 1.

Discretization Direct numerical simulation of transitional flows demands for accurate resolution of smallest flow structures, both spatially and temporal. In this work we use the designated DNS code *NS3D* [1, 2] with 6th-order accurate biased compact differences

$$\begin{aligned}
\frac{1}{5}\frac{\partial q}{\partial \xi}\bigg|_{j-1,+} + \frac{3}{5}\frac{\partial q}{\partial \xi}\bigg|_{j,+} + \frac{1}{5}\frac{\partial q}{\partial \xi}\bigg|_{j+1,+} &= \frac{-q_{j-2} - 19q_{j-1} + 11q_j + 9q_{j+1}}{30\Delta\xi},\\
\frac{1}{5}\frac{\partial q}{\partial \xi}\bigg|_{j-1,-} + \frac{3}{5}\frac{\partial q}{\partial \xi}\bigg|_{j,-} + \frac{1}{5}\frac{\partial q}{\partial \xi}\bigg|_{j+1,-} &= \frac{-9q_{j-1} - 11q_j + 19q_{j+1} + q_{j+2}}{30\Delta\xi},
\end{aligned} \tag{4}$$

used for the convective terms. The subscripts "+" and "–" denote up- and downwind biasing and j the grid point index in some direction ξ, respectively. First derivatives of viscous terms are computed using a standard symmetric compact finite difference stencil

$$\frac{\partial q}{\partial \xi}\bigg|_{j-1} + 3\frac{\partial q}{\partial \xi}\bigg|_{j} + \frac{\partial q}{\partial \xi}\bigg|_{j+1} = \frac{-q_{j-2} - 28q_{j-1} + 28q_{j+1} + q_{j+2}}{12\Delta\xi}. \tag{5}$$

Similarly, second derivatives are obtained directly as

$$2\frac{\partial^2 q}{\partial \xi^2}\bigg|_{j-1} + 11\frac{\partial^2 q}{\partial \xi^2}\bigg|_{j} + 2\frac{\partial^2 q}{\partial \xi^2}\bigg|_{j+1} = \frac{3q_{j-2} + 48q_{j-1} - 102q_j + 48q_{j+1} + 3q_{j+2}}{4\Delta\xi^2}. \tag{6}$$

A standard explicit 4th-order Runge-Kutta method is used for the temporal integration.

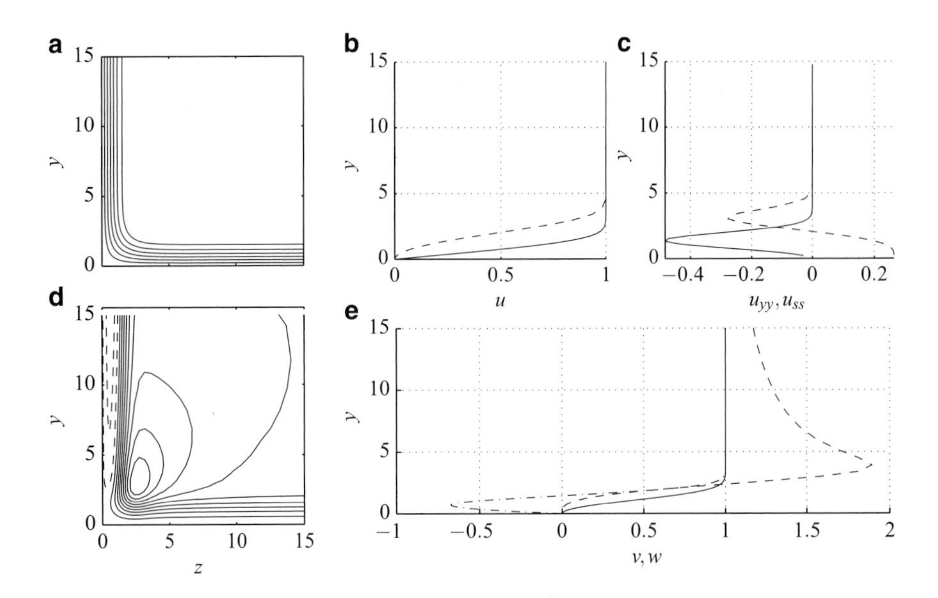

Fig. 2 Self-similar base flow; (**a**) isolines of the streamwise velocity $u(y,z)$; (**b**) streamwise velocity profiles: (*solid line*) far-field, $u(y, z \to \infty)$; (*dashed line*) along the corner bisector, $u(y = z)$; (**c**) second partial derivatives of streamwise velocity profiles: (*solid line*) far-field, $u_{yy}(y, z \to \infty)$; (*dashed line*) along the corner bisector, $u_{ss}(y = z)$; (**d**) isolines of the cross-flow velocity $v(y,z)$: (*dashed line*) negative; (*solid line*) positive; (**e**) cross-flow velocity far-field profiles: (*solid line*) wall normal direction, $v(y, z \to \infty)$; (*dash-dot-dash line*) tangential direction, $w(y, z \to \infty)$; (*dashed line*) along the corner bisector, $v(y = z) = w(y = z)$

Base State and Computational Domain A fully three-dimensional simulation of the transitional flow field is not feasible because of the complex asymptotic behavior of the secondary cross flow, and pressure field. We hence rely on a perturbation formulation that allows us to calculate the perturbation flow field solely upon a fixed steady base state. In the following, perturbation quantities are denoted by $(.)'$ and the base state by $(.)_0$. The implementation of the perturbation formulation is described in detail in [1]. The steady self-similar base state is calculated as a solution to the parabolized Navier-Stokes equation as in [16]. The base flow is depicted in Fig. 2.

The complex nature of the secondary cross-flow becomes apparent in Fig. 2d. Also note that the streamwise velocity component becomes inflectional in the near-corner-region as indicated by the change of sign of the second derivative along the corner bisector coordinate s in Fig. 2c. Inflectional profiles are known to be inviscidly unstable.

The extent and spatial resolution of the computational domain for the DNS and a smaller subdomain used for post-processing are summarized in Table 2. The streamwise domain extent corresponds to a Reynolds number regime of $15 \times 10^4 \leq \mathrm{Re}_x \leq 42.5 \times 10^4$ based on the distance from the leading edge. Simulations at lower, i.e. subcritical, Reynolds numbers are desirable but much more computationally demanding due to the fast boundary layer growth in that region and the associated necessity for a larger computational domain.

Table 2 Computational domain for the direct numerical simulation; subscript (0) and (1) denote start and end, $N_{x,y,z}$ the number of grid points in the respective direction, $x_{p,0}$ and $x_{p,1}$ delimit the streamwise perturbation strip

x_0	Re_0	x_1	Re_1	$x_{p,0}$	$x_{p,1}$	y_1, z_1	N_x	N_y, N_z	Δt
195.68	766.56	554.71	1290.31	197.12	208.61	54	1250	400	0.005
267.49		*396.74*				*22.82*	*225*	*100*	

The heating strip extent corresponds to 30 grid cells, starting at the 15th grid point. Italic numbers in the bottom line refer to the DMD subdomain depicted (as blue numbers) in Fig. 3b

Boundary Conditions Adiabatic no-slip wall boundary conditions are enforced on both walls. The wall pressure is extrapolated from the interior field and the density is calculated from the ideal gas law. Homogeneous Neumann conditions are applied on the inlet and on the far-field boundaries. On the outlet, a subsonic outflow condition [1] is applied. A sponge region is used at the inlet and outlet regions of the computational domain to cancel the fluctuations before reaching the boundaries in order to prevent reflection back into the solution domain. The sponge zone is restricted to the outmost 2.5 % of the inlet and outlet regions and follows a fifth-order polynomial distribution $\sigma(\xi) = \pm\sigma_{max}(1 - 6\xi^5 + 15\xi^4 - 10\xi^3)$ for $\xi \in [0, 1]$, where ξ is the locally scaled distance from the inlet and outlet boundary, respectively, and σ_{max} is the sponge amplitude or gain. A wall heating strip is used to force a harmonic perturbation along the walls. For this purpose, the adiabatic wall boundary condition is locally replaced within some streamwise extend $x_{p,0} \leq x \leq x_{p,1}$. A dipole distribution of the form

$$T' = \mathrm{a}\frac{81}{16}(2\xi)^3\left[3(2\xi)^2 - 7(2\xi) + 4\right]\cos(\omega t) \text{ on } \xi \in [0, 0.5],$$

$$T' = -\mathrm{a}\frac{81}{16}(2 - 2\xi)^3\left[3(2 - 2\xi)^2 - 7(2 - 2\xi) + 4\right]\cos(\omega t) \text{ on } \xi \in [0.5, 1], \quad (7)$$

is used to generate a harmonic perturbation wave of angular frequency ω and an amplitude determined by the amplitude coefficient a. Here again, the auxiliary coordinate $\xi \in [0, 1]$ is then scaled to the desired heating strip extent $x_{p,0} \leq x \leq x_{p,1}$. A mono-frequencial perturbation with $\omega = 0.09$ and $\mathrm{a} = 0.75$ was found to trigger rapid transition due to non-linear interactions and was used to obtain the following results.

3 Results

All results presented are obtained after all initial transients have died out, i.e. taken from a period on the limit cycle. The following three paragraphs are dedicated to the instantaneous coherent flow structures of the transitional flow structures, the mean flow deformation, and the spectral content of the transient flow field.

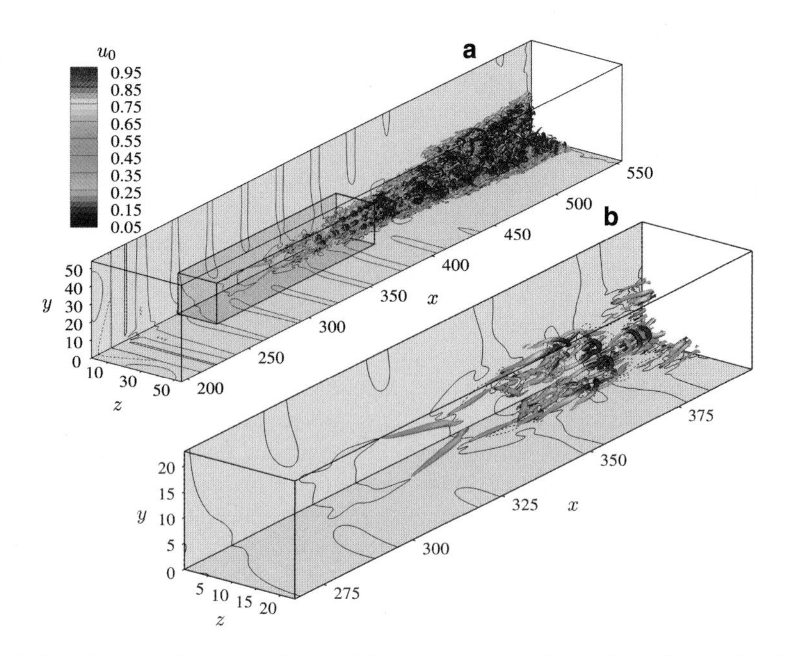

Fig. 3 Snapshot visualization of the transition process; isosurfaces of the λ_2-criterion [6] at $\lambda_2 = -0.01$ colored by the local streamwise velocity of the base state and isocontours of the wall pressure perturbation at $\rho' = 0.002$ (*solid line*) and $\rho' = -0.002$ (*dashed line*): (**a**) entire computational domain and (**b**) enlarged view of the DMD subdomain as listed in the top and bottom row of Table 2, respectively

Coherent Structures of Transitional Flow Instantaneous isosurfaces of the λ_2-criterion are depicted in Fig. 3. It can be seen that turbulence develops in form of a turbulent wedge originating from the near-corner region. A closer look at the flow structures in the initial phase of the transition as depicted in Fig. 3b reveals the emergence of hairpin vortices, typically observed during the transition process, see e.g. [12]. Note that while the perturbation flow field appears symmetric with respect to the corner bisector in the initial stage of the transition process as seen in Fig. 3b, the symmetry is broken further downstream as can be deduced from a closer look at individual flow structures in Fig. 3a. The break of symmetry is likely to stem from small numerical errors and/or the alternating up- and downwind biased compact finite difference scheme used. However, the break of symmetry is an inherent feature of turbulence, and a perturbation-free environment is not experimentally realizable, either.

Mean Flow Deformation The time-averaged streamwise velocity $\overline{\mathbf{u}} = N^{-1}\Sigma_{i=1}^{N}\mathbf{u}_i$ and mean flow deformation $\overline{\mathbf{u}} - \mathbf{u}_0$ field at the beginning, in the middle, and at the end of the post-processing subdomain are depicted in Fig. 4. By comparing with the laminar base state depicted in Fig. 2a it is observed that the flow field at $x = 267.49$ in Fig. 4a still closely resembles the self-similar solution. The velocity isolines in the domain center in Fig. 4b exhibit a distortion in form of a convex bulge as seen

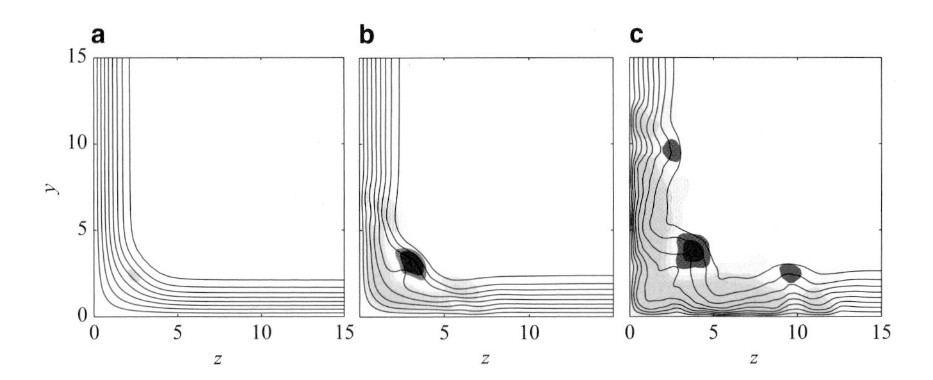

Fig. 4 Contours of the time-averaged streamwise velocity (*solid lines*) and mean flow deformation (*filled*: (*yellow/bright*) positive, (*blue/dark*) negative) in transversal planes at (**a**) $x = 267.49$, (**b**) $x = 332.12$, and (**c**) $x = 396.74$, corresponding to the beginning, the middle, and the end of the DMD subdomain, respectively

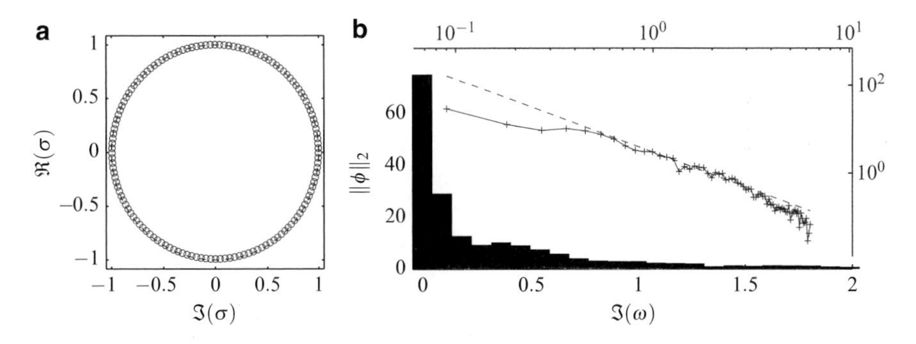

Fig. 5 Dynamic mode decomposition from 140 snapshots over one fundamental period; (**a**) empirical Ritz values (*blue open circles*); (**b**) magnitudes of the Koopman modes on a linear (*black bars*) and a logarithmic scale (*blue solid line*). The *blue dashed line* shows a slope of $\Im(\omega)^{-5/3}$

in experimental studies [17]. A strong mean flow deformation is observed in Fig. 4c for $x = 396.74$ where an additional outward bulge can be seen at $\eta, \zeta \approx 10$ on each of the walls. The latter correspond to the hairpin vortices flanking the bisectorial centered vortex structure seen in Fig. 3.

Spectral Analysis The global stability properties of the transient flow field are best analysed by means of a spectral decomposition technique, i.e. decomposition into global modes of a single frequency. Here, we use the Koopman operator-based dynamic mode decomposition (DMD) introduced by Schmid [15] and analysed in detail by Rowley et al. [13]. The reader is referred to the latter literature for details on the method. The empirical Ritz values are depicted in Fig. 5a in the complex plane. It can be seen that all Ritz values are located on the unit circle, indicating zero temporal growth as expected for a convective problem on its limit cycle. Also, the

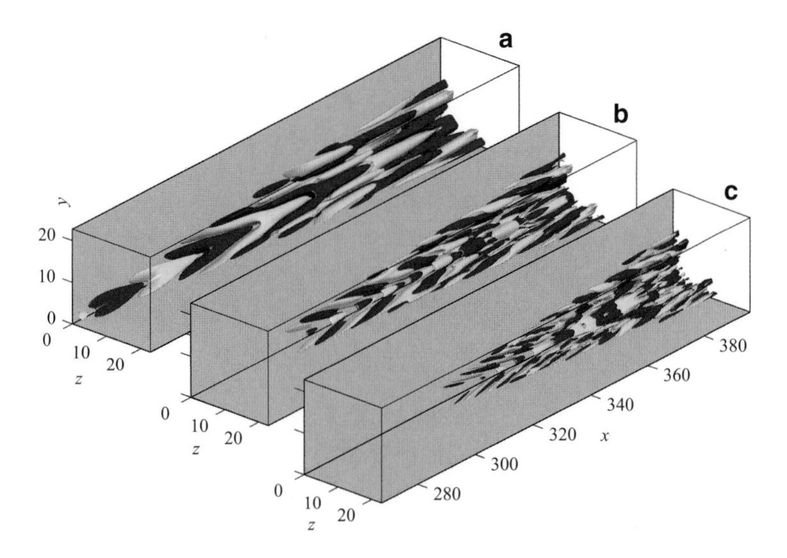

Fig. 6 Dynamic modes from the visualized by isosurfaces of the streamwise perturbation velocity: (**a**) fundamental frequency $\Im(\omega) = 0.09$, (**b**) first higher harmonic $\Im(\omega) = 0.18$, and (**c**), second higher harmonic $\Im(\omega) = 0.27$. Isosurfaces are drawn at $\pm 0.2|u'|_{max}$

values are evenly distributed, meaning that the flow field is clearly decomposed into mono-frequencial modes of integer multiples of the forcing frequency. The modal amplitudes are shown in Fig. 5b as a function of the modal frequency on a linear and a logarithmic scale. The first Koopman mode is found to be the most energetic. This result is not surprising as the first mode embodies the steady component with $\omega = 0$ which is similar but not necessarily equal to the mean flow deformation, cf. Chen et al. [4]. The second most energetic mode is the direct response of the base state to the forcing frequency at $\omega = 0.09$. For higher frequencies, the modal energy distribution is found in good agreement with the $-5/3$ power-law of the inertial subrange of the energy cascade [7].

The modal structures corresponding to the fundamental frequency and the first two higher harmonics are visualised in Fig. 6. The coherent flow structures appear elongated as commonly observed in non-linearly developing flows. It can also be seen that energy is distributed towards higher wave numbers in accordance with the energy cascade. This is clearly indicated by the increasingly fine flow patterns when comparing the modes in the given order of increasing frequency.

4 Performance

Originally the *NS3D* program was designed and optimized for the vector computing architecture of the NEC-SX supercomputers at HLRS Stuttgart. A combination of shared and distributed memory parallelization features are applied to account for

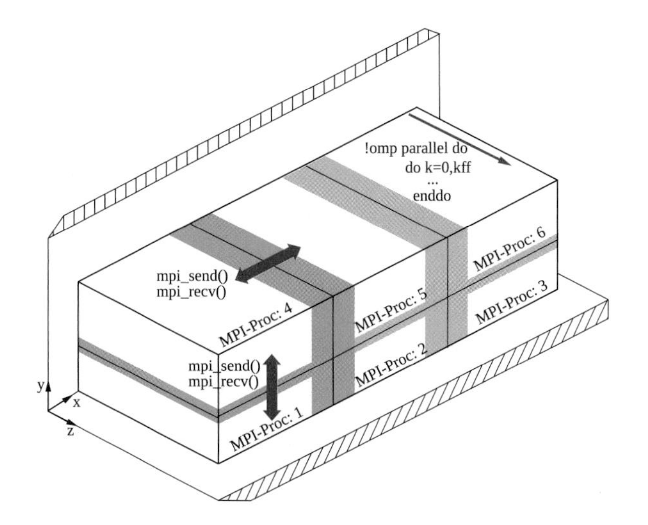

Fig. 7 Domain decomposition in x- and y-direction and parallelization features in *NS3D* program

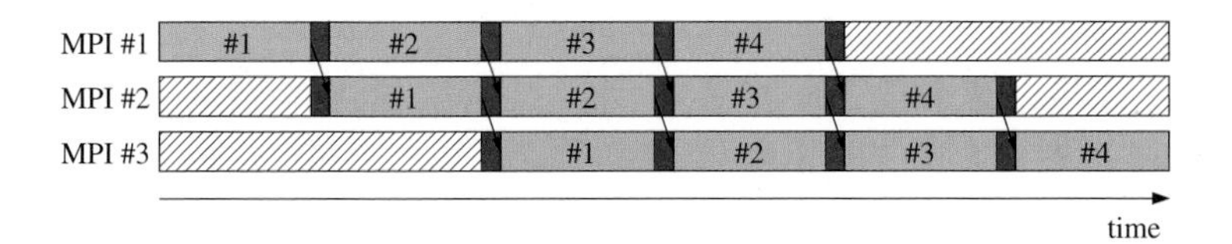

Fig. 8 Pipelined Thomas algorithm: *blue* indicates compute time, *hatched* indicates idle time. Reproduced from [1]

the typically small ratio of CPUs per node to overall number of nodes. A sketch of the layout is shown in Fig. 7. The computational domain is decomposed in blocks of equal number of grid points by splitting the streamwise and wall-normal direction. These blocks communicate via MPI routines [9] to exchange neighbouring data. Additionally the spanwise direction is parallelized by means of openMP directives [10]. Details of the algorithm can be found in [1].

In spanwise direction the classical Thomas algorithm is applied independently on each process in order to efficiently solve the linear system of equations arising from the compact finite difference scheme, i.e. (4)–(6). In streamwise and wall-normal direction however a pipelined version of the Thomas-algorithm [11] is implemented which fully conserves the order of the compact scheme across domain, i.e. process boundaries along the respective direction. Unfortunately this comes at the expense of idle CPU time for subsequent processes as illustrated in Fig. 8. As long as the number of linear systems is sufficiently larger than the number of processes, which has been the case for the NEC-SX vector computers, the benefit of the homogeneous order throughout the whole domain outdoes the computationally costly idling. On scalar architecture computers on the other hand the ratio of CPUs per node to

Table 3 Comparison of CPU time per grid point and time step

Grid	Time steps	Specific CPU time (μs) NEC-SX9	Cray XE6
100 × 100	10,000	2.257	3.051
250 × 1000	10,000	0.335	3.791

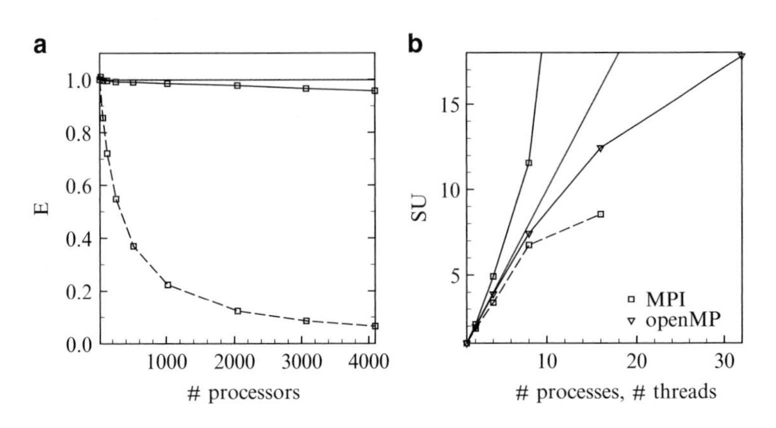

Fig. 9 Parallel scaling of compact FD (*dashed line*) and subdomain compact FD (*solid line*) scheme. (**a**) Efficiency (weak scaling); (**b**) speedup (strong scaling) for openMP and MPI parallelization. Ideal efficiency and speedup are indicated by *blue lines*

overall number of nodes usually is less advantageous for a pipelined computation of the linear systems of equations. One has to compensate the reduced specific CPU times encountered on scaler computers by a larger number of processes. For small scale problems the CPU time per grid point and time step (specific CPU time) on vector parallel computers (NEC-SX9) and massively parallel computers (XE6) are comparable but the NEC SX computers exhibit a large performance boost due to the vectorization for large numbers of grid points as they are usually encountered in direct numerical simulations (cf. Table 3). Taking into account the expansion in z-direction it was opted to use 200 nodes to distribute a total of $2 \cdot 10^8$ grid points on the Cray XE6 supercomputer to simulate the problem at hand. Each node was set to use 32 threads for openMP parallelized tasks. Additionally 20 nodes have been employed to perform online postprocessing routines. Two runs of 22 h each where necessary to reach the limit cycle at which the sampling for all subsequent analysis was undertaken. The specific CPU time for these computations amounted to excruciatingly high 56 μs per grid point and timestep stemming foremost from the use of a pipelined linear systems solver. Figure 9 underlines the massive performance drop of the MPI parallelization in combination with the coupled linear systems on the XE6. The efficiency E is obtained from weak scaling tests on a range of 1–256 MPI processes using 16 CPUs each. Using 4 processes already leads to only 50 % efficiency which keeps on dropping down to 7 % for the largest simulation. Strong scaling tests underline the suboptimal parallelization of the pipelined Thomas algorithm on a scalar computer. The speedup, defined as ratio of sequential compute time to parallel compute time $SU = T_1/T_P$, depicts strong sublinear behaviour starting from 8 processes and onwards. For the sake of completeness the speedup for purely openMP parallelized stimulations is also shown in Fig. 9. Almost ideal speedup can be reached for up to 8 threads which correlates to the size of a NUMA node within a compute node of the Cray XE6 computer. For 9–32 threads still sufficient albeit slightly sublinear speedup can be obtained. In order to significantly improve the parallelization for future simulations a so-called sub-domain compact scheme has been implemented into the *NS3D* program. This scheme decouples the discretization in x- and y-direction such that the linear system of equations can be solved independently on each sub-domain. For that purpose it is necessary to partly replace the left hand side of the compact stencil at domain boundary points by explicit values. In order to sustain the characteristic wave resolution of the compact scheme explicit finite differences of 8th order are used to compute these values. Performing the aforementioned scaling analysis for the new solver routine demonstrates the near-ideal efficiency reached by decoupling the domains as well as a superlinear speedup due to favourable cache effects for the regarded problem size.

5 Conclusions

A direct numerical simulation of the transition in streamwise corner-flow at a technically relevant Mach number of $Ma = 0.8$ was conducted. The transition takes place in form of a turbulent wedge situated in the near-corner region. This result is remarkable in the sense that it is found in accordance with measurements but in opposition to linear theory which does not predict a dominant instability mechanism due to the presence of the corner as compared to the single flat-plate scenario. The turbulent flow structures were found in agreement with other wall-bounded shear flows, and the spectral content of the transitional flow extracted via direct mode decomposition of a snapshot sequence was found to match the theoretically predicted power-law. From the computational side, the DNS code was found to perform inferior in the massive parallel environment of the CRAY XE6 as compared to the vector architectures it was originally designed for. However, a recently developed recast sub-domain compact scheme allowed for a decoupled processing of the routines while maintaining the overall accuracy of its fully implicit precursor. The decoupling led to a significant gain in computational performance for distributed memory parallelization. Scaling tests indicated almost ideal efficiency and superlinear speedup for the new implementation. Scaling tests for the shared memory parallelization showed almost ideal speedup for up to 8 threads as well.

References

1. A. Babucke, Direct numerical simulation of noise-generation mechanisms in the mixing layer of a jet, Ph.D. Thesis, Universität Stuttgart, 2009
2. A. Babucke, M. Kloker, U. Rist, DNS of a plane mixing layer for the investigation of sound generation mechanisms. Comput. Fluids **37**(4), 360–368 (2008)
3. G. Carrier, The boundary layer in a corner. Q. Appl. Math. **4**, 367–370 (1947)
4. K. Chen, J. Tu, C. Rowley, Variants of dynamic mode decomposition: boundary condition, Koopman, and Fourier analyses. J. Nonlinear Sci. **22**(6), 1–29 (2011)
5. I. Galionis, P. Hall, Spatial stability of the incompressible corner flow. Theor. Comput. Fluid Dyn. **19**, 77–113 (2005). doi:10.1007/s00162-004-0153-1
6. J. Jeong, F. Hussain, On the identification of a vortex. J. Fluid Mech. **285**(69), 69–94 (1995)
7. A. Kolmogorov, Dissipation of energy in locally isotropic turbulence. Dokl. Akad. Nauk SSSR **32**(1), 15–17 (1941)
8. A.G. Mikhail, K.N. Ghia, Viscous compressible flow in the boundary region of an axial corner. AIAA J. **16**, 931–939 (1978). doi:10.2514/3.60987
9. MPI—a message-passing interface standard. Technical Report CS-94-230, University of Tennessee, Knoxville (1994)
10. OpenMP: the openmp api specification for parallel programming (2012), http://www.openmp.org
11. A. Povitsky, Parallelization of the pipelined Thomas algorithm. ICASE Report 98-48 (1998)
12. U. Rist, H. Fasel, Direct numerical simulation of controlled transition in a flat-plate boundary layer. J. Fluid Mech. **298**, 211–248 (1995)
13. C. Rowley, I. Mezić, S. Bagheri, P. Schlatter, D. Henningson, Spectral analysis of nonlinear flows. J. Fluid Mech. **641**(1), 115–127 (2009)
14. S.G. Rubin, B. Grossman, Viscous flow along a corner: numerical solution of the corner layer equations. Q. Appl. Math. **29**, 169–186 (1971)
15. P.J. Schmid, Dynamic mode decomposition of numerical and experimental data. J. Fluid Mech. **656**, 5–28 (2010). doi:10.1017/S0022112010001217
16. O. Schmidt, U. Rist, Linear stability of compressible flow in a streamwise corner. J. Fluid Mech. **688**, 569–590 (2011)
17. M. Zamir, Similarity and stability of the laminar boundary layer in a streamwise corner. R. Soc. Lond. Proc. A **377**, 269–288 (1981)

LES of Turbulent Cavitating Shear Layers

Christian Egerer, Stefan Hickel, Steffen Schmidt, and Nikolaus A. Adams

Abstract Preliminary results of large-eddy simulations of turbulent cavitating shear layers are presented. Validation of the computational setup is presented with focus on the extent of the spanwise domain size. We observe characteristic differences in the evolution of the extremal values of the components of the Reynolds-stress tensor at different cavitation numbers.

1 Introduction

Cavitation refers to the formation of vapor cavities in a liquid that undergoes phase change when subjected to low pressures. Coherent structures in turbulent shear flows play an important role in the cavitation process. High negative pressure peaks associated with turbulent eddies are the main source of cavitation in shear flows. Vice versa, cavitation can generate vorticity resulting in a complex mutual interaction, see Arndt [1] for a complete review.

Experimental studies of cavitating shear layers carried out in water tunnels have been published, for example, by O'Hern [2], Iyer and Ceccio [3], or Aeschlimann et al. [4, 5]. Questions related to cavitation and turbulence modulation by phase change have been addressed by the authors. Direct Numerical Simulation (DNS) of cavitating separated flow and cavitating shear layers based on the *incompressible* Navier-Stokes equations have been performed by Kajishima et al. [6] and Okabayashi and Kajishima [7], respectively.

By investigating a spatially developing turbulent shear layer in water under non-cavitating and cavitating conditions, we address questions related to the mutual interaction of turbulence and cavitation, the resulting modulation of turbulence

C. Egerer (✉) · S. Hickel · S. Schmidt · N.A. Adams
Lehrstuhl für Aerodynamik und Strömungsmechanik, Technische Universität München, Boltzmannstr. 15, 85748 Garching bei München, Germany
e-mail: christian.egerer@aer.mw.tum.de

W.E. Nagel et al. (eds.), *High Performance Computing in Science and Engineering '13*, DOI 10.1007/978-3-319-02165-2_24,

due to phase change, and its influence on properties of the shear layer. Within the present project, we plan to add to the rather scarce numerical data available on the aforementioned topics. In particular, effects of compressibility will be investigated.

2 Governing Equations and Numerical Method

We consider the 3-D compressible Navier-Stokes equations for a barotropic fluid written in integral form for an arbitrary control volume Ω with surface $\partial\Omega$

$$\partial_t \int_{\Omega} \mathbf{U} d\Omega + \frac{1}{\Omega} \int_{\partial\Omega} [\mathbf{C}(\mathbf{U}) + \mathbf{D}(\mathbf{U})] \cdot \mathbf{n} dS = 0, \tag{1}$$

where $\mathbf{U} = [\rho, \rho u_i]$ is the vector of conservative variables comprising density ρ and momentum densities ρu_i, $i = 1, 2, 3$; $\mathbf{n}$ denotes the normal vector of the surface increment dS. The total flux across the control volume surface $\partial\Omega$ is split into advection and surface stresses

$$C_i(\mathbf{U}) = [u_i \rho, u_i \rho u_1, u_i \rho u_2, u_i \rho u_3], \qquad \text{and} \tag{2}$$

$$D_i(\mathbf{U}, p) = [0, \delta_{i1} p - \tau_{i1}, \delta_{i2} p - \tau_{i2}, \delta_{i3} p - \tau_{i3}], \tag{3}$$

where $p = p(\rho)$ denotes the pressure and

$$\tau_{ij} = \mu \left(\frac{\partial u_i}{\partial x_j} + \frac{\partial u_j}{\partial x_i} - \frac{2}{3} \delta_{ij} \frac{\partial u_k}{\partial u_k} \right) \tag{4}$$

denotes the viscous stress tensor for a Newtonian fluid with dynamic viscosity μ.

Considering an arbitrary control volume Ω, the solution is represented by the volume average of the vector of conservative variables

$$\overline{\mathbf{U}} = \frac{1}{V} \int_{\Omega} \mathbf{U} dV; \qquad V_{\Omega} = \int_{\Omega} dV. \tag{5}$$

A homogeneous mixture of liquid and vapor is assumed in two phase regions, with $\alpha = \Omega_{\text{v}}/\Omega$ being the vapor volume fraction. Therefore, the mixture or volume averaged density is given by the convex combination of the liquid density ρ_{l} and vapor density ρ_{v}, i.e.,

$$\overline{\rho} = \alpha \rho_{\text{v}} + (1 - \alpha) \rho_{\text{l}}. \tag{6}$$

Assuming that phase change is in thermodynamic equilibrium, i.e., infinitely fast, isentropic and in mechanical equilibrium, the vapor volume fraction can be computed as follows

$$\alpha = \begin{cases} 0 & , \overline{\rho} \geq \rho_{\text{sat,l}} \\ \frac{\rho_{\text{sat,l}} - \overline{\rho}}{\rho_{\text{sat,l}} - \rho_{\text{sat,v}}} & , \overline{\rho} < \rho_{\text{sat,l}} \end{cases}, \tag{7}$$

where $\rho_{\text{sat,l}}$ and $\rho_{\text{sat,v}}$ denote the densities of the vapor and liquid phase at the saturation point. One major advantage of the homogeneous mixture methodology is its capability to reproduce subgrid-scale two-phase regions, e.g., small vapor bubbles, as well as fully resolved vapor structures, e.g., vapor clouds, with the underlying numerical grid. Note, however, that surface tension effects have to be neglected since no interface is reconstructed. Furthermore, no empirically calibrated constants are needed in the computation of mass transfer terms between the liquid and vapor phase. The modeling of two-phase flows using the homogeneous mixture model presented above has proven to be able to accurately describe cavitating flows. A homogeneous model has been successfully applied for the prediction of inertia controlled sheet and cloud cavitation around hydrofoils [8], in micro channels [9], as well as the prediction of erosion sensitive areas in cavitating flows [10].

The set of conservation equations (1) is closed by applying a modified version of Tait's equation of state in pure liquid regions:

$$\overline{p} = (p_{\text{sat}} + B) \left(\frac{\overline{\rho}}{\rho_{\text{sat,l}}} \right)^N - B, \quad \text{if } \alpha = 0. \tag{8}$$

In two phase regions, an isentropic path is followed in the phase diagram to obtain the equilibrium pressure given as

$$\overline{p} = p_{\text{sat}} + C \left(\frac{1}{\rho_{\text{sat,l}}} - \frac{1}{\overline{\rho}} \right), \quad \text{if } 0 < \alpha < 1. \tag{9}$$

We assume that the effective dynamic viscosity of the liquid-vapor mixture satisfies a quadratic law with a maximum in the two-phase region reading [11]

$$\overline{\mu} = (1 - \alpha) \left(1 + \frac{5}{2} \alpha \right) \mu_{\text{sat,l}} + \alpha \mu_{\text{sat,v}}, \tag{10}$$

where $\mu_{\text{sat,l}}$ and $\mu_{\text{sat,v}}$ denote the dynamic viscosities at the saturation point for the liquid and vapor phase, respectively. The properties of water at the reference temperature $T_{\text{ref}} = 293.15\,\text{K}$ and the fitted constants N, B and C of the piecewise defined equation of state (8) and (9) are summarized in Table 1.

The volume-averaged conservative variables, Eq. (5), are advanced in time by a finite volume method. The convective flux, Eq. (2), is discretized by means of the Adaptive Local Deconvolution Method (ALDM) for compressible and barotropic fluids [9, 12]. ALDM is a non-linear discretization method incorporating a subgrid-scale turbulence model. To do so, ALDM reconstructs the primitive variables $\varphi \in [\rho, u_i]$ at cell faces, $\breve{\varphi}^{\pm}_{i\pm1/2}$, by combining Harten-type deconvolution polynomials,

Table 1 Properties of water at $T_{\mathrm{ref}} = 293.15\,\mathrm{K}$

Property	Value	Unit
$\rho_{\mathrm{sat,l}}$	998.1618	kg/m^3
$\rho_{\mathrm{sat,v}}$	0.01731	kg/m^3
$\mu_{\mathrm{sat,l}}$	1.002×10^{-3}	Pa s
$\mu_{\mathrm{sat,v}}$	9.727×10^{-6}	Pa s
p_{sat}	2340.0	Pa
C	1468.54	Pa kg/m^3
N	7.1	–
B	3.06×10^{8}	Pa

$\breve{g}^{\pm}_{kr}$, up to order three non-linearly and solution-adaptively:

$$\breve{\varphi}^{\pm}(x_{i\pm1/2}) = \sum_{k=1}^{3}\sum_{r=0}^{k-1} \omega^{\pm}_{kr}(\gamma_{kr}, \overline{\varphi})\breve{g}^{\pm}_{kr}(x_{i\pm1/2}). \tag{11}$$

The weights $\omega^{\pm}_{kr}$ of the deconvolution polynomials introduce free parameters γ_{kr} which are used to control the truncation error of ALDM. Additionally, a suitable numerical flux function comprising the physical convective flux and a secondary regularization term is used:

$$\breve{C}_{i\pm\frac{1}{2}} = C\left(\frac{\breve{\varphi}^{+}_{i\pm\frac{1}{2}} + \breve{\varphi}^{-}_{i\pm\frac{1}{2}}}{2}\right) - R\left(\varepsilon, \breve{\varphi}^{\pm}, \overline{\varphi}\right) \tag{12}$$

A physically consistent implicit subgrid-scale model is obtained by optimizing the free parameters $\{\gamma_{kr}, \varepsilon\}$ of ALDM so that the effective spectral numerical viscosity matches the eddy viscosity from the Eddy-Damped Quasi-Normal Markovian theory for isotropic turbulence.

The diffusive flux, Eq. (3), is discretized by a linear second order accurate scheme. Time integration is performed by the third order accurate Runge-Kutta method of Gottlieb and Shu [13], which is total-variation diminishing (TVD) for CFL ≤ 1 if the employed spatial discretization is TVD.

3 Computational Setup

The rectangular computational domain comprises a *core domain* of size $L_1 \times L_2 \times L_3 = 10h \times 4h \times 1.2h$ in streamwise (x_1), lateral (x_2), and spanwise (x_3) direction, which is embedded in a coarser *enclosing domain* of size $14h \times 5.6h \times 1.2h$. The enclosing domain is utilized to damp turbulence and acoustic waves.

Two parallel streams of different velocities $U_2 > U_1 > 0$, with a mean convection velocity $U_c = (U_1 + U_2)/2$ and velocity difference $\Delta U = U_2 - U_1$, characterize the shear layer. The initial and inflow boundary condition is defined by the velocity profile

$$u_i(\boldsymbol{x},t) = \begin{bmatrix} U_c + \frac{1}{2}\Delta U \tanh\left(\frac{2x_2}{\delta_{\omega,0}}\right) \\ 0 \\ 0 \end{bmatrix} + u_i'(\boldsymbol{x},t), \tag{13}$$

where the vorticity thickness reads

$$\delta_\omega = \frac{\Delta U}{\left.\frac{\partial u_1}{\partial x_2}\right|_{\max}}. \tag{14}$$

In this paper, we set $\delta_{\omega,0} = 0.01h$ as the initial vorticity thickness. Random time-dependent velocity fluctuations u_i' with an amplitude of $1\,\%$ of ΔU are superimposed on the mean flow to trigger transition.

The mutual interaction of turbulence and phase change is investigated by varying the cavitation number

$$\sigma_c = \frac{2(p_\infty - p_{\text{sat}})}{\rho_\infty(\Delta U)^2}. \tag{15}$$

Since we keep the characteristic velocity difference constant, the variation in σ_c is obtained by imposing the far-field pressure p_∞ at the inflow and lateral boundaries. Note that the far-field density and pressure are directly coupled since we consider a barotropic fluid. At the outflow boundary a convective boundary condition is used. Periodicity is assumed in spanwise directions.

4 Preliminary Results

We present selected results for two cavitation numbers σ_c: one with essentially no cavitation ($\sigma_c = 1.0$), and one strongly cavitating case with cavitation number $\sigma_c = 0.1$. After letting the flow field develop for 10 flow-through times of the core domain, $\Delta t = 10h/U_c$, statistical data have been collected for 10 and 20 flow-through times for $\sigma_c = 1.0$ and $\sigma_c = 0.1$, respectively. Statistics have additionally been averaged in the homogeneous spanwise direction.

4.1 Grid

Preliminary results have been obtained on a grid consisting of approximately 3.88×10^6 cells. The grid spacing in the core domain is $0.025h$ in all directions with the exception of two refined zones at the inlet. The grid spacing is halved in all directions in the first zone; in the second zone the grid spacing is halved another time but only in the lateral direction. The refined inlet zones are necessary to resolve the sharp gradients and ensure a correct growth of instabilities. The enclosing domain comprises three sections where the grid is subsequently coarsed with a ratio of 2:1 with respect to the core domain.

4.2 Computational Aspects

All computations have been run on 10 nodes with 32 parallel tasks per node on Cray XE6 "Hermit" resulting in a total of 320 MPI tasks. The CPU-time per cell and iteration is approximately 26×10^{-6} s. The communication overhead, comprising the exchange of three ghost-cell layers, accounts for roughly 35 % of the total wall time. With the current set-up, approximately 3,000 CPUh are needed for one flow-through time.

4.3 Verification of the Spanwise Domain Size

In order to verify the choice of the spanwise domain size, the spanwise autocorrelation functions

$$R_{ii}(r_3) = \langle u_i(x_3)u_i(x_3 - r_3)\rangle / \langle u_i(x_3)^2\rangle \tag{16}$$

of the velocity components have been evaluated at three positions downstream of the inlet at several time instances for the case with a cavitation number of $\sigma_c = 1.0$. Figure 1 shows the time-averaged autocorrelation functions and indicates that the maximum correlation length for all velocity components and at all downstream positions is equal or less than $r_3 = 0.2h$, which demonstrates that the spanwise domain size of $L_3 = 1.2h$ is sufficient with respect to the periodicity assumption in spanwise direction.

4.4 Coherent Structures

Figure 2 shows coherent structures identified by means of the λ_2-criterion [14] and the corresponding vapor cavities visualized by an iso-surface with $\alpha = 0.1$ at cavitation number $\sigma_c = 0.1$. Initially, spanwise Kelvin-Helmholtz-type vortices are generated. Further downstream, the generation of secondary streamwise instabilities followed by their break-up is observed, completing the transition process to a turbulent shear layer, cf. Fig. 2a. Local low pressure regions associated with turbulent eddies are sources for cavitation. The pairing process in the turbulent shear layers leads to the formation of large spanwise vapor cavities, cf. Fig. 2b.

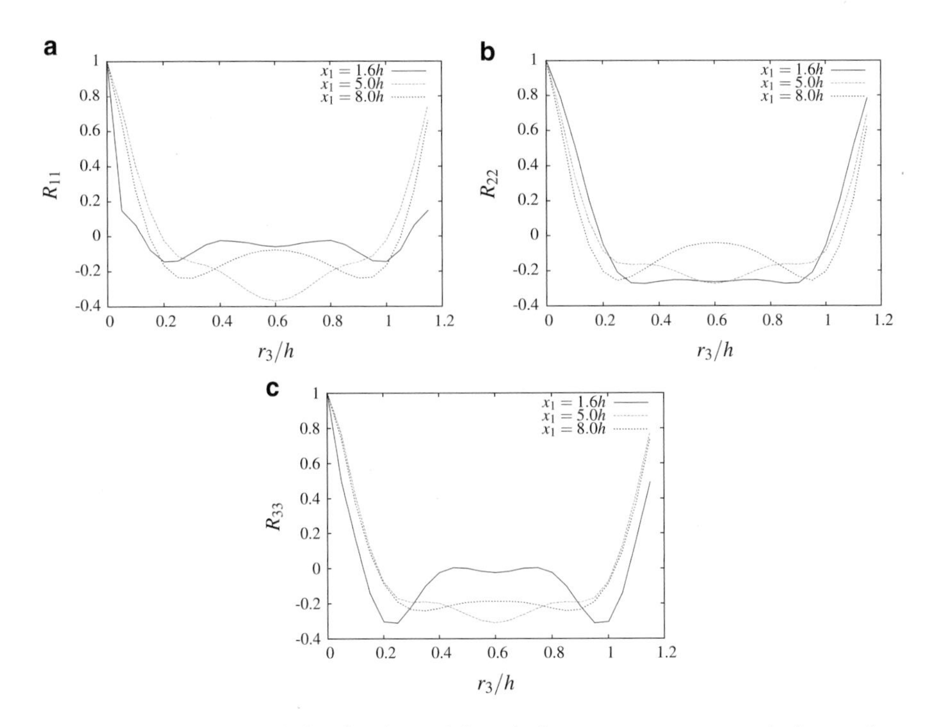

Fig. 1 Spanwise autocorrelation functions of the velocity components u_i at cavitation number $\sigma_c = 1.0$: (**a**) R_{11}, (**b**) R_{22}, and (**c**) R_{33}

4.5 *Reynolds-Stress Tensor*

Figure 3 shows the streamwise evolution of the maximum value in lateral direction of the resolved turbulence kinetic energy (TKE), i.e., $\max(k(x_1 = const., x_2) = \langle u_i' u_i' \rangle/2)$. Differences are observed in the transition region, say $x_1 < 3$, where the cavitation number $\sigma_c = 1.0$ exhibits larger maximum TKE values. However, further downstream the maximum TKE levels decay to approximately the same value.

Further insight is gained by looking at the individual components of the Reynolds-stress tensor, $R_{ij} = \langle u_i' u_j' \rangle$. Figure 4 displays the streamwise evolution of the extremal value in lateral direction of the non-zero components of R_{ij}. Here, one observes that the maximum values of the streamwise component R_{11} are larger for cavitation number $\sigma_c = 0.1$ than for cavitation number $\sigma_c = 1.0$ outside of the transition region, cf. Fig. 4a. On the other hand, the lateral and spanwise components R_{22} and R_{33} are smaller for cavitation number $\sigma_c = 0.1$ than for cavitation number $\sigma_c = 1.0$. In the transition region, the obtained values of R_{22} and R_{33} are approximately twice as high at $\sigma_c = 1.0$ than at $\sigma_c = 0.1$, see Fig. 4c,d. The shear stress R_{12} is more negative for cavitation number $\sigma_c = 0.1$ than for cavitation number $\sigma_c = 1.0$, cf. Fig. 4b.

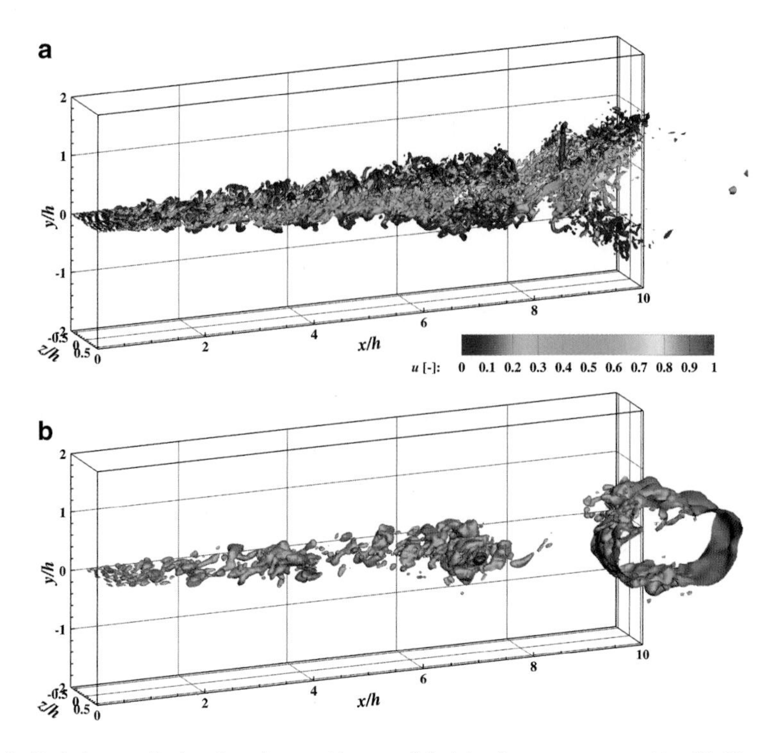

Fig. 2 Turbulent cavitating shear layer with $\sigma_c = 0.1$: (**a**) coherent structures identified by means of the λ_2-criterion, (**b**) iso-surfaces with $\alpha = 0.1$

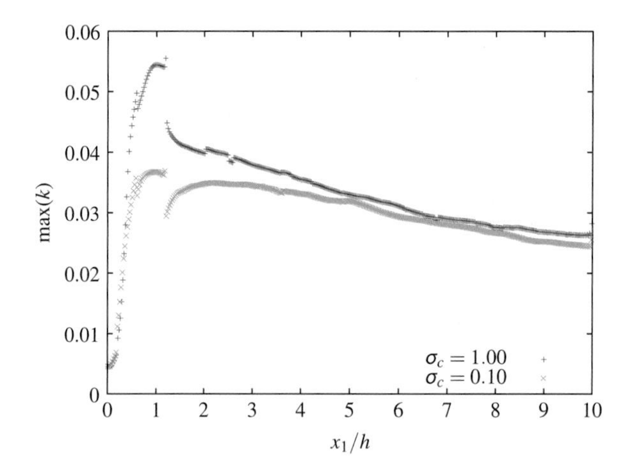

Fig. 3 Streamwise evolution of the maximum value in lateral direction of the turbulence kinetic energy (TKE)

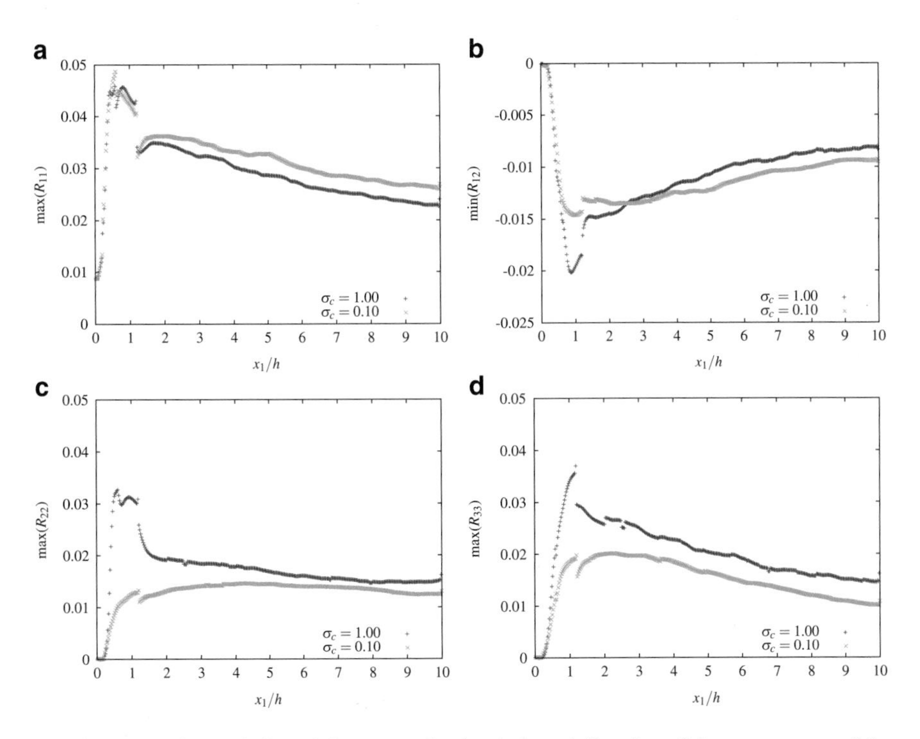

Fig. 4 Streamwise evolution of the extremal value in lateral direction of the components of the Reynolds-stress tensor: (**a**) R_{11}, (**b**) R_{12}, (**c**) R_{22}, and (**d**) R_{33}

The streamwise evolution of the extremal value in lateral direction of the non-zero components of the normalized Reynolds-stress anisotropy tensor

$$b_{ij} = \frac{\langle u_i u_j \rangle}{\langle u_k u_k \rangle} - \frac{1}{3}\delta_{ij}, \tag{17}$$

are shown in Fig. 5. The two different cavitation numbers exhibit a different degree of anisotropy. Differences are most pronounced in the streamwise component b_{11}, see Fig. 5a.

5 Conclusions and Outlook

The preliminary results for LES of turbulent cavitating shear layers presented above show differences in the evolution of the Reynolds-stress tensor and Reynolds-stress anisotropy tensor when the cavitation number is reduced, although the overall level of turbulence measured by the maximum level of turbulence kinetic energy is approximately independent of the investigated cavitation numbers. The cause of these opposing trends needs further investigation.

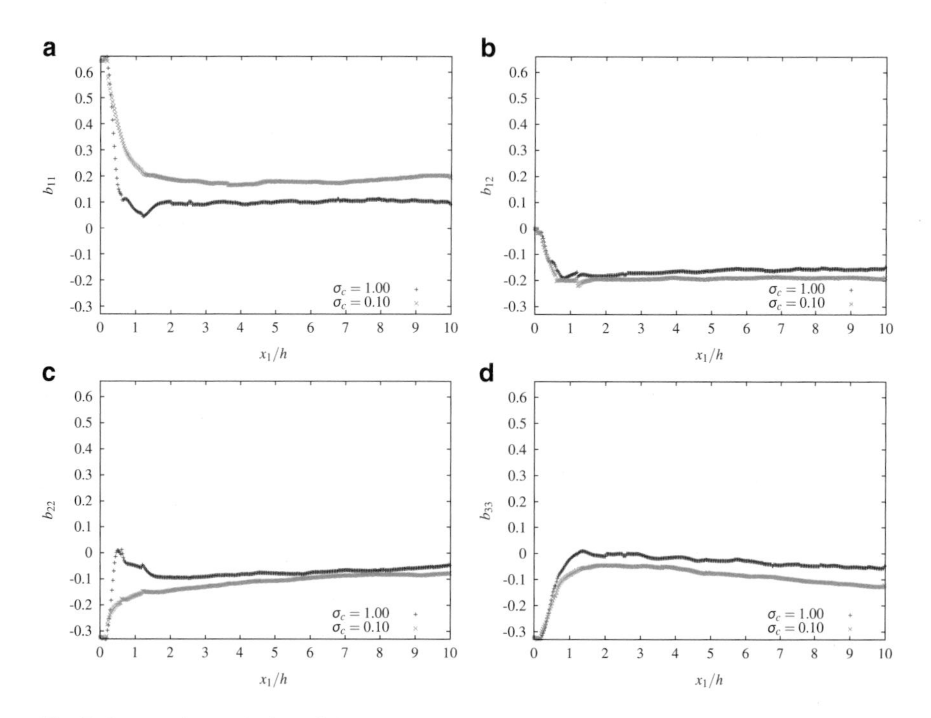

Fig. 5 Streamwise evolution of the extremal value in lateral direction of the components of the normalized Reynolds-stress anisotropy tensor: (**a**) b_{11}, (**b**) b_{12}, (**c**) b_{22}, and (**d**) b_{33}

Simulations on a further refined grid will be conducted and compared to the current results in order to ensure grid independence of our conclusions. Furthermore, the analysis of the Reynolds-stress transport equations is planned in order to identify the reason for the different evolution of the Reynolds-stress tensor.

Acknowledgements This project is supported by the German Research Foundation (DFG) under contract AD186/20-1.

References

1. R.E.A. Arndt, Annu. Rev. Fluid Mech. **34**, 143 (2002)
2. T.J. O'Hern, J. Fluid Mech. **215**, 365 (1990)
3. C.O. Iyer, S.L. Ceccio, Phys. Fluids **14**(10), 3414 (2002)
4. V. Aeschlimann, S. Barre, H. Djeridi, Phys. Fluids **23**, 055105 (2011)
5. V. Aeschlimann, S. Barre, S. Legoupil, Phys. Fluids **23**, 055101 (2011)
6. T. Kajishima, T. Ohta, H. Sakai, K. Okabayashi, in *Fifth International Symposium on Turbulence and Shear Flow Phenomena*, ed. by R. Friedrich, N.A. Adams, J.K. Eaton, J.A.C. Humphrey, N. Kasagi, M.A. Leschziner (2007), pp. 829–834. http://www.tsfp-conference.org

7. K. Okabayashi, T. Kajishima, in *Proceedings of the 7th International Symposium on Cavitation*, Ann Arbor, 2009
8. S.J. Schmidt, M. Thalhamer, G.H. Schnerr, in *Proceedings of the 7th International Symposium on Cavitation*, Ann Arbor, 2009
9. S. Hickel, M. Mihatsch, S.J. Schmidt, in *WIMRC 3rd International Cavitation Forum*. University of Warwick, Coventry, 2011
10. M.S. Mihatsch, S.J. Schmidt, M. Thalhamer, N.A. Adams, in *International Conference on Computational Methods in Marine Engineering*, ed. by L. Eca, E. Oñate, J. García, T. Kvamsdal, P. Bergan, Lisbon, 2011
11. D. Beattie, P. Whalley, Int. J. Multiphase Flow **8**, 83 (1982)
12. S. Hickel, J. Larsson, On implicit turbulence modeling for LES of compressible flows, in *Advances in Turbulence XII*, ed. by B. Eckhardt. Springer, Berlin, 2009), pp. 873–875
13. S. Gottlieb, C.W. Shu, Math. Comput. **67**(221), 73 (1998)
14. J. Jeong, F. Hussain, J. Fluid Mech. **285**, 69 (1995)

Large Eddy Simulation of Cavitating Flows Using a Novel Stochastic Field Formulation

F. Magagnato and J. Dumond

Abstract The basic ideas of the Stochastic Fields method for turbulent reacting flows have been adapted to compressible cavitating flows. A probability density function approach is applied to the vapor mass fraction to simulate vapor bubble size distribution and implemented into our finite volume compressible code. The water-vapor mixture is assumed in homogeneous equilibrium and the vapor mass fraction is described by a set of pure Eulerian transport equations with stochastic source terms.

With this novel technique, major two-phase flow parameters like vapor bubble radius, inter-facial area and volume can be captured. Also the source term non-linearity can be resolved at the sub-grid scale. No Lagrangian solver or equations for bubbles clusters are required leading to a low computational cost and simple implementation. The focus of this work is on the theory of the novel stochastic model and aspects of its implementation. Applications include sheet cavitation.

1 Nomenclature

g	Gravitational acceleration
s	Specific entropy
α	Thermal diffusivity
μ	Dynamic viscosity
ν	Kinematic viscosity

F. Magagnato (✉)
Department of Fluid Machinery, Karlsruhe Institute of Technology, 76131 Karlsruhe, Germany
e-mail: Franco.Magagnato@kit.edu

J. Dumond
Institute for Nuclear and Energy Technologies, Karlsruhe Institute of Technology, 76131 Karlsruhe, Germany

W.E. Nagel et al. (eds.), *High Performance Computing in Science and Engineering '13*, DOI 10.1007/978-3-319-02165-2_25,

2 Introduction

The Stochastic Fields method [1] has been successfully applied to turbulent reacting flows or reacting plumes. In this method, chemical species are described by a set of pure Eulerian transport equations with stochastic source terms. Each field represents an individual realization, so that the set of fields approximates the chemical species Probability Density Function (PDF). By a strict asymptotic theory the equivalence between the evolution of stochastic fields and the PDF transport is proven. Thus, they resolve the species repartition at a sub-grid scale and the highly non-linear source terms are automatically closed. In this paper the Stochastic Field method is used for the first time for two-phase flows. The pure Eulerian fields describing the vapor mass fraction PDF are used to close the highly non-linear mass exchange term between phases. This new concept in two-phase flows is implemented into our compressible code SPARC [2] which can handle large density ratios and Mach numbers as encountered in compressible cavitating flows. Large Eddy Simulations (LES) are performed for turbulence prediction. First numerical results have been compared to experiments of Stutz/Reboud [3] and described below.

3 Mathematical Model

3.1 Two-Phase Homogeneous Equilibrium

In this study a mathematical model based on a locally homogeneous model of a compressible gas-liquid two-phase medium proposed from Okuda/Ikhohagi [4], which belongs to the two-phase models, is used to simulate cavitating flows. Water and vapor are assumed to be in mechanical and thermal equilibrium. Vapor and water evolution is described by the 3D compressible Navier-Stoke vapor-water mixture conservation equations and the vapor mass fraction equation. Cast in integral cartesian form for an arbitrary control volume V with surface ∂V, the system of equations reads:

$$\frac{\partial}{\partial t}\int_V W \mathrm{d}V + \oint_{\partial V} [F - G] \mathrm{d}(\partial V) = \int_V H \mathrm{d}V \tag{1}$$

where

$$W = \begin{Bmatrix} \rho \\ \rho v_x \\ \rho v_y \\ \rho v_z \\ \rho E \\ \rho Y \end{Bmatrix}, F = \begin{Bmatrix} \rho \\ \rho \mathbf{v} v_x + p\mathbf{i} \\ \rho \mathbf{v} v_y + p\mathbf{j} \\ \rho \mathbf{v} v_z + p\mathbf{k} \\ \rho \mathbf{v} E + p\mathbf{v} \\ \rho \mathbf{v} Y \end{Bmatrix}, G = \begin{Bmatrix} 0 \\ \tau_{xi} \\ \tau_{yi} \\ \tau_{zi} \\ \tau_{ij} v_j + q_i \\ 0 \end{Bmatrix}, H = \begin{Bmatrix} 0 \\ 0 \\ 0 \\ 0 \\ 0 \\ S(Y) \end{Bmatrix} \tag{2}$$

and ρ, $\mathbf{v}$, E, p and Y are the density, velocity, total energy per unit mass, pressure and mass vapor fraction of the mixture, respectively. The term τ is the viscous stress tensor and $\mathbf{x} = x\mathbf{i} + y\mathbf{j} + z\mathbf{k}$ is the position vector. S(Y) is the interfacial mass transfer between vapor and water. The modeling of this term will be described later in Sect. 3.3. The pressure is obtained from the equation of state. The total stress tensor τ is a sum of laminar stress tensor plus the turbulent stress tensor according to the Boussinesq assumption. The total heat flux vector $\mathbf{q}$ is obtained from the Fourier law with the constant Prandtl number hypothesis. For the LES the deviatoric part of the sub-grid scale stress tensor is determined with the High Pass Filtered Smagorinsky model of Stolz et al. [5].

3.2 *Equation of State*

Mass and energy balances for an arbitrary control volume indicate that the mixture density is a linear combination of water density ρ and vapor density ρ_g with the void fraction α:

$$\rho = \alpha\rho_g + (1-\alpha)\rho_l \tag{3}$$

and the internal energy a linear combination of water internal energy e_l and vapor internal energy e_g with the mass vapor fraction:

$$e = Ye_v + (1-Y)e_l. \tag{4}$$

Water and vapor internal energy values are extracted from the IAPWS-97 thermodynamic tables and approximated as a function of the temperature.

The water Equation of State (5) has been derived by [6] and the vapor is modeled like an ideal gas (6)

$$p + p_c = \rho_l K(T + T_c) \quad \text{for} \quad Y = 0 \tag{5}$$

$$p = \rho_g RT \quad \text{for} \quad Y = 1 \tag{6}$$

where p and T are the static pressure and temperature, p_c, T_c and K are the pressure, temperature and liquid constants for the liquid state, and R the gas constant.

The gas constant and the constant K have been modified for (5) and (6) to fit vapor and water properties over large pressure and temperature ranges (until at least 75 bar):

$$R(T°C) = 28.8\sqrt{\frac{T}{T_c} \cdot max[T_c - T, 0] + 92} \tag{7}$$

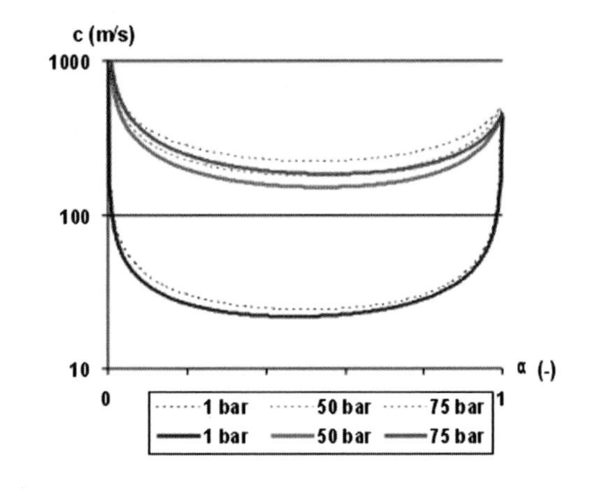

Fig. 1 Mixture sound speed for three pressures 1, 50 and 75 bar. *Dashed lines*: model of Jakobsen; *plain lines*: equation of state used in the simulations

$$K(T°C) = \begin{cases} 0.0016T^2 - 0.0578T + 472.22 \quad \text{J/kg} & \text{if} \quad T < 220°C \\ 0.9893T + 315.61 \quad \text{J/kg} & \text{otherwise} \end{cases} \tag{8}$$

Following the work of Shin et al. [7] and Okuda/Ikohagi [4] the equation of state (9) for a locally homogeneous gas-liquid medium can be rewritten with (3), (5) and (6) in:

$$\rho = \frac{p(p + p_c)}{K(1 - Y)p(T + T_c) + RY(p + p_c)T} \tag{9}$$

The speed of sound of the mixture can be derived from this equation. Heat capacities of water and vapor are required. They are expressed as a function of the temperature to approximate IAPWS-97 values. Mixture speed of sound obtained from this equation of state is displayed exemplary for the three pressures 1, 50 and 75 bar in Fig. 1. The speed of sound obtained with Jakobsen model is also represented in Fig. 1 with dashed lines. This model has been compared to experimental results and show remarkable agreement.

The value for the constants p_c and T_c for water in (6) have been estimated by Tamman [6] to be 1944.61 MPa and 3837 K, respectively.

3.3 Mass Transfer Model

The mass transfer model according to Sone and Sugimoto [8] based on the theory of evaporation/condensation on a plane surface is used:

$$S(Y) = \begin{cases} S(Y)^+ & \text{if} \quad p < p_v \\ S(Y)^- & \text{otherwise} \end{cases} . \tag{10}$$

$$S(Y)^{+} = C_e A\alpha(1-\alpha)\left(\frac{\rho_l}{\rho_g}\right)\frac{p_v^* - p}{\sqrt{2\pi R_g T_g}} \tag{11}$$

$$S(Y)^{-} = C_e A\alpha(1-\alpha)\frac{p_v^* - p}{\sqrt{2\pi R_g T_g}} \tag{12}$$

3.4 Probability Density Function

The time evolution equation for the mass vapor fraction probability density function $P_c \equiv P_c(\phi; \mathbf{x}, t)$ is:

$$\frac{\partial P_c}{\partial t} + U_i \frac{\partial P_c}{\partial x_i} + \frac{\partial}{\partial x_i}\left(\langle u_i' | c = \phi\rangle P_c\right) =$$
$$-\frac{\partial^2}{\partial \phi^2}\left[\gamma \left\langle \frac{\partial c}{\partial x_i}\frac{\partial c}{\partial x_i}\middle| c = \phi \right\rangle P_c\right] + \frac{\partial}{\partial x_i}\left(\gamma \frac{\partial P_c}{\partial x_i}\right) - \frac{\partial}{\partial \phi}\left[S(\phi) P_c\right] \tag{13}$$

Where U is the mean velocity and $\langle u' | c = \phi\rangle$ is the expected value of the fluctuating velocity conditional on the scalar taking the value ϕ. Two open terms needs modeling: the transport due to fluctuating velocity and the molecular mixing term. The first one is modeled by gradient diffusion and the second one with a simple model, the Linear Mean Square Estimation. The transport equation for the PDF becomes with the modeled terms

$$\frac{\partial P_c}{\partial t} + U_i \frac{\partial P_c}{\partial x_i} - \frac{\partial}{\partial x_i}\left(\Gamma \frac{\partial P_c}{\partial x_i}\right) =$$
$$\frac{\partial}{\partial \phi}\left[\frac{\omega_c}{2}(\phi - C) P_c\right] + \frac{\partial}{\partial x_i}\left(\gamma \frac{\partial P_c}{\partial x_i}\right) - \frac{\partial}{\partial \phi}\left[S(\phi) P_c\right] \tag{14}$$

To obviate the difficulty of solving the PDF equation with Lagrangian solvers, the stochastic field method developed by Valino [1] has been extended to the vapor mass fraction.

3.5 Stochastic Fields

The two-phase homogeneous equilibrium formulation has been extended for the Stochastic Field method. Following Valino, P_c is represented by an ensemble of N stochastic fields $\tau^n(\mathbf{x}, t)$, twice differentiable in space at the grid-size length scale:

$$P_c(\phi; \mathbf{x}, t) = \frac{1}{N}\sum_{n=1}^{N} \delta\left[\phi - \tau^n(\mathbf{x}, t)\right] \equiv \langle\, \delta\left[\phi - \tau(\mathbf{x}, t)\right]\rangle \tag{15}$$

Where δ presents the Dirac delta function. The transport equations for these fields can be deduced from the PDF transport (14) by use of stochastic process techniques. Valino proved that (14)

$$d\tau^n = -U_i \frac{\partial \tau^n}{\partial x_i} dt + \frac{\partial}{\partial x_i}\left(\Gamma' \frac{\partial \tau^n}{\partial x_i}\right) dt + (2\Gamma')^{\frac{1}{2}} \frac{\partial \tau^n}{\partial x_i} dW_i' - \frac{\omega}{2}(\tau^n - C)dt + S'(\tau^n)dt \tag{16}$$

gives the rule how the stochastic fields representing a PDF evolve whose (modeled) transport equation is (16).

The equations are filtered and the frequency ω_c is taken to be the inverse of the sub-grid mixing time scale τ_{sgs}:

$$\omega_c = \frac{1}{\tau_{sgs}} = C_d \frac{\mu + \mu_{sgs}}{\rho \Delta^2} \tag{17}$$

with the sub-grid scale constant $C_d = 2$.

Thus, the mass fraction equation has been replaced with a set of N pure Eulerian transport equations with stochastic source terms:

$$\rho d\tau^n = -\rho U_i \frac{\partial \tau^n}{\partial x_i} dt + \frac{\partial}{\partial x_i}\left(\Gamma' \frac{\partial \tau^n}{\partial x_i}\right) dt + \rho \sqrt{\frac{2\Gamma'}{\rho}} \frac{\partial \tau^n}{\partial x_i} dW_i^n$$
$$-\frac{\rho}{2}\frac{(\tau^n - c)}{T_{LES}} dt + S(\tau^n)dt \tag{18}$$

where $n = 1, .., N$; τ^n is the value of mass fraction in field n, U_i is the velocity, Γ^t is the combined molecular and turbulent diffusivity, W_i^n is a Wiener process, independent for each spatial component i but constant in space, c is the local mean of the mass fraction over the N fields, T_{LES} is the constant time used in the LES and $S(\tau^n)$ the source term for cavitation. The last three terms describe the effect of turbulent diffusion in the presence of spatial gradients of vapor mass fraction, micromixing and cavitation, respectively.

All the moments of the scalar field Y can be obtained by averaging over the fields as appropriate. For instance, the mean value of Y is:

$$\overline{Y} = \frac{1}{N}\sum_{n=1}^{N} \tau^n \tag{19}$$

3.6 Numerical Method

The density-based solver SPARC is used for the numerical simulations. It uses a 3D block-structured Finite-Volume-Scheme, the full Multigrid-Method and is parallelized with MPI. For the spatial discretization second up to fourth order accurate cell centered schemes are available. Second order time accurate dual time stepping-scheme is implemented. Approximated Riemann solver and Artificial dissipation schemes can be selected for the flux calculations. Turbulence is predicted with a Reynolds Averaged Navier-Stokes (RANS—Algebraic, linear and non-linear two-equation turbulence models), LES (subgrid-scale models) or Direct Numerical Simulation (DNS). Pre-conditioning yields accurate results for very low as well as high Mach numbers.

In this simulation the Artificial dissipation scheme of Jameson et al. [8] has been used. In contrast to the RANS simulation we scale the amount of artificial dissipation down to a very low value (about 5 %) so that the vortical structure of the flow remains almost unaffected. We stress here again that LES of cavitating flow is not possible without a little amount of artificial dissipation due to the large density variation of the flow field.

Additionally to the above mentioned scheme we pre-condition the Navier-Stokes equations in order to apply a compressible scheme to very low subsonic flows. The method used here is according to Choi and Merkle [9].

4 Results

4.1 Experimental Setup

Experiments of Stutz/Reboud [3] and later Concalves/Patella [10] have been used to validate this novel method. In these experiments, an attached cavitation sheet develops in the venturi type test section CREMHYG (INPG Grenoble). The upper and lower walls of the test section have been designed to reproduce cavitating flows on the blades of space turbopump inducers.

Cross sections are rectangular and their sizes at inlet and throat are $43.7 \times 44\,\text{mm}^2$ and $50 \times 44\,\text{mm}^2$, respectively. Convergence and divergence angles are $4.3°$ and $4°$, respectively. Visualization of cavitation sheet is possible through the transparent walls on the side. The test section is equipped with pressure and temperature sensors as well as double optical probe. The double optical probe is used to evaluate void ratio and velocity fields inside the cavity at five horizontal positions. Figure 2 shows a schematic picture of the venturi profile and the location of measurement sensors.

The computational domain has been extent upstream by six times the channel height and downstream by four times the channel.

A free surface tank imposes the reference pressure in the circuit and the flow rate is imposed by a circulating pump. For the selected operating point, the pressure in the tank is $P = 0.713\,\text{bar}$ and the mass flow rate is $Q = 0.02375\,\text{m}^3/\text{s}$. Experiments indicate that a cavity whose length L ranges between $70\,\text{mm} < L < 85\,\text{mm}$ develops downstream of the contraction in this case.

Numerical analyses from Goncalves et al. [10] among others have been performed to reproduce this attached cavitation sheet. In this numerical investigations

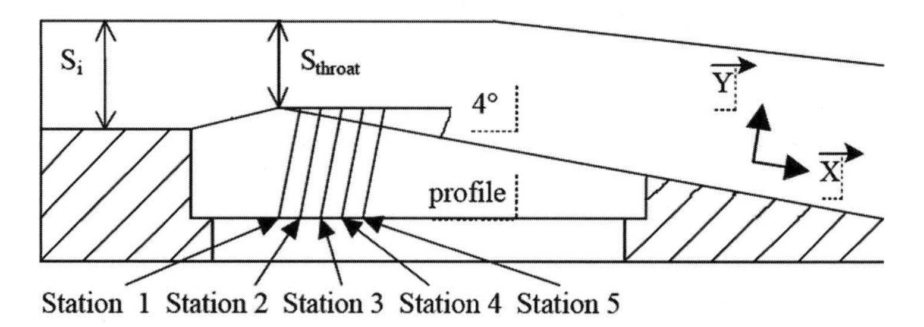

Fig. 2 Schematic picture of the venture profile

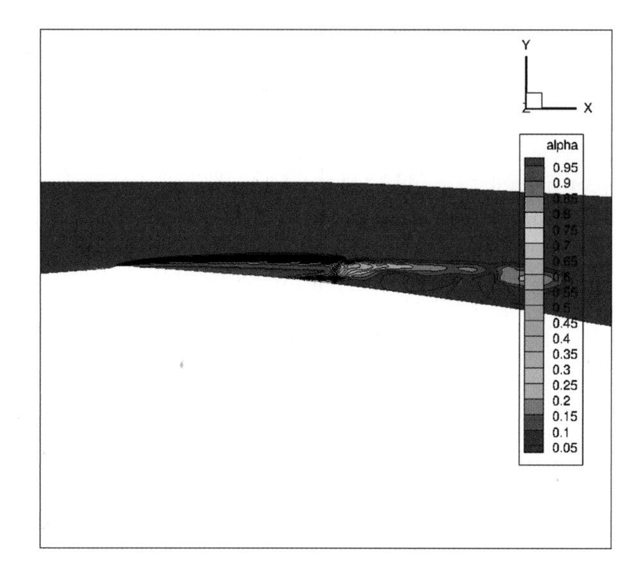

Fig. 3 Void ratio in the symmetry plane

vapor and water are considered to be a compressible continuum. These two-dimensional URANS simulations where done using the Spalart/Allmaras turbulence model as well as the k-ω-SST model from Menter. They obtained acceptable agreement with the experimental findings.

4.2 Numerical Setup

The computational domain has been extent upstream by six times the channel height and downstream by four times the channel height. We are using the full multigrid approach and therefore we were able to investigate three different meshes.

The finest mesh contains about 10 million points, while the coarser meshes uses 1.2 million and 0.16 million points.

The meshes are refined at the contraction and downstream of the contraction to capture accurately turbulent fluctuations and cavitation sheet. The finest mesh

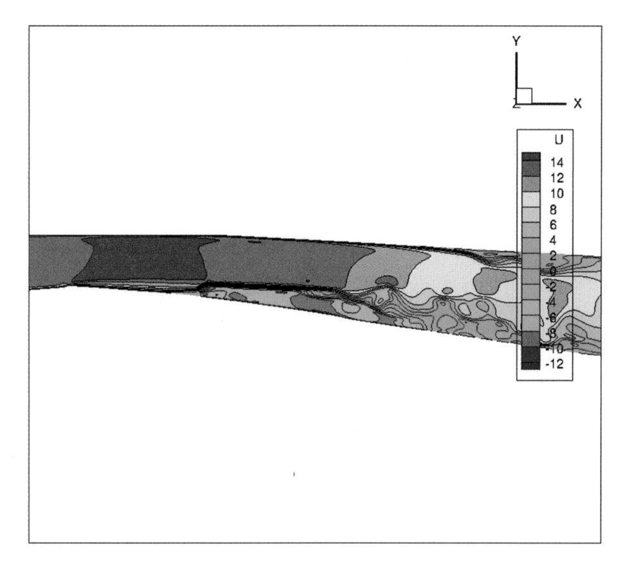

Fig. 4 Stream-wise velocity component in the symmetry plane

has been refined at the walls so that $x+ \approx 20$, $y+ \approx 1$ and $z+ \approx 25$ values corresponds to usual recommended values for LES calculations. The effect of the turbulence at the inlet was investigated by means of the Synthetic Eddy Method (SEM). This method was recently extended for compressible flows by Magagnato et al. [11] and implemented into SPARC. The incoming turbulence level was assumed to be Tu = 10 % of the inlet velocity while the turbulence length scale was chosen to be $L_t = 1$ mm.

At the inlet plane the following values have been set:

$$\mathrm{U_{inlet}} = 10.8\,\mathrm{m/s}$$
$$\mathrm{P_{inlet}} = 35.000\,\mathrm{Pa}$$
$$\mathrm{T_{inlet}} = 293\,\mathrm{K}$$
$$\mathrm{Re_{inlet}} = 2.7 \times 10^6$$

In order to obtain the measured inlet Cavitation number $\sigma_{inlet} = 0.55$ we had to adjust the outlet static pressure since it is not measured in the experiment. Similar to the calculations of Concaves/Patella [10] we were also not able to find a suitable setup in order to adjust σ_{inlet} accurately.

4.3 Calculation

The cavity length in the experiment showed quit stable behaviour [10]. In our calculation we observed a similar behaviour. Also our cavitation length was some 10 % higher (L = 0.088 m) compared to the experiment (see Fig. 3). The velocity profiles as well as the void ration profiles at the five stations have been compared. In contrast to the experiments our predictions showed a separation line somewhat more downstream. While in the experiment the flow separated at station two (x = 0.035 m) we obtained a separation at x = 0.084 m (see Fig. 4).

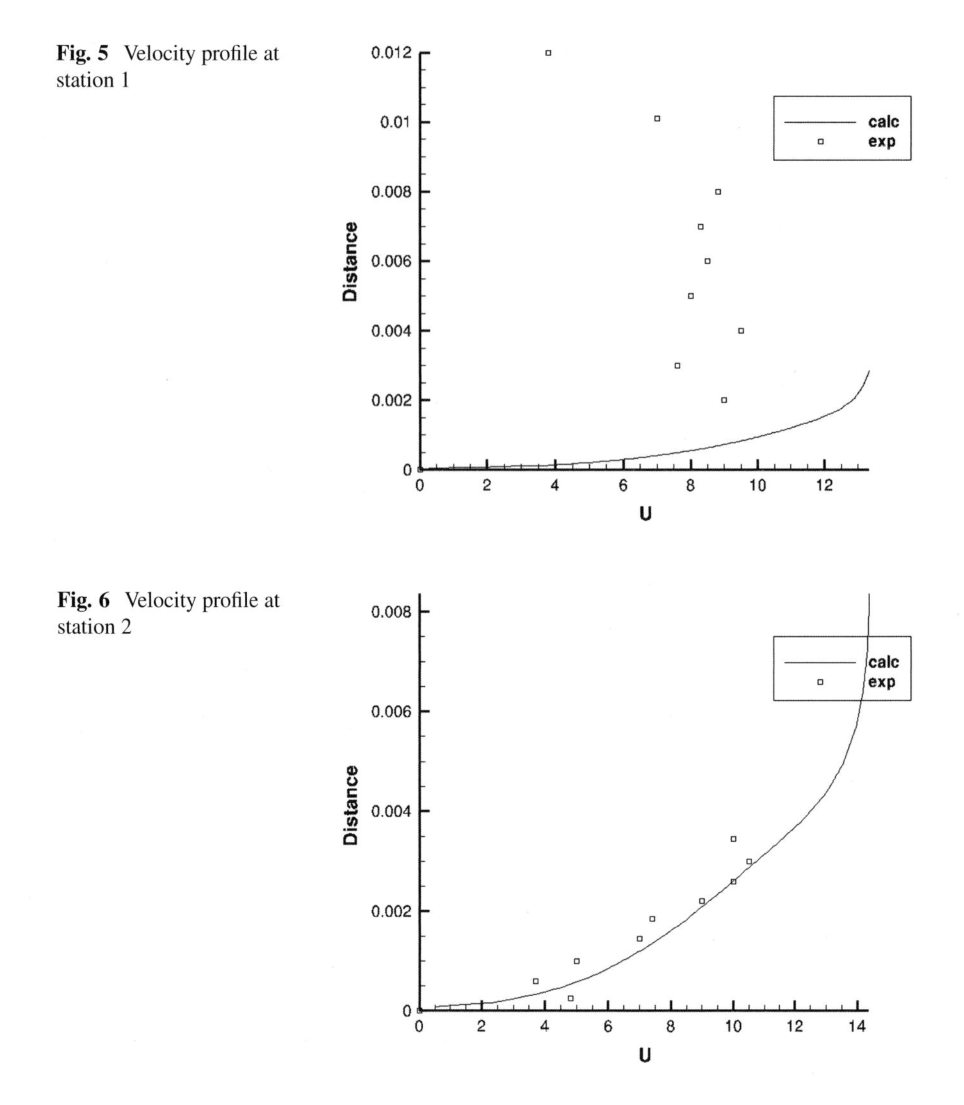

Fig. 5 Velocity profile at station 1

Fig. 6 Velocity profile at station 2

The comparison with the velocity profiles (Fig. 5–14) and the void ratio at the five measured stations reveals the above mentioned slight disagreement.

5 Computational Efficiency

In the computation of these results we have been using up to 932 AMD-Interlagos cores of the CRAY XE6 (Hermit) at HLRS in Stuttgart. The in-house developed code Sparc is parallelized with the MPI-2 software. The computational time for one unsteady calculation in three dimensions using about 10 mio points was about 24 h.

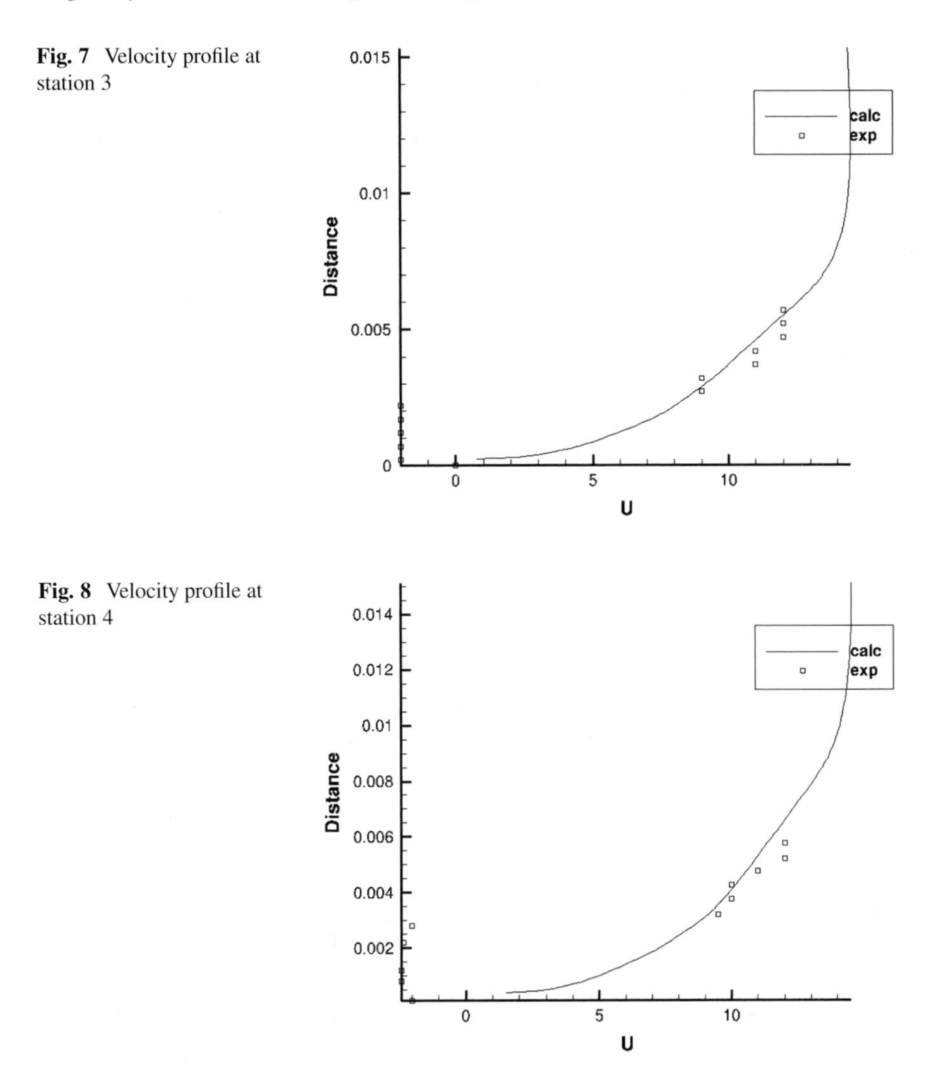

Fig. 7 Velocity profile at station 3

Fig. 8 Velocity profile at station 4

Since we were using more than 10,000 blocks of the finite volume scheme we could efficiently distribute the blocks on these 932 cores with the domain decomposition technique. The load balancing was at about 99 %. The parallel efficiency was also very good. Since in Stuttgart the communication is done with the CRAY GEMINI node-node interconnection the parallel efficiency was close to 95 %. This very good parallel efficiency could only be obtained because we have reduced the amount of output to a minimum. It is desired that the amount of output data should be increased so we need to improve the read/write performance of the code in the near future. From our recent investigations we know that a higher resolution of the computational mesh is required. We think that using about 80 million points in the next phase will be adequate for a well resolved unsteady calculation.

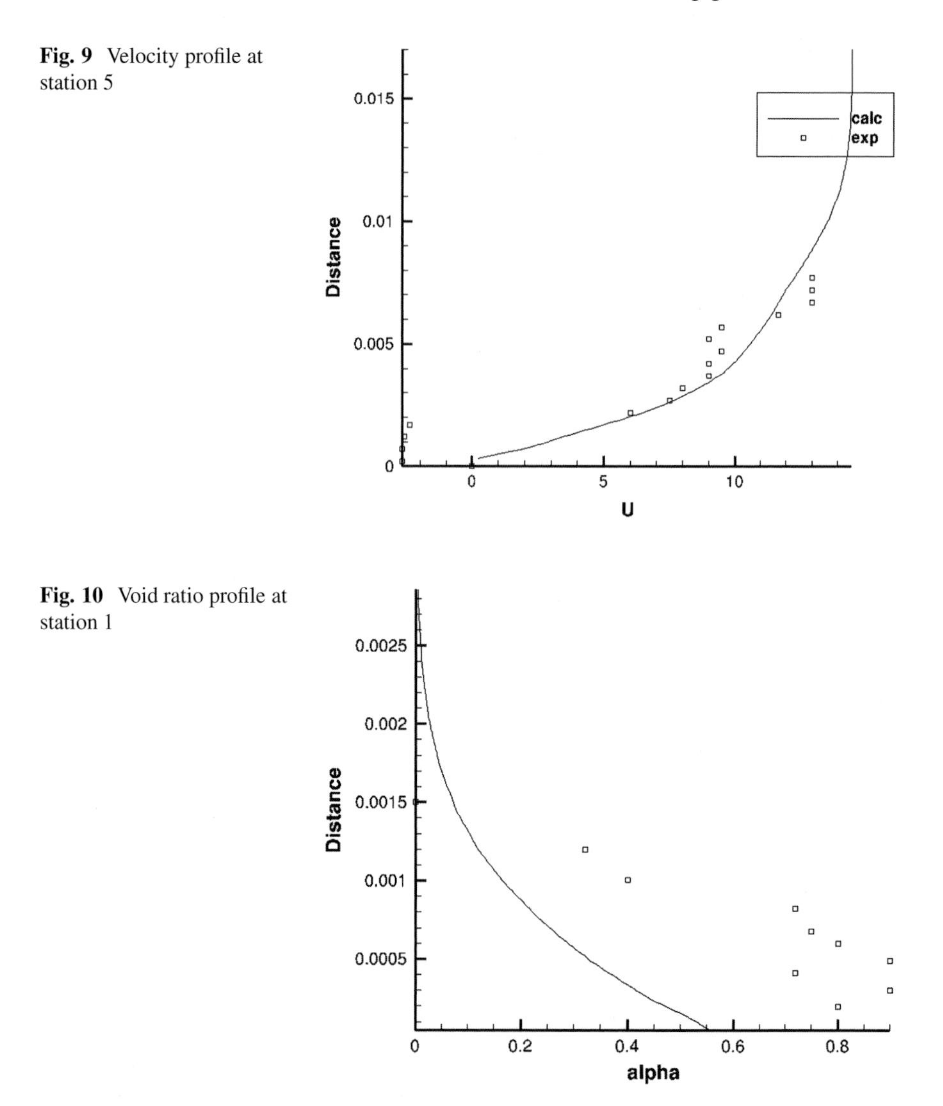

Fig. 9 Velocity profile at station 5

Fig. 10 Void ratio profile at station 1

6 Conclusion

The pure Eulerian fields yield the vapor mass fraction PDF at a low computational cost since no Lagrangian solver or equations for bubbles clusters are required. For our purpose a two-phase flow homogeneous equilibrium model has been selected since compressibility is crucial.

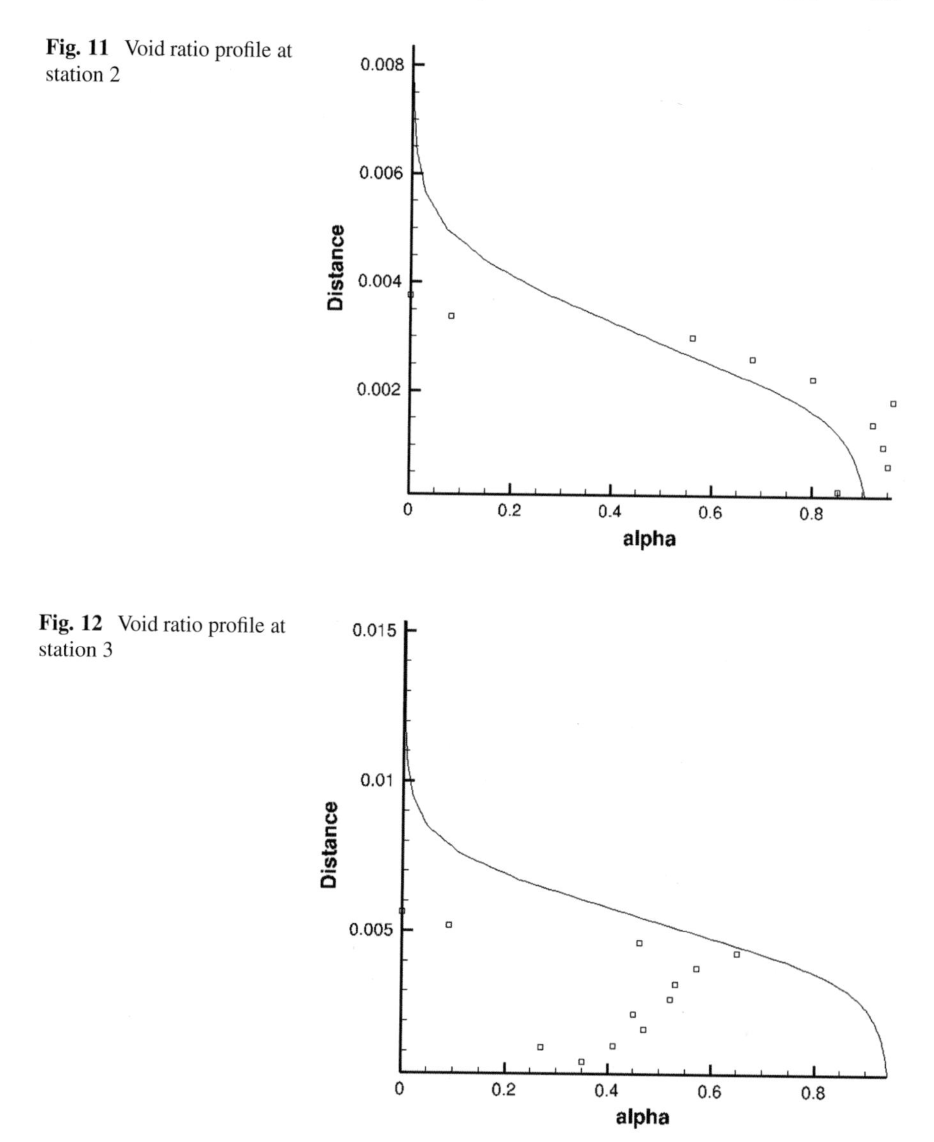

Fig. 11 Void ratio profile at station 2

Fig. 12 Void ratio profile at station 3

The Stochastic Field method implemented in our compressible code has shown encouraging results. Also not fully satisfactory results has been obtained compared with the experiments we feel that also the experiments can't be fully trusted. In the near future we plan to compare this novel approach with a variety of flow configurations in order to fully understand the mechanism of turbulence/two-phase flow interaction.

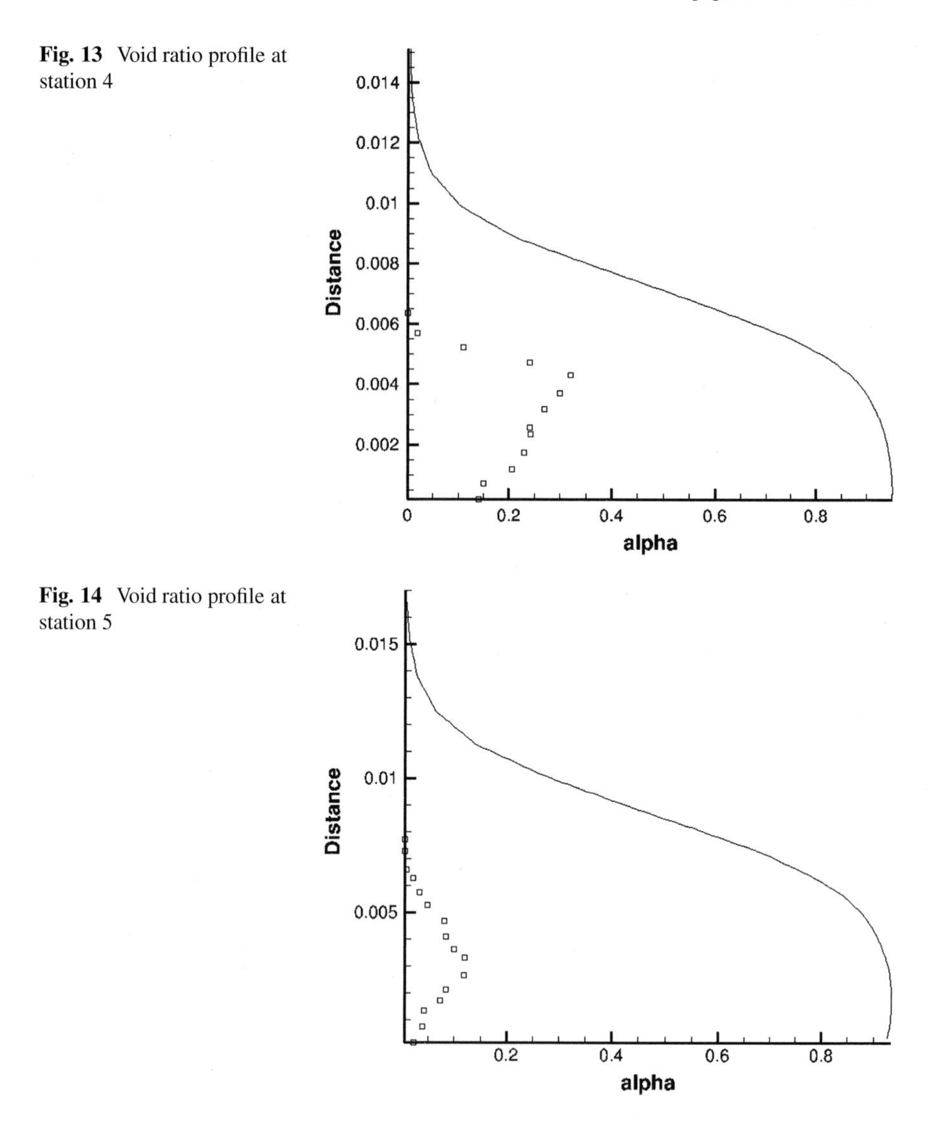

Fig. 13 Void ratio profile at station 4

Fig. 14 Void ratio profile at station 5

Acknowledgements The authors thank Eric Goncalves from University of Grenoble, France for his cooperation related to the experimental setup.

References

1. L. Valino, A field Monte Carlo formulation for calculating the probability density function of a single scalar in a turbulent flow. Flow Turbulence Combust. **60**, 157–172 (1998)
2. F. Magagnato, B. Pritz, M. Gabi, Prediction of the damping characteristics of a combustion chamber by LES, in *The 20th International Symposium On Transport Phenomena*, Victoria, 2009

3. B. Stutz, J.-L. Reboud, Two-phase flow structure of sheet cavitation. Phys. Fluids **9**(12), 3678–3686 (1997)
4. K. Okuda, T. Ikhohagi, Numerical simulation of collapsing behavior of hubble clouds. Trans. JSME B **62**(603), 3792–3797 (1996) (in Japanese)
5. S. Stolz, P. Schlatter, D. Meyer, L. Kleiser, High-pass filtered eddy-viscosity models for LES, in *Direct and Large-Eddy Simulation*, vol. 9, ed. by V.R. Friedrich, B.J. Geurts, O. Metais (Kluwer, Dordrecht, 2003)
6. H.T. Chen, R. Collins, Shock wave propagation past on ocean surface. J. Comput. Phys. **7**, 89–101 (1971)
7. B.R. Shin, Y. Iwata, T. Ikohagi, Numerical simulation of unsteady cavitating flows using a homogeneous equilibrium model. Comput. Mech. **30**, 388–395 (2003)
8. A. Jameson, W. Schmidt, E. Turkel, Numerical simulation of the Euler equations by finite volume method using Runge-Kutta time stepping schemes. AIAA paper 81–1259, in *Proceedings of the 14th Fluid and Plasma Dynamics Conference*, Palo Alto, 1981
9. Y.H. Choi, C.L. Merkle, The application of preconditioning to viscous flows. J. Comput. Phys. **105**, 207–223 (1993)
10. E. Concalves, R.F. Patella, Numerical simulation of cavitating flows with homogeneous models. Comput. Fluids **38**, 1682–1696 (2009)
11. F. Magagnato, B. Pritz, M. Gabi, Inflow conditions for large-eddy simulation of compressible flow in a combustion chamber, in *Proceedings of the 5th International Symposium on Turbulence, Heat and Mass Transfer*, Dubrovnik, 2006

Large-Eddy Simulation of the Particle-Laden Turbulent Flow in a Cyclone Separator

Michael Alletto and Michael Breuer

Abstract A gas cyclone separator represents a classic field of application where turbulent particle-laden flows play a major role. In order to evaluate the performance of a recently developed Euler–Lagrange simulation tool based on the large-eddy simulation (LES) technique and the point-particle approach, this practically relevant flow problem is considered in the present study. As a first step towards a full simulation taking all interactions between the two phases (fluid–particle, particle–fluid and particle–particle) into account, a one-way coupled prediction is carried out. Nevertheless, the entire simulation methodology is described in detail including a sandgrain roughness model and a deterministic collision model. For the latter a performance analysis was carried out demonstrating that even for a high mass loading the computational effort for the collision detection remains below 10 % of the entire CPU-time. The predicted LES results for the cyclone flow are compared with corresponding measurements and a reasonable agreement is found.

1 Introduction

In cyclone separators (see, e.g., Obermair [23] and Obermair et al. [24, 25]) the particle-laden gas flow enters the cyclone body tangentially inducing a highly rotatory flow. Centrifugal forces lead to a migration of the large and thus heavy particles towards the cyclone walls, where they are transported downwards and collected in the dust bin located at the bottom of the apparatus. Turbulent fluctuations lead to a mixing of the particles throughout the cyclone body. Hence, the entrainment of the small and thus light particles from the cyclone walls towards the cyclone axis increases with increasing velocity fluctuations [17]. In the region

M. Alletto · M. Breuer (✉)
Professur für Strömungsmechanik, Helmut-Schmidt-Universität Hamburg, Holstenhofweg 85, Postfach 70 08 22, 22043 Hamburg, Germany
e-mail: breuer@hsu-hh.de; alletto@hsu-hh.de

W.E. Nagel et al. (eds.), *High Performance Computing in Science and Engineering '13*, DOI 10.1007/978-3-319-02165-2_26,

near the axis the air flow is pointing upwards and leaves the cyclone through the vortex finder located at the top of the device. Consequently, the objective of a cyclone to separate particles of a broad spectrum of sizes can be fulfilled. As the representative diameter of the separated particles the cut point is used, i.e., the particle size which is removed with 50 % efficiency from the air stream. The transport and separation mechanism of the particles in a cyclone separator is accompanied with complex flow structures such as curved streamlines, secondary flows and a precessing vortex core already present in the unladen flow. Hence, this type of flow configuration represents a challenging test case for computational approaches treating turbulent flows. Introducing particles further complicates the description of the flow since the interaction between the particles and the turbulent eddies, the particles and the cyclone walls and the particles themselves have to be accurately modeled. Owing to the complicated flow structures and the various interaction mechanisms involved by the disperse phase, it is mandatory to employ an adequate simulation tool in order to reliably predict the pressure loss and the separation efficiency of the cyclone separator. Furthermore, because of the huge number of particles present in such a device, efficient methods to treat the disperse phase are mandatory.

Large-eddy simulations (LES) have already been asserted to be more accurate in treating the flow in cyclone separators than Reynolds-averaged Navier–Stokes (RANS) simulations (see, e.g., [19, 31]). Furthermore, the explorable Reynolds numbers are much higher compared to direct numerical simulations (DNS) because of the coarser grids usually employed and the applicability of wall functions instead of the no-slip boundary condition. Hence, LES seems to be the right modeling approach for the continuous phase in order to achieve reliable results in an affordable time period. Regarding the disperse phase, the Lagrangian point-particle approach was chosen because of its applicability in a wide parameter space and the straightforward way to model particle–particle and particle–wall collisions compared to a description in an Eulerian frame of reference. Hence, the present methodology (LES combined with a point-particle approach) results in a reliable simulation tool which has been successfully tested in the cold flow of a model combustion chamber [1, 6], the conveying of solids in a horizontal pipe [2] and a horizontal channel [12]. In order to further evaluate the present methodology in complex geometries, the flow in a cyclone separator is investigated. This choice is motivated by the still rare investigations of particle-laden flows in cyclone separators using LES combined with a point-particle approach indispensable to gain confidence in the method (for one-way coupled simulations, see, Derksen [16], de Souza et al. [15] and for two-way coupled simulations, see, Derksen et al. [17, 18]). The results shown in Sect. 6 are therefore the first step towards the validation of the present methodology against the experiments of Obermair et al. [24, 25].

2 Governing Equations

In this work the multiphase flow is described using an Euler–Lagrange approach in which the two different phases are solved in two different frames of reference.

The continuous phase is solved in an Eulerian frame of reference. The conservation equations of the filtered quantities used in LES [5] can be extended to take into account the feedback effect of the particles on the fluid (two-way coupling). For that purpose the particle-source-in-cell (PSIC) method described in Crowe et al. [13] is used.

The dispersed phase is solved in a Lagrangian frame of reference. The equation of the translational and rotational motion is given by Newton's second law, where the fluid forces are derived from the displacement of a small rigid sphere in a non-uniform flow [21]. For particles with a density much higher than the carrier fluid only the drag, lift, gravity and buoyancy forces have to be considered. The drag force on the particle is based on the Stokes flow around a sphere, where the corresponding drag coefficient is given by $C_D = 24\alpha/Re_p$ with the correction factor $\alpha = 1 + 0.15\, Re_p^{0.687}$ to extent the validity towards higher particle Reynolds numbers Re_p. The lift force on a particle due to the velocity shear (Saffman force) is calculated as suggested by McLaughlin [22]. The lift force on a particle due to rotation (Magnus force) is calculated as suggested by Crowe et al. [14]. This formulation requires the relative rotation of the particles and thus the angular velocity of the particles.

To account for the rotation of the spherical particles around three Cartesian axes, Newton's second law for the angular momentum is considered. The torque acting on a rotating spherical particle is determined based on the formulation of Rubinow and Keller [29].

For tiny particles with a relaxation time of the same order as the smallest fluid time scales the unresolved scales in LES become important for the particle motion. To consider the effect of the subgrid scales, a simple stochastic model by Pozorski and Apte [26] is applied. It requires the estimation of the subgrid-scale kinetic energy carried out with the help of the scale similarity approach of Bardina et al. [3]. For further details we refer to [1, 2, 6, 9, 12].

3 Numerical Methods

The continuous phase is solved by the computer code $\mathcal{LESOCC}$ (= **L**arge-**E**ddy **S**imulation **O**n **C**urvilinear **C**oordinates [4,5,9]) to integrate the governing equations in space and time. It is based on a 3-D finite-volume method for arbitrary non-orthogonal and block-structured grids. The entire discretization is second-order accurate in space and time. For modeling the non-resolvable subgrid scales the well-known Smagorinsky model [32] with Van Driest damping near solid walls is applied. Alternatively, a dynamic model can be used.

The ordinary differential equation for the particle translational motion is integrated by a fourth-order Runge–Kutta scheme. To avoid time-consuming search algorithms, the second integration to determine the particle position is done in the computational space. Here an explicit relation between the position of the particle and the cell index containing the particle exists [9, 10], which is required to calculate the fluid forces on the particle. Thus a highly efficient particle tracking scheme results allowing to predict the path of millions of particles (see Sect. 6).

The fluid velocity at the particle position is calculated by a Taylor series expansion around the cell center next to the particle [20]. This interpolation was shown to have a weaker filtering effect on the fluid velocity than a trilinear interpolation leading to better results for particles with small relaxation times. The set of three linear ordinary differential equations for the particle angular velocity are solved analytically.

Although not applied in the present case (four-way coupling was not possible because of the prohibitive amount of particles present in the cyclone and two-way coupling is planned in the near future), the entire simulation methodology is intended for high mass loadings and thus allows two-way and four-way coupled simulations. To account for the influence of the particles on the fluid the PSIC method by Crowe et al. [13] is used considering the exchange of momentum between both phases. An additional source term representing the forces exerted by the particles onto the fluid is added to the filtered Navier–Stokes equations. Presently, only the drag force calculated by means of the filtered fluid and particle velocities is considered for the two-way coupling. A smooth source term distribution is achieved by trilinear distributing the contribution of the particles to the eight cell centers surrounding the particle. Any further interactions between the phases are neglected, e.g., possible influences of the particles on the subgrid-scale stresses are not taken into account.

The collisions between particles for a four-way coupled simulation are determined deterministically by a recently developed collision method described in Breuer and Alletto [6]. Based on the technique of uncoupling the calculation of particle trajectories is split into two stages. In the first stage particles are moved based on the equation of motion without inter-particle interactions. In the second stage the occurrence of collisions during the first stage is examined for all particles. If a collision is found, the velocities of the collision pair are replaced by the post-collision velocities without changing their position. The post-collision velocities are calculated by a hard-sphere collision model involving friction between the colliding spheres (see, e.g., [14, 34]). Here changes of the velocity and angular velocity of the two involved particles are modeled by the normal and the tangential restitution coefficients and the static and the dynamic coefficients of friction.

For modeling the particle–wall interactions different options are available: wetted wall, specular wall and rough wall. In case of a wetted wall the particles stick at the wall in case of contact. In case of the specular wall the absolute value of the particle velocity is kept but the sign of the wall-normal velocity component is inverted. In case of a rough wall a recently developed sandgrain roughness model [12] to mimic the rebound behavior of the particles at rough asperities is applied. The model takes into account the momentum loss of the particle during the wall impact by the wall-normal and the tangential restitution coefficients and the static and the dynamic coefficients of friction. The random nature of a rough wall is accounted for by a Gaussian-distributed inclination of the wall-normal vector. The computation of the standard deviation relies on generally used roughness parameters. Furthermore, the model considers the shadow effect leading to a redistribution of the streamwise momentum towards the wall-normal momentum.

4 HPC Strategies

The major time-consuming computations carried out on the Cray XE6 (Hermit) were spent to recompute the flow in a model combustion chamber published in [1, 6, 7] on a finer grid and with improved inflow conditions. The statistics are still not converged and hence the first results of the flow in a cyclone separator are shown in this report. Although these results as a first step towards a full simulation do not consider particle–particle collisions, the collision handling is in general a critical aspect, e.g., in the time consuming computation mentioned before. Hence, a few data about the efficiency of the collision-handling algorithm are provided below.

Concerning the HPC aspects the optimization of the solver for the continuous phase in $\mathcal{LESOCC}$ was already reported in several previous reports, see, e.g., [11, 27].

The HPC strategies concerning the dispersed phase were already published in the reports [7, 8]. Briefly summarized, the present scheme is highly efficient due to the following reasons:

- No CPU-time consuming search algorithm is needed in the present c-space scheme.
- The particle properties are stored in linear arrays which are kept filled even if the particles leave the domain. This guarantees optimal performance.
- The multi-block exchange between blocks for the particle data completely relies on the same arrangement as used for the continuous phase (domain decomposition). The data transfer itself is based on MPI.
- The collision detection procedure is carried out over a small amount of particles contained in a virtual cell which breaks down the computational costs from $O(N_p^2)$ to $O(N_p)$, where N_p denotes the total number of particles.
- As described in detail in [6, 7] several conditions are introduced to further reduce the number of potential colliding particles step by step. At the end only a very few potential colliding particles remain, which have to be evaluated completely.

In order to analyze the performance of the collision-handling algorithm and to investigate the scaling of the computational costs with the number of particles inside the domain, a particle-laden turbulent pipe flow was computed. In order to reduce the performance analysis to the disperse phase, a one-way coupled simulation considering particle–particle collisions but disregarding the two-way coupling was

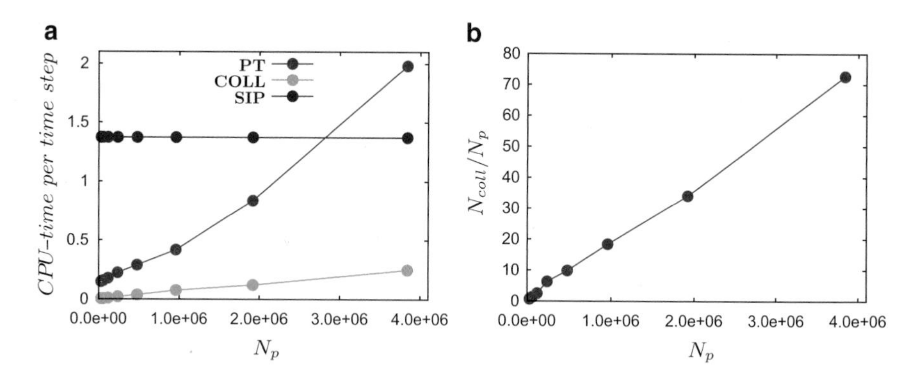

Fig. 1 Performance analysis: (**a**) CPU-time per time step required by the main routines: SIP solver (**SIP**), particle tracking (**PT**) and particle–particle collisions (**COLL**); (**b**) number of collisions found per particle

conducted. By this way the ratio of the computational costs required for the continuous phase to the computational costs required by the disperse phase can be reliably evaluated. In order to obtain realistic conditions the same polydisperse particle distribution as described in Alletto and Breuer [1] and Breuer and Alletto [6] was used and the sandgrain roughness height was assumed to be equal to $k_s = 5\,\mu\text{m}$. In order to keep the computational costs at a reasonable amount the maximum number of particles allowed to be contained in a virtual cell is set equal to 20 for all cases studied. The performance was analyzed for the flow containing in total $N_p = 30{,}000, 60{,}000, 120{,}000, 240{,}000, 480{,}000, 960{,}000, 1{,}920{,}000$, and 3,840,000 particles. The computational domain was discretized by about 3.3 million control volumes divided into 12 blocks. In order to gather reliable statistics, the flow was computed for 10,000 time steps after ensuring fully developed conditions.

Figure 1a displays the CPU-time per time step required for the particle tracking (**PT**), the collision handling (**COLL**) and the SIP solver (**SIP**), which represents the most CPU-time consuming routine of the continuous phase. From Fig. 1a it is evident that the computational costs for the collision handling scale more or less linearly with the number of particles N_p as expected. Since only one-way coupling is applied the computational time required by the SIP solver is constant. Note that the CPU-time required by **PT** increases more than linearly for more than $N_p =$ 960,000 particles located inside the domain. The reason is that for a higher amount of particles (i.e., a higher mass loading) the particles tend to accumulate in the center of the pipe. This leads to a slight load imbalance and hence also to higher CPU-times. That shows that for the test case presented in this section the performance analysis can not be completely separated from the governing physics.

In order to prove that the amount of colliding particles found N_{coll} scales in the proper manner with N_p, Fig. 1b shows the ratio N_{coll}/N_p as a function of the number of particles located inside the domain. It is evident that N_{coll}/N_p scales linearly with N_p which is exactly the dependence assumed in statistical collision models (see, e.g., Sommerfeld [33]).

Furthermore, measurements of the floating-point performance of the CFD code were carried out based on this small test case using three nodes and 12 cores in total. For the case with only 30,000 particles a total performance of about 7 GFlops was measured. With an increasing number of particles the time used for the particle collision detection increases and thus the total performance decreases until about 4.7 GFlops for the case with 3.84 million particles.

5 Description of the Test Case

Obermair et al. [24] carried out LDA measurements of the clean flow in different cyclone geometries. Note that the phase-Doppler anemometry (PDA) measurements of the particle-laden flow can be found in Obermair et al. [25]. Among the various geometries investigated, Obermair et al. [24] found that a cyclone separator with a downcomer tube (the straight tube between the conical part and the dust bin) leads to the best separation efficiency. The geometry of the cyclone separator with the downcomer tube is sketched in Fig. 2a. The mean inlet velocity was $U_{in} = 12.7$ m/s resulting in a Reynolds number of Re = 333,000 based on the mean inlet air velocity at standard conditions and the diameter of the cyclone body $d_b = 400$ mm. The gravity pointed in x-direction of the coordinate system displayed in Fig. 2b, i.e., in the direction of the flow in the inlet section. Due to a better accessibility of the measurement plane a horizontal configuration of the cyclone was chosen since the influence of the gravity was assumed to be negligible compared to the centrifugal forces [23]. In order to compare the results obtained by the present methodology with the experiments of Obermair et al. [24] (unladen flow), one-way coupled simulations were performed. Note that the present one-way coupled simulations have to be seen as the starting point for the two-way coupled simulations planned in the future.

The computational domain employed is depicted in Fig. 2b. The length of the outlet was chosen to reduce the influence of the zero-gradient outlet boundary condition to a minimum. The application of a convective boundary condition was not possible for the present mesh since it produces numerical oscillations near the domain outlet. The length of the inlet was specified in order to ensure a sufficiently long distance between the duct inlet and the cyclone inlet. That allows the steady turbulent velocity profile (no fluctuations, for the mean profile see Reichardt [28]) applied at the inlet boundary to destabilize and develop. This measure leads to a realistic time-dependent velocity field at the inlet of the cyclone body. The computational domain is discretized by 4.4×10^6 control volumes. The maximum wall resolution is $\Delta r^+ = 122, 255, 110, 50.4$ and 247 in the outlet section, the cylindrical part of the cyclone body, the conical part of the cyclone body, the downcomer tube and the bin, respectively. To account for the influence of the unresolved scales the model by Smagorinsky [32] with a constant of $C_s = 0.1$

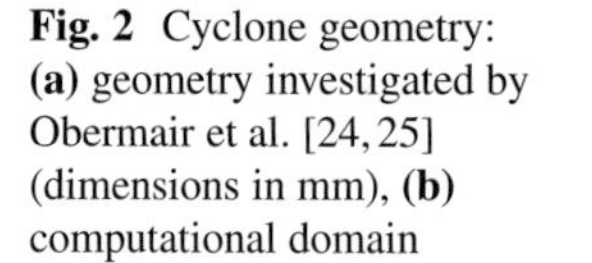

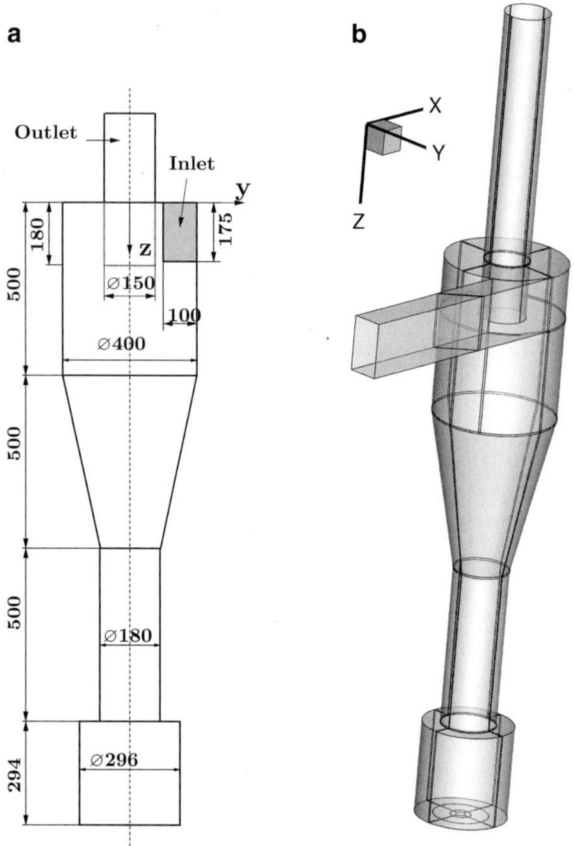

Fig. 2 Cyclone geometry: **(a)** geometry investigated by Obermair et al. [24, 25] (dimensions in mm), **(b)** computational domain

is applied. In order to avoid the resolution of the steep velocity gradients at the walls, the wall model of Schumann [30] is used. The dimensionless time step is $\Delta t = \Delta t^* U_{in}/d_b = 5.2 \times 10^{-5}$.

The particle density is set to $\rho_p = 2,770\,\text{kg/m}^3$ (limestone), i.e., the same material as used in the particle-laden PDA measurements of Obermair et al. [25]. The two-way coupled simulation results are planned to be compared with these measurements in the future. The polydisperse diameter distribution measured by Obermair et al. [25] at the cyclone inlet is approximated by 19 size classes. The particle diameters range between 1.08 and 24.9 μm with a number mean diameter of 1.74 μm. A first rough estimation assuming the same overall mass loading as measured by Obermair et al. [25] (particle-laden flow with $\eta = 0.065\,\%$) with a mean particle diameter of 1.74 μm yields a total number of particles contained in the computational domain of $N_{p,tot} = 1.47 \times 10^{10}$. In order to limit the total number of particles to an affordable number of $N_{p,tot} = 5.0 \times 10^6$, 2,963 particles are grouped into a parcel. That means that to account for the momentum transfer of the particles to the fluid in the two-way coupled simulation planned in the near future, the force exerted by each single tracked parcel is multiplied by the factor 2963. The aerodynamic behavior however, is the same as the single particle. Note that for the present one-way coupled simulation the parcel method has neither an influence on the fluid nor on the particle statistics. At the first stage of this investigation a specular reflection of the particles at the cyclone wall is assumed. In order to avoid an accumulation of the particles in the dust bin and therewith associated unaffordable computational costs, particles hitting the bin wall are removed from the domain. The flow statistics are achieved by averaging the fluid quantities for about 745,000 time steps (about 1.22 s in real time). The particles have been introduced into the computational domain only after it was asserted that the flow field reached a statistically stationary state. Furthermore, in order to ensure that also the particles have reached a statistically stationary state before the start of the averaging, it is checked that the center of mass of all particles in axial direction does not experience any mean displacement. That leads for the particles to a shorter averaging period, i.e., about 240,000 time steps (about 0.39 s in real time).

6 Results

In this section first overall qualitative results of the present simulations compared with the experiments of Obermair et al. [24] are shown. In their measurements Obermair et al. [24] identified two main vortical structures: one is moving towards the dust bin along the cyclone wall and the other vortex is moving towards the outlet near the cyclone axis. Furthermore, a precessing vortex core is identified which rotates with a frequency of 66 Hz around the axis of the downcomer tube. The deviation from an axisymmetric configuration introduced by the inlet leads to a spiral form of the vortex core underlining the multiplicity of complicated flow structures found in cyclone separators.

Figure 3 depicts the contour plot of the time-averaged circumferential fluid velocity of the LES prediction (Fig. 3a) compared with the reference experiment (Fig. 3c). Furthermore, the time-averaged circumferential particle velocity is included for comparison (Fig. 3b). It is evident that the time-averaged circumferential fluid velocity of the LES prediction is qualitatively in good agreement with the measurements of Obermair et al. [24]. However, the magnitude is slightly underpredicted and the deviation of the vortex core (the low velocity region near the cyclone axis) from the cyclone axis is not as pronounced as in the reference experiment. Note that such spirally formed mean vortex cores were also found in other LES simulations (see Derksen [16] and Gronald and Derksen [19]) and also in other experiments (see the comment in Derksen [16]). This indicates that the statistically averaged flow in a cyclone separator is truly three-dimensional due to the deviation of the present geometry from an axisymmetric geometry introduced by a finite number of inlets [16]. The reason for the aforementioned small discrepancies between the flow statistics of the LES and the experiment is probably due to the overprediction of the wall shear stresses by the applied wall model. Complementary simulations applying

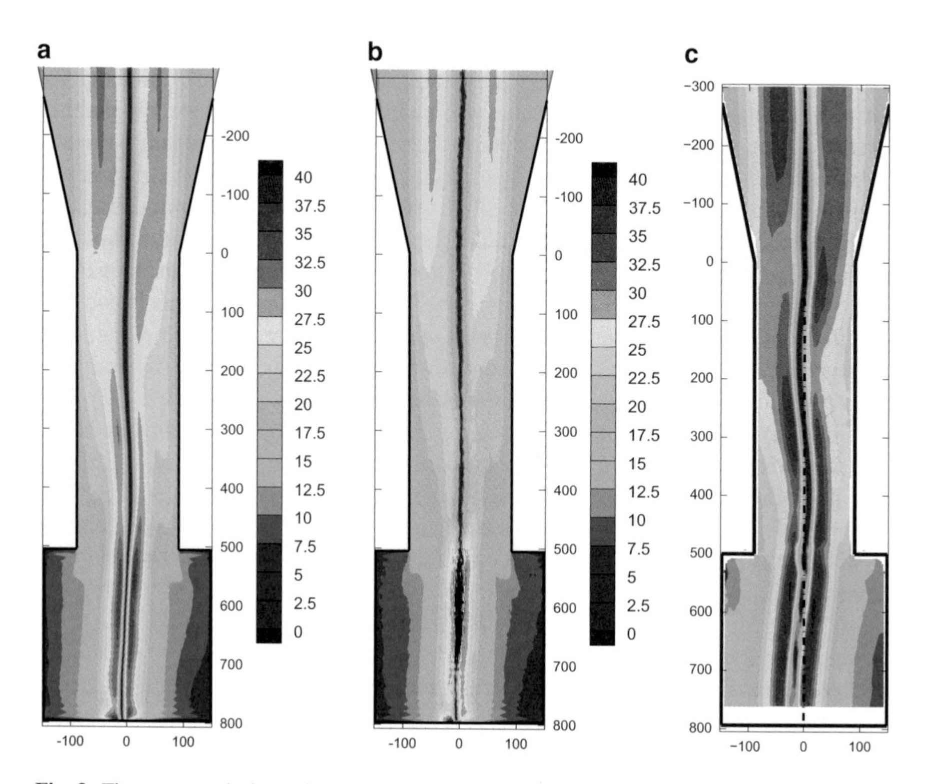

Fig. 3 Time-averaged circumferential velocity (m/s) in the x–z plane (**a**) LES flow; (**b**) LES particles; (**c**) unladen flow experiment [24] (same contour levels)

the no-slip boundary condition at the wall (not shown for the sake of brevity) yield an increase of the circumferential velocity by a factor of 2 compared with the simulation using the wall function of Schumann [30]. This indicates the strong influence of the boundary condition at the wall on the intensity of the precessing vortex. The particle circumferential velocity predicted by LES (Fig. 3b) shows a very similar behavior as the corresponding fluid velocity (Fig. 3a). This seems quite reasonable because of the very small particles which should behave very similar to fluid tracers. Furthermore, due to the specular reflection boundary condition at the wall no momentum loss during the wall impact is expected. The differences to the mean fluid circumferential velocity probably arise due to the different averaging periods.

Figure 4 shows the contour plot of the time-averaged axial velocity of the continuous flow (Fig. 4a) and the particles (Fig. 4b) compared with the reference experiment of Obermair et al. [24] (Fig. 4c) in the x–z plane (see Fig. 2). Positive velocities indicate that the flow is directed towards the dust bin and negative velocities denote that the fluid is transported towards the cyclone outlet. For this statistical quantity the LES prediction is qualitatively and quantitatively in good agreement with the reference experiment in the region close to the wall. Near the

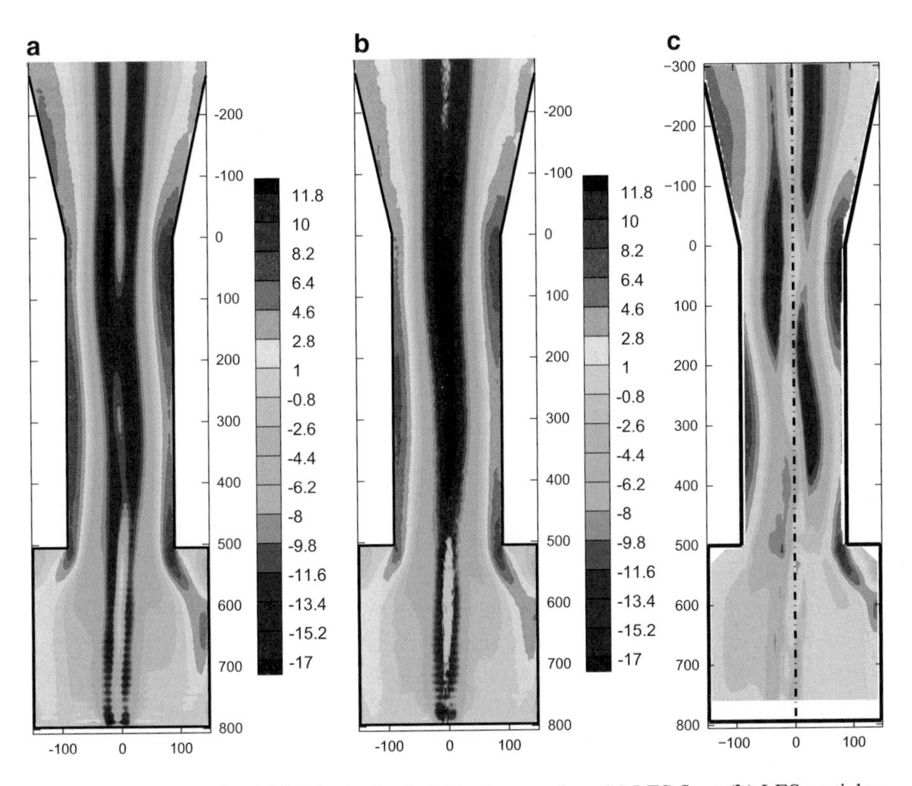

Fig. 4 Time-averaged axial fluid velocity (m/s) in the x–z plane (**a**) LES flow; (**b**) LES particles; (**c**) unladen flow experiment [24] (same contour levels)

axis however, the alternating zones of positive and negative axial velocity are not well captured. Regardless this slight difference between the simulation and the experiment is evident that for both cases the flow is transported towards the dust bin near the wall and towards the outlet in the zone near the axis. As for the mean circumferential velocity also the simulated mean particle axial velocity is very similar to the simulated mean fluid velocity. This is expected for small particles specularly reflected at smooth walls.

The contour plot of the time-averaged RMS-values of the circumferential velocity fluctuations are depicted in Fig. 5 for the same plane as before. It is evident that the LES prediction of the fluid flow (Fig. 5a) is in qualitative agreement with the values measured by Obermair et al. [24]. For both cases the maximum fluctuations coincide with the location of the vortex core (see Fig. 3). However, the amplitudes of the simulated circumferential velocity fluctuations are overpredicted in the region near the axis and underpredicted in the wall region. Note that for the particle statistics (Fig. 5b) very similar observations can be made as for the corresponding fluid statistics. The second moments of the particulate phase (see, also the particle

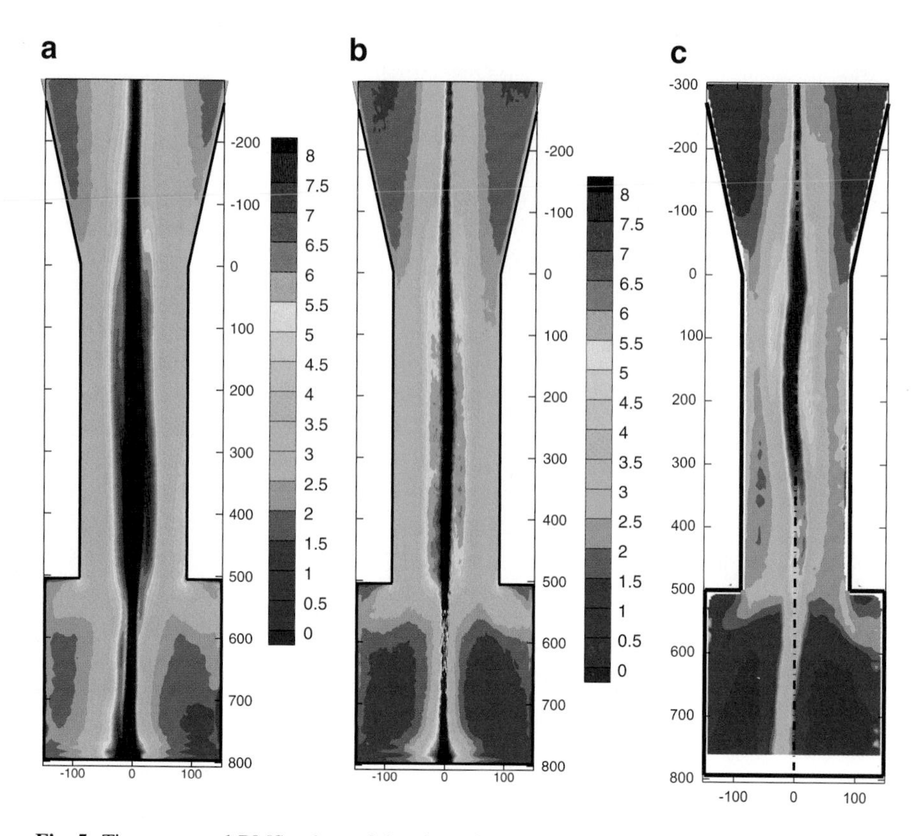

Fig. 5 Time-averaged RMS-values of the circumferential fluid velocity fluctuations (m/s) in the x–z plane (**a**) LES flow; (**b**) LES particles; (**c**) unladen flow experiment [24] (same contour levels)

axial velocity fluctuations in Fig. 6b) are, however, still not fully converged because of the small particle numbers present in the vortex core.

Regarding the agreement with the experiments [24] very similar observations as for the circumferential fluid velocity fluctuations can be found for the time-averaged RMS-values of the axial fluid velocity fluctuations (Fig. 6). Also for this statistical moment the maximum of the fluctuations is observed near the vortex core in accordance with the measurements. The amplitude is, however, overpredicted near the axis and underpredicted in the wall region compared with the measurements of Obermair et al. [24]. Regarding the particle statistics the same observation can be made as for the other statistics shown in this section, i.e., small differences to the continuous phase data can be asserted. Obviously, the particle statistics are not fully converged and need a longer averaging period which has to be ensured in the ongoing study.

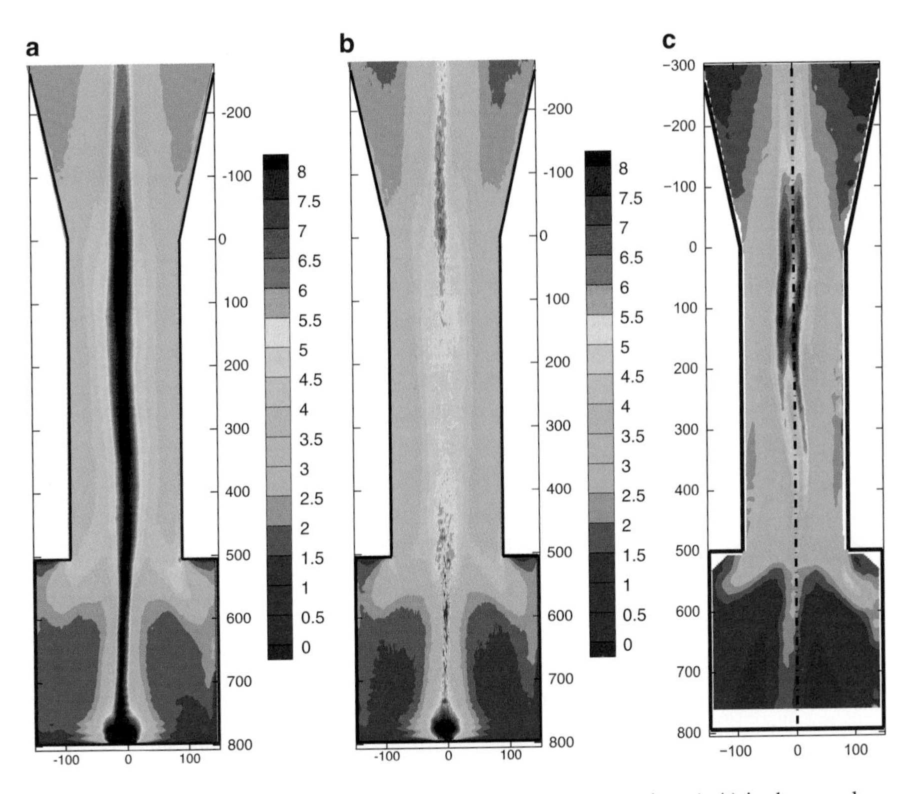

Fig. 6 Time-averaged RMS-values of the axial fluid velocity fluctuations (m/s) in the x–z plane (**a**) LES flow; (**b**) LES particles; (**c**) unladen flow experiment [24] (same contour levels)

7 Conclusions

One-way coupled large-eddy simulations using a point-particle approach are carried out for a cyclone separator flow. The flow data are compared with the unladen PDA measurements of Obermair et al. [24]. Qualitatively good agreement is found with the reference experiments. This underlines that LES is capable to reproduce the complicated flow phenomena found in cyclone separators. In agreement with the experiments of Obermair et al. [24] two main vortical structures are observed: one is moving towards the dust bin near the cyclone walls and the other one moves close to the axis towards the cyclone outlet. Furthermore, the mean vortex core is found to have a helical structure which arises from the deviation of the inlet geometry from an axisymmetric configuration (see Derksen [16]).

Quantitative discrepancies compared with the experiment of Obermair et al. [24] are found in the mean circumferential fluid velocity which in the simulation is lower than in the experiment. The reason is probably that the wall functions applied for the continuous phase overestimate the wall shear stress. The one-way coupled particle statistics are found to be in close agreement with the flow statistics which is reasonably for the small particles considered in the present simulation and the specular reflection at the wall assumed for the particles.

Two-way coupled simulations with particles hitting rough walls are planned for the near future to reproduce more realistically the conditions in real cyclone separators. For particles hitting rough walls a reduction of the circumferential velocity of the particles is expected. By means of the two-way coupling the influence of the expected reduced circumferential velocity of the particulate phase on the carrier phase can be studied. In this way it can be investigated whether the particulate phase leads to substantial changes in the flow structures.

Acknowledgements The time-consuming computations were carried out on the national supercomputer Cray XE6 (Hermit) at the High Performance Computing Center Stuttgart (grant no.: PARTICLE / pfs 12855), which is gratefully acknowledged.

References

1. M. Alletto, M. Breuer, One-way, two-way and four-way coupled LES predictions of a particle-laden turbulent flow at high mass loading downstream of a confined bluff body. Int. J. Multiphase Flow **45**, 70–90 (2012)
2. M. Alletto, M. Breuer, Prediction of turbulent particle-laden flow in horizontal smooth and rough pipes inducing secondary flow. Int. J. Multiphase Flow **55**, 80–98 (2013)
3. J. Bardina, J.H. Ferziger, W.C. Reynolds, Improved subgrid-scale models for large-eddy simulations. AIAA Paper, 80-1357 (1980)
4. M. Breuer, Large-eddy simulation of the sub-critical flow past a circular cylinder: numerical and modeling aspects. Int. J. Numer. Methods Fluids **28**(9), 1281–1302 (1998)
5. M. Breuer, *Direkte Numerische Simulation und Large-Eddy Simulation turbulenter Strömungen auf Hochleistungsrechnern.* Habilitationsschrift, Universität Erlangen–Nürnberg, Berichte aus der Strömungstechnik (Shaker Verlag, Aachen, 2002). ISBN 3-8265-9958-6
6. M. Breuer, M. Alletto, Efficient simulation of particle-laden turbulent flows with high mass loadings using LES. Int. J. Heat Fluid Flow **35**, 2–12 (2012)
7. M. Breuer, M. Alletto, Numerical simulation of particle-laden turbulent flows using LES, in *High Performance Computing in Science and Engineering '11*, ed. by W.E. Nagel, D.B. Kröner, M.M. Resch (Springer, Berlin, 2012), pp. 337–352. ISBN 978-3-642-23868-7
8. M. Breuer, M. Alletto, Effect of wall roughness seen by particles in turbulent channel and pipe flows, in *High Performance Computing in Science and Engineering '12*, ed. by W.E. Nagel, D.B. Kröner, M.M. Resch (Springer, Berlin, 2013), pp. 277–293. ISBN 978-3-642-33373-6
9. M. Breuer, H.T. Baytekin, E.A. Matida, Prediction of aerosol deposition in 90 degrees bends using LES and an efficient Lagrangian tracking method. J. Aerosol Sci. **37**(11), 1407–1428 (2006)
10. M. Breuer, E.A. Matida, A. Delgado, Prediction of aerosol drug deposition using an Eulerian-Lagrangian method based on LES, in *International Conference on Multiphase Flow*, Leipzig, Germany, 9–13 July 2007
11. M. Breuer, P. Lammers, T. Zeiser, G. Hager, G. Wellein, Direct numerical simulation of turbulent flow over dimples—code optimization for NEC SX-8 plus flow results, in *High Performance Computing in Science and Engineering '07, 10th Results and Review Workshop on High Performance Computing in Science and Engineering*, Oct. 04–05, 2007, ed. by W.E. Nagel, D. Kröner, M. Resch, University of Stuttgart, Germany, 2008, pp. 303–318
12. M. Breuer, M. Alletto, F. Langfeldt, Sandgrain roughness model for rough walls within Eulerian-Lagrangian predictions of turbulent flows. Int. J. Multiphase Flow **43**, 157–175 (2012)

13. C.T. Crowe, M.P. Sharma, D.E. Stock, The Particle-Source-In-Cell (PSI-CELL) model for gas-droplet flows. Trans. ASME J. Fluids Eng. **99**, 325–332 (1977)
14. C.T. Crowe, M. Sommerfeld, Y. Tsuji, *Multiphase Flows with Droplets and Particles* (CRC Press, Boca Raton, 1998)
15. F.J. de Souza, R. de Vasconcelos Salvo, D.A. de Moro Martins, Large-eddy simulation of the gas-particle flow in cyclone separators. Sep. Purif. Technol. **94**, 61–70 (2012)
16. J.J. Derksen, Separation performance prediction of a Stairmand high-efficiency cyclone. AIChE J. **49**(6), 1359–1371 (2003)
17. J.J. Derksen, S. Sundaresan, H.E.A. van den Akker, Simulation of mass-loading effects in gas-solid cyclone separators. Powder Technol. **163**, 59–68 (2006)
18. J.J. Derksen, H.E.A. van den Akker, S. Sundaresan, Two-way coupled large-eddy simulations of the gas-solid flow in cyclone separators. AIChE J. **54**(4), 872–885 (2008)
19. G. Gronald, J.J. Derksen, Simulating turbulent swirl flow in a gas cyclone: a comparison of various modeling approaches. Powder Technol. **205**, 160–171 (2011)
20. C. Marchioli, V. Armenio, A. Soldati, Simple and accurate scheme for fluid velocity interpolation for Eulerian-Lagrangian computation of dispersed flow in 3D curvilinear grids. Comput. Fluids **36**, 1187–1198 (2007)
21. M.R. Maxey, J.J. Riley, Equation of motion for a small rigid sphere in a non-uniform flow. Phys. Fluids **26**, 883–889 (1983)
22. J.B. McLaughlin, Inertial migration of a small sphere in linear shear flows. J. Fluid Mech. **224**, 261–274 (1991)
23. S Obermair, Einfluss der Feststoffaustragungsgeometrie auf die Abscheidungsrate und den Druckverlust eines Gaszyklons, Ph.D. thesis, Technische Universität Graz, Austria, 2002
24. S. Obermair, J. Woisetschläger, G. Staudinger, Investigation of the flow pattern in different dust outlet geometries of a gas cyclone by laser Doppler anemometry. Powder Technol. **138**, 239–251 (2003)
25. S. Obermair, C. Gutschi, J. Woisetschläger, G. Staudinger, Flow pattern and agglomeration in the dust outlet of a gas cyclone investigated by phase Doppler anemometry. Powder Technol. **156**, 34–42 (2005)
26. J. Pozorski, S.V. Apte, Filtered particle tracking in isotropic turbulence and stochastic modeling of subgrid-scale dispersion. Int. J. Multiphase Flow **35**, 118–128 (2009)
27. A. Raufeisen, M. Breuer, V. Kumar, T. Botsch, F. Durst, LES and DNS of melt flow and heat transfer in Czochralski crystal growth, in *High Performance Computing in Science and Engineering '06*, ed. by W.E. Nagel, W. Jäger, M. Resch. Transactions of the High Performance Computing Center, Stuttgart (HLRS) 2006 (Springer, Berlin, 2007), pp. 279–291
28. H. Reichardt, Vollständige Darstellung der turbulenten Geschwindigkeitsverteilung in glatten Leitungen. Z. Angew. Math. Mech. **31**, 208–219 (1951)
29. S.I. Rubinow, J.B. Keller, The transverse force on a spinning sphere moving in a viscous fluid. J. Fluid Mech. **11**, 447–459 (1961)
30. U. Schumann, Subgrid-scale model for finite-difference simulations of turbulent flows in plane channels and annuli. J. Comput. Phys. **18**, 376–404 (1975)
31. H. Shalaby, K. Pachler, K. Wozniak, G. Wozniak, Comparative study of the continuous phase flow in a cyclone separator using different turbulence models. Int. J. Numer. Methods Fluids **48**(11), 1175–1197 (2005)
32. J. Smagorinsky, General circulation experiments with the primitive equations. I. The basic experiment. Mon. Weather Rev. **91**, 99–165 (1963)
33. M. Sommerfeld, Validation of a stochastic Lagrangian modelling approach for inter-particle collisions in homogeneous isotropic turbulence. Int. J. Multiphase Flow **27**, 1829–1858 (2001)
34. M. Sommerfeld, B. von Wachem, R. Oliemans, Best practice guidelines for computational fluid dynamics of dispersed multiphase flows, in *SIAMUF, Swedish Industrial Association for Multiphase Flows, ERCOFTAC*, 2008. ISBN 978-91-633-3564-8

Efficient Coupling of an Eulerian Flow Solver with a Lagrangian Particle Solver for the Investigation of Particle Clustering in Turbulence

Christoph Siewert, Matthias Meinke, and Wolfgang Schröder

Abstract Numerical studies show that particles suspended in a turbulent flow tend to cluster due to their inertia (Wang and Maxey, J. Fluid Mech. 256:27–68, 1993; Bec et al., Phys. Rev. Lett. 98:084502, 2007). It was shown by Woittiez et al. (J. Atmos. Sci. 66:1926–1943, 2009) and Onishi et al. (Phys. Fluids 21:125108, 2009) that gravity influences the clustering of small and heavy particles in turbulence. However, these results might be artificially influenced by the periodicity of the used computational domains and also by the turbulence forcing scheme (Rosa et al., J. Phys. Conf. Ser. 318:072016, 2011). In the present study, a new numerical setup to investigate the combined effects of gravity and turbulence on the motion of small and heavy particles is presented, where the turbulence is only forced at the inflow and is advected through the domain by a mean flow velocity. Within a transition region the turbulence develops to a physical state which shares similarities with grid-generated turbulence in wind tunnels. Since the turbulence is decaying in streamwise direction statistical averages can only be performed over small parts of the domain. Hence, a very large number of particles has to be considered to obtain converged statistics compared with the periodic setups of the other numerical studies where averaging can be performed over all particles in the whole domain. This results in the need of a very efficient parallelization strategy. In this study, trajectories of about 43 million small and heavy particles are advanced in time. It is found that specific regions within the turbulent vortices cannot be reached by the particles as a result of the particle vortex interaction. Therewith, the particles tend to cluster outside the vortices. These results are in agreement with the theory of Dávila and Hunt (J. Fluid Mech. 440:117–145, 2001).

C. Siewert (✉) · M. Meinke · W. Schröder
Institute of Aerodynamics, RWTH Aachen University, Wüllnerstraße 5a, 52062 Aachen, Germany
e-mail: c.siewert@aia.rwth-aachen.de

W.E. Nagel et al. (eds.), *High Performance Computing in Science and Engineering '13*, DOI 10.1007/978-3-319-02165-2_27,

1 Introduction

Many investigations of heavy particles transported in turbulent carrier flows have been performed, since this topic is important for several applications, e.g. planetary science, plankton species, spray combustion in diesel engines, volcano ash, dust, spore, and pollen spreading, and droplet growth in clouds. However, for some phenomena physical explanations are still missing. For example in the case of droplet growth in warm clouds the largest scales involved are in the order of the cloud size and the smallest scales are the dust or aerosol particles. The range spans from several kilometers down to a micrometer. This large range of scales is the main reason for still existing gaps in knowledge. In the past mostly theoretical investigations have been performed (e.g. [7]). In numerical simulations of particles in turbulence the dissipation range has to be fully resolved since the particle motions are mainly governed by the smallest scales of turbulence. Thus, only the computational expensive direct numerical simulation method can be used. Only recently the computational power reached a level which enabled these simulations. Thereby the turbulence is typically generated in periodic domains by forcing the largest scales of turbulence. Nevertheless these numerical simulations are restricted to intermediate Taylor-scale Reynolds numbers Re_λ due to the computational effort. So the separation of scales may be an issue. To the authors' knowledge the highest Re_λ concerning particle-laden turbulent flows was reached in [8]. It is found that heavy particles are centrifuged out of vortices and tend to gather in regions of high strain. This effect called "particle clustering" or "preferential concentration" [9].

However, the influence of gravity on the particle phase was neglected in this study. In [10] it is pointed out that the fall velocity of the heavy particle significantly shortens the particle vortex interaction time. Therewith the presence of gravity alters the preferential concentration effect. This is numerically supported by Woittiez et al. [3]. Additionally it is shown by Rosa et al. [5] that the preferential concentration depends on the employed large scale forcing scheme. Therefore, in this study a completely different setup is presented with similarities to grid-generated turbulence in wind tunnels. Synthetic turbulence is forced only at the inflow plane and advected through the domain by a mean flow velocity. Therewith the flow evolves according to the Navier-Stokes equations and a decaying isotropic turbulence can be found. In this flow the combined effects of turbulence and gravity on the motion of small and heavy particles are investigated. However, due to the turbulent decay the flow field is not homogeneous. Hence, statistics can only be gathered at constant streamwise positions. To reach statistical convergence a large number of particle are tracked simultaneously. For that reason an efficient parallelization strategy is developed.

2 Setup

A slice through the domain is shown in Fig. 1. The Reynolds number $Re_L = \frac{UL}{\nu}$ is 80,000 with the domain length L and the kinematic viscosity ν. The domain boundaries are furnished with a sponge layer to avoid spurious oscillations due to

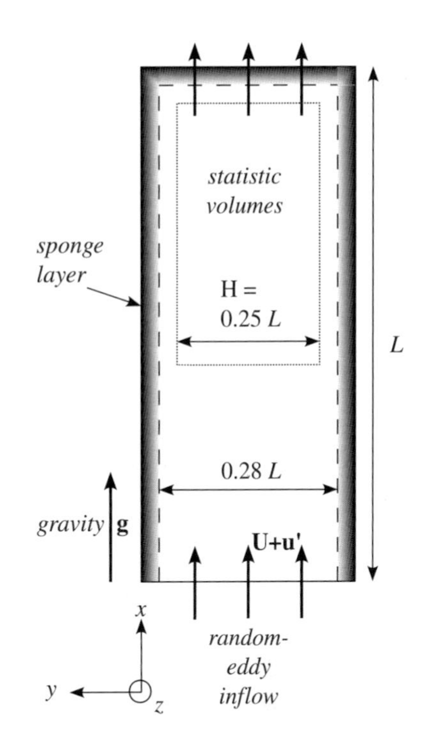

Fig. 1 Sketch of the domain setup. The z and the y layout are the same. At the inflow plane synthetic turbulence is added to the mean advection velocity. The remainder of the domain boundary is furnished with a sponge layer. Gravity points in the mean flow direction. Turbulence and particle statistics are gathered in several statistic volumes at the downstream end of the domain

reflections [11]. At the inflow plane the synthetically generated turbulent fluctuations $\mathbf{u}'$ are added to the mean advection velocity $\mathbf{U}$. The turbulent fluctuations are generated by a method following [12]. The main idea is to generate synthetic fluctuations that match a prescribed Reynolds stress tensor on average. Therefore, the fluctuations consist of a large number of individual Fourier modes, here 200 were used. They are chosen randomly but fulfill the continuity equation. Since there is no mean shear in the flow, turbulence is only produced at the inflow. Downstream the turbulent kinetic energy k decays due to viscous damping. A snapshot is shown in Fig. 2 to provide a qualitative picture of this decay. Thus, the flow field generated in this study shares similarities with grid-generated turbulence in wind tunnel experiments. For grid turbulence it was shown that the decay is of the form: $k(t) = k_0 \left(\frac{t}{t_0}\right)^n$ with $n = -1.3$ [13]. This temporal dependency can be related to a spatial one due to the constant mean flow velocity U by the relationship $t = \frac{x}{U}$. Figure 3 shows the turbulent kinetic energy k over the spatial coordinate x in a logarithmic plot. It can be seen that k is nearly constant at the inflow. This is because the random forcing obeys the momentum equation only on average. However, further downstream k decays with the theoretical decay exponent of -1.3. The size of the transition region can be estimated from this plot. After only $\frac{1}{8}$ of the total domain length a physical state is reached. Re_λ has a value of around 20, weakly depending on the spatial location. However, there is no restriction to scale-up the current method to reach higher turbulent intensities besides the larger

Fig. 2 Isocontours of the vorticity magnitude colored by the streamwise vorticity provide a qualitative picture of strength and sense of rotation of the turbulent vortices

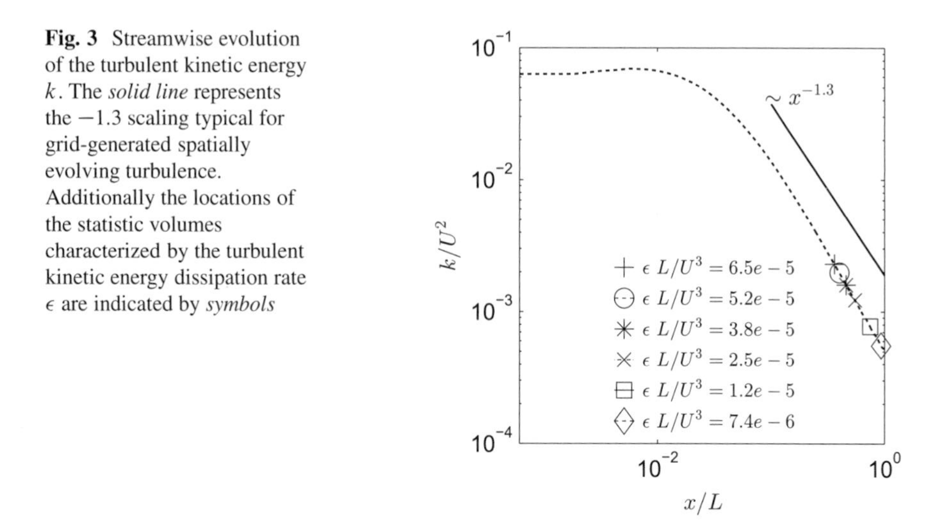

Fig. 3 Streamwise evolution of the turbulent kinetic energy k. The *solid line* represents the -1.3 scaling typical for grid-generated spatially evolving turbulence. Additionally the locations of the statistic volumes characterized by the turbulent kinetic energy dissipation rate ϵ are indicated by *symbols*

computational effort. An advantage of the current method is that the dependency on the dissipation rate can be checked within a single computation (see Fig. 3). The turbulence is spatially decaying and therewith the particles are advected through areas of different turbulence intensities. More details can be found in [14].

2.1 Particle Phase

The particles considered here are small (ratio of the particle radius to the Kolmogorov length $\frac{a_p}{\eta_K} \ll 1$) and heavy (ratio of the particle density and the fluid density $\frac{\rho_p}{\rho_f} = 843$). Within this limit the equation of motion [15] can be simplified to

$$\frac{d\mathbf{v}_p}{dt} = \frac{f_D}{\tau_p}\left[\mathbf{u}_f\left(\mathbf{x}_p\right) - \mathbf{v}_p\right] + \mathbf{g} \ , \tag{1}$$

with $\mathbf{v_p}$ the particle velocity, $\mathbf{u_f}$ the fluid velocity at the particle position $\mathbf{x_p}$ interpolated by a tricubic least square method, the gravity force $\mathbf{g}$, a non-linear Stokes drag correction f_D and the particle response time $\tau_p = \frac{\rho_p}{\rho_f}\frac{2a_p^2}{9\nu}$. A total number of about 45 million particles in ten different radii classes are advanced simultaneously in the turbulent flow as discussed above. Despite the high number of particles the volume loading is small enough to assume that the particle phase does not influence the flow phase. This procedure is called one-way coupling. The particles are released at the inflow boundary into the flow. After a interval the particle motions are independent of the initial release conditions. The turbulence and particle statistics are gathered in different statistic volumes downstream of the initial transition region, see Fig. 3. The vertical extent of the downstream volumes is chosen small enough to ensure that the statistics are unaffected by the presence of the sponge layers.

2.2 Particle Characterization

The ratio of the particle response time and the flow time scale is called Stokes number St. Because of the small particle sizes the Kolmogorov time scale is the appropriate time scale [3]. Particles with a very small inertia respectively response time follow the flow similar to fluid tracers. In the opposite limit of very large inertia the particles are hardly affected by the turbulence. If the particle and turbulent time scale match, particles are driven out of the vortices and gather in regions of high strain [16]. This particle clustering is measured by a so called radial distribution function (RDF) g_{11} [3]. It compares the actual number of particle pairs at contact with the number expected for a normal distribution of the particles. In the presence of gravity the behavior discussed above is altered. Without surrounding flow or in laminar flow the heavy particles would fall with their so-called terminal velocity $v_t = \tau_p \frac{g}{f_D}$ (derived from (1)). The relative importance of gravity and turbulence can be evaluated by the non-dimensional settling velocity Sv which is the ratio of the terminal velocity and the Kolmogorov velocity scale $Sv = \frac{v_t}{u_\eta}$ [10]. If the ratio is high, the particles sediment faster through the vortex than its turnover time. In this case gravity dominates the motion of the particles compared to turbulence due to the small particle vortex interaction times. However, gravity and turbulent acceleration are not simply superposed as (1) might suggest. Dávila and Hunt [6] suggest to rescale the Stokes number by the non-dimensional settling velocity resulting in the particle Froude number $F_p = St\ Sv^2$. They predict that for $F_p \sim 1$

the average turbulent falling velocity is higher than the laminar one because the particles preferentially pass vortices on their downward motion side. Additionally, they state that due to their inertia particles cannot reach certain areas of the vortices.

3 Parallelization

The simulations for this study were performed on the Cray XE6 (Hermit) supercomputer at HLR Stuttgart. It consists of 113,664 compute cores on 3,552 compute nodes. The nodes are connected by the high speed Cray Gemini network. For disk storage a Lustre parallel file system is available.

To start a simulation only a small amount of data has to be provided besides the solver itself. The bounding box of the computational domain has to be provided in STL format along with a text input file specifying all input parameters of the simulation. At first a computational grid is generated in parallel. The computational domain is covered by a single Cartesian cell which is subsequent refined by iteratively splitting each cell into eight equal sized child cells until the desired grid resolution is reached. Since a DNS is intended, the domain is discretized into cubic cells with a side length corresponding to the Kolmogorov length η_K, leading to a total number of about 53 million cells. The resulting grid is stored into a single file on disk. Thereby the parallel netcdf library is used [17]. It can handle the storage of several hundred gigabytes of data in binary form automatically while using the advantages of the Lustre parallel file system. Afterwards the solver can be started. The flow solver used here is a general purpose solver developed and successfully applied at the Institute of Aerodynamics at RWTH Aachen [11]. It solves the integral form of the compressible Navier-Stokes equations with a finite volume method. The flow simulated here is essentially incompressible, hence the Mach number is set to $Ma = 0.1$. The computational domain is automatically divided into load balanced blocks during the solver startup. This can be achieved by exploiting the hierarchical grid structure. During the parallel grid generation the cells have been sorted after the Hilbert curve which is a space filling curve [18]. Since it self-similar with respect to changing resolution, each cell in grid file is followed by its entire offspring. All processors read in only the upper part of the cell hierarchy and divide the grid on these cells weighted by their number of offspring. Additional weights can be easily introduced accounting for the different amount of workload each individual cell contributes. For example the inflow boundary condition used in this study needs considerably more computing time than the other boundary conditions. Also the number of particles in each cell can be used to further balance the computational load. After the division on the upper part of the cell hierarchy each processor reads in only the offspring of its part of the hierarchy. This is very efficient since the information that each processors needs are located continuously on disk. Additionally this strategy leads to a very efficient communication topology since the usage of the space filling curve garanties locality. A block contains only neighboring cells which leads to a high volume to surface ratio of each block. This is desirable

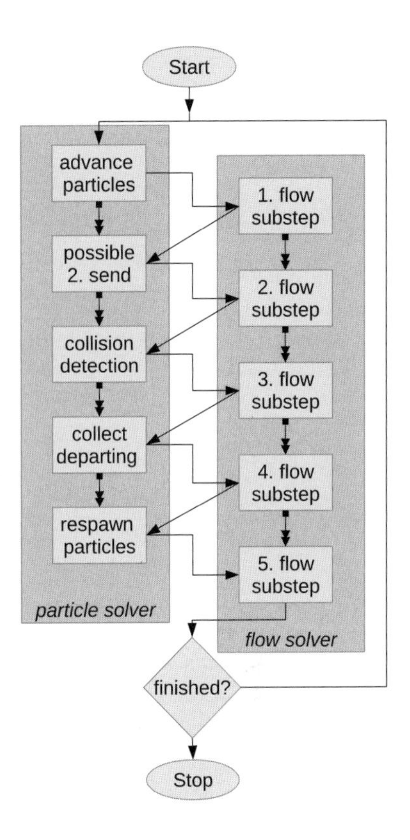

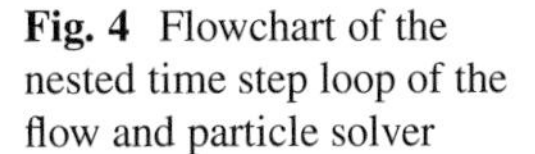
Fig. 4 Flowchart of the nested time step loop of the flow and particle solver

because it keeps the number respectively the size of the information which has to be communicated from the boundary to the neighboring blocks at a minimum, such that only a limited amount of network bandwidth is necessary. Furthermore, adjacent blocks are handled preferentially by neighboring processors. This leads to short communication latencies and also strengthens the low need of network bandwidth. To store the data received from the neighboring blocks layers of halo cells are automatically added at the block borders during the start up of the solver. The data is communicated via the Message Passing Interface (MPI) [19]. All neighbor communications are written to be non-blocking which means that they are independent of the order of the incoming messages. This is a necessary prerequisite to reach high scalability in a massive parallel computation.

To further decrease the communication waiting time the flow and the particle solver communications are nested. The proceeding is visualized in Fig. 4. This nesting is required since the high number of individual communications would otherwise restrict the scalability of the complete solver. The flow solver uses a five step Runge–Kutta method for time integration. After each of the five substeps the flow variables at the borders of the block have to be communicated to the neighboring blocks. In the meantime the particle solver can operate. First the particles are advanced in time. To enable parallel collision detection the particles

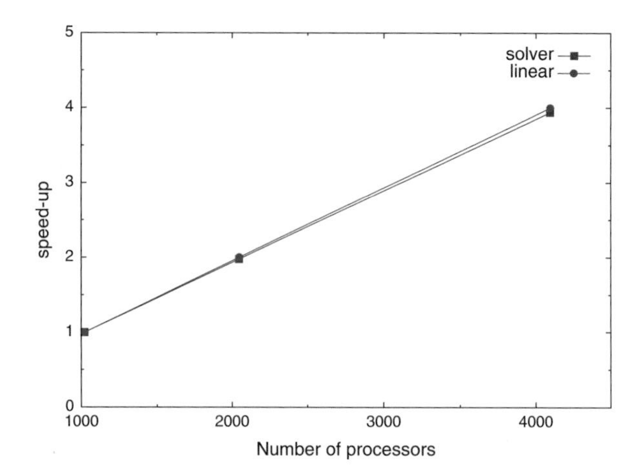

Fig. 5 Strong scaling from 1,024 up to 4,096 cores

at the borders of the block are send to the neighboring blocks. Meanwhile, the first substep of the flow solver can take place. The number of particles to be send to the neighboring domains is unknown and varies in time. Since dynamical memory allocation has to be avoided during run time due to the limited availability of random access memory, the send and receive buffers are allocated with a constant size during the solver start up. Due to the prior unknown spatial clustering of the particles no reasonable sized buffer can be allocated which is large enough for all required particle communications during a long simulation run. Only a size large enough to cover most of all particle communications can be estimated. In the rest of the cases additional communications have to take place. During that the second flow solver substep can be performed. Afterwards each block can locally search for particle collisions and process them. Since the particle positions can be affected by the collisions, they must be communicated to the neighboring blocks. As only dense particulated flows are considered particle collisions are rare events. Hence, only a small subset of the particles has to be transferred and in this case the exchange buffers are large enough so that only one communication is needed. After the third flow solver substep all particles outside of the block, i.e., within the halo cells of the block are deleted. To keep the total number of particles statistically stationary, particles should be reintroduced at the inflow if they have left the computational domain. For that reason they are send to the processor which handles the particle respawn. Meanwhile, the fourth flow solver substep takes place. The new particle respawn positions are chosen from a normal distribution and the particles are scattered to the corresponding blocks sharing the inflow. The timestep is completed by the last flow solver substep. Hence, the communication time of the flow as well as the particle solver can be successfully hidden by executing the two solvers alternately. Thus, an excellent parallel performance can be achieved. This is demonstrated in Fig. 5. It shows a strong scaling on the massive parallel system

Table 1 Computational details of typical simulation

Number	Run description	No. cores	Walltime
1	Grid generation	512	10 min
2	Flow solver	512	23 std.
3	Flow + particle solver	1,024	23.5 std.
4–6	Statistics	1,024	23.5 std.
7–15	Post-processing	32–512	0.5–12 std.

from 1,024 to 4,096 cores. In this range the speed up nearly follows the ideal linear scaling.

A typical simulation within this study is summarized in Table 1. The grid is generated within 10 min on 512 cores. Afterwards the flow solver is run for the rest of the maximal 24 h compute time which can be allocate within Hermit queuing system. Since one-way coupling is assumed the presence or absence of the particle does not change the flow solution. So the simulation is started with the flow solver alone. After an initial period of $8 \times \frac{L}{U}$, the flow field is independent of its initial condition. The flow variables are stored on disk in a restart file. Then the solver is restarted, this time which both the flow and the particle solver. Thus, a higher number of cores is allocate to account for the higher memory consumption. After another $2 \times \frac{L}{U}$ the particles are independent of their initial release conditions and statistics are gathered over $20 \times \frac{L}{U}$. It should be pointed out that the whole simulation is in principle independent of number of processors. However, a trade-off has to be found due to the limited available memory, the total duration of the simulation and queue waiting time.

4 Results

Figure 6 shows qualitatively the preferential concentration effect by displaying the particle positions in a slice. In (a) the particle Froude number is much smaller than unity resulting in a nearly homogeneous particle distribution. For a particle Froude number of unity (b) many regions without particles and as a result regions with particle clusters can be found.

The particle clustering effect is depicted quantitatively in Fig. 7 by means of the RDF values over the particle Stokes number. It can be seen that the particles in the considered size range generally tend to cluster ($g_{11} > 1$). From small to big particles the clustering effect gets stronger and attains its maximum value for $F_p \sim 1$ and decays after that. In contrast the position of the maximum value in [20] is at $St \sim 0.6$. At this value the strongest effect was found in [8] neglecting the influence of gravity. This is surprising because of the otherwise good quantitative agreement with the data of [20]. The differences might be originated in the spatial decay of the turbulence in this study and/or the periodicity of the domain and the large scale forcing scheme in the studies reported in the literature. This study supports the

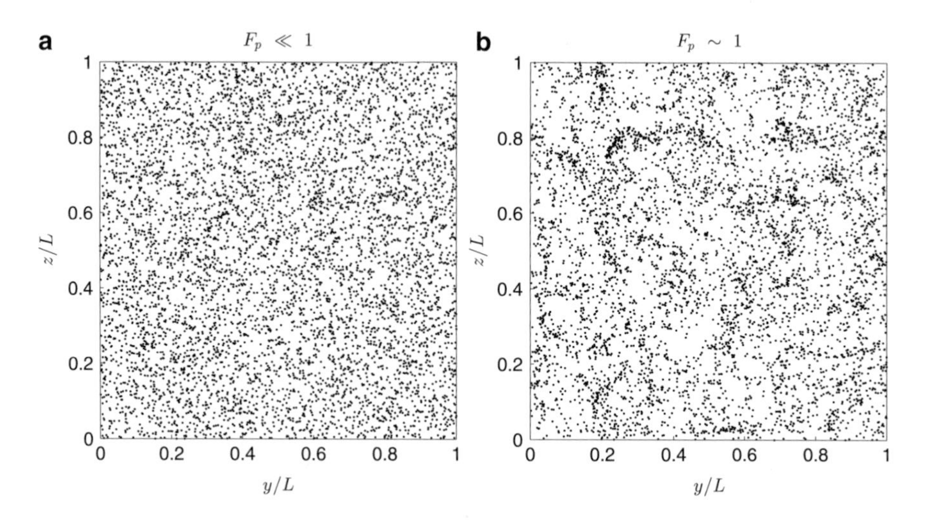

Fig. 6 Visualization of the preferential concentration effect. The particle positions are displayed in a slice perpendicular to the streamwise direction. (**a**) $F_p \ll 1$ and (**b**) $F_p \sim 1$

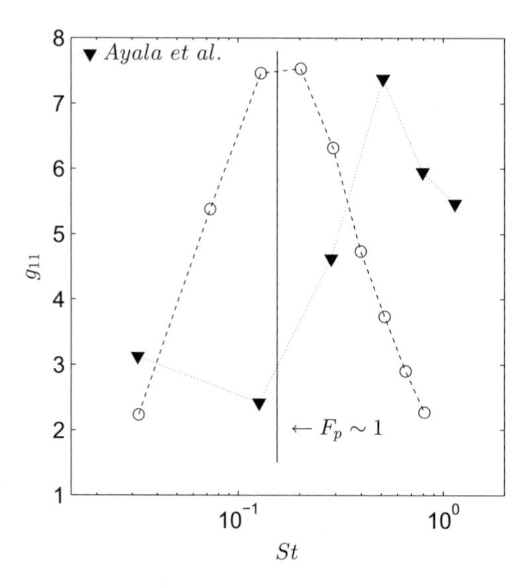

Fig. 7 RDF g_{11} over the particle Stoke number. The *filled symbols* represent data from [20]

theory of [6] and highlights the importance of the particle Froude number. However, based on the existing data it remains unclear what causes this difference and which result is more physically correct. Therefore, it is planned to numerically simulate a laboratory experiment to answer this question.

5 Conclusion

A new numerical setup is presented to investigate the combined effects of gravity and turbulence on the motion of small and heavy particles. The turbulence is only forced at the inflow and is advected through the domain by a mean flow velocity. Within a transition region the turbulence develops to a physical state which shares similarities with the grid-generated turbulence in wind tunnel experiments. In this flow trajectories of about 43 million small and heavy particles are advanced in time. This was possible due to an efficient parallelization of the solver. To avoid communication wait times the flow and particle solver are called alternately. The outcome of this study is that the particle clustering is strongest for $F_p \sim 1$. This is in contrast to other numerical simulations in the literature. It remains unclear which result is more physically correct. Since the effect of particle clustering is very important, e.g., for the collision rates of cloud droplets and therewith the formation of rain, it is planned to numerically reproduce windtunnel experiments to quantitatively compare experimental and numerical results.

Acknowledgements The funding of this project under grant number SCHR 309/39 by the Deutsche Forschungsgemeinschaft is gratefully acknowledged. The authors thank HLR Stuttgart for the provided computational resources.

References

1. L.-P. Wang, M.R. Maxey, Settling velocity and concentration distribution of heavy particles in homogeneous isotropic turbulence. J. Fluid Mech. **256**, 27–68 (1993)
2. J. Bec, L. Biferale, M. Cencini, A. Lanotte, S. Musacchio, F. Toschi, Heavy particle concentration in turbulence at dissipative and inertial scales. Phys. Rev. Lett. **98**, 084502 (2007)
3. E.J.P. Woittiez, H.J.J. Jonker, L.M. Portela, On the combined effects of turbulence and gravity on droplet collisions in clouds: a numerical study. J. Atmos. Sci. **66**, 1926–1943 (2009)
4. R. Onishi, K. Takahashi, S. Komori, Influence of gravity on collisions of monodispersed droplets in homogeneous isotropic turbulence. Phys. Fluids **21**, 125108 (2009)
5. B. Rosa, H. Parishani, O. Ayala, L.-P. Wang, W.W. Grabowski, Kinematic and dynamic pair collision statistics of sedimenting inertial particles relevant to warm rain initiation. J. Phys. Conf. Ser. **318**, 072016 (2011)
6. J. Dávila, J.C.R. Hunt, Settling of small particles near vortices and in turbulence. J. Fluid Mech. **440**, 117–145 (2001)
7. P.G. Saffman, J.S. Turner, On the collision of drops in turbulent clouds. J. Fluid Mech. **1**, 16–30 (1956)
8. J. Bec, L. Biferale, M. Cencini, A.S. Lanotte, F. Toschi, Intermittency in the velocity distribution of heavy particles in turbulence. J. Fluid Mech. **646**, 527–536 (2010)
9. K.D. Squires, J.L. Eaton, Preferential concentration of particles by turbulence. Phys. Fluids **3A**, 1169–1178 (1991)
10. W.W. Grabowski, P. Vaillancourt, Comments on "preferential concentration of cloud droplets by turbulence: effects on the early evolution of cumulus cloud droplet spectra". J. Atmos. Sci. **56**, 1433–1436 (1999)
11. D. Hartmann, M. Meinke, W. Schröder, An adaptive multilevel multigrid formulation for Cartesian hierarchical grid methods. Comput. Fluids **37**, 1103–1125 (2008)

12. P. Batten, U. Goldberg, S. Chakravarthy, Interfacing statistical turbulence closures with large-eddy simulation. AIAA J. **42**(3), 485–492 (2004)
13. S.B. Pope, *Turbulent Flows* (Cambrigde University Press, Cambrigde, 2000)
14. R.P.J. Kunnen, C. Siewert, M. Meinke, W. Schröder, K. Beheng, Numerically determined geometric collision kernels in spatially evolving isotropic turbulence relevant for droplets in clouds. Atmos. Res. **127**, 8–21 (2013)
15. M.R. Maxey, J.J. Riley, Equation of motion for a small rigid sphere in a nonuniform flow. Phys. Fluids **26**, 883–889 (1983)
16. S. Sundaram, L.R. Collins, Numerical considerations in simulating a turbulent suspension of finite-volume particles. J. Comput. Phys. **124**, 337–350 (1996)
17. J. Li, W. keng Liao, A. Choudhary, R. Ross, R. Thakur, R. Latham, A. Siegel, B. Gallagher, M. Zingale, Parallel netcdf: a high-performance scientific i/o interface, in *Proceedings of Supercomputing*, 2003
18. D. Hilbert, Über die stetige Abbildung einer Linie auf ein Flächenstück. Math. Ann. **38**, 459–460 (1891)
19. Message Passing Interface Forum, *MPI: A Message-Passing Interface Standard, Version 2.2* (High Performance Computing Center Stuttgart (HLRS), Stuttgart, 2009)
20. O. Ayala, B. Rosa, L.-P. Wang, W.W. Grabowski, Effects of turbulence on the geometric collision rate of sedimenting droplets. Part 1. Results from direct numerical simulation. New J. Phys. **10**, 075015 (2008)

Lagrangian Simulation of a Fluid with Solid Particle Loading Performed on Supercomputers

Florian Beck, Florian Fleissner, and Peter Eberhard

Abstract The simulation of a fluid with particle loading is a challenging task. In this contribution two Lagrangian simulation methods are coupled to handle the challenges when simulating such scenarios. The Smoothed Particle Hydrodynamics method is used for modeling the fluid. The particles within the fluid are simulated with the Discrete Element Method, which is also applied for the boundary geometry. These methods are coupled to model a fluid and solid particles in one common simulation. In this work Lagrangian methods are employed because the natural handling of the solid-fluid interface.

The simulation of a fluid with particle loading requires a large numbers of particles. With an increasing number of particles the computational effort rises. In the presented examples, the number of particles is up to 50 million. Therefore, the calculation is done on supercomputers.

First, in this work the simulation framework for complex fluid particles simulation is presented. Then, the employed simulation methods are introduced and some simulations for verifying the framework are discussed. Afterwards simulations of a fluid with particles as loading and complex boundaries and the results are presented.

F. Beck (✉) · F. Fleissner · P. Eberhard
Institute of Engineering and Computational Mechanics, University of Stuttgart, Pfaffenwaldring 9, 70569 Stuttgart, Germany
e-mail: florian.beck@itm.uni-stuttgart.de; florian.fleissner@itm.uni-stuttgart.de; peter.eberhard@itm.uni-stuttgart.de

W.E. Nagel et al. (eds.), *High Performance Computing in Science and Engineering '13*, DOI 10.1007/978-3-319-02165-2_28,

1 Introduction

The coupling of two simulation methods can be used to merge their advantages in one common simulation. In doing this it is possible to simulate complex systems which are consisting, e.g., of a fluid with solid particles. In this context, one possibility is to model the solid particles with the Discrete Element Method (DEM). Several studies about such fluid-particle systems have been conducted in the past, a review can be found in [25]. Different techniques have been developed to investigate these systems, a summary of the recent developments is summarized in [24]. Several approaches exist for the modeling of fluid-particle systems, which can be classified as continuum or discrete approaches [24]. When using a discrete approach on a microscopic scale, it is a common way to couple a classical computational fluid dynamics (CFD) method with the DEM method.

Instead of this Lagrangian-Eulerian coupling, it is another possibility to model the fluid also with a mesh-less method. In this work the Smoothed Particle Hydrodynamics (SPH) method is used for the simulation of the fluid. There are different approaches for the Lagrangian-Lagrangian coupling, e.g. [19] or [9]. The advantage of the application of two mesh-less methods is a very natural way of handling the fluid-solid interface. Also the description of the free surface of the fluid is easier to handle with a mesh-less method like the SPH method than with some classical CFD methods. This advantage is of particular importance when considering the impact of the fluid to complex geometries.

For the simulation of a complex system with particle methods millions of DEM and SPH particles are necessary [24]. To handle large numbers of particles the simulation has to be parallelized. In this work parallel simulations on a supercomputer of the High Performance Computing Center Stuttgart (HLRS) with our simulation framework are presented. The parallel simulation on the supercomputer allows the efficient simulation of a fluid with particle loading. Besides the simulation of a free jet with particle loading some numerical experiments for verifying the parallel simulations on the supercomputer are presented first. The simulation of a fluid with particle loading is an interesting step forward for the later investigation of abrasive wear due to solid particles.

Therefore, this work is divided into three parts. In the first part the SPH and the DEM method are introduced briefly. Then the coupling of the two simulation techniques as well as the boundary interface treatment is discussed. Afterwards the simulation framework itself and the parallelization are presented. In the next part some numerical experiments are described. The numerical experiments are carried out to verify the parallel simulations with our framework on the used supercomputer. In the last part the simulation of a fluid with particle loading is presented. In this scenario the impact of a liquid free jet with solid particle loading to a turbine blade is simulated. This is a preliminary numerical experiment for upcoming wear analyses of fluids with particle loading. At the end a conclusion of this work is given.

2 Simulation Method

2.1 Smoothed Particle Hydrodynamics

The SPH method is applied for modeling the fluid. The fluid is described by the Navier-Stokes (N-S) equations. Due to its mesh-less character, the SPH method is a suitable approach for describing a fluid with free surfaces and the changing interface between the fluid and a solid.

Originally SPH was introduced in [7, 15] to investigate different phenomena in astrophysics. In the past decades there are several fields of applications to which the method was applied, like free surfaces [16, 17, 23], fluid-solid interaction [1, 10] and multi-phase flow [2, 8]. Commonly the weakly compressible SPH method or the truly incompressible SPH method is used. An overview about the SPH method and its extensions can be found in [13].

The equations for conserving the momentum (1) and for the conservation of mass (2) of the fluid discretized with the SPH method are

$$\rho\left(\frac{\partial \mathbf{v}}{\partial t}+\mathbf{v}\cdot\nabla\mathbf{v}\right)=-\nabla p+\mu\nabla^2\mathbf{v}+\mathbf{f}\,, \tag{1}$$

$$\nabla\cdot\mathbf{v}=0\,. \tag{2}$$

In these equations $\mathbf{f}$ are the body forces acting on the fluid, p is the pressure, $\mathbf{v}$ the velocity, ρ the density and μ the viscosity of the fluid. To obtain the SPH formulation of the N-S equations basically two steps are required. The first step is the kernel approximation and the second one is referred to as particle approximation [13]. The continuum, which is described with the N-S equations, is discretized into so called particles during these two steps and at these particles any quantity is described with the functions

$$A\left(\mathbf{r}_a\right)=\sum_b m_b\frac{A_b}{\rho_b}W\left(\mathbf{r}_{ab},h\right)\,, \tag{3}$$

$$\nabla A(\mathbf{r}_a)=\sum_b m_b\frac{A_b}{\rho_b}\nabla_a W\left(\mathbf{r}_{ab},h\right)\,. \tag{4}$$

In these functions m_b is the mass, ρ_b the density, $\mathbf{r}_b$ the position of a particle and h the smoothing length. There are several kernel functions W_{ab} which can be applied in these functions, depending on the simulation scenario. In this work the classical Gaussian kernel function

$$W(\mathbf{r}_{ab},h)=\left(\frac{1}{\sqrt{(\pi h^2)^d}}\right)e^{-\mathbf{r}_{ab}^2/h^2} \tag{5}$$

is used for the simulation of the fluid with particle loading. The solid particles inside the fluid are not modeled with the SPH method, but with the DEM.

2.2 Discrete Element Method

The particles, here the loading of the fluid, are modeled with the Discrete Element Method (DEM) [3]. The Granular media is discretized with particles which are interacting. It is possible to use this method also for bulk material that consists of free particles which have not a permanent contact like in [4]. It is also possible to couple several particles with inner-particle bonds to form a multi-particle body. In this way it is possible to model natural failure [5]. The dynamics of the particles can be described with the Newton-Euler equation, [21], which yield

$$m_i \mathbf{a}_i = \mathbf{f}_i , \tag{6}$$

$$\mathbf{I}_i \cdot \dot{\boldsymbol{\omega}}_i + \boldsymbol{\omega}_i \times \mathbf{I}_i \cdot \boldsymbol{\omega}_i = \mathbf{l}_i . \tag{7}$$

Here m_i is the mass, $\mathbf{I}_i$ the inertia tensor and $\mathbf{f}_i$ and $\mathbf{l}_i$ are forces and torques. The translational acceleration is $\mathbf{a}_i$ and the angular velocity $\boldsymbol{\omega}_i$. In the following only the translational part is taken into account and particle rotations are ignored.

There exist several contact models for calculating the force between two particles [24]. In this work a Hertzian contact is used. For calculating the force one can use

$$F_{ij} = K_{ij} \delta_{ij}^{\frac{3}{2}} + d \dot{\delta}_{ij} \tag{8}$$

with

$$K_{ij} = \frac{4}{3\pi(h_i + h_j)} \left(\frac{R_i R_j}{R_i + R_j} \right)^{\frac{1}{2}} ,$$

$$h_j = \frac{1 - \nu_j^2}{\pi E_j} .$$

In (8) R_j is the radius of a particle, E_j Young's modulus of the granular material, ν_j the Poisson number, d a damping parameter and δ_{ij} the overlap of two particles. This contact model was introduced in [11].

2.3 DEM-SPH Coupling

In this section the coupling of the DEM and SPH method is described. Depending on the ratio of size of the SPH particles and the DEM particles, there are different techniques for coupling the DEM and the SPH method, see e.g. [19] for a much larger DEM particle than SPH particle. In [9] a coupling technique for small DEM particles is applied. The coupling which is used in this work is similar to the one described in [19].

Every time step in the particle simulation involves a neighborhood search. All adjacent particles, which are interacting, are determined during the neighborhood search. If a contact between an SPH and a DEM particle is detected, a contact force is calculated. This force then is applied to the DEM and the SPH particle. The equation of motion for the DEM particle and the equation for conserving the momentum of the fluid are

$$m_{\mathrm{i}}\mathbf{a}_{\mathrm{i}} = \mathbf{f}_{\mathrm{i}} + \mathbf{f}_c \,, \tag{9}$$

$$\rho\left(\frac{\partial \mathbf{v}}{\partial t} + \mathbf{v}\cdot\nabla\mathbf{v}\right) = -\nabla p + \mu\nabla^2\mathbf{v} + \mathbf{f} + \mathbf{f}_c \,. \tag{10}$$

In these equations $\mathbf{f}_c$ is the volumetric interaction force between a DEM particle and the fluid.

In this work we use a force model which we have already successful applied to other simulations [12]. In the contact model we calculate a boundary force similar to [17]. We take the distance and velocity difference with different parameters into account. For the contact of any particle with the boundary we are using the same contact model.

3 Simulation Code

3.1 Pasimodo

Pasimodo is the particle simulation framework developed at the ITM "www.itm.uni-stuttgart.de/research/pasimodo/pasimodo_en.php". It is completely written in C++ and has a modular design. Basically it consists of a core and so called plugins which are shared libraries which are dynamically loaded. The core controls the overall simulation sequence. For a typical particle based simulation, a simulation loop cycle consists of three phases. In the first one the neighborhood search is performed. After the contacts between the particles are determined, the forces are calculated in the second phase. In the last phase the particles are integrated, e.g. their state variables are updated. Every component is implemented as plugin, but it is also possible to merge several components in one plugin. The necessary components for a DEM simulation are particles, an interaction model and an integrator. Also the SPH method is implemented as plugin. Of course it is possible to combine several plugins into one simulation.

3.2 Parallel Implementation

In this section some details about the parallelization of the software package Pasimodo are introduced. Currently there are two different possibilities for a parallel simulation with Pasimodo. The first one is used for systems with shared memory and the second one is used for distributed memory. The distributed memory version uses the Message Passing Interface (MPI) for the interprocess communication. In this work it is applied for the simulation on supercomputers.

In contrast to grid-based methods, where the neighbors of any grid-point and thus the system topology are known and fixed, in particle methods the neighbors are varying and the particles can move throughout the whole simulation domain. Of course there are several other challenging points in particle simulations. For an efficient parallel simulation, a load-balancing algorithm is implemented in Pasimodo "www.itm.uni-stuttgart.de/research/pasimodo/pasimodo_en.php" which deals with these challenges. The aim of this algorithm is to minimize idle time of processors and to prevent excessive communication.

Of course it would be easy to make a static distribution of particles to a processor before the simulation, but this would be very inefficient. The particles are not restricted in their motion, therefore an efficient approach has to deal with dynamically varying domains. This is achieved by combining a load balancing algorithm combined with the orthogonal recursive bisection ORB [20] decomposition scheme. For the dynamic adaption of the domain decomposition we are using a PI-controller. In this way a reduced running time is achieved resulting from the mentioned complex load-balancing algorithm.

4 Numerical Experiments

In this section some numerical experiments are described. These simulations are used to verify that Pasimodo is working correctly on the NEC Nehalem Cluster, a supercomputer of the HLRS. Also the result of the investigation about the maximum number of particles, which can currently handled with Pasimodo, is shown. For the numerical experiments several quantities of the simulation were compared. The simulation methods which are used for the verification examples are the DEM method, the SPH method for fluids and the SPH method for solid state simulations.

4.1 Breaking Dam 2D

First a commonly used example in the SPH community, a breaking dam, is simulated. In this example a fluid sloshes from the left to the right in a virtual tank

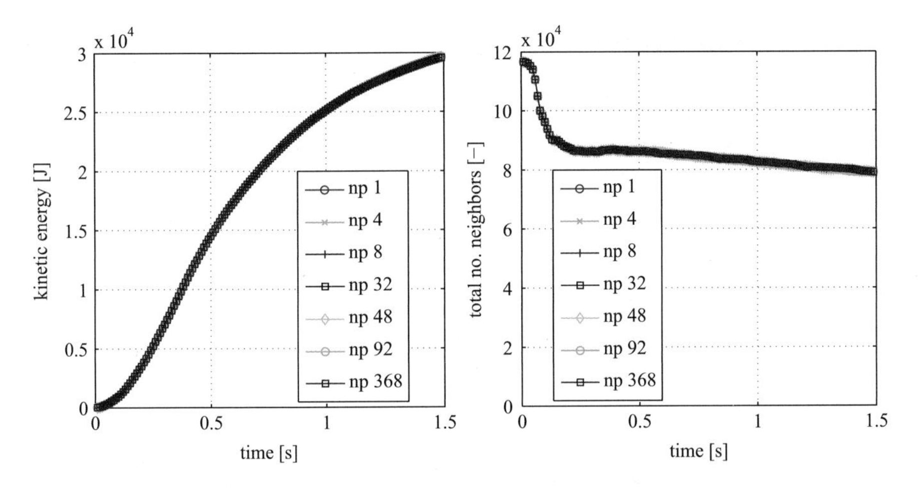

Fig. 1 Comparsion of the kinetic energy and total number of neighbors of simulations made with different numbers of MPI processes (np)

which is open to the right. There are also some modified versions where the tank is closed to the right and the fluid sloshes back to the left. In the following some details about this example are given.

In this simulation the SPH method is used. The example is based on the simulation described in [17]. Instead of the spline kernel a Gaussian kernel is used with the coefficients for the two dimensional case. The extensions XSPH, artificial viscosity and a repulsive force are used. The simulation is done without real viscosity like in [14]. The shear and bulk viscosity is taken into account with the artificial viscosity. The coefficients are chosen to $\beta = 0$ and $\alpha = 0.01$ as in [17]. For the boundary conditions a repulsive force based on a modified Lenard Jones potential is applied. This force is calculated between boundary particles and the SPH particles.

In Fig. 1 the kinetic energy and the total number of neighbors are shown. This simulation scenario was calculated on the NEC Nehalem cluster. In order to verify the parallel implementation, the simulation was performed using different numbers of MPI processes. The curves of the different runs, which are shown in Fig. 1, are identical. If there were differences it would be an indication for a problem of the parallelization. Also the progress of this simulation can be seen in these curves. At the beginning the fluid is at rest and afterwards accelerated due to gravity. Therefore, the kinetic energy has to increase.

In Fig. 2 four different states of the simulation are shown. The decrease of the total number of neighboring particles which correspond to the number of interactions can also be seen in this figure. The next numerical experiment is similar to this two dimensional case.

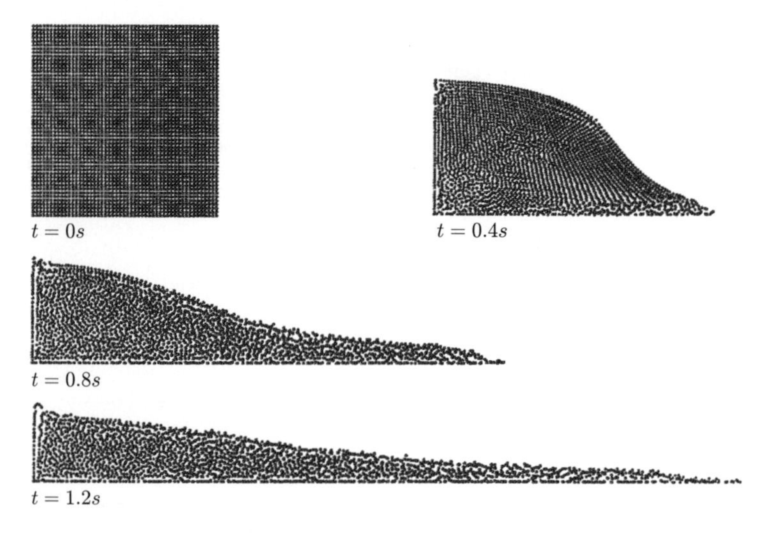

Fig. 2 Four different states of the simulation of a 2D breaking dam

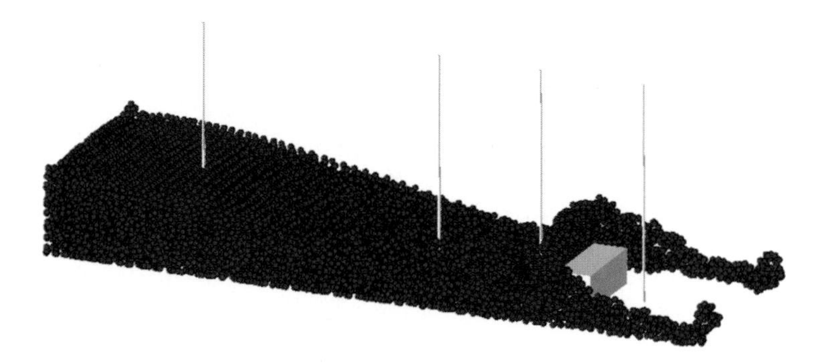

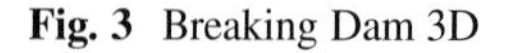

Fig. 3 Breaking Dam 3D

4.2 Breaking Dam 3D

The next numerical experiment is the simulation of a three dimensional breaking dam in a closed tank, see Fig. 3. The impact of the fluid on the obstacle and the back wall of the tank can be seen. The tank itself is not shown in this figure. The four small rods are fixed points for measuring the height of the fluid during the simulation.This example is commonly used to verify the SPH method. This simulation is originally used to simulate green water loadings. These are large amounts of water which damage, e.g. ship housings [6, 18]. An advantage of the SPH method in contrast to some classical grid based methods is the more natural description of the free surface. The simpler description of the free surface and not only but also the interface

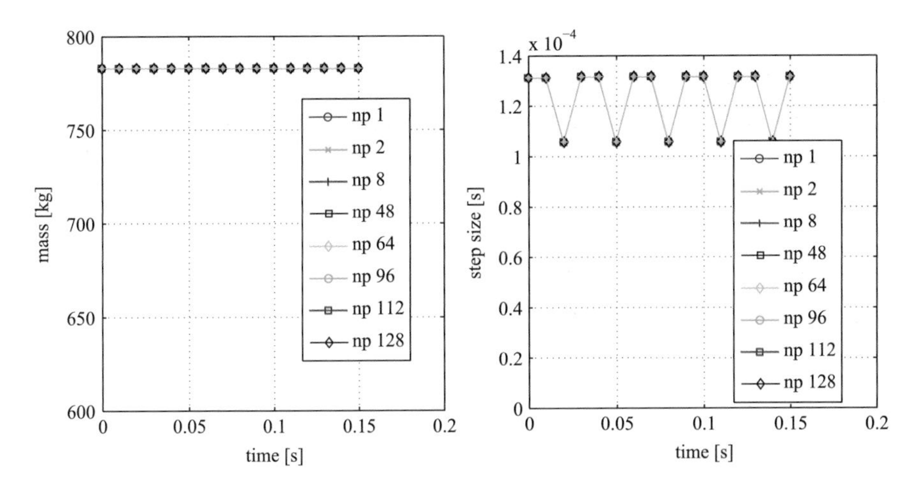

Fig. 4 Comparsion of the mass and the time step size of simulations for different numbers of MPI processes

between the fluid and solid bodies is due to the mesh-less character of the SPH method.

In Fig. 4 the mass and the time step size of this simulation are shown. It is important that the mass in the simulations with the different number of MPI processes is constant. Also the time step size of the simulation should not depend on the number of MPI processes, as it can be seen in this figure.

4.3 *Tensile Test*

With the SPH method it is not only possible to simulate fluids but also to simulate solid bodies. One interesting topic is the simulation of the solid body in machining processes [22]. The next numerical experiment is therefore the simulation of a tensile test with the SPH method.

The tensile test is commonly used in material science to obtain material parameters or to select a material for an application. During the test several properties can be measured. The properties are the tensile strength, elongation and diameter reduction. From these properties it is possible to determine other material properties. In Fig. 5 the deformation during the tensile test is shown.

For the simulation of a solid body with the SPH method a strain tensor is computed instead of the viscous forces in fluids. For a more detailed description see [22]. In Fig. 6 the number of iteration steps and the total number of interactions are shown. This numerical experiment was also performed with different numbers of MPI processes. The number of iterations and the number of interactions may not depend on the number of processes and this can clearly be seen.

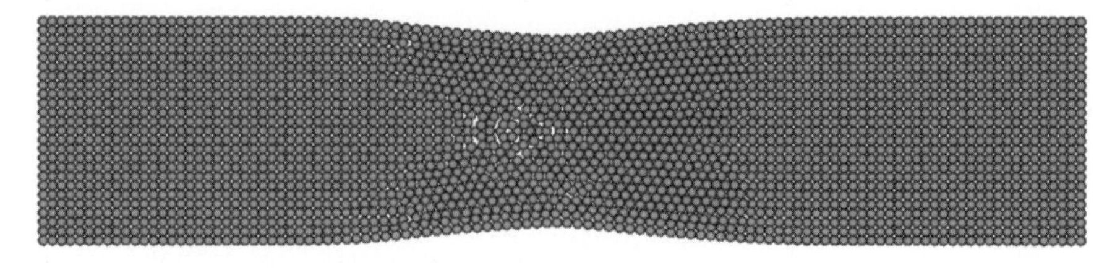

Fig. 5 Deformation of the solid body during the tensile test

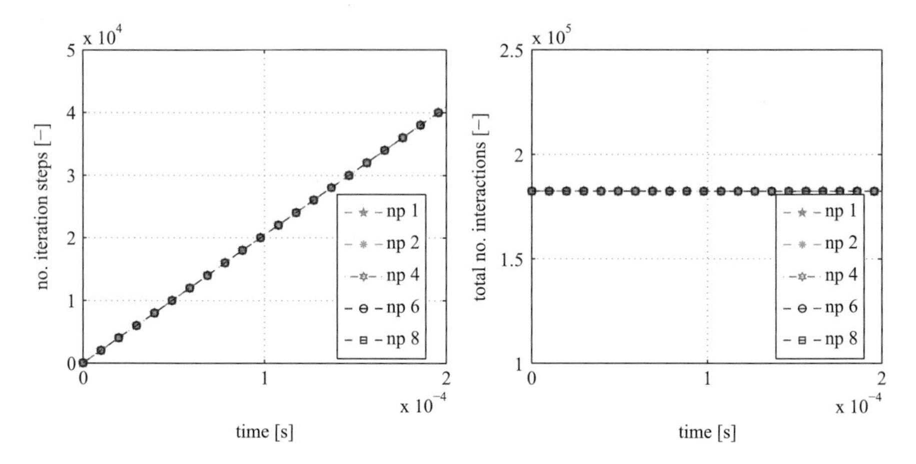

Fig. 6 Comparsion of the number of iteration steps and the total number of interaction of simulations made with different numbers of MPI processes

4.4 *Maximum Particle Number Using DEM*

An interesting point in particle simulations is the maximum number of particles which can be treated. The next numerical experiment is carried out in order to get an idea about the maximum number of particles which can currently be handled by Pasimodo on this specific supercomputer.

In this simulation the DEM method is used. To investigate the maximum number of particles the following scenario is set up. There is a box consisting of boundary particles and within the box are DEM particles. As contact law the Hertzian contact law is used. The particles are placed with a small initial distance and have varying radii. Due to gravity the particles are then falling down inside the box.

For the analysis of the maximum number of particles which can be handled with Pasimodo. We are using the maximum number of nodes, which can be used with our account at the HLRS. The maximum number of nodes of our account is limited to 257 nodes on the NEC Nehalem supercomputer. Therefore, only the maximum number of particles up to that maximum number of nodes is investigated.

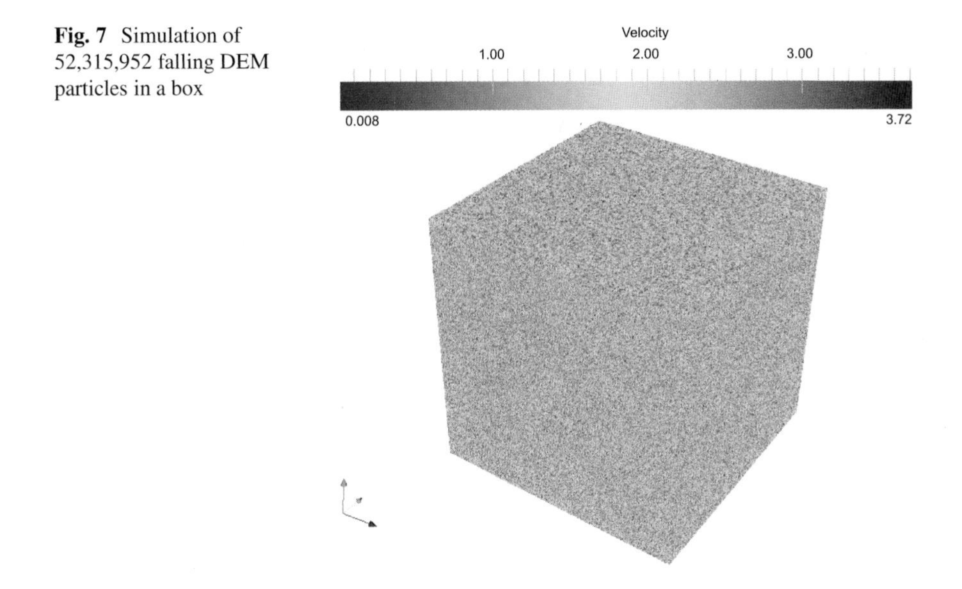

Fig. 7 Simulation of 52,315,952 falling DEM particles in a box

The numerical experiment was calculated on nodes with different properties. One special node with 144 GB memory was used. The other 256 nodes are nodes of the property Nehalem without special characteristics offering 12 GB of memory each. The maximum number of particles to that number of nodes was 52,315,952. In Fig. 7 is a screenshot of the simulation.

5 Free Jet with Particle Loading

In this example the SPH and the DEM method are coupled in one simulation for a free jet with particle loading. For this numerical experiment only mesh-less methods are applied. The advantage of mesh-less methods is the more natural handling of the interface of the fluid-solid interaction and the boundaries compared to mesh-based methods. The SPH method is used to discretize the fluid and the DEM for modeling the particle loading of the fluid. The boundary geometry is represented through boundary triangles from CAD data.

A free jet with particle loading is simulated which is shown in Fig. 8. For this example the free jet is discretized with 805,408 SPH particles. For the loading 702 DEM particles are used. For this simulation different force laws are applied between the different components of the simulations. A contact law based on [17] is used for contact between a DEM particle and an SPH particle, as well as for the contact between the particles and the boundary geometry. Additionally the force calculation within the particle methods takes place. The simulation of the experiment was performed on 36 nodes. This reasonable number was determined

Fig. 8 Simulation of a free jet with particle loading

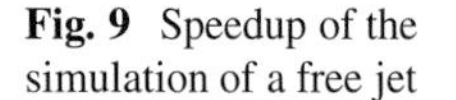

Fig. 9 Speedup of the simulation of a free jet

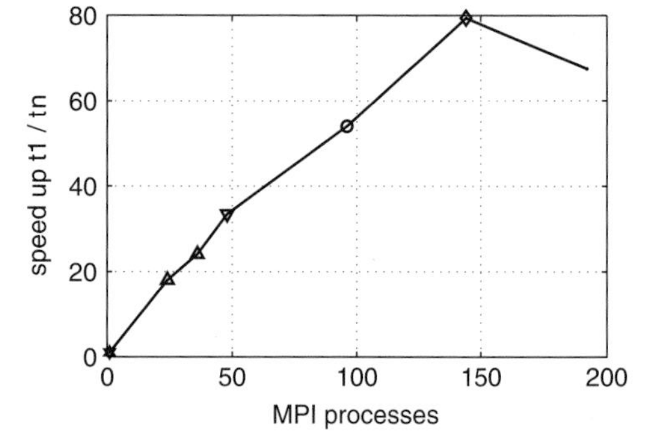

with some pre-simulations. In those pre-simulations different numbers of processes were applied. The speedup of the simulation is shown in Fig. 9. The best result was obtained with 144 processes on 36 nodes. Here, the simulation was made on 36 nodes with 12 GB memory and two quad cores Intel Xeon X5560 each. The whole simulation time was 10,368 CPU hours. The number of processes per node was 4, which is fixed due to the restricted available memory of a single node.

The speedup of the simulation decreases with further increasing number processes, which can be explained with the particle number per process. In this example there are around 5,000 particles per process. On each process the particles are organized in several groups. This is necessary to avoid taking all particles of a process into account for the neighborhood search between two processes. With an increasing number of processes the number of particles per node is decreasing. With a decreasing number of particles more particles have to be transferred between groups and processes. The transfer of particles requires more communication, which in turn slows down the simulation. It is quite obvious that an optimal setup requires

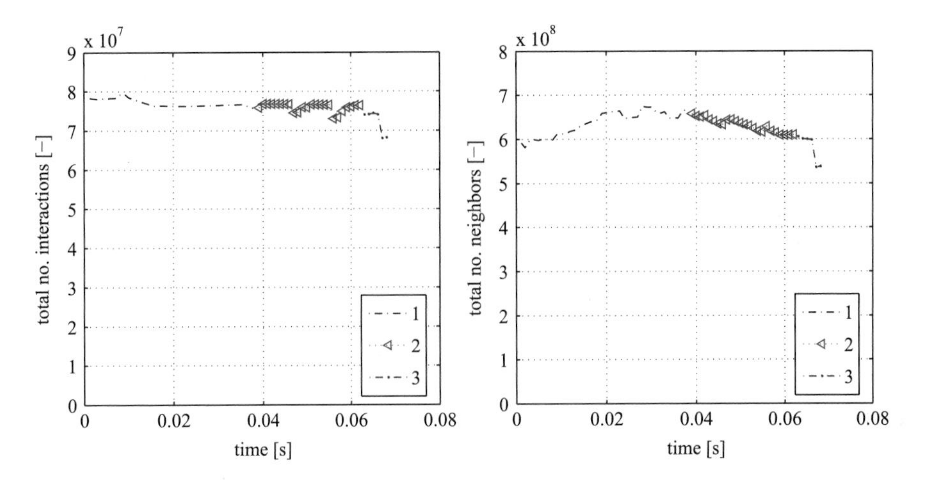

Fig. 10 Total number of interactions and total number of neighbors of the simulation of the free jet

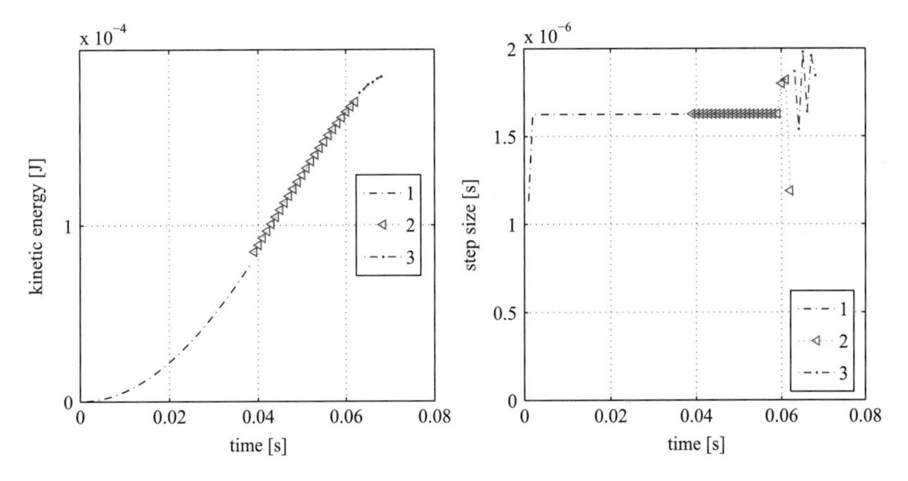

Fig. 11 Kinetic energy and step size of the simulation of the free jet

a lot of knowledge and experience—not only about the numerical behavior, but also about the available hardware and distribution strategy. Due to the available walltime of 24 h, the simulation was interrupted and continued three times. A different color/ line type is used for each simulation in the following plots. In Fig. 10 the total number of interactions and the total number of neighbors are shown. It can be seen that the numbers decrease, after the impact of the jet on the geometry. This is due to the simulation methods. For the calculation of the force between the SPH particles and the geometry fewer particles are taken into account than for computing the force between SPH particles. In Fig. 11 the kinetic energy and the

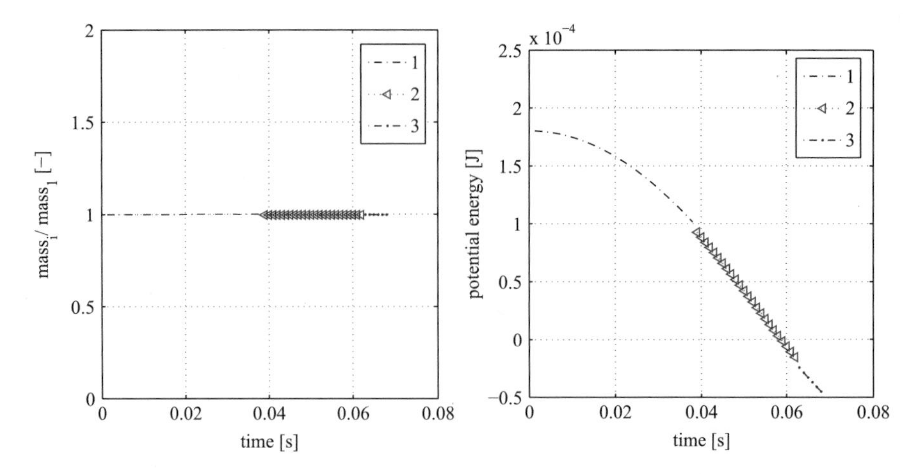

Fig. 12 Mass of particles and potential energy of the simulation of the free jet

step size of the simulation are shown. The kinetic energy increases first, because in this preliminary numerical experiment the jet is in the beginning at rest. After the impact the increment of the kinetic energy decreases, because the free jet is decelerated during the contact with the geometry. The step size is varying more and more with increasing boundary contact of the SPH method. A challenging point of this simulation is to suspend and continue the simulation due to the restricted walltime. In Fig. 12 the mass referred to the initial simulation and the potential energy of the simulation are shown. It can be seen, that the mass in the simulation is constant. If the mass would vary between the simulation periods the continuing simulation would be inaccurate. Also the decrease of the potential energy of the different simulation parts fits together. The free jet at different times is shown in Fig. 13. The particle loading of the free jet at different times of the simulation is shown in Fig. 14. In the beginning there is no contact of the free jet and the geometry. Then, there is the impact of the free jet to the geometry. After the impact the free jet spreads on the geometry.

6 Conclusion

In this work numerical experiments are presented which were performed on the NEC Nehalem Cluster of the HLRS. The numerical experiments have shown that the parallel implementation works well and scales nicely on the used supercomputer. The simulation results of the parallel simulations are consistent with the serial results. Given to the limitation of resources of our account, it was possible to investigate up to 50 million particles. After this, the primary objective of this work, a coupled simulation, was presented and the speedup was analyzed. The simulations

Fig. 13 Simulation of the free jet with particle loading

have shown that a lot of knowledge of the hardware and the simulation software is necessary to choose parameters for a good speedup. The simulation of the jet shows that without the capability of a parallel simulation with many MPI processes, the required calculation time would not be acceptable.

Fig. 14 Particle loading of the free jet

Acknowledgements The research leading to the presented results has received funding from the German Research Foundation (DFG) under the program SFB 716 "A5 Dynamic simulation of systems with large particle numbers", subproject "Simulation of abrasive damage processes using hybrid smoothed particle hydrodynamics". Also, the support of the colleagues within the SFB 716, especially of D2, M.Sc. Amer Wafai and Dr.-Ing. Rainer Keller has been very helpful. This financial support by the DFG is highly appreciated.

References

1. J. Campbell, R. Vignejevic, L. Libersky, A contact algorithm for smoothed particle hydrodynamics. Comput. Meth. Appl. Mech. Eng. **184**, 49–65 (2000)
2. A. Colagrossi, M. Landrini, Numerical simulation of interfacial flows by smoothed particle hydrodynamics. J. Comput. Phys. **191**, 448–475 (2003)
3. P.A. Cundall, O.D.L. Strack, A discrete numerical model for granular assemblies. Géotechnique **29**(1), 47–65 (1979)
4. C. Ergenzinger, R. Seifried, P. Eberhard, A discrete element model to describe failure of strong rock in uniaxial compression. Granul. Matter **13**(4), 341–364 (2011)
5. C. Ergenzinger, R. Seifried, P. Eberhard, A discrete element approach to model breakable railway ballast. J. Comput. Nonlinear Dynam. **7**(4), 041,010–1–8 (2012)
6. G. Ersdal, A. Kvitrud (eds.), Green water on Norwegian production ships. Proceedings of the 10th ISOPE Conference, Seattle, 2000
7. R.A. Gingold, J.J. Monaghan, Smoothed particle hydrodynamics: Theory and application to non-spherical stars. Mon. Not. Roy. Astron. Soc. **181**, 375–389 (1977)
8. X. Hu, N. Adams, A multi-phase SPH method for macroscopic and mesoscopic flows. J. Comput. Phys. **213**, 844–861 (2005)
9. Y. Jun Huang, O.J. Nydal, Coupling of discrete-element method and smoothed particle hydrodynamics for liquid-solid flows. Theor. Appl. Mech. Lett. **2**(1), 012002 (2012)
10. S. Kulasegaram, J. Bonet, R.W. Lewis, M. Profit, A variational formulation based contact algorithm for rigid boundaries in two-dimensional SPH applications. Comput. Mech. **33**, 316–325 (2004)
11. H.M. Lankarani, P.E. Nikravesh, A contact force model with hysteresis damping for impact analysis of multibody systems. J. Mech. Des. **112**, 369–376 (1990)
12. A. Lehnart, F. Fleissner, P. Eberhard, Simulating tank vehicles with sloshing liquids load. Simpack News **9**, 10–12 (2010)
13. M. Liu, G. Liu, Smoothed particle hydrodynamics (SPH): an overview and recent developments. Arch. Comput. Meth. Eng. **17**, 25–76 (2010)
14. E. Lo, S. Shao, Simulation of Near-Shore Solitary Wave Mechanics by an Incompressible SPH Method. Appl. Ocean Res. **24**, 275–286 (2002)
15. L.B. Lucy, A numerical approach to the testing of the fission hypothesis. Astron. J. **82**(12), 1013–1024 (1977)
16. J. Monaghan, A. Kocharyan, SPH Simulation of multi-flow. Comput. Phys. Commun. **87**, 225–235 (1995)
17. J.J. Monaghan, Simulating free surface flows with SPH. J. Comput. Phys. **110**, 399–406 (1994)
18. W. Morris, J. Millar, B. Buchner (eds.), Green water susceptibility of North Sea FPSO/FSUs, Proceedings of the 15th Conference on Floating Production Systems, London, 2000
19. A.V. Potapov, M.L. Hunt, C.S. Campbell, Liquid solid flows using smoothed particle hydrodynamics and the discrete element method. Powder Tech. **116**(2–3), 204–213 (2001)
20. J. Salmon, Parallel hierarchical N-Body methods. Ph.D. thesis, Caltech University, Pasadana, 1990
21. W. Schiehlen, P. Eberhard, *Technische Dynamik (in German)* (Teubner, Wiesbaden, 2011)
22. F. Spreng, P. Eberhard, F. Fleissner, An approach for the coupled simulation of machining processes using multibody system and smoothed particle hydrodynamics algorithms. Theor. Appl. Mech. Lett. **3**(1), 013,005–1–7 (2013)
23. H. Takeda, S. Miyama, M. Sekiya, Numerical simulation of viscous flow by smoothed particle hydrodynamics. Progr. Theor. Phys. **92**(5), 939–960 (1994)
24. H. Zhu, Z. Zhou, R. Yang, A. Yu, Discrete particle simulation of particulate systems: Theoretical developments. Chem. Eng. Sci. **62**(13), 3378–3396 (2007)
25. H. Zhu, Z. Zhou, R. Yang, A. Yu, Discrete particle simulation of particulate systems: A review of major applications and findings. Chem. Eng. Sci. **63**(23), 5728–5770 (2008)

Computation of Helicopter Phenomena Using a Higher Order Method

Ulrich Kowarsch, Constantin Oehrle, Martin Hollands, Manuel Keßler, and Ewald Krämer

Abstract The enhancement of the structured Computational Fluid Dynamics solver FLOWer in the field of flux computation and its advantage to numerical helicopter simulations is presented. The improvement includes the replacement of the second order spatial scheme with a fifth order scheme for flow state reconstruction and the implementation of different Riemann solvers for flux computation. Aim of the implementation is to reduce the numerical dissipation and to achieve a high vortex preservation essential for numerical investigations of helicopter flows, such as rotor-fuselage interaction and noise emission of the rotor due to blade vortex interactions. For these phenomena which are sensitive to the rotor wake preservation, an investigation with the second order and the fifth order scheme is performed to compare the numerical results as well as the computational performance. The results show significant improvements in the rotor wake conservation, especially in case of an acoustic evaluation. It is shown that the Riemann solver has a high influence to the vortex conservation with low additional computational cost.

1 Introduction

The helicopter flow is characterized by a highly unsteady flow field dominated by blade tip vortices. With the aim to investigate the influence of the rotor wake to the surrounding area, a conservation of these vortices is required in terms of low numerical dissipation. The current second order scheme has shortcomings to conserve the rotor wake over long distances, therefore a further development of the numerical solver is required. To overcome this deficit the spatial second order Jameson Schmidt Turkel (JST) scheme is replaced with the fifth order spatial

---

U. Kowarsch (✉) · C. Oehrle · M. Hollands · M. Keßler · E. Krämer
Institute of Aerodynamics and Gas Dynamics, University of Stuttgart,
Pfaffenwaldring 21, D-70569 Stuttgart, Germany
e-mail: kowarsch@iag.uni-stuttgart.de

W.E. Nagel et al. (eds.), *High Performance Computing in Science and Engineering '13*, DOI 10.1007/978-3-319-02165-2_29,

Weighted Essentially Non-Oscillatory (WENO) scheme for flux reconstruction. In addition a more detailed Riemann solver is used to achieve a more precise flux computation over the cell boundaries. This new method is tested for cases which are sensitive to the rotor wake preservation as the rotor-fuselage interaction in the fast forward flight, called tail-shake phenomenon, and for an acoustic investigation of an isolated rotor in descent flight causing blade-vortex-interaction noise. The following chapters present the WENO method and its implementation into the structured Computational Fluid Dynamics (CFD) solver FLOWer with focus to achieve high performance on the HLRS cluster platform CRAY XE6 Hermit.

2 Numerical Method

The implementation of the WENO scheme is made in the structured finite volume flow solver FLOWer [12], developed by the German Aerospace Center (DLR). The code discretizes the unsteady Reynolds-averaged Navier Stokes equations with a spatial second order central difference JST scheme for flux computation according to [7], as already mentioned. The time discretization is achieved by merging the governing differential equation in space with the implicit dual time-stepping approach according to Jameson [8], which transforms each time step into a steady-state problem. The steady-state problems can then be merged with a conventional time stepping scheme. In case of FLOWer a Runge-Kutta scheme is used. To support an efficient computation, convergence accelerators like multigrid and residual smoothing are implemented in the code. Essential for helicopter flows, fluxes due to grid movements are taken into account using an Arbitrary Lagrangian Eulerian (ALE) approach. In addition, the Chimera technique for overset grids enables relative movements between grids which allows for example the simulation of rotor-fuselage configurations. An efficient computation is achieved by a multi-block structure of the grid to enable parallel computing.

3 Motivation

With the requirement to achieve a more precise preservation of the rotor wake and the propagation of small disturbances over a wide range, the FLOWer code, using the 2nd order JST scheme, approaches its limits. The only degree of freedom is to increase the amount of grid cells to achieve a more precise solution. Hence this leads to significant higher memory requirements for the solution data and the evaluation becomes more elaborate. Another possibility is the implementation of a higher order scheme requiring a significant lower amount of grid cells to achieve a comparable accuracy.

High order schemes exhibit a lower numerical dissipation for unsteady flows, compared to state of the art 2nd or 3rd order schemes. Especially the WENO

(Weighted Essentially Non-Oscillatory) scheme, as one kind of higher order schemes, has been quite successful in the past years for flows with discontinuities and complex flow structures in smooth regions [1]. Therefore this scheme is chosen for the flow state reconstruction in the FLOWer code to improve the numerical quality of the solution.

4 Weighted Essentially Non-Oscialltory Scheme

Based on the well known ENO (Essentially Non-Osciалltory) scheme [5], the WENO scheme was introduced by Liu et al.[13] as a modification of the robust and widely used ENO scheme. The fundamental idea of the ENO scheme is to compute polynomial stencils of order r, based on the surrounding cell condition (cf. Fig. 1). On this basis the flow state is reconstructed. To suppress an oscillation of the solution, the stencil for the reconstruction with the lowest oscillation, measured by a smoothness indicator, is chosen. The WENO scheme extends this concept by introducing a nonlinear weight for each computed stencil depending on its smoothness indicator. The consideration of all stencils results in a higher order of $2r - 1$ compared to the ENO scheme which is of order r. At discontinuities or high gradients the weights cause a reduction in order to remain stable and lead to a rth-order ENO scheme. The flow state reconstruction using WENO is employed for the left and right handed flow state at the cell boundary. The right handed state is computed in the same manner as Fig. 1, just mirrored at the considered cell boundary (marked red) to achieve a symmetrical reconstruction. The intercell flux has to be computed out of the two reconstructed states by the consideration of the Riemann problem (see Sect. 5). Besides the lower numerical dissipation and dispersion than schemes of lower order, the WENO reconstruction has the advantage to be parameter free in contrast to e.g. the JST scheme. For more details see [9]. The implementation into the FLOWer code is presented in more detail in [3].

The scheme is adapted to three dimensional problems by a separate reconstruction in each grid dimension according to Fig. 1.

With the implementation of the scheme for equidistant grids, it has to be noted, that geometrical growth of cells cause a formal error in the state reconstruction. Therefore, a study has been made by implementing the WENO scheme under consideration of the geometrical growth of the cells according to [17]. However, this method showed an unprofitable additional expense in computational effort by about factor 12 compared to the equidistant WENO scheme. For this reason, the meshing strategy is to generate a mesh topology with mainly equidistant cartesian cells. This can be realized by embedding structure adapted meshes into a cartesian background mesh using the overlapping grids method implemented in FLOWer. In addition, by using mostly cartesian meshes the highest numerical quality is guaranteed for the numerical solver.

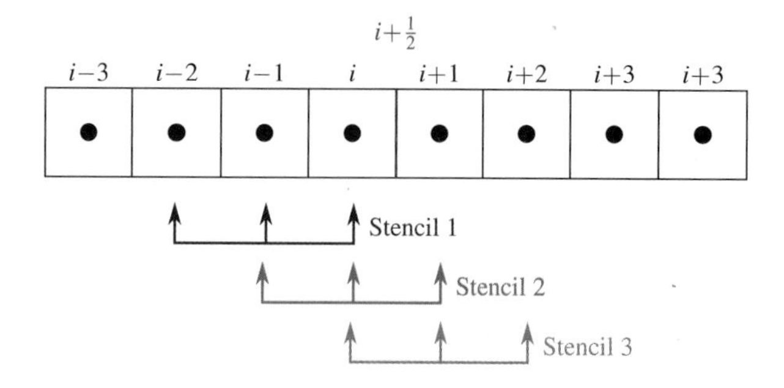

Fig. 1 Stencils for 5th order WENO or 3rd order ENO left handed flux reconstruction at the cell boundary $i + \frac{1}{2}$

5 Riemann Solver

Finding a solution for a system a hyperbolic differential equation, as in case of the Euler equations, requires a consideration of the Riemann problem which is formulated to

$$U_t + F(U)_x = 0\,, U(x,0) = \begin{Bmatrix} U_L \text{ if } x < 0, \\ U_R \text{ if } x > 0. \end{Bmatrix} \quad (1)$$

The structure of the solution is characterized by the eigenvalues λ_i of the system which represent characteristic waves constituting discontinuities in U propagating with the speed λ_i. In case of the Euler solution three eigenvalues are found which physically map the shock wave, rarefaction fan and contact discontinuities/shear waves. The Godunov method transfers this problem into the discrete space and formulates a consideration of the Riemann problem for the intercell flux at each cell boundary. One possibility to avoid the solution of the Riemann problem is the usage of artificial viscosity which is used by the JST scheme. The biggest drawback of this computationally efficient method is the diffusion and dissipation trough the missing reproduction of discontinuities. Solving the Riemann problem to conserve the waves is getting of interest for vortex preservation due to high gradients over the vortex cross-section. Comparable to the state reconstruction, the solution of the Riemann problem is reduced to a one dimension consideration of the flux over a intercell boundary. Therefore the left and the right sided state at the considered boundary is taken into account, which are reconstructed using the WENO scheme.

5.1 *Lax-Friedrich Riemann Solver*

One of the simplest Riemann solvers is the Lax-Friedrich method which is also known as the Rusanov solver. The method estimates the fastest present wave speed

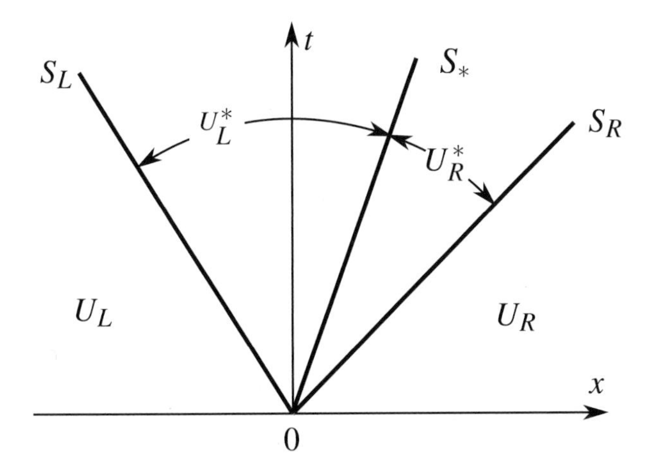

Fig. 2 Structure of the HLLC Riemann solution in the x-t plane for the Euler equations according to [15]

λ_{max} at the cell boundary and computes the flux according to [15]

$$f(u) = \frac{1}{2}\left[(f(u_l) + f(u_r)) + \lambda_{max}(u_r - u_l)\right], \tag{2}$$

denoting u_l as the left and u_r as the right sided state of the cell boundary. Since only the fastest wave speed is estimated, the Lax-Friedrich solver is an one-wave model which is the simplest type of an incomplete Riemann solver. The standard WENO implementation is done under the usage of the Lax-Friedrich Riemann solver, which is denoted as WENO LF scheme further on in this report.

5.2 *HLLC Riemann Solver*

The HLLC (Harten-Lax-van Leer-Contact) scheme is an approximate Riemann solver which estimates all occurring wave speeds of the Euler equation. In addition to the Lax-Friedrich scheme, the HLLC scheme approximates the speed of the slowest wave and the contact and shear wave, denoted as the *middle* wave, which divides the wave fan resulting in an additional state (see Fig. 2). The fastest and slowest wave speeds are estimated using the Roe averaged eigenvalues according to [15]. The shear wave speed is estimated using a pressure-velocity based wave speed estimation introduced by Toro [15]. Advantage of the HLLC solver is the estimation of all eigenvalues without any required knowledge about the flux-Jacobian matrix. Especially the preservation of vortices profits from the HLLC method owing to the estimation of shear waves and lower dissipation.

The WENO scheme in combination with the HLLC Riemann solver is denoted as WENO HLLC further on.

Table 1 Computation effort and additional expenditure for the different flux computation schemes examined on different systems

Scheme	i7 $\left[\frac{s}{iter}\right]$	i7	Hermit $\left[\frac{s}{iter}\right]$	Hermit
JST	0.2294		0.5384	
WENO LF	0.6198	270.2 %	1.3958	259.3 %
WENO HLLC	0.7752	337.9 %	2.0798	386.3 %

5.3 Performance Tests

The initial performance tests were executed on a Intel i7-2600 with up to 3.4GHz CPU (denoted by i7) and FLOWer compiled with the ifortran compiler. A computation of a three dimensional block was performed on a single process with a constant flow condition. The simulation was conducted for the JST, WENO LF, and WENO HLLC scheme as described in Sects. 4 and 5. To analyze possible benefits of the cray gfortran compiler for the WENO loop, the investigation has been conducted on the CRAY XE6 HERMIT (AMD Interlagos at 2.3GHz) as well and is listed in Table 1 together with the investigation on the i7 platform. The additional computational cost of the WENO schemes was calculated by means of the JST performance. The evaluation shows a markedly additional effort of a simulation using the WENO schemes by a factor of 3. Therefore it is required to consider the actual benefit of the scheme under the same required computational resources, which is conducted as part of the validation case in Sect. 6.

In addition, the different computational effort requires a discussion about the necessity of a WENO computation in all regions of the mesh. In case of large simulations it is inefficient to use the computationally intensive WENO scheme for regions without relevance. Therefore blocks in the concerning areas are marked for a second order JST computation. Hence this requires a precise reconsideration of the parallelization algorithm due to the different computational efforts of the schemes. In order to achieve an equal distribution of the workload for the processes estimated by the assigned block sizes, blocks marked for WENO computation are scaled by the additional expense factor found in Table 1. The HPC performance of the FLOWer code on the HLRS CRAY XE6 HERMIT has been described in [2]. The new implementation does not change this characteristics due to the consideration of the additional work load in case of a WENO computation in the parallelization routine. No further scaling studies have been carried out.

6 Validation Case

For the investigation of the three dimensional behavior, the propagation of a wing tip vortex is computed with the different schemes. The investigation with the different schemes is limited to the cartesian background mesh (cf. Fig. 3), while the computation in the wing mesh is performed with the standard JST scheme.

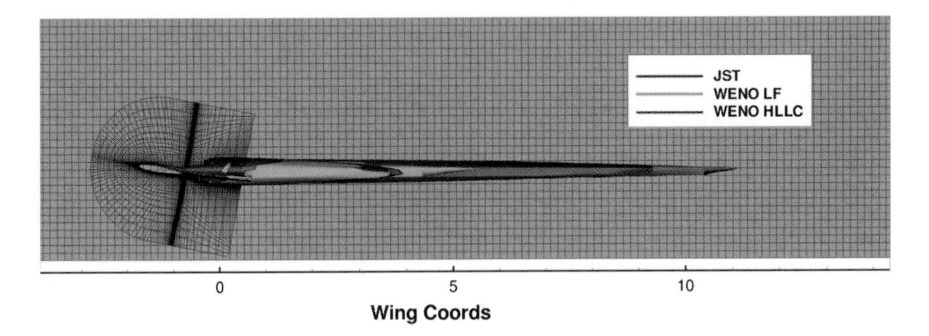

Fig. 3 λ_2 iso-surface plot of a wing tip simulation for different flux computation schemes in the cartesian background mesh

To examine the benefit to the physical accuracy of the higher order scheme at the same computational effort, the resolution of the meshes the WENO schemes are applied to are scaled according to the findings of Table 1. This is achieved by a coarsening of a three dimensional mesh by factor $\sqrt[3]{3.25} \approx 1.5$ (average of additional computational effort of WENO schemes) in each coordinate direction for a comparable computational expense. Adapted to this test case, the background mesh is coarsened by factor 1.5 in case of the WENO schemes. The visualization of the vortex is realized with the λ_2-criterion, which is a reliable indicator for vortices. Figure 3 shows the λ_2 iso-surface for the computation with the different schemes over the vortex age measured in multiples of the wing chord length. The origin of the axis is placed at the chimera boundary, where the solution is transferred from the wing mesh to the background mesh. The WENO LF scheme is found to be slightly better than the JST scheme despite the better mesh resolutions in case of the JST scheme. Significant improvement on the vortex conservation is achieved using the WENO scheme in combination with the HLLC Riemann solver where a nearly 170 % higher preservation of the λ_2 level is achieved compared to the JST scheme on the finer grid.

The same characteristic is found in the minimal pressure of the vortex core (cf. Fig. 4). The same initial pressure level is handed over by the chimera interpolation onto the background mesh. In the following the pressure increases until the scheme resolvable level is reached. At this point, the WENO LF and WENO HLLC scheme show already a lower pressure level which is preserved over the vortex age. All schemes show approximately the same growth gradient of the pressure level which is presumably characterized by the physical dissipation of the vortex. They just differ in a constant offset between the curves.

It has to be denoted that the slightly different computational effort between WENO and WENO HLLC has not been considered in the mesh resolution. Anyway it is not expected that this difference significantly displaces the results of this investigation.

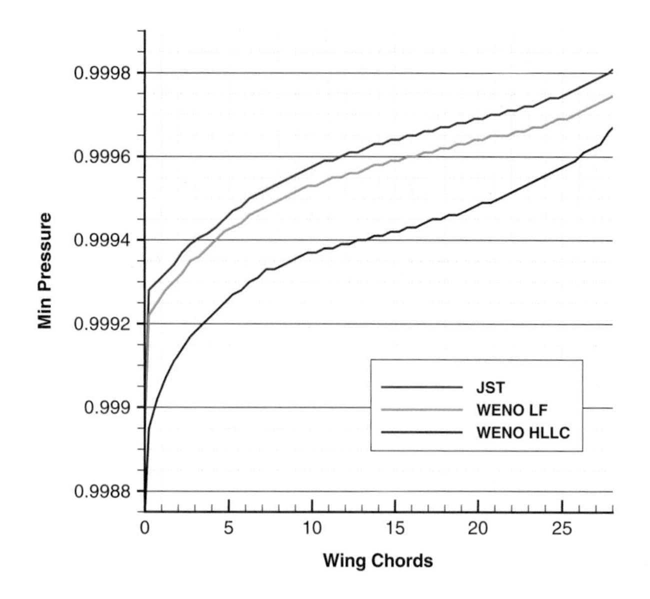

Fig. 4 Minimal pressure progression over the age of the wing vortex scaled by the distance to the wing in chord length

7 Helicopter Simulations

The following section describes the simulation and results of helicopter phenomena using the WENO LF and WENO HLLC scheme. In addition, JST simulations are performed to evaluate the efficiency and performance of the newly implemented schemes for large sized simulations.

7.1 Tail Shake

One up-to date and still unsolved issue of CFD simulations in the field of helicopters is the rotor-tail interaction, also called tail shake phenomenon [16]. This phenomenon occurs in fast forward flight, where the rotor wake interacts with the tail boom of the fuselage and excites a high lateral vibration. Since the cause of the phenomenon is not fully understood yet, multiple well known helicopter prototypes showed a high tail shake vibration level during the first test flights, e.g. the Boeing AH-64D Longbow Apache [6] and the Eurocopter EC135 [10]. Present solutions to tackle the tail shake problem are expensive wind tunnel test campaigns which are based on a geometry variation of the rotor head and turbine fairing. To replace these wind tunnel tests with CFD simulations, the high numerical dissipation is found to be a significant obstruction. Therefore a project supported by the Deutsche

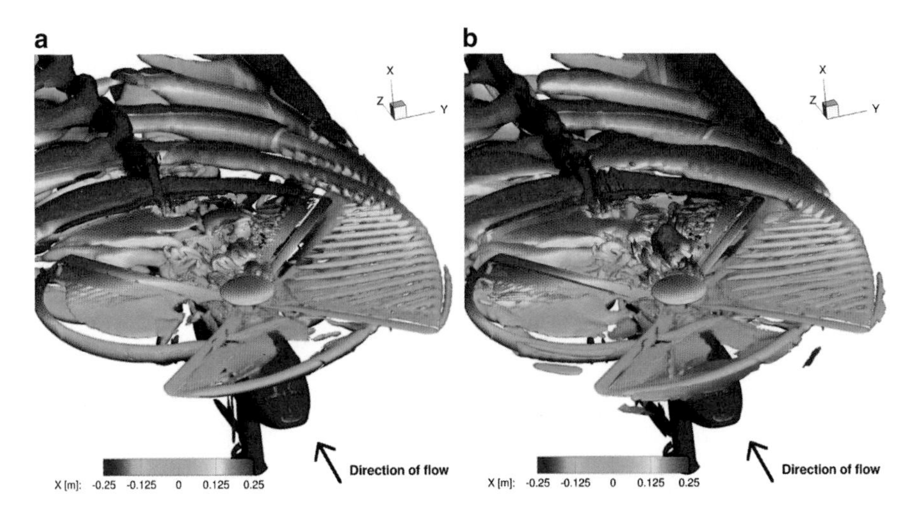

Fig. 5 Comparison of the λ_2 vortex topology for a snapshot of the flow field (14. rotor revolution, $\Psi = 45.0°$). (**a**) 5th order WENO LF scheme; (**b**) 2nd order JST scheme

Forschungs Gesellschaft (DFG) has been conducted at the IAG to investigate the advantage of the higher vortex preservation due to the WENO computation to the tail shake phenomenon. The simulation is limited to computations using the WENO LF and JST scheme.

Figure 5 shows the λ_2 iso-surface of the simulation based on the wind tunnel test campaign GoAHEAD [14], providing extensive measurement data e.g. forces, pressure sensors, and PIV. The higher vortex preservation due to WENO LF is considerable, as found in the λ_2 surface. Especially at the front area of the helicopter and around the helicopter rotor hub, the more discrete mapping of the vortices can be seen. A further advantage is found in the lower diffusion of the vortices especially in the front area of the rotor. The JST scheme shows in comparison to the WENO LF computation more vortex fragments and smearing.

This higher preservation of the rotor wake has also a positive influence on the pressure distribution on the tail boom by comparison with the experimental data as in case of a JST computation, which is reviewed in detail in [11].

8 Aeroacoustic Evaluation of an Isolated Helicopter Rotor

The aeroacoustic study of a helicopter main rotor requires high resolution grids. The wavelength of the highest frequency relevant for acoustic evaluation should be resolved with about twenty grid cells for good acoustic results. In this study, a grid with about 52 million cells is used to resolve frequencies up to 1200 Hz. To achieve good parallelization, the mesh setup consists of 496 blocks (Fig. 6). The

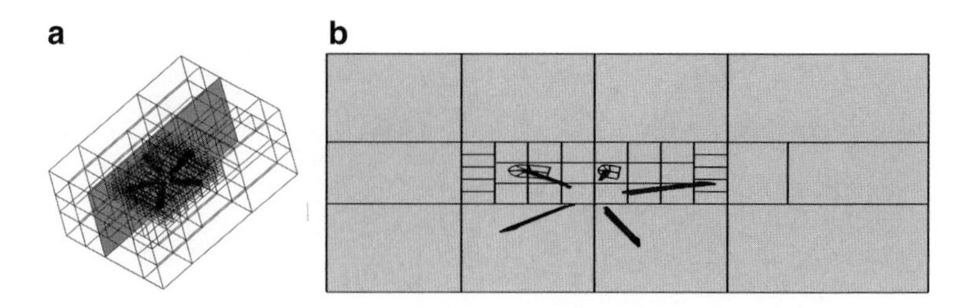

Fig. 6 Computational mesh setup. (**a**) Mesh overview; (**b**) Slice with rotor blades

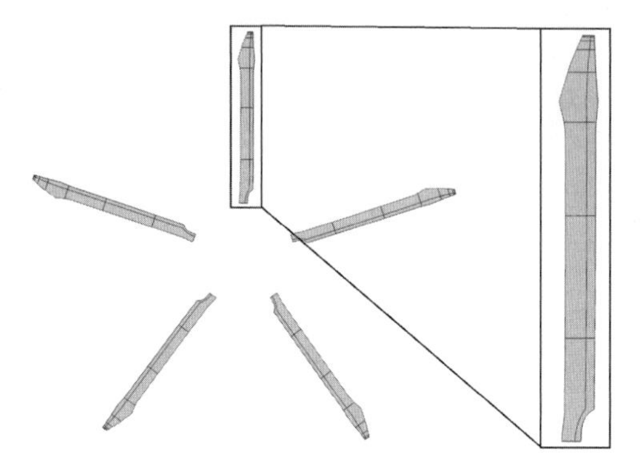

Fig. 7 Rotor with close-up view of rotor blade

setup consists of six independent grids: Five blade grids with about 6.5 million cells and a background grid with finer mesh resolution in proximity to the rotor.

In the context of this study a CFD simulation of an isolated five-bladed helicopter rotor (see Fig. 7) in a six degree descent flight with 65 knots was performed on the HLRS CRAY XE6 HERMIT. This flight case is relevant for Blade Vortex Interaction (BVI) Noise. Due to the lift created by the rotor blades, strong blade tip vortices are formed (Fig. 8). The down-wash of the rotor forces these vortices under the rotor plane, yet in descent flight, the vortices pass back through the rotor plane and thus interact with the blades. This causes fluctuations of rotor loads and generates loud and impulsive noise. A BVI situation is shown in Fig. 8. The vortex of each blade is marked with different colors and the areas of parallel interaction of the rotor blades with the blade tip vortices is indicated. The governing parameters of BVI noise are the strength, the size, and position of the interacting vortices. Thus for the simulation of the BVI noise, a good conservation of the blade tip vortices during the numerical procedure is very important. Therefore the newly implemented WENO schemes are expected to increase the quality of the solution. Their influence, compared to the JST scheme, on the acoustic results are evaluated in this investigation.

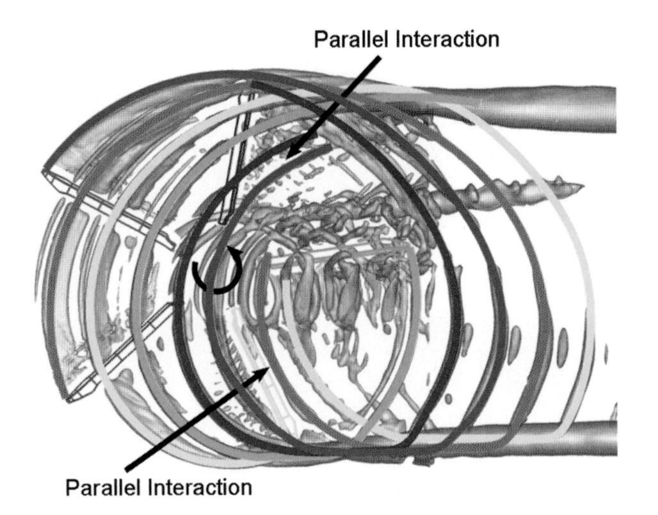

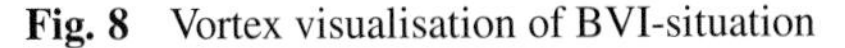
Fig. 8 Vortex visualisation of BVI-situation

8.1 *Computational Effort*

The computations using WENO LF and WENO HLLC were limited to the inner zone of the background mesh around the rotor. For the other mesh parts including the blade grids, the JST scheme is used to reduce the computational effort. Therefore a total of about 13 million cells, corresponding to 25 % of all cells are calculated with the WENO LF and WENO HLLC scheme. Because of a different blocking of the WENO HLLC scheme, only WENO LF and JST are compared. The number of cells for the WENO HLLC computation differs by just about 0.1 %. A similar cell size in proximity of the rotor is assured to reduce mesh influence on the results to a minimum. Using the same amount of 496 CPUs the computation for 144 time steps with 50 sub-iterations using the WENO LF scheme required 19.8 h or 9,832 CPU hours while the JST scheme required 18.6 h or 9,229 CPU hours. Therefore the WENO LF computation consumed about 6.5 % more CPU time. Thus for one rotor revolution resolved with 720 time steps about 3,015 CPU hours more are required for the WENO LF scheme.

8.2 *Vortex Conservation*

Figure 9 shows the vortices for the three numerical schemes, using the iso-surface of the λ_2-criterion. Between the WENO HLLC scheme and the JST scheme a significant difference is found. A higher vortex conservation is clearly observable and the vortices are generally more intense. The difference between the WENO LF scheme and the JST scheme is minor, compared to the differences to the WENO HLLC, as found in the validation case (cf. Sect. 6).

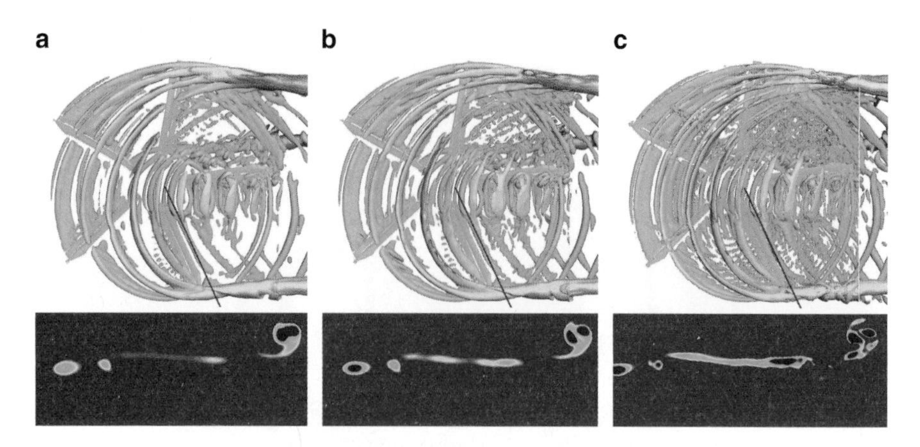

Fig. 9 Vortex visualization. (**a**) JST scheme; (**b**) WENO LF scheme; (**c**) WENO HLLC scheme

The figures underneath the iso-surfaces show slices of the flow field at the gray lines. The position is placed on the retreating blade side in front of a rotor blade. Due to the counterclockwise rotation, in one of the following time steps the vortex and the blade are interacting. The slices show that the vortices of the WENO HLLC scheme are stronger and more distinct than they are for the other two schemes. The WENO LF scheme shows less distinct vortices than WENO HLLC, but the vortex cores are still more concentrated than for the JST scheme. Additionally, the structure of the vortex is resolved more in detail in case of the WENO HLLC scheme, which is significant considering the blade tip vortex.

8.3 Load Distribution

As already stated, the BVI affects the loads of the rotor blades (cf. Fig. 10). According to the five-bladed rotor a 72°-periodic load distribution is found. In each period, small oscillations in the thrust coefficient caused by the interaction of the blade with the vortices are present. These areas are shown in detail in Fig. 10b. The amplitude of the oscillations are higher for the WENO HLLC scheme compared to the WENO LF and the JST scheme. This effect can directly be related to the better conservation of the vortices by the WENO HLLC scheme.

8.4 Acoustic Results

With the Ffowcs Williams Hawkings solver ACCO (Acoustic Coupling), a code developed at the Institute of Aerodynamics and Gasdynamics (IAG), the acoustic sources on defined integration surfaces around the rotor blades are determined and afterwards the sound propagation to distant observers is computed. This allows to determine a noise footprint of the rotor.

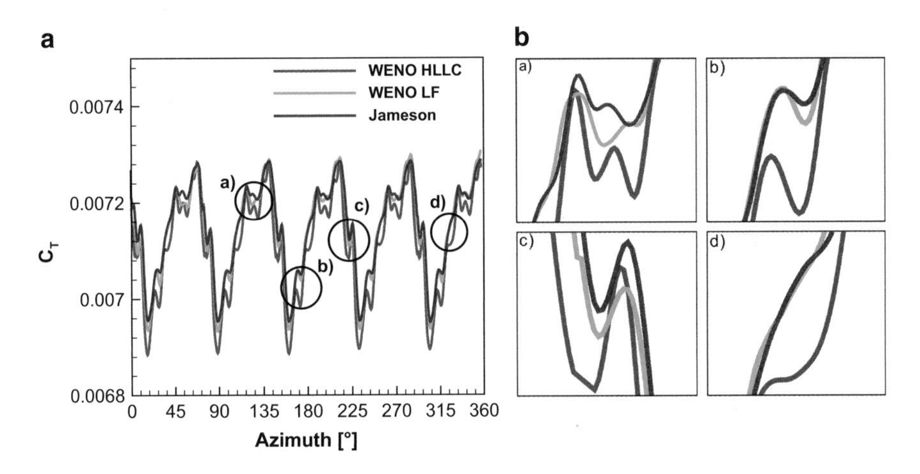

Fig. 10 Load Distribution. (**a**) Thrust coefficient over azimuth; (**b**) Close up of oscillations

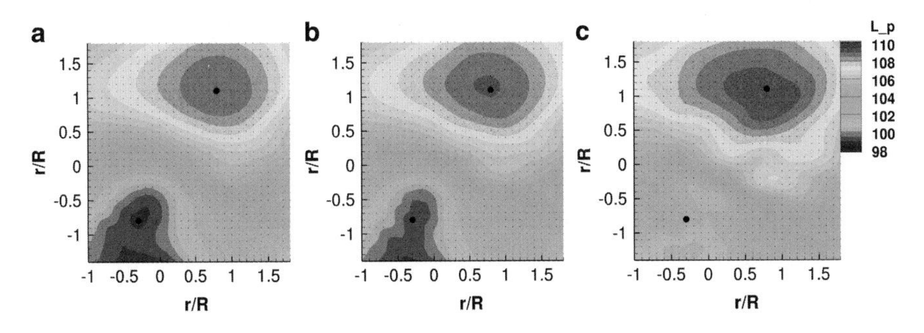

Fig. 11 Noise footprints below the rotor (sound pressure level L_p in dB). (**a**) JST scheme; (**b**) WENO LF scheme; (**c**) WENO HLLC scheme

For the three schemes the noise footprints ten meters underneath the rotor are shown in Fig. 11. The two axes are scaled according to the radius of the rotor. The characteristic noise emission is qualitatively the same for the three schemes. The noise emission is more intense for the WENO HLLC scheme than for the WENO LF and for the JST scheme. Especially on the retreating blade side, the difference between the WENO HLLC scheme and the other two schemes is obvious. At the two black squares the sound pressure levels on the advancing blade side are: JST 109.0 dB, WENO LF: 109.1 dB and WENO HLLC: 109.2 dB and on the retreating blade side: JST: 99.0 dB, WENO LF: 99.3 dB and WENO HLLC: 101.6 dB.

For the aircraft certification, the helicopter noise emission during a flyover is recorded by a row of microphones. Figure 12 shows this flyover schematic. The microphones are placed in a line perpendicular to the flight path of the helicopter with 75 m distance between the microphones. Afterwards the sound pressure level values recorded by the microphones are converted to the so called effective perceived noise level (EPNL). For each microphone one EPNL value is computed.

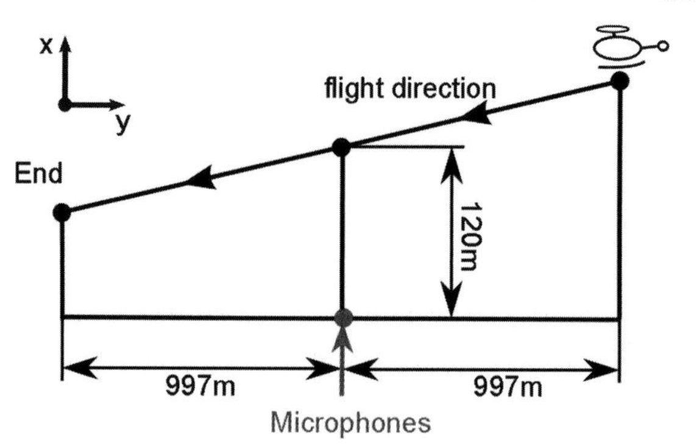

Fig. 12 Flyover schematic

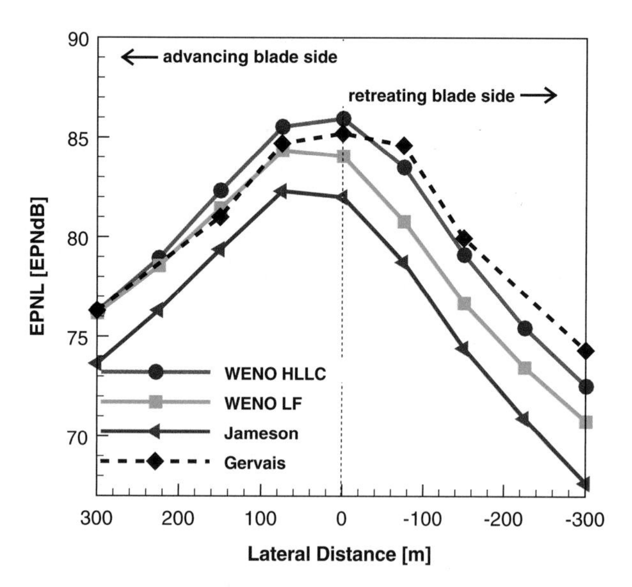

Fig. 13 EPNL diagram

The EPNL values for each microphone and for the three schemes are shown in Fig. 13. The WENO HLLC scheme results in about 3–5 EPNdB louder values during the flyover than the JST scheme. Especially on the retreating side, this difference is remarkable. The difference between the WENO scheme and the JST scheme is about 1.5–2.5 EPNdB. The dotted black line in this figure represents a similar acoustic computation of Gervais et al. from Eurocopter [4]. Their results are shifted to the value at a lateral distance of 300 m (advancing side) of the WENO HLLC scheme, as only relative values are indicated in their paper.

The EPNL value of the Eurocopter EC135 helicopter during the flyover is 84 EPN*dB*. This value is calculated as a logarithmic average of the three microphones at $\pm 75\,m$ and $0\,m$ lateral distance. The comparable value in this study accounts for 82.5 EPN*dB* (WENO HLLC scheme). This shows, that the WENO HLLC setup gives promising results and the additional computational effort can be justified by the benefits in the acoustic results for this flight case.

9 Conclusion

The present paper shows significant advantages using the higher order WENO reconstruction method newly implemented into the CFD code FLOWer. Especially in combination with the HLLC Riemann solver a higher vortex conservation is achieved. Even if a higher computational effort is required, a computation with as much as 70 % less grid cells compared to a 2nd order scheme still shows a more precise solution. The two presented applications to different helicopter flight situations show obvious benefits of the higher accuracy of the fifth order scheme. Remarkable results are achieved in the computation of the acoustic noise emission being the outcome of a more detailed vortex topology and a good preservation of vortex strength.

Acknowledgements We greatly acknowledge the provision of supercomputing time and technical support by the High Performance Computing Center Stuttgart (HLRS) for our project HELISIM.

References

1. D.D. Boyd Jr., HART-II Acoustic Predictions using a Coupled CFD/CSD Method. *Proceedings AHS Annual Forum*, Grapevine, Texas, 2009
2. R.E. Busch, M.S. Wurst, M. Keßler, E. Krämer, Computational aeroacoustics with higher order methods. In: W.E. Nagel, D.H. Kröner, M. Resch (eds) *High Performance Computing in Science and Engineering '12* (Springer, Berlin, Heidelberg, 2012), pp. 239–253
3. D. Flad, Implementation of a higher order method for flux calculation in FLOWer, Studienarbeit, Institute of Aerodynamics and Gas Dynamics, University of Stuttgart, Germany, 2011
4. M. Gervais, V. Gareton, Analysis of main rotor noise reduction due to novel planform design. *The Blue^{TM}, 37th European Rotorcraft Forum*, Gallarate, Italy, 2011
5. A. Harten, B. Engquist, S. Osher, S. Chakravarthy, Uniformly high order essentially non-oscillatory schemes. J. Comput. Phys. **71**, 231–303 (1987)
6. A.A. Hassan, T. Thompson, E.P.N. Duque, J. Melton, Resolution of tail buffet phenomenon for AH-64D^{TM} longbow apace. *Proceedings of the 53rd Annual Forum of the American Helicopter Society*, Virginia, USA, 1997
7. A. Jameson, W. Schmidt, E. Turkel, Numerical solution of the Euler equations by finite volume methods using Runge-Kutta time-stepping schemes. *14th AIAA Fluid and Plasma Dynamic Conference*, Palo Alto, California, USA, 1981
8. A. Jameson, Time dependent calculations using multigrid, with applications to unsteady flows past airfoils and wings. *Proceedings of the 10th AIAA Computational Fluid Dynamics Conference*, Honolulu, Hawaii, USA, 1991

9. G.-S. Jiang, C.-W.Shu, Efficient implementation of weighted ENO schemes. J. Comput. Phys. **126**, 202–228 (1996)
10. K. Kampa, B. Enenkl, G. Polz, G. Roth, Aeromechanic aspects in the design of the EC135. *Proceedings of the 23rd European Rotorcraft Forum*, Dresden, Germany, 1997
11. U. Kowarsch, F. Bensing, M. Keßler, E. Krämer, Untersuchungen zur Rotornachlauf-Rumpf-Interaktion mit einem hybriden Strömungslöser, Final report of the DFG-project KR 2959/1, Stuttgart, Germany, 2012
12. N. Kroll, B. Eisfeld, H.M. Bleeke, The Navier-Stokes code FLOWer. Notes Numer. Fluid Mech. **71**, 58–68 (1999)
13. X.-D. Liu, S. Osher, T. Chan, Weighted essentially non-oscillatory schemes. J. Comput. Phys. **115**, 200–212 (1994)
14. K. Pahlke, The GOAHEAD project. *Proceedings of the 33rd European Rotorcraft Forum*, Kazan, Russia, 2007
15. E.F. Toro, *Riemann Solvers and Numerical Methods for Fluid Dynamics* (Springer, Berlin, 1997)
16. P.G. de Waard, M. Trouvï£¡, Tail shake vibration, NLR-TP-99505, NLR, The Netherlands, 1999
17. R. Wang, H. Feng, R.S. Spiteri, Observations on the fifth-order WENO method with non-uniform meshes. Appl. Math. Comput. **196**, 433–447 (2008)

Unsteady CFD Simulation of the NASA Common Research Model in Low Speed Stall

Philipp P. Gansel, Sebastian A. Illi, Stephan Krimmer, Thorsten Lutz, and Ewald Krämer

Abstract In advance of cryogenic time-resolved PIV measurements in the European Transonic Windtunnel ETW unsteady CFD simulations of the NASA Common Research Model have been carried out. In the observed high Reynolds number and low Mach number flow regime a pressure induced boundary layer separation occurs on the main wing at high angles of attack. In steady RANS simulations the development of the separation over the wing surface was investigated and found to begin on the mid-board sections. For analysis of the unsteadiness of the wake flow and the impact on the htp an URANS calculation was conducted for a high angle of attack case. Calculated spectra of surface and wake pressures are presented. The presence of a spectral gap in the URANS simulations is discussed on the basis of the resolved fluctuations' spectra and a modelled turbulent energy spectrum derived from turbulence model entities. Finally an overview of computational resources spent and computational performance achieved within the SCBOPT project is given.

1 Introduction

At the edge of the flight envelope at low speeds large boundary layer separations occur at the main wing of transport-type aircraft. The flow separation not only leads to a loss of lift, but also typically shows unsteady behaviour and thus produces an unsteady flow in the wake. While the large scale turbulent wake can influence the empennage flow, its boundary layer state, the control surface efficiency or cause structural excitation through aerodynamic loads, the fluid mechanical processes of formation, development and transport of the turbulent wake are not well understood

P.P. Gansel (✉) · S.A. Illi · S. Krimmer · T. Lutz · E. Krämer
Institute of Aerodynamics and Gas Dynamics, University of Stuttgart, Pfaffenwaldring 21, D-70569 Stuttgart, Germany
e-mail: gansel@iag.uni-stuttgart.de

W.E. Nagel et al. (eds.), *High Performance Computing in Science and Engineering '13*, DOI 10.1007/978-3-319-02165-2_30,

and described yet. Still, they are a challenging test case for numerical as well as experimental investigations.

In order to investigate the effects present in propagation of the large scale turbulent wake structures and the influence on the tailplane aerodynamics a campaign of cryogenic wind tunnel tests is planned to be conducted in the European Transonic Windtunnel (ETW) in Cologne [12]. Low-speed stall as well as high-speed buffet conditions are to be measured, of which only the former case is object to the presented work due to the different physics in the transonic flow regime. Along with standard measurement of integral forces and surface pressures some more sophisticated techniques are applied as an optical dynamic deformation measurement system and time-resolved particle image velocimetry (TR-PIV) in the wake field.

The tests within the ESWI^{RP} project (European Strategic Wind Tunnels Improved Research Potential) are funded by the European Commission in the 7th framework program. The sub-proposal entitled "Time-Resolved Wake Measurements of Separated Wing Flow & Wall Interference Investigations" was submitted by a consortium of eight universities and research facilities from six European countries. The University of Stuttgart (Institute of Aerodynamics and Gas Dynamics) initiated the proposal and performs preparatory computational fluid dynamics (CFD) studies to define the test matrix [12].

The aim of the calculations is to determine adequate freestream conditions and parameters for the wind tunnel tests and derive the spatial and temporal resolution of the TR-PIV system required to measure the occurrent turbulent fluctuations in the wake. Also the positioning of the PIV measurement planes in the wake was discussed based on the numerical results. After the test campaign the comparison with the experimental data will provide a validation of numerical methods for a complex separated flow and geometry case. In the presented simulations the approach of the Unsteady Reynolds-Averaged Navier-Stokes equations (URANS) was applied.

Apart from stalling airfoils and tip vortex or engine jet interactions few investigations on turbulent wake of separated aircraft wing flow have been published. Havas and Jenaro-Rabadan showed the effect of main wing stall on the unsteady rolling moment of the empennage and presented a method to model the unsteady loads based on wind tunnel tests [9]. A similar empirical model for horizontal tailplane (htp) yaw and rolling moments was developed by Whitney et al. based on URANS results [23]. In the HINVA research program [17] in-flight stall tests are conducted with the DLR ATRA A320 aircraft. During the experiments the full-scale transport aircraft is also equipped with unsteady pressure transducers at the main wing and the htp to provide a validation basis for unsteady numerical simulations and a possibility to analyse the wake interference.

2 Geometry and Test Case

The NASA Common Research Model (CRM) is used as geometry in this study. It is a generic configuration representative for current commercial transport aircraft [21]. The model has a wide-body fuselage and a supercritical transonic wing. The

Table 1 CFD model geometric data [21]

Reference wing area	S	383.690 m^2	htp reference area	S_{htp}	92.903 m^2
Reference chord	c	7.005 m	htp reference chord	c_{htp}	4.691 m
Wing span	b	58.763 m	htp span	b_{htp}	21.336 m
Aspect ratio	Λ	9.0	htp aspect ratio	Λ_{htp}	4.90
Taper ratio	λ	0.275	htp taper ratio	λ_{htp}	0.35
Sweep angle	$\varphi_{c/4}$	35°	htp sweep angle	$\varphi_{c/4,\text{htp}}$	37°

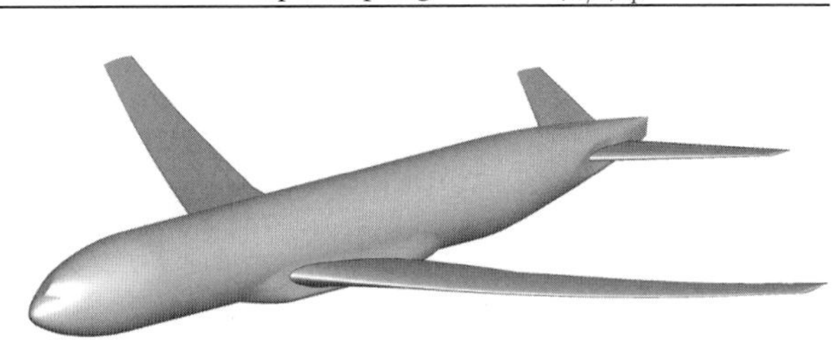

Fig. 1 NASA CRM geometry including wing, body, belly fairing and htp

design point is a cruise flight at a Mach number of $Ma = 0.85$, a Reynolds number of $Re = 40 \times 10^6$ and a lift coefficient of $C_L = 0.5$. The geometric parameters are listed in Table 1.

The CRM was designed as a test case for CFD code validation and comparison of various grid generators and flow solvers in the AIAA Drag Prediction Workshop (DPW) series [22]. Since its geometry is freely available (e.g on [1, 4]) it appears to be suitable for the ESWI^{RP} tests, where the obtained experimental data shall be published likewise as validation basis for numerical methods. Some transonic and partially cryogenic wind tunnel tests with the model have already been conducted and published [2, 15, 16], however none with low-speed condition is among them. A 80 % down-scaled CRM wind tunnel model has been built by JAXA for their low Reynolds number tests [8, 20].

The CRM geometry is available in different configurations. In the presented work a simple setting is used consisting of wing, body, belly fairing and htp at 0° incidence, see Fig. 1. A clean wing is utilized to avoid additional interference effects with the pylon and the engine nacelle. In the studies low-speed freestream conditions of $Ma = 0.25$ and $Re = 17.2 \times 10^6$ are set in accordance to the planned wind tunnel tests where the Reynolds number was chosen to fit in the ETW pressure and temperature envelopes.

3 Numerical Methods

A short overview of relevant fundamentals as well as applied numerics and settings is given in the following sections.

3.1 Governing Equations

To simulate the turbulent high-Reynolds number flow the compressible unsteady RANS equations are solved in the implemented integral form

$$\frac{\partial}{\partial t}\int_{\Omega}\mathbf{U}\mathrm{d}\Omega + \oint_{\partial\Omega}(\mathbf{F}_\mathrm{c} - \mathbf{F}_\mathrm{v})\,\mathrm{d}S = \mathbf{0}, \tag{1}$$

where the right-hand sided source term in this case is equal to zero (cf. [3]). They describe the equality of the change of conservative variables

$$\mathbf{U} = (\rho\ \rho u\ \rho v\ \rho w\ \rho E)^\top \tag{2}$$

in a control volume Ω and the integral convective and viscous fluxes $\mathbf{F}_\mathrm{c}$ and $\mathbf{F}_\mathrm{v}$ over its bounding surface $\partial\Omega$ with the normal vector $\mathbf{n} = (n_x\ n_y\ n_z)^\top$. The flux vectors are

$$\mathbf{F}_\mathrm{c} = \begin{pmatrix} \rho\,\mathbf{V}\cdot\mathbf{n} \\ \rho u\,\mathbf{V}\cdot\mathbf{n} + n_x p \\ \rho v\,\mathbf{V}\cdot\mathbf{n} + n_y p \\ \rho w\,\mathbf{V}\cdot\mathbf{n} + n_z p \\ \rho(E + p/\rho)\,\mathbf{V}\cdot\mathbf{n} \end{pmatrix} \tag{3}$$

and

$$\mathbf{F}_\mathrm{v} = \begin{pmatrix} 0 \\ n_x\sigma_{xx} + n_y\sigma_{xy} + n_z\sigma_{xz} \\ n_x\sigma_{yx} + n_y\sigma_{yy} + n_z\sigma_{yz} \\ n_x\sigma_{zx} + n_y\sigma_{zy} + n_z\sigma_{zz} \\ \mathbf{n}\cdot\sigma\times\mathbf{V} + \lambda\nabla\cdot T \end{pmatrix}. \tag{4}$$

Please note that the velocities $\mathbf{V} = (u\ v\ w)^\top$, density ρ, pressure p, temperature T and specific total energy E are considered being the Favre- or Reynolds-averaged (for p and ρ) entities. The tensor σ here denotes the sum of (laminar) viscous stresses and the turbulent Reynolds stresses. Utilizing the hypotheses of Stokes and Boussinesq it is set proportional to the rate-of-strain tensor yielding

$$\sigma_{ij} = (\mu_\mathrm{L} + \mu_\mathrm{T})\left(\frac{\partial V_i}{\partial x_j} + \frac{\partial V_j}{\partial x_i}\right) - \frac{2}{3}(\nabla\cdot\mathbf{V} + \rho k_\mathrm{T})\,\delta_{ij}. \tag{5}$$

The laminar dynamic viscosity is given by Sutherland's formula

$$\mu_\mathrm{L} = \mu_0\left(\frac{T}{T_0}\right)^{3/2}\frac{T_0 + C_\mathrm{S}}{T + C_\mathrm{S}} \tag{6}$$

with its constants $\mu_0 = 1.1716 \times 10^{-5}$ Pa s, $T_0 = 273$ K and $C_S = 110.4$ K, while the turbulent viscosity μ_T is determined by the turbulence model. Accordingly, the heat-flux vector in (4) is composed using laminar and turbulent thermal conductivity

$$\lambda = \lambda_L + \lambda_T = c_p \left(\frac{\mu_L}{Pr_L} + \frac{\mu_T}{Pr_T} \right) \tag{7}$$

with constant laminar and turbulent Prandtl numbers (assumption for gases) of 0.72 and 0.9, respectively.

For the modelling of the turbulent viscosity the one-equation model of Spalart and Allmaras [19] is used. The term containing the turbulent kinetic energy k_T in the Reynolds stresses in (5) is therein neglected.

The system is finally closed by the thermal (used to calculate the temperature $T = p/(\rho R)$) and caloric equations of state which for a perfect gas ad up to

$$p = \rho\,(\gamma - 1)\left(E - \mathbf{V}^2/2\right). \tag{8}$$

3.2 *Computational Grid*

Due to the absence of a side-slip angle the simulations are run on a half model grid with a symmetry boundary condition on the mid plane. In the case of stalling wing flow under subsonic conditions this is not a completely insignificant simplification. Since in reality—but also in CFD calculations on unstructured grids due to their more perturbing (numerical) character—boundary layer separation generally appears on one wing first and then influences the opposite wing. Still the benefit of half the number of grid points is reason enough to use this method, but one should bear in mind the possible effects.

The hybrid computational mesh with an overall cell count of 19.3 million is built of prisms and hexahedra in the boundary layer grid and of tetrahedra in the remaining flow domain. Pyramids are inserted to connect tetrahedra to quadrilateral cell faces. Unlike the usual way to estimate the required boundary layer mesh height and resolution not only the reference chord length (of the wing) was employed. Rather a couple of patches have been placed on the surface and local boundary layer mesh parameters have been determined for each (Table 2) based on an individual travelled length of the fluid. Figure 2 shows the transition between patches of different resolution solved with stripes of hexahedral cells. Constrained by the turbulence model the first wall normal cell size is adjusted to achieve non-dimensional y_1^+ values of 1 or beneath.

To provide a sufficient high-quality resolution of the flow in the wing-body and htp-body junctions structured hexahedral blocks are put in allowing the two different boundary layer meshes to intersect (see Fig. 3a). The effects of such grid design on the flow solution was discussed in [6]. Because of the sweep angles of leading and trailing edge and the strong curvature of the fuselage another structured O-type

Table 2 Boundary layer mesh parameters

Patch	Layers	First cell size	Growing factor	Total height
A	55	45.21 μm	1.142	467.61 mm
B	72	19.74 μm	1.171	906.78 mm
C	72	13.06 μm	1.171	1,612.39 mm
D	72	16.51 μm	1.171	2,162.56 mm
E	50	7.569 μm	1.171	470.41 mm
F	50	18.03 μm	1.171	278.69 mm
G	50	0.1334 μm	1.159	137.34 mm
H	50	7.074 μm	1.171	109.29 mm
I	50	13.61 μm	1.159	140.21 mm
J	50	5.842 μm	1.159	59.94 mm

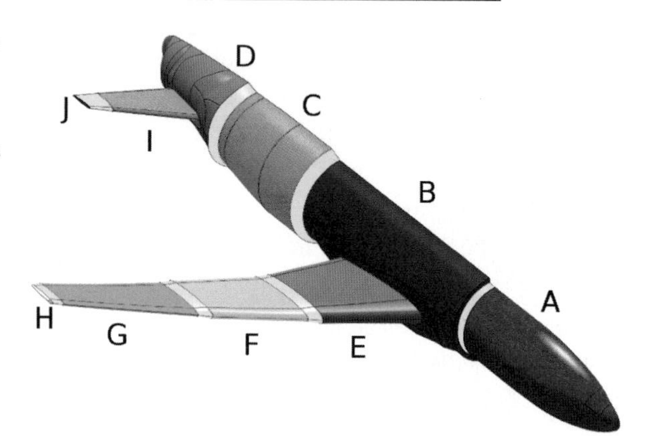

Fig. 2 Boundary layer mesh patches with different parameters (Table 2); the *yellow stripes* mark the structured intersection blocks

block, shown in Fig. 3b, is wrapped around the junction block to ensure the prisms to be extruded perpendicularly to the body wall and improve their quality. The surface grid generally consists of triangles except at the wing-body and htp-body junctions, the transition blocks with changing boundary layer grid entities, the afterbody, the leading and trailing edges and at the tips of wing and htp. In these regions hexahedra build up the boundary layer mesh (see examples in Fig. 3c and d).

The farfield boundary condition is placed on a hemisphere with a radius of 25 times the wing span around the geometry. Simulations with a special wake grid with higher resolution [7] is object to further investigations.

3.3 *Solver and Settings*

The flow has been calculated with the DLR TAU code [18], a second order accurate unstructured finite volume solver for CFD. Central JST discretization scheme was used with artificial matrix dissipation for stabilization. The convergence was accelerated by residual smoothing and a 5w geometrical multigrid cycle. For the unsteady simulation at stalled conditions the cycle was reduced to 2v for stability

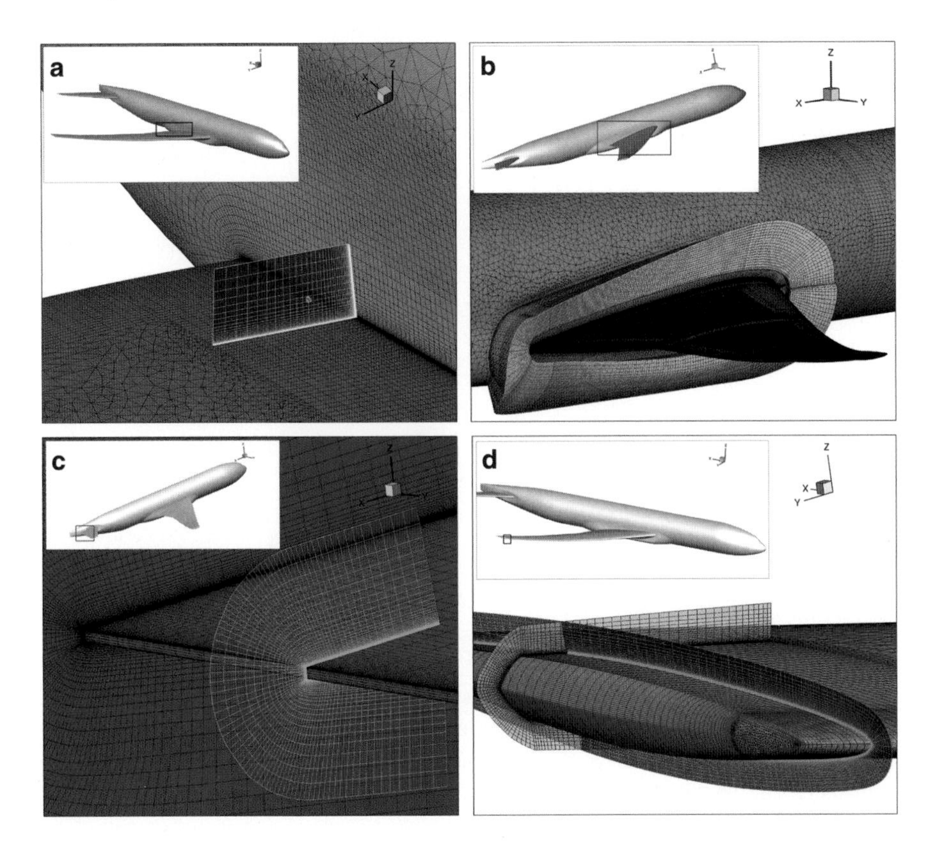

Fig. 3 Structured boundary layer mesh (**a**) in the wing-body junction (**b**) in the transition block to maintain wall-normal prism extrusion on the fuselage (**c**) at the htp trailing edge (**d**) at the wing tip

reasons. A physical time step size of $0.01 \times c/U_\infty = 0.9$ ms is prescribed in the dual time stepping scheme. After reaching statistical convergence of the mean integral force coefficients the last 4,000 time steps of the simulation were evaluated.

All simulations were conducted fully turbulent with the eddy-viscosity model of Spalart and Allmaras [19]. Further investigations with more sophisticated turbulence modelling as Reynolds stress models and utilizing Detached Eddy Simulations (DES) are planned for higher accuracy and because of the known turbulence model dependency of the corner flow and separation in the wing-body junction [11].

4 Results

The main focus of the present studies is the unsteady behaviour of the flow in the wake and its spectral content. Nevertheless, some basic results of the steady alpha sweep simulations are shown first to give an overview of the stall mechanism of this particular wing.

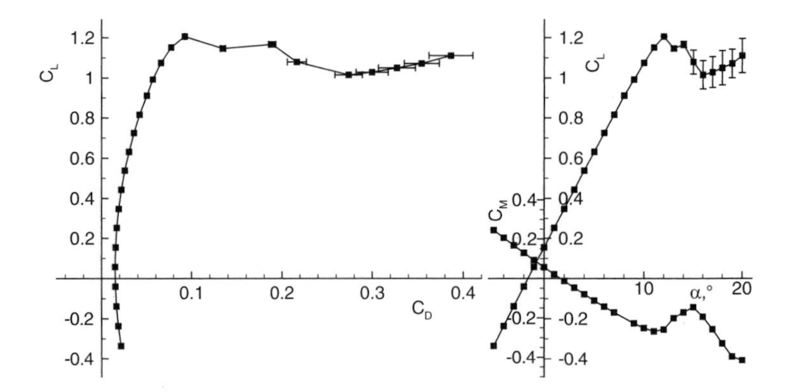

Fig. 4 Lift, drag and pitching moment polars of the CRM calculated using RANS ($Ma = 0.25$, $Re = 17.2 \times 10^6$)

4.1 RANS Polar Simulation

The steady RANS calculations were conducted for angles of attack between $-5°$ and $20°$. Up to $12°$ the simulations show an excellent convergence with a density residual norm of 10^{-7}. At this incidence the maximum lift coefficient of $C_L = 1.206$ is reached (see Fig. 4). A quite high maximum lift-to-drag ratio of 20.9 is achieved at $\alpha = 4°$. At angles of attack above $12°$ the lift collapses while the drag increases extremely due to the stalled wing. In this α range the steady simulations do not converge any more but end up with a constant oscillation. The mean values of this oscillation are plotted in Fig. 4 with the standard deviation as error bars. Note that this oscillation is not identical to the physical oscillation of the flow at these conditions, because the steady equations are used. The reference point of the displayed pitching moment is located at $x = 33.678$ m, $y = 11.906$ m, $z = 4.520$ m.

Under the investigated freestream conditions which differ considerably from the CRM design point, the wing shows an undesired stall behaviour. The boundary layer separation first occurs at the mid wing sections which is indicated by the loss of lift in Fig. 5. At $14°$ the lift is still increasing on the inboard wing up to 30 % of the half span. Also near the wing tip there is a big area with attached flow. A flow separation beginning at the outboard wing causes higher rolling moments (if unsymmetrical), induces higher unsteady wing bending loads, can affect the aileron functionality and therefore is assessed poor.

Since the present study is aimed at the interference of the htp flow with the unsteady wake behind a stalling wing an angle of attack was chosen for the unsteady simulations that maintains boundary layer separation at the inboard wing. At $19°$ it is the case in a wide range of the htp span which covers the innermost 36 % of the wing span.

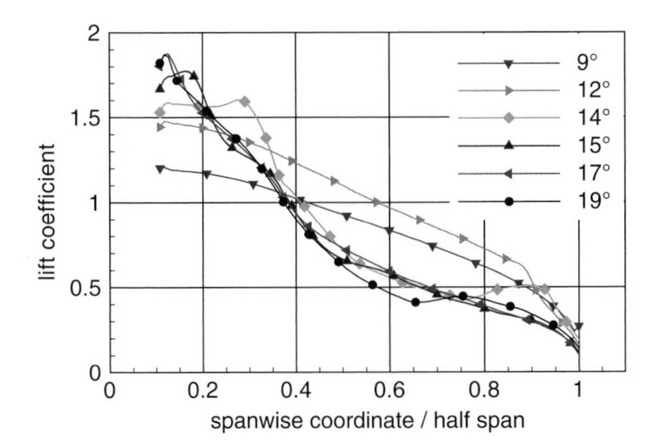

Fig. 5 Spanwise lift distribution on the main wing for selected angles of attack ($Ma = 0.25$, $Re = 17.2 \times 10^6$)

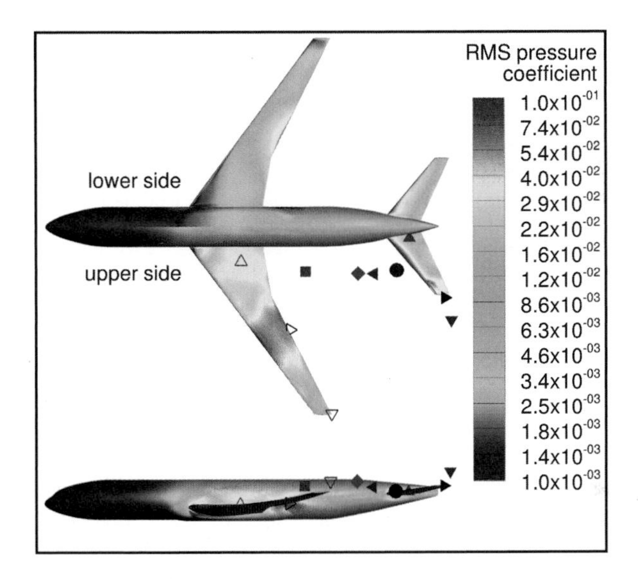

Fig. 6 Pressure fluctuations on the surface and locations of analysed points in the wake ($Ma = 0.25$, $Re = 17.2 \times 10^6$, $\alpha = 19°$)

4.2 URANS Results and Spectral Wake Analysis

Due to the big amount of grid points it is not possible to store the whole flow field in every time step for spectral analysis. This is only done for the geometry surfaces and distinct points in the surrounding mesh. Figure 6 shows the location of the points selected for closer study. Furthermore the rms value of the pressure coefficient c_p is plotted on the CRM surface. Although the boundary layer is separated on the whole

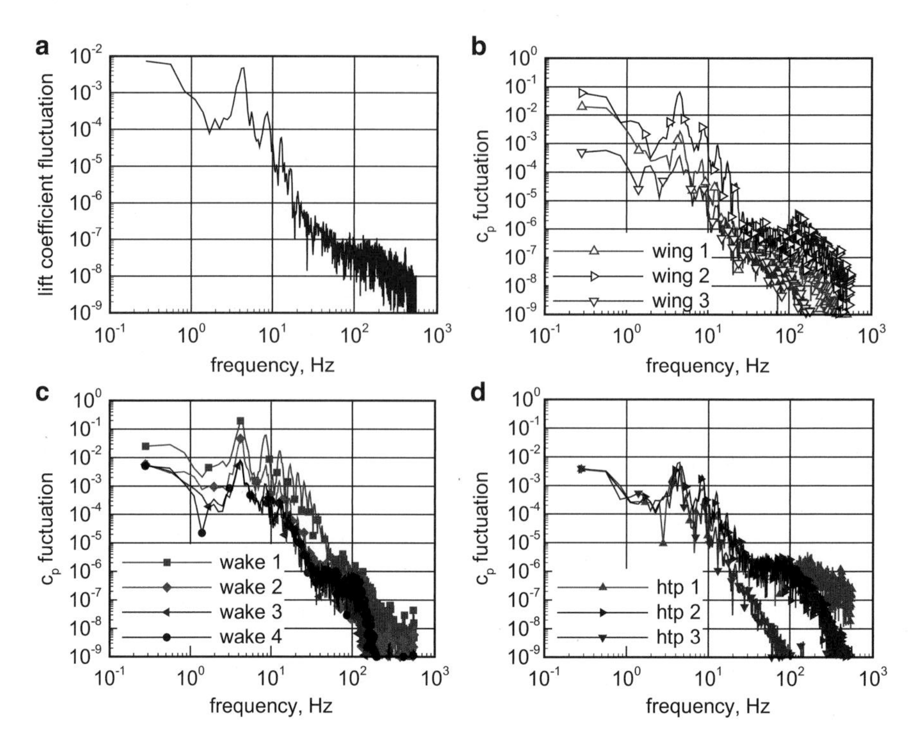

Fig. 7 Turbulent spectra of (**a**) lift coefficient (**b**) pressure coefficient at the wing (**c**) pressure coefficient in the wake (**d**) pressure coefficient at the htp ($Ma = 0.25$, $Re = 17.2 \times 10^6$, $\alpha = 19°$)

wing, a spot in the middle of the wing provides significantly high fluctuations. The flow unsteadiness of the wing stall is transported downstream and also reaches the htp. Here the fluctuations are highest on the upper side close to the leading edge due to the sensitivity of the suction peak to small variations in the angle of attack.

The spectral analysis of the simulation is summarized in Fig. 7. The lift coefficient spectrum is dominated by a peak of 0.005 at a frequency around 4.3 Hz (see Fig. 7a). This frequency and its higher harmonics arise from large coherent eddies in the separated flow above the wing and can be found in the whole wake and htp flow. Figure 7b depicts the pressure spectra close to the wing surface at three spanwise positions. According to the rms level in Fig. 6 the strongest fluctuations occur in the middle of the wing. The lowest amplitudes are found at the wing tip, where also the spectrum in the large-scale range differs from the other positions. In the wake Fig. 7c documents the decay of amplitudes downstream for all frequencies. Only at the last two positions the amplitude of the peak frequency is almost the same. The higher harmonics seem to underlie a stronger damping than the fundamental frequency from position wake 1 to 4. This indicates a smoothing of the fluctuations in the wake at higher frequencies. In the vicinity of the htp the pressure spectra are very homogeneous along the spanwise coordinate, even in a certain distance outside of the tip (position htp 3 in Fig. 7d). There is no significant shift in the peak frequency along the wake. The amplitude at this frequency has decreased about one order of magnitude compared to the wing positions. However, the investigated spectra are recorded at medium chordwise positions of the htp and near the suction peak at the htp leading edge the amplitudes are considerably higher.

4.3 On the Validity of the URANS Approach

Because the resolved flow unsteadiness in URANS cannot cover the whole range of turbulent motions, the small scale turbulence has to be modelled. It is not evident a priori how the URANS approach distinguishes between which turbulent fluctuations to resolve in the Reynolds-averaged flow variables and which the turbulence model shall cover or where to draw the line between these two ranges. To depict the turbulence-modelled part of fluctuations a model spectrum of turbulent energy is utilized [14]. The energy spectrum

$$
\begin{aligned}
E(\kappa) &= C\varepsilon^{2/3}\kappa^{-5/3} f_L(\kappa L) f_\eta(\kappa\eta)\,, \\
f_L(\kappa L) &= \left(\frac{\kappa L}{\left[(\kappa L)^2 + c_L\right]^{1/2}}\right)^{5/3+p_0}\,, \\
f_\eta(\kappa\eta) &= \exp\left(-\beta\left\{\left[(\kappa\eta)^4 + c_\eta^4\right]^{1/4} - c_\eta\right\}\right)
\end{aligned}
\tag{9}
$$

is given as function of the wave number κ. The flow-dependent turbulent dissipation rate ε, turbulence length scale L and Kolmogorov length scale η are constant for each particular time step and position. Further model constants are $C = 1.5, c_L = 6.78, p_0 = 2, \beta = 5.2$, and $c_\eta = 0.40$. The non-dimensional functions f_L and f_η are introduced to model the shape of the spectrum in the energy-containing large scale range and the dissipation range, respectively. The length scales $L = k_{\mathrm{T}}^{3/2}/\varepsilon$ and $\eta = (\nu_{\mathrm{L}}^3/\varepsilon)^{1/4}$ are solely defined by the turbulence model quantities ε and k_{T} and the laminar kinematic viscosity $\nu_{\mathrm{L}} = \mu_{\mathrm{L}}/\rho$. When using a two-equation turbulence model, especially a k-ε model, these quantities can be extracted directly from the flow field solution. In our case, the Spalart-Allmaras model delivers only one turbulence quantity which is the turbulent viscosity ν_{T}. The required values of ε and k_{T} can be estimated using the norm of the rate-of-strain tensor $S_{ij} = \left(\partial V_i/\partial x_j + \partial V_j/\partial x_i\right)/2 - (\nabla \cdot \mathbf{V})\,\delta_{ij}/3$ as proposed in [5, 13]. The kinetic energy is

$$
k_{\mathrm{T}} = \frac{\nu_{\mathrm{T}}\sqrt{2S_{ij}S_{ij}}}{\sqrt{C_\mu}}
\tag{10}
$$

with the constant $C_\mu = 0.09$. The dissipation rate can now be obtained from

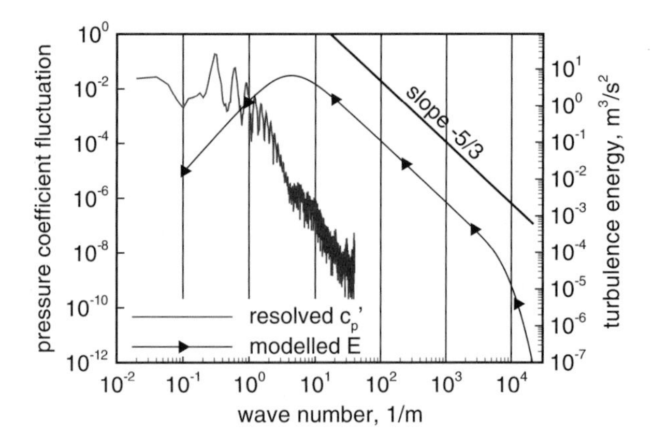

Fig. 8 Spectral content of resolved pressure fluctuation and energy spectrum function based on turbulence model entities in the wake flow (position wake 1 marked with a *red square* in Fig. 6)

$$\varepsilon = \frac{C_\mu k_\mathrm{T}^2}{\nu_\mathrm{T}} \,. \tag{11}$$

In Fig. 8 the turbulent energy spectrum at the first evaluation point in the wake is plotted as well as the c_p fluctuation spectrum (note the different ordinate scaling). For this the pressure spectrum was transformed from frequency into wave number space using Taylor's hypothesis of frozen turbulence which is applicable at the considered high freestream and local velocity. An additional line illustrates the slope of $-5/3$ to identify the inertial sub-range of the spectra. The maximum of E is found at a wave number one order of magnitude higher than the one corresponding to the peak in resolved fluctuations and in a region where they are already decaying. In the frequency range of high resolved pressure fluctuations the modelled spectrum, thus the turbulence model contributes only a small part of turbulence. The spectrum of resolved fluctuations in turn has no additional peak in the vicinity of the maximum of E. This indicates that the URANS approach is feasible here and a spectral gap between the resolved large scale flow motions and the modelled turbulence is maintained.

5 Computational Resources

A scaling test for a comparable mesh with 42 million points was performed to distinguish the amount of domains used for the present simulations. For the solution process each domain is given to one computational node to run the simulation on. Therefore the amount of cells in each domain and the computational effort per node is reduced by splitting the mesh into an increasing number of domains. However, the scaling of the DLR TAU code is limited, because the amount of domain

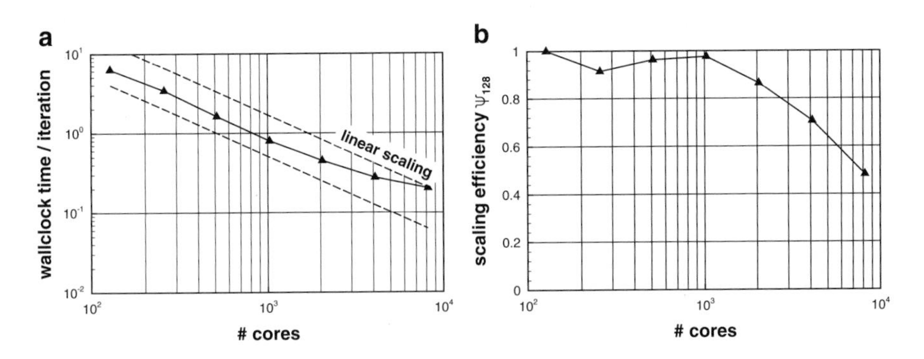

Fig. 9 TAU scaling on CRAY XE 6 Hermit, 42 million points, LUSGS, central differences (**a**) time per iteration (**b**) scaling efficiency

surfaces and hence the amount of communication between the nodes increases while decomposing into more domains. Also the amount of ghost cells which are needed to calculate the gradients at the boundaries is rising and reducing the parallel performance.

For optimal performance the graph partitioner Chaco [10] was linked to the DLR TAU code. To perform scaling tests the mesh was split into a large number of domains ranging from 128 up to 8,192. The lower limit of 128 domains was set by the limited memory per node, the upper limit of 8,192 by the drop in scaling performance at this point.

In Fig. 9a it is shown, that the code performs close to linear scaling in a range of 128 to 2,048 cores. To highlight this result the scaling efficiency

$$\Psi_{128} = \frac{\text{CPU time with 128 cores}}{\text{CPU time with } n \text{ cores}} \tag{12}$$

was defined. In Fig. 9b it can be seen that up to 1,024 cores the scaling efficiency Ψ_{128} is above 90 %. By further increasing the amount of cores the scaling efficiency is still acceptable with 82 % using 2,048 cores and around 70 % for 4,096 cores.

6 Conclusions

Steady and unsteady RANS simulations were conducted in high Reynolds number, low Mach number flow regime to investigate the low-speed stall behaviour of the CRM. The focus is on the development of the unsteady wake flow behind the boundary layer separation on the wing and its influence on the htp. The simulations precede time-resolved PIV measurements of the wake flow in the ETW.

A high quality grid was constructed for the simulations with the DLR TAU code. It includes a boundary layer mesh adapted to the local parameters in several patches with different wall normal extend and resolution. This ensures the correct prediction of the boundary layer on all surfaces, especially the fuselage tail with very high boundary layer thicknesses. Special attention was also paid to the wing-body and htp-body junctions where structured hexahedral blocks maintain good resolution of both boundary layers.

In steady polar simulations a maximum lift coefficient of 1.206 was found at 12°. For higher angles of attack the boundary layer separation initially starts at mid-board sections and then spreads over the whole wing. For high incidences the calculations reveal a very unsteady flow character.

At the considered angle of attack of 19° in the unsteady simulation the spectra of the lift coefficient as well as the pressure coefficients at the wing, in the wake and near the htp show a peak frequency of 4.3 Hz. The highest fluctuations are observed in the middle of the wing. The amplitudes decay as the wake develops downstream while the decrease is stronger for high frequencies. At the htp no significant variation is found in the spectra along the spanwise direction. There is no shift in the peak frequency during the propagation of the turbulent wake.

The validity of the URANS approach for this kind of application with an unsteady and turbulent wake flow was shown by comparison of the resolved fluctuation spectrum with a turbulent energy spectrum model based on local turbulence model data. The peaks of the two spectra have a satisfactory distance. Since there is no second peak in the resolved spectrum at frequencies with the highest modelled turbulent energy, a spectral gap is maintained and the URANS method is applicable in this case.

A scaling test clarified the ability of the DLR TAU code to perform well on a large number of parallel domains. Using a graph partitioner the calculations achieved good results close to linear scaling up to 1,024 cores. Running on 4,096 CPUs they still score a scaling efficiency of 70 %.

The wake flow is object to further investigations where the influence of mesh refinement and other eddy-viscosity turbulence models as well as hybrid RANS-LES (large eddy simulation) methods will be studied. Regarding the movability of the model in the planned wind tunnel tests and the optical access for the PIV measurement system preliminary CFD simulations with lower angles of attack in the stall region of the lift polar should be considered.

References

1. 5th AIAA CFD Drag Prediction Workshop, http://aaac.larc.nasa.gov/tsab/cfdlarc/aiaa-dpw. Accessed 21 Mar 2013
2. S. Balakrishna, M.J. Acheson, Analysis of the NASA Common Research Model dynamic data, in *49th AIAA Aerospace Sciences Meeting including the New Horizons Forum and Aerospace Exposition, 2011–1127*, AIAA, Orlando, Florida, 2011
3. J. Blazek, *Computational Fluid Dynamics: Principles and Applications*, 2nd edn. (Elsevier, Amsterdam, 2005)
4. Common Reserach Model, http://commonresearchmodel.larc.nasa.gov. Accessed 21 Mar 2013

5. E. Fares, W. Schröder, A general one-equation turbulence model for free shear and wall-bounded flows. Flow Turbul Combust **73**(3–4), 187–215 (2005)
6. P.P. Gansel, P. Dürr, M. Baumann, T. Lutz, E. Krämer, Influence of meshing on flow simulation in the wing-body junction of transport aircraft, in *18. DGLR-Symposium of STAB*, Stuttgart, Germany, 2012
7. P.P. Gansel, S.A. Illi, T. Lutz, E. Krämer, Numerical simulation of low speed stall and analysis of turbulent wake spectra. in *15th International Conference on Modelling Fluid Flow*, vol. 1, ed. by J. Vad, pp. 199–206, Budapest, Hungary, 2012
8. A. Hashimoto, T. Aoyama, M. Kohzai, K. Yamamoto, Transonic wind tunnel simulation with porous wall and support devices, in *27th AIAA Aerodynamic Measurement Technology and Ground Testing Conference*, Chicago, Illinois, 2010
9. J. Havas, G. Jenaro Rabadan, Prediction of horizontal tail plane buffeting loads (2009). IFASD 2009-128
10. B. Hendrickson, R. Leland, The chaco users guide, version 2.0. Tech. Rep. SAND94-2692, Sandia National Laboratories, Albuquerque, 1994
11. A. Klein, S.A. Illi, K. Nübler, T. Lutz, E. Krämer, Wall effects and corner separations for subsonic and transonic flow regimes, in *High Performance Computing in Science and Engineering '11*, ed. by W.E. Nagel, D.B. Kröner, M.M. Resch, pp. 393–407 (Springer, Berlin, Heidelberg, 2012)
12. T. Lutz, P.P. Gansel, J.L. Godard, A. Grobushin, R. Konrath, J. Quest, S.M.B. Rivers, Going for experimental and numerical unsteady wake analyses combined with wall interference assessment by using the NASA CRM model in ETW, in *51th AIAA Aerospace Sciences Meeting, New Horizons Forum and Aerospace Exposition*, Grapevine, Texas, 2013
13. Y. Nagano, C.Q. Pei, H. Hattori, A new low-Reynolds-number one-equation model of turbulence. Flow Turbul. Combust. **63**(1–4), 135–151 (2000)
14. S.B. Pope, *Turbulent Flows*. (Cambridge University Press, Cambridge, 2011)
15. M.B. Rivers, C.A. Hunter, R.L. Campbell, Further investigation of the support system effects and wing twist on the NASA Common Research Model, in *30th AIAA Applied Aerodynamics Conference, 2012–3209*, AIAA, New Orleans, Louisiana, 2012
16. S.M.B. Rivers, A. Dittberner, Experimental investigations of the NASA Common Research Model in the NASA Langley National Transonic Facility and NASA Ames 11-ft transonic wind tunnel, in *49th AIAA Aerospace Sciences Meeting including the New Horizons Forum and Aerospace Exposition, 2011–1126*, AIAA, Orlando, Florida, 2011
17. R. Rudnik, D. Reckzeh, J. Quest, HINVA—High lift INflight VAlidation—project overview and status, in *50th AIAA Aerospace Sciences Meeting including the New Horizons Forum and Aerospace Exposition, 2012–0106*, AIAA, Nashville, Tennessee, 2012
18. D. Schwamborn, T. Gerhold, R. Heinrich, The DLR TAU-code: Recent applications in research and industry, in *ECCOMAS CFD 2006 Conference*, Egmond aan Zee, The Netherlands, 2006
19. P.R. Spalart, S.R. Allmaras, A one-equation turbulence model for aerodynamic flows (1992). AIAA 92-0439
20. M. Ueno, T. Kohzai, S. Koga, H. Kato, K. Nakakita, N. Sudani, 80 % scaled NASA Common Research Model wind tunnel test of JAXA at relatively low Reynolds number, in *51st AIAA Aerospace Sciences Meeting including the New Horizons Forum and Aerospace Exposition, 2013–0493*, AIAA, Grapevine, Texas, 2013
21. J.C. Vassberg, M.A. DeHaan, S.M.B. Rivers, R.A. Wahls, Development of a Common Research Model for applied CFD validation studies (2008). AIAA 2008-6919
22. J.C. Vassberg, E.N. Tinico, M. Mani, B. Rider, T. Zickuhr, D.W. Levy, O.P. Brodersen, B. Eisfeld, S. Crippa, R.A. Wahls, J.H. Morrison, D.J. Mavriplis, M. Murayama, Summary of the fourth AIAA CFD drag prediction workshop (2010). AIAA 2010-4547
23. M.J. Whitney, T.J. Seitz, E.L. Blades, Low-speed-stall tail buffet loads estimation using unsteady CFD (2009). IFASD 2009-127

Flow Simulation of a Francis Turbine Using the SAS Turbulence Model

Timo Krappel, Albert Ruprecht, and Stefan Riedelbauch

Abstract The operation of Francis turbines in part load conditions causes high fluctuations and dynamic loads in the turbine and the efficiency of the draft tube decreases strongly. For a better understanding of the various instantaneous phenomena, flow simulations of the complete Francis turbine are done with particular focus on the draft tube flow.

The SAS (Scale Adaptive Simulation) model is used as turbulence model. Depending on the grid resolution and the time step size, this model allows the resolution of smaller turbulent structures compared to RANS turbulence models.

Flow phenomena in a Francis turbine require long physical time to convect through the machine. Therefore, long computational time with many processors is necessary to conduct such a flow simulation.

1 Introduction

Due to the liberalisation of the power market and due to an increasing amount of renewable energy, power plants have to operate in a wider range. The operation of Francis turbines in part load conditions causes high fluctuations and dynamic loads in the turbine and the efficiency of the draft tube decreases strongly. The dynamic load can even lead to structural damage.

Flow simulations of the complete Francis turbine are done. This allows for analysing and better understanding of the various instantaneous flow phenomena. The draft tube flow is very complex, because it is decelerated and rotating. These flow attributes are very difficult being captured by numerical flow simulations, respectively by the chosen turbulence model. The SAS (Scale Adaptive Simulation)

T. Krappel (✉) · A. Ruprecht · S. Riedelbauch
Institute of Fluid Mechanics and Hydraulic Machinery, Pfaffenwaldring 10, 70550 Stuttgart, Germany
e-mail: timo.krappel@ihs.uni-stuttgart.de

W.E. Nagel et al. (eds.), *High Performance Computing in Science and Engineering '13*, DOI 10.1007/978-3-319-02165-2_31,

model is chosen in this study, because similar studies of flow simulations of hydraulic turbines [5, 11] show, that RANS turbulence models do not have the capability to predict the complex flow as correctly as a SAS model.

2 Numerical Method

2.1 *Flow Solver*

For the flow simulation of the Francis turbine two different CFD codes are used: the commercial Ansys CFX version 14.5 and the open source OpenFOAM® version 1.6-ext. Both CFD codes are able to handle the rotation of the turbine and to couple different meshes by an interface, e.g. [1]. They are also using finite-volume method for discretisation.

2.2 *Governing Equations*

Laminar and turbulent flow can be described using Navier-Stokes equations. For incompressible and isothermal flow, they read:

$$\rho \frac{\partial U_i}{\partial t} + \rho U_j \frac{\partial U_i}{\partial x_j} + \frac{\partial P}{\partial x_i} - \frac{\partial \tau_{ij}}{\partial x_i} = 0\,. \tag{1}$$

For turbulence closure the Boussinesq approximation is used with the eddy viscosity formulation of the SST model [8].

2.3 *SAS-SST Turbulence Model*

The SAS approach [3, 9] enables the unsteady SST RANS turbulence model [8] to operate in SRS (Scale Resolving Simulation) mode [10]. This is achieved by introducing a new quantity, namely Q_{SAS}, into the ω-equation of the SST model [2], as described in (2).

$$\begin{aligned} &\frac{\partial \rho\omega}{\partial t} + \nabla\cdot(\rho U \omega) = \alpha \frac{\omega}{k} P_k - \rho\beta\omega^2 + Q_{SAS} \\ &+\nabla\cdot\left[\left(\mu + \frac{\mu_t}{\sigma_\omega}\right)\nabla\omega\right] + (1 - F_1)\frac{2\rho}{\sigma_{\omega 2}\omega}\nabla k \nabla\omega \end{aligned} \tag{2}$$

Q_{SAS} is defined as:

$$Q_{SAS} = \max\left[\rho\zeta_2\kappa S^2\left(\frac{L}{L_{vK}}\right)^2 - C\frac{2\rho k}{\sigma_\Phi}\max\left(\frac{|\nabla\omega|^2}{\omega^2}, \frac{|\nabla k|^2}{k^2}\right), 0\right], \quad (3)$$

containing the turbulent length scale L and the von Karman length scale L_{vK}. Based on the theory of Rotta [12], L_{vK} describes the second derivative of the velocity field:

$$L_{vK} = \kappa\left|\frac{\overline{U}'}{\overline{U}''}\right|$$

$$\overline{U}'' = \sqrt{\frac{\partial^2\overline{U}_i}{\partial x_k^2}\frac{\partial^2\overline{U}_i}{\partial x_j^2}} \quad \overline{U}' = S = \sqrt{2S_{ij}S_{ij}} \quad S_{ij} = \frac{1}{2}\left(\frac{\partial\overline{U}_i}{\partial x_j} + \frac{\partial\overline{U}_j}{\partial x_i}\right) \quad (4)$$

This procedure reduces the turbulent viscosity and therefore enables the resolution of unsteady structures in the simulation. Compared to the SST model, smaller structures can be generated and the turbulence cascade goes down to grid limit. As the SAS model does not provide sufficient damping of the smallest turbulent scales at grid limit, a limiter from a LES model is applied to the turbulent viscosity. This ensures a proper dissipation of turbulence. An advantage of the SAS model it has the ability to operate in RANS mode, if grid resolution and time step is not sufficient.

2.4 Computational Setup

For temporal discretisation a second order backward differencing scheme and for spatial discretisation a bounded second order central differencing scheme [4] was used and for turbulence quantities first order schemes were applied for temporal and spatial discretisation.

The different components of the Francis machine are (in streamwise direction): spiral case, stay vanes and guide vanes, runner blades and elbow draft tube with expansion tank (Fig. 1). As the model has "test rig size", the overall length is roughly 2.5 m with a height of 1.5 m (this is a scaling factor of up to 25 compared to the Francis turbines used in real power plants). The Reynolds number based on the spiral case inlet diameter equates $Re = 3 \cdot 10^5$.

Two different grids of hexahedral type are evaluated for the simulation of the Francis turbine: one with a size of $\approx$16 million elements (16M) and a refined one with a size of $\approx$40 million elements (40M) (see Table 1). Although the grid size is quite large, wall resolution is still away from resolving the boundary layer. This is one of the main challenges, as wall portion is very high for turbine simulations. As it is essential to have a Courant-number smaller than one to resolve turbulent structures to small scales, the time steps size corresponds to 2° and 0.5° of runner revolution for the coarser (16M) respectively for the finer mesh (40M).

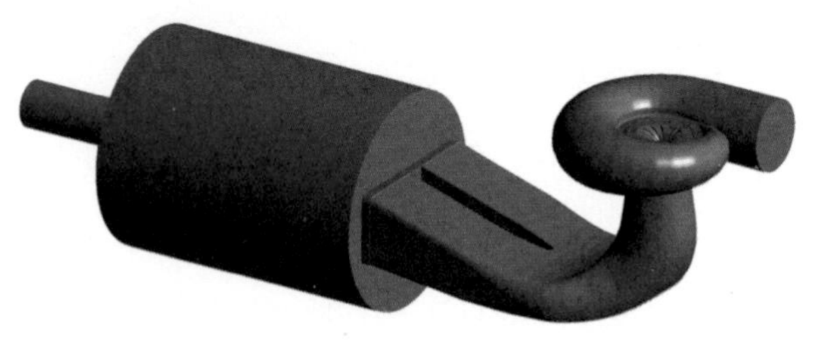

Fig. 1 Hydraulic contour of Francis machine used for the simulations

Table 1 Description of grid size and wall resolution for different domains

	16M		40M	
Domain	No. of elements	$y^+_{average}$	No. of elements	$y^+_{average}$
Spiral case	1,019 k	28.0	4,643 k	21.0
Stay/guide vanes	3,708 k	17.5	8,561 k	12.4
Runner	3,778 k	22.2	9,582 k	16.3
Draft tube with exp. tank	8,091 k	15.6	18,204 k	8.9
Total	16.2M		40.1M	

The simulations are done with steady state inlet boundary conditions from a fully developed pipe flow.

3 Results

A flow simulation of a complete Francis machine with all main components, e.g. spiral case, stay vanes and guide vanes, runner and elbow draft tube with expansion tank, needs very long physical time until the flow is convected through all parts. The convective time for one through flow equals roughly 20 runner revolutions. For the evaluation of time averaged quantities, even more simulation time is necessary: 20 runner revolutions for global data and 40 runner revolutions for time averaged velocity profiles in the draft tube. As the flow simulations using OpenFOAM-1.6-ext have not reached above mentioned simulation time, the flow is mainly analyzed in detail for the results obtained with CFX.

3.1 Global Machine Data

One main criterion for validation of the flow in Francis turbines are the hydraulic losses of different components. In particular because it is very feasible to compare this quantity with experimental data.

Interestingly, the losses summarised over all components of the Francis turbine are almost the same for both grid sizes. For the finer grid the losses are only

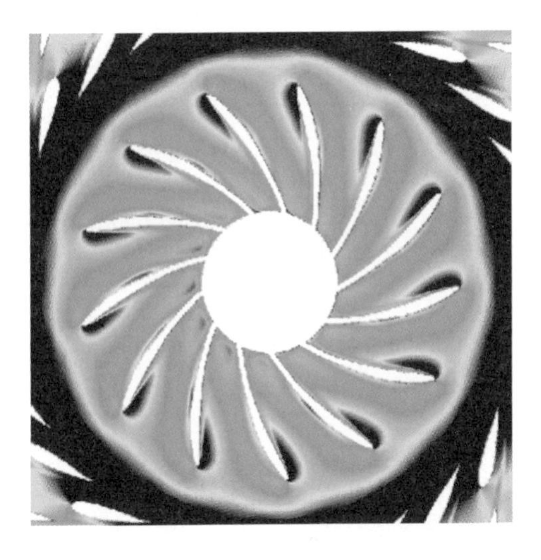

Fig. 2 Relative velocity in a cutting plane through the runner

0.07 % smaller than with the coarser grid (compared to $\approx$11 % in total based on the available hydraulic head). The difference for the first components in stream wise direction spiral case and stay vanes and guide vanes is almost negligible. In this region the flow is very well predicted by the RANS part of the SAS model and therefore there is no further mesh dependency, which could occur by the LES part of the SAS model. For the runner domain the simulation with the coarser mesh shows 0.31 % higher losses than with the finer mesh. In contrast to the draft tube flow, where the finer grid predicts 0.22 % more losses. This indicates, that mesh independency has not been reached yet.

3.2 Typical Flow Phenomena

Francis turbines usually operate with constant rotational speed. If, for mass flow reduction, the guide vanes are somewhat closed, at runner inlet the flow is misaligned (Fig. 2). Flow is guided more to direction of suction side of runner, than to pressure side. This is an explanation for mesh dependency in runner domain, already ascertained in Sect. 3.1.

Another typical phenomenon is the so called vortex rope, which occurs at the hub of runner outlet. It has a helical shape and is propagated into draft tubes straight cone (Fig. 3). Most of mass flow through the runner is pushed towards shroud, respectively wall of draft tube cone. This leads to an eddy water region in the centre of the cone, which is visualised by an isosurface of pressure in Fig. 3.

In the draft tubes elbow the vortex rope is breaking down as the through flow area increases and dissipates to turbulence (Fig. 4). Due to the flow direction in the elbow, there is almost no through flow in the left channel of the straight diffuser. In the right channel of the straight diffuser some more turbulent structures collapse.

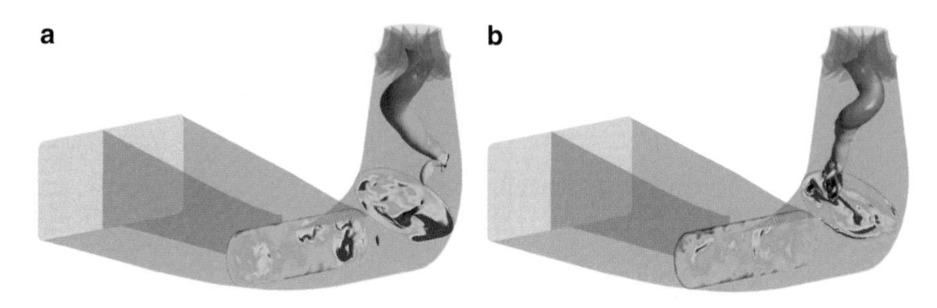

Fig. 3 Isosurfaces of pressure (500 Pa), coloured by viscosity ratio 0–150. (**a**) 16M; (**b**) 40M

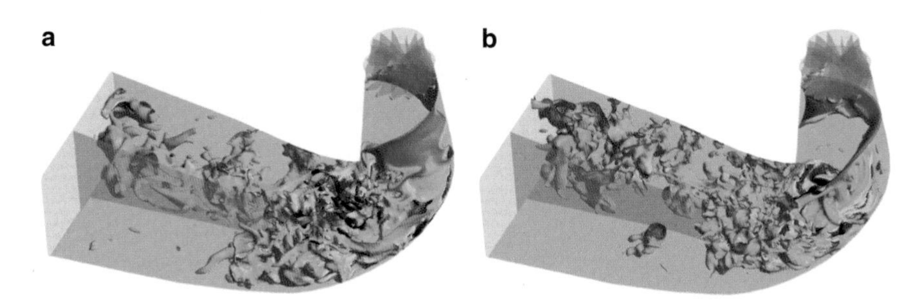

Fig. 4 Isosurfaces of pressure (1300 Pa), coloured by viscosity ratio 0–150. (**a**) 16M; (**b**) 40M

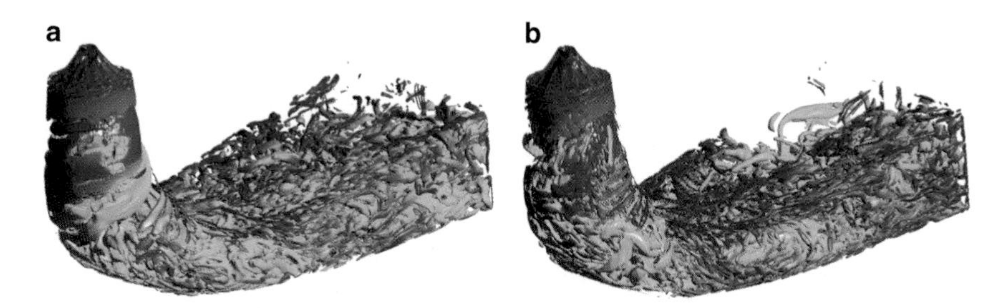

Fig. 5 Isosurfaces of velocity invariant $Q = 10$, coloured by viscosity ratio 0–300. (**a**) 16M; (**b**) 40M

In the cone larger vortex structures of the vortex rope can be seen having higher values of turbulent viscosity (Fig. 5). These structures, originated by the vortex rope, too, collapse to smaller structures in the straight diffuser.

The theoretical aspect, that the SAS turbulence model is able to resolve smaller turbulent structures, if grid and time resolution is sufficient, is discussed on the basis of the results depicted in Figs. 3–5, coarse grid each in (a) and fine grid each in (b).

First at all, it seems, that the vortex rope is conducted further into the draft tube for finer grids (Fig. 3). For coarser grids the vortex rope is dissipated earlier, as

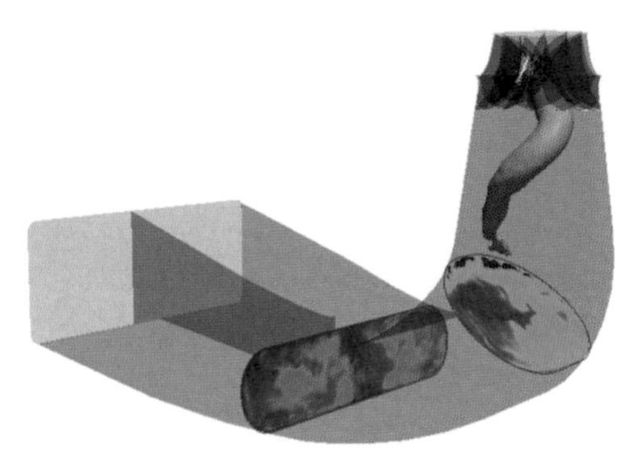

Fig. 6 Isosurfaces of pressure, coloured by viscosity ratio 0–150

turbulent viscosity is higher, especially at the end of the vortex rope and in the elbow. Higher grid resolution and smaller time step lead in general to a reduced turbulent viscosity. This enables the resolution of smaller turbulent structures (Figs. 4 and 5).

Flow simulation of a Francis turbine with OpenFOAM-1.6-ext using hybrid RANS-LES turbulence models has shown good results in [7]. The results depicted in Fig. 6 show a well developed vortex rope. The eddy viscosity seems to tend to somewhat lower values compared to the results with CFX.

4 Computational Resources

The flow simulations were performed on the CRAY XE6 (HERMIT) installed at the HLRS Stuttgart. This cluster contains 3,552 compute nodes with 32 cores per node with 32 GB and 64 GB memory. Data between the partitions is exchanged via MPI (message passing interface). For domain decomposition METIS library [6] is used for both codes.

For both used CFD codes a strong scaling analysis has been performed to determine the parallel performance for the flow simulations with the finer grid (40M) (Fig. 7). As the computational resources for CFX are limited, this study has been performed up to 256 cores. The results in Fig. 7a show, that the speedup of the simulation with CFX is observable away from "ideal". For 128 cores it is 1.63/2 and for 256 cores 2.29/4. In contrast, the speedup of the simulation with OpenFOAM-1.6-ext performs quite well up to 256 cores (Fig. 7b). With 512 cores the speedup decreases to 5.64/8 and with 1,028 cores further decreases to 7.73/16.

Based on the results of Fig. 7, the chosen number of cores for the simulations with CFX is 192 and for OpenFOAM-1.6-ext 768. This leads to simulation time of 176 days for CFX and 113 days for OpenFOAM-1.6-ext.

Within OpenFOAM-1.6-ext some pre- and post-processing steps have to be done in serial, such as partitioning, reconstruction of partitioned cases or interpolation of data for initialisation of results. This requires a lot of memory and has therefore to be done on pre- & post-processing nodes (as 64 GB of memory are not sufficient).

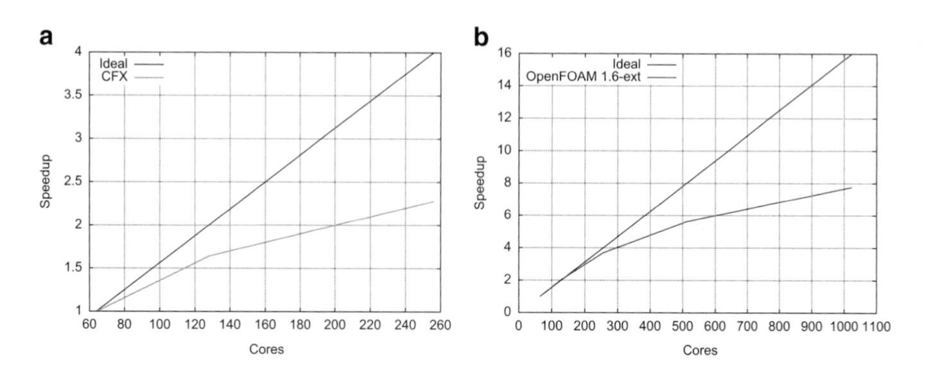

Fig. 7 Speedup based on simulations with 64 cores for simulations with the finer grid (40M). (**a**) Simulations with CFX on 64, 128 and 256 cores; (**b**) Simulations with OpenFOAM-1.6-ext on 64, 128, 256, 512 and 1,028 cores

The size of one result including transient statistics is 22.2 GB for CFX and 7.1 GB without and 15.4 GB with grid for OpenFOAM-1.6-ext each with the larger grid (40M).

5 Conclusion and Outlook

Simulations of the turbulent flow in a complete Francis turbine at part load conditions were performed using a commercial and an open source CFD code. For turbulence modelling the hybrid RANS-LES SAS model was applied. To investigate the influence of grid resolution and time step size on resolving small turbulent structures, two different grids were used with 16 and 40 million elements.

The hydraulic losses differ slightly in the runner and the draft tube domain. At runner inlet the flow is misaligned due to more closed guide vanes at constant rotational speed. At runner outlet a helical vortex rope occurs at the hub and propagates into the draft tube cone. In the elbow the vortex rope breaks down and dissipates to turbulent structures.

The SAS turbulence model enables a flow simulation down to turbulent structures. For finer grids and smaller time steps, the resolved turbulent structures are getting smaller. Also the vortex rope is better developed, as turbulent viscosity is decreased.

The flow simulations require high computational effort. To obtain turbulent statistics, several months of computational time is necessary. Strong scaling speedup tests show some limitations at higher amount of cores.

Future investigation will focus on some more numerical details, influence of turbulent inlet conditions, as well as boundary layer resolution. Also other hybrid RANS-LES methods, such as IDDES (Improved Delayed DES) [13] or wall-modelling approaches will be applied. Ideally, these hybrid RANS-LES models are compared with fully resolved Large Eddy simulations.

Currently, a test rig is under construction at the laboratory of IHS. For physical validation of the simulations flow field, a comparison with measurements will be done. The velocity field will be measured with laser Doppler anemometry and the pressure pulsations with piezo-resistive transducers.

References

1. M. Beaudoin, H. Jasak, Development of a generalized grid interface for turbomachinery simulations with OpenFOAM. Open source CFD international conference, Milano, Italy, 2008
2. Y. Egorov, F.R. Menter, Development and application of SST-SAS turbulence model in the DESIDER project. *Advances in Hybrid RANS-LES Modelling*. Notes on Numerical Fluid Mechanics and Multidisciplinary Design, Vol. 97 (Springer, Berlin, Heidelberg, 2008), pp. 261–270
3. Y. Egorov, F.R. Menter, D. Cokljat, The scale-adaptive simulation method for unsteady turbulent flow predictions. Part 2: Application to aerodynamic flows. J. Flow Turbul. Combust. **85**(1), 139–165 (2010)
4. H. Jasak, H.G. Weller, A.D. Gosman, High resolution NVD differencing scheme for arbitrarily unstructured meshes. Int. J. Numer. Meth. Fluid. **31**, 431–449 (1999)
5. D. Jošt, A. Škerlavaj, A. Lipej, Numerical flow simulation and efficiency prediction for axial turbines by advanced turbulence models. 26th IAHR Symposium on Hydraulic Machinery and Systems, Beijing, China, 2012
6. G. Karypis, V. Kumar, A fast and high quality multilevel scheme for partitioning irregular graphs. SIAM J. Sci. Comput. **20**, 359–392 (1998)
7. T. Krappel, I. Buntić-Ogor, O. Kirschner, A. Ruprecht, S. Riedelbauch, Numerical simulation of the vortex rope in a pump turbine. 7th OpenFOAM Workshop, Darmstadt, 25–28 June 2012
8. F.R. Menter, Two-equation eddy-viscosity turbulence models for engineering applications. AIAA J. **32**(8), 269–289 (1994)
9. F.R. Menter, Y. Egorov, The scale-adaptive simulation method for unsteady turbulent flow predictions. Part 1: Theory and model description. J. Flow Turbul. Combust. **85**(1), 113–138 (2010)
10. F.R. Menter, J. Schütze, M. Gritskevich, Global vs. zonal approaches in hybrid RANS-LES turbulence modelling. *Progress in Hybrid RANS-LES Modelling*. Notes on Numerical Fluid Mechanics and Multidisciplinary Design, Vol. 117 (Springer, Berlin, Heidelberg, 2012), pp. 15–28
11. A.D. Neto, R. Jester-Zuerker, A. Jung, M. Maiwald, Evaluation of a Francis turbine draft tube flow at part load using hybrid RANS-LES turbulence modelling. 26th IAHR Symposium on Hydraulic Machinery and Systems, Beijing, China, 2012
12. J.C. Rotta, *Turbulente Strömungen* (BG Teubner, Stuttgart, 1972)
13. M.L. Shur, P.R. Spalart, M.K. Strelets, A.K. Travin, A hybrid RANS-LES approach with delayed-DES and wall-modelled LES capabilities. Int. J. Heat Fluid Flow **29**, 1638–1649 (2008)

3D CFD Simulation of Twin Entry Turbochargers in an Engine Environment

Benjamin Boose, Hans-Jürgen Berner, and Michael Bargende

Abstract In this long term project the behaviour of a twin scroll turbocharger in an engine environment with pulsating boundary conditions is investigated. A complete three dimensional turbocharger fluid model is built including exhaust manifold, turbine and compressor housing as well as the complete runners. The turbine and the compressor side is coupled by a momentum equilibrium (and) the instantaneous rotational speed of the turbocharger is calculated for each timestep. This model is capable of analyzing both, the transient pulse efficiency, which is generated by the working cycle and the operating behavior, especially the overflow losses in a twinscroll turbine caused by pressure pulses.

1 Introduction

The increasing demands on the reduction of fuel consumption of passenger cars and the reduction of climate relevant emissions e.g. CO_2, lead to a higher focus on smaller turbocharged engines, the so called downsizing effect. Downsizing means the increase of engine efficiency as well as the reduction of emissions by reducing the engine displacement volume while maintaining constant driving performance by the use of a turbocharger [1]. The combination of direct injection and variable camshaft timing together with turbocharging has become the key technology today to reduce fuel consumption of gasoline and diesel engines and to improve efficiency. Reduced friction and moving the relevant operating points to higher loads near the efficiency sweet spot lead to lower specific fuel consumption compared to natural aspirated engines. The energy of exhaust gas exiting the engine should be harnessed to the most efficient extend in order to achieve rapid responding behavior

B. Boose (✉)
Institut für Verbrennungsmotoren und Kraftfahrwesen, Pfaffenwaldring 12, 70569 Stuttgart, Germany
e-mail: benjamin.boose@ivk.uni-stuttgart.de

W.E. Nagel et al. (eds.), *High Performance Computing in Science and Engineering '13*, DOI 10.1007/978-3-319-02165-2_32,

and a high low-end-torque. For this reason the interaction between the engine and the turbocharger needs to be investigated especially due to pulsation. The turbine manufacturer measures turbine maps on a stationary hot gas test bench. But when flow pulsation takes place, the turbine mainly exceeds outside its stationary maintained operating map range as the flow will be highly pulsatile which leads to a varying efficiency during the engine cycle. Knowledge of this transient behavior is necessary for further procedure.

With this in mind, the goal of early research into turbochargers was to quantify the unsteady performance and operating behavior of turbines. Palfreyman and Botas [2] used an explizit rotating turbine runner mesh in their 3D CFD calculations. The comparison between the unsteady turbine efficiency with the efficiency measured on a stationary test bench showed a slightly lower efficiency for the quasi-steady experiment. Lam and Robert [3] came almost to the same conclusion by using a static runner mesh with a multiple reference frame. Winkler et al. [4] investigated the unsteady performance of a twin entry turbine in an engine environment. A significant interaction between the two turbine scrolls was discovered. This interaction influenced the isentropic efficiency. Also Copeland et al. [5] as well as Müller et al. [6] calculated the unsteady performance of a twin entry turbine and compared the results to a steady flow analysis. A discrepancy between steady state and pulse efficiency of the turbine has been proved in their work. To characterize the flow conditions inside a twin scroll turbine was the goal of Brinkert et al. [7]. The authors took measurements with dissimilar pressure admission on the scrolls and introduced parameters to describe the appeared results which were further analyzed by three dimensional CFD calculations.

2 Theoretical Background

Within this work a qualitative prediction on the efficiency and behaviour of a twin entry turbine under pulsating conditions will be made. This chapter contains a short overview of the most important equations and physics behind turbine efficiency and velocity ratio. The efficiencies used in this work are based on isentropic adiabatic conditions. The states 3* and 4* describe the turbine housing inlet and the diffusor outlet. State 3* is divided into scroll 1 and scroll 2 in case of a twin entry turbine. The points right in front of the turbine runner and behind it are named 3 and 4. The total to total isentropic turbine efficiency $\eta_{tt,34}$ is the relation between total enthalpy drop Δh_t and the isentropic total enthalpy drop Δh_{ts} over the turbine, which is calculated by (1) [1] using total temperature T_t, total pressure p_t and the specific heat capacity c_p of the exhaust gas. The coefficient of expansion is here named γ. The total to static efficiency is to be calculated analogically with respect to different indices.

$$\eta_{tt,34} = \frac{\Delta h_t}{\Delta h_{ts}} = \frac{h_{t,3} - h_{t,4}}{h_{t,3} - h_{t,4s}} = \frac{1 - \frac{T_{t,4}}{T_{t,3}}}{1 - \frac{p_{t,4}}{p_{t,3}}^{\frac{\gamma-1}{\gamma}}} \tag{1}$$

The turbine blade speed ratio S, presented in (2), defines the ratio of the rotor inlet tip speed u and the isentropic velocity c_s, which is the velocity the flow would obtain if expanded in an ideal nozzle [1].

$$S = \frac{u}{c_s} = \frac{u}{\sqrt{2 \cdot c_p \cdot T_{t,3} \left(1 - \frac{p_{t,4}}{p_{t,3}}^{\frac{\gamma-1}{\gamma}}\right)}} \tag{2}$$

3 Model Building

The turbocharger modeled in this work originates from a four cylinder Otto engine with direct injection and 1,998 cm^3 volume of displacement. The engine produces 194 kW power and delivers a maximum torque of 353 Nm with a boost pressure of 2,320 mbar. The valve overlap in accordance to the twin entry turbine leads to a high low-end-torque reached by scavenging the residual gas out of the cylinder. The twin scroll turbocharger with asymmetric scroll alignment is connected to the engine by a four in two exhaust runner (see Fig. 1). Combined are the two outer cylinders 1 and 4 and the two inner cylinders 2 and 3 to one entry of the turbine and are merged immediately upstream the runner.

3.1 Computational Mesh

The resulting flow model from the introduced components is shown in Fig. 1. The computational domain extends from the inlet to the exhaust runner over the turbine including the internal wastegate up to the entrance of the catalitic converter on the exhaust side. Air-sidely the compressor is modeled from the short suction pipe up to the end of the volute. The computational mesh of the complete flow model consists of 1.75 million nodes. In Table 1 the rounded node numbers of all components can be found. The turbine runner as well as the compressor runner are discretized by hexaeder elements, the housings and the exhaust runner are meshed by tetraeder elements. The tip clear between blade and shroud is resolved by five layers. Figure 2 shows the computational mesh of a turbine and a compressor blade passage.

3.2 Physics

For solving the problem the software Ansys CFX in version 14.0 has been used for the reason of its high reputation in the turbomachinery field. There were no experimental data available for a validation. Relevant properties used in this model can be found in Table 2. The specific enthalpy of the exhaust gas is temperature dependent and derives from a look-up table. For the rotating turbine and the

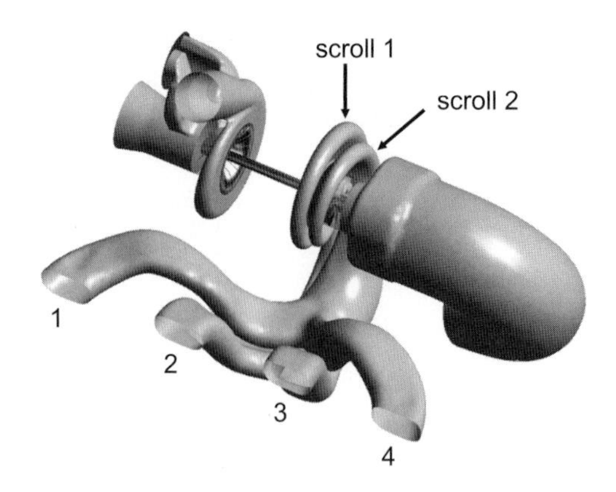

Fig. 1 Geometry of the three dimensional CFD model including the exhaust runner plus turbine and the compressor side. Cylinder 1+4 are connected to the outer, more bent scroll 2 and cylinder 2+3 are connected to the inner scroll 1

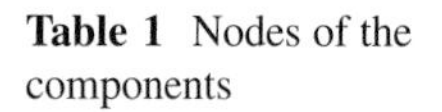

Table 1 Nodes of the components

Component	Number of nodes
Turbine runner blade passage	82,000
Compressor runner blade passage	76,000
Turbine housing	250,000
Compressor housing	190,000
Exhaust runner	125,000
Diffusor exhaust	75,000
Sum	About 1,750,000

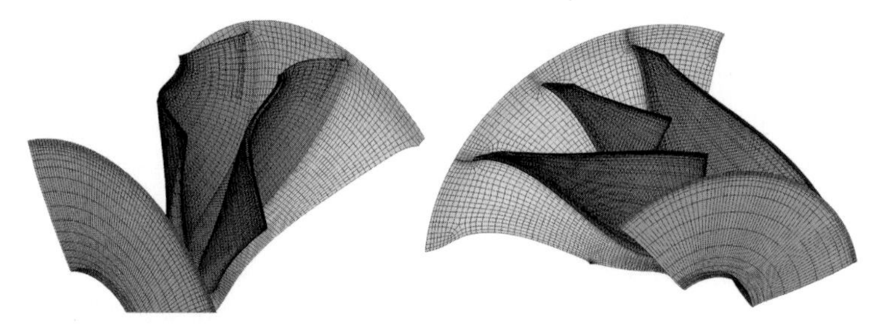

Fig. 2 Computational mesh of the turbine runner blade passage on the *left side* and the compressor runner blade passage on the *right side*. The turbine runner consists of eight passages and the compressor runner consists of six passages. The tip clear is coloured in *red*

compressor runner an explicit mesh motion is used. The stationary and the dynamic meshes are connected every timestep by a general grid interface. All walls in this case are treated as adiabatic. This complete model is independent from turbine or compressor maps. The rotational speed of the rotor is calculated for each timestep by a momentum equilibrium between the turbine and the compressor side including the bearing friction with (3)–(5). The momentum difference $M_{difference}$ is obtained by the sum of the momentum that the turbine delivers $M_{turbine}$, the momentum that the

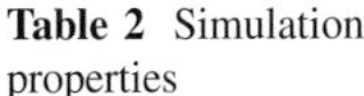

Table 2 Simulation properties

Property	Value
Heat transfer	Total energy incl. viscous work term
Turbulence model	Shear stress transport
Time step	$1.0 \cdot 10^{-6}$ [s]
Coefficient loops	4
Advection scheme	High resolution
Transient scheme	Second order backward Euler

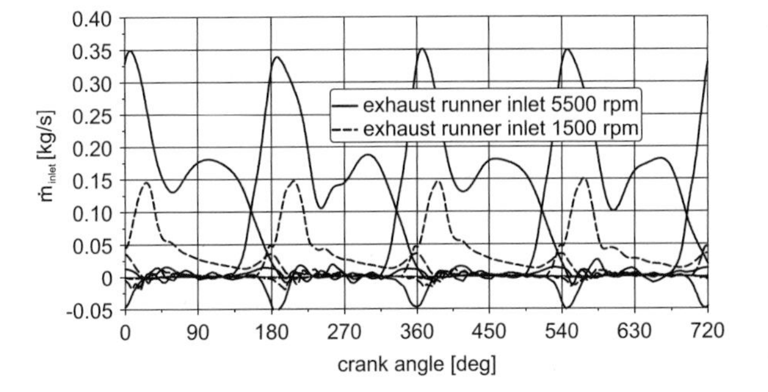

Fig. 3 Inlet boundary conditions for the four exhaust runners. Massflow over crank angle for the engine operating points 1,500 rpm and 5,500 rpm full load

compressor transfers to the fresh air $M_{compressor}$ and the friction momentum $M_{friction}$ that is lost in the bearings. With this momentum difference an angular acceleration $\dot{\omega}$ is calculated and used to set the rotational speed for the next time step. The moment of inertia of the rotor J_{TC} in (4) amounts $2,30 \cdot 10^{-5}\ kg \cdot m^2$. The friction torque is a rotation speed dependent quantity. Axial movement of the rotor and variations in oil temperature are not taken into account.

$$M_{difference} = M_{turbine} + M_{compressor} + M_{friction} \tag{3}$$

$$\dot{\omega} = \frac{M_{difference}}{J_{TC}} \tag{4}$$

$$\omega\,(t+1) = \omega\,(t) + \dot{\omega} \cdot \Delta t \tag{5}$$

3.3 Boundary Conditions

The boundary conditions used in this work come from a one dimensional fluid model of the complete engine that delivers time-resolved data for massflow, temperature and pressure for both inlets and outlets of the turbocharger model. The operating point analyzed in this paper is full load at 5,500 engine rpm at the engine maximum power. The massflow $\dot{m}_{inlet}$ over crank angle for two operating points at the exhaust runner inlets is shown in Fig. 3. The compressor inlet boundary

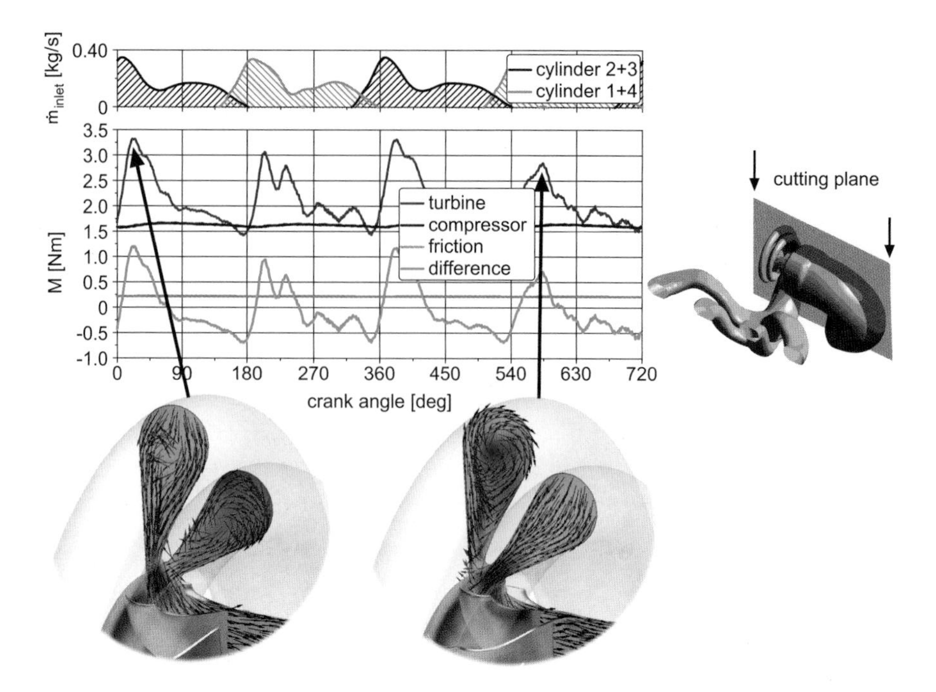

Fig. 4 Resulting torques on turbine and compressor runner over crank angle. The difference torque which accelerates or decelerates the rotor is coloured in *green*. Above, the inlet massflows of the four exhaust runners are plotted in *black* for scroll 1 and in *grey* for scroll 2. The cutting plane through the turbine is shown at two peak points of torque. The *left picture* shows the scroll 1 active at 38 °CA the *right one* shows scroll 2 active at 585 °CA. The cutting planes are coloured with the tangentially projected velocity streamline vectors

condition is ambient pressure at 300 K and it produces boost pressure against a static pressure of 0.4 bar upto 1.4 bar, depending on the engine operating point.

4 Simulation Results

In Fig. 4 the torque produced by the pressure forces of the exhaust gas on the turbine blades is displayed over crank angle. When substracting the torque the compressor transfers to the fresh air and the torque lost in the bearings from the turbine torque the difference torque remains, which is responsible for the dynamic change in the turbocharger's rotational speed. As can be seen, the torque peaks from scroll 1 are higher than those from scroll 2. This can be explained by the either differently angled scrolls since a higher angle results in higher incidence losses.

The difference torque results in rotational speed fluctuations of the turbocharger that are plotted over crank angle in Fig. 5. At the engine operating point of 5,500 rpm the turbocharger speed varies by 3,500 rpm with higher peaks for scroll 1 that

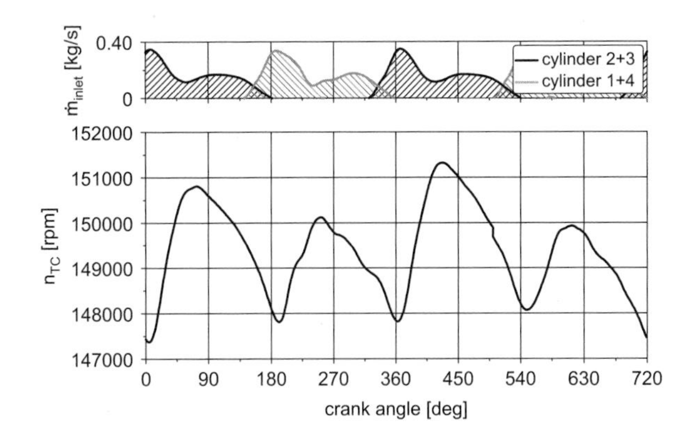

Fig. 5 Rotational speed of the turbocharger n_{TC} at 5,500 engine rpm. Above, the inlet massflows into the four exhaust runners are plotted in *black* for scroll 1 and in *grey* for scroll 2

delivers the higher torque to the rotor shaft. Further the phase shift of 57 °CA in pressure pulse between the exhaust runner inlet and the resulting rotor accelerating can be observed.

4.1 Efficiency

With respect to (1) the adiabatic total to static turbine efficiency $\eta_{ts,3^*4^*}$ for each scroll over the complete turbine is plotted in Fig. 6. The diagram shows that the peaks for scroll 1 are about 10 % higher than for scroll 2. Due of the time shift between inlet and outlet conditions of the turbine, efficiencies over 1 can occur. An efficiency below 0 means an inactivity of the scroll or flowback into the exhaust runner. The averaged efficiency over four working cycles for scroll 1 is 53.1 % and 44.6 % for scroll 2. Further turbine blade speed ratio, that is plotted in Fig. 7, varies during the pulse cycle. The ratio rises to an optimum value of 0.7 when pressure pulse decreases and rapidly drops to 0.5 when again pressure pulse increases. It is quite visible that the turbine is mainly operating away from the optimum value of 0.7, further information can be found in [1].

4.2 Transient Effects

Transient effects occur when the turbine works under pulsating conditions. The turbine charges when pressure pulse increases and discharges when pulse decreases. In Fig. 8 the instantaneous shaft power is plotted over the turbine runner inlet massflow $\dot{m}_3$. It can be observed that the turbine is charging from 0 to 30 °CA. After peak massflow, the turbine is discharging until 180 °CA. The shaft power,

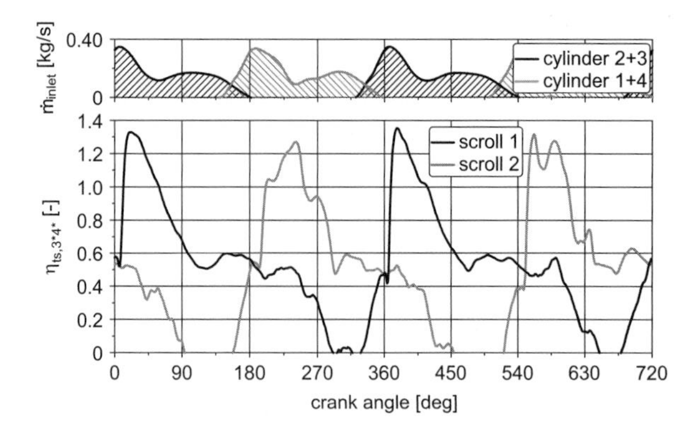

Fig. 6 Total to static isentropic turbine efficiency $\eta_{ts,3*4*}$ at 5,500 engine rpm. Above, the inlet massflows of the four exhaust runners are plotted in *black* for scroll 1 and in *grey* for scroll 2

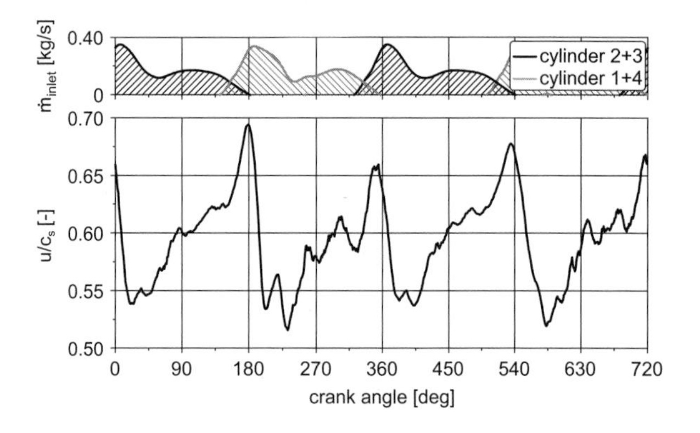

Fig. 7 Isentropic turbine velocity ratio $^{u}/_{c_s}$ at 5,500 engine rpm. Above, the inlet massflows into the four exhaust runners are pictured in *black* for scroll 1 and in *grey* for scroll 2

generated from the turbine, is up to 40 % higher for the discharging phase for the same turbine inlet massflow. So it can be assumed that there is a mass storage in the turbine, since different shaft powers for the same massflow are delivered. This mass storage influences the operating behaviour and the turbine can not be treated as quasi-stationary.

4.3 Overflow Losses

The cutting plane in Fig. 9 shows the tangentially projected streamline vectors. An overflow from the active scroll 1 to the opposite one can be observed. Two circumferential control surfaces $sf1$ and $sf2$ in Fig. 9 are defined in order to

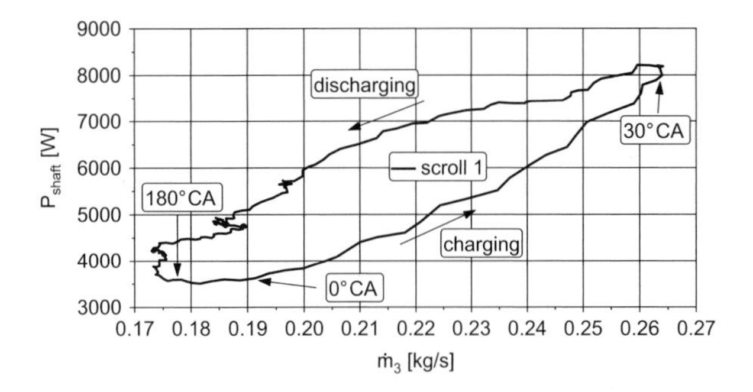

Fig. 8 Instantaneous shaft power over turbine runner inlet massflow $\dot{m}_3$ for scroll 1

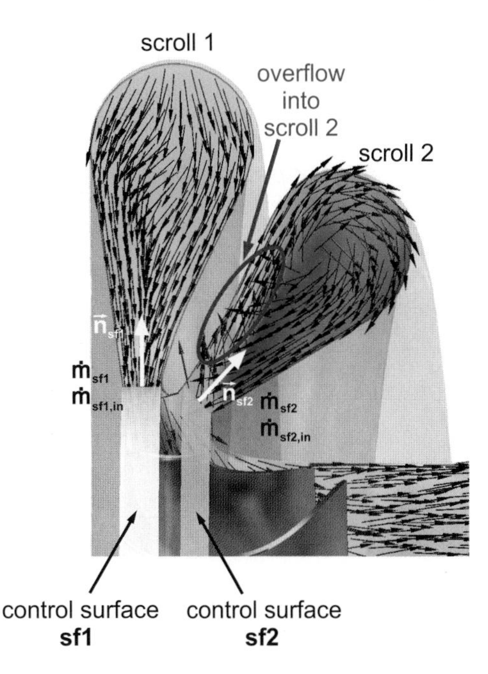

Fig. 9 Twin scroll specific overflow factor. Cutting plane through the turbine is coloured with the velocity and shows tangentially projected streamline vectors. An overflow from scroll 1 in scroll 2 is shown. For each scroll a circumferential control surface has been defined. Its normal vectors are n_{sf1} and n_{sf2}. While m_{sf} describes the complete massflow through the control surface, $m_{sf,in}$ accounts for the massflow only in surface normal direction which is the massflow from the active scroll into the inactive one

calculate the massflow through each scroll. By the use of a surface normal vector n_{sf} an overflow massflow $\dot{m}_{sf,in}$ to the inactive scroll can be calculated by means of (6) and (7). To quantify the overflow losses, by means of the massflows $\dot{m}_{sf}$ and $\dot{m}_{sf,in}$, a specific number, namely the scroll overflow factor SOF is introduced and calculated for each scroll. The overflow determined by $\dot{m}_{sf,in}$ is considered as a loss

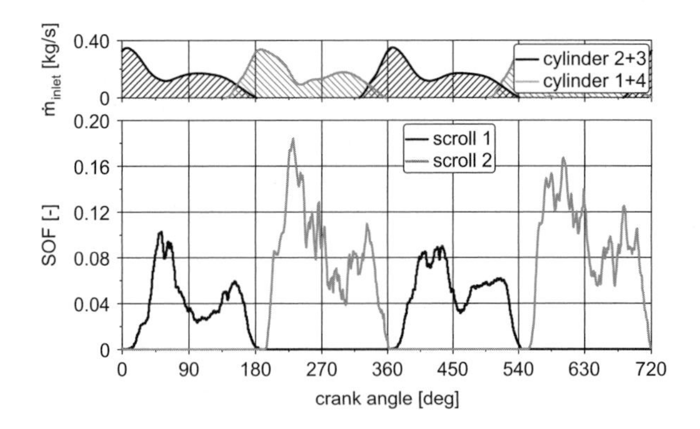

Fig. 10 Scroll overflow factor *SOF* for each scroll over crank angle for engine operating at 5,500 rpm. Above, the inlet massflows into the four exhaust runners are plotted in *black* for scroll 1 and in *grey* for scroll 2

of usable exhaust gas enthalpy since the overflown mass mainly flows back into the exhaust runner and cannot be converted into shaft power.

$$v_{normal} = \mathbf{n}_{sf} \cdot \mathbf{v} \tag{6}$$

$$\dot{m}_{sf,in} = \int_{sf} \rho \cdot v_{normal} dA \tag{7}$$

$$SOF_{scroll1} = 1 - \frac{\dot{m}_{sf1} - \dot{m}_{sf2,in}}{\dot{m}_{sf1}} \tag{8}$$

$$SOF_{scroll2} = 1 - \frac{\dot{m}_{sf2} - \dot{m}_{sf1,in}}{\dot{m}_{sf2}} \tag{9}$$

In Fig. 10 the scroll overflow factor is shown over crank angle for each scroll. A higher *SOF* for scroll 2 can be observed. This behaviour matches the lower efficiency of the angled scroll shown in Fig. 6, which means also that there is a higher backflow into the exhaust runner through scroll 1.

4.4 Backflow

In Fig. 11 the inlet and outlet massflows of the turbine housing are visualized with normally projected streamline vectors. It can be observed that the overflown mass from scroll 2 in the opposite one flows through the volute back into the exhaust runner. The backflow through scroll 1 is higher due to the higher overflow of the opposite scroll 2. The problem of this backflow is the negative influence on the gas exchange, especially at low engine speeds during valve overlap phase.

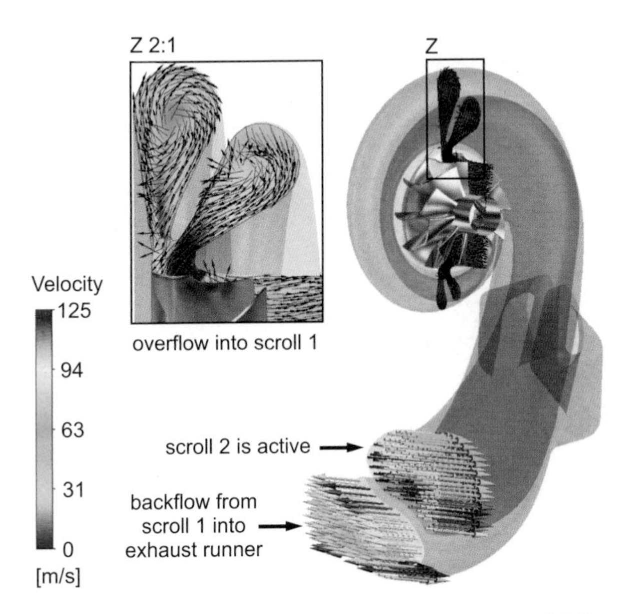

Fig. 11 Mechanism of exhaust gas backflow into the exhaust runner at 5,500 engine rpm. The cutting plane through the turbine volute shows the tangentially projected streamline vectors and an overflow from the active scroll 1 into the inactive one can clearly be seen. This overflown mass flows back into the exhaust runner

5 CPU Time

The Nec Nehalem Cluster of the *Höchstleistungsrechenzentrum Stuttgart* is used to solve this problem. Its CPUs are Intel Xeon (X5560) Quad Cores with 2.8 GHz and 12 GB memory. The node to node interconnection runs through an Inifiniband ethernet. In Fig. 12 the time per iteration in seconds is shown from eight CPUs in use up to 256 CPUs on the left side. On the right side there is shown the job time in days to run four engine cycles. The usage of 64 or 128 CPUs gives the best compromise, the calculation takes around six to seven days, starting off from a good initial solution.

6 Conclusions

In this work the turbine behaviour under pulsating conditions was investigated. The instantaneous efficiency curve for the twin entry turbine for four workings cycles was plotted, the peak efficiency occurs at 390 °CA for scroll 1. Time-averaged the scroll 1 shows a 7 % higher efficiency over the whole cycle than the opposite more angled scroll 2. In addition, an overflow from the active to the inactive scroll was detected. With respect to the efficiency curve the more angled scroll has a

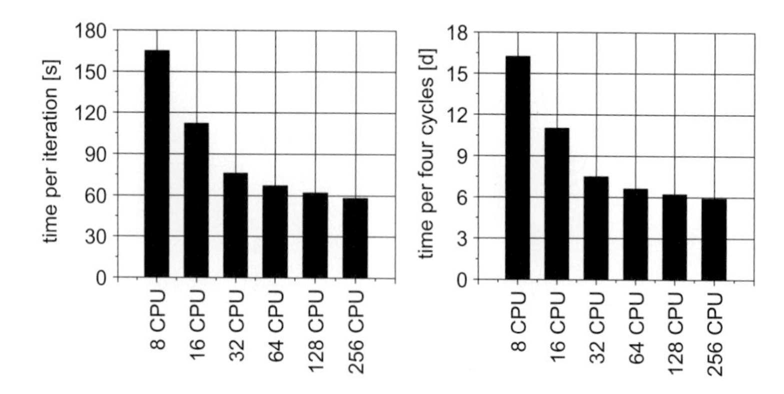

Fig. 12 Job time in seconds to converge one iteration on the *left* and in days to converge four engine cycles on the *right side*

higher overflown massflow and hence a higher loss than scroll 1. A backflow of the overflown mass into the exhaust runner is a result of this overflow effect. In future the turbine efficiency and overflow factor will be investigated for low engine speeds. Additionally, different twin entry turbine designs, e.g. the dual volute will be compared to the standard twin scroll turbine. A design for delivering the best overall performance at low and high engine speeds, concerning efficiency, scroll overflow and resulting backflow is sought.

References

1. N.C. Baines, Fundamentals of turbocharging. *Concepts NREC*, Vermont, 2004
2. D. Palfreyman, R.F. Martinez-Botas, The pulsating flow field in a mixed flow turbocharger turbine: an experimental and computational study. ASME paper GT2004-53143 (2004)
3. J.K.-W. Lam, Q.D.H. Roberts, Flow modelling of a turbocharger turbine under pulsating flow. Turbocharging and Turbochargers, I Mech E (2002)
4. N. Winkler, H.-E. Angstrom, Instantaneous on-engine twin-entry turbine efficiency calculations on a diesel engine. SAE paper 2005-01-3887 (2005)
5. C.D. Copeland, R. Martinez-Botas, M. Seiler, Unsteady performance of a double entry turbocharger turbine with a comparison to steady flow conditions, in *Proceedings of the ASME Turbo Expo 2008: Power for Land, Sea and Air*, GT2008-50827, Berlin, 2008
6. M. Müller, T. Streule, S. Sumser, G. Hertweck, A. Knauss, A. Knüspert, A. Nolte, W. Schmid, The asymmetric twin scroll turbine for daimler heavy duty engines, 13. Aufladetechnische Konferenz Dresden (2008)
7. N. Brinkert, S. Sumser, A. Schulz, S. Weber, K. Fiesweger, H. Bauer, Understanding the twin scroll turbine—flow similarity, in *ASME Turbo Expo*, Vancouver, 2011

Numerical Simulation of Flow with Volume Condensation in a Model Containment

Jing Zhang and Eckart Laurien

Abstract One severe accident scenario is a leak in the primary circuit of a Pressurized Water Reactor (PWR) resulting in steam injection into the containment. In this case, wall and volume condensation can occur in the containment. The condensation phenomena are of interest for safety considerations. This paper presents a newly developed volume condensation model with a two-phase flow for the ANSYS CFX code. A two-fluid model with a continuous gas phase consisting of a steam-air mixture and a dispersed liquid phase composed of water droplets in thermal and mechanical non-equilibrium is applied, so that both phases are modeled with separate temperatures and velocities. The motion of the droplet due to gravitational force is considered. Volume condensation is modeled as a sink of mass and source of energy at the droplet interfaces. In order to focus on the volume condensation, the new model is applied to a simple two-dimensional steady state test case. The volume condensation model is validated with a condensation experiment TH2, which was performed in the German THAI facility.

1 Introduction

During a severe accident (e.g. Loss of Coolant Accident), steam (water vapor) can be released through a leak in the primary circuit into the containment of a PWR, see Fig. 1. Condensation can occur on the containment walls and within its volume, referred to as "volume condensation". If the steam partial pressure within the mixture is locally higher than the steam saturation pressure, the condensation into small droplets must be considered.

J. Zhang (✉) · E. Laurien
Institute of Nuclear Technology and Energy Systems, University of Stuttgart, Pfaffenwaldring 31, 70569 Stuttgart, Germany
e-mail: jing.zhang@ike.uni-stuttgart.de; eckart.laurien@ike.uni-stuttgart.de

W.E. Nagel et al. (eds.), *High Performance Computing in Science and Engineering '13*, DOI 10.1007/978-3-319-02165-2_33,

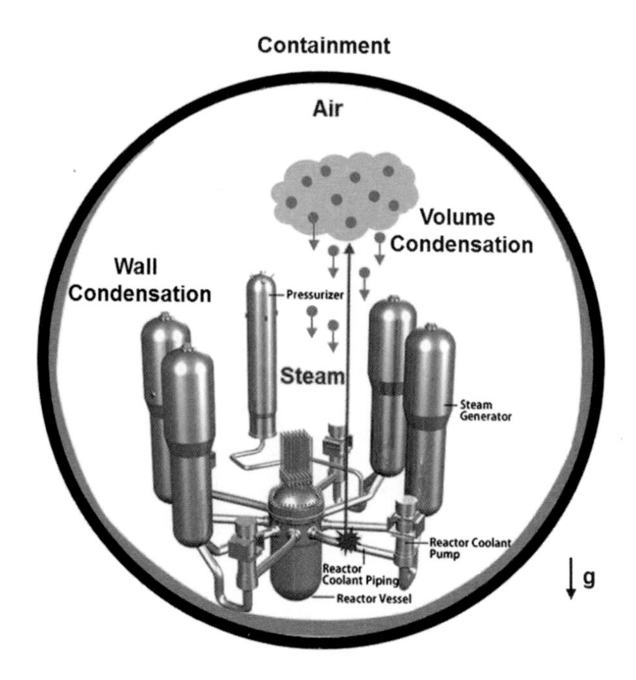

Fig. 1 Condensation inside containment

Wall condensation and volume condensation have been modeled by different authors using a thermodynamic equilibrium model with a single-phase flow with the Computational Fluid Dynamics (CFD) codes CFX [1–5], Fluent [6] or GASFLOW [7]. However, a slip velocity between the gas and droplet phase, as it may occur in the containment, has so far not been taken into account. Thus, phenomena such as downward transport of droplets due to gravity, droplet deposition on surfaces as well as wetting fog could not be simulated. For steam and air mixtures, this may have an effect on the spatial distribution of liquid sources. For other mixtures, which include hydrogen, this may have an effect on the formation of explosive mixtures. It is therefore necessary to develop numerical models, which are able to predict volume condensation in mechanical non-equilibrium. Such models must involve the possibility of a relative (slip) velocity between the gas and the liquid phases.

In general, condensation of steam in presence of non-condensable gases, i.e. nitrogen, oxygen and hydrogen (in an experiment helium), is referred to as "dewing" and the condensate as "dew". The equilibrium thermodynamics of air (79 % nitrogen, 21 % oxygen) and water in mechanical and thermal equilibrium is described by the Mollier diagram [8, 9]. A slip velocity or a temperature difference between the gas and liquid phases is not included in this theory. Consequently, the size and number of droplets or nucleation sites is irrelevant within this description. These assumptions are only valid for extremely small droplets.

In a multi-dimensional approach of a CFD code, two-phase flows in non-equilibrium can be described by the two-fluid model. In this model, both phases, including the dispersed droplet phase, are regarded as inter-penetrating fluids, which can move independently. The temporally averaged mass, momentum and energy balance of each phase is described by a set of conservation equations. These equations can be derived from the general Navier-Stokes equations by phasic averaging, which result in averaged Reynolds stresses, turbulent heat fluxes, as well as exchange terms of mass, momentum and energy. These terms must be modeled by two-phase turbulence models and phase-exchange models.

The aim of the present work is to develop and validate a phase exchange model for volume condensation for a two-phase flow using the CFD code ANSYS CFX 14.0. Using a two-dimensional steady-state test case, the flow for the development and test of the model is significantly simplified and the simulation is focused on the volume condensation process.

In order to study the flow with condensation under severe accident conditions, the TH2 experiment has been performed in the THAI experimental facility (THAI = **T**hermal hydraulics, **H**ydrogen, **A**erosols, **I**odine) by Becker Technologies. The THAI facility is a mock-up of reactor containment and consists of a closed cylindrical vessel, in which steam can be injected. In order to validate the volume condensation model, a two-dimensional computational domain is used due to the rotational quasi-symmetry of the TH2 geometry.

2 Computational Model

2.1 Two-Fluid Model

The volume condensation is modeled on the basis of the two-fluid model. The gas phase (index G) is modeled as a continuous mixture of steam and air (79 % nitrogen, 21 % oxygen), and the liquid phase (index L) is treated as a disperse phase of droplets. It is assumed that small droplets are already present as nucleation sites, i.e. heterogeneous nucleation is modeled by assuming a given droplet number density and initial droplet diameter. The volume fraction of each phase is denoted by $\alpha_k (k = L, G)$, where k is the phase index. Both phases may have different velocities $\overrightarrow{\overline{u}}^k$ and temperatures $\overline{T}^k$ (the overbar refers to temporal phasic averaging), thus allowing for thermal and/or mechanical non-equilibrium, e.g. wet fog. The source-sink term Γ_k of the mass transport equation and the source-sink term E_k of the energy transport equations are implemented for the volume condensation model in CFX. The two-fluid model describes the time dependent mass, momentum and energy conservation for each phase [10, 11] in a three-dimensional coordinate system $x_m (m = 1, 2, 3)$. The mass transport is described by the phase continuity equations:

$$\rho_k \{ \frac{\partial \alpha_k}{\partial t} + \nabla(\alpha_k \overrightarrow{\overline{u}}^k_m)\} = \Gamma_k \tag{1}$$

where ρ_k is the density of each phase. The momentum balance is described by the momentum equations of each phase, which reads for the coordinate direction m:

$$\rho_k\{\frac{\partial(\alpha_k \overline{u}_m^k)}{\partial t} + \nabla(\alpha_k \overrightarrow{\overline{u}}^k \overline{u}_m^k)\}$$
$$= -\frac{\partial(\alpha_k p)}{\partial x_m} + \nabla[\alpha_k(\overline{\overline{\tau}}^k + \overline{\overline{\tau}}^{Re,k})]_m + \overline{u}_m^k \Gamma_k + \alpha_k \rho_k g_m + M_{k,m} \quad (2)$$

in which p is the pressure, $\overline{\overline{\tau}}^k$ and $\overline{\overline{\tau}}^{Re,k}$ are the molecular stresses and the turbulent (Reynolds) stresses of each phase, g_m is the gravity acceleration and $M_{k,m}$ is the momentum exchange term. The energy equations in CFX are expressed by enthalpy equations, using $\overline{h}^k$ for each phase.

$$\rho_k\{\frac{\partial(\alpha_k \overline{h}^k)}{\partial t} + \nabla(\alpha_k \overrightarrow{\overline{u}}^k \overline{h}^k)\} = \nabla[\alpha_k(\overline{q}^k + \overline{q}^{Re,k})] + \overline{h}^k \Gamma_k + E_k \quad (3)$$

Here $\overline{q}^k$ and $\overline{q}^{Re,k}$ are the molecular and turbulent heat fluxes. The sum of the volume fractions α_k of both phases is 1.

$$\alpha_G + \alpha_L = 1 \quad (4)$$

Considering the mass balance, a mass that leaves the gas phase must be added to the liquid phase. The sum of the mass transfer rate per unit volume Γ_k is equal to zero.

$$\Gamma_G + \Gamma_L = 0 \quad (5)$$

The interfacial momentum transfer force $M_{k,m}$ consists primarily of the drag force $M_{k,m}^D$. Other forces such as the lift force, virtual mass force and Basset force are also present but are assumed to be negligible in the present analysis.

$$M_{k,m} = M_{k,m}^D \quad (6)$$

The drag force per unit volume on the continuous gas phase is the negative drag force on the liquid phase. To model the exchange terms of mass, momentum and energy for the droplet flow, two limiting cases are considered: First, the droplet diameter is assumed to be given as a constant value d. Then the momentum exchange term results to:

$$M_{G,m}^D = -M_{L,m}^D = \alpha_L c_D \frac{3\rho_G}{4d} |\overrightarrow{\overline{u}}^L - \overrightarrow{\overline{u}}^G| (\overrightarrow{\overline{u}}^L - \overrightarrow{\overline{u}}^G) \quad (7)$$

Second, the droplet number density (number of droplets per unit volume) is given as a constant value n. Then the momentum exchange term is given as:

$$M_{G,m}^{D} = -M_{L,m}^{D} = n c_D \frac{\pi \rho_G}{8} (\frac{6\alpha_L}{n\pi})^{\frac{2}{3}} |\vec{u}^L - \vec{u}^G| (\vec{u}^L - \vec{u}^G) \tag{8}$$

These cases are regarded as limiting cases of a "real" flow with both variable droplet diameter and number density. In the real situation the droplets may have a size distribution rather that a constant size and the number density may vary as a function of droplet sub-cooling and other nucleation parameters. An implementation of such physical models into a CFD code will only be possible in the near future, when experiences with a two-fluid formulation and a nucleation model will be available. Therefore, the present formulation may be regarded only as a first step towards such a complete and physically accurate description.

The drag coefficient c_D is given by the Schiller-Naumann correlation for a droplet:

$$c_D = \frac{24}{Re}(1 + 0.15 Re^{0.687}) \tag{9}$$

The Reynolds number is defined as:

$$Re = \frac{\rho_G |\vec{u}^L - \vec{u}^G| d}{\mu_G} \tag{10}$$

where ρ_G is the gas mixture density, d is the droplet diameter and μ_G is the dynamic viscosity of the gas mixture.

2.2 *Volume Condensation Model*

The volume condensation model is based on the two-fluid model. In order to understand the volume condensation process between small droplets and the surrounding gas phase, the interactions are described first between a single droplet and the gas mixture of condensable steam and non-condensable air, as shown in Fig. 2.

The droplet is assumed to be spherical with a diameter d. The liquid droplet is in thermodynamic equilibrium at its surface and has a uniform temperature $\overline{T}^L$. The gas phase at the edge of the thermal boundary layer around the droplet has a temperature $\overline{T}^G$. The partial pressure of steam directly at the interface must be equal to the saturation pressure corresponding to the droplet temperature $p_{H_2O,sat}(\overline{T}^L)$ [12]. The molar fraction at the interface $y_{H_2O,sat}$ is assumed to be in equilibrium and is equal to the ratio of the saturation pressure to the static pressure p.

$$y_{H_2O,sat} = \frac{p_{H_2O,sat}(\overline{T}^L)}{p} \tag{11}$$

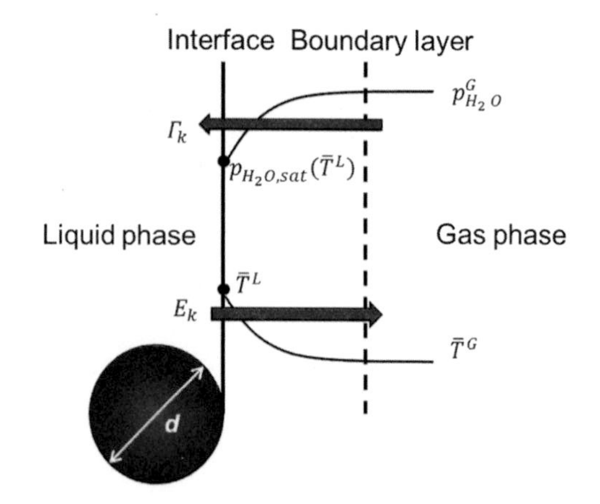

Fig. 2 Pressure and temperature profiles for a condensing droplet in the gas mixture

The mass fraction at the interface $c_{H_2O,sat}$ is defined as follows:

$$c_{H_2O,sat} = y_{H_2O,sat} \frac{M_{H_2O}}{M_G} \tag{12}$$

where M_G is the molar mass of the gas mixture with steam and non-condensable gas. The non-condensable gas consists of at least one component. In this present work, the non-condensable gas is air, but the formulation can easily be modified to take account of other non-condensable gases, e.g. individual fractions of nitrogen, oxygen and hydrogen.

$$M_G = c_{H_2O,sat} M_{H_2O} + (1 - c_{H_2O,sat}) M_{Air} \tag{13}$$

The molar mass of water M_{H_2O} is 18 g/mol and of air M_{Air} is 29 g/mol. The gradient of concentration $c_{H_2O,G} - c_{H_2O,sat}$ leads to mass transport across the diffusion boundary layer. If the steam mass fraction in the gas is higher than the steam saturation concentration, the volume condensation rate is calculated from the following simple expression:

$$\Gamma_G = -\rho_G \beta A (c_{H_2O,G} - c_{H_2O,sat}) = -\Gamma_L \tag{14}$$

The mass transfer coefficient β due to the flow around a sphere is given by the correlation of Ranz-Marshall [13] for the Sherwood number Sh.

$$Sh = \frac{\beta d}{D_{H_2O,Air}} = 2 + 0.6 Re^{\frac{1}{2}} Sc^{\frac{1}{3}} \tag{15}$$

where $D_{H_2O,Air}$ is the diffusion coefficient of the steam-air mixture. The Schmidt number is expressed as:

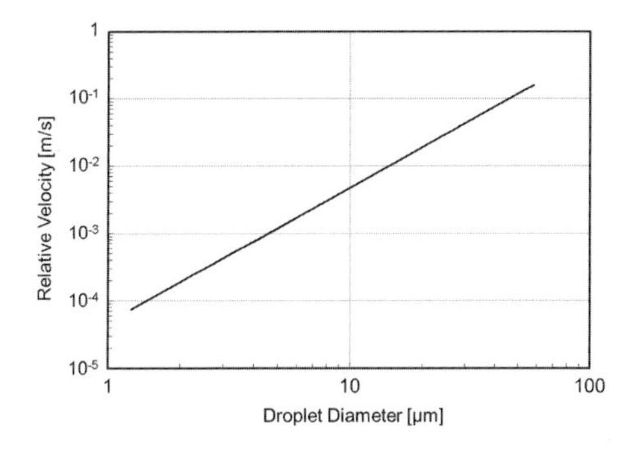

Fig. 3 Relative velocity between droplet and surrounding gas

$$Sc = \frac{\mu_G}{\rho_G D_{H_2O,Air}} \tag{16}$$

In (14) the interface area density of all droplets is defined as follows.

$$A = \pi n d^2 = \frac{6\alpha_L}{d} \tag{17}$$

During the volume condensation heat is extracted from the droplets into the atmosphere. The energy source due to the release of latent heat Δh_{LG} is determined from the expression for the volume condensation as:

$$E_G = -\Gamma_G \Delta h_{LG}, \quad E_L = 0 \tag{18}$$

In order to estimate the magnitude of the effect of the relative velocity between the phases, we have computed the terminal velocity of droplet deposition according the Hadamard-Rybczynski expression of creeping motion [14, 15]. In (10) the relative velocity $|\vec{u}^L - \vec{u}^G| = \Delta\vec{u}$ is affected by the buoyancy force and is given by:

$$\Delta\vec{u} = \frac{2g(\frac{d}{2})^2(\rho_L - \rho_G)}{3\mu_G} \cdot \frac{1 + \frac{\mu_G}{\mu_L}}{2 + 3\frac{\mu_G}{\mu_L}} \tag{19}$$

Here, μ_G and μ_L are dynamic viscosities of gas and liquid. Figure 3 shows the relative velocity $\Delta\vec{u}$ between the droplet and the surrounding gas as a function the droplet diameter from the above expression.

Because of the similar phenomena of the volume condensation and cloud formation, the size of cloud is taken into account. A typical cloud droplet is on the order of 20 μm [16]. The droplet with a diameter of 20 μm has relative velocity of around 0.02 m/s, see Fig. 3. It can be assumed, that a safety-related flow and condensation event within the reactor containment occurs within a time interval in the order of 500 s and has an extension of several meters, estimated 10 m. Then a droplet estimated velocity of 0.02 m/s must be considered by the theory, because it takes only 500 s for the droplet to move through the entire condensation region. Therefore, we consider a droplet size of 20 μm, for which a relative velocity must be taken into account. Of course, this estimation is somewhat arbitrary, but it may help as a guideline for model selection.

3 Simple Test Case

3.1 Geometry and Boundary Conditions

As an initial step, a two dimensional steady state test case was designed to simplify the flow condition and perform the volume condensation model. The geometry is an open channel of 5.0 m height and 0.2 m width. Two inlets have a width of 0.1 m, (see Fig. 4). The saturated steam-air mixture flows with a higher temperature of 50 °C through inlet 1 and with a lower temperature of 25 °C through inlet 2 into the channel. Both flows have the same mass flow rate of 10 g/m^{3}s. The mixture leaves the channel through the outlet 3. The outlet is modeled as a nozzle to ensure a smooth outflow. The front and back area are modeled using symmetry conditions. All remaining faces are adiabatic walls with no slip condition. With this test case it is assumed that the pressure in the channel is constant of 1 bar and the droplet number density is a constant of 10^9 m^{-3}. The initial droplet volume fraction is 10^{-5}.

3.2 Numerical Setup

A steady state simulation was performed with ANSYS CFX 14.0 using the coupled double precision solver. The SST turbulence model and buoyancy model were applied for the calculation. A first order scheme was used for turbulence model equations. The second order scheme was applied for other transport equations. The root mean square (RMS) criterion was set to 10^{-4}. The grid was generated using the program ICEM CFD by ANSYS and has 22,088 grid points. The simulation was carried out on the NEC Nehalem Cluster at the High Performance Computing Center Stuttgart (HLRS). The test case has been simulated on one node with four processors. The steady state calculation needed a CPU time of 7.3×10^4 s.

3.3 Results

By using the newly developed model the volume condensation could be calculated and the preliminary simulation for the steady state case is discussed, as shown in Fig. 5. After the mixing of two wet air flows, the volume condensation happens

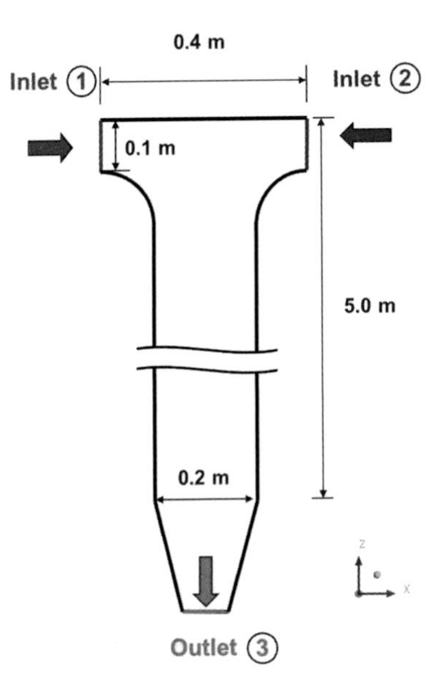

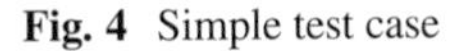
Fig. 4 Simple test case

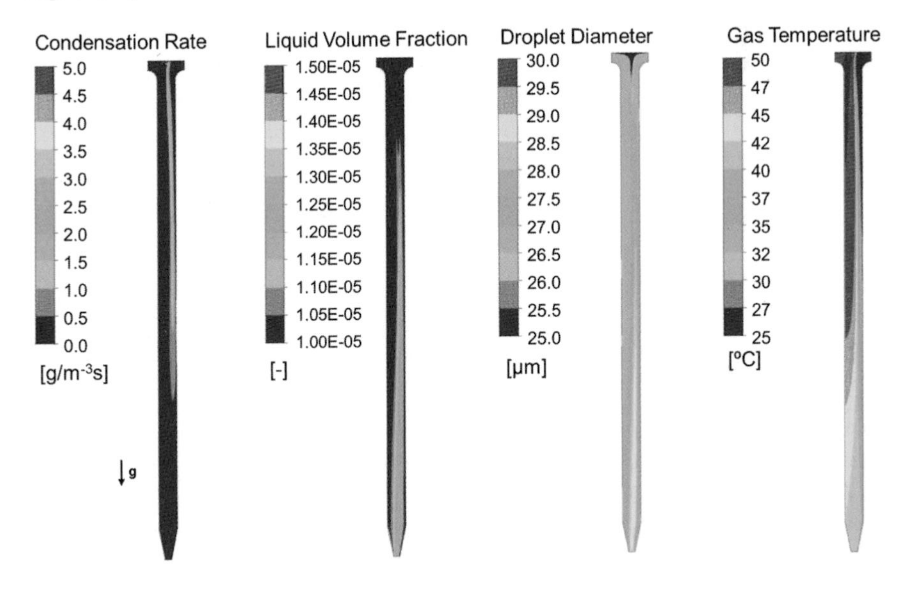

Fig. 5 Results for test case

primarily in the upper channel. At the end of the channel, the condensation rate is almost zero. The volume condensation causes an increase of the liquid volume fraction in the domain, so that the outflow has higher liquid volume fraction. With a constant droplet number density, steam condenses at the droplet interfaces and

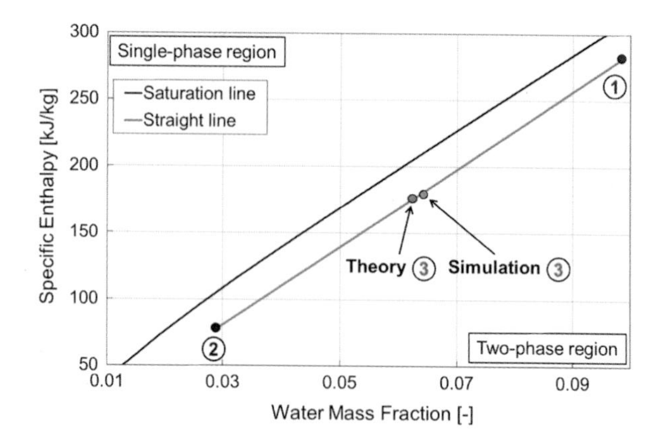

Fig. 6 Comparison of simulation result and theory in Mollier diagram

the droplet diameter grows. Caused by the gravitational force, the gas phase and liquid phase have different velocities and bigger droplets fall faster. At the outlet, the change in gas temperature is negligibly small, so that the fluid at the outlet is supposed to be in the equilibrium condition.

The equilibrium state is shown in the developed Mollier diagram (Fig. 6), in which the specific enthalpy is plotted over the water mass fraction (consisting of vapour and droplets). The saturation line separates the single-phase and two-phase regions in the diagram. Because of the mass and energy conservation equations, the mixing point 3 must be on the straight line between inlet points 1 and 2. The result of the specific enthalpy shows a good agreement to the theory.

4 TH2 Experiment

4.1 Geometry and Boundary Conditions

In order to validate the volume condensation model, a two-dimensional computational domain is used due to the rotational quasi-symmetry of the TH2 geometry [17]. The TH2 experiment investigates the flow with steam condensing inside a pressure vessel. The main component of the THAI facility is a cylindrical vessel with a volume of $60\,\mathrm{m}^3$ and was performed by Becker Technologies. The height of the vessel is 9.2 m with a diameter of 3.16 m. An open inner cylinder with a diameter of 1.4 m is installed inside the vessel. In the TH2 experiment colder air is initially present in the vessel. The TH2 experiment was divided into six phases. For the simulation only the first $3,600\,s$ of the first phase are used. During the first phase, hot steam is injected through an upper inlet into the vessel. The injection opening is near the center of the vessel and located at 6.7 m, see Fig. 7. The volume condensation can occur in the upper vessel. The wall condensation can also occur

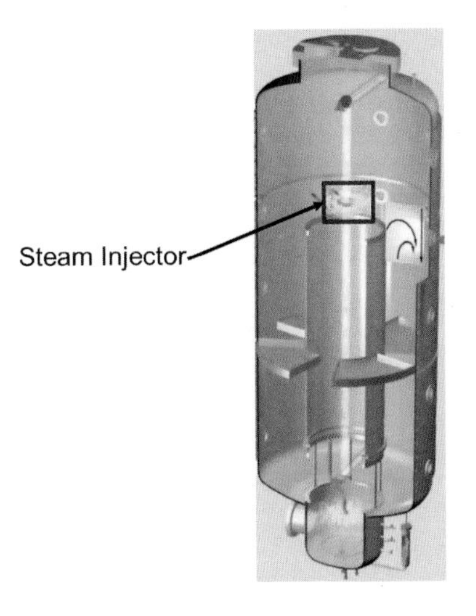

Fig. 7 THAI vessel

on the colder wall. The heat transfer coefficient from the wall to ambience and the ambient temperature were set to constant value. The values of the mass flow rates and injection temperatures were taken from experimental data.

4.2 Numerical Setup

The transient simulations for the TH2 test case were performed using the coupled double precision solver. The SST turbulence model and the buoyancy model were used for all simulations. For mass, momentum and energy conservation equations the second order scheme was applied. A first order scheme was chosen for the turbulence model. For the convergence RMS criterion was set to 5×10^{-4}.

4.3 Results

The first calculation is a single-phase simulation and focuses on the natural convection flow of air and steam without condensation. The second calculation is also a single-phase simulation but performed within the CFX implemented wall condensation model, which was developed by ANSYS [18]. The wall condensation model was activated at the fluid domain interface on the fluid side. Steam condensation on the walls is modeled as a sink of mass and energy. The liquid film was not modeled. The third calculation is a two-phase simulation including the wall condensation model and the newly developed volume condensation model. In this calculation the droplets are present in the vessel with a constant droplet diameter

Table 1 Position of monitor points

Name	Vertical elevation [m]	Radius [m]
MP25	7.7	0.7
MP12	1.2	0.35

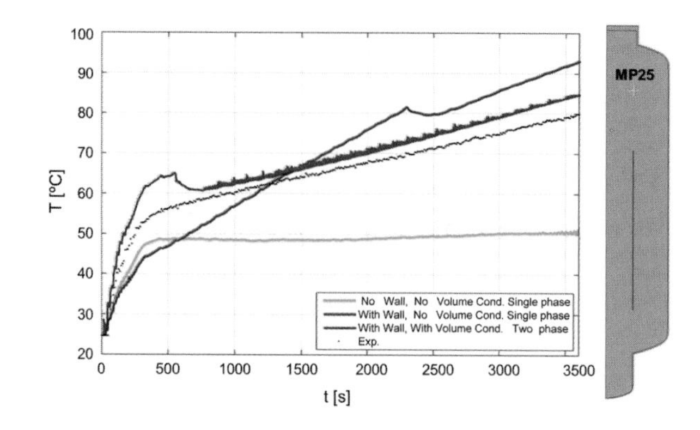

Fig. 8 Temperature comparison between three calculated values and experiment at MP25

of 100 μm and an initial droplet volume fraction of 10^{-5}. For the presentation of results, two monitor points (MP) were chosen, which represent two main conditions. Their positions are provided in Table 1.

As shown in Fig. 8, the calculated temperatures are compared with the experiment during the first 3,600 s at MP25. The simulation without any condensation model shows a poor agreement with the experiment. The calculated temperature is consistently underpredicted after 400 s. While the steam without condensation mixes quickly with the cold atmosphere, the steam is more efficiently distributed in the vessel. The simulation with only the wall condensation model shows a temperature difference of 10 K between the calculated and measured temperature. The simulated temperature is lower during the first 2,300 s and higher after 2,300 s than the measured temperature.

The calculated temperature, which is simulated with the wall condensation model and the newly developed volume condensation model, increases rapidly in first 600 s, because the injected steam condenses directly in the upper region of the vessel and the heat is released into the ambience. After contact with the wall, the steam condenses not only in the volume but also at the wall of the vessel. After 800 s the calculated temperature achieves a good agreement with the experiment.

The calculated temperatures and the experiment are plotted in Fig. 9 at MP12. Without any condensation model, the hot steam is distributed easily to the lower region of the containment, so that the calculated temperature shows an overprediction compared to the experiment. With the usage of the condensation models, the temperatures are in good agreement with the experiment. Because of the condensation only appeared in the upper vessel, the temperature changes are small in the lower vessel. The simulation results with the wall and volume condensation model show a better agreement with the experiment, than solely with the wall condensation model.

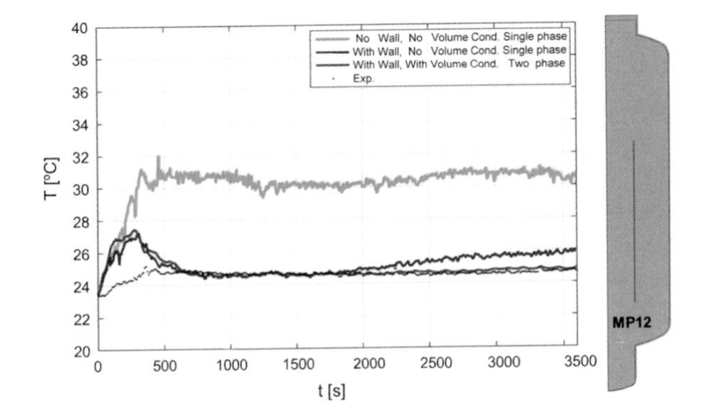

Fig. 9 Comparison of experiment with three calculated temperatures at MP12

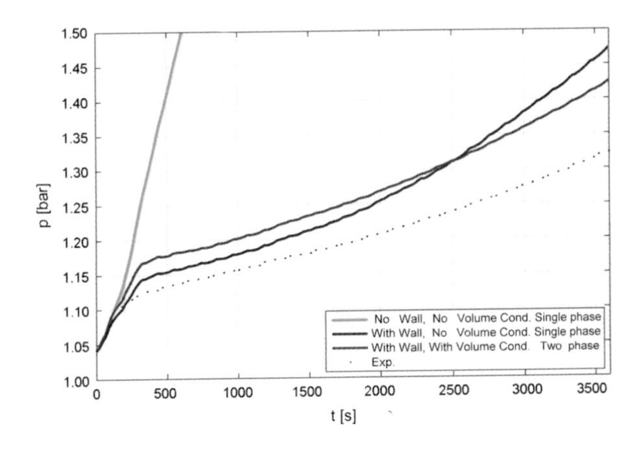

Fig. 10 Comparison of calculated pressures with experiment in THAI vessel

Figure 10 shows the measured and simulated pressures in the THAI vessel. Without any condensation model the released steam could not be condensed in the vessel, so that the calculated pressure is much higher than the measured pressure. Because of the pressure drop with the steam condensation, the calculations of the pressure with condensation models show a better agreement with the experiment data. Here the simulation result of the pressure with the wall and volume condensation model shows a best agreement with the experiment.

During the first 400 s the steam condenses mostly in the volume of the vessel. To investigate the influence of droplet diameter, the simulations were carried out with the wall and volume condensation model. For these simulations the initial droplet volume fraction is assumed to be 10^{-5}. Four different droplet diameters (e.g. 50, 100, 200 and 300 μm) are given as input parameters. Figure 11 shows the comparison of the pressures between the experiment and calculated values with different droplet diameters. Obviously, smaller droplet diameters lead to lower pressures. While the smaller droplets have larger interfaces of droplet, the steam

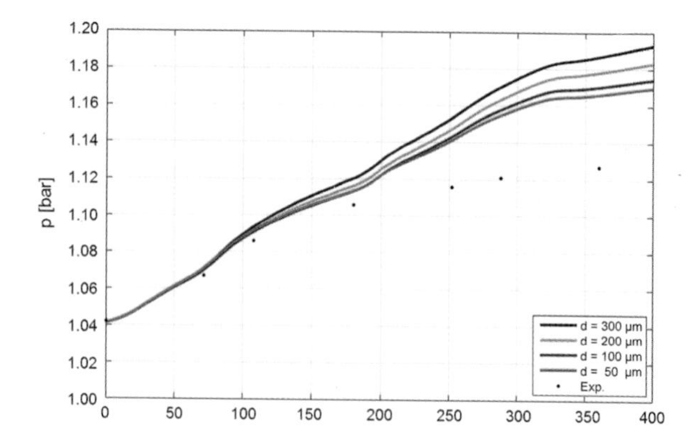

Fig. 11 Comparison of the pressures between experiment and calculated values with different droplet diameters

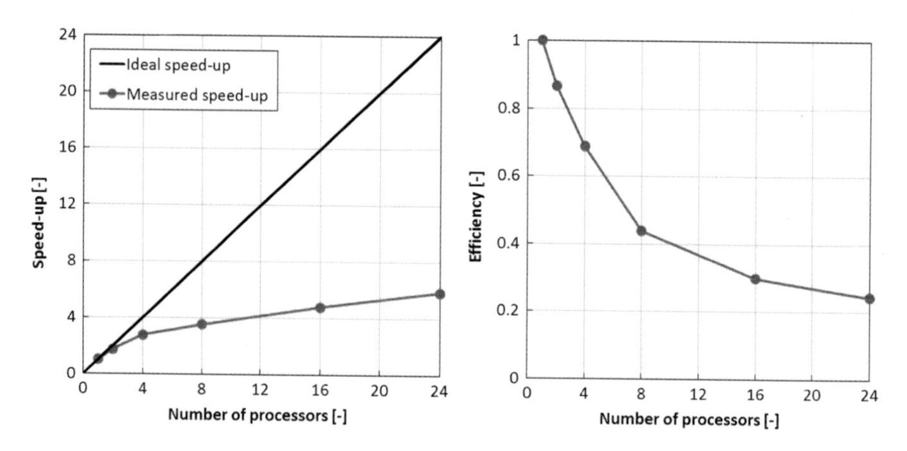

Fig. 12 Speed-up (*left*) and efficiency (*right*) with the increase number of processors

could be condensed faster at these droplet interfaces. The calculated pressure with the smallest droplet diameter of 50 μm shows the best agreement with the experiment.

4.4 Computational Performance

To estimate the efficiency on the NEC Nehalem Cluster at the HLRS, the single-phase simulations were carried out only with the wall condensation model for up to 24 processors. The mesh for all scaling tests contains approximately 0.1 million grid points. In order to analyse the performance of the parallel program ANSYS CFX, Fig. 12 shows the speed-up and efficiency of the simulations with the increase number of the processors. The reference for the speed-up is the computation time on a single processor. If the numbers of the processor are more than one, the measured speed-up could not reach the ideal speed-up. The efficiencies for two and four processors are above circa 70 %. Parallelisation on more processors makes more communications, which has further influence on the efficiency. For future three-dimensional simulations with a model-containment mesh, which consists of about 10 million grid points, we plan to ensure the efficiency on up to 112 processors at the HLRS.

5 Conclusions

The physical formulation of newly developed volume condensation model was presented in this paper. It was based on the two-fluid model, which simulates a continuous gas mixture and a dispersed droplet. The thermal and mechanical non-equilibrium were considered for the temperatures and velocities between both phases. The sink term of mass and source term of energy were modeled for the volume condensation. The SST turbulence model and buoyancy model were used for the simulations. Firstly, the volume condensation model was performed for a two-dimensional steady state test case. Based on a Mollier diagram, the simulation results have been verified and shown a good agreement with the theory. Then the volume condensation model was tested in a two-dimensional test case of the TH2 experiment. The two-phase simulation results, using the wall and volume condensation model, were compared with the single-phase simulations without any condensation models, solely with wall condensation model, as well as with the experiment during the first 3,600 s. The transient temperature and pressure profiles using the wall and volume condensation model have shown to give the best agreement with the experiment. To investigate the sensitivity of the results, the influence of the droplet diameter was discussed during the first 400 s.

Acknowledgements This work was supported by the German Federal Ministry of Economics and Technology (BMWi) on the basis of a decision by the German Bundestag, project number 1501414. The simulations were performed on the national supercomputer NEC Nehalem Cluster at the High Performance Computing Center Stuttgart (HLRS) under the grant number TurboCon/12843.

References

1. M. Babić, I. Kljenak, B. Mavko, Simulation of atmosphere mixing and stratification in the ThAI experimental facility with a CFD code. *International Conference Nuclear Energy for New Europe*, Bled, Slovenia, 2005
2. I. Kljenak, M. Babić, B. Mavko, I. Bajsić, Modeling of conainment atmosphere mixing and stratification experiment using a CFD approach. Nucl. Eng. Des. **236**, 1682–1692 (2006)
3. M. Babić, I. Kljenak, B. Mavko, Prediction of light gas distribution in experimental containment facilities using the CFX4 code. Nucl. Eng. Des. **238**, 538–550 (2008)

4. J.M. Martín-Valdepeñas, M.A. Jiménez, F. Martín-Fuertes, J.A. Fernández, Improvements in a CFD code for analysis of hydrogen behavior within containments. Nucl. Eng. Des. **237**, 627–647 (2007)
5. M. Houkema, N.B. Siccama, Validation of the CFX4 CFD code for containment thermal-hydraulics. Nucl. Eng. Des. **238**, 590–599 (2008)
6. L. Vyskocil, J. Schmid, J. Macek, CFD simulation of air-steam flow with condensation. Computational Fluid Dynamics (CFD) in Nuclear Reactor Safety (NRS), Korea, 2012
7. P. Royl, J.R. Travis, J. Kim, GASFLOW a computational fluid dynamics code for gases aerosols, and combustion, vol. 3, Karlsruhe, 2008
8. R. Mollier, Ein neues Diagramm für Dampfluftgemische. Z. VDI **67**, 869–872 (1923)
9. R. Mollier, Das ix-Diagramm für Dampfluftgemische. Z. VDI **73**, 1009–1013 (1929)
10. M. Ishii, K. Mishima, Two-fluid model and hydrodynamic constitutive relations. Nucl. Eng. Des. **82**, 107–126 (1984)
11. M. Ishii, *Thermo-fluid Dynamics of Two-phase Flow* (Springer, New York, 2006)
12. N. Kolev, *Multiphase Flow Dynamics 2* (Springer, Berlin, 2005)
13. W.E. Ranz, W.R. Marshall, Evaporation from drops. Chem. Eng. Progr. **48**, Part I, 141–146, Part II, 173–180 (1952)
14. R. Clift, J.R. Grace, M.E. Weber, *Bubbles, Drops, and Particles* (Academic Press, New York, 1978)
15. C.T. Crowe, M. Sommerfeld, Y. Tsuji, *Multiphase Flows with Droplets and Particles* (CRC Press, Boca Raton, 1998)
16. H.R. Pruppacher, J.D. Klett, *Microphysics of Clouds and Precipitation* (Springer, Dordrecht, 2010)
17. T. Kanzleiter et al., Final Report, Experimental Facility and Program for the Investigation of Open Questions on Fission Product Behaviour in the Containment. Report No. 1501218 (2003)
18. G. Zschaeck, T. Frank, A.D. Burns, CFD modelling and validation of wall condensation in the presence of non-condensable gas. Computational Fluid Dynamics (CFD) in Nuclear Reactor Safety (NRS), Korea, 2012

Highly Efficient Integrated Simulation of Electro-membrane Processes for Desalination of Sea Water

Kannan Masilamani, Harald Klimach, and Sabine Roller

Abstract Electrodialysis is an efficient process for sea water desalination. In this process, sea water flows through the channel with a complex spacer structure that separates the ion exchange membranes. The multi-species lattice Boltzmann model for liquid mixture modeling has been chosen to study the transport phenomena of ionized liquids in the spacer filled channel. This model is implemented in the highly scalable simulation framework APES based on octree meshes. In this paper, the performance and scalability of our implementation on the Cray XE6 system Hermit in Stuttgart are presented. The performance analysis is performed on periodic cubic domains with different compilers and communication patterns. The scalability of our solver with spacer structure for single-fluid LBM and multi-species LBM with three species mixture are presented. The spacer structure is scaled from single spacer element to full length of laboratory experiment spacer.

1 Introduction

Scarcity of drinking water is a growing problem in many regions of the world. This problem motivates the development of new efficient sea water desalination technologies. There are numerous sea water desalination process techniques developed

K. Masilamani (✉)
Siemens AG, Corporate Technology, CT RTC ENC ENT-DE, Günther-Scharowsky-Str. 1, 91058 Erlangen, Germany

Simulationstechnik und wissenschaftliches Rechnen, University of Siegen, Hölderlinstr. 3, 57076 Siegen, Germany
e-mail: kannan.masilamani@uni-siegen.de

H. Klimach · S. Roller
Simulationstechnik und wissenschaftliches Rechnen, University of Siegen, Hölderlinstr. 3, 57076 Siegen, Germany
e-mail: harald.klimach@uni-siegen.de; sabine.roller@uni-siegen.de

W.E. Nagel et al. (eds.), *High Performance Computing in Science and Engineering '13*, DOI 10.1007/978-3-319-02165-2_34,

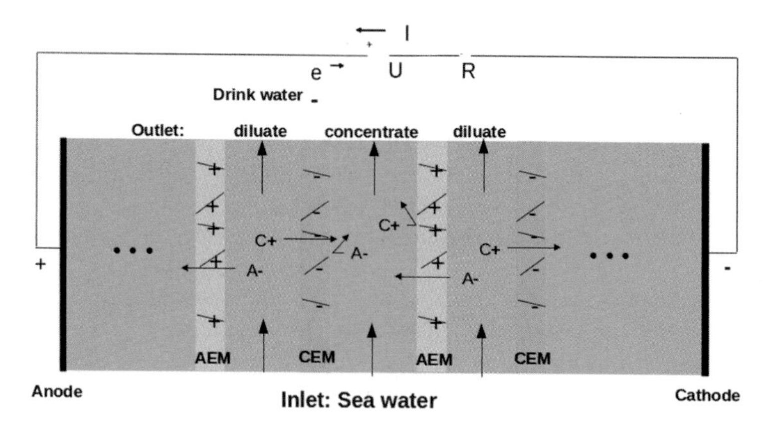

Fig. 1 Simplified structure of a desalination stack in two dimensions. A periodic arrangement of anion-exchange-membrane (AEM), diluate channel, cation-exchange-membrane (CEM) and concentrate channel is embedded between anode and cathode. The sea water is fed in both channels. After removing the ions, drink water can be extracted at the end of the diluate channel

to supply drinking water to mankind. However, energy and environmental impact are additional constraints and for the mass production of drinking water an energy and cost efficient technique is required. Electrodialysis is a promising technology in this area, which uses ion exchange membranes to separate ionic species from sea water under the influence of an applied electric field. This technique is already applied in practice, the underlying physical and chemical process are not understood in the detail that is necessary for optimizations of the devices. Numerical simulation together with large HPC systems are used to study these underlying physical phenomena in this project.

A schematic illustration of an electrodialysis stack is shown in Fig. 1. It consists of anion exchange membranes (AEM) and cation exchange membranes (CEM) arranged alternatively. At the ends of the stack anodes and cathodes are used to induce an electrical field. When an electric potential is applied, anions and cations start to separate from sea water. AEM and CEM are selective permeable membranes, they allow either anions or cations to pass through respectively. In diluate channels, ions are removed, while they are collected in concentrate channels. At the outflow of the stack, desalinated water is extracted from the diluate channels, this water is fed again into the stack several times until the desired concentration level is obtained.

Ion exchange membranes (AEM and CEM) in the stack are separated by a complex structure called spacer. This spacer structure is shown in Fig. 2. It acts as a mechanical stabilizer and also induces turbulence. The effective transport of ions through the membranes and also the total energy consumption of the stack is mainly influenced by the geometry of the spacer.

In order to study the physical phenomena in the spacer filled channel, a suitable numerical method is required. The Lattice Boltzmann Method (LBM) is chosen as our numerical scheme due to its advantage of easily integrating complex geometries

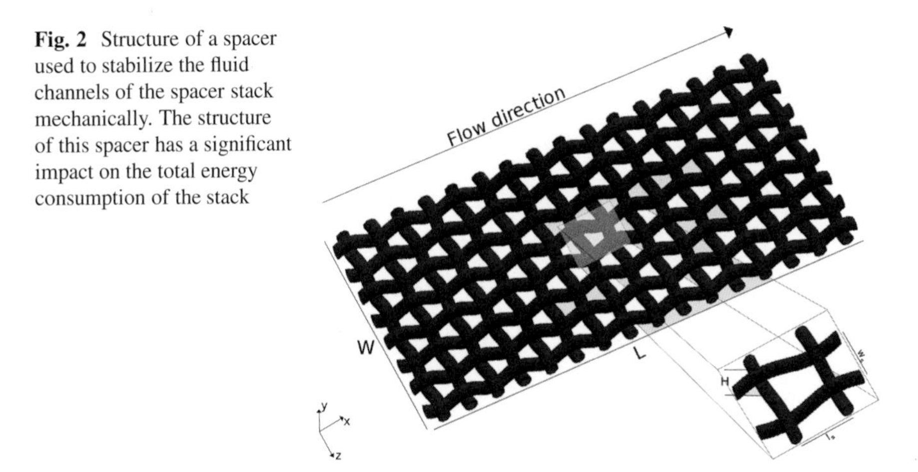

Fig. 2 Structure of a spacer used to stabilize the fluid channels of the spacer stack mechanically. The structure of this spacer has a significant impact on the total energy consumption of the stack

and its performance on large scale HPC systems [1]. This method is used to simulate the fluid flow in the spacer filled channel. For the mass transport of ionic species this method is extended to the multi-species LBM for liquid mixture [2]. The multi-species LBM model recovers the Maxwell-Stefan equations for diffusive fluxes and the Navier-Stokes equations for ions and mixture transport in the macroscopic regime. This solver is embedded in our highly scalable parallel framework named APES (Adaptable Poly-Engineering Simulator), which is based on an octree data structure [3].

In this paper we focus on the performance of our implementation of the numerical schemes on the Cray system Hermit. The paper is organized as follows: In Sect. 2 we describe the Lattice Boltzmann Method and the multi-species LBM for liquid mixtures. In Sect. 3 we briefly describe the APES framework. Section 4 is concerned with performance and scalability of both numerical schemes. Finally, in Sect. 4 we summarize our work and present an outlook to future work on this topic.

2 Numerical Scheme

In this section the chosen numerical scheme the Lattice Boltzmann Method for flow simulation and the multi-species Lattice Boltzmann Method for ions and mass transport simulation are briefly presented.

2.1 The Lattice Boltzmann Method

The Lattice Boltzmann Method (LBM) represents the incompressible Navier-Stokes equations in the macroscopic level in the limit of small Mach and Knudsen numbers and its derived from Boltzmann equation [4]. The LBM equation describes the time evolution of particle distribution function f_i on the mesoscopic level on the uniform grid with fixed set of discrete velocities. The most popular LBM equation with BGK [5] collision operator is given as

$$\underbrace{f_i(\mathbf{x} + \mathbf{e_i}\Delta t, t + \Delta t) = \underbrace{f_i(\mathbf{x}, t) - \omega(f_i(\mathbf{x}, t) - f_i^{eq}(\mathbf{x}, t))}_{\text{Collision}}}_{\text{Streaming}} \tag{1}$$

where $f_i(\mathbf{x}, t)$ is the particle distribution function defined at position $\mathbf{x}$ and at time t. $\mathbf{e_i}$ is set of discrete velocity vectors and the index $i = 1 \ldots Q$ where Q is the number of velocity directions which is dependent on the velocity model. The velocity model are chosen for our simulation is $D2Q19$ where D stands for dimension. The collision term is defined by BGK operator, it defines the relaxation of particle distribution f_i towards thermodynamic equilibrium f_i^{eq} with single relaxation parameter ω. The relaxation parameter is related to lattice viscosity by $\omega = 1/(3\nu + 0.5)$. The spatial and temporal step size in lattice units are $\Delta x = \Delta t = 1$, i.e a particles travel only one lattice per time step.

Equation (1) is solved for each time step through two step procedure: streaming and collision. In streaming step, the distribution function f_i are streaming to its neighboring cells in the corresponding velocity direction. In collision step, the distribution functions are relaxed towards its equilibrium f_i^{eq}, this is the cell local step whereas the streaming step requires neighbor information. The thermodynamic equilibrium distribution for incompressible model has following form

$$f_i^{eq}(\mathbf{x}, t) = w_i \rho \left(1 + \frac{\mathbf{e_i} \cdot \mathbf{u}}{c_s^2} + \frac{(\mathbf{e_i} \cdot \mathbf{u})^2}{2c_s^4} - \frac{\mathbf{u}^2}{2c_s^2}\right) \tag{2}$$

where ρ and $\mathbf{u}$ are macroscopic density and velocity. c_s is the speed of sound and in lattice unit it is given by $c_s = \frac{c}{\sqrt{3}}$ and c is the lattice speed $c = \Delta x / \Delta t$. w_i is weighting factor for each discrete directions. The macroscopic density and velocity can be computed from distribution function f_i from the following equations

$$\rho = \sum_{i=1}^{n} f_i(\mathbf{x}, t) \tag{3}$$

$$\mathbf{u} = \frac{1}{\rho} \sum_{i=1}^{n} f_i(\mathbf{x}, t)\mathbf{e_i}. \tag{4}$$

The macroscopic pressure p can be computed from density ρ from the relation

$$p = c_s(\rho_0 + \rho) \tag{5}$$

where ρ_0 is the reference density.

2.2 The Multi-species LBM for Liquid Mixture

In multi-species LBM, (1) has to be solved to each individual species k and the interaction between the species is achieved by changing the velocity in the equilibrium function [2]. Now, for each species k (1) can be written as

$$\underbrace{f_i^k(\mathbf{x} + \mathbf{e_i}\Delta t, t + \Delta t) = \underbrace{f_i^k(\mathbf{x}, t) - \omega^k (f_i^k(\mathbf{x}, t) - f_i^{eq,*,k}(\mathbf{x}, t))}_{\text{Collision}}}_{\text{Streaming}}. \tag{6}$$

$f^{eq,*,k}$ is the modified equilibrium distribution function for species k and it has the following form

$$f_i^{eq,*,k}(\mathbf{x}, t) = w_i \rho^k \left(s_i^k + \frac{\mathbf{e_i} \cdot \mathbf{u^{*,k}}}{c_s^2} + \frac{(\mathbf{e_i} \cdot \mathbf{u^{*,k}})^2}{2c_s^4} - \frac{\mathbf{u^{*,k}} \cdot \mathbf{u^{*,k}}}{2c_s^2} \right) \tag{7}$$

Here, we choose $s_0^k = (3 - 2\phi_k)$ and $s_{i \neq 0}^k$ for $D3Q19$ model and $\phi_k = \frac{min_l(m_l}{m_k} \leq 1$. where ρ_k is the macroscopic density of species k and $\mathbf{u^{*,k}}$ is the modified macroscopic velocity of species k which is given as

$$\mathbf{u^{k,*}} = \mathbf{u^k} + \sum_l \frac{B_{k,l}}{B} \phi_k \chi_l (\mathbf{u_l} - \mathbf{u_k}) \tag{8}$$

where B_{k_l} is the Maxwell-Stefan diffusion resistance coefficient between species k and l. B is the free parameter fixed during initialization. $\chi_k = n_k/n$ is the molar fraction of species k and $n_k = \rho_k/\rho$ and n are number density of species k and total number density. $rho = \sum_k \rho_k$ is the mixture density. Equation (8) defines the interaction between species which corresponds to Maxwell-Stefan equations in the macroscopic regime as

$$\nabla \chi_k = \frac{1}{n} \sum_{l \neq k} \frac{1}{D_{k,l}} (\chi_k J_l - \chi_l J_k) \tag{9}$$

For further details on the multi-species LBM for liquid mixture modeling refer to [2]. Here, we can notice that multi-species LBM can be solved in a similar fashion like single-fluid LBM 2.1 with stream and collision step. Only overhead is in the computation of collision term which is cell local operation. However, this additional computational cost (i.e floating point operations) has an advantage in efficient usage of compute cores which is presented in Fig. 9.

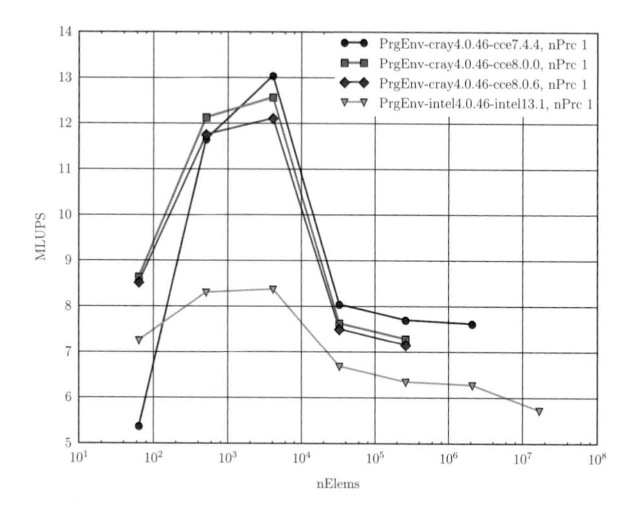

Fig. 3 Comparision of single core performance of different compiler

3 Apes Framework

The Adaptable Poly-Engineering Simulator (APES) is a highly scalable solver framework that can be used to simulate different physics with numerical schemes based on meshes [3]. Its central component is an octree based element-wise data structure (*TreElM*) [6], which is implemented as a library and used in the other parts of the framework. *APES* covers all necessary steps in mesh based simulations from mesh generation over dedicated solvers to a post-processing tool. All tools are based on the same data structure ensuring that the scalability of the simulation is maintained across all steps and allowing the deployment on large scale parallel systems. The Lattice Boltzmann Method described in the previous section is implemented in this framework as a dedicated solver named *Musubi*. It is capable of deploying various stencil layouts and specialized kernels. The handling of the meshes is focused on complex geometries and uses an unstructured management to deal with sparse trees. Though the unstructured approach imposes some overhead on the computation it is beneficial for non-trivial geometries. More detailed information on the solver itself is provided in [7].

4 Performance and Scalability

This section describes the performance of Musubi on the Cray XE6 system Hermit at HLRS. The Hermit system provides of 3,552 computing nodes with AMD Interlagos on two sockets where each socket has 16 cores resulting in 32 cores per node. For our performance analysis, up to 1,024 computing nodes or 32,768 cores are used. Only MPI parallelism is considered here. Both, intranode and internode performance are measured i.e performance within a single node (up to 32 cores with as many MPI processes) and between multiple nodes. Two different test cases are chosen for performance and scaling analysis. The first one is an academic setup with a fully periodic cubic domain. And the second one is a channel with complex spacer geometry. While the first one provides a well defined environment for scaling with ideally self-similar problems for weak scaling, the second resembles the actual production runs for the considered application.

Usually, the performance of the machine for numerical methods is given in floating point operations per second, i.e GFLOPs (Giga floating point operations per second). However, for Lattice-Boltzmann codes the measurement of the lattice updates per second is commonly used to compare the performance. As the number of floating point operations per lattice update is fixed this can be directly translated into GFLOPs. This number of lattice updates per second (LUPs) will be used in the following presentation, as it provides a direct estimation on how long a given simulation, that requires a certain number of lattice nodes and time step updates will take.

We represent the behavior of the code in terms of performance per execution unit, that is per node or per core, to get a clearer impression of the performance independent of the number of used execution units. An ideal parallel execution is expected to just replicate the serial behavior on each execution unit. However, the execution performance is influenced by cache usage, non-computational implementation overheads, vector lengths, communication times and so on. The performance maps in Figs. 6 and 10 shows the achieved performance per node over the problem size per node for various total number of nodes. The performance map combines both, weak scaling and strong scaling. Weak scaling can be measured by the vertical comparison of points between different lines for the node count i.e. fixing the number of elements per node. The closer the points are located to each other the better the weak scaling. Strong scaling on the other hand is not as easily seen in the performance map, but can be derived by moving to the left when increasing the number of nodes. This reduces the number of elements per process with increasing node counts, as required by strong scaling with a fixed overall problem size.

4.1 Periodic Testcase

We choose a fully periodic cubic simulation domain as a test case in which the problem size and communication surface on each core are exactly the same. A refinement by one level increases number of elements by a factor of 8, due to a doubling of element counts in each direction. This is in some sense a worst case scenario for the MPI communication, since there are no obstacles anywhere, that could reduce the communication surfaces. In our analysis, problem sizes from 64 elements up to 1 billion elements are covered. At first, the performance of Musubi

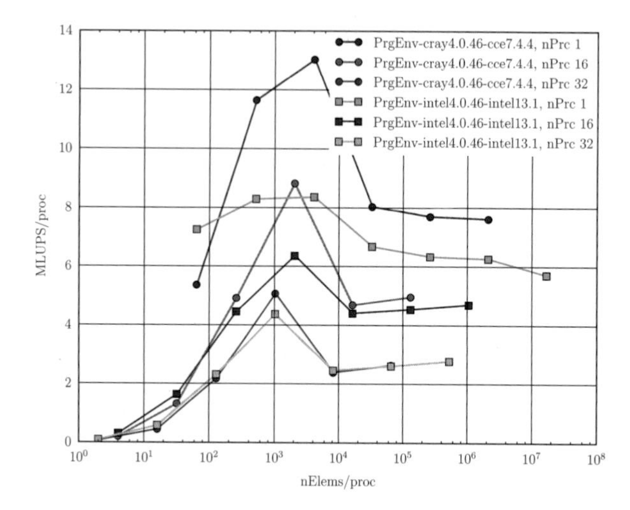

Fig. 4 Comparision of intranode performance of Cray vs Intel compiler

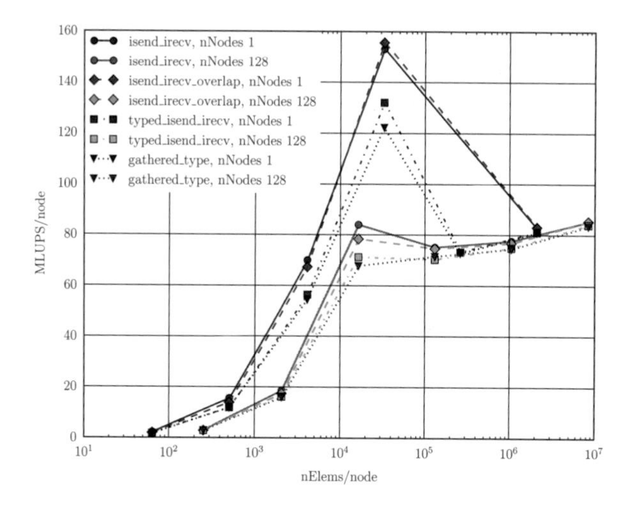

Fig. 5 Performance of different communication patterns on nNodes 1 and 128

with different compilers and different communication patterns are investigated to find the most suitable compiler and communication pattern for the application on the Hermit system.

The single core performance of the application with different compilers is shown in Fig. 3. Various Cray compiler versions in the fixed Cray programming environment *PrgEnv-cray/4.0.46* and the Intel compiler in version *13.1* are compared. The measurements clearly show, that for Musubi on Hermit, the Cray compiler is clearly better suited than the Intel compiler and especially the oldest deployed Cray compiler *cce/7.4.4* provides the highest performance for our solver. Figure 4 shows the intranode performance map for the Cray *cce/7.4.4* and the Intel compiler. The performance is given per core, where one MPI process is used per core.

The overall performance can be obtained by multiplying the shown performance with the number of participating cores. For small problem size i.e at the left of the curves, the performance is low due to non-computation overheads, especially in parallel executions a high communication to computation ratio gets visible in this domain. These overheads diminish the advantages of small problems fitting into the cache, however a peak from the cache usage is still clearly visible at around 1,000 elements per core. To the right end the out of cache performance can be seen. The comparison between Cray and Intel compiler shows, that the difference between them is reduced with higher number of processes, however the Cray compiler seems to maintain a slight advantage in the cache region. Furthermore, the analysis shows, that it is advantageous to make use of all cores on a single node, though the gain from 16 cores per node to 32 cores per node is not large.

In terms of memory usage, the executable produced by the Intel compiler is capable of fitting 16 Million elements on a single node, while the Cray compiler fits only 2 Million elements on a single node. This seems to be due to different treatments of arrays of derived data types by the two Fortran compilers.

For our production simulations and in the following analysis, the Cray compiler is used for its better performance. As the intranode analysis showed and advantage in full node performance when using all 32 cores with one MPI process each, this mode of operation is used.

Our next investigation is concerned with the performance of different communication patterns in Musubi for a single node with 32 processes and on 128 nodes to account for the usage of the network. There are several communication patterns available in Musubi and configurable at runtime. The most basic one is the *isend_irecv* pattern, where buffers are used explicitly to gather information from the actual simulation data and these buffers are than exchanged by non-blocking point to point communications. In this pattern the steps are to first copy the data from their original location to contiguous send buffers, then post all receives and sends, wait on all communications to finish and finally copy the data in the received buffers to their destination. A slight variation of this pattern is offered by *isend_irecv_overlap*, where the copying of data from and to the buffers is overlapped with the communication (with different processes). Finally, there are two modes that do not make use of explicit buffers in the application for the communication but instead use MPI data types to describe the location of data to communicate in the solver arrays directly. *Typed_isend_irecv* describes the data type with individual entries for every number to exchange, while *gathered_type* refers to a data type where contiguous parts are gathered and described as a block in the MPI data type.

The performance of the different communication patterns are given in Fig. 5. It shows that the difference between communication patterns has its maximum in the region where problem sizes fit into the cache. On a single full node *isend_irecv* and *isend_irecv_overlap* exhibit almost the same behavior and are better than the communication patterns deploying MPI data types. The *isend_irecv* also yields the highest performance on 128 nodes, though the distance between all communication

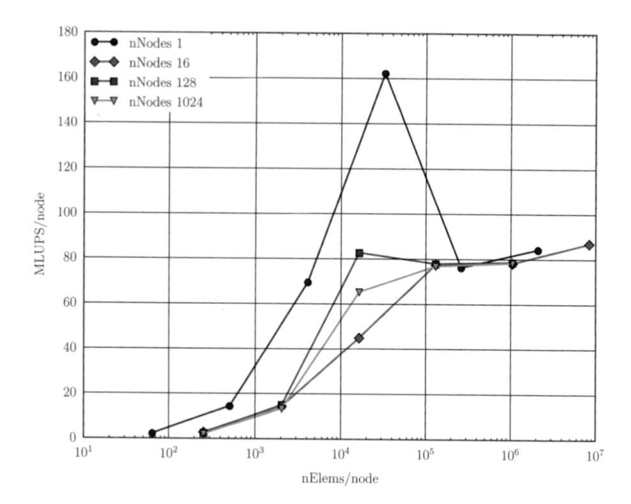

Fig. 6 Internode performance map of LBM model for periodic testcase

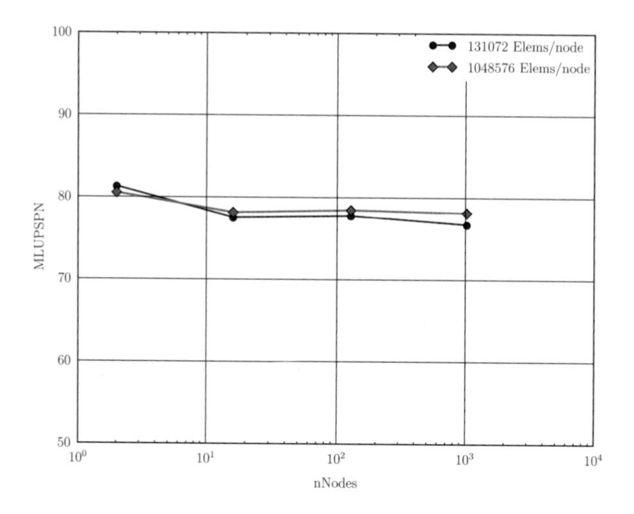

Fig. 7 Weak scaling for different number of nodes for two different number of elements per node for periodic test case

patterns gets smaller in this internode setting. Due to this behavior we stick to the *isend_irecv* communication pattern for production runs and the following performance measurements.

The performance map in Fig. 6 depicts the performance in terms of million lattice updates per second (MLUPs) per node over the problem size per node for various node counts. Compared to the single node performance, communication now involves network transfers, which increases the time spent on communication. This effect can be seen clearly for small problem sizes per node with a large ratio of communication to computation and low computing efficiency. The larger the

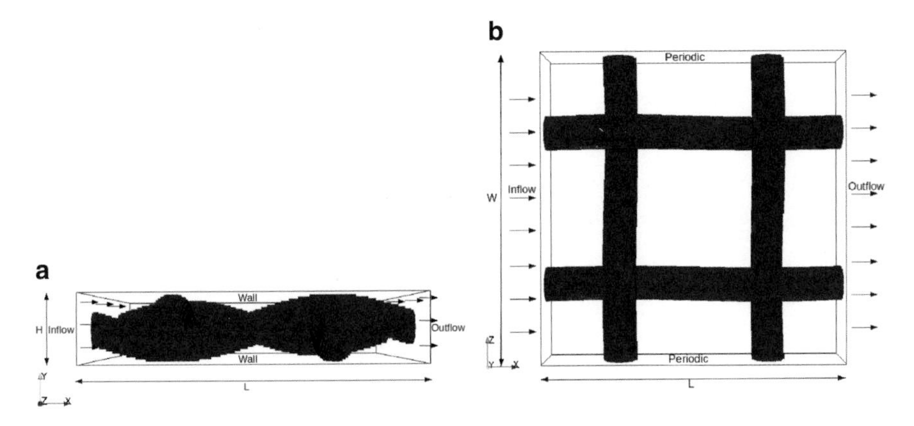

Fig. 8 Schematic layout of the simulation setup with interwoven spacer geometry. (**a**) Side view of the channel; (**b**) Top view of the channel

problem size per node, the smaller this ratio becomes. For problem sizes fitting into the cache, weak scaling is not so good but for larger problem sizes weak scaling is nearly optimal. Figure 7 shows a dedicated plot for weak scaling with approximately 131,000 elements per node and 1 million elements per node. The points on the line correspond to (from left to right), $2, 16, 128$ and 1,024 computing nodes on the Hermit system. Ideal weak scaling would be a straight line which is almost achieved for these problem sizes per node.

4.2 Complex Spacer Geometry Testcase

Our production runs require a complex geometry given by mesh structured spacer, that holds the two membranes for the electrodialysis apart. The fluid flows horizontally along the filaments of the mesh between the two membranes as shown in Fig. 8. To find the optimal point of operation for the production runs, a scaling analysis was done with this complex geometry in addition to the simple setup without any boundary conditions discussed in the previous section. In the following we'll refer to the domain with two filaments covering enclosing one free space section as a single spacer element. It has a length of $0.2\,cm$ and is discretized by 66,000 cells. For the investigation of different problem sizes, this single element is repeated up to a total length of $20\,cm$ with roughly 6 million cells. In the width of this channel slice, a periodic boundary is assumed. This setup covers all relevant production settings, hence all effects that influence the production performance are included in the analysis.

Our solver framework ensures nearly perfect balancing of the computational load (neglecting communication costs), with at most a difference of one in the cell count between any two partitions. However, due to the irregular domain and

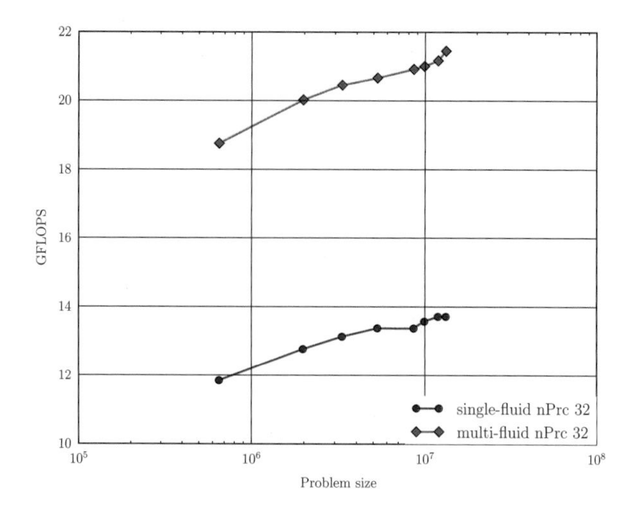

Fig. 9 Full node performance of single fluid and multi-species LBM model with spacer structure

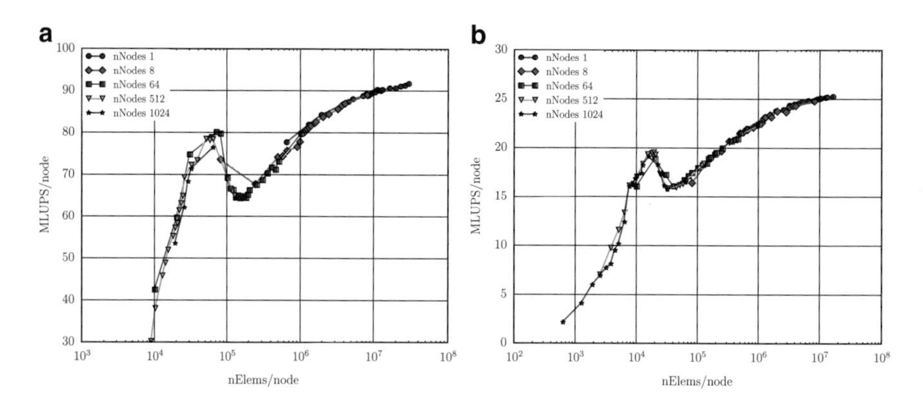

Fig. 10 Internode performance map with spacer structure. (**a**) Single-fluid LBM model; (**b**) Multi-species LBM model

the large number of walls in the spacer filled channel, the communication surface between different processes might vary drastically, resulting in a large imbalance of communication costs. This might offer an explanation for the rather dependency of the performance on the problem size as observed in Fig. 10, even in the out of memory region.

First, the performance runs of the single-fluid and multi-species LBM are performed using a single node with 32 cores, depicted in Fig. 9. In both models a D3Q19 layout with a BGK-like collision operator is used. Streaming and collision steps are solved at each time step for both models. However, the multi-species LBM model used in this work requires the solution of an additional cell local linear equation system, increasing the required number of floating point operations per lattice update. The single fluid solver requires only about 150 floating point operations, while a simulation with 3 species requires with 850 operations more than just a factor of 3 more operations. This factor increases more than linearly with the number of species, as the equation system to be solved increases with the number of species.

In our LBM solver, instead of communicating all probability density functions on the communication surface between the processors, only the required links of probability density functions are communicated. This reduces the MPI buffer size and bandwidth driven communication times. The amount of data is depending on the number of species. Thus also the data to be exchanged via MPI is increased with larger number of species. In the simulation of a ternary mixture the MPI buffer sizes are therefore by a factor of three larger than for single fluid simulations. Full node performance of both LBM models over different problem sizes is shown in Fig. 9 in terms of GFLOPS. Since the number of elements with a single spacer element i.e. 66,000 is above the cache size, it is not possible to show the effect of the cache for this single node execution. The Hermit system has a theoretical peak performance of 294.4 GFLOPS per node. A sustained performance of roughly 7.2 % has been achieved with multi-species LBM and 4.2 % with single-fluid LBM on a single node. This reveals a better utilization of the Interlagos processors by the multi-species simulation, due to the increased number of operations per byte.

Figure 10 shows the performance of the single fluid LBM (Fig. 10a) and the multi-species LBM (Fig. 10a) implementation for different problem sizes and number of computing nodes. Both models show similar profiles, though at different absolute performance levels. It should also be noted, that with the single-fluid LBM nearly 80,000 cells can be put on a single node, while with three species this figure is reduced to around 20,000 elements per node. A small cache effect peak is observed in both cases between 10^4 and 10^5 cells per node, however it is outperformed by sufficiently large problem sizes in the out of cache region.

The steep slope in the out of cache region in both models is supposedly due to load imbalances in the communication times, caused by the walls of the spacer geometry. As can be seen in the periodic test case in Fig. 6, the performance is much more independent of the problem size in the out of cache region when there is no complex geometry. The main difference between the two setups is the exactly balanced communication effort in the one case and a less than optimal balancing for the complex spacer geometry. This can be resolved using a dynamic load balancing algorithm to redistribute the simulation domain on each processor at runtime according to the actual load, including the communication times.

From Fig. 10, it can be seen that weak scaling works fine on different process counts for all problems, that fit into memory down to the cache-sized problems, where the communication gets dominant and the performance per node drops down with a steeper slope.

An explicit plot for weak scaling for two different problem sizes per node is shown in Fig. 11. One with approximately 21,000 elements per node which fit into the cache for multi-species LBM and another with approximately 63,000 elements

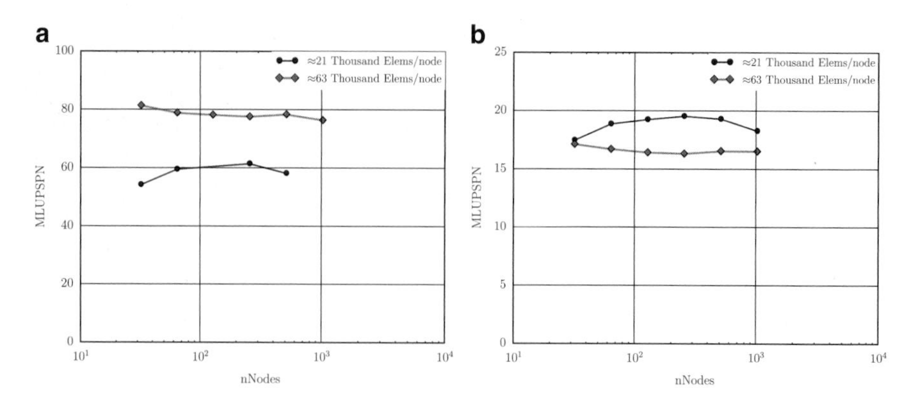

Fig. 11 Weak scaling for two different number of elements per node with spacer structure for different number of nodes. Ideal weak scaling is a straight line. (**a**) Single-fluid LBM model; (**b**) Multi-species LBM model

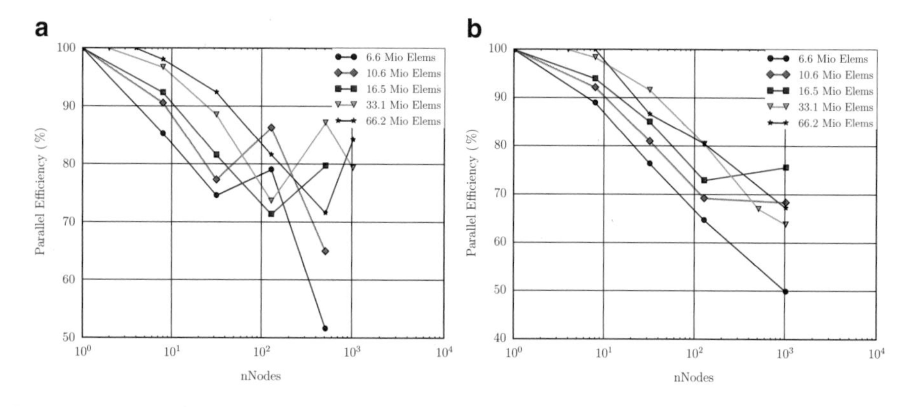

Fig. 12 Strong scaling parallel efficiency for different problem sizes with spacer structure. (**a**) Single-fluid LBM model; (**b**) Multi-species LBM model

per node which fits into the cache for single-fluid LBM. In both cases, weak scaling is almost perfect with only a small drop in the performance for larger counts of compute nodes.

A dedicated graph for strong scaling is shown in Fig. 12 with number of nodes on the horizontal axis and parallel efficiency (%) in the vertical axis. Here, testcases with problem sizes of roughly 6.6, 10.6, 16.5 and 66.2 million elements are used. In both LBM models, with a problem size of 6.6 million elements, the performance drops from 1 node to 1,024 nodes because communication dominates computation. The peak in the plots are due to caching effects as also obserable in the performance plots. For single-fluid LBM 66.2 million elements fit into cache on 1,024 nodes and with three species in the multi-species LBM this figure reduces to 33.1 million elements. For the testcase with full laboratory scale spacer length of 20 cm and 66.2 million elements, a minimum of 8 nodes is required to fit the problem into memory

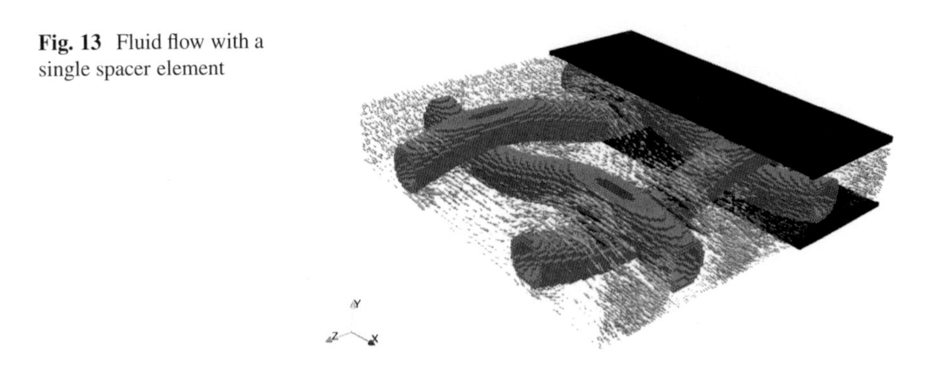

Fig. 13 Fluid flow with a single spacer element

for multi-species and 4 nodes for single-fluid LBM. The performance efficiency of multi-species LBM for large problem sizes with 1,024 nodes is less than for smaller problem sizes because it did not fit into the cache and is at the local performance minimum immediately after leaving the cache instead. Hence, to fit the problem with full spacer length into cache, compute nodes of roughly 2 or 3 times more than 1,024 nodes are required. Otherwise, the problem size per node should be increased to gain higher performance levels in the region where the influence of the communication is negligible.

4.3 Production Run and Complete Toolchain

For the actual production runs, a mesh with 520 million elements covering the spacer geometry was generated in serial on the Hermit pre- and post processing node with 128 GB memory and it took around 3 h. The single-fluid LBM model is used to simulate flow on this problem using 128 computing nodes on Hermit (4,096 cores). Number of time steps used for this simulation is 225,000 (which represents physical 0.25 s) and the complete simulation took around 4 h. For post-processing 8 Hermit nodes are used and it consumed around 2 h. Figure 13 shows the flow distribution in the channel with a spacer structure.

5 Summary and Outlook

In this paper, we briefly described our implementation of the Lattice Boltzmann method for incompressible flows and its extension to liquid mixture modeling to enable the simulation of electrodialysis for sea water desalination. The performance and scalability of the implementation was presented for an academic simulation domain without boundaries and a complex geometry resembling the actual flow channel in the desalination simulation. We showed the results for different compilers and communication patterns on the Hermit system and highlighted the well suited behavior of the code for large scale runs with a high sustained performance. The analysis showed the capability to solve full scale simulations and the largest production runs conducted up to now made use of more than 500 million cells.

The main focus of our ongoing work is put on the influence of the spacer configuration on the flow. One important aspect in this context are also the boundary conditions and higher order boundary conditions might have to be deployed. The simulated pressure drop will be validated against laboratory results. For multi-species LBM for liquid mixture the special boundary condition to treat diffusive fluxes on membrane interfaces will be investigated. Further scaling improvements might be achieved by accounting for the required communication times in the balancing.

Acknowledgements This work was funded by the German Federal Ministry of Education and Research (Bundesministerium für Bildung und Forschung, BMBF) in the framework of the HPC software initiative in the project HISEEM.

References

1. J. Bernsdorf, Lattice-Boltzmann simulation of reactive flow through a porous media catalyst. In *Recent Progress in CFD, No.08-07 JSAE Symposium*, 2007
2. J. Zudrop, S. Roller, A. Pietro, A Lattice Boltzmann Scheme for electrolytes. Phys. Rev. E, submitted 2013
3. S. Roller, J. Bernsdorf, H. Klimach, M. Hasert, D. Harlacher, M. Cakircali, S. Zimny, K. Masilamani, L. Didinger, J. Zudrop, An adaptable simulation framework based on a linearized octree. *High Performance Computing on Vector Systems: Proceedings of the High Performance Computing Center Stuttgart* (Springer, Berlin, Heidelberg, 2011), pp. 93–105
4. S. Chen, G.D. Doolean, Lattice Boltzmann method for fluid flows. Ann. Rev. Fluid Mech. **30**, 329–364 (1998)
5. P.L. Bhatnagar, E.P. Gross, M. Krook, A model collision processes in gases. I. Small amplitude processes in charged and neutral one-component systems. Phys. Rev. **94**, 511–525 (1954)
6. H. Klimach, M. Hasert, J. Zudrop, S. Roller (2012) Distributed octree mesh infrastructure for flow simulations. European Congress on Computational Methods in Applied Sciences and Engineering, pp. 1–15
7. M. Hasert, K. Masilamani, S. Zimny, H. Klimach, J. Qi, J. Bernsdorf, S. Roller, Complex Fluid Simulations with the Parallel Tree-based Lattice Boltzmann Solver Musubi. Journal of Computational Science (submitted 2013)

Part VI
Transport and Climate

Christoph Kottmeier

The modelling of large natural systems like the ocean and the atmosphere, and in particular their coupled spatio-temporal evolution, has become a major challenge for HPC computing in the last decades. The rapid progress being achieved in terms of computer generations with much higher CPU power (and storage capacities) could be easily "balanced" by reducing the grid spacing of such models. CPU requirements for finer mesh sizes and the CFL-criterion limit the rapid transition to direct grid-scale resolution of important processes like convection, cloud processes, and flow over topography.

Several projects focus on regionalizations of global climate models. They are characterized by large storage requirements and CPU-times *HRCM*: *"High Resolution Climate Modelling with the CCLM Regional Model"* (KIT Karlsruhe), *WRFCLIM "High resolution climate predictions and short-range forecasts to improve the process understanding and representation of land-surface interactions in the WRF model in Southwest Germany"* (University of Hohenheim) and *"High resolution climate Change Information for the Lower Mekong River Basin of Southeast Asia"* (KIT Garmisch-Partenkirchen). The latter project aims at giving quantitative information to assess changes in future land-use and agricultural productivity in Central Vietnam. HRCM addresses the needs of hydrology and flood management in medium size river catchments of basic land-atmosphere exchange studies and climate predictability in Europe and Africa. The studies for Germany in WRFCLIM and HRCM provide assessments of the capabilities of regional climate models in simulating the observed climate of the last decades and estimate the future changes in various regions on Earth. Regional climate change projections of this kind are used more and more to enable estimates of climate change consequences in various economic and social sectors. The approach on downscaling global climate

C. Kottmeier
Institut für Meteorologie und Klimaforschung, Karlsruher Institut für Technologie, Wolfgang-Gaede-Str. 1, 76131 Karlsruhe, Germany
e-mail: christoph.kottmeier@imk.uka.de

with an ensemble of regional models becomes increasingly common, causing additional CPU load and better statistical certainty in the climate change signals.

A new approach in climate modeling is addressed by *RUCACI "Reducing the uncertainties of climate projections: High-resolution climate modeling of aerosol and climate interactions on the regional scale using COSMO-ART"* (KIT Karlsruhe). The objective is to provide more sophisticated treatment of aerosols, which interact with clouds and thus affect the radiation budget both by direct and indirect (cloud related) effects. The model system COSMO-ART fulfills these requirements and will be applied for the first time for a decadal downscaling run.

The project *MIPAS* (KIT Karlsruhe) on processing the data from the satellite Michelson interferometer MIPAS on the ENVISAT also reflects the high importance of the HLRS and SSC computing facilities for most visible research programs in actual research in meteorology and oceanography. The retrieval of altitude-resolved profiles of various trace species of the atmosphere from MIPAS spectra is a very challenging HPC task. The actual focus of analysis is on studying the chemistry and dynamics of the middle atmosphere in response to solar proton events in the Arctic and Antarctic regions.

The project *AGULHAS "The Agulhas System as a Key Region of the Global Oceanic Circulation"* (GEOMAR, Kiel) applies the ocean general circulation model NEMO to study the leakage of Indian Ocean water to the Atlantic as a feature of the Agulhas current south of Africa. After implementation of the model, various sensitivity studies have been performed and first estimates of the climate change signal on the leakage have been obtained.

High Resolution Climate Modeling with the CCLM Regional Model

H.-J. Panitz, G. Fosser, R. Sasse, K. Sedlmeier, S. Mieruch, M. Breil, H. Feldmann, and G. Schädler

Abstract Using the CRAY XE-6 at the HLRS high performance computing facilities provides the possibility to study various aspects of the regional climate employing the regional climate model COSMO-CLM. The research activities of the group "Regional Climate and Water Cycle" at the KIT focus on the regional atmospheric water cycle and, especially, on extremes and different goals are pursued in the individual research projects. Different regions and orographies are studied using different resolutions from 50 to 3 km. Furthermore, different time spans are investigated and computational capacities from 2 to 500 node-hours per year (Wall Clock Time) are required. The analyses comprise decadal climate simulations of Germany, Europe and Africa to assess regional decadal climate predictability. Further, climate projections are carried out for Baden-Württemberg (Germany) and novel ensemble generating techniques are implemented to better describe the involved uncertainties. High resolution (3 km) experiments are performed for Baden-Württemberg to study extremes and the effects of climate change on soil erosion. Moreover, the possibilities of adaption to climate change for Baden-Württemberg are analysed, with focus on extremes and combination of extremes (such as dry soil and extreme precipitation).

1 Introduction

The regional climate simulations efforts carried out at the Institute for Meteorology and Climate Research (IMK) of Karlsruhe Institute of Technology (KIT) investigate the regional climate of Central Europe/Germany and of Africa by means of the

H.-J. Panitz (✉) · G. Fosser · R. Sasse · K. Sedlmeier · S. Mieruch · M. Breil · H. Feldmann · G. Schädler
Institut für Meteorologie und Klimaforschung, Karlsruher Institut für Technologie (KIT), Karlsruhe, Germany
e-mail: hans-juergen.panitz@kit.edu

W.E. Nagel et al. (eds.), *High Performance Computing in Science and Engineering '13*, DOI 10.1007/978-3-319-02165-2_35,

climate version of the COSMO model (CCLM). The spatial resolutions vary from about 50 to 3 km, depending on the specific goals of different national and international research programs and projects, namely MiKlip, REKLIM, KLIWA, and KLIMOPASS. We want to study the processes and interactions involved in the regional hydrological cycle, to assess the impact of climate change on soil erosion and extreme weather events, and to investigate the possibility of decadal climate prediction for Europe and also for the Monsoon region in Africa. Ensemble simulations are essential for these investigations. The ensemble will be built by different methods: usage of different driving models and other boundary conditions like sea surface temperature, land cover changes, greenhouse gas concentrations, changes of grid size and physical parameterizations, and perturbation of physical parameters of the model. The results will be compared with climate observations; this permits an assessment of the quality of the model results. Furthermore, the ensemble approach allows determining the uncertainty of climate projections and predictions, which is essential for all impact studies.

2 The CCLM Model

The regional climate model (RCM) CCLM is the climate version of the operational weather forecast model COSMO (Consortium for Small-scale Modeling) of the German Weather Service (DWD). It is a three-dimensional non-hydrostatic model which means that spatial resolutions below 10 km (which is considered the limit for hydrostatic models) are possible. The model solves prognostic equations for wind, pressure, air temperature, different phases of atmospheric water, soil temperature and soil water content.

Further details on COSMO and its application as a RCM can be found in [1, 2], on the web-page of the COSMO consortium (http://www.cosmo-model.org), and in [3–5].

3 Regional Climate Simulations Using the HLRS Facilities

3.1 The MiKlip Program

The climate system shows variability over a large range of temporal scales. This variability constitutes a "noise" with respect to climate change studies, which try for instance to determine the effects of the anthropogenic green house gas forcings. Variations of these trends arise from slow variations within the coupled ocean/atmosphere/land earth system. The superposition of these variations increases or reduces the climate trends during certain periods. Among the main drivers are decadal-to-multidecadal variations in the heat transport by ocean currents. For the North-Atlantic sector the Atlantic Multidecadal Oscillation (AMO) is a key driver for climate in North-America, Europe and Africa [6]. From the study of these modulations a potential to predict the climate up to decadal time scales arises. To exploit this potential—which could also be of great use to economical and governmental planning purposes—the BMBF has initiated the large research program MiKlip (Mittelfristige Klimaprognosen). Within this program a decadal ensemble prediction system will be established. IMK coordinates a module of MiKlip, which is dedicated to the regionalization of such decadal forecasts. The institute participates in three projects within this MiKlip module. Next to the coordination project of the module (REGIO PREDICT)—which also has the task to assess the skill of the ensemble of the regional decadal hindcast experiments—the other two aim at regional decadal predictions of climate for Europe (project DecReg) and the West African monsoon region (project DEPARTURE), respectively. MiKlip is organized into three development stages. During each stage simulations with an enhanced decadal prediction system will be provided. During the first stage a so called Baseline(0) Ensemble using the global climate model MPI-ESM has been established, by using simple initialization and ensemble techniques. In further versions improvements will be implemented. This baseline ensemble consists of initialized 10-year hindcast simulations for the period 1960–2010 with a low resolutions version of the model (MPI-ESM-LR, $\approx$1.87° resolution).

3.1.1 Regional Decadal Prediction for Europe (REGIO_PREDICT)

The regional baseline ensemble for Europe is a coordinated effort of several partners. IMK, DWD (German Weather Service) and the University Frankfurt perform decadal simulations with CCLM, whereas the max Planck Institute for Meteorology (MPI), Hamburg contributes hindcasts with the RCM REMO to the multi- model ensemble. The simulation domain covers Europe with a resolution of about 25 km. Five decades covering the hindcast period 1960–2010 (1961–1970, ..., 2001–2010) have been selected to study the basic performance and added value of regional downscaling. The ensemble size is ten realizations for each decade. This accounts for 50 simulations alone with CCLM plus several reference simulations (driven by re-analysis or un-initialized climate simulations). A detailed analysis of this ensemble is now analysed. During the current second stage various studies are needed to increase the model performance and to establish sufficient regional ensemble spread. Based on the analysis and the experiments an updated regional (prototype) ensemble will be generated.

3.1.2 Decadal Regional Predictability (DecReg)

The main goals of DecReg comprise the characterization of temporal and spatial variability, the initialization of the regional model, the coupling between the global and regional model and the postprocessing of the simulations as well as the assessment of an added value (if it exists) of the regional hindcasts compared to

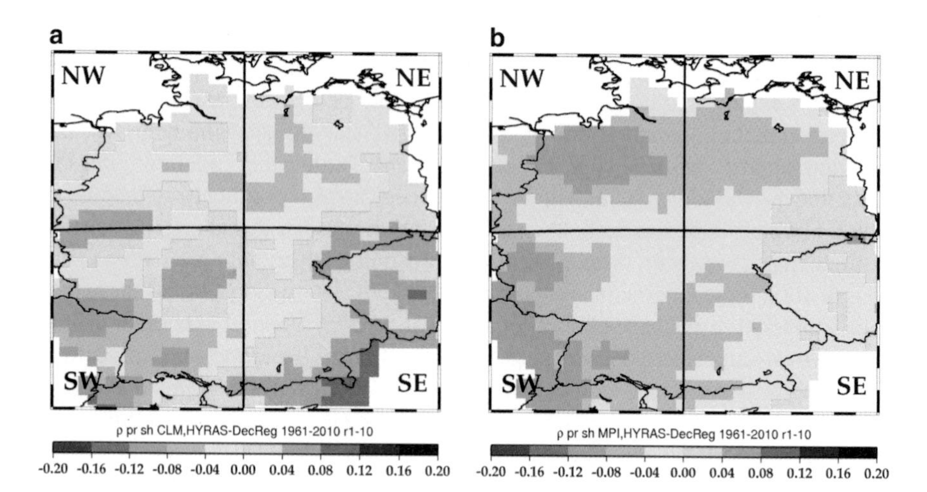

Fig. 1 Average correlation coefficient (ρ) of summer-half precipitation from 1961 to 2010 over the ten member ensemble. The reference is a merged observational dataset comprising of HYRAS and DecReg data. **(a)** Correlation of the regional model CCLM vs. observations. **(b)** Correlation of the global model MPI vs. observations (interpolated to the regional grid)

the global predictions. A special interest applies to the appearance and distribution of extremes such as heavy precipitation and droughts. The pivotal point is the generation of the regional hindcast ensemble by the MiKlip projects DecReg (KIT, Goethe University Frankfurt), LACEPS (DWD) and RedClip (MPI, Hamburg) and the pivotal question is, how to achieve a regional forecast skill and added value. To this end, the DecReg strategy covers the development of high quality gridded observational validation datasets as reference (DWD). To achieve a forecast skill, crucial questions concern the coupling of the global and regional model, which is analysed by the partners from the Johannes Gutenberg University Mainz, and the soil initialization (including data assimilation) in CCLM, which is considered by the partners from Frankfurt. Here in Karlsruhe we analyse the regional ensemble with respect to different combinations of metrics, variables, temporal and spatial resolutions, postprocessing etc. to achieve forecast skill. Finally, the aim of the partners within DecReg is to define a strategy how to maximize predictability and added value on the regional scale. To assess the potential added value of the downscaling approach with respect to the global simulations, we have analysed summer half-year sums of precipitation. Figure 1a shows the mean correlation coefficient between the ten member CCLM ensemble and an observational dataset consisting of HYRAS data from 1961 to 2000 and data generated by the DWD partners from DecReg (2001–2010). Figure 1b shows the correlation coefficient between MPI-ESM (interpolated to the regional grid) and the observations. As can be seen, the correlations are weak in both cases. However, coherent structures are present for the CCLM correlation at mountainous regions, i.e. the Alps, the Ore Mountains, the Bergisches Land and the Odenwald. Thus, although the correlation

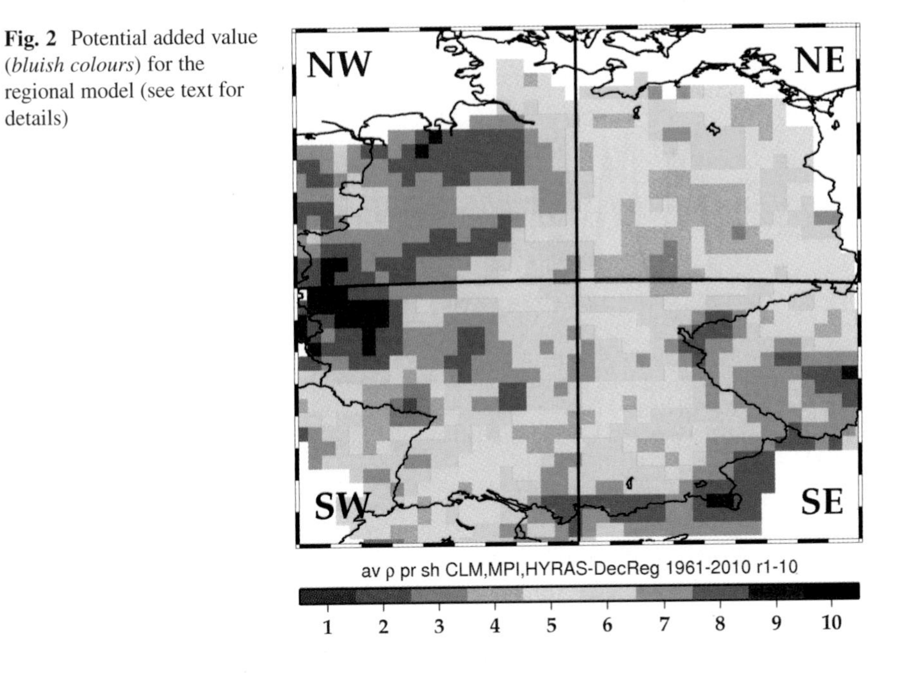

Fig. 2 Potential added value (*bluish colours*) for the regional model (see text for details)

is weak the regional model outperforms the global model in this regions. To quantify that, we have counted the cases when the regional model shows larger correlations than the global model among the ten realisations, which is shown in Fig. 2. Here, the bluish colours denote the potential added value. In line with Fig. 1 the CCLM shows added value in the above mentioned mountainous regions and, additionally, the analysis reveals that added value is gained in the Black Forest as well, and in a large region in the North-West part of Germany.

3.1.3 Decadal Climate Predictability in West Africa (DEPARTURE)

Within DEPARTURE simulations using CCLM contribute to a multi RCM ensemble. The overall aim of DEPARTURE is assessing the decadal climate predictability in the West African monsoon region and the Atlantic region of tropical cyclogenesis. The decadal forecast skill will be assessed for decadal hindcast periods, namely the decades 1966–1975, 2001–2010, and 2005–2015, by downscaling decadal MPI-ESM-LR (ECHAM6/MPIOM) simulations and, as far as possible, the comparison of the results with appropriate observations. According to current knowledge, West African climate may be characterized by high decadal predictability because it is governed by processes with long-term memory such as the state of the oceans, land-cover and soil characteristics, greenhouse gas and aerosol concentrations. Sensitivity studies with different lower boundary conditions and with different Surface-Vegetation-Atmosphere-Transfer (SVAT) models coupled to CCLM, via the OASIS coupler, allow assessing the effect of soil and vegetation processes on

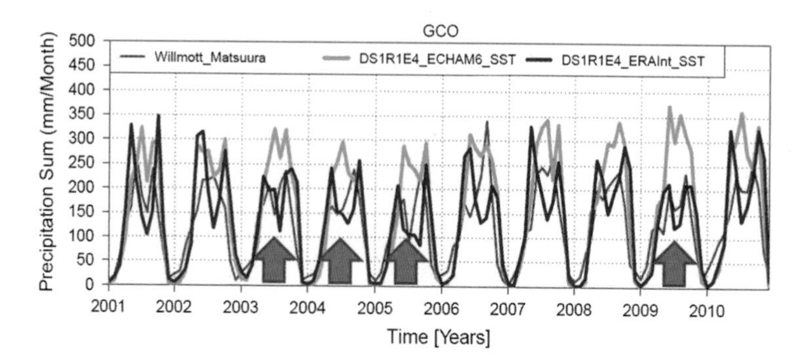

Fig. 3 Monthly sums of total precipitation for decade 2001–2010 in the coastal Gulf of Guinea region (GCO). *Green arrows* indicate those years showing the highest improvement of results due to the use of more realistic SST. Willmott-Matsuura (UDEL) denotes an observational climate data set [10]

decadal climate predictability in the African monsoon region. In addition, time-dependent distributions of the Aerosol Optical Depth (AOD) will be used which take into account the aerosol forcing.

The DEPARTURE model domain is an extension of the CORDEX Africa domain [7] to the West. Using a horizontal grid-spacing of 0.44° it has as size of 275 grid-points from West to East and 207 grid-points from South to North. The model configuration corresponds to that of the CORDEX Africa simulations [7].

At first, a climate simulation, driven by ERA-Interim reanalyses [8] had been carried out for the period 1979–2010. One purpose of this simulation was to provide so-called balanced soil conditions, which can be used as initial conditions for the actual decadal calculations in order to avoid a long spin-up time.

Several sensitivity runs have been carried out for the decade 2001–2010. As forcing data the realization r1i1p1 of the corresponding MPI-ESM-LR (=ECHAM6) decadal simulation has been used [9].

The intention of one series of sensitivity runs was to investigate the impact of different strategies of soil and vegetation initialization on key parameters like the temperature in 2 m height (T_2m) and the total precipitation (TOT_PREC) over Africa. For this purpose, the reference initial conditions representing an unbalanced soil and being derived from the MPI-ESM-LR data, had been substituted by initial conditions derived from the ERA-Interim driven CCLM climate simulation, representing a balanced soil. This sensitivity study showed that these changes in the initial conditions change the results, even years after the initialization. However, in comparison to climate observations, these changes do not improve or deteriorate the model results very much (not shown).

Sensitivity to SST was found for precipitation in the coastal areas of the African monsoon region. The simulation results could be improved for some years of the decade, indicated by the green arrows in Fig. 3 for the Gulf of Guinea region (GCO), when using SST from ERA-Interim reanalysis instead of the SST from the MPI-ESM-LR decadal run (Fig. 4), which has been proved to be not very unrealistic (Paeth et al., University Würzburg, pers. comm.).

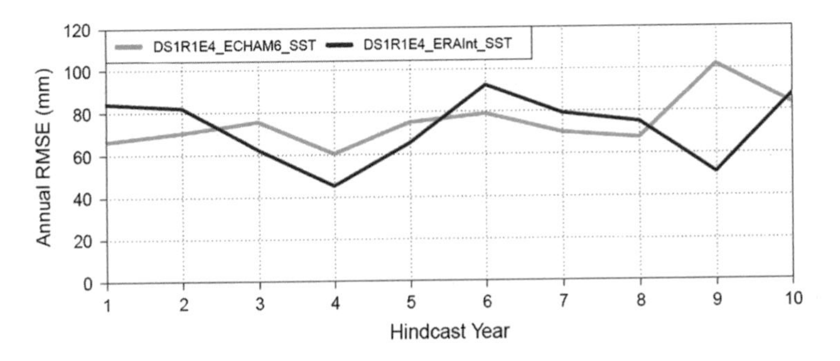

Fig. 4 Sensitivity of CCLM result for total precipitation (TOT_PREC) to SST boundary condition. Annual RMSE of total precipitation with respect to UDEL [10] climate observations

3.2 *The Helmholtz Climate Initiative (REKLIM)*

Reliable knowledge of the variability and changes in regional climate, and in particular extreme climate, is crucial for the development and implementation of appropriate adaptation and mitigation strategies. Within the Helmholtz Network REKLIM (Regional Climate Change), an ensemble of high-resolution climate simulations is generated for past and future decades in order to assess the predictability of regional climate change and to account for its uncertainties. Such uncertainties can arise from the positioning of synoptic systems in the large-scale atmospheric forcing, which is derived from low-resolution global climate models (GCMs) and drives the regional climate model (RCM) at its lateral boundaries. To capture this kind of uncertainty, an ensemble of RCM simulations is created by introducing small shifts to the large-scale atmospheric conditions [11–13].

The RCM simulations are performed using CCLM. For obtaining high-resolution simulations, a double nesting strategy is applied. During the first nesting step, CCLM simulations for Europe at 50-km horizontal resolution are driven by initial and boundary conditions from the GCM ECHAM6. During the second nesting step, the 50-km CCLM data drive CCLM runs for Germany at 7-km horizontal resolution. The simulations span the periods 1968–2000 and 2018–2050, each including 3 years spin-up. For generating the CCLM ensemble, the atmospheric fields from ECHAM6 are shifted with respect to topography to the North (prefix N), South (prefix S), West (prefix W), and East (prefix E) by 50 km (suffix 2), respectively. Thus, in addition to the reference (i.e., non-shifted forcing), four driving datasets are created by means of Atmospheric Forcing Shifting (AFS).

On the CRAY XE-6 at the HLRS facilities, the computing requirements amount to approx. 96 node-hours (i.e., 4 node-days) per simulation year at 50-km resolution using $118 \times 110 \times 40$ grid points (Table 2). With a domain size of $165 \times 200 \times 40$ grid points, the 7-km CCLM runs require approx. 384 node-hours (i.e., 16 node-days) per simulation year and, thus, significantly more computing resources than the 50-km simulations. The necessary storage capacity per simulation year is in the

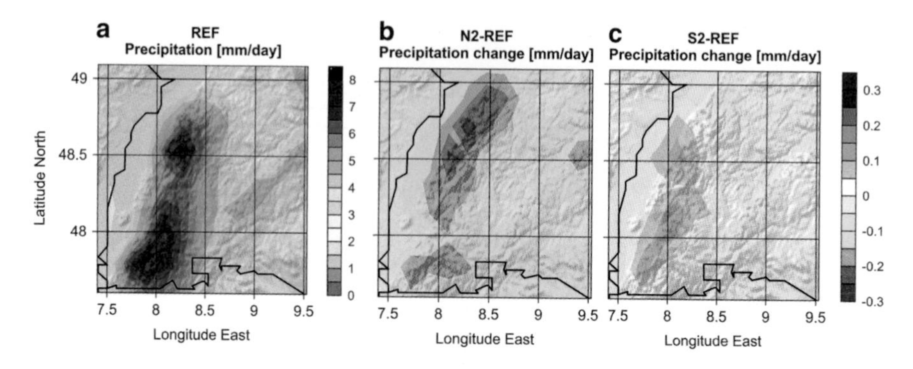

Fig. 5 Mean spatial distribution of daily precipitation (mm/day) for the period 1971–2000 in (**a**) REF and its changes in the (**b**) N2 and (**c**) S2 shifting scenario

order of 0.05 Tb for the 50-km CCLM runs. For the 7-km CCLM simulations, the storage capacity is again considerably higher and amounts to approx. 0.4 Tb.

The previous report focused on the effect of AFS on 50-km CCLM simulations from 1980 to 1984 [12]. In this paper, the precipitation changes caused by AFS are investigated for the period 1971–2000 at 7-km resolution in order to study the variability of heavy precipitation in more detail. For this purpose, the shifting scenarios N2 and S2 are compared to the reference simulation (REF). On average, the mean annual precipitation over Baden-Württemberg is approx. 1,584 mm/year in REF, 1,540 mm/year in the N2 scenario, and 1,586 mm/year in the S2 scenario. This means, most precipitation is produced in the S2 scenario and in REF whereas AFS causes the mean annual precipitation to decrease in the N2 scenario. Looking at the mean spatial precipitation distribution (Fig. 5), the effect of AFS is particularly evident over complex terrain since small shifts of synoptic systems affect the interaction of air masses with orography and, thus, the precipitation initiation and intensity [11, 12]. For instance, in the N2 scenario (Fig. 5b), precipitation decreases by up to 0.3 mm/day over the western slope of the Black Forest. This amounts to approx. 3.5 % of the mean daily precipitation in REF (Fig. 5a). In contrast, the S2 scenario shows precipitation increases by up to 0.2 mm/day (i.e., up to approx. 2.4 % more precipitation than in REF) over the western Black Forest (Fig. 5c).

During single precipitation events, the spatial variability can be more heterogeneous and, thus, the difference between the model experiments REF, N2, and S2 can be considerably larger (Fig. 6). This is demonstrated for the heaviest precipitation event in Baden-Württemberg that occurred in REF on 03 August 1992 with a mean intensity of 53.8 mm/day. On average, the shifting scenarios produce less precipitation with intensities in the order of 46.8 mm/day (N2 scenario) and 37.4 mm/day (S2 scenario). However, the local precipitation amounts differ significantly from the spatial averages (Fig. 6). In REF, precipitation occurs with intensities of up to 110 mm/day, particularly in the southern Black Forest (Fig. 6a). Furthermore, the shift of the large-scale atmospheric forcing to the North by 50 km leads to precipitation increases of up to 20 mm/day in the northern Black Forest and eastern

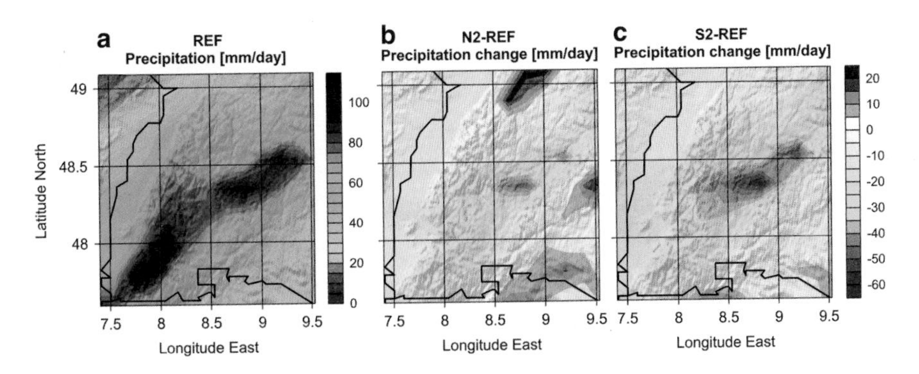

Fig. 6 Spatial distribution of daily precipitation (mm/day) in (**a**) REF and its changes in the (**b**) N2 and (**c**) S2 shifting scenario on 03 August 1992. This date corresponds to the precipitation event with a 30-year return period in REF

Swabian Jura (Fig. 6b). This corresponds to approx. 40 % more precipitation than in REF. However, precipitation is generally reduced in the N2 scenario. The shifting scenario S2 is also mainly related to precipitation decreases on 03 August 1992 in comparison to REF (Fig. 6c). This is especially distinct over the Swabian Jura where changes of up to −60 mm/day occur (i.e., approx. 75 % less precipitation). Consequently, AFS has a significant impact on the local intensity during heavy precipitation events and local rainfall can vary considerably in comparison to REF.

Figure 7 illustrates changes in daily rainfall over Baden-Württemberg depending on the precipitation return periods for the three model experiments [14]. For instance, in REF, rainfall in the order of 53.8 mm/day occurs once every 30 years. It is important to note that the event with a 30-year return period does not need to appear on the same date in the three model experiments. In REF and N2, the event with a 30-year return period is on 03 August 1992. However, in the S2 scenario, the heaviest precipitation event in Baden-Württemberg takes place on 28 July 1973 with a mean intensity of about 45.1 mm/day (Fig. 7). Thus, for the 30-year return period, the mean rainfall varies by up to 8.7 mm/day between REF, N2, and S2. For the 15-year return period, the return value (i.e., the precipitation intensity occurring once every 15 years) is in the range of 41.2 mm/day (S2) and 49.9 mm/day (REF). Again, the maximum spread of the return values between REF, N2, and S2 is about 8.7 mm/day. It decreases with reduced return periods (Fig. 8) and, e.g., amounts to 6 mm/day for 10-year return periods, 5.1 mm/day for 7.5-year return periods etc. For events with return periods between 10 and 1.5 years, the return values are highest in the N2 scenario. As known from Fig. 6, the local rainfall can differ much more between REF and the shifting scenarios than indicated by the spatial averages and, thus, accordingly, the local return values can be considerably different between the three model runs depending on the precipitation return periods.

In conclusion, ensemble generation by means of AFS is appropriate in order to capture uncertainties arising from the positioning of synoptic systems in the large-scale atmospheric forcing, which have not been considered in RCM ensembles so far

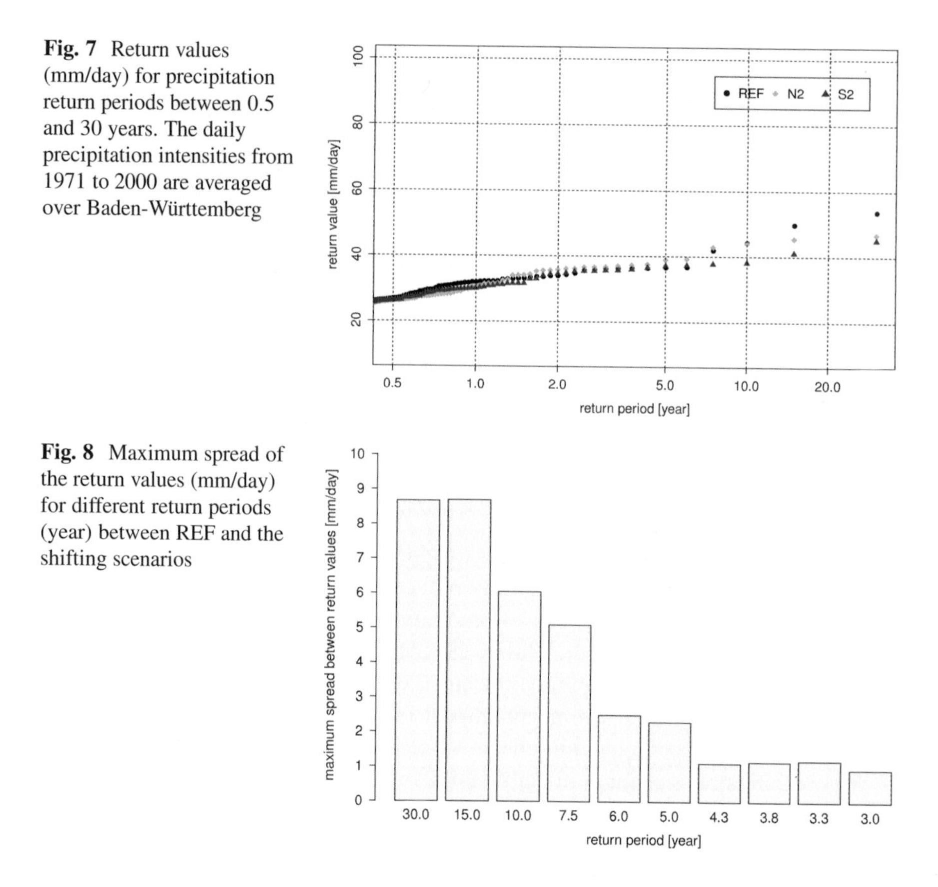

Fig. 7 Return values (mm/day) for precipitation return periods between 0.5 and 30 years. The daily precipitation intensities from 1971 to 2000 are averaged over Baden-Württemberg

Fig. 8 Maximum spread of the return values (mm/day) for different return periods (year) between REF and the shifting scenarios

[11]. AFS is particularly valuable for sampling the statistics of extreme values and, thus, for assessing the variability and changes in extreme climate in a more reliable way. In future, the CRAY XE-6 at HLRS will be further used for completing the high-resolution CCLM ensemble with 7-km simulations for the period 2018–2050 and, moreover, for investigating future changes in extreme climate.

3.3 *Soil Erosion Due To Precipitation in the Context of Climate Change (KLIWA)*

For the near future of the south of Germany, recent projections for winter indicate a widespread increase of 5–10 % in mean precipitation coupled with a slight decrease and more variability in summer [15]. The intensity of extreme events is expected to increase in all seasons, but for summer (JJA), the situation is more complex and highly depending on the specific area under investigation. In general, a decrease in the number of wet days is expected in the future leading to dryer

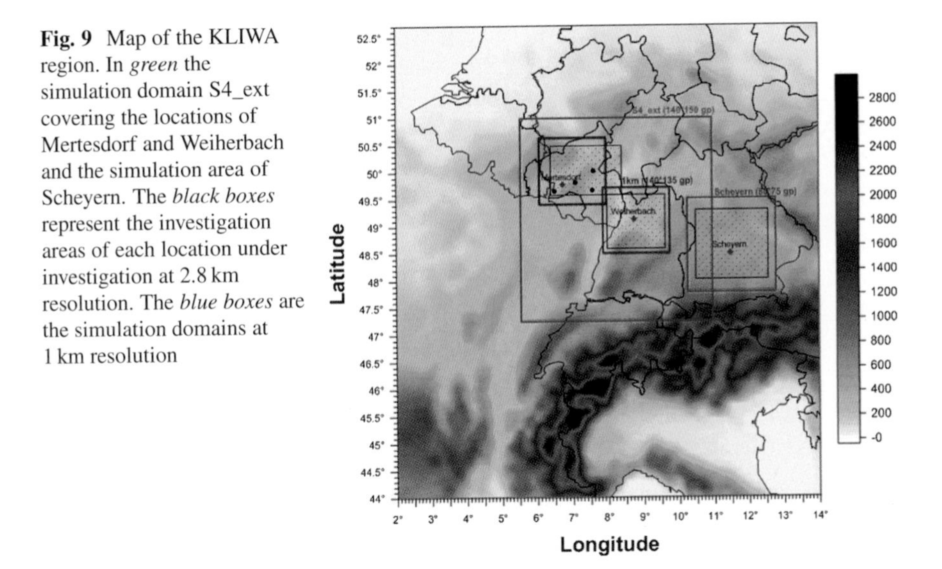

Fig. 9 Map of the KLIWA region. In *green* the simulation domain S4_ext covering the locations of Mertesdorf and Weiherbach and the simulation area of Scheyern. The *black boxes* represent the investigation areas of each location under investigation at 2.8 km resolution. The *blue boxes* are the simulation domains at 1 km resolution

summers [16]. This situation is likely to lead to an enhancement of erosion and thus creates economical loss especially for agricultural areas in Southern Germany. In this context, the KLIWA project "Bodenabtrag durch Wassererosion in Folge von Klimaveränderungen" aims to assess the impact of climate change on soil erosion for selected locations with high erosion risk (namely Weiherbach in Baden-Württemberg, Mertesdorf in Rheinland-Pfalz, and Scheyern in Bayern). Since this phenomenon acts on a scale of few hundreds meters, precipitation data in very high spatial and temporal resolution are necessary as input for erosion models. To reach the final resolution of 1 km COSMO-CLM is employed to downscale global climate projection to regional scale using a multiple nest strategy. This procedure consists in using the results from the previous nest at coarser resolution to force the simulations with higher resolution. So, global model output are dynamically downscaled to the 50 km nest, which covers most of Europe. The next finer nest at 7 km comprises the Germany and the near surroundings while the 2.8 km nest concentrates on the state of Baden-Württemberg (140×116 grid points) in southwestern Germany (Fig. 9). One kilometer resolution is used to simulate only to certain episodes in specific areas because of the constraints in simulation time requirements and in memory storage capacity. All simulations use the same CCLM version, consistent model setup and ECHAM5 [17] as forcing data for the recent past (1971–2000) and the near future (2021–2050). Long-term simulations at both 50 and 7 km resolution were already available thanks to the CEDIM project "Flood Hazard in a Changing Climate" [7, 18]).

This report presents the first results on the analysis of the climate change signal for the area of Weiherbach at 2.8 km resolution. Figure 10a compares the probability distribution of different daily precipitation intensities for the recent past and near future in summer months (JJA). Future summers show a light increase

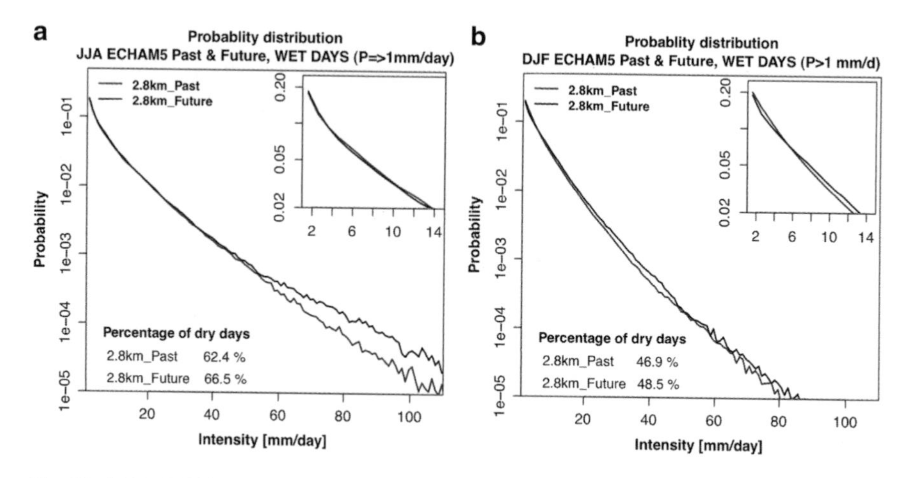

Fig. 10 (**a**) Probability distribution of daily precipitation in JJA for the recent past (*pink*) and near future (*violet*). Percentage of dry days for the past and the future in text. (**b**) Probability distribution of daily precipitation in DJF for the recent past (*pink*) and near future (*violet*). Percentage of dry days for the past and the future in text

Table 1 Changes in the maximum number of consecutive dry days and number of dry periods between past and future for the Weiherbach area at 2.8 km resolution

Maximum number of consecutive dry days (P < 1 mm)	17 ÷ 41 Mean 23	22 ÷ 67 Mean 30
Number of dry periods (dry period = 5 consecutive dry days)	67 ÷ 122 Mean 92.7	83 ÷ 125 Mean 103.3

for precipitation below 3 mm/day followed by a decrease for intensities below 25 mm/day and then a strong enhancement of precipitation above 50 mm/day. The percentage of dry days, defined as day with a daily precipitation below 1 mm/day, will increase of 4 % in the near future. Moreover, the maximum number of consecutive dry days will in average increase from 23 to 30 with maxima peaks in the plain area up to 67 days without rain. The number of dry periods, defined as period of more than 5 consecutive dry days, will also increase (Table 1). Therefore, Weiherbach area will experience in future summers longer dry periods associated with rainy periods with higher amount of precipitation. This trend seems to be confirmed also on hourly base with an increase in the percentage of dry hours as well as in the extreme precipitation, which could reach maxima of 140 mm/h, event never experienced in the past. For future winters, the percentage of dry days will also raise of almost 2 %. Moreover, Fig. 10b shows an increase (decrease) of intensities above (below) 4 mm/day, stronger than in summer but limited to intensities below 50 mm/day. Above this threshold, in fact, the differences between past and future reduce considerably. On hourly base, precipitation intensities above 1 mm/h will systematically increase of about 2 mm/h in comparison with the past. Nevertheless, it is important to notice that the "extreme" event in winter reach a maximum of 24 mm/h that is not so extreme compared with the summer maximum of 140 mm/h.

The analysis conducted up to now leads to the conclusion that in a near future the Weiherbach area will suffer in summer from longer dry periods associated with days characterized by higher intensities, especially in terms of extreme, in comparison with past observations. In winter, the trend is confirmed with longer dry period and a general increase of medium range intensities on daily base and of high intensities on hourly base. These findings are in good agreement with previous studies at 7 km resolution [15, 16]. Higher resolutions shows the same climate change signal but allows a better spatial and temporal representation of the precipitation pattern.

3.4 Climate Change and Exemplary Adaptation in Baden-Württemberg (Klimopass)

Regional measures of adaptation to a changing climate require high resolution climate data concerning the possible changes of climate variables and statistics of practical interest, especially extremes and combinations of extremes. Furthermore, there is an increasing awareness of the importance of assessing the uncertainty of the data. For this purpose, ensemble simulations are a suitable method as has been shown, for example, in the ENSEMBLES project (http://ensembles-eu.metoffice.com). However, the resolution of such simulations (25 km) is still too low for regional impact studies.

The research program KLIMOPASS ("Klimawandel und modellhafte Anpassung in Baden-Württemberg", "Climate change and exemplary adaptation in Baden Württemberg") was initiated to investigate climate change and its impacts as well as the possibilities of adaptation measures for the state of Baden-Württemberg in Germany. The contribution of IMK-TRO to the Klimopass project is to provide an ensemble of high-resolution regional climate simulations (with a horizontal resolution of 7 km), to analyze the change of different climate variables and their statistics as well as giving a quantitative estimate of the robustness of the data based on ensemble simulations. This information can then be used as input for impact studies of other projects within and outside KLIMOPASS. A first set of data has been provided for TU Dortmund, Fachgebiet Energieeffizientes Bauen.

Our regional high resolution Ensemble will consist of COSMO-CLM runs using different global driving models (e.g. ECHAM6, CNRM, EC Earth) and different CCLM setups (e.g. usage of an alternative soil vegetation model). Model runs using the atmospheric forcing shifting (AFS) (described in Sect. 3.2) will also be part of the ensemble. The time periods simulated are the recent past (1968–2000) and the near future (2018–2050). Transient runs (1968–2050) are planned for selected ensemble members.

For the 7 km-resolution simulations a twofold nesting strategy is used; in a first step the data of a global model are downscaled to 50 km resolution and used as

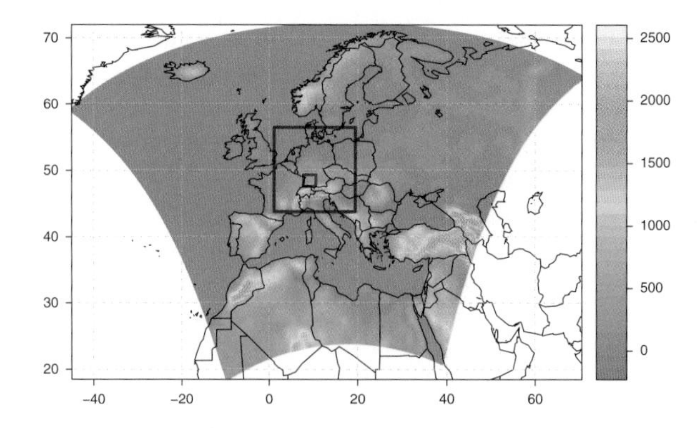

Fig. 11 Topography (m) of CCLM simulation domain. *Whole area*: 50 km domain; *blue area*: 7 km domain; *red area*: investigation area Baden Württemberg

forcing data for the CCLM regional model at that resolution. In a second step, the CCLM model results are downscaled to 7 km resolution and used to force the high-resolution simulation. The 50 km model domain covers Europe and parts of northern Africa (Fig. 11, whole area). A horizontal resolution of 50 km equals a horizontal grid spacing of 0.44° and the domain covers 118 grid points from West to East and 110 grid-points from South to North. The 7 km model domain (Fig. 11, blue box) covers central Europe. The horizontal grid spacing is 0.625° and the simulation area expands from 165 grid points from West to East and 200 grid-points from South to North. The main investigation area is the state of Baden-Württemberg in Germany (Fig. 11, red box)

Simulations for a first ensemble member, with ECHAM6 as a global driving model have been carried out for both recent past (1971–2000) and near future (2021–2050), simulations with for a second global driving model, CNRM are in progress for the recent past.

The analysis will focus on the variables precipitation, 2-m temperature, wind and radiation as well as water balance and soil moisture and their statistics. Besides calculating mean values a special focus will be placed on extremes (e.g. duration of heat and dry spells, extreme wind or precipitation). Due to the increased amount of data which results from using the ensemble method, a broader statistical data basis is created which also allows the examination of combined extremes (such as simultaneous occurrence of dry and hot spells or dry soil and extreme precipitation). These are of great interest for impact studies e.g. agriculture and forestry.

4 Computational Requirements

Regional climate modeling requires a large amount of computer resources, depending on the number of grid-points considered, the numerical time-step necessary to receive stable numerical solutions, and the length of the simulation period.

Table 2 Computational requirements

Research program	Project	Domain size (no. of grid-points)	Horizontal grid spacing in km	Numerical time-step (s)	No. of nodes used	WCT (nh/y)	No. of simulated years
MiKlip	DecReg	232 × 226 × 40	25	240	64	384	500
	DEPARTURE	275 × 207 × 35	50	240	64	200	100
REKLIM	Regional climate	118 × 110 × 40	50	360	64	96	260
	Ensembles	165 × 200 × 40	7	60	64	384	130
		165 × 200 × 40	7	60	64	384	60
KLIWA	Erosion	140 × 150 × 40	2.8	25	64	704	60
		140 × 135 × 40	1	10	64	960	2
KLIMO-PASS	RegEns	118 × 110 × 40	50	360	64	96	105
	BaWü	165 × 200 × 40	7	60	64	384	105

The following table (Table 2) shows typical computational needs that the different projects have. The times, i.e. the Wall Clock Time, are given in node-hours per year (WCT (nh/y)). Thus, they are scaled to one simulation year and denote the times one CRAY-XE6 node needs in order to simulate 1 year if all of its cores are used. Overall, the storage capacity amounts to 2–3 TB per 10 years simulation time including the data input, output and post-processing.

5 Future Work

Within the MiKlip projects REGIO_PREDICT, DecReg and DEPARTURE, the generation of the baseline 1 CCLM ensemble will be initiated in using the MPI-ESM baseline 1 global decadal predictions. In general, the computational effort is similar to the generation of the baseline 0. However, the now on-going research activities are aimed to find strategies how to improve the baseline 1 ensemble over the baseline 0 beyond using MPI-ESM baseline 1. These efforts comprise e.g. the soil initialization and ensemble perturbations of the simulations, the coupling of the global and regional models, the number of ensemble members, the resolution and the temporal initialization. The plans for the project REKLIM are, amongst others, the completion of the high-resolution CCLM ensemble with 7-km simulations for the period 2018–2050 and the statistical analysis of future changes in extreme climate. High resolution experiments down to 1 km are planned as well for the project KLIWA. The project Klimopass continues the simulations to generate a multi-model ensemble to achieve a statistical basis to analyse extremes and combinations of extremes together with their uncertainties.

References

1. M. Baldauf, A. Seifert, J. Förstner, D. Majewski, M. Raschendorfer, Operational convective-scale numerical weather prediction with the COSMO model: description and sensitivities. Mon. Weather Rev. **139**, 3887–3905 (2011). doi:10.1175/MWR-D-10-05013.1
2. G. Doms, U. Schättler, A description of the nonhydrostatic regional model LM, Part I: dynamics and numerics. COSMO Newslett. **2**, 225–235 (2002)
3. C. Meissner, G. Schädler, Modelling the regional climate of Southwest Germany: sensitivity to simulation setup, in *High Performance Computing in Science and Engineering '07*, ed. by W.E. Nagel, D. Kröner, M. Resch (Springer, Berlin, 2007). ISBN 978-3-540-74738-3
4. B. Rockel, A. Will, A. Hense, Regional climate modelling with COSMO-CLM (CCLM). Meteorol. Z. **17**(4) (2008). ISSN 0941-2948 (special issue)
5. C. Meissner, G. Schädler, H.-J. Panitz, H. Feldmann, C. Kottmeier, High resolution sensitivity studies with the regional climate model COSMO-CLM. Meteorol. Z. **18**, 543–557 (2009). doi:10.1127/0941-2948/20090400
6. R.T. Sutton, D.L.R. Hodson, Atlantic ocean forcing of North American and European summer climate. Science **309**, 115–118 (2005). doi:10.1126/science.1109496
7. H.-J. Panitz, P. Berg, G. Schädler, G. Fosser, Modelling regional climate change for Germany and Africa, in *High Performance Computing in Science and Engineering '11*, ed. by W.E. Nagel, D. Kröner, M. Resch (Springer, Berlin, 2012), pp. 503–512. doi:10.1007/978-3-642-23869-7
8. A. Simmons, S. Uppala, D. Dee, S. Kobayashiera, New ECMWF reanalysis products from 1989 onwards. ECMWF Newslett. **110**, 25–35 (2006) (Winter 2006/07)
9. W.A. Müller, J. Baehr, H. Haak, J.H. Jungclaus, J. Kröger, D. Matei, D. Notz, H. Pohlmann, J.S. von Storch, J. Marotzke, Forecast skill of multi-year seasonal means in the decadal prediction system of the Max-Planck Institute for Meteorology. Geophys. Res. Lett. **39**, L22707 (2012). doi:10.1029/2012/GL053326
10. C.J. Willmott, K. Matsuura, D.R. Legates, Global air temperature and precipitation: regridded monthly and annual climatologies (version 2.01) (1998). Available online at http://climate.geog.udel.edu/~climate/
11. R. Sasse, G. Schädler Generation of regional climate ensembles using atmospheric forcing shifting. Int. J. Climatol. (2013 submitted)
12. H.-J. Panitz, G. Fosser, R. Sasse, A. Sehlinger, H. Feldmann, G. Schädler, Modelling near future regional climate change for Germany and Africa, in *High Performance Computing in Science and Engineering '12*, ed. by W.E. Nagel, D. Kröner, M. Resch (2013)
13. I. Schlüter, G. Schädler, Sensitivity of heavy precipitation forecasts to small modifications of large-scale weather patterns for the Elbe river. J. Hydrometeorol. **11**, 770–780 (2010). doi:10.1175/2010JHM1186.1
14. B. Früh, H. Feldmann, H.-J. Panitz, G. Schädler, D. Jacob, P. Lorenz, K. Keuler, Determination of precipitation return values in complex terrain and their evaluation. J. Climate **23**, 2257–2274 (2010). doi:10.1175/2009JCLI2685.1
15. H. Feldmann, G. Schädler, H.-J. Panitz, C. Kottmeier, Near future changes of extreme precipitation over complex terrain in Central Europe derived from high resolution RCM ensemble simulations. Int. J. Climatol. (2012). doi:10.1002/joc.3564
16. S. Wagner, P. Berg, G. Schädler, H. Kunstmann, High resolution regional climate model simulations for Germany, Part II: projected climate changes. Climate Dyn. (2012). doi:10.1007/s00382-012-1510-1
17. G. Roeckner, G. Baeuml, L. Bonaventura, R. Brokopf, M. Esch, M. Giorgetta, S. Hagemann, I. Kirchner, L. Kornblueh, E. Manzini, A. Rhodin, U. Schlese, U. Schulzweida, A. Tompkins The atmospheric general circulation model ECHAM 5, Part I: model description. Technical Report 349, Max-Planck-Institut für Meteorologie, Hamburg (2003)

18. G. Schädler, P. Berg, D. Düthmann, H. Feldmann, J. Ihringer, H. Kunstmann, J. Liebert, B. Merz, I. Ott, S. Wagner, Flood hazards in a changing climate. Project Report, p. 83. Centre for Disaster Management and Rsik Reduction Technology (CEDIM) (2012), http://www.cedim.de/download/FloodHazardsinaChangingClimate.pdf

High-Resolution Climate Predictions and Short-Range Forecasts to Improve the Process Understanding and the Representation of Land-Surface Interactions in the WRF Model in Southwest Germany (WRFCLIM)

Kirsten Warrach-Sagi, Hans-Stefan Bauer, Oliver Branch, Josipa Milovac, Thomas Schwitalla, and Volker Wulfmeyer

1 Introduction and Motivation

The application of numerical modeling for climate projections is an important task in scientific research since they are the most promising means to gain insight in possible future climate changes. The quality of the prepared global projections has been continuously improved in recent years, enabled by more powerful supercomputers as well as advanced numerical and physical schemes (e.g. [1–3]). During the last two decades, various regional climate models (RCM) have been developed and applied for simulating the present and future climate of Europe. First of all, the performance of the RCMs to successfully reproduce the observed regional climate characteristics within the last decades was extensively assessed. Within the EU projects ENSEMBLES and PRUDENCE, ensemble simulations of RCMs forced with ERA-40 reanalysis data were executed and analyzed with a grid resolution of the order of 50 km [4, 5]. It was found that these models were able to reproduce the pattern of temperature distributions reasonably well but a large variability was found with respect to the simulation of precipitation. The performance of the RCMs was strongly dependent on the quality of the boundary forcing, namely if precipitation was due to large-scale synoptic events. Additionally, summertime precipitation was subject of significant systematic errors as models with coarse grid resolution have difficulties to simulate convective events. This resulted in deficiencies with respect to simulations of the spatial distribution and the diurnal cycle of precipitation. Correspondingly, the 50-km resolution RCMs were hardly capable of simulating the statistics of extreme events such as flash floods. The results of ENSEMBLES

K. Warrach-Sagi (✉) · H.-S. Bauer · O. Branch · J. Milovac · T. Schwitalla · V. Wulfmeyer
Institute of Physics and Meteorology, University of Hohenheim, Garbenstrasse 30, 70599, Stuttgart, Germany
e-mail: Kirsten.Warrach-Sagi@uni-hohenheim.de

W.E. Nagel et al. (eds.), *High Performance Computing in Science and Engineering '13*, DOI 10.1007/978-3-319-02165-2_36,

and PRUDENCE are in accordance with a variety of studies of RCMs at the order of 25–50 km (e.g. [6–9]).

Due to these reasons, RCMs with higher grid resolution of 10–20 km were developed and extensively verified, e.g. in southwest Germany [10, 11]. While still inaccuracies of the coarse forcing data were transferred to the results, these simulations indicated a gain from high resolution due to better resolution of orographic effects. These include an improved simulation of the spatial distribution of precipitation and wet day frequency and extreme values of precipitation. However, three major systematic errors remained: the windward-lee effect [10], phase errors in the diurnal cycle of precipitation [12], and precipitation return values, especially on longer return periods [11].

In order to provide high-resolution ensembles and comparisons of regional climate simulations, the World Climate Research Program (WCRP) initiated the **CO**ordinated **R**egional climate **D**ownscaling **EX**periment (**CORDEX**). CORDEX is performed in preparation of the fifth assessment report of the Intergovernmental Panel on Climate Change (IPCC AR5; [13]). Within WRFCLIM a verification run for Europe was performed with the Weather Research and Forecasting (WRF) model [2] for a 20-year period (1989–2009) on the NEC Nehalem Cluster at a grid resolution of approx. 12 km with CORDEX simulation requirements. Precipitation and soil moisture results of this simulation were analysed [14, 15] and due to the results in the beginning of 2012 the simulation was repeated with an updated version of WRF on the CRAY X6 at HLRS. This simulation was completed in April 2012 and is currently under evaluation e.g. within the EURO-CORDEX (www.euro-cordex.net) ensemble of regional climate models (e.g. [16]). The regional climate projection runs (1950–2100) will be forced with the latest global climate model runs (CMIP5 data), which currently become available to the regional climate modeling community.

Within this context the WRFCLIM objectives are to

- provide high resolution (12 km) climatological data to scientific community
- support of quality assessment and interpretation of the regional climate projections for Europe through the contribution to the climate projection ensemble for Europe (EURO-CORDEX) for the next IPCC report

Generally, three forcing conditions for the formation of precipitation in mid-latitude terrains with complex orography and land-surface heterogeneity can be distinguished [17]: (1) strongly-forced conditions, e.g., due to the presence of a surface front; (2) weakly-forced conditions, no surface front but upper-level instabilities, e.g., due to the presence of an upper-level trough and advection of potential vorticity; (3) air mass convection: no strongly-forced or weakly forced conditions, likelihood of convection is increased equivalent to potential temperature advection in combination with thermally-induced slope and valley flows. The importance of surface forcing increases from (1) to (3) but, during summertime, the interaction of all three forcing mechanisms is responsible for the development of specific convergence lines resulting in a complex distribution of precipitation [17–20]. The results demonstrated that a substantial increase of forecast skill can be achieved if the models are operated at convection-permitting resolution without the need of the parameterization of deep convection. The first downscaling experiment to the convection permitting scale within WRFCLIM on the NEC Nehalem Cluster [14] shows, that regional climate modeling benefits from this resolution, too, as similar systematic errors are present such as the windward-lee-effect in orographic terrain.

The natural deviation of boundaries from global models from the true state due to model physics and assumed initial conditions, inconsistent physics between global and regional models, and the poor representation of orography and the heterogeneity of land-surface-vegetation properties in RCMs at 10–50 km resolution result in a large gap in our knowledge concerning regional impacts of climate change (IPCC 2007, [21]) due to a reduced skill of regional simulations of feedback processes between the land-surface and the atmospheric boundary layer as well as of clouds and precipitation development. Doherty et al. [21], who summarized the research needs that shall follow IPCC AR4 and the new Strategic Plan of the World Weather Research Program [22] come to consistent conclusions concerning the promotion of research in two areas: (1) high-resolution, advanced mesoscale atmospheric ensemble modeling, and (2) high-resolution variational data assimilation, e.g., for testing and improving regional climate models in weather forecast mode. The scope of the WRFCLIM project, is to investigate in detail the performance of regional climate projections with WRF in the frame of EURO-CORDEX at 12 km down to simulations at the convection permitting scale within the DFG funded Research Unit 1695.[1]

The main objectives of the CP simulations of WRFCLIM are as follows:

- Replace the convection parameterization by the dynamical simulation of the convection chain to better resolve the processes of the specific location and actual weather situation physically.
- Gain of an improved spatial distribution and diurnal cycle of precipitation through better landsurface-atmosphere feedback simulation and a more realistic representation of orography to support the interpretation of the 12 km climate simulations for local applications e.g. inhydrological and agricultural management.

Special attention will be paid to the land-surface-vegetation-atmosphere feedback processes. Further, the model will be validated with high-resolution case studies applying advanced data assimilation to improve the process understanding over a wide range of temporal scales. This will also address whether the model is able to reasonably represent extreme events.

---

[1]RU 1695 Agricultural Landscapes under Climate Change Processes and Feedbacks on a Regional Scale: An objective is the development and verification of a convection-permitting regional climate model system based on WRF-NOAH including an advanced representation of land-surface-vegetation-atmosphere feedback processes with emphasis on the water and energy cycling between croplands, atmospheric boundary layer and the free atmosphere.

2 WRF Simulations at HLRS

2.1 Technical Description

All simulations within this project are performed on the CRAY XE6 at the HLRS with the Weather and Research Forecast (WRF) model. The model is applied in Europe with focus on Germany and in Israel. WRF offers the choice of various physical parameterizations to describe subgrid processes like e.g. cloud formation and radiation transport. In WRFCLIM it is usually applied with the land surface model NOAH [23, 24], the Morrison two-moment microphysics scheme [25], the YSU atmospheric boundary layer parameterization [26], the Kain-Fritsch-Eta convection scheme [27] in case of grid cells larger than 4 km and the CAM shortwave and longwave radiation schemes [28]. Experiments with the other five boundary layer parameterizations, two other radiation schemes and a new land surface model are also part of the current simulations.

Within this project convection permitting WRF simulations are performed to further downscale the climate data for applications in hydrology and agriculture and an improved representation of feedback processes between the land surface and the atmosphere. Two pre-studies were carried out since June 2012 to test the model configuration including the choice of physical parameterizations under different climate conditions, where and when measurements for verification are available, one in Germany and one in Israel with a horizontal resolution of approx. 2 km. For Germany the sensitivity tests were performed with 50 vertical levels and for Israel with 92 vertical levels.

Further the whole year of 2012 was simulated with WRF for Germany on the convection permitting resolution of 3 km with 92 vertical levels to resolve the atmospheric boundary layer in an optimal way for an improved representation e.g. of convection processes. The model domain was selected to account for the predominant westerly winds in central Europe and the importance of the eastern Atlantic for the development of European weather. Figure 1 shows the domain configuration. The outer domain (black rectangle) was applied for the simulation of the whole year 2012. The inner domain (red rectangle) will be applied for single case studies of interesting weather situations in the future.

To improve the forecast of an atmospheric model on the short-range and climatological periods, it is necessary to start from an analysis which is as accurate as possible with respect to the observations and the model background. To merge observations with the existing model background, a data assimilation technique is required. For our investigations we decided to apply the three-dimensional variational data assimilation (3DVAR) technique with the WRF model in a Rapid Update Cycle (RUC) configuration. This allows the data assimilation step in arbitrary time intervals (currently between 1 and 3 h). The aim is that the observations and the model form a balanced background to reduce spin-up effects especially of precipitation. The simulations at 3.6 km horizontal resolution with 50 vertical levels proved to be too coarse in the data assimilation experiment. A second experiment

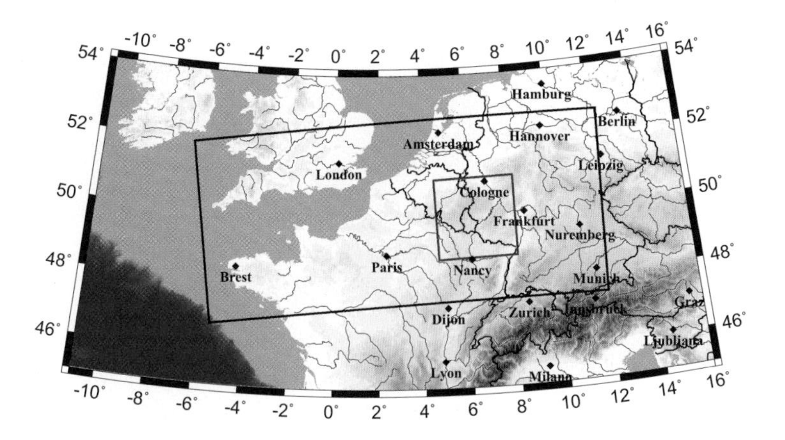

Fig. 1 Domain configuration of the 2012 simulation. The outer domain is 425 × 189 grid cells at 3 km horizontal resolution and 92 vertical levels up to 50 hPa. The inner domain (*red rectangle*) consists of 250 × 250 grid cells at 1 km horizontal resolution applying the same 92 vertical levels

Table 1 Technical details of the WRF simulations performed from June 2012 to April 2013

Project	Number of CPUs	Grid resolution (km)	Simulation period	Number of grid cells	Δt (s)	Number of simulations	CPUh
RU 1695	384	2	4–6 Aug 2007	301 × 301 × 50	18	5	
RU 1695	1,088	2	5–11 July 2010	501 × 501 × 50	12	26	
RU 1695	1,088	2	5–11 Dec 2010	501 × 501 × 50	12	26	
RU 1695	1,088	2	7–13 Sep 2009	501 × 501 × 50	12	44	
RU 1695	1,088	2	7–13 Sep 2009	501 × 501 × 91	12	3	1,399,700
Impact	640	3.6	7 days in summer 2007	550 × 550 × 50	15	35	313,504
Impact	640	3	July 2012	691 × 682 × 57	15	62	61,984
Impact	360	3	Jan–Dec 2012	425 × 189 × 92	12	1	112,000
Impact	864	2	May 01–Aug 31, 2012	444 × 444 × 91	12	7	2,566,336

was set up on a convection permitting horizontal resolution of approx. 3 km with 57 levels over central Europe. Table 1 summarizes some technical details of the WRF simulations performed from June 2012 to date (15th April 2013).

2.2 *Results of Convection-Permitting Simulations*

2.2.1 Sensitivity on Parameterization Schemes

When setting-up a model for use in particular region, the foremost issue is the determination of the appropriate model configuration. Different regions experience

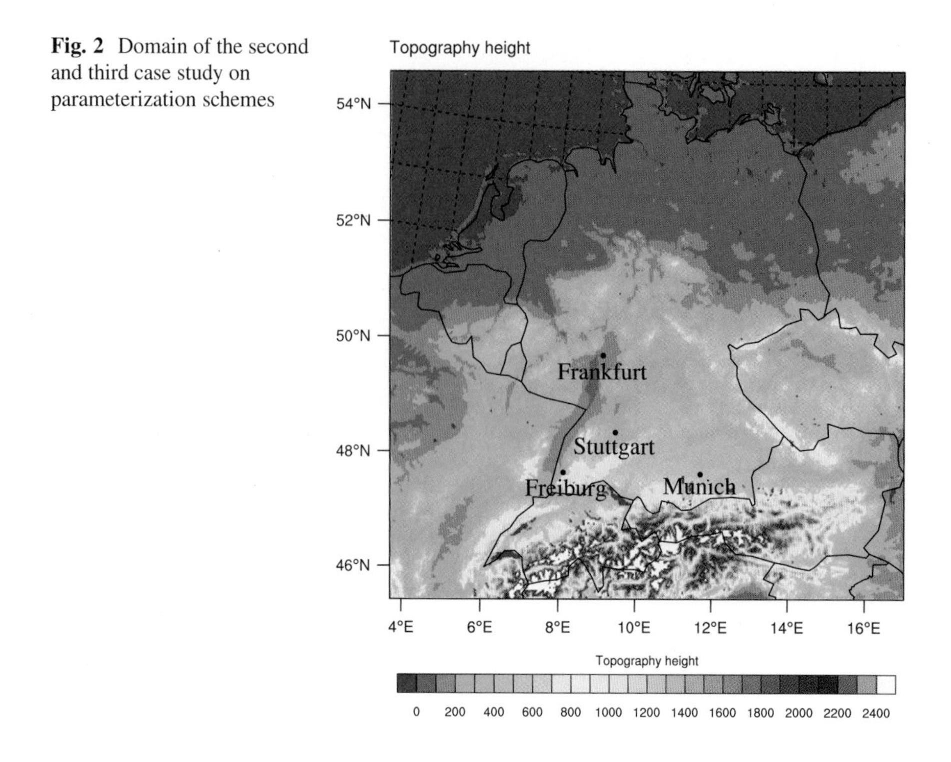

Fig. 2 Domain of the second and third case study on parameterization schemes

varied conditions, and the optional setup of a model is spatially and temporarily dependent. One of the most important aspects in configuring a model is to select the parameterizations to be used and therefore sensitivity tests are an unavoidable part in improving a model performance. The WRF model provides multiple parameterization schemes and allows their evaluation within the same model core.

In the first case study WRF was applied for Southern Germany (Fig. 1, red frame). Sensitivity studies were performed with two boundary layer schemes (Mellor-Yamada-Janjic—MYJ and Yonsei University–YSU) and two land surface models (NOAH and NOAH-Multi Physics NOAM-MP). NOAH-MP provides 12 switches that can be set by user and it has been just recently implemented into WRF (April, 2012). However, default settings provided by the WRF community were used in this case study. The impact of the different parameterizations used in the WRF model can be significant, e.g. the difference in latent heat flux between two configurations that differ only in the setting of boundary layer schemes or land surface models can be more than $100\,\mathrm{W\,m^{-2}}$ or more than 30 % of the total latent heat flux.

In the second case study the WRF model was applied on a domain covering the whole of Germany (Fig. 2). Sensitivity studies were performed with six boundary layer schemes (MYJ, YSU, the Asymmetric Convective Model version 2—ACM2, Total Energy Mass Flux—TEMF, Quasi-Normal Scale Elimination—QNSE, Mellor-Yamada-Nakanishi-Niino level 2.5—MYNN 2.5) two land surface models (NOAH and NOAH-MP) and two radiation transfer schemes (the Rapid Radiative Transfer Model for General circulation models—RRTMG and New

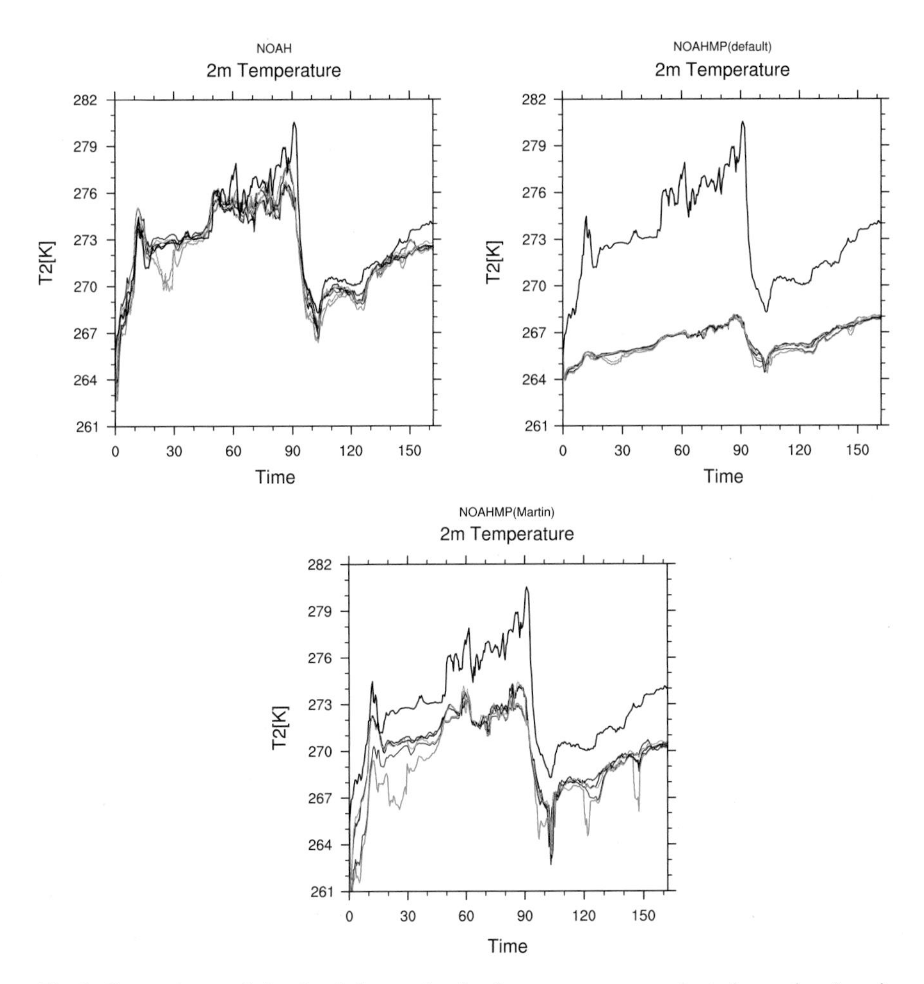

Fig. 3 Comparisons of simulated time series for 2 m temperature against observation (x axis denotes time in [h] since the start of the simulations and y axis denotes temperature at 2 m height in [K]). *Black lines* represents observations, while six *colored lines* show the results simulated with WRF show configured with four different boundary layer schemes and two radiation schemes (MYJ (*red*), MYJ+New Goddard (*green*), YSU (*blue*), YSU+New Goddard (*green*), QNSE (*cyan*), MYNN 2.5 (*purple*))

Goddard scheme) for a week in summer and in winter 2010. NOAH-MP was used in two different settings: the default one and the version adjusted for Germany. Simulated results were compared with measurements from Eddy Covariance (EC) stations in southern Germany Fig. 3 shows exemplary the importance and impact of the switches in NOAH-MP on the final results. The winter simulation has a strong cold bias when using WRF with NOAH-MP, but the results are significantly improved when the adjusted NOAH-MP is applied. The ACM2 and TEMF boundary layer schemes are instable when running in combination with NOAH-MP, especially during the winter period.

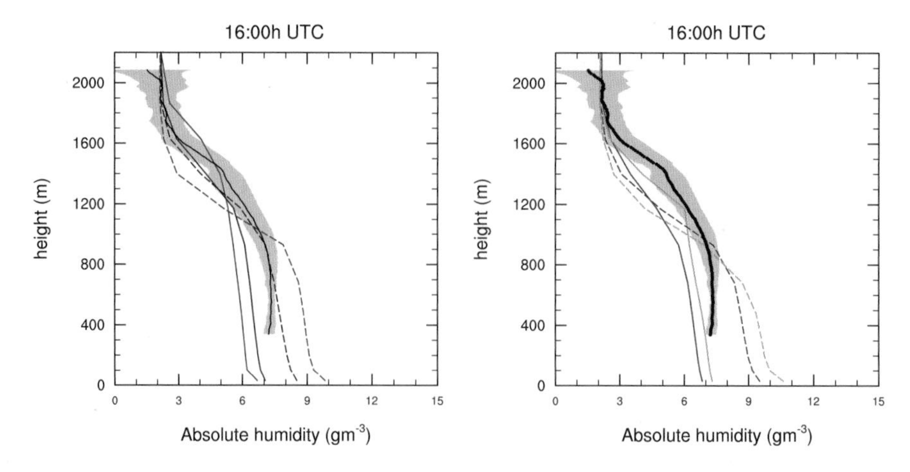

Fig. 4 Vertical profiles of absolute humidity on September 8th 2009 at 16:00 UTC. *Black lines* represent DIAL measurements, *blue* YSU boundary layer scheme, *red* MYJ boundary layer scheme, *orange* QNSE and *purple lines* represent MYNN 2.5 boundary layer scheme. *Solid lines* denote Noah land-surface model, while Noah-MP is represented in *dashed lines*

In the third case study on the domain of Germany (Fig. 2) various combinations of four boundary layer schemes (MYJ,YSU,MYNN 2.5, QNSE) and two land surface models (NOAH, NOAH-MP), including different combinations of switches in NOAH-MP were analyzed. The resulting ensemble of 44 simulations were compared against measurements of absolute humidity from the DIAL (Differential Absorbtion Lidar) of the University of Hohenheim in western Germany in September 2009.

Both, the boundary layer schemes and land surface models, impact the simulated vertical absolute humidity profiles. The impact of the land surface models extends also to higher altitudes, even up to the entrainment zone at the top of the boundary layer (Fig. 4). Results also showed that the WRF model is more sensitive on the choice of land surface model then on the boundary layer scheme.

2.2.2 Sensitivity on Land Use in a Dry Hot Climate

Geoengineering increasingly becomes an issue in a changing climate, namely in an expected hotter and drier climate. One practice is to change land cover. One method is changing the land use through e.g. plantations. However, in a large scale style this potentially impacts the weather and climate. In Israel we have access to measurements of a real land use change experiment. A sensitivity study is carried out with WRF in Israel (Fig. 5) to verify the model under such conditions and to study the impact of large scale plantations on the weather.

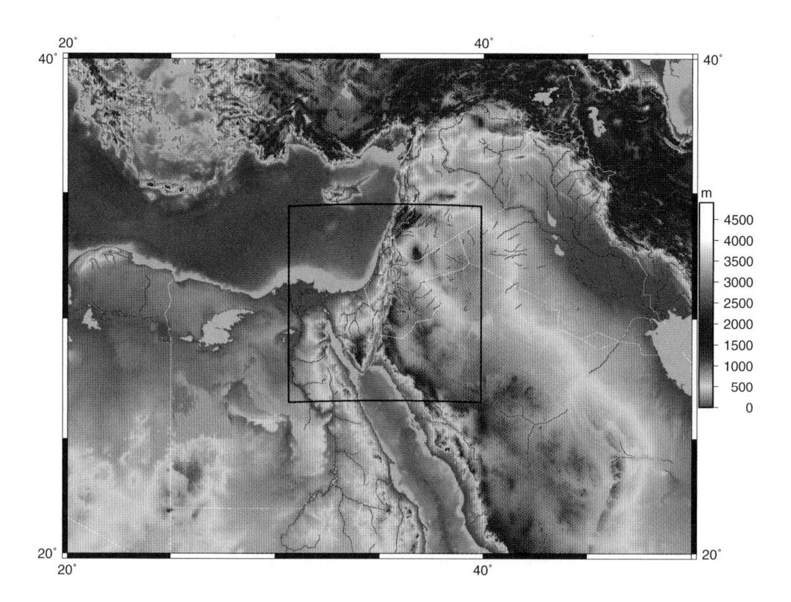

Fig. 5 Topographic map (2 km × 2 km horizontal resolution) of the region of interest, at the eastern end of the Mediterranean. The model domain (approx. 888 × 888 km) is marked in the centre with a *black line*. Care was taken to include synoptic features such as the NNE sea airflow into Israel, and also to avoid strong features at the boundaries, such as orography

Validation data includes surface measurements of vapor mixing ratio (Q2), temperature (T2), wind speed and direction (WS) and net surface radiation from three meteorological stations (a desert surface and two plantations). The stations lie within 1 km of each other and are named respectively—Desert, Jatropha and Jojoba.

Comparison of 2 m temperatures, 2 m humidity mixing ratio, net surface radiation and wind speeds of the Control model and desert observations (Obs) are shown in Fig. 6. The daytime temperatures are reproduced extremely well by the model. However, during the night time there is a cold bias of around 2 °C. Q2 (a measure of humidity) shows a dry bias of between 1 and 2 g kg^{-1} which occurs during daytime and night time. It should be noted here that the Q2 variability is high however. If we compare the humidity to the original 6-hourly 12 km forcing data, there is also a dry bias over the same area especially in the later part of the day which may account for some of this Q2 bias. NetRad and WS values are also reproduced quite accurately by the model with only small differences. WS are fractionally higher in the model, which can be expected, given the difference between the 10 m model height and measurement height of 5 m. The variability of the observations are well represented by the model for all variables.

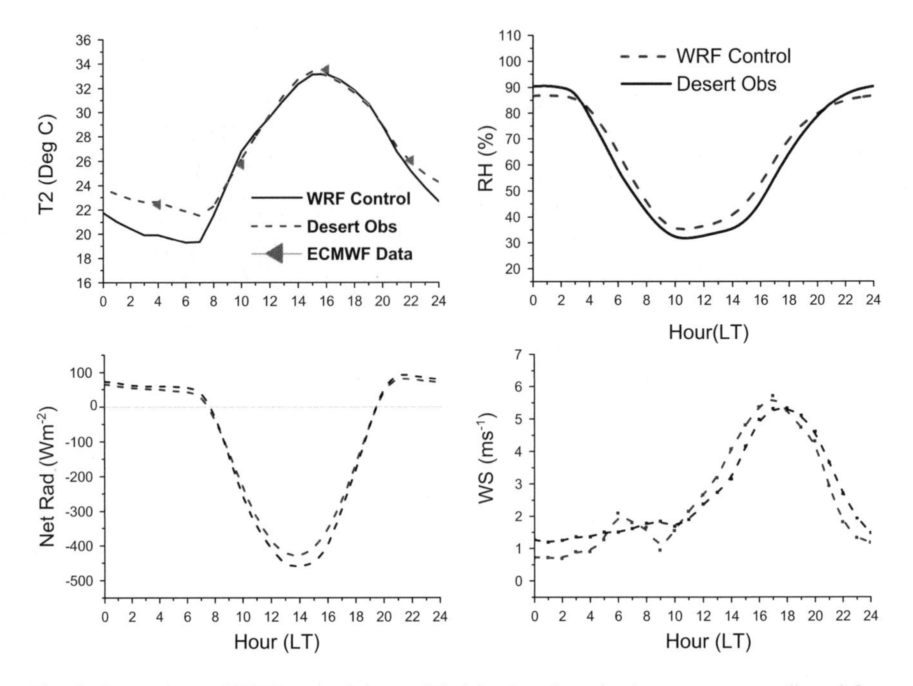

Fig. 6 Comparison of WRF control (unmodified land surface data) mean summer diurnal 2 m vapor mixing ratio (Q2), 2 m temperature (T2), net surface radiation (Net Rad) and 10 m wind speeds with equivalent observations from the desert site. WRF variables were averaged over a 25 grid cell box centred at the desert site geographical coordinates. Plotted also are the 6-hourly downscaled Q2 and T2 values from ECMWF. Note that WRF wind speeds are at 10 m while the observations are at 5 m

The same variables from the vegetation simulations were compared with the vegetation observations from the Jatropha and Jojoba plantations and are shown in Fig. 7.

We can see that the daytime 2 m temperatures are not modelled over the vegetation as well as over the desert because there is a cold bias Cof between 2 and 4 °C. The Jatropha observations are closer to the model runs than the Jojoba data. At night time this bias is increased to around 4 °C. There is also a large constant bias in the Q2 (humidity), up to 4 g kg^{-1} even more so than the Control run. Wind speeds are reasonable given the difference in height between the model (10 m) and measurements (5 m) and the net surface radiation is also quite well represented.

Overall, the model performs better over the desert than over the simulated vegetation in respect to the relevant measurements but the model has large biases when simulating vegetation, particularly temperature and humidity. Some of this bias could be accounted for by the bias in the forcing data but not all.

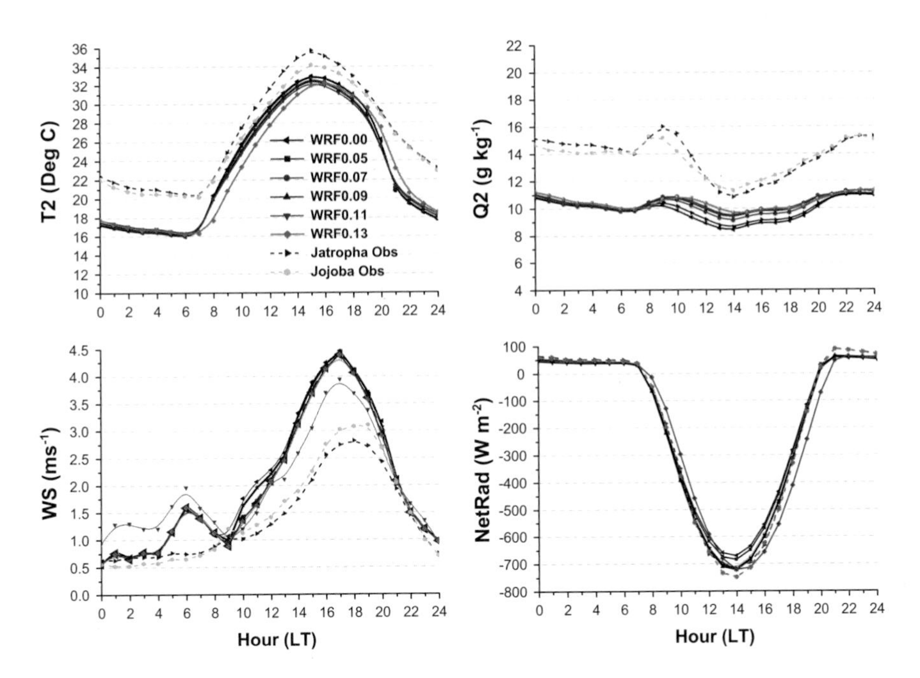

Fig. 7 WRF summer mean 2 m temperature, vapour mixing ratio, wind speed and net radiation values averaged over a 25 grid cell box centred on the equivalent plantation measurement sites, compared with mean summer Jatropha and Jojoba plantation measurements. WRF curves are marked with *solid lines* and observations are *dashed*

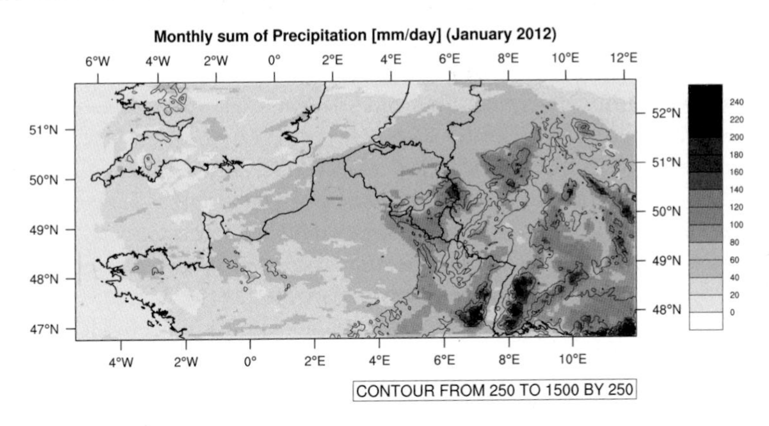

Fig. 8 Example result of the 2012 simulation showing the monthly precipitation for January 2012

2.2.3 Convection Permitting Simulation of 2012 in Germany

The analysis of the simulation is work in progress. A first impression is given with Fig. 8, showing the monthly sum of precipitation from January 2012. The simulation will be the basis for an assessment of the high resolution climatological conditions in southwestern Germany.

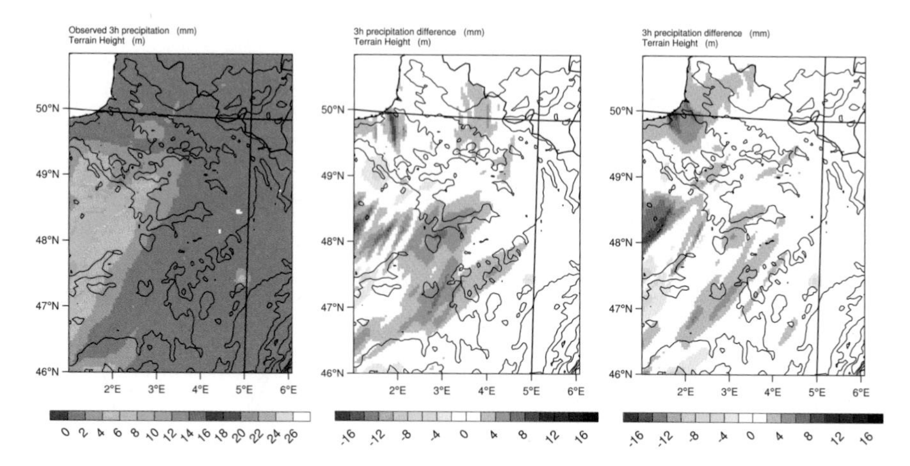

Fig. 9 Absolute 3-h precipitation sum (*left*). The *middle* and *right* panel show the precipitation difference for the simulation without (*middle*) and with radar radial velocities (*right*) with respect to the observation

2.2.4 Data Assimilation (3DVAR) with the WRF Model with a Rapid Update Cycle (RUC)

With the three-dimensional variational data assimilation (3DVAR) technique a simulated field is modified by minimizing a cost function which describes the differences between observations and this simulated field. The so gained analysis is then used as initial field for the next forecast with the model. This variational technique allows to use measured data like GPS derived Zenith Total Delay (GPS-ZTD), doppler precipitation radar data and satellite brightness temperatures for an improved analysis.

As the results with the assimilation of GPS-ZTD show promising results with respect to precipitation [20], the next step is to adjust the three-dimensional precipitation and wind fields by using Doppler precipitation radar data.

The assimilation of radar radial velocities helps to improve convergence lines and thus to improve the short-range precipitation forecast (see Fig. 9).

The influence of the additional assimilation of radar reflectivities on short-range precipitation forecast is not yet clear and needs further investigation. We also discovered that our vertical resolution in the boundary layer is too coarse so that we increased the number of vertical levels from 50 to 57 with a simultaneous increase in the horizontal resolution to 3 km to better represent orographic features (Fig. 10).

This new model domain requires a new background error covariance matrix for data assimilation which required 62 forecasts of 24 h each to estimate a climatological model variance and this is work in progress.

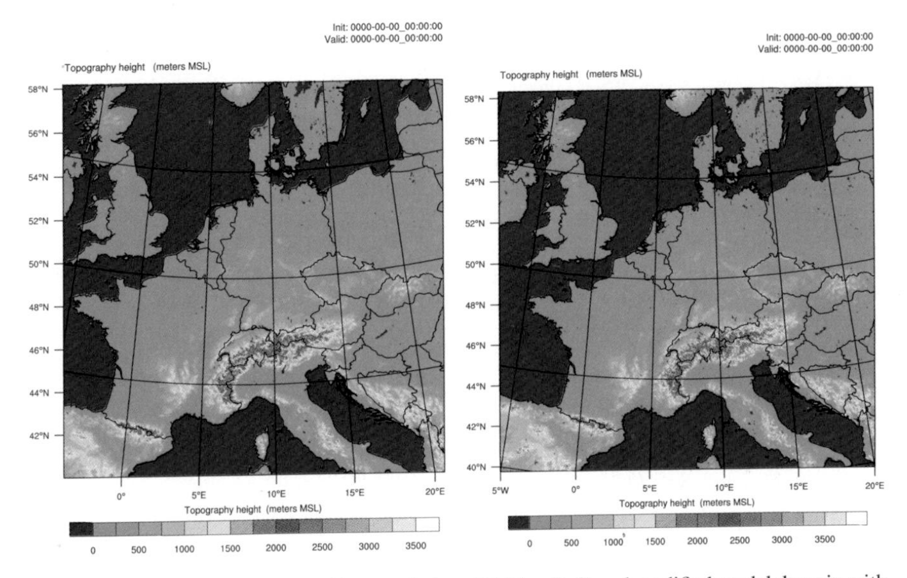

Fig. 10 Original model domain with a resolution of 3.6 km (*left*) and modified model domain with a resolution of 3 km

References

1. T. Thuburn, J. Comp. Phys. **227**, 3715 (2008)
2. W.C. Skamarock, J.B. Klemp, J. Dudhia, D. Gill, D.M. Barker, M.G. Duda, X.Y. Huang, W. Wang, J.G. Powers, A description of the advanced research WRF version 3. NCAR Technical Note TN-475+STR, NCAR, Boulder (2008)
3. H. Morrison, A. Gettelman, J. Climate **21**, 3642 (2008)
4. J.H. Christensen, O.B. Christensen, Clim. Change **81**, 7 (2007)
5. J.H. Christensen, T.R. Carter, M. Rummukainen, G. Amanatidis, Clim. Change **81**, 1 (2007)
6. M. Beniston, D.B. Stephenson, O.B. Christensen, C.A.T. Ferro, C. Frei, S. Goyette, K. Halsnaes, T. Holt, K. Jylhü, B. Koffi, J. Palutikoff, R. Schöll, T. Semmler, K. Woth, Clim. Change **81**, 71 (2007)
7. M. Déqué, D.P. Rowell, D. Lüthi, F. Giorgi, J.H. Christensen, B. Rockel, D. Jacob, E. Kjellström, M. De Castro, B. Van den Hurk, Clim. Change **81**, 53 (2007)
8. D. Jacob, L. Bähring, O.B. Christensen, J.H. Christensen, S. Hagemann, M. Hirschi, E. Kjellström, G. Lenderik, B. Rockel, C. Schär, S.I. Seneviratne, S. Somot, A. Van Ulden, B. Van der Hurk, Clim. Change **81**, 31 (2007)
9. E.B. Jaeger, I. Anders, D. Lüthi, B. Rockel, C. Schär, S.I. Seneviratne, Meteorol. Z. **17**, 349 (2008)
10. H. Feldmann, B. Früh, G. Schädler, H.J. Panitz, D. Keuler, K. Jacob, P. Lorenz, Meteorol. Z. **17**, 455 (2008)
11. B. Früh, H. Feldmann, H.J. Panitz, G. Schädler, D. Jacob, P. Porenz, K. Keuler, J. Climate **23**, 2257–2274 (2010). doi:10.1175/2009JCLI2685.1
12. P. Brockhaus, D. Lüthi, C. Schär, Meteorol. Z. **17**(4), 433 (2008)
13. F. Giorgi, C. Jones, G. Asrar, Addressing climate information needs at the regional level: the CORDEX framework. Technical Report WMO Bulletin 58, World Meteorological Organization (WMO) (2009)

14. K. Warrach-Sagi, T. Schwitalla, V. Wulfmeyer, H.S. Bauer, Climate Dyn. (2013). doi:10.1007/s00382-013-1727-7
15. P. Greve, Evaluating soil water content in a WRF-NOAH downscaling experiment, Diploma thesis, IFM-Geomar Kiel, Institute of Physics and Meteorology, University of Hohenheim, 2012
16. R.Vautard, A. Gobiet, D. Jacob, M. Belda, A. Colette, M. Deque, J. Fernandez, L.M. Garcia-Diez, K. Goergen, I. Guettler, T. Halenka, K. Keuler, S. Kotlarski, G. Nikulin, M. Patarcic, M. Suklitsch, C. Teichmann, K. Warrach-Sagi, V. Wulfmeyer, P. Yiou, Climate Dyn. **20**, 30 (2013). doi:10.1007/s00382-013-1714-z
17. V. Wulfmeyer, A. Behrendt, C. Kottmeier, U. Corsmeier, C. Barthlott, G.C. Craig, M. Hagen, D. Althausen, F. Aoshima, M. Arpagaus, H.S. Bauer, L. Bennett, A. Blyth, C. Brandau, C. Champollion, S. Crewell, G. Dick, P. Di Girolamo, M. Dorninger, Y. Dufournet, R. Eigenmall, R. Engelmann, C. Flamant, T. Foken, T. Gorgas, M. Grzeschik, J. Handwerker, C. Hauck, H. Höller, W. Junkermann, N. Kalthoff, C. Kiemle, M. König, L. Krauss, C.N. Long, F. Madonna, S. Mobbs, B. Neiniger, S. Pal, G. Peters, G. Pigeon, E. Richard, M.W. Rotach, H. Russchenberg, T. Schwitalla, V. Smith, R. Steinacker, J. Trentmann, D.D. Turner, J. Van Baelen, S. Vogt, H. Volkert, T. Weckwerth, H. Wernli, A. Wieser, M. Wirth, Q. J. R. Meteorol. Soc. **137**, 3 (2011). doi:10.1002/qj.752
18. A. Behrendt, S. Pal, M. Bender, A. Blyth, U. Crasmeier, J. Cuesta, G. Dick, M. Dorninger, C. Flamant, P. Di Girolamo, T. Gorgas, Y. Huang, N. Kalthoff, S. Khodayar, H. Mannstein, K. Träumer, A. Wieser, V. Wulfmeyer, Q. J. R. Meteorol. Soc. **137**, 81 (2011). doi:10.1002/qj.758
19. U. Corsmeier, N. Kalthoff, Ch. Barthlott, A. Behrendt, P. Di Girolamo, M. Dorninger, F. Aoshima, J. Handwerker, Ch. Kottmeier, H. Mahlke, St. Mobbs, G. Vaughan, J. Wickert, V. Wulfmeyer, Driving processes for deep convection over complex terrain: a multi-scale analysis of observations from COPS-IOP 9c. COPS Spec. Issue Q. J. R. Meteorol. Soc. **137**, 137–155 (2011). doi:10.1002/qj.754
20. T. Schwitalla, H.S. Bauer, V. Wulfmeyer, F. Aoshima, Q. J. R. Meteorol. Soc. **137**, 156 (2011). doi:10.1002/qj.721
21. S.J. Doherty, S. Bojinski, A. Henderson-Sellers, K. Noone, D. Goodrich, N.L. Bindoff, J.A. Church, K.A. Hibbard, T.R. Karl, L. Kaifez-Bogataj, A.H. Lynch, D.E. Parker, I.C. Prentice, V. Ramaswamy, F.W. Saunders, M.S. Smith, K. Steffen, T.F. Stocker, P.W. Throne, K.E. Trenberth, M.M. Verstraete, F.W. Zwies, Bull. Am. Meteorol. Soc. **90**, 497 (2009)
22. WMO (2010). Available online: http://www.wmo.int/pages/prog/arep/wwrp/new/documents/final_WWRP_SP_6_oct.pdf
23. F. Chen, J. Dudhia, Mon. Weather Rev. **129**, 569 (2001)
24. F. Chen, J. Dudhia, Mon. Weather Rev. **129**, 587 (2001)
25. H. Morrison, G. Thompson, V. Tatarskii, Mon. Weather Rev. **137**, 991 (2009)
26. S.Y. Hong, Y. Noh, J. Dudhia, Mon. Weather Rev. **134**, 2318 (2006)
27. J.S. Kain, J. Appl. Meteorol. **43**, 170 (2004)
28. W.D. Collins, P.J. Rasch, B.A. Boville, J.R. Mc Caa, D.L. Williamson, J.T. Kiehl, B. Briegleb, C. Bitz, S.J. Lin, M. Zhang, Y. Dai, Description of the ncar community atmosphere model (cam 3.0). NCAR Technical Note NCAR/TN-464+STR, NCAR, Boulder (2004), 226 pp.

High Resolution Climate Change Information for the Lower Mekong River Basin of Southeast Asia

Patrick Laux, Van Tan Phan, Tran Thuc, and Harald Kunstmann

Abstract Regional climate projections are derived for Southeast Asia with the goal to estimate future agricultural productivity and to derive adaptive land use strategies for the future. Therefore, the regional non-hydrostatic *Weather Research and Forecasting* (WRF) model is used to dynamically downscale large-scale coupled atmosphere–ocean general circulation model information. WRF is driven by ECHAM5 data for the control period 1960–2000 as well as for the period 2001–2050 using the two different SRES scenario A1B and B1. In addition to these long-term climate simulations, a 30-year WRF simulation using ERA40 reanalysis data (1971–2000) is performed to validate the performance of the WRF simulations for this region and to enable calibration of climate impact models. In total, around 1.7 Mio CPUh were used to finalize the climate simulations at the Steinbuch Centre for Computing (KIT, SCC). Trend analysis of the past reveals significant positive trends up to 0.04 K/year for the northern part of the Lower Mekong River Basin. Except for relatively small regions, precipitation shows positive trends

P. Laux (✉) · H. Kunstmann
Karlsruhe Institute of Technology (KIT), Institute for Meteorology and Climate Research, Atmospheric Environmental Research (IMK-IFU), Kreuzeckbahnstrasse 19, 82467 Garmisch-Partenkirchen, Germany
e-mail: patrick.laux@kit.edu; harald.kunstmann@kit.edu

V.T. Phan
Hanoi University of Science, Faculty of Hydrology, Meteorology and Oceanography, 334 Nguyen Trai, Thanh Xuan, Hanoi, Vietnam
e-mail: tanpv@vnu.edu.vn

T. Thuc
Ministry of Natural Resources and Environment (MONRE), Institute of Meteorology, Hydrology and Environment (IMHEN), Nguyen Chi Thanh street, Hanoi, Vietnam
e-mail: thuc@netnam.vn

W.E. Nagel et al. (eds.), *High Performance Computing in Science and Engineering '13*, DOI 10.1007/978-3-319-02165-2_37,

with a magnitude of approximately +10 mm/year. Temperature, in general, is expected to be increased in the future following both the A1B and the B1 scenario. The magnitude of increase, however, strongly depends on the scenario. For the period 2021–2050 the magnitude ranges between +3.5 K to +1.5 K on average for the A1B and the B1 scenario, respectively. For precipitation, however, the signal is not that clear. While for the B1 scenario an increase of precipitation is expected for almost the whole basin, both positive and negative signals are found for A1B.

1 Introduction

Climate change and climate variability are of major concern for the environment and people's well-being in SE Asia. Increasing frequency and severities of extreme events like floods, droughts, hurricanes but also increasing temperatures, sea level rise and salt water intrusion in the coastal areas are expected to have dramatic consequences for agricultural productivity and thus food security in this region (AR4, IPCC). These challenges demand for informed stakeholders and a land management strategies to increase the resilience of the ecosystems [5]. Scientifically sound land management strategies, however, require reliable and high resolution climate information for the past, present and future.

Especially for poorly gauged regions such as Southeast Asia [10] where statistical downscaling is no option, regional climate models (RCMs) are indispensable tools to dynamically downscale global climate simulations obtained by general circulation models (GCMs). The horizontal resolution of GCMs is typically not finer than 100 km, while state-of-the-art RCMs are able increase the resolution to less than 10 km. For this study transient (1961–2050) weather Research and Forecast (WRF) simulations are performed, driven by ECHAM5 [8] and the A1B and B1 scenario using a nested approach. This finally results in hydrometeorological information with high spatiotemporal detail. Before applying dynamical downscaling of future climate projections (by using a RCM), simulations driven by perfect boundary conditions should be conducted to detect systematic biases of the RCM, which are primarily caused by internal model dynamics and physics [1, 7]. The perfect boundary conditions are provided by reanalysis data, as e.g. provided by the National Centers for Environmental Prediction/National Center for Atmospheric Research (NCEP/NCAR) or the European Centre for Medium-Range Weather Forecasts (ECMWF). Here the ERA40 reanalysis product [11] of ECMWF is chosen for SE Asia due to better performance for the summer season when compared gainst gridded temperature and precipitation data (please see [5] for more details).

2 WRF Model Configurations

WRF is a next-generation mesoscale numerical weather prediction system designed to serve both operational forecasting and atmospheric research needs. The WRF Software Framework (WSF) provides the infrastructure that accommodates the dynamics solvers, physics packages that interface with the solvers, programs for initialization, WRF-Var, and WRF-Chem. There are two dynamics solvers in the WSF. The one applied in this project is the Advanced Research WRF (ARW) solver which was primarily developed at NCAR (National Centre for Atmospheric Research, USA). The ARW dynamics solver integrates the compressible, non-hydrostatic Euler equations. The equations are cast in flux form using variables that have conservation properties. The equations are formulated using a terrain-following mass vertical coordinate. The flux form equations in Cartesian space are extended to include the effects of moisture in the atmosphere and projections to the sphere.

For the temporal model discretization the ARW solver uses a time-split integration scheme. Slow or low-frequency meteorologically significant modes are integrated using a third-order Runge-Kutta (RK3) time integration scheme, while the high-frequency acoustic modes are integrated over smaller time steps to maintain numerical stability. The horizontally propagating acoustic modes (including the external mode present in the mass-coordinate equations using a constant-pressure upper boundary condition) and gravity waves are integrated using a forward-backward time integration scheme, and vertically propagating acoustic modes and buoyancy oscillations are integrated using a vertically implicit scheme (using the acoustic time step). The time-split integration for the flux-form equations is described and analyzed in [4].

The spatial discretization in the ARW solver uses a C grid staggering for the variables. That is, normal velocities are staggered one-half grid length from the thermodynamic variables. The grid lengths Δx and Δy are constants in the model formulation; changes in the physical grid lengths associated with the various projections to the sphere are accounted for using map factors. The vertical grid length $\Delta 2\eta$ is not a fixed constant; it is specified in the initialization. The user is free to specify the η values of the model levels subject to the constraint that $\eta = 1$ at the surface, $\eta = 0$ at the model top, and η decreases monotonically between the surface and model top. Climate simulations with a target resolution of 5 km for Central Vietnam for 1960–2050 are performed. For this purpose, WRF is nested in the general circulation model ECHAM5 using the following setup and location of the domains (Fig. 1, source: [5]). The same settings are used for the downscaling of ERA40 reanalysis data (1971–2000). In this study it is focussed on the LMRB in Southeast Asia, which is fully included in Domain 2.

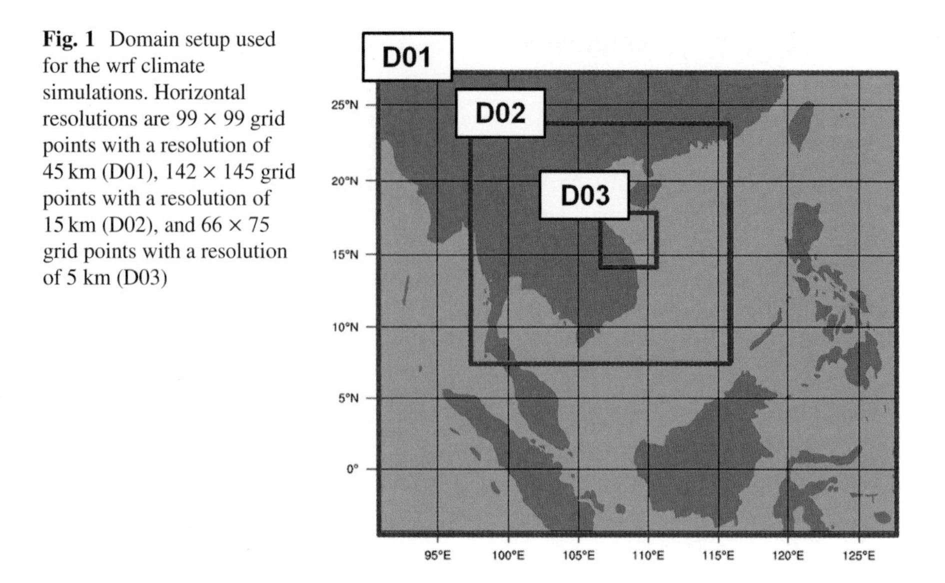

Fig. 1 Domain setup used for the wrf climate simulations. Horizontal resolutions are 99 × 99 grid points with a resolution of 45 km (D01), 142 × 145 grid points with a resolution of 15 km (D02), and 66 × 75 grid points with a resolution of 5 km (D03)

3 Validation of WRF Simulation Results

In a previous study it could be demonstrated that the spatial patterns of temperature could be simulated with a relatively high accuracy if compared to gridded observation data [5]. Here, it is analysed how the precipitation patterns evolve in time and how the patterns of Domain 2 correspond to the gridded observation data APHRODITE [12]. In Fig. 2 the mean spatial precipitation distribution precipitation of WRF-ERA40 and APHRODITE (1971–2000) are compared. It can be seen that the major precipitation regimes (caused by the two different monsunal systems) can be reproduced by the wrf simulations. Partly high biases in the magnitude of precipitation can be observed. It still remains unclear if these are real biases or if these are artifacts and origin from the low density of the observation network in Southeast Asia.

4 Climate Trends

In this section, the trends of the dynamically downscaled ERA40 reanalysis data are calculated on an annual basis. The slope of each trend is estimated using the non-parametric Sens slope method [9]. A positive value of the Sens slope indicates an upward trend (i.e. increasing with time), whereas a negative value indicates a downward trend (i.e. decreasing with time). The rank-based non-parametric Mann-Kendall [3, 6] statistical test is used to assess the significance of trends in hydro-meteorological data in this study. The trends are tested at $\alpha = 0.05$ and

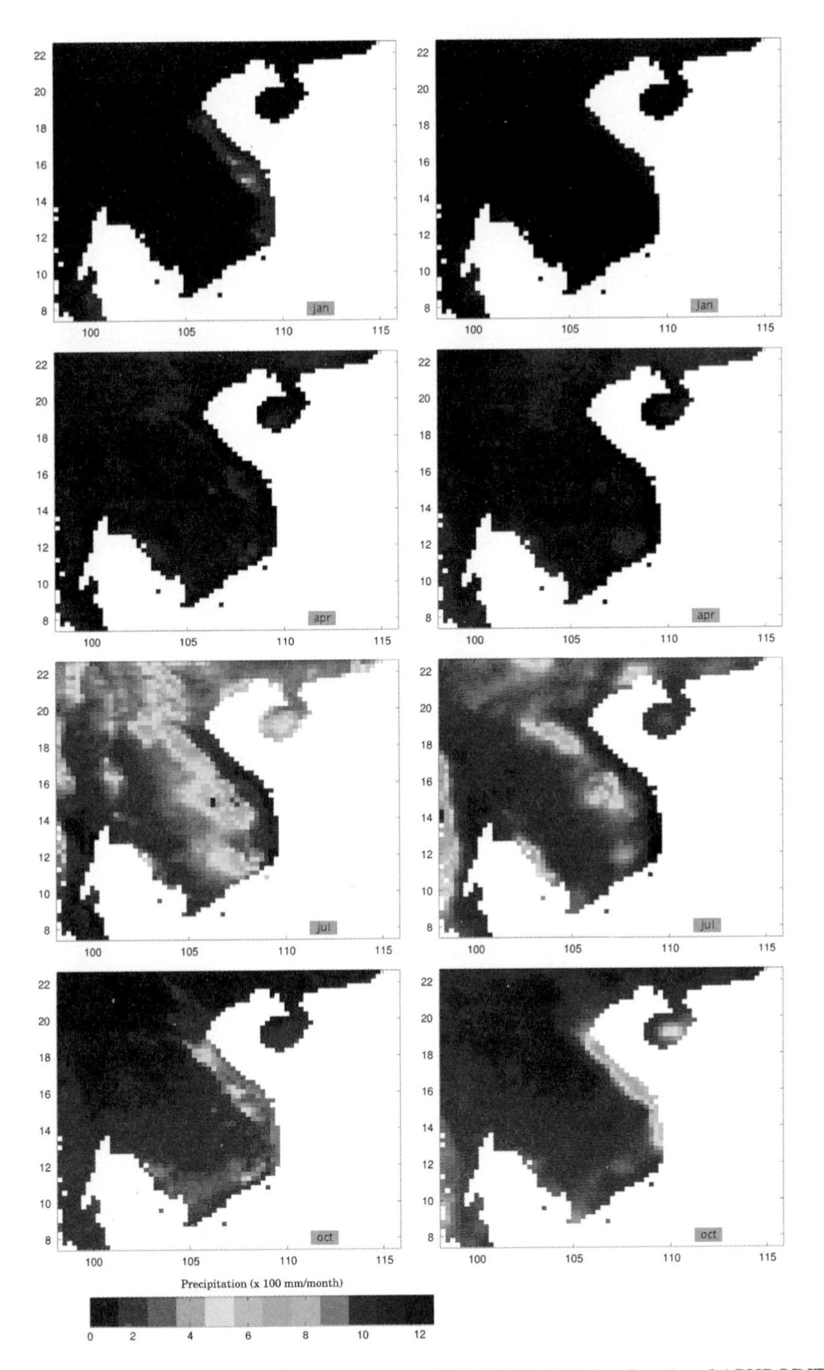

Fig. 2 1971–2000 average precipitation amounts of WRF-ERA40 (*left column*) and APHRODITE (*right column*) for January, April, July and October. The WRF-ERA40 precipitation amount is upscaled and regridded from 15 km resolution to 0.25 resolution using bilinear interpolation method

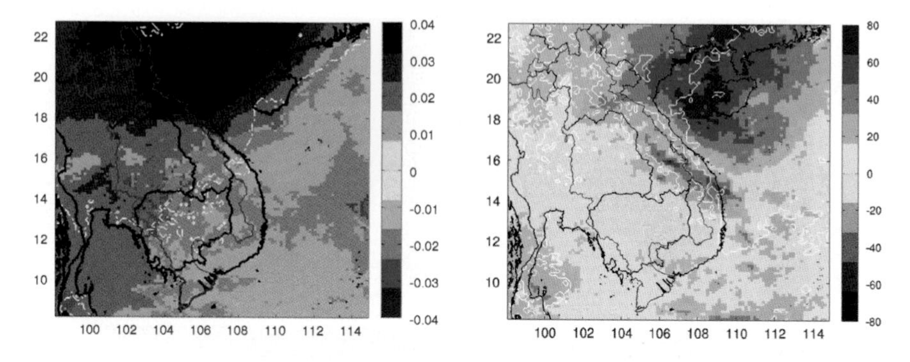

Fig. 3 Temperature trends [K/year] (*left*) and rainfall trends [mm/year] (*right*) as obtained by the WRF simulations for D02 driven by ERA40 reanalysis data. *White dashed (solid) line* indicates significance at $\alpha = 0.01$ (0.05)

0.01, indicating significant results and highly significant results, respectively. The approach is chosen because it makes no assumption about the distribution of the data. It proves therefore robust to the effect of extreme values and outliers [2]. As can be seen from Fig. 3, the trends of mean temperature in the Lower Mekong River Basin (LMRB) can reach values up to +0.04 k/year. This corresponds to a warming up to 1.2 K during 1971–2000. In the southern part of the basin, there prevail regions with slightly negative trends (> 0.01). Except for these regions, the trends are found to be very significant. For precipitation, the trends in the LMRB are dominantly positive range between 0 and +50 mm/year, however, relatively small areas are found to be significant. On average the trends in the LMRB are about +10 mm/year (300 mm for the 1971–2000 period).

5 Climate Change Signal

The climate change signal is assessed for precipitation and temperature by calculating the differences of two different climate projections, i.e. between ECHAM5-A1B and ECHAM5-B1, and the ECHAM5 control period for 1971–2000 (ECHAM5-CTRL). For precipitation, the long-term amounts are considered whereas the long-term mean values are calculated for temperature. The future projections (2001–2050) are subdivided into two different time slices (2001–2030 and 2021–2050) for further considerations. Figure 4 illustrates the results for precipitation. For the A1B scenario one can see regions with expected precipitation increases (blue colored regions) and decreases (red values) for both time slices. The period 2021–2050 shows a strong precipitation decrease for the northern part of the LMRB and for central Vietnam. There, the expected decrease amounts 600–800 mm. Following the B1 emission scenario, almost the whole basin has to expect precipitation increase in the order of +200 mm on average. Temperature is very likely to be increased in the

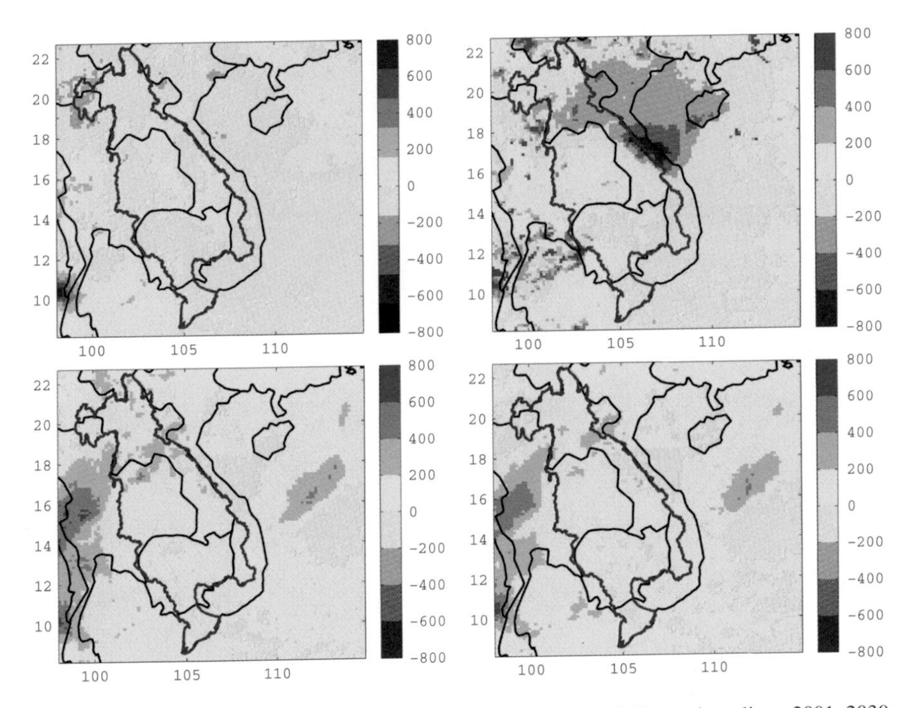

Fig. 4 Expected climate change signal for precipitation using two different time slices, 2001–2030 (*left column*) and 2021–2050 (*right column*) and two different emission scenarions, ECHAM5-A1B (*top row*) and ECHAM5-B1 (*bottom row*), given as differences from the 1971–2000 long-term mean values of the control period (ECHAM5-CTRL)

long run. The magnitude of increase, however, is strongly depending on the scenario. For the 2021–2050 period, the differences range between averagely +3.5 K to +1.5 K for A1b and B1, respectively. Only for the near future (2001–2020 period) slightly negative temperatures are simulated following the B1 scenario (Fig. 5).

6 Outlook

More than 30 TB of data are produced which will be analyzed in the future. Besides the instantaneous variables, retained in a six hourly and hourly resolution, additional meteorological surface variables containing the magnitude and timing of the actual minimum and maximum values of a day, i.e. precipitation, temperature, wind speed, and water vapour mixing ratio are retained. This will allow for detailed analysis of extreme events later on.

The results are currently post-processed and will be provided to the LUCCi project consortium, but results can be provided on request to the climate impact

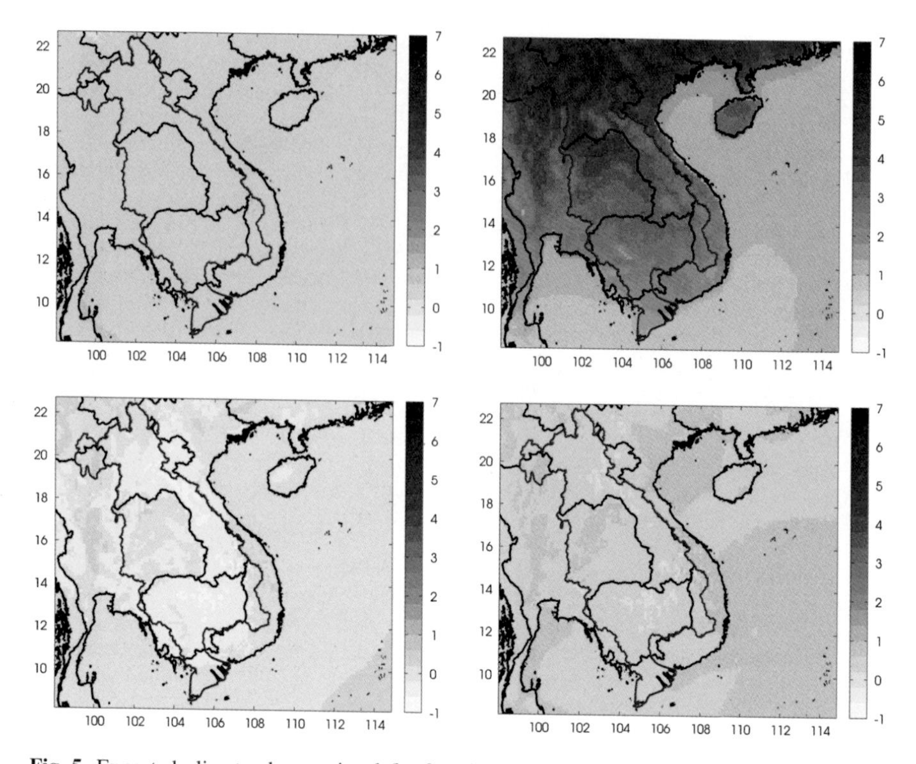

Fig. 5 Expected climate change signal for 2m-air-temperature using two different time slices, 2001–2030 (*left column*) and 2021–2050 (*right column*) and two different emission scenarions, ECHAM5-A1B (*top row*) and ECHAM5-B1 (*bottom row*), given as differences from the 1971–2000 long-term mean values of the control period (ECHAM5-CTRL)

community as well. Currently, the interest of local stakeholder are collected and analyzed, and tailor-made information can be provided on request.

Acknowledgements This research is funded by the Federal Ministry of Education and Research (research project: Land Use and Climate Change Interactions in Central Vietnam (LUCCi), reference number 01LL0908C). The provision of CPU and storage resources at Karlruhe Institute of Technology (KIT), Steinbuch Centre for Computing (SCC) and Karlruhe Institute of Technology (KIT), Institute of Meteorology and Climate Research (IMK-IFU) is highly acknowledged.

References

1. F. Giorgi et al. (2001) Regional climate information—evaluation and projections, in *Climate Change 2001:The Scientific Basis. Conribution of Working Group I to the Third Assessment Report of the Intergovernmental Panel on Climate Change*, ed. by J.T. Houghton, Y. Ding, D.J. Griggs, M. Noguer, P.J. van der Linden, X. Dai, K. Maskell, C.A. Johnson (Cambridge University Press, Cambridge, 2001), pp. 583–638

2. D.R. Helsel, R.M. Hirsch, *Statistical Methods in Water Resources* (Elsevier, Amsterdam, 2002), p. 522
3. M.G. Kendall (1975) *Rank Correlation Methods* (Charles Griffin, London, 1975), p. 272
4. J.B. Klemp, W.C. Skamarock, J. Dudhia, Conservative split-explicit time integration methods for the compressible nonhydrostatic equations. Mon. Weather Rev. **135**, 2897–2913 (2007)
5. P. Laux, V.T. Phan, Ch. Lorenz, T. Thuc, L. Ribbe, H. Kunstmann, Setting up regional climate simulations for Southeast Asia, in *Proceedings of High Performance Computing in Science and Engineering'12*, ed. by W.E. Nagel, D.B. Kröner, M.M. Resch (Springer, Berlin, 2013), pp. 391–406
6. H.B. Mann, Nonparametric tests against trend. Econometrica **13**(3), 245–259 (1945)
7. K. Prömmel, B. Geyer, J.M. Jones, M. Widmann, Evaluation of the skill and the added value of renalysis driven regional simulation for Alpine temperature. Int. J. Climatol. **30**, 760–773 (2010)
8. E. Roeckner et al., The atmospheric general circulation model ECHAM5. Part I: Model description. Rep. 349, Max Planck Institute for Meteorology, 127 pp., (2003) [Available from MPI for Meteorology, Bundesstr. 53, 20146 Hamburg, Germany.]
9. P.K. Sen, Estimates of the regression coefficient based on Kendall's Tau. J. Am. Stat. Assoc. **63**(324), 1379–1389 (1968)
10. M. Souvignet, P. Laux, J. Freer, H. Cloke, D.Q. Thinh, T. Thuc, J. Cullmann, A. Nauditt, W.A. Flgel, H. Kunstmann, L. Ribbe, Recent climatic trends and linkages to river discharge in Central Vietnam. Hydrol. Process. (2013). doi:10.1002/hyp.9693
11. S.M. Uppala et al., The ERA-40 re-analysis. Q. J. R. Meteorol. Soc. **131**, 2961–3012 (2005). doi:10.1256/qj.04.176
12. A. Yatagai et al., APHRODITE: Constructing a long-term daily gridded precipitation dataset for Asia based on a dense network of rain gauges. Bull. Am. Meteorol. Soc. (2012). doi:10.1175/bams-d-11-00122.1

Reducing the Uncertainties of Climate Projections: High-Resolution Climate Modeling of Aerosol and Climate Interactions on the Regional Scale Using COSMO-ART

Implementation and Testing of COSMO-ART on HERMIT Cluster at HLRS

Bernhard Vogel, Andrew Ferrone, and Tobias Schad

Abstract Aim of this project is to investigate the impact of aerosol in high resolution climate runs. Aerosols and their interactions with radiation and clouds represents in the moment one of the major uncertainties in our understanding of climate system as they can described only roughly in coarse resolution global models. The online coupled comprehensive chemistry model system COSMO-ART already showed in several case studies the potential of closing this gap. In order to apply it on decadal climate time scales the use of high performance computing becomes a necessity. We propose to perform decadal climate simulations on two domains. A first one covers Europe at a horizontal resolution of 14 km for a period from 1995 to 2005. The second domains covers Germany at a horizontal resolution of 2.8 km for the same period and it will be forced at the boundaries by the results of the first domain. On the European domain only the interactions of aerosols and radiation will be represented, whereas in the high-resolution German domain, the simulated aerosol can additionally act as cloud condensation and ice nuclei. The results will be obtained will be unprecedented and lead to a better understanding of changes in the regional climate of Europe and Germany. Up to now model implementation, some technical testing and scaling results could be performed.

B. Vogel (✉) · T. Schad
KIT, Karlsruhe, Germany
e-mail: bernhard.vogel@kit.edu; tobias.schad@kit.edu

A. Ferrone
Public Research Centre - Gabriel Lippmann, Belvaux, Luxembourg
e-mail: ferrone@lippmann.lu

W.E. Nagel et al. (eds.), *High Performance Computing in Science and Engineering '13*, DOI 10.1007/978-3-319-02165-2_38,

1 Model Implementation

We managed to successfully compile the model system COSMO-ART on CRAY XE-6. The source code is written in Fortran F90 and F77. We used the Cray Fortran compiler which also includes MPI by default.

The following optimization is used: -Os (This was recommended by H.-J. Panitz, as higher optimization levels may lead to problems in connection with Intel compilers).

For compilation of COSMO-ART following libraries were needed: NETCDF (Version 4.2.0), GRIB (DWD), which were available on CRAY XE-6.

2 Technical Testing

Furthermore some technical testing was done. We successfully tested the model chain from preparing global model data with int2lm (pre-processor for model input data) towards initial- and boundary data for our regional climate model COSMO-ART.

In comparison with regional climate model COSMO-CLM COSMO-ART needs additionally input data: global model data for chemistry and aerosol boundary conditions, anthropogenic emissions and land use data.

There still has to be done some technical testing regarding the independence of results of processors used, and the independence of the results on restarts.

3 Scaling Results

To better understand the behaviour of COSMO-ART on CRAY XE-6 we performed a scaling test. For a model domain of $332 \times 328 \times 40$ grid-points (European domain) we performed model runs over a simulation period of 1 day. We were beginning with a total amount of processors of 128 and were increasing them stepwise up to 2,704 processors. A saturation of CPU time at a processor amount of 1,920 can be seen (see Fig. 1). If we declare our run with 128 processors as a reference run, the speed-up of the run with 1,920 processors is about seven. With more processors than 1,920 speed-up is increasing in smaller steps. We assume the efficient processor amount for the chosen model domain of $332 \times 328 \times 40$ grid-points is 1,920 processors. Overall, the storage capacity for output of 1 day of simulation is 19 GB. So far COSMO-ART was used for process studies covering time intervals of several days. This required a huge amount of output i.e. each variable at each hour. In the future

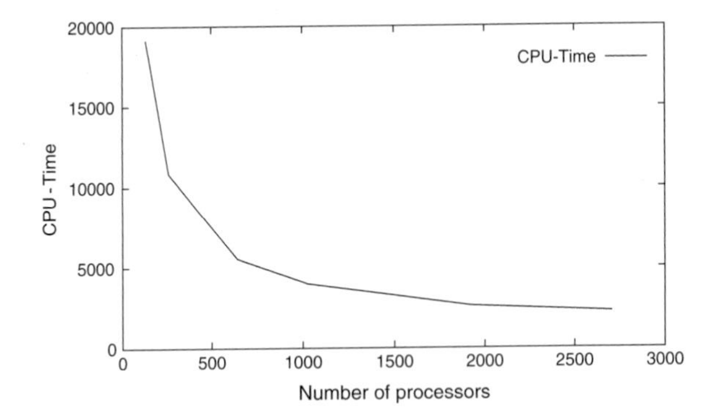

Fig. 1 CPU-time as a function of processors

we are aiming on decadal climate runs. Therefore, we are currently working on a strategy to reduce the output of the model runs, avoiding the risk that we are losing important information.

Global Long-Term MIPAS Data Processing

M. Kiefer, F. Friederich, B. Funke, A. Linden, M. López-Puertas, and T. von Clarmann

Abstract The Michelson Interferometer for Passive Atmospheric Sounding (MIPAS) was a Fourier transform mid-infrared limb scanning high resolution spectrometer which allowed for simultaneous measurements of more than 30 atmospheric trace species related to atmospheric chemistry and global change. MIPAS on Envisat was operated by ESA since mid of 2002 through April 8th 2012, when, due to a power supply failure, the platform and all onboard instruments were lost. At the Institute for Meteorology and Climate Research (IMK), MIPAS spectra are used for retrieval of altitude-resolved profiles of abundances of trace species of the atmosphere. The vertical profiles processed at IMK are available to external data users since many years. The trace gas distributions are used for the assessment of e.g. stratospheric ozone chemistry, stratospheric cloud physics and heterogeneous chemistry, stratospheric exchange processed with troposphere and mesosphere, intercontinental transport of pollutants in the upper troposphere, effects of solar proton events on stratospheric chemistry, thermosphere temperature, and chemistry-climate models. In the reporting period 2012/2013 most CPU time of MIPAS data processing on the XC4000 supercomputer was used for retrievals including computationally expensive NLTE calculations. This comprises the kinetic temperature for several special high altitude observation modes of MIPAS, NO, CO and NO_2 volume mixing ratios. Most of the processing work was done in close

---

M. Kiefer · F. Friederich · A. Linden · T. von Clarmann
Institut für Meteorologie und Klimaforschung, Karlsruhe Institute of Technology, Karlsruhe, Germany
e-mail: michael.kiefer@kit.edu; felix.friederich@kit.edu; andrea.linden@kit.edu; thomas.clarmann@kit.edu

B. Funke · M. López-Puertas
Instituto de Astrofísica de Andalucía, Granada, Spain
e-mail: bernd@iaa.es; puertas@iaa.es

W.E. Nagel et al. (eds.), *High Performance Computing in Science and Engineering '13*, DOI 10.1007/978-3-319-02165-2_39,

collaboration with the Instituto de Astrofísica de Andalucía (IAA) in Granada, Spain. Two examples of scientific exploitation of this kind of data will be given:

1. Determination of the NO_x-production rate and the NO_x-lifetime during the solar proton event in October–November 2003.
2. The Impact of two solar proton events in 2012 on chemistry and dynamics of the middle atmosphere at the northern and southern polar caps have been examined using MIPAS data of temperature and several species.

1 The MIPAS/Envisat Mission

The Michelson Interferometer for Passive Atmospheric Sounding (MIPAS) was a space-borne cooled Fourier transform spectrometer which operated in the mid-infrared spectral region 4.15–14.6 μm. The design spectral resolution was 0.035 cm^{-1}. It measured thermal emission spectra of the Earth's limb, whereby variation of the tangent altitude provided altitude-resolved information [1, 2].

MIPAS was part of the core-payload of ESA's research satellite Envisat. Envisat was a sun-synchronous polar orbiter which passed the equator at 10:00 a.m. local time in southward direction 14.4 times a day. The MIPAS mission intended to improve the understanding of the composition of the Earth's atmosphere by measurement of 4D distributions of more than 30 trace species relevant to atmospheric chemistry and climate change. Operation of satellite and instrument and level-1b (L1b) data processing (Fourier transformation, phase correction, calibration) have been performed by the European Space Agency (ESA). MIPAS was operational in its original, high spectral resolution specification from June 2002 to 26 March 2004 and provided about 1,000 limb scans a day, each consisting of 17 times 60,000 spectral radiance measurements. In summary 4 TB of data are available for scientific analysis from this period. Due to an instrument problem the measurements could not be resumed before the beginning of 2005. The spectral resolution had to be reduced to about 40 % of the initial value since then. After recovery the duty cycle was initially lowered to approximately 50 %, however, due to the very good instrument conditions, the full 100 % duty cycle was reached again soon. Just weeks after celebrating its tenth year in orbit, communication with the Envisat satellite was suddenly lost on 8th April. Following fruitless attempts to re-establish contact, and the investigation of failure scenarios, the end of the mission has being declared by ESA.

2 Data Analysis

Retrieval of atmospheric constituents abundances requires the inverse solution of the radiative transfer equation. Measured radiances are compared to radiative transfer calculations and residuals are minimized in a least squares sense by adjustment of the constituents abundances which are fed into a forward model. The Retrieval Control Program (RCP), together with the forward model, is the core of the processing chain.

The forward model, the Karlsruhe Optimized and Precise Radiative Transfer Algorithm (KOPRA) [7] is a computationally optimized line-by-line model which simulates radiative transfer through the Earth's atmosphere under consideration of all relevant physics: the spectral transitions of all involved molecules, atmospheric refraction, line-coupling and non-local thermodynamic equilibrium (NLTE). Along with the spectral radiances, KOPRA also provides the Jacobian matrices, i.e. the sensitivities of the spectral radiances to changes in the atmospheric state parameters.

Atmospheric state parameters are retrieved by constrained multi-parameter non-linear least-squares fitting of simulated and measured spectral radiances. The inversion is regularized with a Tikhonov-type constraint which minimizes the first order finite differences of adjacent profile values [8].

Instead of simultaneous retrieval of all target parameters from a limb sequence, the retrieval is decomposed in a sequence of analysis steps. First, spectral shift is corrected, then the instrument pointing is corrected along with a retrieval of temperature, using CO_2 transitions. Then the dominant emitters in the infrared spectrum are analyzed one after the other (H_2O, O_3, HNO_3...), each in a dedicated spectral region where the spectrum contains maximum information on the target species but least interference by non-target species. Finally, the minor contributors are analyzed, whereby pre-determined information on the major contributors is used. The analysis is done limb sequence by limb sequence, i.e. limb scans are processed independently from each other by separate calls of the RCP. Typically some ten thousands to hundred thousand of limb sequences are processed for several species in one project.

3 Computational Considerations

3.1 *Processing System Overview*

The retrieval of atmospheric trace constituents is performed for each geolocation separately. A geolocation is determined by the mean values of the geographic coordinates and of the times of the spectral radiance measurements of the corresponding limb scans. ESA delivers L1b data as binary files with one file containing spectra plus additional information for all geolocations of one Envisat orbit.

The processing steps consist of:

- preprocessing with eight single steps performed by eight different programs, always performed at IMK;
- transfer of preprocessed data from IMK to the XC4000

- retrieval (core processing step) with the Retrieval Control Program (RCP) on the XC4000 supercomputer;
- postprocessing on the XC4000 to generate some elementary result diagnostics and corresponding plots.
- transfer of result and diagnostics data from XC4000 to IMK.

A detailed description of the above steps, as well as of the adaptation of the retrieval software to the XC4000 supercomputer, already has been given in [6]. A new concept for the MIPAS data processing was developed together with W. Augustin of the scientific supercomputing group of the Karlsruhe University's computing center, and subsequently implemented in the first half of 2008. Details of this enhanced processing system have been given in the report of 2008–2009.

3.2 *Processor Usage*

The increased performance of the processing system since mid 2008 has made the XC4000 a major contributor to the total amount of data processed at IMK during the year 2008 and the beginning of 2009. However, the in-house computing capabilities at IMK have considerably increased since then and, as a consequence, the focus of data processing on the XC4000 shifted towards the species, which require the most computing time per geolocation, namely those which require costly non-local thermodynamic equilibrium (NLTE) calculations during the retrieval.

As in the reporting period before, in 2012–2013 the XC4000 has mainly been used to process CO, NO, NO_2, and the kinetic temperature in the upper atmosphere above 60 km. All these retrievals require NLTE calculations.

Since the retrieval of one species at one geolocation is a task completely independent from retrievals at other geolocations, there is no overhead like e.g. message or data passing between parallel threads of a retrieval job. This allows to use as many XC4000 processors as available for MIPAS data processing.

For the reporting period 2012–2013 the lowest two frames of Fig. 1 show the number of projects (upper frame) and the accumulated number of projects (lower frame) over the number of simultaneously used processors.

A compilation of the data for two other reporting periods (2008–2009 and 2011–2012) is shown in Fig. 1, too. Approximately 130 new projects were processed during the current reporting period. Peak numbers of 800 simultaneously used processors could be achieved for approx. 70 projects. The steady increase in those projects which used 600/800 processors (depending on whether three or four processors per node were activated) over time, from 2008 to 2013, might be an indication that the overall workload (other users) on the XC4000 has decreased in the last reporting periods, compared to e.g. 2008–2009.

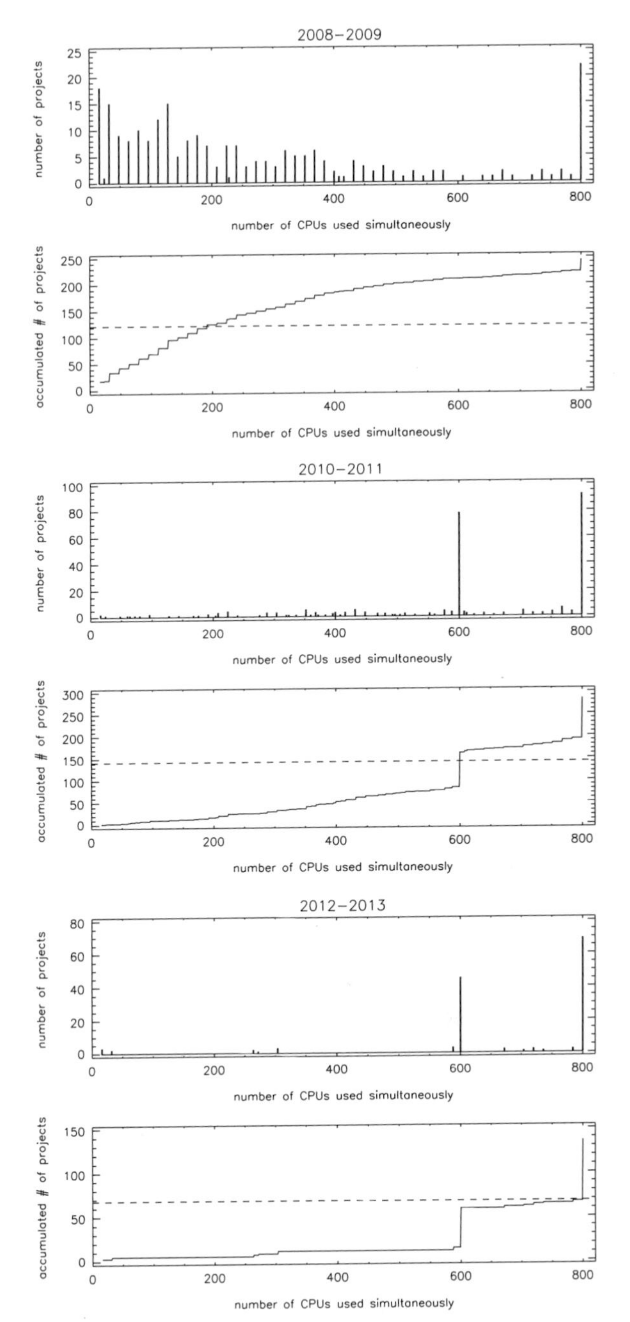

Fig. 1 Frequency and accumulated frequency of peak numbers of simultaneously used processors per computation project. *Dashed line* marks the value of half the number of projects. Data of three reporting periods (plot titles) are shown

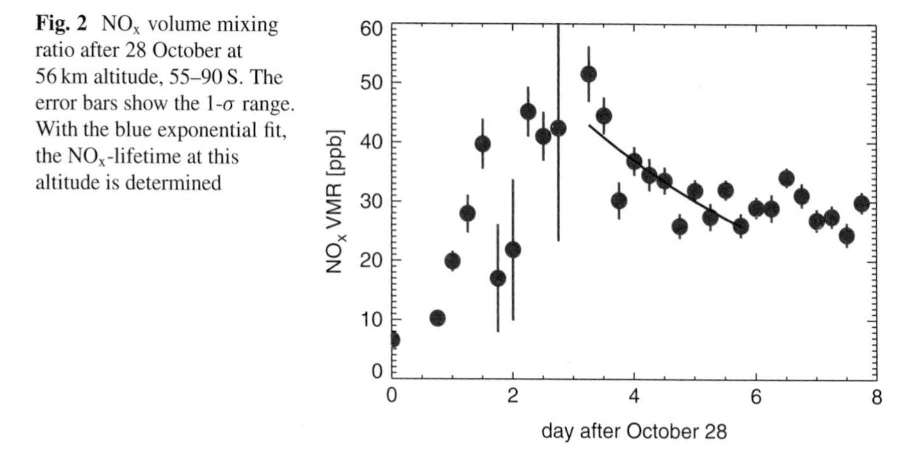

Fig. 2 NO_x volume mixing ratio after 28 October at 56 km altitude, 55–90 S. The error bars show the 1-σ range. With the blue exponential fit, the NO_x-lifetime at this altitude is determined

4 Two Examples of Scientific Exploitation of MIPAS Data

4.1 Determination of the NO_x-Production Rate and the NO_x-Lifetime During the Solar Proton Event in October–November 2003

During a solar proton event (SPE) solar electrons, protons, and more massive ions intrude in polar latitudes of the Earth's atmosphere guided by geomagnetic field lines. The particles can penetrate down into the mesosphere and upper stratosphere ($\geq$40 km) leading to excitation, ionization, dissociative ionization, and dissociation of N_2. As a result, NO_x ($NO + NO_2$) is produced through N_2 dissociation and subsequent (ion-)chemical reactions. In October/November 2003 a strong SPE occurred, which enhanced NO_x many times over its background volume mixing ratio (vmr). Comparisons between measured and modelled NO_x showed an overestimation of all models between 2 and 0.5 hPa at high latitudes for that SPE [5].

The aim of this study [3] is to determine the lifetime and the effective production rate of NO_x, in order to provide model diagnostics. Measurements of NO and NO_2 by MIPAS are used to determine the atmospheric content of NO_x at altitudes of 44–62 km in October/November 2003. The temporal decrease of NO_x after the SPE-triggered maximum defines the altitude-dependent lifetime of NO_x. It is controlled by photolysis, transport and mixing, and decreases with altitude (from $\sim$20 days at 44 km to $\sim$2 days at 62 km).

Figure 2 shows an example at 56 km altitude, averaged over all latitudes higher than 55 S. The red points show the 6-h average of the measured NO_x volume mixing ratio, the error bars show the 1-sigma range. By means of an exponential function (blue fit), the NO_x-lifetime is determined.

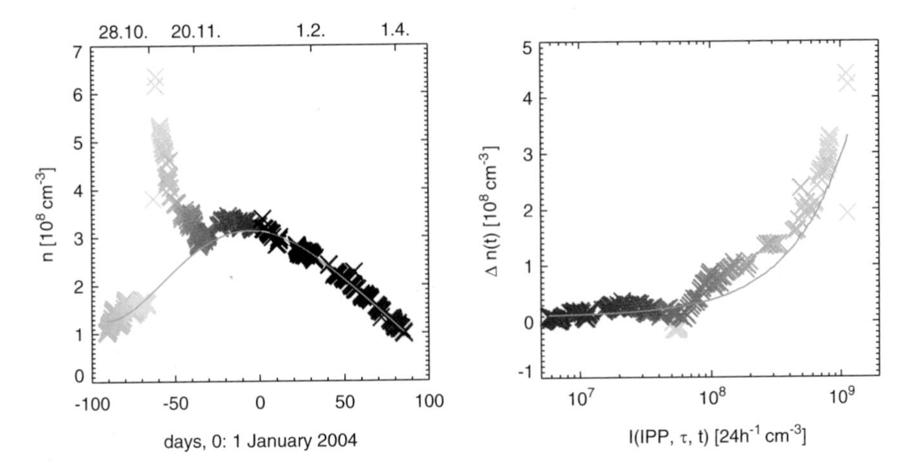

Fig. 3 *Left*: NO_x-number density in 50 km altitude, averaged over all geomagnetic latitudes higher than 55 S, the *green curve* shows the "backround" NO_x-number density. *Right*: NO_x-enhancement compared to accumulated ionisation rates. The gradient of the green fit is the effective NO_x-production rate

By considering NO_x-enhancement, NO_x-lifetime, and ionization rates of electrons, protons, and alpha particles of the 3-D model AIMOS (Atmospheric Ionization Module OSnabrück), effective NO_x-production rates are calculated.

Figure 3 (left) shows the NO_x-number density in 50 km altitude, averaged over the southern polar cap (higher than 55 S geomagnetic latitude). The color code is time dependent. Green and yellow are the number densities before the SPE; orange and red during and shortly after the SPE; black, violet, and blue from 2004 on. The green fit shows the "backround" NO_x-number density. Figure 3 (right) shows the relation between accumulated ionization rates weighted with the NO_x lifetime (logarithmic scale), and the NO_x enhancements (linear scale). The gradient of the resulting green linear fit is the effective NO_x-production rate. It is mainly triggered by the following two competing reactions, including both N as reactant:

$$N + O_2 \longrightarrow NO + O \quad (1)$$

$$N + NO \longrightarrow N_2 + O \quad (2)$$

A comparison with a box model shows that the determined NO_x-production rates are too low, which likely is a hint that the ionization rates are overestimated.

4.2 MIPAS Observation of Solar Proton Events in 2012

As discussed in the preceding section, solar particles can ionize the atmosphere down to stratospheric altitudes. Hence polar HO_x, nitrogen and chlorine chemistry

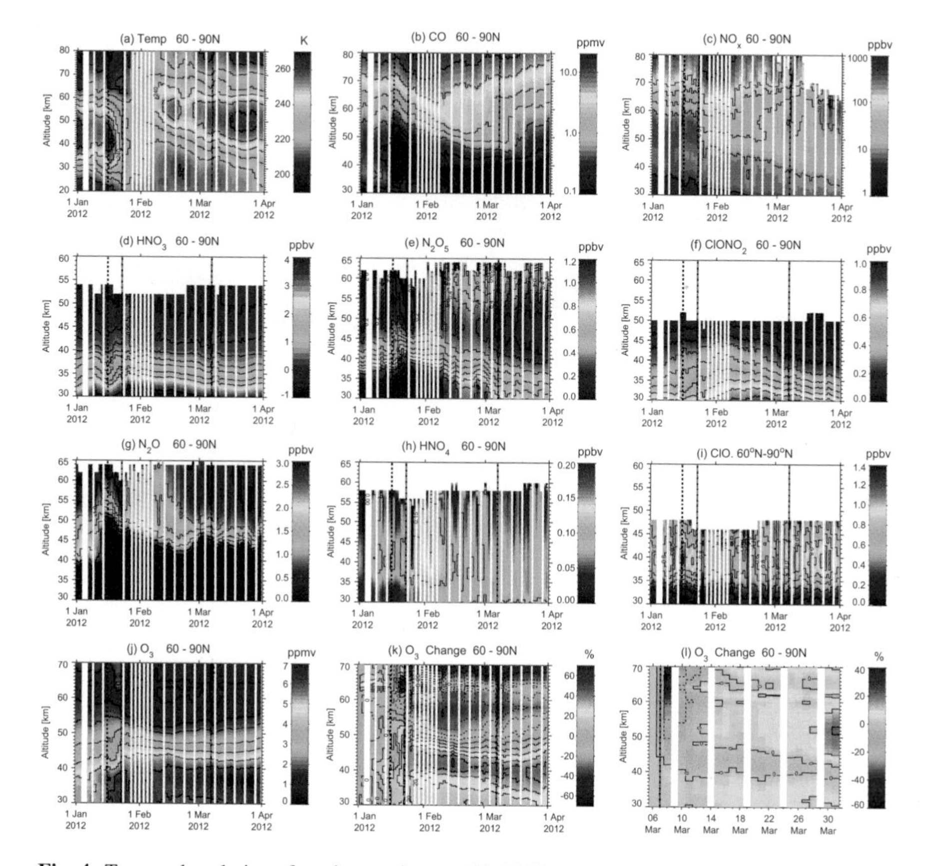

Fig. 4 Temporal evolution of northern polar cap (60–90 N) mean temperature and trace gas mixing ratios as a function of altitude. The onset of the stratospheric warming is marked by a *black dotted vertical line*, while the SPEs are indicated by *dashed-dotted lines*. Ozone changes shown in (**k**) and (**i**) are calculated with respect to the profile of the first day plotted

can be affected, which in consequence can affect ozone concentrations [4]. During two solar proton events (SPE) in 2012, namely on 23–30 January and 7–15 March MIPAS monitored atmospheric temperature and, amongst other species, O_3, CO, CH_4, NO_2, NO, HNO_3, N_2O_5, HNO_4, $ClONO_2$, N_2O, and ClO. A more detailed discussion of the involved chemical and dynamical processes can be found in [9].

Over the northern polar cap sudden stratospheric warming (SSW) events along with mesospheric cooling occurred during January (Fig. 4a). Using CO as a dynamics tracer it can be deduced from Fig. 4b that mesospheric CO-rich air continues to move down which indicates continued polar winter conditions. In the upper stratosphere lower CO mixing ratios hint at air of subpolar stratospheric origin mixed towards polar latitudes during the sudden warming event.

Nitrogen chemistry during the SPEs is characterized by abrupt NO_x increase at altitudes of approx. 50–70 km (Fig. 4c) and by abrupt enhancements of HNO_3

mixing ratios above approximately 45 km during the January SPE and above 35 km during the March SPE (Fig. 4d), which is attributed to a recombination reaction of H^+ and NO_3^- ions on water cluster ions. $ClONO_2$ shows a weak positive response to the January SPE in the upper stratosphere (Fig. 4f), while the longterm development of $ClONO_2$ resembles that of temperature, which is an indication of the major influence of vortex dynamics on $ClONO_2$. For ClO, another chlorine species observed by MIPAS, no change was observed which exceeds the observed day-to-day variability (Fig. 4i). Near 65 km altitude the formation of the tertiary ozone maximum is clearly seen in Fig. 4j, k. After an interruption during the SSW this maximum is reformed. Two days after the January SPE, about a third of the mesospheric ozone is destroyed. This ozone loss coincides with the phase of excess HNO_4 (Fig. 4h) and is attributed to SPE-triggered HO_x chemistry. Mesospheric ozone concentrations, however, recover within a few days. Immediately after the March SPE, ozone loss by up to 60 % is observed in the mesosphere, and again ozone recovers within a few days (Fig. 4l). Around 55 km NO_x-related ozone loss starts a couple of days after the January SPE; the pattern of ozone-depleted air coincides in latitude and time with the NO_x enhancements.

At the southern polar cap the dynamic conditions are quite different: January through April 2012 the southern polar stratosphere and mesosphere are characterized by continuous stratopause cooling (Fig. 5a). A pronounced immediate response of NO_x to both SPEs is observed down to the stratopause (Fig. 5b). Since there is no subsidence of air over the southern polar cap, and since the NO_x increase is observed simultaneously over a wide range of altitudes, it is attributed to SPE-induced in-situ production. Increases of 2, 5, 10, 20, and 30 ppbv at altitudes of 52, 56, 59, 63 and 70 km are observed during the January SPE, and 2, 5, 10, 20, 30, 35 ppbv at altitudes of 47, 50, 53, 60, 63 and 66 km after the March SPE. As in the Northern Hemisphere (NH), the NO_x increase persists several weeks. In January, these increases are weaker and shorter-lived than in the NH which is attributed to photochemical losses of NO in the SH upper stratosphere and above. In March, the NO_x enhancements are similar for both hemispheres.

A short term (1–3 days, depending on altitude and date) negative response of ozone is noticeable after both SPEs (Fig. 5c). At around 55 km altitude, the seasonal increase of ozone is overcompensated by ozone loss due to the enhanced NO_x mixing ratios.

5 Conclusions

For the reporting period 2012–2013 the XC4000 supercomputer again has proven to be a highly valuable tool for MIPAS data processing. Processing of the species NO, CO, NO_2 (all require expensive NLTE calculations) for the whole MIPAS mission, up to the end in April 2012, could be completed. Additionally kinetic temperature retrievals for MIPAS' MA, UA, and NLC measurement modes have been completed.

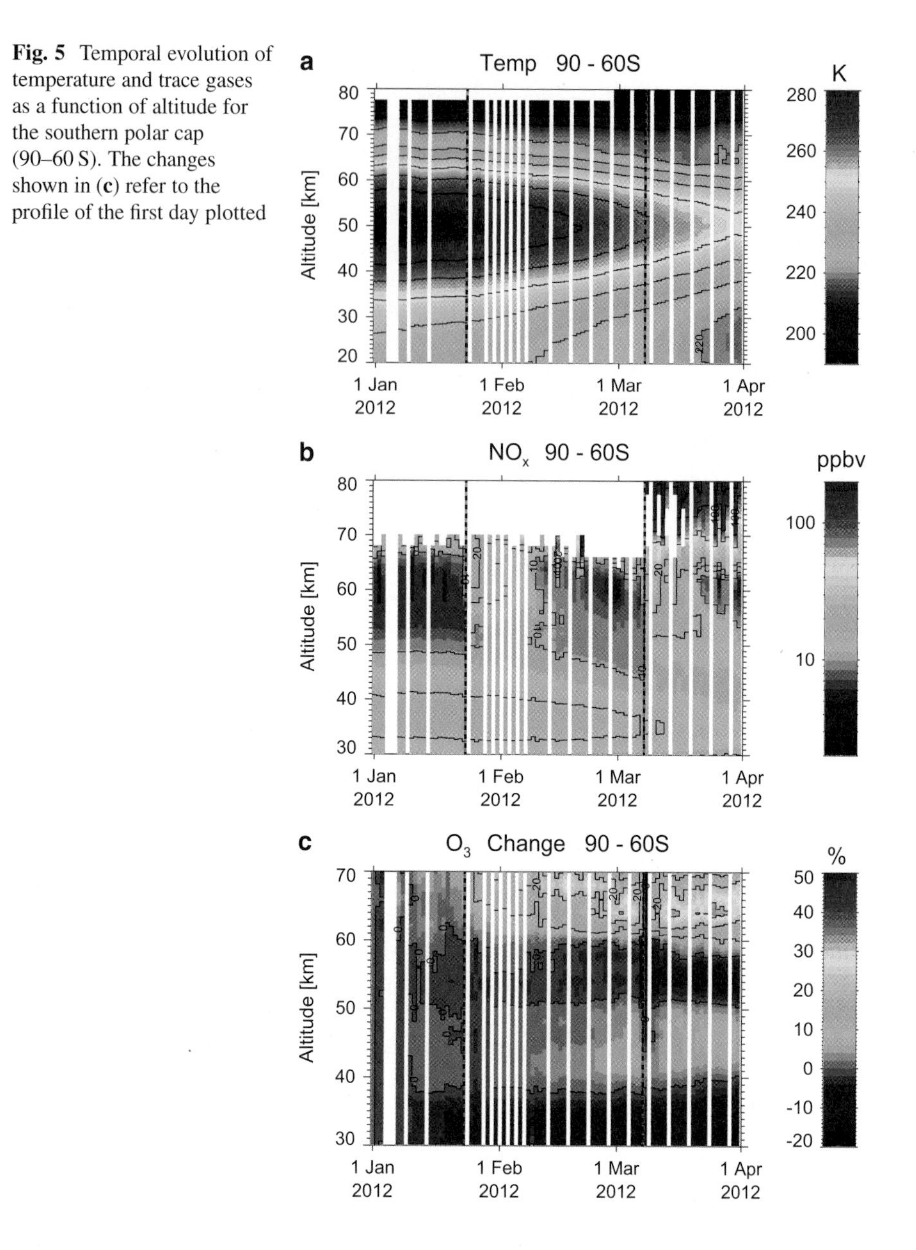

Fig. 5 Temporal evolution of temperature and trace gases as a function of altitude for the southern polar cap (90–60 S). The changes shown in (**c**) refer to the profile of the first day plotted

Two scientific applications of the data have been illustrated: First, the production rate and lifetime of NO_x during the solar proton event in October–November 2003 have been examined using MIPAS NO and NO_2 data. Second, the impact of two solar proton events in 2012 on the chemistry in the middle and upper atmosphere over the northern and southern polar caps has been studied with MIPAS retrievals of temperature, O_3, CO, CH_4, NO_2, NO, HNO_3, N_2O_5, HNO_4, $ClONO_2$, N_2O, and ClO.

References

1. H. Fischer, C. Blom, H. Oelhaf, B. Carli, M. Carlotti, L. Delbouille, D. Ehhalt, J.-M. Flaud, I. Isaksen, M. López-Puertas, C.T. McElroy, R. Zander, in *Envisat-MIPAS, an Instrument for Atmospheric Chemistry and Climate Research*, ed. by C. Readings, R.A. Harris. European Space Agency-Report SP-1229 (ESA Publications Division, ESTEC, Noordwijk, 2000)
2. H. Fischer, M. Birk, C. Blom, B. Carli, M. Carlotti, T. von Clarmann, L. Delbouille, A. Dudhia, D. Ehhalt, M. Endemann, J.M. Flaud, R. Gessner, A. Kleinert, R. Koopmann, J. Langen, M. López-Puertas, P. Mosner, H. Nett, H. Oelhaf, G. Perron, J. Remedios, M. Ridolfi, G. Stiller, R. Zander, MIPAS: an instrument for atmospheric and climate research. Atmos. Chem. Phys. **8**, 2151–2188 (2008)
3. F. Friederich, T. von Clarmann, B. Funke, H. Nieder, J. Orphal, M. Sinnhuber, G.P. Stiller, J.M. Wissing, Lifetime and production rate of NO_x in the upper stratosphere and lower mesosphere in the polar spring/summer after the solar proton event in October–November 2003. Atmos. Chem. Phys. **13**(5), 2531–2539 (2013)
4. B. Funke, M. López-Puertas, M. García-Comas, D. Bermejo-Pantaleón, G.P. Stiller, T. von Clarmann, The impact of energetic particle precipitation on the Earth's atmosphere, in *Astrophysics and Space Science Proceedings* (Springer, Berlin, 2010), pp. 181–189
5. B. Funke, A. Baumgaertner, M. Calisto, T. Egorova, C.H. Jackman, J. Kieser, A. Krivolutsky, M. López-Puertas, D.R. Marsh, T. Reddmann, E. Rozanov, S.-M. Salmi, M. Sinnhuber, G.P. Stiller, P.T. Verronen, S. Versick, T. von Clarmann, T.Y. Vyushkova, N. Wieters, J.M. Wissing, Composition changes after the "Halloween" solar proton event: the High Energy Particle Precipitation in the Atmosphere (HEPPA) model versus MIPAS data intercomparison study. Atmos. Chem. Phys. **11**(17), 9089–9139 (2011)
6. M. Kiefer, U. Grabowski, H. Fischer, Global long-term MIPAS processing, in *High Performance Computing in Science and Engineering '07, Transactions of the High Performance Computing Center, Stuttgart (HLRS) 2007*, ed. by W.E. Nagel, D. Kröner, M. Resch (Springer, Berlin, 2008), pp. 519–532 (Höchstleistungsrechenzentrum Stuttgart)
7. G.P. Stiller (ed.), *The Karlsruhe Optimized and Precise Radiative Transfer Algorithm (KOPRA), Wissenschaftliche Berichte*, vol. FZKA 6487. (Forschungszentrum, Karlsruhe, 2000)
8. T. von Clarmann, N. Glatthor, U. Grabowski, M. Höpfner, S. Kellmann, M. Kiefer, A. Linden, G. Mengistu Tsidu, M. Milz, T. Steck, G.P. Stiller, D.Y. Wang, H. Fischer, B. Funke, S. Gil-López, M. López-Puertas, Retrieval of temperature and tangent altitude pointing from limb emission spectra recorded from space by the Michelson Interferometer for Passive Atmospheric Sounding (MIPAS). J. Geophys. Res. **108**(D23), 4736 (2003). doi:10.1029/2003JD003602
9. T. von Clarmann, B. Funke, M. López-Puertas, S. Kellmann, A. Linden, G.P. Stiller, C.H. Jackman, V.L. Harvey, The solar proton events in 2012 as observed by MIPAS. Geophys. Res. Lett. 40, 10–15 (2013). doi:10.1002/grl.50119

The Agulhas System in a Global Context

Jonathan V. Durgadoo, Christina Roth, and Arne Biastoch

1 Introduction

The Agulhas is a convoluted and multifarious system [1]. It consists of a western boundary current, the Agulhas Current, which is arguably one of the most prominent current systems of the Southern Hemisphere (Fig. 1). The Agulhas Current, roughly on par with its Northern Hemisphere counterpart, the Gulf Stream, carries vast amount of heat and salt towards the pole [2]. Unique to the Agulhas, is the fact that the current abruptly retroflects (turns back on itself) south of the African continent, and approximately 75 % of its water transport returns into the Indian Ocean [3]. About 15 Sv (1 Sv = 1 million cubic m/s) of water enters the South Atlantic [4] through what is commonly referred to as "Agulhas leakage" [5]. Large parts of this water subsequently flow across the Equator and become part of the Gulf Stream system.

Recent scientific efforts have highlighted the climatic relevance of Agulhas leakage [2, 6, 7]. In particular the injection of heat and salt into the Atlantic circulation makes leakage an alleged key player in the process of water transformation that takes place in the North Atlantic. In the course of a warming climate, on the one hand, leakage has been increasing [7] and on the other hand, Arctic sea-ice is decreasing and Greenland melting is enhancing [8]. Here, it should be noted that the Agulhas density contribution to the Atlantic circulation is of opposing sign to that of subolar/subarctic freshening (e.g., Greenland melt water). Because of this dichotomy, understanding the back-bone processes that modulate and expedite leakage across the Equator is of current relevance in the oceanographic and climatic communities.

J.V. Durgadoo (✉) · C. Roth · A. Biastoch
GEOMAR Helmholtz-Zentrum für Ozeanforschung Kiel, Düsternbrooker Weg 20, 24105 Kiel, Germany
e-mail: jdurgadoo@geomar.de

W.E. Nagel et al. (eds.), *High Performance Computing in Science and Engineering '13*, DOI 10.1007/978-3-319-02165-2_40,

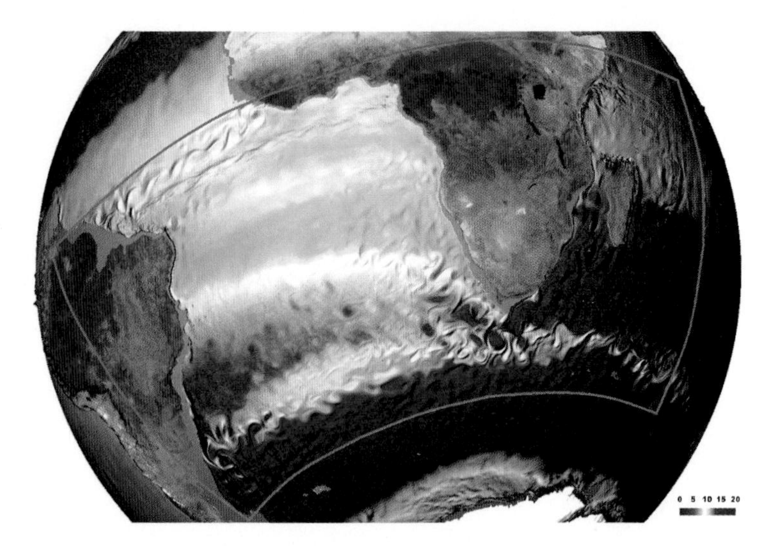

Fig. 1 The INALT01 configuration. Mid-depth temperature (*colour-scale*, °C) and velocity (gradients) show the path of the Agulhas Current flowing along the east coast of South Africa, the retroflection south of the continent, and large rings being exported into the South Atlantic. The regional nested component of INALT01 is bounded by the *thick grey lines* [9]

Measuring Agulhas leakage is no simple task, owing to the complex retroflection regime [10]. Part of what is termed leakage is through the intermittent expulsion of large ring-like features at the retroflection. Other sources are through so-called filaments and cyclones. Obtaining direct observations using either ship-borne or autonomous equipment is logistically not feasible because of the vast monitoring area, large number of variables, and time coverage required in order to understand the various processes and assess the impact of leakage. Employing ocean general circulation models overcomes these limitations by providing coherent sets of data, and by lending themselves to multiple realisations (sensitivity experiments).

In this report we highlight the relevant features of the model configuration we implemented, briefly present the major scientific achievements, and finally outline the basis for on-going work, including the challenges we are facing with the CRAY XE6 system.

2 Model Description

In constructing a model configuration capable of capturing the complexities of the Agulhas system, several issues have to be considered. Such a configuration has to have sufficient spatial and temporal resolution. Many features of the Agulhas system are of the order of 10–100 km in size and lasting 10–60 days. The areal coverage has

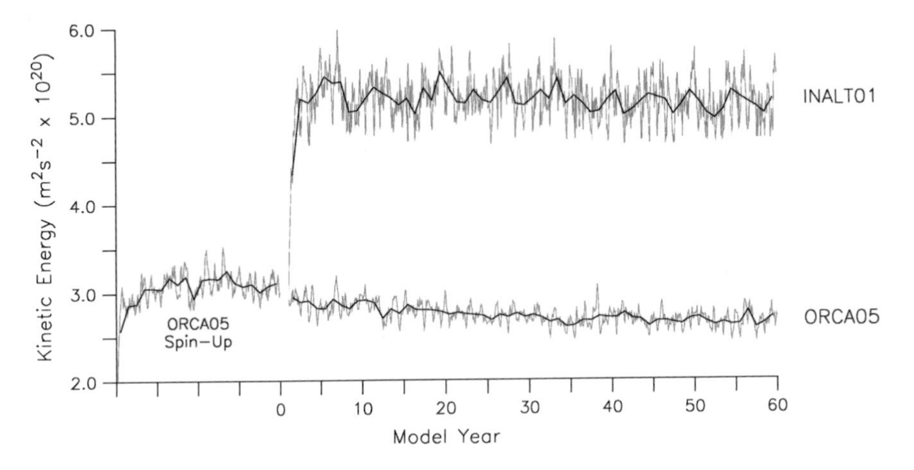

Fig. 2 Volume integrated (10°W–60°E; 10°S–45°S) kinetic energy per unit mass (m^2/s^2) with annual values (*thick lines*) overlaying monthly values (*grey*). Following a 20 year spin-up, reference experiments were performed for all configurations. Adapted from [15]

to be global in order to assess large-scale impact. An ocean-only model also requires coherent atmospheric forcing fields for a sufficiently long time period. Lastly the need of running multiple realisations requires a configuration that is efficiently optimised.

The INALT01 configuration (Fig. 1), based on the NEMO (Nucleus for European Modelling of the Ocean, v3.1.1, [11]) model code, was built with the above-mentioned criteria in mind. INALT01 is a nested configuration consisting of two components embedded within each other [9]. The first component is a coarse (half degree horizontal) resolution, and is global. It successfully represents the large-scale circulation. The second component refines the horizontal resolution five-fold and extends a regional domain comprising of the greater Agulhas region as well as the south Atlantic. The resulting horizontal resolution over the Agulhas is of the order of 9 km. Both components communicate with each other, effectively providing a method of assessing the global impact of the Agulhas system. This nesting approach results in a configuration that has over 62 million combined grid points, and that ran on 16 processors on HLRS, defunct NEC SX-9. A realisation of the configuration required 12–15 h per model-year, and produced 9 TB of raw data for a 60-year simulation (timespan of the available forcing field). Furthermore, in oder to assess the impact of the Agulhas mesoscale on the large-scale circulation, a stand-alone ORCA05 configuration is used. This configuration is identical to INALT01's first component. Both configurations ran in parallel (Fig. 2). An adaptation of the model system on to the new Cray XE6 took place in 2012, and is ongoing.

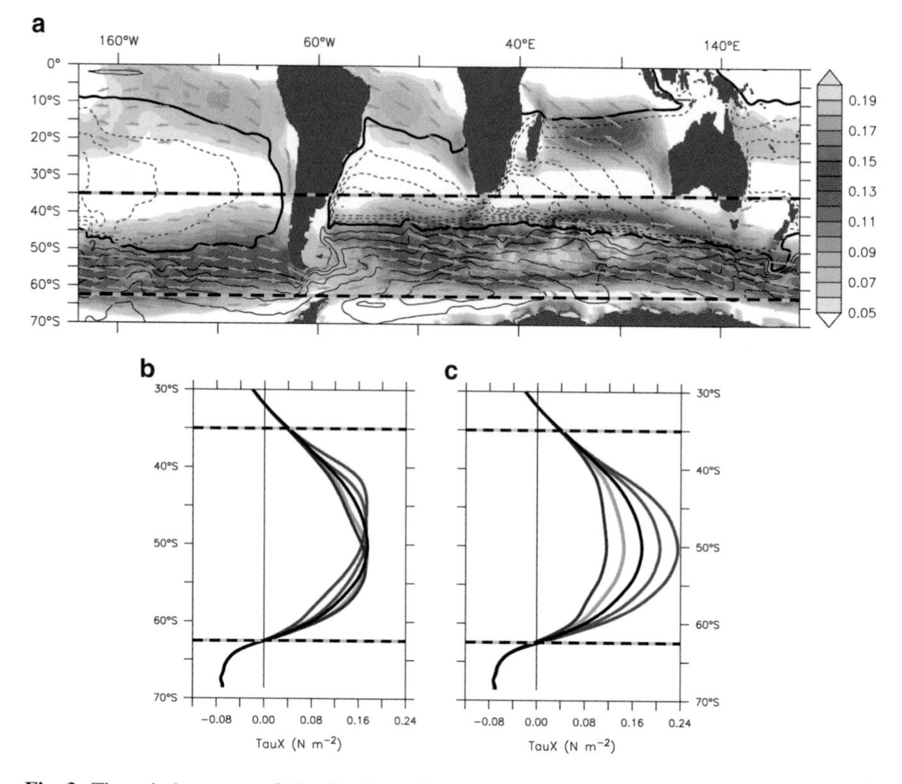

Fig. 3 The wind system of the Southern Hemisphere. (**a**) Wind stress (*colour-scale*, N/m^2) and barotropic stream-function (contours) showing the relative position of the wind patterns and oceanic features respectively. (**b** and **c**) Longitudinal mean wind stress (*black thick line*) and examples of sensitivity experiments altering the easterly (**b**) westerly (**c**) winds (*coloured lines*) [15]

3 Controls of the Southern Hemisphere winds on Agulhas leakage

Theoretical arguments propose that leakage ought to largely depend on the Southern Hemisphere winds [2, 5]. The Agulhas system is sandwiched between two major wind belts (Fig. 3a), namely the easterly trades between the Equator and about 30°S and the westerlies (roughly 30°S–60°S). In a series of experiments (Fig. 3b, c), each wind belt was individually, gradually, and coherently altered, and the sensitivity of Agulhas leakage was evaluated.

Results suggest that the trade winds, which modulate the Agulhas Current transport, play a minor role in determining the amount of water that flows into the Atlantic (Fig. 4a) [12]. In the cases where the trades were gradually increased, enhanced recirculation was observed, with the western boundary current system becoming more energetic. This, however, had no subsequent impact on leakage.

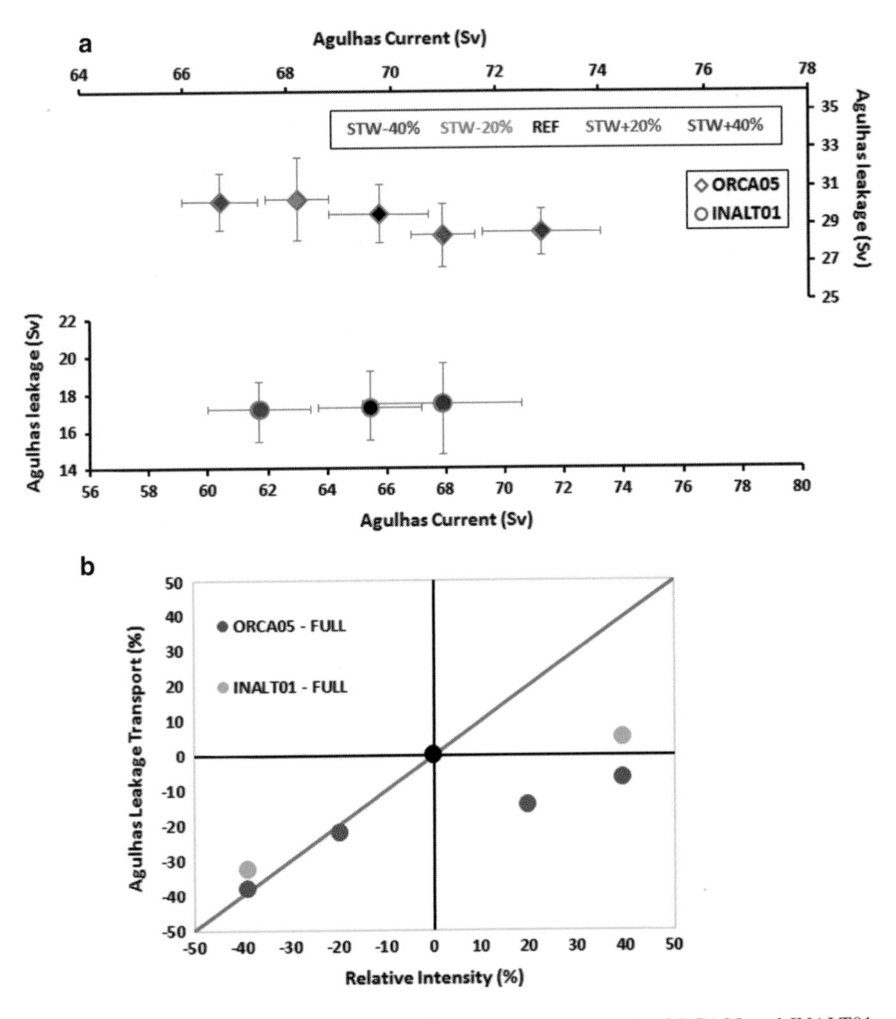

Fig. 4 (**a**) Agulhas leakage versus Agulhas Current transport (Sv) for ORCA05 and INALT01. Each *point* represents a 20 year average, and error bars represent detrended inter-annual standard-deviations [12]. (**b**) Change in Agulhas leakage (%) versus change in intensity (%) of the Southern Hemisphere westerlies. Reference values (*black dot*) are set at the origin for ORCA05 and INALT01 and each *dot* represents a decade average [15]

In lieu, varying the intensity of the westerly belt gave rise to significant changes in leakage (Fig. 4b) [9]. In addition a preferential dependency on the westerlies strength was found. Leakage responded proportionally to the westerlies intensity up to a certain point. Beyond this point, interactions (also stimulated by the strengthened westerlies) between the Antarctic Circumpolar Current and the Agulhas system became important.

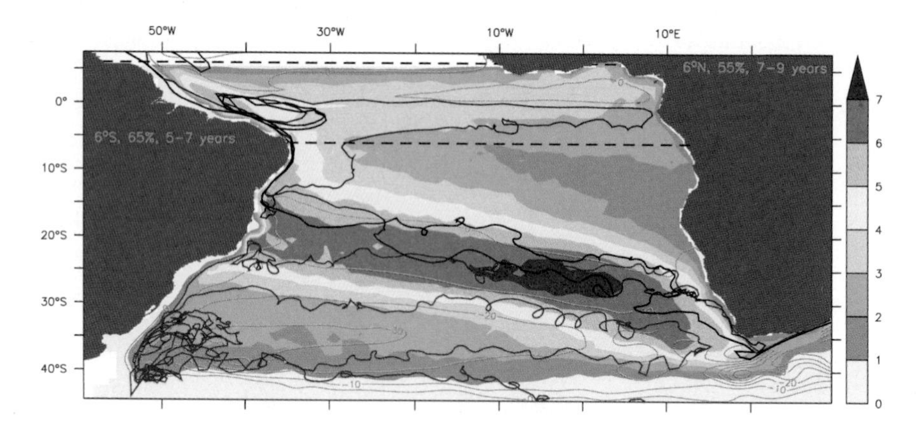

Fig. 5 Spreading of Agulhas leakage into the Atlantic based on the Lagrangian analysis of 40,000 virtual floats, representing a mean leakage transport of 15.3 Sv. *Background colors* show the total number of float counts per 1 degree grid cell. *Boxes* contain the calculated mean volume transport of Agulhas leakage through marked sections in the Atlantic, its fraction of the mean total leakage transport, and the most probable transit times. Three exemplary float trajectories are overlaid to illustrate the possible pathways [15]

4 Impact, fate and consequences of Agulhas leakage

Recent work suggests that changes in the Southern Hemisphere winds have led to an increase in Agulhas leakage and a subsequent salinification of the Atlantic [7, 9]. Climate model projections for the twenty-first century especially predict a progressive southward migration and intensification of the westerlies. The potential effects on the ocean circulation of such an anthropogenic trend in wind stress were studied with a sensitivity experiment utilizing the high-resolution ocean model [14]. The model suggests an increase of 4.5 Sv in Agulhas leakage in response to the wind changes. The change in leakage is reflected in a concomitant change in the transport of the South Atlantic subtropical gyre, but leads only to a small increase in the Atlantic Meridional Overturning Circulation. A main effect of the increasing inflow of Indian Ocean waters with potential long-term ramifications for the overall Atlantic circulation is the salinification and densification of upper-thermocline waters in the South Atlantic, which extends into the North Atlantic within the first decades.

To explicitly study the timescales and pathways of Agulhas leakage, a Lagrangian analysis with virtual floats was used [13]. Consistent among different model resolutions, typical timescales for the advective fate into the subtropical North Atlantic fall between one and two decades (Fig. 5). They are associated with a relative direct path via the South Equatorial Current, the North Brazil current, and Florida Current.

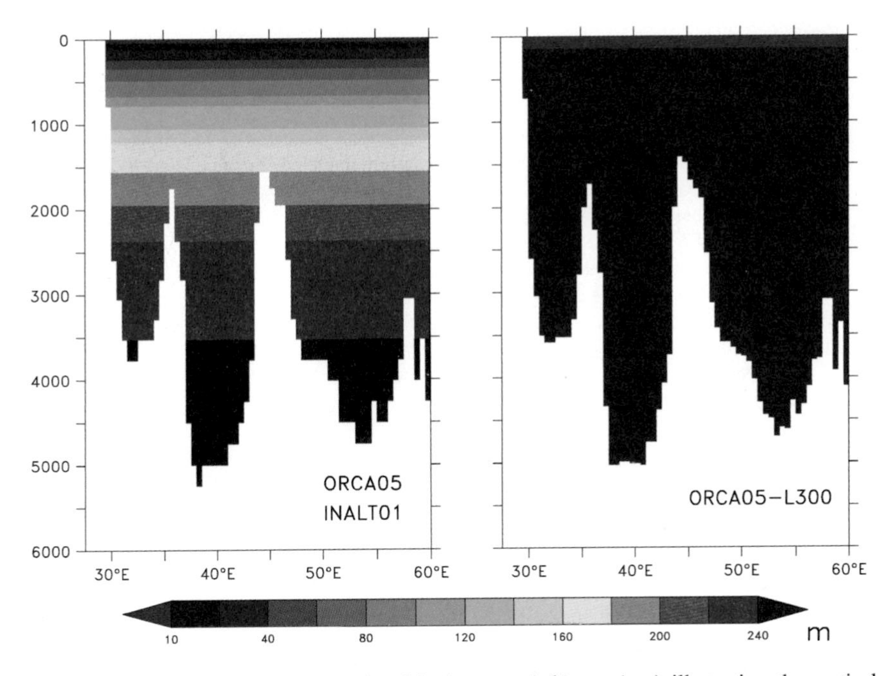

Fig. 6 Layer thickness (m) and associated bathymetry (*white regions*) illustrating the vertical representation of 46 z-level ORCA05 and INALT01 (*left panel*) and the new enhanced vertical refinement of ORCA05-L300 (*right panel*)

5 Outlook and Challenges

The experience gained from constructing a configuration with adequate horizontal resolution has been essential. ORCA05 and INALT01 represent the vertical with 46 z-levels, with thickness ranging between 6 and 250 m. Many of the regional questions pertaining to the Agulhas system require one further step: a refinement of the vertical. Currently, a version of ORCA05, the ORCA05-L300 is being tested with increased resolution in the vertical. In this configuration, 300 z-levels are used, and the level thickness is set to a maximum of 20 m (Fig. 6).

On-going work seeks to build the next generation of the INALT configuration, INALT20, twentieth degree (5 km) horizontal resolution. This new configuration promises to be suitable for both regional and large-scale global studies. This will be possible through either a double-nesting strategy or doubling the resolution of the existing configuration, in addition to the implementation of the vertical refinement.

References

1. J.R.E. Lutjeharms, *The Agulhas Current* (Springer, Berlin, 2006)
2. L.M. Beal, W.P.M. de Ruijter, A. Biastoch, R. Zahn, members of SCOR/WCRP/IAPSO Working Group 136, On the role of the Agulhas system in global climate. Nature **472**, 429–436 (2011). doi:10.1038/nature09983
3. J.R.E. Lutjeharms, R.C. van Ballegooyen, The retroflection of the Agulhas Current. J. Phys. Oceanogr. **18**, 1570–1583 (1988)
4. P.L. Richardson, Agulhas leakage into the Atlantic estimated with subsurface floats and surface drifters. Deep–Sea Res. I **54**, 1361–1389 (2007)
5. W.P.M. De Ruijter, P.J. van Leeuwen, J.R.E. Lutjeharms, Generation and evolution of Natal Pulses, solitary meanders in the Agulhas Current. J. Phys. Oceanogr. **29**, 3043–3055 (1999)
6. A. Biastoch, C.W. Böning, J.R.E. Lutjeharms, Agulhas leakage dynamics affects decadal variability in Atlantic overturning circulation. Nature **456**, 489–492 (2008)
7. A. Biastoch, C.W. Böning, F.U. Schwarzkopf, J.R.E. Lutjeharms, Increase in Agulhas leakage due to poleward shift of the Southern Hemisphere westerlies. Nature **462**, 495–498 (2009)
8. A.Hu, G.A. Meehl, W. Han, J. Yin, Transient response of the MOC and climate to potential melting of the Greenland Ice Sheet in the 21st century. Geophys. Res. Lett. **36**(10), L10707 (2009)
9. J.V. Durgadoo, B.R. Loveday, C.J.C. Reason, P. Penven, A. Biastoch, Agulhas leakage predominantly responds to the Southern Hemisphere Westerlies. J. Phys. Oceanogr. **43**, 2113–2131 (2013). doi:http://dx.doi.org/10.1175/JPO-D-13-047.1
10. O. Boebel, J. Lutjeharms, C. Schmid, W. Zenk, T. Rossby, C. Barron, The Cape Cauldron: a regime of turbulent inter-ocean exchange. Deep–Sea Res. II **50**, 57–86 (2003)
11. G. Madec, NEMO ocean engine. Technical report, Note du Pole de modelisation, Institut Pierre Simon Laplace (IPSL), France, 2008
12. B.R. Loveday, J.V. Durgadoo, C.J.C. Reason, A. Biastoch, P. Penven, Decoupling of the Agulhas leakage from the Agulhas current. J. Phys. Oceanogr. under review (2013)
13. S. Rühs, J.V. Durgadoo, E. Behrens, A. Biastoch, Advective timescales and pathways of Agulhas leakage. Geophys. Res. Lett. **40**, 3997–4000 (2013). doi:10.1002/grl.50782
14. A. Biastoch, C.W. Böning, Anthropogenic impact on Agulhas leakage. Geophys. Res. Lett. **40** (2013). doi:10.1002/grl.50243
15. J.V. Durgadoo, Controls and impact of Agulhas leakage. Ph.D. Thesis (2013). http://macau.uni-kiel.de/receive/dissertation_diss_00013313?lang=en

Part VII
Miscellaneous Topics

Wolfgang Schröder, Willi Jäger, and Hans-Joachim Bungartz

The Miscellaneous Topics section documents not only the breadth but also the bounty numerical simulation. It supports research in many other fields than just the topics like fluid dynamics, aerodynamics, structure mechanics, and so forth. The following five articles show that today's computational approaches are by far neither complete from a numerical nor from a modeling point of view. However, the physically correct focus will lead to reliable prediction methods to substantiate new theories.

The report of the University of Hohenheim emphasizes the superiority of the optimal allocation method of economic capital. The discussion restricts itself to an introductory model case providing optimal conditions in the form of an informed central management of the model bank. First, a detailed description of certain alternative allocations is given. Subsequently, a benchmark study investigates the performance differences between the optimal and the alternative allocation schemes.

The origin of life is addressed in the contribution of the Friedrich-Schiller-University in Jena. That is the endogenic dynamics of Earth and Mars are discussed. A self-consistent model of plate tectonics is presented, followed by a new model of episodic growth of the continental crust which is of interest since the existence of the continental crust with epicontinental seas, lagoons and ponds is directly linked to the origin of life.

W. Schröder (✉)
Aerodynamisches Institut, RWTH Aachen, Wüllnerstr. 5a, 52062 Aachen, Germany
e-mail: office@aia.rwth-aachen.de

W. Jäger
Interdisziplinäres Zentrum für Wissenschaftliches Rechnen, Universität Heidelberg, Im Neuenheimer Feld 368, 69120 Heidelberg, Germany
e-mail: jaeger@iwr.uni-heidelberg.de

H.-J. Bungartz
Fakultät für Informatik, Technische Universität München, Boltzmannstrae 3, 85748 Garching, Germany
e-mail: bungartz@in.tum.de

The history of terrestrial planets is touched in the contribution of the Institute of Planetology in Münster, the Technical University Berlin, and the DLR Institute of Planetary Research in Berlin. The fractional crystallization of a magma ocean leads to a chemical stratification characterized by a progressive enrichment in heavy elements from the core-mantle boundary to the surface. This results in an unstable configuration that causes the overturn of the mantle and the subsequent formation of a stable chemical layering. Assuming scaling parameters appropriate for Mars, simulations of 2D thermo-chemical convection in a Cartesian geometry and in a 2D cylindrical geometry are performed.

The chair of Thermodynamics and Energy Engineering of the University of Paderborn uses molecular simulations to compute surface tension of oxygen and nitrogen, to determine a large scale thermodynamic data set, and to analyze vapor-liquid equilibria (VLE) of binary mixtures. As to the binary mixtures two new molecular force field models for cyanogen and cyanogen chloride are developed and the simulated phase equilibria are compared to experimental VLE data.

The Technical University of Kaiserslautern and the University of Paderborn use massively parallel molecular dynamics simulation to analyze systems containing electrolytes, vapor-liquid interfaces, and biomolecules in contact with water-oil interfaces. Novel molecular models of alkali halide salts are employed for the simulation of electrolytes in aqueous solution. The enzymatically catalyzed hydroxylation of oleic acid is investigated by molecular dynamics simulation taking the internal degrees of freedom of the macromolecules into account.

Modeling and simulation of cellular processes are providing better understanding of the basic principles of living systems, of single cells, cell systems, tissue and organs. The complexity and high dimensionality arising in these systems are demanding on one side advanced computational technology, on the other also a proper modeling adapted to the potential of new hard and software. Focal adhesive kinase (FAK) is in the focus of a project located in Heidelberg. This protein complex is an important factor in signaling pathways, involved in many major human diseases such as cancer, diabetes and autoimmune disorders. The main result achieved so far states that interaction PIP_2, a phospho-lipid component of cell membranes, with FAK may induce a partial opening of its structures, but cannot induce an active state of FAK. The project demonstrates that high-performance computing is offering highly valuable tools for quantitative life sciences.

The project ASIL-KA studies the scalability properties of a special parallel direct solver for Finite Element problems. Starting from two reference scenarios for the Maxwell's equations (addressing cavity and eigenvalue problems, respectively), results concerning convergence behavior and scalability on Hermite for up to 4,096 cores are presented.

Optimal Versus Alternative Economic Capital Allocation in Banking

H.-P. Burghof and J. Müller

Abstract Burghof and Müller (High Performance Computing in Science and Engineering '11. Springer, Heidelberg, 2012) already describes a basic model of economic capital allocation in banking. Furthermore, Burghof and Müller (High Performance Computing in Science and Engineering '12. Springer, Heidelberg, 2013) addresses the model's heuristic solving algorithm in the form of threshold accepting by focusing on the algorithm's parameterization. The current report finally focusses on emphasizing the superiority of the optimal method of economic capital allocation. Thereto, the report restricts itself to an introductory model case providing optimal conditions in the form of an informed central management of the model bank.

1 Introduction

In the present introductory model case the management knows the different probabilities of success of the model bank's business units. On the basis of this information, the management can realize economic capital allocations which perfectly satisfy the demands of the bank. Potential bias arising from an uninformed bank management remains disregarded during the present report.

In order to investigate the superiority of an optimal economic capital allocation, the present report at first provides a detailed description of certain alternative allocations methods. Subsequently, a benchmark study investigates on the performance differences between the optimal and the alternative allocation schemes. The study subdivides into an in-sample- and an out-of-sample part. Furthermore, the report

---

H.-P. Burghof · J. Müller (✉)
Lehrstuhl für Bankwirtschaft und Finanzdienstleistungen, Universität Hohenheim, 70599 Stuttgart, Germany
e-mail: Burghof@uni-hohenheim.de; Jan.Mueller@uni-hohenheim.de

W.E. Nagel et al. (eds.), *High Performance Computing in Science and Engineering '13*, DOI 10.1007/978-3-319-02165-2_41,

provides technical information on the typical computations of the present research approach. The final paragraph concludes.

2 Alternative Allocation Schemes for Benchmarking Purposes

The alternative allocation schemes represent the expected return, the uniform and the random method.[1] The first alternative assigns economic capital proportional to the respective business unit's expected return. The allocation scheme completely disregards business units exhibiting negative expected returns leaving these units without any capital assignments. In contrast, the second alternative provides each successful business unit with the same amount of economic capital while the third one allocates the capital completely randomly among the successful units. Since the first alternative disregards business units exhibiting negative return expectations, the second and third alternative behave identic to that extent. Otherwise, their lack in competitiveness compared to the first alternative would be considerable from the start.

Finally, there are countless further potential allocation methods. The present approach, however, focusses on the very obvious alternatives to outright optimization since the investigation of countless allocation methods with only marginal differences appears useless. Besides the optimal allocation, during the present analyses the expected return method represents the most sophisticated allocation scheme. On the one side, this method considers the expected returns of the business units and provides the most successful units with the highest shares of economic capital. On the other side, this allocation scheme still creates relatively balanced economic capital allocations implicitly inducing considerable diversification effects. Finally, the method regularly achieves very high expected returns outperforming the allocations of the uniform and random method. Therefore, the expected return method represents the most serious competitor for the optimal allocation method.

An important means to establish a basic competitiveness between the alternative allocation methods and the optimal method represents the scaling of the alternative limit allocations according to the scaling algorithm from Fig. 1.

During the first line of the algorithm, each alternative allocation method executes its individual allocation scheme resulting in a vector of value at risk limits $\mathbf{vl}$ representing the respective economic capital allocation. Potential schemes here are the mentioned expected return, the uniform and the random scheme. The rest of the scaling algorithm then establishes the maximum use of the available investment capital c_{bank} (first while-loop) and economic capital ec (second while-

[1] See Burghof and Müller [2, 3] for information on the optimal allocation method. Burghof and Müller [3] also provides a quick introduction to the expected return and the uniform method.

Fig. 1 Scaling algorithm for economic capital allocations **vl** in order to induce maximum feasible use of the available investment capital c_{bank} and economic capital ec

1: compute **vl** according to the respective alternative allocation method
2: initialize $\mathbf{vl}^{\text{init}} = \mathbf{vl}$, $\varphi = 1$, $\varphi_{\min} = 0$ and $\varphi_{\max} = 2$, *iteration* = 0
3: **while** max $c_{\text{bank},i} \neq c_{\text{bank}}$ **do**
4: *iteration* + 1
5: **for** $i = 1$ to m **do**
6: compute $c_{\text{bank},i}$
7: **end for**
8: **if** max $c_{\text{bank},i} < c_{\text{bank}}$ **then** increase φ (Algo. 2), $\mathbf{vl} = \varphi \cdot \mathbf{vl}^{\text{init}}$
9: **else if** max $c_{\text{bank},i} > c_{\text{bank}}$ **then** decrease φ (Algo. 2), $\mathbf{vl} = \varphi \cdot \mathbf{vl}^{\text{init}}$
10: **end while**
11: *iteration* = 0, *token* = 1
12: **while** 1 = 1 **do**
13: *iteration* + 1, *count* = 0
14: **for** $i = 1$ to m **do**
15: compute $pl_{\text{bank},i}$
16: **if** $pl_{\text{bank},i} < -vl_{\text{bank}}$ **then** *count* + 1
17: **end for**
18: **if** *token* = 1 **and** $count \leq m \cdot \alpha$ **then** leave loop **else** *token* = 0
19: **if** $count < m \cdot \alpha$ **then** increase φ (Algo. 2), $\mathbf{vl} = \varphi \cdot \mathbf{vl}^{\text{init}}$
20: **else if** $count > m \cdot \alpha$ **then** decrease φ (Algo. 2), $\mathbf{vl} = \varphi \cdot \mathbf{vl}^{\text{init}}$
21: **else if** $count = m \cdot \alpha$ **then** leave loop
22: **end while**

Fig. 2 Binary search algorithm for adjusting the scaling factor φ

1: **if** increase φ **then** $\varphi_{\min} = \varphi$ **else** $\varphi_{\max} = \varphi$
2: **if** iteration > 100 **and** increase φ **then** $\varphi_{\max} + 1$, *iteration* = 0
3: **if** iteration > 100 **and** decrease φ **then** $\varphi_{\min} = 0$, *iteration* = 0
4: $\varphi = \varphi_{\min} + 0.5(\varphi_{\max} - \varphi_{\min})$

loop). Without this scaling, the alternative allocation methods would not represent serious alternatives to the optimal allocation method.

Key variable in context with the scaling algorithm represents the factor φ responsible for the scaling of **vl** while $\varphi_{\min}$ and $\varphi_{\max}$ determine the factor's lower and upper bound. Depending on the interim results, the scaling algorithm adjusts φ according to the binary search algorithm from Fig. 2 whose description follows below.[2] Thereby, after a certain number of iterations, both algorithms together establish an allocation **vl** featuring the maximum use of the available investment capital c_{bank} and economic capital ec.

An important point of the scaling algorithm from Fig. 1 represents the initialization of $\mathbf{vl}^{\text{init}}$ in line 2. Variable $\mathbf{vl}^{\text{init}}$ represents the initial limit allocation constantly serving as a starting point during the scaling computations in order to create a new version of **vl** according to $\mathbf{vl} = \varphi \times \mathbf{vl}^{\text{init}}$.

The first while-loop of the scaling algorithm addresses the maximum usage of the available investment capital c_{bank}. This maximum usage arrives as soon as the maximum invested capital max $c_{\text{bank},i}$ of $1, \ldots, i, \ldots, m$ simulation iterations matches c_{bank}. This state at the same time determines the loop's exit according to line 3. Before reaching this state, lines 8 and 9 adjust the scaling factor φ on the basis of the binary search algorithm from Fig. 2.

[2]See e.g. Knuth [6, pp. 409–426] for further information on binary search algorithms.

Depending on whether the decision variables **vl** require an increase or a decrease the binary search algorithm shifts the lower or upper bound of factor φ in the form of $\varphi_{\min} = \varphi$ or $\varphi_{\max} = \varphi$ according to line 1. The shift causes line 4 to adjust factor φ for the next iteration of the first while-loop of the scaling algorithm from Fig. 1. If 100 repetitions can not establish $\max c_{\text{bank},i}$ matching c_{bank} either the upper bound $\varphi_{\max}$ is too low or the lower bound $\varphi_{\min}$ is too high. Then line 2 of the binary search algorithm increases the current $\varphi_{\max}$ by one, or line 3 sets $\varphi_{\min}$ back to zero.

The second while loop of the scaling algorithm from Fig. 1 establishes the maximum usage of the available economic capital ec. Thereto, line 15 computes the bank's profit and loss $pl_{\text{bank},i}$ of each simulation iteration i. Variable $count$ saves the respective values' violations of the bank's risk limit vl_{bank} according to line 16. If the number of violations fits the current confidence level β, the scaling algorithm ends via line 18, the maximum usage of the investment capital c_{bank} provided. The latter holds true for the case of $token = 1$. The feasible maximum number of risk limit violations follows from the current confidence level β in the form of $m \times (1 - \beta) = m \times \alpha$.

A very important variable during the second while-loop of the scaling algorithm represents the variable $token$. As long as $token = 1$ holds true, the current economic capital allocation **vl** induces the maximum use of the available investment capital c_{bank}. Further increases of **vl** would then violate the budget constraint while decreases would inevitably cause the invested capital to fall below its feasible maximum. Consequently, according to line 18, the scaling algorithm ends in this state, the satisfaction of the VAR-constraint provided.

If the scaling algorithm does not end according to line 18 while $token = 1$ holds true the current **vl** obviously causes VAR-constraint violations. In this case, the algorithm from Fig. 1 continues scaling **vl** according to the lines 19 and 20 and finally ends via line 21. Instead of newly initializing φ, $\varphi_{\min}$ and $\varphi_{\max}$ after the first while-loop, the scaling algorithm continues on the basis of the current values.

For reasons of competitiveness, the different alternative allocation methods also use, besides the identic scaling algorithm, an identic risk assessment. Since the optimal allocation method assesses risk on the basis of a comprehensive trading simulation, the alternative allocation methods do the same. This risk assessment contrasts with the classic risk assessment known from portfolio management. The classic risk assessment commonly exclusively uses returns from financial instruments held by the bank in seemingly constant portfolios, without considering trading and its consequences in the form of unstable correlations.

Finally, all allocation methods could also use this classic risk assessment. However, which portfolio constellation should the risk assessment assume in this case? Within the present model, the decision makers act autonomously. As a result, the bank's portfolio constellation varies constantly by default. Consequently, in order to reliably prevent the economic capital allocations to regularly induce violations of the confidence level, the classic risk assessment had to assume the worst-case portfolio constellation. The worst case describes the state where all decision makers build positions of the same trading direction. The present analysis, however, disregards this save but rather inefficient approach of the classic risk assessment and exclusively applies the risk assessment on the basis of m trading simulations.

3 In-Sample Generation of Optimal and Alternative Economic Capital Allocations

In order to investigate on the performances of the different allocation methods, each method undergoes tests on the basis of one and the same model bank. Thereto, the bank simulates the upcoming trading day $m = 20\text{k}$ times and then uses the resulting sample data to develop the ex-ante economic capital allocation by the respective allocation method. Analyses concerning the allocations' performances follow. The computations on the basis of this sample data represent the in-sample computations. In contrast, the subsequent paragraph investigates on the robustness of the in-sample findings by using a newly drawn data sample representing the out-of-sample case. To keep the report short, we do not provide any further methodical details concerning the simulations of the trading days here.

For comparative purposes, the configuration of the model bank should at least enable the two most promising allocation methods, the optimal and the expected return method, to induce exactly the same risk VAR_{bank} and expected invested capital $\text{E}(c_{\text{bank}})$. However, the inflexibility of the alternative allocation methods excludes the creation of an appropriate level playing field for any model bank configuration. Main setscrews concerning the bank's configuration consist in the available economic capital ec and investment capital c_{bank}. Even particular configurations can only establish a perfect level playing field between two allocation methods at once. The present analyses tolerate less perfect level playing fields between the optimal and the less sophisticated alternatives in the form of the uniform and random method. Nevertheless, the analyses' results still provide useful indications concerning the fundamental performances of these two alternative allocation methods.

The current analysis uses a model bank configuration of $ec = 662$ and $c_{\text{bank}} = 150\text{k}$. Furthermore, the bank consists of $n_{\text{units}} = 50$ business units each exhibiting a different probability of success p_j with $1, \ldots, j, \ldots, n_{\text{units}}$. The units' average probability of success lies at 0.55.

The following diagrams display the economic capital allocations resulting from each allocation method.

Note that Fig. 3 arranges the economic capital limits of each allocation according to the expected returns of the respective business unit. Therefore, the order of the units stays the same throughout the diagrams always starting with the most successful unit on the left. The allocation of the optimal method clearly differs from the allocation of the expected return method. The optimal method does not follow exactly the expected returns of the business units. Furthermore, the simple uniform method provides each business unit with the same amount of economic capital. Since in the present case every business unit exhibits positive expected returns no

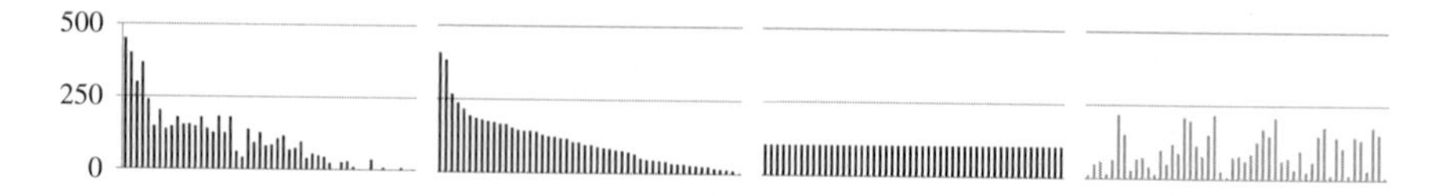

Fig. 3 Limit allocations according to the optimal, expected return, uniform and random method

Table 1 In-sample results of the optimal, expected return, uniform and random method using $m = 20$k simulations

	Optimal		Expected return		Uniform		Random	
$\mu_{\text{bank}}/\sigma_{\mu\text{ bank}}$	396	494	370	464	222	348	212	382
$\text{E}(c_{\text{tank}})/\sigma_{c\text{ tank}}$	149,157	256	149,253	220	149,393	171	141,923	6,522
VAR_{bank}/β	662	0.99	662	0.99	566	0.995	654	0.9906
$ES_{\text{bank}}/\sigma_{ES}$	896	225	842	213	703	146	818	173
$\Sigma_{\text{vl}}/d_{\text{vl}}$	5,376	8.13	5,515	8.34	5,227	9.23	4,965	7.59
$\text{E}(RORAC_{ec})/$ $E(RORAC_{VAR})$	0.599	0.599	0.560	0.560	0.336	0.392	0.320	0.324

unit remains disregarded by the allocation. The allocation of the random method naturally appears mixed-up. Its grey color indicates that the random allocation takes a different form for each of the $1, \ldots, i, \ldots, m$ simulations.

The data below provides the in-sample performances of the different allocations.

The optimal method maximizes the expected profit μ_{bank} of the model bank. Its expected profit outperforms that of the direct competitor in the form of the expected return method by 7 %. On the other hand, the expected profit of the optimal method also exhibits the strongest standard deviation $\sigma_{\mu_{\text{bank}}}$. This finding, however, is less relevant since the central risk measure represents value at risk (line 3 of Table 1). The expected profits μ_{bank} of the uniform and random method can not compete with those of the optimal and the expected return method.

Concerning the invested capital c_{bank} the random method induces the lowest usage rate with an average invested capital of $\text{E}(c_{\text{bank}}) = 141{,}923$. The allocations of the random method naturally cause the invested capital to deviate considerably according to $\sigma_{c_{\text{bank}}} = 6{,}522$.

The optimal and the expected return method manage to induce the maximum risk level and fully use the available economic capital according to $VAR_{\text{bank}} = ec = 662$. In contrast, the uniform and the random method both cause much lower risk levels. As a consequence, the actual confidence levels β of these methods lie above those of the optimal and expected return method.

Table 1 additionally provides the risk measure expected shortfall ES_{bank}. In this context, the optimal allocation method causes higher values than the expected return method. The optimal method's allocation exhibits a lower degree of natural diversification. As a consequence, its expected shortfall ES_{bank} exceeds that of the expected return method's allocation. However, since the value at risk VAR_{bank} represents the decisive risk measure in the present case, the ES_{bank}-assessment just serves as a form of plausibility check.

Variable Σ_{vl} simply sums up the economic capital limits. In case of the optimal method the limits add up to $\Sigma_{vl} = 5{,}376$ representing $d_{vl} = 8.13$ times the actually available amount of economic capital $ec = 662$. The completely autonomous decision making of the business units concerning the taking of long and short positions results in uncorrelated decisions and uncorrelated financial instruments' returns. As a consequence, strong diversification effects enable these immense Σ_{vl}- and d_{vl}-values.

Furthermore, $\mathrm{E}(RORAC_{ec})$ and $\mathrm{E}(RORAC_{VAR})$ measure the rate of expected profit μ_{bank} to available economic capital ec and to actually used economic capital in the form of VAR_{bank}. In context with the efficient allocation of the available economic capital in particular $\mathrm{E}(RORAC_{ec})$ represents the relevant measure. In the present case the optimal and the expected return method fully use the available resources. As a consequence, each method for itself exhibits the same values for $\mathrm{E}(RORAC_{ec})$ and $\mathrm{E}(RORAC_{VAR})$. However, the optimal method clearly outperforms the expected return method concerning both values.

Finally, the optimal method maximizes the expected profit μ_{bank} of the model bank. At the same time, the optimal method induces only the same risk VAR_{bank} as the direct competitor in the form of the expected return method. Therefore, the in-sample computations clearly confirm the optimal method's superiority.

4 Out-of-Sample Back Testing of Optimal and Alternative Economic Capital Allocations' Performances

The previous paragraph describes the generation of different economic capital allocations relying on one particular sample of simulation data. These in-sample computations reveal the optimal allocation method to outperform the alternative allocation methods. In order to test the robustness of this finding, now the allocation methods undergo tests on the basis of a newly drawn sample of simulation data on the basis of different random numbers.

The histograms of Fig. 4 reveal the distributions of the profits and losses pl_{bank} of the model bank under use of the new data sample.

The distributions of the optimal and the expected return method exhibit means lying distinctly right from those of the uniform and the random method. Hence, also under the use of the new data sample the sophisticated allocation methods induce much higher expected profits μ_{bank} than the less sophisticated allocation methods. However, the histograms hardly allow identifying any differences between the optimal and the expected return method. Compared to the expected return method, the optimal method appears to reallocate probability mass from the center of its distribution to the right tail which would be advantageous. In context with Fig. 4 "TA" denotes the optimal allocation method. The abbreviation stands for "threshold accepting" representing the underlying heuristic optimization algorithm.

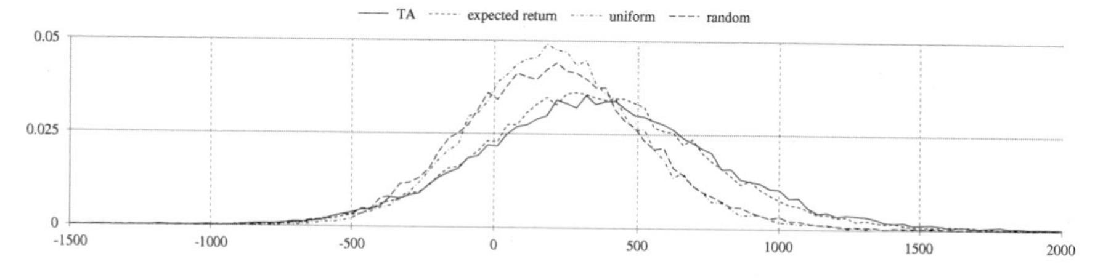

Fig. 4 Histograms of the allocation methods' out-of-sample results concerning pl_{bank} using $m =$ 20k simulations

Table 2 Out-of-sample results of the optimal, expected return, uniform and random method using $m = 20$k simulations

	Optimal		Expected return		Uniform		Random	
$\mu_{\text{bank}}/\sigma_{\mu\ \text{bank}}$	356	439	333	412	201	311	193	341
$\text{E}(c_{\text{tank}})/\sigma_{c\ \text{bank}}$	148,804	212	148,880	192	148,794	153	141,354	6,474
VAR_{bank}/β	628	0.992	592	0.993	493	0.997	589	0.994
$ES_{\text{bank}}/\sigma_{ES}$	823	194	772	182	632	147	733	152
$\Sigma_{\text{vl}}/d_{\text{vl}}$	5,376	8.56	5,515	9.32	5,227	10.59	4,965	8.42
$\text{E}(RORAC_{ec})/$ $E(RORAC_{VAR})$	0.538	0.566	0.504	0.563	0.304	0.408	0.292	0.327

Comparing the allocation methods in detail requires a comprehensive analysis of their performances according to the data from Table 2. The data reveals the optimal method to outperform the expected return method in expected profit μ_{bank} by 7 % again. Both methods almost exhibit the same average invested capital $\text{E}(c_{\text{bank}})$. With the new data sample, however, the optimal method induces a higher risk level VAR_{bank}. Consequently, the optimal method implements a higher and more efficient usage rate of the available economic capital ec. Unsurprisingly, the optimal method also again exhibits the highest expected shortfall ES_{bank}. The variables Σ_{vl} and d_{vl} remain more or less unchanged. The optimal method also outperforms the alternative methods concerning $\text{E}(RORAC_{ec})$ and $\text{E}(RORAC_{VAR})$. The uniform and the random method still do not represent serious alternatives to the sophisticated methods. Finally, the out-of-sample computations confirm the optimal method's superiority from the in-sample case.

5 Technical Information on Typical Computations

In context with the present research approach the use of parallel computing has considerable benefits. The central computations of the present research approach consist in the threshold accepting algorithm for heuristic optimization. The threshold

Table 3 Technical information on typical computations with the NEC Nehalem cluster

Computation	A	B	C
Model cases	400	1	1
Business units	50	50	200
Restarts	24,000	100	100
Rounds	10	10	10
Steps	2,000	2,000	2,000
Sequential	–	3.71 h	4.07 h
Parallel	5.08 h	4.16 min	6.43 min
Nodes	120	13	13
Processors	960	101	101
Servants	959	100	100

accepting algorithm falls into the category of trajectory methods.[3] Search algorithms of this category use one or more independent search processes moving through the respective solution space on their own. The independence of the single search processes causes the whole computational efforts to exhibit a high degree of granularity enabling the effective use of parallel computing. This also holds true for trajectory methods in the form of threshold accepting.

In contrast, the single search processes of e.g. population based methods interact with each other.[4] This interaction reduces the independence of these processes and at the same time the effectiveness of parallel computing. Furthermore, population based methods not necessarily generate better solutions.

The data from Table 3 emphasizes the benefits of using parallel computing in context with the present research approach.[5] Type A computations enable improving the threshold accepting implementation's effectiveness. Burghof and Müller [3] addresses such computations in detail. Finally, type A computations test different variants of the algorithmic design and/or different variants of the algorithm's parameterization. Table 3 uses the term "model cases" instead of variants. The tests commonly use a model bank consisting of 50 business units. Further explanations concerning the remaining lines of Table 3 provide Burghof and Müller [2, 3]. Type A computations use a large number of restarts. As a consequence, the realization of precise parameterization by a sequential execution of a type A computation would take much too long.

Type B and C computations investigate on the superiority of the optimal allocation method of economic capital compared to alternative allocation methods. The present paragraphs 3 and 4 use similar computations. The exemplary

[3]See Gilli and Winker [5] for an overview concerning categories of heuristic optimization algorithms. See also Maringer [7, pp. 38–76] for a slightly different alternative introduction to heuristic optimization.

[4]See e.g. Gilli and Winker [5].

[5]Differences compared to the data from previous annual reports result from the different development stages of the research.

computations B and C differ in the number of business units of the respectively underlying model bank. The present research approach requires many reruns of such computations. On the one side, the high rate of required reruns results from development purposes. On the other side, establishing a perfect level playing field between two different allocation methods can become complicated and require several attempts trying different model bank configurations. Without parallel computing, these processes would become extremely inconvenient or even impossible.

Obviously, parallel computing plays a vital role concerning the realization of the present research approach. The use of dynamic load balancing minimizes negative effects of parallel computing in the form of parallel overhead. For programming, the research approach uses C++. The program parts addressing parallel computing use the Intel Message Passing Interface (IMPI).

6 Conclusions

The report shows the superiority of optimal economic capital allocation within the model. The optimal allocation method successfully anticipates the decentralized autonomous decision making of the business units. Consequently, within the model, the optimal allocation method overcomes the difficulties arising from unstable correlations.

In contrast to the present report, the whole research approach also investigates the superiority of the optimal allocation method for less optimal conditions. Thereto, an uninformed bank management replaces the informed bank management. In contrast to the informed management, the uninformed requires Bayesian learning in order to estimate the probabilities of success of the model bank's business units.[6] In this case, the learning intensity determines the precision of the resulting information. The analyses reveal the optimal method to outperform the alternative methods also under relatively low learning intensities and imprecise information respectively. The superiority of the optimal method even holds true for further increases of the uncertainty through the additional consideration of herding.[7] In case of herding, a business unit potentially quits taking decisions autonomously as soon as the imitation of the colleagues' trading decisions appears more promising from a rational perspective.

The research approach describes the situation of optimal economic capital allocation under comprehensive consideration of decentralized and autonomous decision making. This contrasts to the understanding of common risk management concepts which rather completely neglect behavioral aspects. Admittedly, the objective

---

[6] See Burghof and Müller [1, pp. 205–210] for a previous version of the modeling of a Bayesian learning bank management.

[7] See Burghof and Sinha [4] for an investigation of herding in context with VAR-limit-systems without the consideration of optimization.

consideration of behavioral aspects within the economic capital management of banks appears hardly tractable. Nevertheless, the significance of behavioral aspects in context with risk assessment matters suggests the rethinking of the established concepts.

References

1. H.-P. Burghof, J. Müller, Allocation of economic capital in banking: a simulation approach, in *The VaR Modeling Handbook*, ed. by G.N. Gregoriou (McGraw-Hill, New York u. a., 2009)
2. H.-P. Burghof, J. Müller, Allocation of economic capital in banking: a simulation approach, in *High Performance Computing in Science and Engineering '11*, ed. by W.E. Nagel, D.B. Kröner, M. Resch (Springer, Heidelberg u. a., 2012)
3. H.-P. Burghof, J. Müller, Parameterization of threshold accepting: the case of economic capital allocation, in *High Performance Computing in Science and Engineering '12*, ed. by W.E. Nagel, D.B. Kröner, M. Resch (Springer, Heidelberg u. a., 2013)
4. H.-P. Burghof, T. Sinha, Capital allocation with value-at-risk – the case of informed traders and herding. J. Risk **7**(4), 47–73 (2005)
5. M. Gilli, P. Winker, Heuristic optimization methods in econometrics, in *Hand-Book of Computational Econometrics*, ed. by D.A. Belsley, E.J. Kontoghiorghes (Wiley, Chichester, 2009)
6. D. Knuth, *The Art of Computer Programming 3* (Addison-Wesley Longman, Amsterdam, 1997)
7. D. Maringer, *Portfolio Management with Heuristic Optimization*, ed. by H. Amman, B. Rustem (Springer, Dordrecht, 2005)

Geodynamic Mantle Modeling and Its Relation to Origin and Preservation of Life

Uwe Walzer and Roland Hendel

Abstract Section 1 refers to hypotheses on the origin of life. These different hypotheses require distinct geodynamic and structural-geology prerequisites. For example, in case of chemoautotrophic metabolism-first hypotheses, a plate-tectonic mechanism is necessary that contains sites of reducing volcanic exhalations. It was shown that the mass extinctions of biological species are influenced by the convection-differentiation mechanism of the endogenic evolution of the Earth's mantle. Especially LIP-producing eruptions appear to be the principal reason for mass extinction events. Occasionally, bolide impacts cause an extinction event in an ecologically stressed, LIP-generated situation. Section 2 reports on our efforts pertaining to the self-consistent modeling of plate tectonics. To facilitate plate-like motions, two conditions are required, namely a low-viscosity asthenosphere and a deviation from the purely viscous constitutive equation of the lithosphere. Our modeling results show that already relatively simple additional assumptions in a 3-D spherical-shell model of the Earth's mantle produce oceanic lithospheric plates moving along the Earth's surface and changing their shape and size as a function of time. Section 3 describes a new model of episodic growth of continental crust (CC). In the case of genetics-first hypotheses or of metabolism-first hypotheses with solar irradiation or lightning energy supply, the existence of CC with epicontinental seas, lagoons and ponds is directly determining for the origin of life. Furthermore, the most important sources of nutrients originate from the upper CC. We put a water-concentration dependent solidus model of mantle peridotite into a 3-D spherical-shell, dynamic mantle model with chemical differentiation that redistributes the heat-producing elements. As a result, we obtain a set of temporal

U. Walzer (✉) · R. Hendel
Institut für Geowissenschaften, Friedrich-Schiller-Universität, Humboldtstr. 11, 07743 Jena, Germany
e-mail: u.walzer@uni-jena.de

W.E. Nagel et al. (eds.), *High Performance Computing in Science and Engineering '13*, DOI 10.1007/978-3-319-02165-2_42,

distributions of CC growth episodes that show a certain temporal invariance for a variation of the melting-criterion parameter, f_3. The laterally averaged surface heat flow density qob, the Urey number Ur, and the kinetic creep energy E_{kin} show temporally sinusoidal components superposing a monotonously decreasing curve. Section 4 discusses partly unknown distributions of physical quantities, the knowledge of which is necessary for the computation of a dynamic Martian convection-differentiation system. It is ambiguous whether the early strong Martian magnetic dipole was generated by the more effective core cooling due to a plate-tectonic mode of solid-state convection in the Martian mantle during the first 500 Ma. Section 5 outlines the numerical progress in the advancement of the Terra code that was achieved by cooperation of the international group of Terra developers.

1 What Has Life to Do with the Endogenic Dynamics of Earth and Mars?

1.1 Introduction: Definition of Life

It is difficult to define life and the biological species more precisely. *Barbieri* [5] specified life as follows. Life is a metabolizing material informational system with the ability of self-reproduction with changes (evolution), that requires energy and suitable environment. *Trifonov* [122] investigated 123 definitions of life and found a "lowest common denominator" as follows: Life is self-reproduction with variations. An overview of the problem is given by [45]. The definition by *Ruiz-Mirazo et al.* [100] is somewhat more specific: Living beings are autonomous systems with open-ended evolution capacities. They must have a semi-permeable active boundary, an energy transduction apparatus and, at least, two types of functionally interdependent macromolecular components. The latter is required to articulate a phenotype-genotype decoupling.

With regard to the species, the definition of *Mayr* [77] is customary: Species are groups of interbreeding natural populations which are reproductively isolated from other such groups. *Van Regenmortel* [124] objects that interspecies hybridization is quite common, especially in the plant kingdom and that many Archean and Proterozoic organisms did not reproduce by sexual means. Furthermore, we want to emphasize that biologists and paleontologists use differing definitions of the term species.

As it seems, reproduction is the most conspicuous feature of life. Animate beings carry their blueprint in the form of a genetic code which enables them to produce individuals that are *in principle* of the same type. Because living beings are objects of an extremely high degree of order and on account of the second law of thermodynamics, their existence is only possible by metabolism, i.e. they are materially and energetically open systems. A further consequence of the second law

Table 1 The five large Phanerocoic mass extinction events. Data acc. to [6], * after [105]

Event	Final age (Ma)	Duration (Ma)	Loss of genera	Estimated loss of species
Ordovician	443	3.3 to 1.9	57 %	86 %
Devonian	359	29 to 2	35 %	75 %
Permian	251	2.8 to 0.16	56 %	96 %
Triasic	200	8.3 to 0.6	47 %	80 %
Cretaceous	65	≤ 2.5	40 % (60 % *)	76 %

is the fact that at least all more complex living beings grow old and finally decease. A further complexity is hidden in the words *"in principle"*. Paleontologically, we observe a very slow, gradual alteration of the fossil's shapes whereas the extinction of a species is or seems to be abrupt.

1.2 Extinctions

In the Phanerozoic, i.e. in the last 570 Ma, we observe six large mass extinctions where up to 19 families of marine invertebrates and vertebrates died off during 1 Ma [94]. If, relating to extinctions, the word "large" is defined by a loss of over 75 % of estimated species then we determine five large mass extinction events during the Phanerozoic [6]. Cf. Table 1.

After every mass extinction, other families and species relatively quickly occupied the free living environment. At first sight, it looks like a discontinuity in the development of life.

What is the cause of the large mass extinction events? *The first explanation*. Independent of the present problem, it is evident that the terrestrial magnetic dipole and the magnetosphere serves as a protective shield for the genetic material. Therefore it is an essential prior condition for the continuity of life because the majority of the highly energetic, electrically charged particles of the solar wind and of the cosmic "radiation" are deflected. That is why it is understandable that one of the first proposals of an explanation of the mass extinction events was the assumption that the magnetic dipole reversals cause the mass extinctions. If we investigate the observations in detail we find, however, that, e.g., the large mass extinction at the Cretaceous-Paleocene Boundary (KPB) does *not* coincide with any magnetic dipole reversal. The temporal distribution of the Phanerozoic dipole reversals is now comprehensively known. It totally differs from the distribution of Phanerozoic mass extinctions. A magnetic dipole reversal takes only a period of 2,000–5,000 a. During this process, the magnetic dipole moment declines to 20–30 % of the original amount, however, the quadrupole and octupole contributions rise in this time span. Evidently, the magnetic protection of life is provided during these time intervals so that this first explanation fails.

The second explanation. Alvarez et al. [2] report on iridium increases of about 30, 160, and 20 times in deep-sea limestones exposed in Umbria, Denmark, and New Zealand. On the other hand, the upper continental crust is depleted in platinum metals. Because these iridium anomalies arose precisely at the time of KPB, *Alvarez et al.* [2] proposed that the impact of an asteroid had caused both the KPB mass extinction event and the iridium anomaly. *Schulte et al.* [105] confirm that the Cretaceous-Paleogene extinction coincided with the Chicxulub impact and that a global perturbation of the δ^{13}C curve and a drop of carbonate sedimentation in the marine realm simultaneously took place. However, they emphasize that these relatively sudden variations occurred within the time of Deccan flood basalt volcanism but not at its beginning. They and also *Miller et al.* [82] conclude that the Chicxulub impact triggered the mass extinction. *Renne et al.* [95] show ^{40}Ar/^{39}Ar data and establish synchrony between KPB and the corresponding mass extinction to within 32 ka. However, they indicate that the global climate instability began with six abrupt temperature drops already $\sim$1 Ma *before* the Chicxulub impact [140]. Therefore it is evident that the ecosystem was under critical stress already *before* the Chicxulub event. *Johnson and Hickey* [57] investigated the loss of flora around the KPB in North Dakota and showed that many species died off conspicuously *before* the KPB. Also the decrease of the number of Antarctic invertebrate species distinctly begins *before* the KPB [143]. *Benest and Froeschlé* [9] systematically compared the marine extinction events from the Lower Cretaceous until the present time with the age estimates of large impacts. They found a coincidence only for the KPB and for the Eocene-Oligocene Boundary but else not at all. In India, the dinosaurs and some fish and frog species died off somewhat *before* the KPB. Damage cannot appear before an impact if the impact is the only cause. On the other hand, there are also big impacts without any extinction event. Therefore, the suspicion raised that impacts only amplify the effect of another main mechanism.

The third explanation. Courtillot et al. [24] and *Archibald* [4] proposed that large eruptions of continental and oceanic flood basalts are the chief cause of mass extinction events. The giant volcanic eruptions not only produce the large igneous provinces (LIPs) [21] but also rise to a reduction of the solar irradiation into the atmosphere. The latter effect is caused by the volcanic change of the chemical composition of the atmosphere, particularly by aerosols and SO_2 gas emissions. Sulfur dioxide forms sulfate aerosol particles that reflect the incoming solar radiation [65]. Thereby a global cooling is caused [140] which effectuates a growth of the polar ice caps. Thus large areas of the epicontinental seas will become lowland. Already *Smith* [109] showed that the extinctions at the Triassic-Jurassic Boundary and at KPB are connected with worldwide regressions of epicontinental seas.

According to [19, 46, 64, 65], studies of the Deccan Volcanic Province revealed three major volcanic phases: Phase-1 in C30n at 67.4 Ma followed by a 2 Ma period of quiescence, the main Phase-2 in C29r just before the KPB, and the late Phase-3 in the early Danian (C29r/C29n). Phase-2 generates $\sim$ 80 % of the 3,500 m thick Deccan lava pile. *Keller* [64] believes that the catastophic effects of the Chicxulub impact have been overestimated and that the KPB mass extinction is essentially caused the volcanic main Phase-2 of the Deccan lava and gas eruptions.

The end-Triassic extinction at 201.4 Ma is tied to a sharp negative spike in $\delta^{13}C$. *Whiteside et al.* [138] analysed the $\delta^{13}C$ of n-C_{25}-n-C_{31} n-alkanes and of wood. *Schoene et al.* [103] confirm this result. Furthermore, *Schoene et al.*[103] and *Deenen et al.*[30] found rapid sea-level fluctuations and a global cooling. They confirm that the Triassic-Jurassic Boundary and the end-Triassic mass extinction correlate with the onset of flood volcanism in the Central Atlantic Magmatic Province to <150 ka.

A majority of authors [38, 59] [and some of their quotations] link the Siberian flood basalts through production of large amounts of sulfur with the end-Permian mass extinction. *Shen et al.* [107] report that U-Pb dating reveals an end-Permian extinction peak at 252.28 ± 0.08 Ma and that the negative $\delta^{13}C$ excursion lasted $\leq$ 20,000 a. Admittedly, they [107] concluded that a massive release of thermogenic CO_2 and CH_4 is a plausible explanation for this sudden collapse of the marine and terrestrial ecosystems. Also in this case, the flood basalt eruption is the proper cause.

The 260 Ma-old Emeishan volcanism in Southwest China and interbedded Middle Permian carbonates contain a record of the Guadalupian mass extinction connected with the extinction of 56 % of plant species in the North China Block [14, 139]. This extinction predates a major negative $\delta^{13}C$ excursion. There is a clear temporal link that suggests that the flood basalt eruptions triggered the Guadalupian extinction [14, 139].

LIPs are obviously considered a relevant cause for mass extinctions and justifiably so. It is disputed yet, what thereby evoked additional mechanisms could amplify the direct effect, e.g. a haline euxinic acidic thermal transgression [66]. There are indications that occasionally bolide impacts can destabilize an already strained state. Apart from that, bolide impacts often do not have any essential influence on the variation of the number of species. Independent of the extinctions, we observe that the longer lasting major orogenic intervals are connected with the main glaciations which effect life otherwise.

Apparently, mutation and recombination of genes always act on the populations. They produce a very slow alteration of a species: Related populations in successive layers often show moving changes, especially the microfossils. However, the connection of orders, families and genera is often obscure [37]. Evidently, extinctions clear the way for the resettlement of living environments which formerly were dominated by the extinct species. It is increasingly obvious that *the mass extinctions of biological species are influenced by the endogene evolution of the Earth's mantle, especially by LIP-producing eruptions.*

1.3 Origin of Life

In the strict sense we know neither how nor where life came into existence [15]. The genetic information is stored by the DNA (desoxyribonucleic acid) which will be transcoded into RNA (ribonucleic acid) and then translated into proteins [36]. In reference to life, we want to limit ourselves to the well-known DNA-RNA-protein triad and exclude transeunt speculations. The origin of life divides into two stages. The first stage gives rise to first replicable molecules, probably to RNA [29]. Amino acids, sugars, and other molecules of life can be generated in the laboratory. There are several suggestions where these prebiotic reactions could take place in the Archean. The second stage is how organic molecules form a protocell, i.e. a system with proteins, nucleic acids, and cell membranes. This second stage contains the unresolved root of the matter. The two stages are connected by the question, what originated first, DNA or proteins? Genetics-first hypotheses are widely spread and connected with the opinion that the synthesis of RNA nucleotides and oligonucleotides is fundamental (*Orgel* [90–92]). As precursors of RNA, several alternatives to RNA have been proposed, namely peptide nucleic acid (PNA) and threose nucleic acid. Possibly RNA was directly synthesized. This process could have facilitated by a catalytic system on the surface of minerals [92]. *Johnston et al.* [58] found ribozymes which are able to produce complementary copies of RNA molecules.

As a metabolism-first hypothesis we mention the papers by *de Duve* [26–28]. He proposes a protometabolism in a thioester world and supposes catalysis on multimers which stem from thioesters. It is a question of a sophisticated chain of reactions based on thioesters and Fe-S compounds which result in a complete protometabolism using energy supply by lightnings or ultraviolet (UV) irradiation. Already *Miller and Urey* [83] experimentally showed that aldehydes and amino acids are produced in an anoxic atmosphere of H_2O, CH_4, NH_3, H_2, CO_2, CO and N_2 by electrical sparks (lightnings). These experiments depend on external energy sources. For this configuration, Archean "warm little ponds" and lightnings on continental sites have been proposed.

A second metabolism-first hypothesis has been suggested by *Wächtershäuser* [126–129]. He assumes that chemoautotrophic microorganisms played the most ancient role in the biosphere. Therefore he proposes a chemoautotrophic origin of life in a volcanic iron-sulphur world. In this case, the energy has not an indirectly solar origin but is released at the mouths of churning deep-sea vents, e.g. at the black smokers which are bound to the Earth-embracing system of plate boundaries of the oceanic lithospheric plates. So, in the case of the Wächtershäuser hypothesis, the existence of oceans and plate tectonics is necessary for the emergence of life on Earth and perhaps also on Mars if there were an ocean and plate tectonics in the first about 500 Ma. In spite of the lower luminosity of the early Sun and the larger distance of Mars from the Sun, the early Mars had a *fluid* ocean because of an atmosphere of higher density [31, 39–41]. An important framework requirement for life is fluid water. Because of the lower atmospheric pressure, this temperature window for *present-day* Mars is between 273 K and 283 K. The inorganic nutrients of life are molecules such as H_2, N_2, H_2O, H_2S, NH_3, CH_4, CO, CO_2, HCN and P_4O_{10} [129]. These molecules emerge as volcanic exhalations from the planet's mantle. (O_2 and O_3 are considered to be virtually absent.) The listed multitude of molecules catalytically reacted at the electrically positively charged surfaces of metallic sulfides, e.g. FeS, NiS, ZnS, (Fe,Ni)S. Generally expressed, transition metal centers with sulphido, carbonyl and other ligands were catalytically active and promoted the growth of organic superstructures [129]. *Wächtershäuser* [129] believes to be able to explain also the cellularization and the emergence of the genetic machinery. He also delineates the track from chemoautotrophic life to Bacteria, Archea and Eukarya. *Kundell* [70] demonstrated by computational analysis what special proto-nucleic acid could come into consideration as a pioneer organism.

Both genetics-first hypotheses and metabolism-first hypotheses encounter difficulties. *Vasas et al.* [125] indicate that, in case of metabolism-first hypotheses, replication of compositional information is so inaccurate that fitter compositional genomes cannot be maintained by selection and, therefore, the system lacks evolvability which, however, is an essential feature of life.

In case of genetics-first hypotheses and of metabolism-first hypotheses with solar UV radiation or lightnings as an energy source, the following geological conditions have to be fulfilled: It is imperative that there are continents with shallow epicontinental seas. Furthermore shallow and nearly or totally closed small lagoons or ponds with a primordial soup of organic compounds are preconditions. *In case of chemoautotrophic metabolism-first hypotheses, in addition to continents, a plate-tectonic mechanism is necessary which exhibits sites of reducing volcanic exhalations.* In this case solar energy is irrelevant, but only in the early stages. For all types of origin-of-life hypotheses, sources of nutrients are necessary. *Only three rock types serve as a source of nutrient [76], namely (a) tonalite-trondhjemite-granodiorite (TTG) complexes of the Archean upper continental crust (UCC), (b) carbonatite magmatic rocks enriched in U and Th, and (c) primordial continents with KREEP basalts which are rich in potassium (K), rare earth elements (REE) and phosphorus (P).* The last one is essential for life because it is a component of DNA, RNA and phospholipids that form the cell membranes.

1.4 Preservation of Life

The preservation of life depends on many geochemical conditions. The cycles of water, carbon, sulfur and phosphorus are very important. It makes sense to include the endogenous parts of these cycles which include the silicate mantles of Earth and Mars. The endogenous water cycle has a bearing on both the effective shear viscosity and thus on mantle solid-state convection [104] and on the location- and time-dependent solidus and thus on chemical differentiation of the mantle [131]. Phosphorus has a restrictive effect on the biological productivity of the Earth [42]. In Sect. 1.2, we describe how geodynamics influences the large mass extinction events and the preservation of life. On the other hand, the products of life's metabolic processes have a profound effect on the chemistry of the Earth which possibly affects the mechanics of tectonic and magmatic evolution of the mantle [108]. As a survey on the specific problems of the chemistry of the Earth, we recommend [18], for the chemistry of Mars [33, 79, 88, 121].

The existence of an ocean is a necessary but not sufficient condition for plate tectonics which, in turn, is the most effective mechanism for a stronger cooling of the planet's iron core. If the core's ferrous alloy is not or not completely frozen then a sufficiently large cooling of the planet drives the hydromagnetic convection of the fluid part of the core. From the geomagnetic secular variation, it has been concluded that the magnitudes of the flow velocities in the metallic fluid of the Earth's outer core (OC) amount to 10 and 30 km/a. So the OC viscosity is between 1 and 100 Pa·s. Therefore, rather small eddies are expected which are oriented along the Earth's axis of rotation by the Coriolis force. The mechanism can be explained by the theory of averaged fields. The terrestrial magnetic induction field determines the magnetosphere which is the essential protection system of life. *Therefore plate tectonics is twofold or threefold important for life, first by the volatile cycles of the mantle, second by the magnetosphere and third by the black smokers if Wächtershäuser [129] is right. In the case of genetics-first hypotheses or of metabolism-first hypotheses with solar irradiation or lightning energy supply, the existence of continents with epicontinental seas, lagoons and ponds are directly determining for the origin of life. Because of the sources of nutrients, the origin and growth of the continental crust (CC) is relevant for life, in either case [76, 131].*

2 Self-Consistent Modeling of Plate Tectonics

It was a brilliant performance to recognize that a multitude of geological, kinematic, geochemical, isotopic and geophysical observations can be systematized by plate tectonics. But this systematization does not mean physical comprehension. Therefore, some pioneering papers derive the plate movements by self-consistent numerical modeling (*Trompert and Hansen* [123], *Tackley [116, 117], Richards et al.* [96], *Bercovici and Karato* [11], *Bercovici and Ricard* [12]). This means that the plates are not artificially prescribed at the upper boundary of the model but they develop from the system of equations of mantle convection. In the following, we concentrate on our own papers because this is a statement of accounts on our use of supercomputing facilities at SCC Karlsruhe and HLRS Stuttgart.

We converted the balance equations of momentum, energy, and mass for our purpose [132]. For example, the energy conservation was rewritten in such a way that the Grüneisen parameter, γ, explicitly occurs several times (cf. (33), (34), (38) of [132] and (6) of [130]). Using the Vashchenko-Zubarev equation [55], we determine γ directly from seismic observations, i.e. from the Preliminary Reference Earth Model (PREM, [35]). In this way, it is not necessary to use mineralogical mantle models to determine γ. To estimate the buoyancy, we need the gravity, g, and the thermal expansivity, α, as a function of radius, r. We use the parameterized form of gravity from PREM. As for the expansivity, we adopt a modified version of the model by *Chopelas and Boehler* [20]. Using

$$\gamma_{th} = \frac{\alpha \cdot K_T}{c_v \cdot \rho} = \frac{\alpha \cdot K_S}{c_p \cdot \rho} \tag{1}$$

taking the adiabatic bulk modulus, K_S, and the density, ρ, from PREM, α from [20], and equating the thermodynamic Grüneisen parameter, γ_{th}, with the Vashchenko-Zubarev gamma, we are able to determine c_p, the specific heat at constant pressure.

The temperature and pressure dependence of the shear viscosity, η, is, however, a good deal more critical for the solution of the balance equations. Both the *energy balance*

$$\frac{\partial T}{\partial t} = -\frac{\partial (T v_j)}{\partial x_j} - (\gamma - 1) T \frac{\partial v_j}{\partial x_j} + \frac{1}{\rho c_v} \left[\tau_{ik} \frac{\partial v_i}{\partial x_k} + \frac{\partial}{\partial x_j} \left(k \frac{\partial}{\partial x_j} T \right) + \mathcal{Q} \right] \tag{2}$$

and the *momentum balance*

$$0 = -\frac{\partial}{\partial x_i}(P - P_r) + (\rho - \rho_r) g_i(r) + \frac{\partial}{\partial x_k} \tau_{ik} \tag{3}$$

include the deviatoric stress tensor, τ_{ik}, where

$$\tau_{ik} = \eta \left(\frac{\partial v_i}{\partial x_k} + \frac{\partial v_k}{\partial x_i} - \frac{2}{3} \frac{\partial v_j}{\partial x_j} \delta_{ik} \right) \tag{4}$$

and the viscosity, η,

$$\eta = \eta(r, \theta, \phi, t) = 10^{r_n} \cdot \frac{\exp(c\, \overline{T_m / T_{av}})}{\exp(c\, \overline{T_m / T_{st}})} \cdot \eta_3(r) \cdot \\ \cdot \exp\left[c_t \cdot T_m \left(\frac{1}{T} - \frac{1}{T_{av}} \right) \right] \tag{5}$$

K_T is the isothermal bulk modulus, c_v specific heat at constant volume, T temperature, t time, x_j position vector, v_j velocity vector of solid-state creeping, k thermal conductivity, $\mathcal{Q}$ heat generation rate per unit volume, P pressure, g_i gravity vector, θ colatitude, ϕ longitude, r_n viscosity-level parameter, c_t and c=7 are parameters (see [130]), T_m melting temperature, T_{av} laterally averaged temperature, T_{st} starting temperature, η_3 radial viscosity profile (see [130]), the indices r at P_r and ρ_r refer to the adiabatic reference state.

From the postglacial uplift of continents, e.g. Fennoscandia and Laurentia, it is evident that the asthenospheric viscosity is 10^{21} Pa·s. Therefore and on account of the profile η_3, we choose $\eta_3 = 3.45 \cdot 10^{20}$ Pa·s at 367 km depth [130]. Geophysical multi-layer models often assume a certain number of layers where every layer has a constant viscosity. *Sabadini and Vermeersen* [101] give an excellent survey of applications of normal mode relaxation theory to the derivation of terrestrial viscosity profiles. We [133] introduced a new viscosity profile of the Earth's mantle and supplemented [135] it to introduce a numerically feasible approach to take account of the mantle's secular cooling and the slow rising of the viscosity profile using the second factor on the right-hand side of (5). Some ideas of [135] are listed as follows.

(a) A chemically homogeneous layer of the mantle cannot have a constant viscosity because of the pressure dependence of activation enthalpy. Therefore, the viscosity mostly rises with pressure. A compensation or even overcompensation by the temperature dependence is virtually possible only near the core-mantle boundary (CMB) because of the extremely high temperature gradient in the D" layer.
(b) Viscosity discontinuities occur in the chemically homogeneous parts of the mantle only at phase boundaries, because P and T do not jump.
(c) At 410, 520 and 660 km depth, the lattice of olivine changes toward denser packings of atoms. Therefore it would be inconsistent to assume that there only the seismic velocities, v_s and v_p, and the density, ρ, jump but not activation energy and activation volume. Because the activation enthalpy is in the exponent of the viscosity function, we expect appreciable viscosity jumps. Therefore we had considerable numerical problems with the Terra code and encouraged corresponding efforts (*Müller* [87], *Köstler* [69]).
(d) Other authors often use only the temperature dependence of viscosity to numerically produce an ocean lithosphere. But the real ocean lithosphere evolves not only from the temperature dependence but also by dehydration, other devolatilizations and not least by chemical differentiation. The latter one generates a basaltic and gabbroic ocean crust, underneath a harzburgitic layer and under it a lherzolithic layer. Therefore the lithosphere-asthenosphere boundary (LAB) corresponds to a *chemical* jump and a viscous jump. That is why we introduced a stiff lithosphere in our viscosity profile, $\eta_3(r)$. For numerical reasons, however, we had to replace the η-jump by a steep viscosity gradient. It can be shown by numerical experiments that *it is not possible to obtain self-consistent plates near the surface of a spherical shell if we use a purely viscous constitutive equation.* Therefore we additionally introduced a viscoplastic yield stress, σ_y. For the uppermost 285 km, an effective viscosity, η_{eff}, was implemented where

$$\eta_{eff} = \min\left[\eta(P,T), \frac{\sigma_y}{2\dot{\varepsilon}}\right], \quad (6)$$

The second invariant of the strain rate tensor is denoted by $\dot{\varepsilon}$. *The second essential condition for the appearance of self-consistent plates is the low-viscosity asthenosphere* in η_3. Using these principles we [133] calculated Fig. 1. Further developments of the model are demonstrated in [130]. Figure 2 shows an improved plate-like behavior.

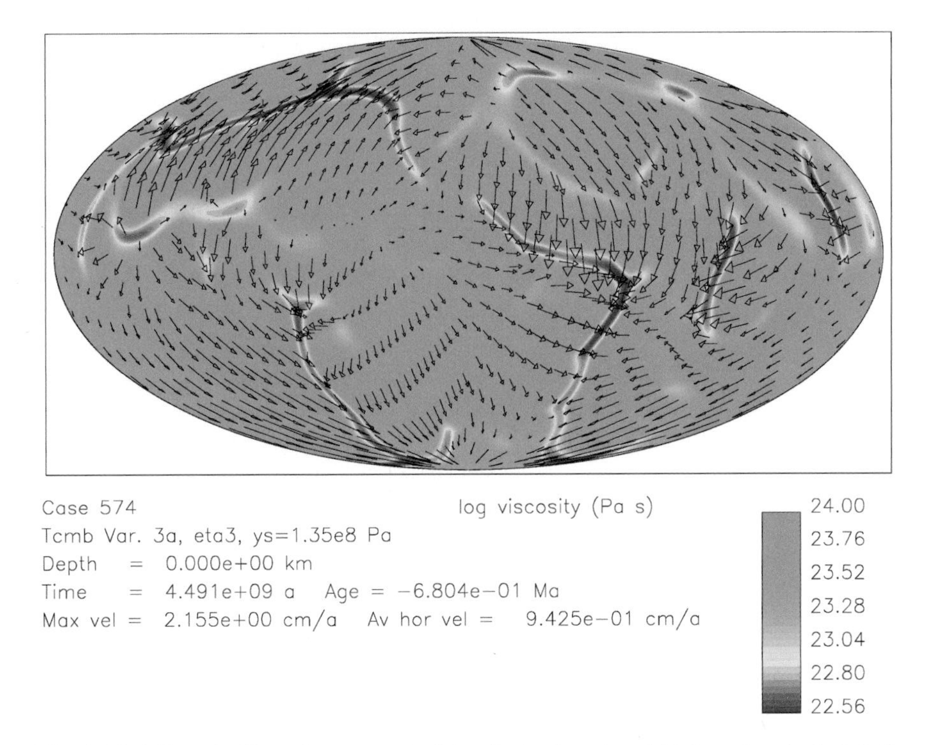

Fig. 1 Modeling of self-consistent oceanic lithospheric plates in a 3-D dynamic spherical shell according to [133]. This equal-area projection shows a log-viscosity (Pa· s) distribution with a yield stress of 135 MPa in the lithosphere. This solution type was obtained in a larger Ra-σ_y area by parameter variation. *Ra* denotes the Rayleigh number. *Arrows* show the plate-like array of creeping velocities of the lithospheric rocks

3 Modeling of Continental Growth

The formation of basaltic oceanic crust occurs by a single-stage melting of the depleted mantle (DM) near mid-ocean ridges (MOR). According to [53], 30–80 % of the Earth's mantle are depleted (DM) in incompatible elements (e.g. U, Th, K, REE), according to [10], 30–60 %. The low-viscosity asthenosphere is preferably composed of DM. This fact is verified by observations at MOR. However, the production of continental crust (CC) is considerably more complex. Furthermore, the chemical composition of the lower continental crust (LCC) is not accurately known [98]. It is assumedly andesitic or at the boundary between andesite and basaltic andesite, i.e. at about 56.5 wt % SiO_2 [118–120]. A minimum of two stages of chemical differentiation are necessary to produce granodiorite-dominated upper continental crust (UCC) [118]. *Davidson and Arculus* [25] present five conceptional models for generating CC where particularly Model V appears to be realistic. According to the last model, CC essentially evolves from oceanic island arc crust (OIAC). For instance, the crustal structure of the Izu-Bonin island arc [115]

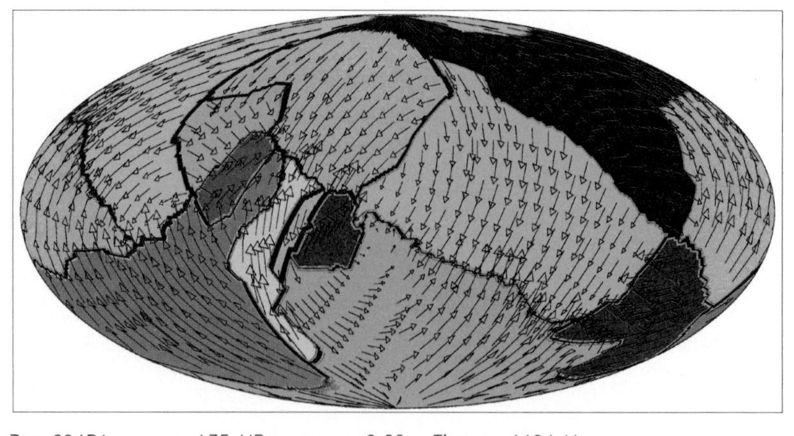

Fig. 2 Improved lithospheric plates (*colors*) and creep velocities (*arrows*) for a 3-D convection model [130] with chemical differentiation and continents which are not shown here. For decreasing r_n, the viscosity difference between oceanic lithosphere and asthenosphere grows and the number of lithospheric plates decreases if the other parameters are fixed

may be considered as a prototype of OIAC. After the orogenic shortening, an ultramafic cumulate resides beneath the seismological Moho. This cumulate ranges down to the genetic Moho. After continuing orogenic thickening, this cumulate and the subjacent lithospheric mantle will be removed by thermal erosion through convection in the space of sub-arc wedge or by delamination through a Rayleigh-Taylor instability. A presumably somewhat smaller part of the CC-growth emerges from the accretion of oceanic plateau basalts which result from increased plume activity. Meanwhile we have got evidence that the juvenile plateau basalt proportion of the total basalt production of an orogenically active period was considerably larger than that of the present time. Furthermore, these orogenic events took place episodically [21]. We [131] did not include the specific CC-differentiation with full details into our global convection model but in a simplified way. We designed a new dynamic 3-D spherical-shell convection model of the Earth's mantle and incorporated a novel solidus, T_{sol}, [72] of peridotite which is not only a function of pressure, P, but also of water concentration. Partial melting in major volumes occurs only when

$$T > f_3 \cdot T_m \tag{7}$$

applies or when the total water abundance exceeds the water solubility [72, 81]. The quantity f_3 is a parameter that is somewhat smaller than or equal to 1 and that we vary in the different Terra runs. Because plenty of water escapes during every chemical differentiation event, the solidus will be enhanced for some time after the differentiation event in the differentiation volume. We use the concept of chemical mantle reservoirs considerably different than in previous times [52, 53]. We do not imagine sharp boundaries between the reservoirs but we define the time-dependent chemical composition of a specific mantle location by the mixing ratio of the conventional reservoirs using the reservoir concentrations of the incompatible elements by *McCulloch and Bennett* [78]. We use the three chemical mantle reservoirs (CC, DM, PM) with different abundances [131] of heat-producing elements U, Th, and K. The segregation processes quickly change the distribution of these elements; the solid-state mantle convection slowly modifies the distribution. Because the heat generation rate per unit volume, Q, depends on the temporally and spatially variable U, Th and K abundances, there is a feedback on convection and thermal evolution of the mantle. Zircon and sandstone age estimates result in *observed* peaks of frequency at 2697, 1824, 1435, 1047, 594, 432, and 174 Ma. *These episodes are reproduced by run 498* (cf. Fig. 3, first panel). Our model [131] is based on the numerical solution of the balance equations of energy, momentum, moment of momentum, and mass in a spherical shell which represents the Earth's mantle. Furthermore, we use equations which guarantee the conservation of the four sums of number of atoms of the pairs ^{238}U-^{206}Pb, ^{235}U-^{207}Pb, ^{232}Th-^{208}Pb, and ^{40}K-^{40}Ar. In the present model of continental evolution, we replaced the viscosity profile $\eta_3(r)$ [130, 133] of (5) by the newly developed $\eta_4(r)$ [131] which resembles the viscosity profile by *Mitrovica and Forte* [84] although the derivation of it was completely different. The radial dependence of viscosity, $\eta_4(r)$, is based on some solid-state physics considerations. We took the mean of the relative viscosity between LAB and a depth, h, of 1,250 km. The absolute value of this mean was defined as 10^{21} Pa·s, i.e. the observed Haskell value. Furthermore, the Grüneisen parameter, γ, is important for solid-state geophysics [3, 60, 93, 113, 114]. Cf. (2) and the Gilvarry-Lindemann equation [134], (2.8). In [130, 133], we utilized the Vashchenko-Zubarev gamma, γ_{VZ}. Because, however, the shear modes contribute an essential part to γ, we [131] replace γ_{VZ} by the acoustic gamma, γ_a, where

$$\gamma_a = \frac{1}{6}\frac{K_T}{K_S + (4/3)\mu}\left[\left(\frac{\partial K_S}{\partial P}\right)_T + \frac{4}{3}\left(\frac{\partial \mu}{\partial P}\right)_T\right] + \frac{1}{3}\frac{K_T}{\mu}\left(\frac{\partial \mu}{\partial P}\right)_T - \frac{1}{6} \tag{8}$$

The quantity K_S denotes the adiabatic bulk modulus and μ the shear modulus. For the depth interval between 771 and 2,741 km, we [131] use the observed seismic PREM values and (8) to determine γ_a. For h <771 km and in the thin D" layer immediately above the CMB, the observed dK/dP and $d\mu/dP$ of PREM lead to physically implausible depth variations of γ. Therefore, we employ the gamma estimates of [113] for the latter depth ranges. The combined Grüneisen parameter is called extended acoustic gamma, γ_{ax}, that finally has been used in [131]. Furthermore, we [131] derived improved profiles of $\alpha(r)$ and $c_v(r)$. Because we now adopt the water-abundance dependent solidus [72], our viscosity function [131] also depends on the water concentration (cf. (5)), in contrast to [130, 133]. As a result, there are magmatically quiescent time intervals between the active periods (Fig. 3, first panel). Furthermore, *qob* (Fig. 3, second panel), *Ur* (Fig. 3, third panel), and E_{kin} (Fig. 3, fifth panel) distinctly show temporally sinusoidal components which are superimposed to slowly decreasing curves. However, the volumetrically averaged mantle temperature, T_{mean}, *does not change rapidly and is monotonously decreasing* (cf. Fig. 3, fourth panel). This result corroborates a principal conclusion of *Gurnis and Davies* [50]. Therefore, we can dismiss catastrophic mechanisms that *simultaneously* incorporate the *whole* mantle.

In our model, the number, size, form, distribution and angular velocity of the evolving continents is *not* prescribed or constrained. These quantities are a result of the dynamics of the numerical system. The only exception is the rule that if a terrane, e.g. an oceanic plateau, touches a continent, it has to be united with this continent. This regulation simulates the geological accretion of terranes. Figure 4 shows a typical continental distribution for the geological present time. Of course, we can show the continental distribution also for earlier instants of time. To estimate whether we obtained a *realistic* solution, we developed the model continents of the modeled present time into spherical harmonics. Because the spherical-harmonics coefficients, A_n^m and B_n^m, depend on the poles of the grid of parallels and meridians, i.e. on a human addition, we have to form pole-independent functions h_n of A_n^m and B_n^m. Then we compare the computed h_n with the h_n of the real, present-day distribution of continents. In this way, we found a realistic Ra-σ_y area. If we draw nearer to this special Ra-σ_y area in a wider Ra-σ_y space then also the other magnitudes, e.g. *qob*, converge to realistic values which are known from other investigations.

4 Considerations on a Numerical Model of the Thermal Evolution of Martian Mantle

Today it is clear that the Earth always had, exclusive of the earliest stages, fluid oceans, continents with sediment-producing running water and an atmosphere which initially was CO_2-rich and nearly anoxic. The animate beings themselves essentially contributed to the decrease of the CO_2 concentration in the terrestrial atmosphere by the sedimentation of limestone. This decrease diminished the greenhouse effect and balanced the luminosity increase of the sun. Therefore, the outlined mechanism works like a temperature control device. Hydrous mineral phases and the thereby produced diminished melting temperature in the upper layers of the Earth's oceanic lithosphere as well as the existence of an asthenosphere are essential preconditions for the mechanism of terrestrial plate tectonics. Plate tectonics, on the other hand, cools the Earth essentially more effective than a one-plate planet like the present-day Mars. This effective cooling, the *fluid* metallic core and a sufficiently large rotational speed are necessary for the sufficiently large magnetic dipole field of the Earth and the life-protective magnetosphere.

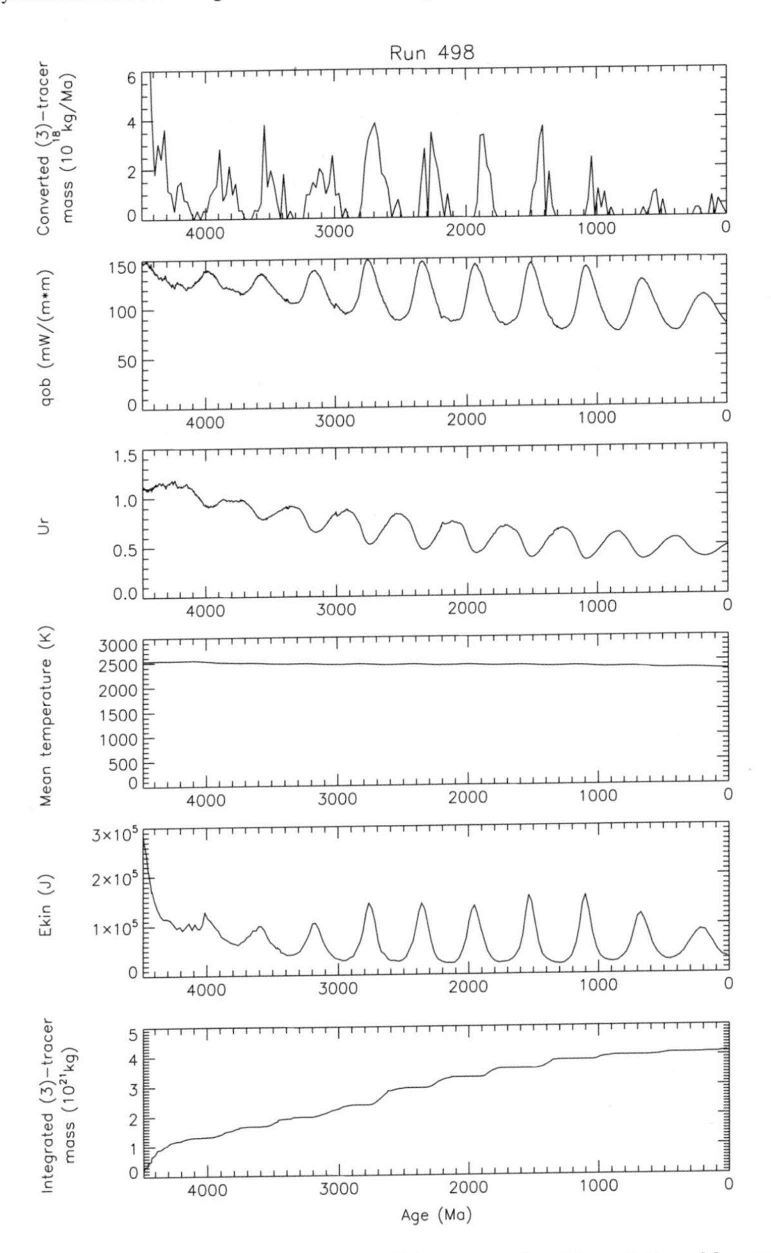

Fig. 3 The Earth's evolution curves of run 489. This case was found by variation of four parameters [131], namely, Rayleigh number, Ra, viscoplastic yield stress, σ_y, thermal conductivity, k, and melting-criterion parameter, f_3. For this run, the viscosity-level parameter is $r_n = 0.5$ (This equates to about Ra= 10^8), $\sigma_y = 120$ MPa, $k = 5.0$ W/(m·K), and $f_3 = 0.995$. The first panel shows the episodic, juvenile CC-magmatic activity, the second one the laterally averaged surface heat flow density, qob, and the third one the Urey number, Ur. The fourth panel displays the volumetrically averaged mantle temperature, T_{mean}, the fifth one the kinetic energy, E_{kin}, of the solid-state convection of the mantle. The sixth panel, finally, represents the CC cumulative growth

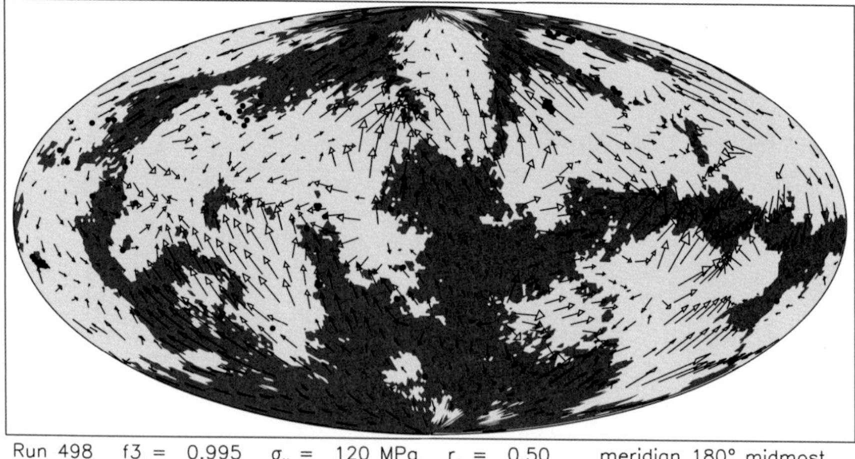

Fig. 4 Present-day distribution of the terrestrial continents (*red*), oceanic lithosphere (*yellow*) and oceanic plateaus (*black dots*) for run 498. The quantities r_n=0.5, σ_y=120 MPa, k=5.0 W/(m· K), and f_3=0.995 are kept constant. The *arrows* show the present-day creep velocities at the surface

Is it possible to transfer this overall picture to the early Mars? On the northern Martian hemisphere, shore lines were found that point to an early ocean that covered about 42 % of the surface. It is notable that the Martian soils in the Meridiani Planum and in the Gusev Crater show a good correlation between the concentration of phosphates and those of sulphates and chlorides. The outcome of this is that a homogenization took place in a large water reservoir like an ocean [48]. Even if an early Martian ocean existed we cannot directly conclude that there was early plate tectonics.

Large empty fluviatile valleys have been observed on Mars. It is not clear, however, whether they represent previous rivers or streams due to episodic huge mud eruptions. At the Martian poles, water ice exists even in the present time. Numerous structures in mid-latitude regions of Mars point to previous glaciers with classical moraine structure. *Kargel* [61] supposes that the present-day Martian crust contains dikes of water ice, hydratized sulphates, clathrates etc. in a configuration similar to pegmatite dikes in the Earth's UCC. Also the shergottites, nakhlites, and Chassigny (SNC meteorites) are sign of a wet Martian mantle [79]. On the other hand, we cannot find any mid-ocean ridge system at the bottom of the ancient Martian ocean because it is covered with sediments. *Acuña et al.* [1] discovered that, in the first 500 Ma, Mars had a magnetic dynamo that disappeared before the Hellas and Argyre impacts that happened 4,000 Ma ago. The intensity of the remanent field of the Martian crust is surprisingly high. Mars Global Surveyor revealed magnetic inductions $\geq$ 1,500 nT in an altitude of 120 km. There are appreciable contributions to the Martian magnetic induction field for spherical harmonics degrees n $\geq$ 14 that are rather small on Earth. It is an unresolved problem whether the early

Martian dynamo was particularly powerful or whether a quadrupolar dynamo is more probable [54, 71, 137]. ASPERA-3 results of Mars Express [74] suggest that the solar wind could cause the losing of ocean water and other volatiles. Therefore, the dynamo switching off would not only annihilate potential Martian life but also dry out the Martian ocean and heavily diminish the mass of Martian atmosphere.

In the strongly cratered highlands of the southern Martian hemisphere, there are intensive, EW striking anomalies of the radial and northern components of the magnetic induction field that have been associated with the terrestrial mid-ocean magnetic anomalies [22, 23, 112]. But the extent of these Martian anomalies is considerably larger. Furthermore, we would expect them at the bottom of the previous ocean on the northern hemisphere. Therefore, the problem of the Martian plate tectonics is unresolved.

Before we can design and compute a dynamic 3-D spherical-shell Mars model, we need a structural model analogous to PREM. This is a difficult task because Mars is virtually seismically not explored. Furthermore, the problem of the chemical model is not resolved. We know only the mass, moment of inertia, radius, tidal Love number k_2, the very well explored Martian surface and the SNC meteorites. But the SNCs are differentiation products that come from the crust and are *not* representative of the Martian mantle as a whole. *Sohl and Spohn* [111] derive two end-member structural models where they assume a bulk chondritic ratio Fe/Si=1.71 for the first model. For the second model, however, they hypothesize a maximum value $C = 0.366 * M_p r_p^2$ of the polar moment of inertia factor. They derive the density, ρ, and other physical parameters as a function of radius, r. Using a chondritic basic assumption, *Walzer et al.* [136] estimated ρ, the isothermal bulk modulus K_r, the thermal expansivity α, and the Grüneisen parameter γ as a function of depth of the Martian mantle. *Rivoldini et al.* [97] derive the density and other profiles for Mars where they use five different chemical bulk compositions as a presupposition, namely DW84 [33], LF97 [73], EH45 [102], EH70 [102] and MM03 [85]. The existing uncertainities of the Martian structural models are also illustrated by the fact that even up-to-date values of the tidal Love number range between 0.12 ± 0.004 [75] and 0.236 ± 0.058 [110] where a *probably* realistic value is at 0.159 ± 0.009 [68].

Modern *parameterized* convection models of Mars were presented by [16, 44, 51, 86, 106], where the model by [44] predicts that the Martian mantle must have been degassed more extensively (>80 %) than previously thought. *Ogawa and Yanagisawa* [89] computed a *dynamical* 2-D Cartesian-box convection model of the Martian mantle. They present the evolution of temperature, internal heating, composition, and water content in 2-D boxes. The estimated plume magmatism is sufficiently strong to generate a large crustal growth and dehydration in the early Martian history. *Ruedas et al.* [99] give an account of another *dynamical* model, in this case in a 2-D spherical annulus geometry. To avoid the difficulties with the first 500 Ma of Martian history, only the thermochemical evolution of Martian mantle over the past 4 Ga was computed. The authors of [99] and [136] independently concluded that, of all the above named chemical Martian mantle models, the *Dreibus-Wänke* [33] model tends to explain observations best.

A future dynamic Mars model should solve the full set of balance equations of solid-state creeping in a 3-D spherical shell like the procedure for the Earth's mantle summarized in Sects. 2 and 3. The first 500 Ma of Martian evolution should not be included. But for Mars, that involves some difficulties:

(a) We have no seismic model corresponding to PREM. Therefore, we should derive a structural model analogous to *Sohl and Spohn* [111] or *Gudkova and Zharkov* [49]. Hence, a hypothetical seismic model and the radius dependence of some other physical quantities should be estimated using a chemical Martian mantle model. So, the most fundamental decision concerns the chemical model. *Dreibus and Wänke* [34] propose 40 % oxidized components of C1 carbonaceous chondritic material and 60 % heavily reduced material (component A). In contrast, *Sanloup et al.* [102] assume 45 % enstatite meteorites EH and 55 % ordinary chondrites of type H because they want to explain the isotope ratios of oxygen. However, they generate other difficulties with their proposal. Of course, every realistic structural model has to produce the correct values of the observed mass, moment of inertia, radius and tidal Love number, i.e. correct within the error limits. A further constraint is the conclusion, derived from satellite observations [142], that the CMB of Mars separates a solid upper domain from a liquid lower one. Since the Martian iron core is not frozen out, we have to look for another explanation of the missing present-day magnetic dipole field of major intensity.
(b) The estimation of the effective shear viscosity of solid-state creep in the Martian mantle as a function of radius is yet considerably more difficult than for the Earth's mantle. In addition, we do not know a fixed point analogous to the Earth's Haskell value. Because of the formulas of Sects. 2 and 3, it would be important to determine the temperature-pressure phase diagram of the Martian-mantle material. The resultant phase boundaries with jumps of activation energy and activation volume are relevant to determine the expected essential viscosity discontinuities. Furthermore, the pressure dependence of the solidus, T_{sol}, can be derived from the phase diagram. Moreover, there is a connection between the phase diagram and the Grüneisen parameter, γ, and the thermal expansivity, α. In the dynamic model, we need also the radius dependence of γ and α.
(c) The structural model of (a) cannot be derived independently of (b) because the acoustic gamma, γ_a, depends on K, dK/dP, μ, $d\mu/dP$. Furthermore, the density, ρ, depends on K.
(d) One could consider to adopt phase diagrams for a hydrous peridotite system [56, 62, 63] because the early Martian mantle probably had a higher water concentration than the Earth's mantle. But this is not so simple because the Earth's mantle contains a principal chemical component of about 8 % FeO whereas the Martian mantle probably contains 18 % FeO [13, 34].
(e) We have to take into account that, also for Mars, the lithosphere does not simply develop in consequence of the temperature dependence of viscosity but also due to chemical differentiation and owing to the relation between water abundance and water solubility [81] as a function of depth.

(f) We propose that the viscosity distributions of Earth [131] and Mars are also time-dependent because the water abundance is also temporally variable. If the early Martian mantle lost plenty of water then the curves of water solubility and water concentrations lose their two intersection points [81]. Therefore, the early Martian asthenosphere vanishes. But the asthenosphere is a prerequisite of plate tectonics.
(g) Also the Martian lithosphere cannot resist shear stress in any order. Therefore, we should introduce a viscoplastic yield stress.
(h) *The items (a) to (g) could possibly result in a model with early plate tectonics an Mars, early stronger core cooling and, consequently, a strong early Martian magnetosphere that was life-protective if life was existent.* In any case, we know that Mars had a strong magnetic dipole field in the first 500 Ma.
(i) In contrast to Earth, the Martian mantle does not have a uniform $^{182}W/^{184}W$ ratio [43, 67, 80]. Therefore, Mars has a heterogeneous mantle [80]. That is why we expect that considerable deviations from the spherical-shell model are necessary.
(j) Mars has only a basaltic crust, no felsic igneous rocks [80].

5 Numerical Improvements

We obtained the numerical solutions of the system of balance equations of convection in a spherical shell using a three-dimensional finite-element discretization, a fast multigrid solver and the second-order Runge-Kutta procedure. The mesh is generated by projection of a regular icosahedron onto a sphere to divide the spherical surface into twenty spherical triangles or ten spherical diamonds. A dyadic mesh refinement procedure connects the midpoints of each side of a triangle with a great circle such that each triangle is subdivided into four smaller triangles. Successive grid refinements generate an almost uniform triangular discretization of the spherical surface of the desired resolution. Corresponding mesh points of spherical surfaces at different depths are connected by radial lines. The radial distribution of the different spherical-surface networks is so that the volumes of the cells are nearly equal. The details of the Terra code are given by *Baumgardner* [7, 8], *Bunge et al.* [17] and *Yang* [141]. For many runs that we needed for variation of parameters, we used a mesh with 1351746 nodes. Some runs were made with 10649730 nodes to check the convergence of the lower resolution runs. The result is that the laterally averaged surface heat flow, the Urey number, the Rayleigh number and the Nusselt number as functions of time show hardly any discernable differences ($<0.5\,\%$). The code was benchmarked for constant-viscosity convection by *Bunge et al.* [17] with numerical results of *Glatzmeier* [47] for Nusselt numbers, peak temperatures, and peak velocities. A good agreement ($\leq 1.5\,\%$) was found.

In 2009, an international group of Terra developers decided to set up a community svn-repository for further code development, supplemented by trac, a web-based project management and bug-tracking tool, and automated compile/test cycles using BuildBot. From then on the group worked on a common code base using automated tests for every revision of the code. The group enhanced the code to increase global resolution and maximum number of MPI processes. A Ruby test framework was integrated [87] into the automated BuildBot tests, and a finite-element inf-sup stabilization using pressure-polynomial projections [32] was implemented. An efficient preconditioner for the variable-viscosity Stokes system [69] has been developed. The code was restructured to use language features of Fortran95 and Fortran2003. Using doxygen, an automated code documentation has been integrated. The free-slip boundary conditions and propagator matrix benchmark tests have been investigated. Peter Bollada and Rhodri Davies showed that adding boundary terms to the right hand side of the momentum equation reduces some sort of errors while other kinds of error still exist. They will continue to figure out the exact cause of that behavior and work on fixing this.

After fruitful discussions about possibilities to proper formulate the variable viscosity operator in Terra, Peter Bollada (Leeds), John Baumgardner (San Diego, USA) and Christoph Köstler (Jena) figured out in which way the code has to be changed to apply a physically consistent A-operator using cell-averaged viscosities. The most significant code change is the switch from nodal based to triangle based operator parts on the sphere. The viscosity-weighted summation over triangular integrals is then done in the application of the operator. We expect that the cost for applying $A*u$ will be doubled but a consistent formulation on all grid-levels could pay off for this, especially if we get a better convergence rate of the multigrid algorithm. J. Baumgardner found small missing terms in the current variable-viscosity formulation of Terra and outlined several approaches to pull viscosity out of the derivatives. He made further progress in scaling the momentum tensor and in using local spherical coordinates.

We are going to further improve the parallelization of the particle tracking routines in Terra. Compared to previous Terra versions, there is an extra need for communication among several MPI-processes to figure out connected regions of partial melting in the mantle from which incompatible elements are extracted and transported to the surface. A similar communication is required to define the extent of continents. With the high number of tracers, it is crucial to compress the required global information locally before it is exchanged among neighboring processors. R. Hendel will continue to reduce the communication overhead for tracking globally connected regions, so that the scalability of the particle routines will be extended to 500 and more processors.

Acknowledgements We gratefully acknowledge stimulating discussions with J. Baumgardner, H.-P. Bunge, P. Bollada, S. Brune, H. Davies, R. Davies, C. Köstler, M. Mohr, M. Müller, J. Weismüller etc. in the group of Terra developers. This paper benefited from geochemical and geological discussions with L. Viereck and J. Kley. We thank the Steinbuch Center for Computing, Karlsruhe, for supply of computational time under grant sphshell. This work was partly supported by the Deutsche Forschungsgemeinschaft under grant KL 495/16-1.

References

1. M.N. Acuña, J.E.P. Connerney, N.F. Ness, et al., Global distribution of crustal magnetization discovered by the Mars Global Surveyor MAG/ER experiment. Science **284**, 790–793 (1999)
2. L. Alvarez, W. Alvarez, F. Azaro, H. Michel, Extraterrestrial cause for the Cretaceous-Tertiary extinction. Science **208**, 1095–1108 (1980)
3. O.L. Anderson, *Equations of State of Solids for Geophysics and Ceramic Science* (Oxford University Press, New York, 1995)
4. J.D. Archibald, *Dinosaur Extinction and the End of an Era: What the Fossils say* (Columbia University Press, New York, 1996)
5. M. Barbieri, The organic codes. An introduction to semantic biology. Genet. Mol. Biol. **26**, 105–106 (2003)
6. A.D. Barnosky, N. Matzke, S. Tomiya, G.O. Wogan, B. Swartz, T.B. Quental, C. Marshall, J.L. McGuire, E.L. Lindsey, K.C. Maguire, et al., Has the Earth's sixth mass extinction already arrived? Nature **471**(7336), 51–57 (2011)
7. J.R. Baumgardner, A three-dimensional finite element model for mantle convection. PhD Thesis, University of California, Los Angeles, 1983
8. J.R. Baumgardner, Three-dimensional treatment of convective flow in the Earth's mantle. J. Stat. Phys. **39**, 501–511 (1985)
9. D. Benest, C. Froeschlé, *Impacts on Earth.* Lecture Notes in Physics (Springer, New York, 1998)
10. V.C. Bennett, Compositional evolution of the mantle. In *Treatise on Geochemistry, Vol.2: The Mantle and the Core*, ed. by R.W. Carlson (Elsevier, Amsterdam, 2003), pp. 493–519
11. D. Bercovici, S.-I. Karato, Whole-mantle convection and the transition-zone water filter. Nature **425**, 39–44 (2003)
12. D. Bercovici, Y. Ricard, Tectonic plate generation and two-phase damage: Void growth versus grain size reduction. J. Geophys. Res. **110**, B03401 (2005). doi:10.1029/2004JB003181
13. C.M. Bertka, Y. Fei, Implications of Mars pathfinder data for the accretion history of the terrestrial planets. Science **281**, 1838–1840 (1998)
14. D.P. Bond, J. Hilton, P.B. Wignall, J.R. Ali, L.G. Stevens, Y. Sun, X. Lai, The Middle Permian (Capitanian) mass extinction on land and in the oceans. Earth Sci. Rev. **102**(1), 100–116 (2010)
15. O. Botta, The chemistry of the origins of life. In *Astrobiology: Future Perspectives*, ed. by P. Ehrenfreund et al. Astrophysics and Space Science Library, vol. 305 (Kluwer (Springer Netherlands), Dordrecht, 2004), pp. 359–392
16. D. Breuer, *Thermal evolution, crustal growth and magnetic field history of Mars* (Habilitation, W. W. Univ. Münster, 2003)
17. H.-P. Bunge, M.A. Richards, J.R. Baumgardner, A sensitivity study of three-dimensional spherical mantle convection at 10^8 Rayleigh number: Effects of depth-dependent viscosity, heating mode and an endothermic phase change. J. Geophys. Res. **102**, 11991–12007 (1997)
18. R.W. Carlson. The mantle and the core. In *Treatise on Geochemistry*, vol. 2 (Elsevier, Amsterdam, 2003)
19. A.-L. Chenet, X. Quidelleur, F. Fluteau, V. Courtillot, S. Bajpai, ^{40}K–^{40}Ar dating of the Main Deccan large igneous province: Further evidence of KTB age and short duration. Earth Planet Sci. Lett. **263**(1), 1–15 (2007)
20. A. Chopelas, R. Boehler, Thermal expansivity in the lower mantle. Geophys. Res. Lett. **19**, 1983–1986 (1992)
21. M.F. Coffin, O. Eldholm, Large igneous provinces: Crustal structure, dimensions and external consequences. Rev. Geophys. **32**, 1–36 (1994)
22. J.E.P. Connerney, M.H. Acuña, P.J. Wasilewski, et al., Magnetic lineations in the ancient crust of Mars. Science **284**, 794–798 (1999)

23. J.E.P. Connerney, M.H. Acuña, P.J. Wasilewski, G. Kletetschka, N.F. Ness, H. Rème, R.P. Lin, D.L. Mitchell, The global magnetic field of Mars and implications for crustal evolution. Geophys. Res. Lett. **28**(21), 4015–4018 (2001)
24. V. Courtillot, J. Besse, D. Vandamme, R. Montigny, J.-J. Jaeger, H. Cappetta, Deccan flood basalts at the Cretaceous/Tertiary boundary? Earth Planet. Sci. Lett. **80**(3), 361–374 (1986)
25. J.P. Davidson, R.J. Arculus, The significance of Phanerozoic arc magmatism in generating continental crust. In *Evolution and Differentiation of the Continental Crust*, ed. by M. Brown, T. Rushmer (Cambridge University Press, Cambridge, 2006), pp. 135–172
26. C. de Duve, *Blueprint for a cell: The Nature and Origin of Life* (Neil Patterson, Burlington, NC, 1991)
27. C. de Duve, The beginnings of life on Earth. American Scientist **83**(5), 428–437 (1995)
28. C. de Duve, Clues from present-day biology: the thioester world. In *The Molecular Origins of Life*, ed. by A. Brack (Cambridge University Press, Cambridge, 1998), pp. 219–236
29. C. de Duve, Life as a cosmic imperative? Phil. Trans. Math. Phys. Eng. Sci. **369**(1936), 620–623 (2011)
30. M.H. Deenen, M. Ruhl, N.R. Bonis, W. Krijgsman, W.M. Kuerschner, M. Reitsma, M. Van Bergen, A new chronology for the end-Triassic mass extinction. Earth Planet. Sci. Lett. **291**(1), 113–125 (2010)
31. J.M. Dohm, J.C. Ferris, V.R. Baker, R.C. Anderson, T.M. Hare, R.G. Strom, N.G. Barlow, K.L. Tanaka, J.E. Klemaszewski, D.H. Scott, Ancient drainage basin of the Tharsis region, Mars: Potential source for outflow channel systems and putative oceans or paleolakes. J. Geophys. Res. **106**, 32943–32958 (2001). doi:10.1029/2000JE001468
32. C. Dohrmann, P. Bochev, A stabilized finite element method for the Stokes problem based on polynomial pressure projections. Int. J. Num. Meth. Fluid. **46**, 183–201 (2004)
33. G. Dreibus, H. Wänke, Mars, a volatile-rich planet. Meteoritics **20**, 367–381 (1985)
34. G. Dreibus, H. Wänke. Supply and loss of volatile constituents during the accretion of terrestrial planets. In *Origin and Evolution of Planetary and Satellite Atmospheres*, ed. by S.K. Attreya, J.B. Pollack, M.S. Matthews (Univ. Arizona Press, Tucson, 1989), pp. 268–288
35. A.M. Dziewonski, D.L. Anderson, Preliminary reference Earth model. Phys. Earth Planet. Int. **25**, 297–356 (1981)
36. M. Eigen, *Steps Towards Life: A Perspective on Evolution* (Oxford University Press, Oxford, 1992)
37. H.K. Erben, *Evolution: Eine Übersicht sieben Jahrzehnte nach Ernst Haeckel*. Haeckel-Bücherei, vol. 1 (Enke, Stuttgart, 1990)
38. D.H. Erwin, *Extinction: How Life on Earth Nearly Ended 250 Million Years ago* (Princeton University Press, Princeton, NJ, 2006)
39. A.G. Fairén, J.M. Dohm, Age and origin of the lowlands of Mars. Icarus **168**, 277–284 (2004). doi:10.1016/j.icarus.2003.11.025
40. A.G. Fairén, J.M. Dohm, V.R. Baker, M.A. de Pablo, J. Ruiz, J.C. Ferris, R.C. Anderson, Episodic flood inundations of the northern plains of Mars. Icarus **165**, 53–67 (2003) doi:10.1016/S0019-1035(03)00144-1
41. A.G. Fairén, D. Fernández-Remolar, J.M. Dohm, V.R. Baker, R. Amils, Inhibition of carbonate synthesis in acidic oceans on early Mars. Nature **431**, 423–426 (2004) doi:10.1038/nature02911
42. G.M. Filippelli, The global phosphorus cycle. In *Phosphates: Geochemical, Geobiological, and Materials Importance*, ed. by M.L. Kohn, J. Rakovan, J.M. Hughes. Reviews in Mineralogy and Geochemistry, vol. 48 (Mineralogical Society of America, Washington, DC, 2002), pp. 367–381
43. N.C. Foley, M. Wadhwa, L.E. Borg, P.E. Janney, R. Hines, T.L. Grove, The early differentiation history of Mars from ^{182}W-^{142}Nd isotope systematics in the SNC meteorites. Geochim. Cosmochim. Acta **69**(18), 4557–4571 (2005). doi:10.1016/j.gca.2005.05.009
44. A.A. Fraeman, J. Korenaga, The influence of mantle melting on the evolution of Mars. Icarus **210**(1), 43–57 (2010)

45. J. Gayon, C. Malaterre, M. Morange, F. Raulin-Cerceau, S. Tirard, Special issue: Definitions of life. Orig. Life Evol. Biosph. **40**, 119–244 (2010)
46. B. Gertsch, G. Keller, T. Adatte, R. Garg, V. Prasad, Z. Berner, D. Fleitmann, Environmental effects of Deccan volcanism across the Cretaceous–Tertiary transition in Meghalaya, India. Earth Planet. Sci. Lett. **310**(3), 272–285 (2011)
47. G.A. Glatzmaier, Numerical simulations of mantle convection: Time-dependent, three-dimensional, compressible, spherical shell. Geophys. Astrophys. Fluid Dynam. **43**, 223–264 (1988)
48. J.P. Greenwood, R.E. Blake, Evidence for an acidic ocean on Mars from phosphorus geochemistry of martian soils and rocks. Geology **34**, 953–956 (2006)
49. T.V. Gudkova, V.N. Zharkov, Mars: Interior structure and excitation of free oscillations. Phys. Earth Planet. Int. **142**, 1–22 (2004)
50. M. Gurnis, G.F. Davies, Apparent episodic crustal growth arising from a smoothly evolving mantle. Geology **14**, 396–399 (1986)
51. S.A. Hauck, R.J. Phillips, Thermal and crustal evolution of Mars. J. Geophys. Res. **107**(E7), 5052 (2002)
52. A.W. Hofmann, Chemical differentiation of the Earth: The relationship between mantle, continental crust and oceanic crust. Earth Planet. Sci. Lett. **90**, 297–314 (1988)
53. A.W. Hofmann, Sampling mantle heterogeneity through oceanic basalts: Isotopes and trace elements. In *Treatise on Geochemistry, Vol.2: The Mantle and the Core*, ed. by R.W. Carlson (Elsevier, Amsterdam, 2003), pp. 61–101
54. L.L. Hood, East-west trending magnetic anomalies in the southern hemisphere of Mars: Modeling analysis and interpretation. In *37th annual Lunar and Planetary Science Conference*, pp. 2203, LPI, 2006
55. R.D. Irvine, F.D. Stacey, Pressure dependence of the thermal Grüneisen parameter, with application to the lower mantle and outer core. Phys. Earth Planet. Int. **11**, 157–165 (1975)
56. H. Iwamori, Phase relations of peridotites under H_2O-saturated conditions and ability of subducting plates for transportation of H_2O. Earth Planet. Sci. Lett. **227**, 57–71 (2004)
57. K.R. Johnson, L.J. Hickey, Megafloral change across the Cretaceous/Tertiary boundary in the northern Great Plains and Rocky Mountains, USA. In *Global Catastrophes in Earth History; an Interdisciplinary Conference on Impact, Volcanism and Mass Mortality*, ed. by V.L. Sharpton, P.D. Ward, vol. 247 (Geological Society of America Special Paper, Boulder, Colorado, 1990), pp. 433–444
58. W.K. Johnston, P.J. Unrau, M.S. Lawrence, M.E. Glasner, D.P. Bartel, RNA-catalyzed RNA polymerization: Accurate and general RNA-templated primer extension. Science **292**, 1319–1325 (2001)
59. S.L. Kamo, G.K. Czamanske, Y. Amelin, V.A. Fedorenko, D. Davis, V. Trofimov, Rapid eruption of Siberian flood-volcanic rocks and evidence for coincidence with the Permian–Triassic boundary and mass extinction at 251 ma. Earth Planet. Sci. Lett. **214**(1), 75–91 (2003)
60. S.-I. Karato, *Deformation of Earth Materials: An Introduction to the Rheology of Solid Earth.* (Cambridge University Press, Cambridge, 2008)
61. J.S. Kargel, *Mars: A Warmer, Wetter Planet.* (Springer, Berlin, 2004)
62. T. Kawamoto, Hydrous phase stability and partial melt chemistry in H_2O-saturated KLB-1 peridotite up to the uppermost lower mantle conditions. Phys. Earth Planet. Int. **143**, 387–395 (2004)
63. T. Kawamoto, Hydrous phases and water transport in subducting slabs. Rev. Mineral. Geochem. **62**, 273–289 (2006)
64. G. Keller, The Cretaceous–Tertiary mass extinction, Chicxulub impact, and Deccan volcanism. In *Earth and Life* (Springer, Berlin, 2012), pp. 759–793
65. G. Keller, A. Sahni, S. Bajpai, Deccan volcanism, the KT mass extinction and dinosaurs. J. Biosci. **34**(5), 709–728 (2009)

66. D.L. Kidder, T.R. Worsley, Phanerozoic Large Igneous Provinces (LIPs), HEATT (Haline Euxinic Acidic Thermal Transgression) episodes, and mass extinctions. Palaeogeogr. Palaeoclimatol. Palaeoecol. **295**(1), 162–191 (2010)
67. T. Kleine, K. Mezger, C. Münker, H. Palme, A. Bischoff, ^{182}Hf-^{182}W isotope systematics of chondrites, eucrites, and Martian meteorites: Chronology of core formation and early mantle differentiation in Vesta and Mars. Geochim. Cosmochim. Acta **68**, 2935–2946 (2004). doi:10.1016/j.gca.2004.01.009
68. A.S. Konopliv, S.W. Asmar, S.M. Foiles, Ö. Karatekin, D.C. Nunes, S.E. Smrekar, C.F. Yoder, M.T. Zuber, Mars high resolution gravity fields from MRO, Mars seasonal gravity, and other dynamical parameters. Icarus **211**, 401–428 (2011). doi:10.1016/j.icarus.2010.10.004
69. C. Köstler, Iterative solvers for modeling mantle convection with strongly varying viscosity. PhD Thesis, Friedrich-Schiller-Univ. Jena, http://www.geodyn.uni-jena.de, 2011
70. F.A. Kundell, A suggested pioneer organism for the Wächtershäuser origin of life hypothesis. Orig. Life Evol. Biosph. **41**(2), 175–198 (2011)
71. B. Langlais, M.E.P. Purucker, M. Mandea, The crustal magnetic field of Mars. J. Geophys. Res. 109, E02008 (2004)
72. K.D. Litasov, Physicochemical conditions for melting in the Earth's mantle containing a C-O-H fluid (from experimental data). Russ. Geol. Geophys. **52**, 475–492 (2011)
73. K. Lodders, B. Fegley, An oxygen isotope model for the composition of Mars. Icarus **126**, 373–394 (1997)
74. R. Lundin et al., Solar wind-induced atmospheric erosion at Mars: First results from ASPERA-3 on Mars Express. Science **305**, 1933–1936 (2004)
75. J.C. Marty, G. Balmino, J. Duron, P. Rosenblatt, S. Le Maistre, A. Rivoldini, V. Dehant, T. van Hoolst, Martian gravity field model and its time variations from MGS and ODYSSEY data. Planet. Space Sci. **57**, 350–363 (2009). doi:10.1016/j.pss.2009.01.004
76. S. Maruyama, M. Ikoma, H. Genda, K. Hirose, T. Yokoyama, M. Santosh, The naked planet Earth: Most essential pre-requisite for the origin and evolution of life. Geosci. Front. **4**(2), 141–165 (2013). ISSN 1674-9871. doi:10.1016/j.gsf.2012.11.001 URL http://www.sciencedirect.com/science/article/pii/S1674987112001272.
77. E. Mayr, *Populations, Species, and Evolution: An Abridgment of Animal Species and Evolution* (Havard University Press, Cambridge, MA, 1970)
78. M.T. McCulloch, V.C. Bennett, Progressive growth of the Earth's continental crust and depleted mantle: Geochemical constraints. Geochim. Cosmochim. Acta **58**, 4717–4738 (1994)
79. H.Y.J. McSween, Mars. In *Treatise on Geochemistry, Vol. 1, Meteorites, Comets and Planets*, ed. H.D. Holland, K.K. Turekian (Elsevier, Amsterdam, 2003), pp. 601–621
80. K. Mezger, V. Debaille, T. Kleine, Core formation and mantle differentiation on Mars. Space Sci. Rev. **174**(1–4), 27–48 (2013)
81. K. Mierdel, H. Keppler, J.R. Smyth, F. Langenhorst, Water solubility in aluminous orthopyroxene and the origin of the Earth's asthenosphere. Science **315**, 364–368 (2007). doi:10.1126/science.1135422
82. K.G. Miller, R.M. Sherrell, J.V. Browning, M.P. Field, W. Gallagher, R.K. Olsson, P.J. Sugarman, S. Tuorto, H. Wahyudi, Relationship between mass extinction and iridium across the Cretaceous-Paleogene boundary in New Jersey. Geology **38**(10), 867–870 (2010)
83. S.L. Miller, H.C. Urey, Organic compound synthesis on the primitive Earth. Science **130**, 245–251 (1959)
84. J.X. Mitrovica, and A.M. Forte, A new inference of mantle viscosity based upon joint inversion of convection and glacial isostatic adjustment data. Earth Planet. Sci. Lett. **225**(1), 177–189 (2004)
85. R.K. Mohapatra, S.V.S. Murty, Precursors of Mars: Constraints from nitrogen and oxygen isotopic compositions of Martian meteorites. Meteoritics Planet. Sci. **38**(2), 225–241 (2003). doi:10.1111/j.1945-5100.2003.tb00261.x
86. A. Morschhauser, M. Grott, D. Breuer, Crustal recycling, mantle dehydration, and the thermal evolution of Mars. Icarus **212**(2), 541–558 (2011). doi:10.1016/j.icarus.2010.12.028

87. M. Müller, Towards a robust Terra code. PhD Thesis, Friedrich-Schiller-Univ. Jena, http://www.geodyn.uni-jena.de, 2008
88. L.E. Nyquist, D.D. Bogard, C.-Y. Shih, A. Greshake, D. Stöffler, O. Eugster, Ages and geological histories of Martian meteorites. Space Sci. Rev. **96**, 105–164 (2001)
89. M. Ogawa, T. Yanagisawa, Two-dimensional numerical studies on the effects of water on Martian mantle evolution induced by magmatism and solid-state mantle convection. J. Geophys. Res. Planet. (1991–2012) **117**, E06004 (2012). doi:10.1029/2012JE004054
90. L.E. Orgel, The origin of life—a review of facts and speculations. Trends Biochem. Sci. **23**, 491–495 (1998)
91. L.E. Orgel, Self organizing biochemical cycles. Proc. Natl. Acad. Sci. USA **97**, 12503–12507 (2000)
92. L.E. Orgel, Prebiotic chemistry and the origin of the RNA world. Crit. Rev. Biochem. Mol. Biol. **39**(2), 99–123 (2004)
93. J.P. Poirier, *Introduction to the Physics of the Earth's Interior* (Cambridge University Press, Cambridge, 2000)
94. D.M. Raup, J.J. Sepkowski, Mass extinction in the marine fossil record. Science **215**, 1501–1503 (1982)
95. P.R. Renne, A.L. Deino, F.J. Hilgen, K.F. Kuiper, D.F. Mark, W.S. Mitchell, L.E. Morgan, R. Mundil, J. Smit, Time scales of critical events around the Cretaceous-Paleogene boundary. Science **339**(6120), 684–687 (2013)
96. M.A. Richards, W.-S. Yang, J.R. Baumgardner, H.-P. Bunge, Role of a low-viscosity zone in stabilizing plate tectonics: Implications for comparative terrestrial planetology. Geochem. Geophys. Geosys. **3**, 1040 (2001). doi:10.1029/2000GC000115
97. A. Rivoldini, T. van Hoolst, O. Verhoeven, A. Mocquet, V. Dehant. Geodesy constraints on the interior structure and composition of Mars. Icarus **213**(2), 451–472 (2011)
98. R.L. Rudnick, D.M. Fountain, Nature and composition of the continental crust: A lower crustal perspective. Rev. Geophys. **33**(3), 267–309 (1995). doi:10.1029/95RG01302
99. T. Ruedas, P.J. Tackley, S.C. Solomon, Thermal and compositional evolution of the Martian mantle: Effects of phase transitions and melting. Phys. Earth Planet. In. **216**, 32–58 (2013). doi:10.1016/j.pepi.2012.12.002.
100. K. Ruiz-Mirazo, J. Peretó, A. Moreno, A universal definition of life: autonomy and open-ended evolution. Orig. Life Evol. Biosph. **34**(3), 323–346 (2004)
101. R. Sabadini, B. Vermeersen, *Global Dynamics of the Earth* (Kluwer Academic Publishers, Dordrecht, 2004)
102. C. Sanloup, A. Jambon, P. Gillet, A simple chondritic model of Mars. Phys. Earth Planet. In. **112**, 43–54 (1999)
103. B. Schoene, J. Guex, A. Bartolini, U. Schaltegger, T.J. Blackburn, Correlating the end-Triassic mass extinction and flood basalt volcanism at the 100 ka level. Geology **38**(5), 387–390 (2010)
104. G. Schubert, D.L. Turcotte, T.R. Olson, *Mantle Convection in the Earth and Planets* (Cambridge University Press, Cambridge, 2001)
105. P. Schulte, L. Alegret, I. Arenillas, J.A. Arz, P.J. Barton, P.R. Bown, T.J. Bralower, G.L. Christeson, P. Claeys, C.S. Cockell, et al., The Chicxulub asteroid impact and mass extinction at the Cretaceous-Paleogene boundary. Science **327**(5970), 1214–1218 (2010)
106. S. Schumacher, D. Breuer, Influence of a variable thermal conductivity on the thermochemical evolution of Mars. J. Geophys. Res. **111**, E02006 (2006)
107. S.-Z. Shen, J.L. Crowley, Y. Wang, S.A. Bowring, D.H. Erwin, P.M. Sadler, C.-Q. Cao, D.H. Rothman, C.M. Henderson, J. Ramezani, et al., Calibrating the end-Permian mass extinction. Science **334**(6061), 1367–1372 (2011)
108. N.H. Sleep, D.K. Bird, E. Pope, Paleontology of Earth's mantle. Ann. Rev. Earth Planet. Sci. **40**, 277–300 (2012)
109. A.B. Smith, *Systematics and the Fossile Record: Documenting Evolutionary Patterns* (Blackwell Sci. Publ., Oxford, 1994)

110. D.E. Smith, M.T. Zuber, M.H. Torrence, P.J. Dunn, G.A. Neumann, F.G. Lemoine, S.K. Fricke, Time variations of Mars' gravitational field and seasonal changes in the masses of the polar ice caps. J. Geophys. Res. Planet. (1991–2012) **114**, E05002 (2009). doi:10.1029/2008JE003267
111. F. Sohl, T. Spohn, The interior structure of Mars: Implications from SNC meteorites. J. Geophys. Res. **102**(E1), 1613–1635 (1997)
112. K.F. Sprenke, L.L. Baker, Magnetization, paleomagnetic poles, and polar wander on Mars. Icarus **147**, 26–34 (2000). doi:10.1006/icar.2000.6439
113. F.D. Stacey, P.M. Davis, *Physics of the Earth*, 4th edn. (Cambridge University Press, Cambridge, 2009)
114. L. Stixrude, C. Lithgow-Bertelloni, Influence of phase transformations on lateral heterogeneity and dynamics in Earth's mantle. Earth Planet. Sci. Lett. **263**, 45–55 (2007)
115. K. Suyehiro, N. Takahashi, Y. Ariie, Y. Yokoi, R. Hino, M. Shinohara, T. Kanazawa, N. Hirata, H. Tokuyama, A. Taira, Continental crust, crustal underplating, and low-Q upper mantle beneath an oceanic island arc. Science **272**(5260), 390–392 (1996)
116. P.J. Tackley, Self-consistent generation of tectonic plates in time-dependent, three-dimensional mantle convection simulations. Part 1. Pseudoplastic yielding. Geochem. Geophys. Geosys. **1**, 2000GC000036 (2000a). doi:10.1029/2000GC000036
117. P.J. Tackley, Self-consistent generation of tectonic plates in time-dependent, three-dimensional mantle convection simulations. Part2. Strain weakening and asthenosphere. Geochem. Geophys. Geosys. **1**, 2000GC000043 (2000b). doi:10.1029/2000GC000043
118. S.R. Taylor, S.M. McLennan, *The Continental Crust: Its Composition and Evolution* (Blackwell Scientific, Oxford, 1985)
119. S.R. Taylor, S.M. McLennan, The geochemical evolution of the continental crust. Rev. Geophys. **33**, 241–265 (1995)
120. S.R. Taylor, S.M. McLennan, *Planetary Crusts: Their Composition, Origin and Evolution* (Cambridge University Press, Cambridge, 2009)
121. A.H. Treimann, The Nakhlite meteorites: Augite-rich igneous rocks from Mars. Chemie der Erde **65**, 203–270 (2004)
122. E.N. Trifonov, Vocabulary of definitions of life suggests a definition. J. Biomol. Struct. Dynam. **29**(2), 259–266 (2011)
123. R. Trompert, U. Hansen, Mantle convection simulations with rheologies that generate plate-like behavior. Nature **395**, 686–689 (1998). doi:10.1038/27185
124. M.H. Van Regenmortel, Logical puzzles and scientific controversies: the nature of species, viruses and living organisms. Syst. Appl. Microbiol. **33**(1), 1–6 (2010)
125. V. Vasas, E. Szathmáry, M. Santos, Lack of evolvability in self-sustaining autocatalytic networks constraints metabolism-first scenarios for the origin of life. Proc. Natl. Acad. Sci. **107**(4), 1470–1475 (2010)
126. G. Wächtershäuser, Before enzymes and templates: Theory of surface metabolism. Microbiol. Rev. **52**(4), 452–484 (1988)
127. G. Wächtershäuser, The case for the chemoautotrophic origin of life in an iron-sulfur world. Orig. Life Evol. Biosph. **20**(2), 173–176 (1990)
128. G. Wächtershäuser, Origin of life in an iron-sulfur world. In *The Molecular Origins of Life*, ed. by A. Brack (Cambridge University Press, Cambridge, 1998), pp. 206–218
129. G. Wächtershäuser, From volcanic origins of chemoautotrophic life to Bacteria, Archaea and Eukarya. Phil. Trans. R. Soc. Biol. Sci. **361**(1474), 1787–1808 (2006)
130. U. Walzer, R. Hendel, Mantle convection and evolution with growing continents. J. Geophys. Res. **113**, B09405 (2008). doi:10.1029/2007JB005459
131. U. Walzer, R. Hendel, Real episodic growth of continental crust or artifact of preservation? A 3-D geodynamic model. J. Geophys. Res. Solid Earth **118** (2013). doi:10.1002/jgrb.50150
132. U. Walzer, R. Hendel, J. Baumgardner, Viscosity stratification and a 3D compressible spherical shell model of mantle evolution. In *High Perf. Comp. Sci. Engng. '03*, ed. by E. Krause, W. Jäger, M. Resch (Springer, Berlin, 2003), pp. 27–67

133. U. Walzer, R. Hendel, J. Baumgardner, The effects of a variation of the radial viscosity profile on mantle evolution. Tectonophysics **384**, 55–90 (2004)
134. U. Walzer, R. Hendel, J. Baumgardner, Toward a thermochemical model of the evolution of the Earth's mantle. In *High Perf. Comp. Sci. Engng. '04*, ed. by E. Krause, W. Jäger, M. Resch (Springer, Berlin, 2005), pp. 395–454. doi:10.1007/3-540-26589-9_38
135. U. Walzer, R. Hendel, J. Baumgardner, Plateness of the oceanic lithosphere and the thermal evolution of the Earth's mantle. In *High Perf. Comp. Sci. Engng. '05*, ed. by W.E. Nagel, W. Jäger, M. Resch (Springer, Berlin, 2006), pp. 289–304
136. U. Walzer, T. Burghardt, R. Hendel, J. Kley, Towards a dynamical model of the Mars' evolution. In *High Perf. Comp. Sci. Engng. Stuttgart '09*, ed. by W.E. Nagel, D.B. Kröner, M.M. Resch (Springer, Berlin, 2010), pp. 485–510
137. A.K.W. Whaler, and M.E.P. Purucker, Martian magnetization-preliminary models. Leading Edge **22**(8), 763–765 (2003)
138. J.H. Whiteside, P.E. Olsen, T. Eglinton, M.E. Brookfield, R.N. Sambrotto, Compound-specific carbon isotopes from Earth's largest flood basalt eruptions directly linked to the end-Triassic mass extinction. Proc. Natl. Acad. Sci. **107**(15), 6721–6725 (2010). doi:10.1073/pnas.1001706107
139. P.B. Wignall, Y. Sun, D.P. Bond, G. Izon, R.J. Newton, S. Védrine, M. Widdowson, J.R. Ali, X. Lai, H. Jiang, et al., Volcanism, mass extinction, and carbon isotope fluctuations in the Middle Permian of China. Science **324**(5931), 1179–1182 (2009)
140. P. Wilf, K.R. Johnson, B.T. Huber, Correlated terrestrial and marine evidence for global climate changes before mass extinction at the Cretaceous-Paleogene boundary. Proc. Natl. Acad. Sci. **100**(2), 599–604 (2003)
141. W.-S. Yang, Variable viscosity thermal convection at infinite Prandtl number in a thick spherical shell. PhD Thesis, University of Illinois, Urbana-Champaign, 1997
142. C.F. Yoder, A.S. Konopliv, D.N. Yuan, E.M. Standish, W.M. Folkner, Fluid core size of Mars from detection of the solar tide. Science **300**, 299–303 (2003)
143. W.J. Zinsmeister, R.M. Feldmann, M.O. Woodburne, D.H. Elliot, Latest Cretaceous/earliest Tertiary transition on Seymour Island, Antarctica. J. Paleontol. **63**, 731–738 (1989)

Magma Ocean Cumulate Overturn and Its Implications for the Thermo-chemical Evolution of Mars

Ana-Catalina Plesa, Nicola Tosi, and Doris Breuer

Abstract Early in the history of terrestrial planets, the fractional crystallization of primordial magma oceans may have led to the formation of large scale chemical heterogeneities. These may have been preserved over the entire planetary evolution as suggested for Mars by the isotopic analysis of the so-called SNC meteorites. The fractional crystallization of a magma ocean leads to a chemical stratification characterized by a progressive enrichment in heavy elements from the core-mantle boundary to the surface. This results in an unstable configuration that causes the overturn of the mantle and the subsequent formation of a stable chemical layering. Assuming scaling parameters appropriate for Mars, we first performed simulations of 2D thermo-chemical convection in Cartesian geometry with the numerical code YACC. We ran a large set of simulations spanning a wide parameter space, by varying systematically the buoyancy ratio B, which measures the relative importance of chemical to thermal buoyancy, in order to understand the basic physics governing the magma ocean cumulate overturn and its consequence on mantle dynamics. Moreover, we derived scaling laws that relate the time over which chemical heterogeneities can be preserved (mixing time) and the critical yield stress (maximal yield stress that allows the lithosphere to undergo brittle failure) to the buoyancy ratio. We have found that the mixing time increases exponentially with B, while the critical yield stress shows a linear dependence.

A.-C. Plesa
WWU, Institute of Planetology, Muenster and German Aerospace Center, Institute of Planetary Research, Berlin
e-mail: a_ples01@uni-muenster.de

N. Tosi
Technical University Berlin, Department of Planetary Geodesy, Berlin
e-mail: nic.tosi@gmail.com

D. Breuer
German Aerospace Center, Institute of Planetary Research, Berlin
e-mail: doris.breuer@dlr.de

W.E. Nagel et al. (eds.), *High Performance Computing in Science and Engineering '13*, DOI 10.1007/978-3-319-02165-2_43,

We investigated then Mars early thermo-chemical evolution using the code GAIA in a 2D cylindrical geometry and assuming a detailed magma ocean crystallization sequence as obtained from geochemical modeling. A stagnant lid forms rapidly because of the strong temperature dependence of the viscosity. This immobile layer at the top of the mantle prevents the uppermost dense cumulates to sink, even when allowing for a plastic yielding mechanism. The convection pattern below this dense stagnant lid is dominated by small-scale structures caused by perturbations in the chemical component. Therefore, large-scale volcanic features observed over Mars surface cannot be reproduced. Assuming that the stagnant lid will break, the inefficient heat transport due to the stable density gradient and the entire amount of heat sources above the core-mantle-boundary (CMB) lead to a strong increase of the temperature to values that exceed the liquidus. We conclude that a fractionated global and deep magma ocean is difficult to reconcile with observations. Other scenarios like shallow or hemispherical magma ocean or even another freezing mechanism, which would reduce the strength of chemical gradient need to be considered.

1 Introduction

Thermal and chemical buoyancy are the most important dynamical processes in planetary mantles shaping their interior structure, influencing the heat transport and modelling the surface through tectonics and volcanism. Beside the Moon and Earth, Mars is the third body from which samples in form of meteorites are available. The so called SNC meteorites, named after the places where they were found (Shergotites, Nakhlites and Chassignites), reveal an important aspect of the thermo-chemical evolution of Mars. The large variation in lithophile radiogenic isotope systems such as ^{87}Rb$-^{86}$Sr, ^{147,146}Sm$-^{143,142}$Nd, and ^{176}Lu$-^{176}$Hf revealed by the analysis of these basaltic cumulates samples, suggest the existence of three to four distinct chemical reservoirs that have been preserved over the entire thermo-chemical evolution of Mars. Moreover, geochemical investigations of these samples offer constraints on their age and formation mechanisms. For example, for the incompatible trace-element-depleted shergottite source, located in the upper Martian mantle, the differentiation age has been estimated at $\sim$ 4.535 Ga [2], while the enriched end-member formed around 4.465 to 4.457 Ga. The nakhlites which have ϵ^{142}Nd similar to depleted shergottites but a higher ϵ^{182}W than all shergottites are best explained by a 4-stage model. In a first stage, the core segregation stage, tungsten is enriched in the core while Hf remains preferentially in the mantle. The second stage is characterized by the crystallization of garnet or majoritic garnet and explains the high values of ϵ^{182}W obtained for nakhlites. In a third stage, garnet or majoritic garnet segregation takes place to generate the observed decoupling of ϵ^{182}W and the shergottite-like ϵ^{142}Nd of nakhlites. The nakhlite crystallization takes place in stage 4 [4].

To explain the complex thermo-chemical processes needed to form such reservoirs and also their young age, most geochemical studies argue that fractional crystallization of a global magma ocean may reproduce the isotopic characteristic of the SNCs. In such scenario, the large amount of heat present in the early stage of the planetary evolution can cause melting of a significant part or even of the entire mantle of a terrestrial body producing a liquid magma ocean. The subsequent cooling of the mantle causes the magma ocean to freeze from the core-mantle boundary to the surface due to the steeper slope of the mantle adiabat compared to the slope of the solidus. Freezing of a magma ocean is a highly complex process which has been investigated by several authors e.g. [7, 25]. In [25] two main mechanisms for the freezing of a magma ocean are described: equilibrium crystallization and fractional crystallization, depending on the size of crystals formed during the freezing of the magma ocean and their settling velocity. In the equilibrium case, the magma ocean crystallizes without differentiating, i.e. the crystal size and the settling velocity are so small that the magma ocean freezes before crystal-melt separation can take place. In this case a homogeneous mantle is formed after the freezing of the magma ocean. In the case of fractional crystallization, instead, the residual liquid evolves producing denser cumulates as the crystallization proceeds, largely due to iron enrichment in the evolving magma ocean liquid [6]. After the magma ocean fractional crystallization has completed, the mantle is gravitationally unstable and hence prone to overturn.

Most mantle dynamics simulations assume no early magma ocean or an equilibrium crystallization, and therefore a homogeneous mantle as initial condition e.g. [12, 26]. A few studies have focused on the effects of a magma ocean fractional crystallization and the subsequent overturn on the thermal evolution of Mars [4,6,7]. However, the simulations presented in the above mentioned papers make very strong assumptions on the domain geometry and mantle rheology.

In this study we perform numerical simulations to investigate the consequences of the fractional crystallization of a whole mantle magma ocean on the subsequent thermo-chemical evolution of Mars.

2 Models and Methods

We model thermo-chemical mantle convection by solving the conservation equations of mass, momentum, energy and composition e.g., [23]. These equations are scaled using the thickness of the mantle D as length scale, the thermal diffusivity κ as time scale, the temperature drop across the mantle ΔT as temperature scale and the chemical density contrast $\Delta\rho$ as compositional scale. Assuming a Boussinesq fluid with Newtonian rheology and infinite Prandtl number, as appropriate for highly viscous media with negligible inertia, the non-dimensional equations of thermo-chemical convection read e.g. [5]:

$$\nabla \cdot \mathbf{u} = 0 \tag{1}$$

$$\nabla \cdot \left[\eta(\nabla \mathbf{u} + (\nabla \mathbf{u})^T)\right] + RaT\mathbf{e}_r - Ra_C C - \nabla p = 0 \tag{2}$$

$$\frac{\partial T}{\partial t} + \mathbf{u} \cdot \nabla T - \nabla^2 T - \frac{Ra_Q}{Ra} = 0 \tag{3}$$

$$\frac{\partial C}{\partial t} + \mathbf{u} \cdot \nabla C = 0 \tag{4}$$

where η is the viscosity, p the dynamic pressure, $\mathbf{u}$ the velocity, t the time, T the temperature, C the composition that can be translated as density and $\mathbf{e_r}$ is the unit vector in radial direction. The variables in the above equations are non-dimensionalized using the relationships to physical properties presented e.g. in [1]. In (2), Ra and Ra_C are the thermal and compositional Rayleigh numbers respectively, which are defined as:

$$Ra = \frac{\rho g \alpha \Delta T D^3}{\kappa \eta_{ref}} \tag{5}$$

$$Ra_C = \frac{\Delta \rho g D^3}{\kappa \eta_{ref}}, \tag{6}$$

where ρ is the reference density, $\Delta\rho$ the density contrast, g the gravitational acceleration, α the thermal expansivity, ΔT the temperature contrast between outer and inner boundaries, D the thickness of the mantle, κ the thermal diffusivity, and η_{ref} the reference viscosity. In (3), Ra_Q represents the Rayleigh number due to the internal heat sources and is given by:

$$Ra_Q = \frac{\rho^2 g \alpha Q_m D^5}{\kappa k \eta_{ref}}, \tag{7}$$

where Q_m is the mantle radioactive heat production and k is the thermal conductivity.

The viscosity is calculated according to the Arrhenius law for diffusion creep [11]. The non-dimensional formulation of the Arrhenius viscosity law for temperature dependent viscosity is given by [22]:

$$\eta(T) = \exp\left(\frac{E}{T + T_{surf}} - \frac{E}{T_{ref} + T_{surf}}\right) \tag{8}$$

where E is the activation energy, T_{surf} the surface temperature and T_{ref} the reference temperature. Due to the strongly temperature-dependent viscosity, a stagnant lid will rapidly form on top of the convecting mantle. Additionally we use a pseudo-plastic approach such that the stagnant lid undergoes plastic yielding if the convective

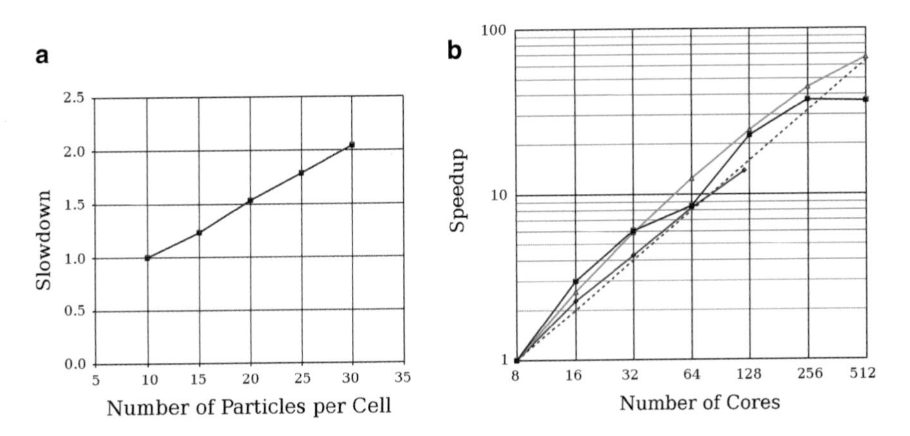

Fig. 1 (**a**) Slowdown factor when increasing the number of particles from 10 to 30 per computational cell and using a grid with $6 \cdot 10^5$ computational points; (**b**) Speedup factor obtained when using a grid with $6 \cdot 10^5$ computational points and 30 particles per cell: PFCluster1 (*red*), HLRN (*blue*), HLRS Cray Xe6 Hermit (*green*)

stresses are high enough to overcome the imposed yield stress σ_y. An effective viscosity can thus be defined as follows:

$$\frac{1}{\eta_{eff}} = \frac{1}{\eta(T)} + \frac{1}{\frac{\sigma_y}{2\dot{\varepsilon}}} \quad \text{with} \quad \sigma_y = \sigma_0 + z\frac{\partial \sigma}{\partial z} \tag{9}$$

where z is the depth and $\dot{\varepsilon}$ is the second invariant of the strain rate tensor e.g. [28].

We treat thermo-chemical convection by using a particle in cell method (PIC) e.g. [20, 27, 31]. To account for the advection of different chemical components (4), the motion of massless particles is first calculated from the velocity field computed on a fixed mesh. The compositional field is then interpolated from the distribution and composition of the particles. The method has the advantage over classical grid-based methods of being essentially free of numerical diffusion and to enable to naturally advect an arbitrary large number of different compositional fields by solving only one equation.

The performance for the Gaia code [9, 10, 19] was tested using the particle in cell method (PIC) for the advection of different chemical components.

Figure 1 shows performance tests when keeping the mesh fixed. First we calculate the slowdown factor when increasing the number of particles per computational cell. Doubling the number of particles per cell, the computational time increases by a factor of 1.5. Second the code performance is shown when using the same number of particles per computational cell and increasing the number of processors used. Using a $6 \cdot 10^5$ computational nodes grid, a good performance was reached when using up to 512 cores.

3 Results and Discussion

In all our models, if not otherwise specified, we use the Arrhenius viscosity law with an activation energy $E = 300$ kJ/mol and a surface temperature $T_{surf} = 250$ K. The initial temperature is set at the solidus apart from the uppermost region, where the temperature rapidly decreases to a cold surface temperature of 250 K. We therefore assume that a steam atmosphere which may have been formed due to efficient degassing during the magma ocean crystallization phase, has been rapidly lost as a consequence of the enhanced atmospheric loss processes during the early stage of the solar system. This is also supported by a recent study [15] in which the evolution of a magma ocean coupled with a convective-radiative atmospheric model has been investigated. They showed that for a distance from the sun equal to that of Mars the surface temperature drops to $\sim$ 500 K in about 1 Ma and is expected to decrease steadily as a consequence of water vapor condensation [15].

3.1 *Magma Ocean Cumulate Overturn: A Numerical Parameter Study*

In the first part of the project, we have investigated the basic physics governing the magma ocean cumulate overturn process. For a detailed description of the parameters used and the obtained results we refer the reader to [30]. Assuming scaling parameters appropriate for Mars, we first performed simulations of 2D thermo-chemical convection in Cartesian geometry with the numerical code YACC (Yet Another Convection Code, https://code.google.com/p/yacc-convection/) [13, 29]. We investigated systems heated either solely from below or from within by varying systematically the buoyancy ratio $B = Ra_C/Ra$, which measures the relative importance of chemical to thermal buoyancy, and the mantle rheology, by considering systems with constant, strongly temperature-dependent (using viscosity contrasts higher than 10 orders of magnitude) and pseudo-plastic viscosity. Since here we aim at deriving scaling laws based on the buoyancy ratio B and the thermal Rayleigh number Ra, in this part of the project we used steady-state simulations, i.e. no cooling boundary conditions and the amount of radiogenic heat sources is kept constant throughout the simulation.

The initial profiles for the temperature and composition (density) are shown in Fig. 2. Since the composition distribution in the mantle after the fractional crystallization of the initial magma ocean is poorly constrained, we use here for simplicity a linear unstable profile. This kind of profile has been widely used in numerical studies involving chemical heterogeneities and density distribution in a fractionally crystallized magma-ocean [8, 32].

Using systems with constant and strongly temperature-dependent viscosity, we have derived scaling laws that relate the time over which chemical heterogeneities can be preserved (mixing time) to the buoyancy ratio. After a first episode of mantle

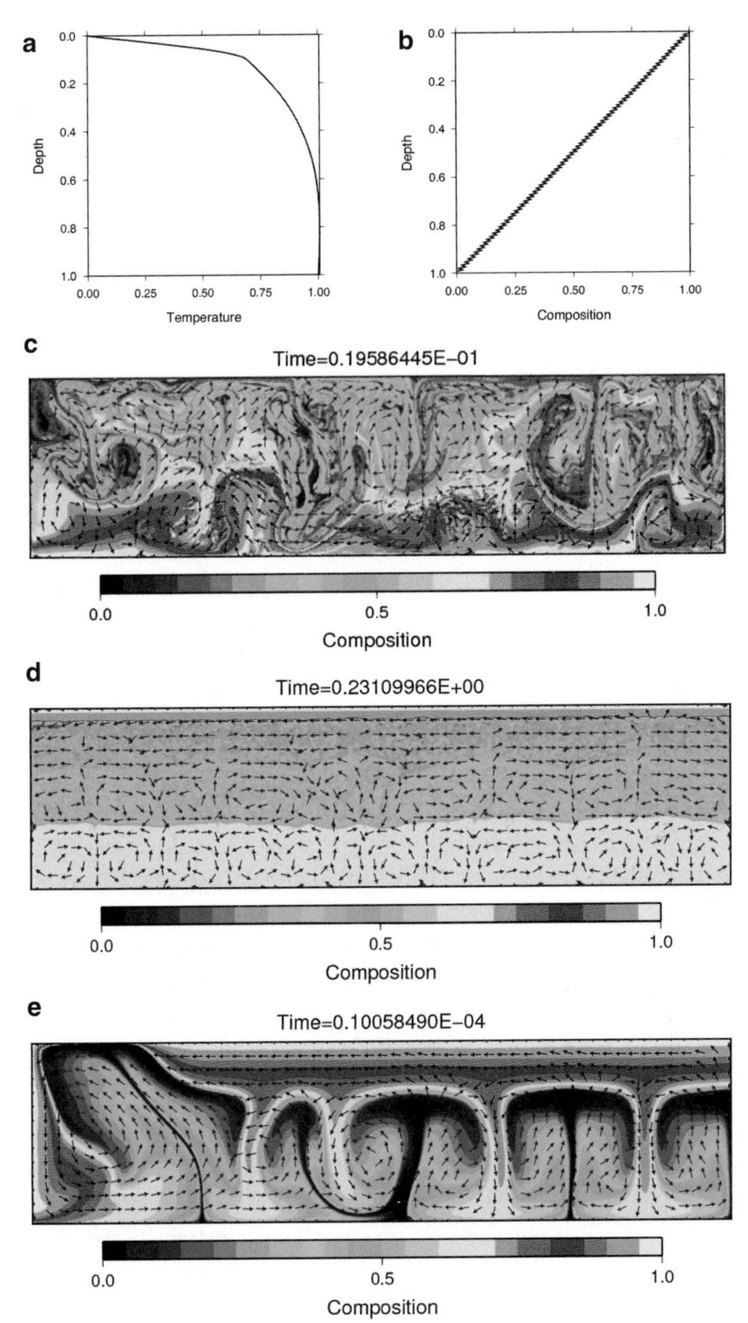

Fig. 2 Initial (**a**) temperature and (**b**) composition profiles. Snapshots for $B = 0.8$ and $Ra = 1e6$ for purely bottom heated systems showing: (**c**) constant viscosity, (**d**) temperature-dependent viscosity and (**e**) pseudo-plastic rheology cases

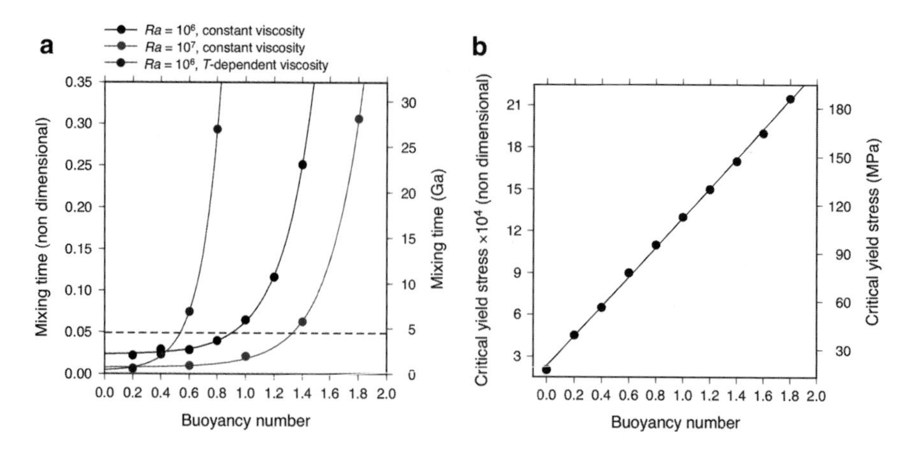

Fig. 3 Varying the buoyancy ratio, (**a**) the mixing time follows an exponential curve, while (**b**) the critical yield stress increases linearly with B

overturn which involves the whole mantle in the constant viscosity case and occurs only below the stagnant lid when the viscosity is strongly affected by temperature, a stable compositional gradient is reached. The mixing time over which the chemical heterogeneities can be preserved increases exponentially with B (Fig. 3). Increasing the Rayleigh number but keeping B constant will only cause a shift in the exponential function (compare red and black curves in Fig. 3a). Interestingly, for small values of B, when using a strong temperature-dependent viscosity, chemical heterogeneities will be erased faster compared to the constant viscosity case. This is due to the higher convective vigor, since the temperature decreases the viscosity by about two orders of magnitude. Therefore, thermally driven convection entirely removes the initial chemical gradient, and the system behaves as a purely thermal convection system. Increasing B, the picture changes, and the mixing time highly increases for the strongly temperature-dependent viscosity cases. The reason is that the temperature first evolves to a nearly conductive state—the chemical gradient being preserved. Small perturbations in both the thermal and chemical profile eventually increase causing the entire mixing of the mantle, albeit at a later time than in the constant viscosity case. For the temperature-dependent cases and $B > 1$, the chemical gradient dominates the convection and small or even no amount of material can be transported between different mantle layers. In purely internally heated cases this fact is even more enhanced, since instabilities that drive convection and favor mixing will only form at the top of the domain (in comparison, for a bottom heated system convection is driven from both top and bottom boundary layer instabilities).

Furthermore, we have investigated the behavior of systems when using a pseudo-plastic rheology. To determine the critical yield stress (maximal yield stress that allows the lithosphere to undergo brittle failure), we have used a constant yield stress value (in (9) $\partial\sigma/\partial z = 0$). The critical yield stress linearly increases with B (Fig. 3b), independently of the reference viscosity and therefore the Rayleigh number.

However, it should be noted, that if the Rayleigh number changes due to the geometry (i.e. a thicker mantle), then the yield stress will increase with increasing Rayleigh number. The value of the critical yield stress does not depend on the heating mode, i.e. for both bottom and purely internally heated systems the same scaling values have been found. Therefore the principal agent, which determines if the lithosphere undergoes plastic deformation is the buoyancy value B. Realistic yield stress values are on the order of several hundreds MPa [14], meaning that for small B values the lithosphere will most likely remain immobile (stagnant-lid convection).

3.2 *Magma Ocean Cumulate Overturn: Application to Mars*

In the second part of the project we use a detailed magma ocean crystallization sequence as obtained from geochemical modeling to investigate the cumulate overturn and its consequences on the subsequent thermo-chemical evolution of Mars. Compared to the study presented in the first part of the project, we use decaying radiogenic heat sources and cooling boundary conditions at the CMB. For a detailed description of the parameters used and the obtained results we refer the reader to [21]. Due to the lithophile nature of the radiogenic heat sources, they will preferentially be enriched in the late stage of the liquid magma ocean and therefore a high amount will be present in the near-surface layers. Following [7], we set the entire amount of heat sources in the uppermost 50 km. In previous studies e.g. [4, 7], where only the cumulate overturn has been investigated, the density profile which is established after the fractional crystallization of the magma ocean has been calculated at the solidus temperature. Since we are interested in the subsequent evolution, we recalculate the density profile at a reference temperature of 1 °C using the values from [6]. Moreover, we also account for an exothermic phase transition from olivine to γ-olivine and garnet and pyroxene to majorite present at about 14 GPa in the Martian mantle. Although this phase transition has been discussed in a previous work [6], it has not been dynamically modeled but assumed to have already taken place and therefore the density in the lower part of the mantle has been adjusted to a density which otherwise would only be present in the upper part of the mantle. We model the exothermic phase transition, such that material from the lower part of mantle becomes about 300 kg/m^3 lighter when rising above the phase transition depth, while material from the upper mantle becomes up to 500 kg/m^3 heavier when undergoing the phase transition. Due to the exothermic behavior, cold downwellings undergo the phase transition at a shallower depth than hot upwelling material. Additionally we use the extended Boussinesq approximation (EBA) which changes the energy equation (3) as follows:

$$\frac{\partial T}{\partial t} + \mathbf{u} \cdot \nabla T - Di(T + T_{surf})\mathbf{u}_r - \nabla^2 T - \frac{Di}{Ra}\Phi - H = 0 \qquad (10)$$

where Di is the dissipation number which depends on the thermal expansivity α, the gravity acceleration g, the mantle thickness D and the heat capacity of the mantle c_p:

$$Di = \frac{\alpha g D}{c_p} \tag{11}$$

The viscous dissipation, Φ, depends on the viscosity and the second invariant of the strain rate tensor as follows:

$$\Phi = 2\eta\dot{\varepsilon}^2 \tag{12}$$

Compared to previously published studies, we use a strongly temperature-dependent viscosity (8), which results in a viscosity contrast higher than 10 orders of magnitude, while previous studies used viscosity contrast of 3–5 orders of magnitude. Additionally we use a pseudo-plastic rheology with a depth-dependent yield stress ((9) with $\sigma_0 = 10^8$ Pa and $\partial\sigma/\partial z = 160$ Pa/m). Even though the lithosphere is allowed to undergo plastic yielding if the convective stresses overcome the depth-dependent yield stress and the buoyancy ratio $B = 3.3$ is high enough such that a realistic yield stress value may be used, this does not take place when using realistic rheological parameters (i.e. $E = 300$ kJ/mol and $T_{surf} = 250$ K) for the Arrhenius viscosity law. Responsible for this fact is the initial density profile, which in this case, apart from the unstable density gradient in the upper and lower part of the mantle, is overall stable. Indeed, using a linear profile as in the previous part of the project, the upper dense layer rapidly overturns. In this case both the hot upwellings and the return flow caused by sinking material from the upper layers, produce high enough convective stresses such that plastic deformation of the lithosphere takes place (Fig. 4). Note that in the linear density case, the chemical phase-transition has not been accounted for.

In previous studies e.g. [4, 7], a small viscosity contrast has been used, to reduce computational costs. However, this has major effects on the overturn efficiency. Using a low activation energy and a high surface temperature or imposing a fixed viscosity contrast by using a linearized form of the Arrhenius law (i.e. Frank-Kamenetskii approximation) both result in a small viscosity contrast and allow for surface mobilization. We have used the latter, to investigate the effects of the cumulate overturn on the interior dynamics, when all mantle layers are involved in the overturn process. When the uppermost layer sinks, it brings the densest material and the entire amount of heat sources to the CMB. Figure 5b shows the whole mantle overturn with the phase transition depth colored in black and white. Material from the upper part of the mantle becomes denser while material from the lower part of the mantle becomes lighter when crossing this depth. Due to the stable layering of the chemical component which suppresses efficient heat transport and, additionally, the high amount of heat sources concentrated above the CMB, the lower mantle heats up such that the temperature increases to values

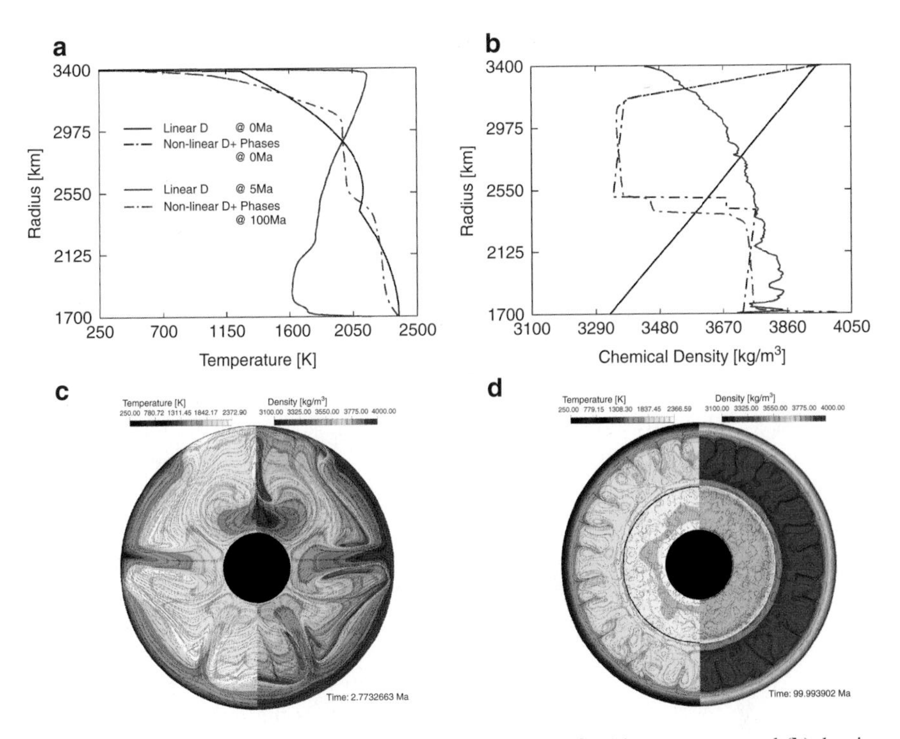

Fig. 4 Initial (*black*) and after the overturn (*magenta*) profiles for (**a**) temperature and (**b**) density when using an initial linear density profile (*full lines*) or a detailed magma ocean crystallization sequence with an additional exothermic phase transition (*dashed-dotted lines*). Further, temperature and density distribution for (**c**) the linear profile and (**d**) the detailed crystallization sequence

that exceed the liquidus (Fig. 5d)—the temperature above which the silicate mantle is entirely molten. The melt produced at the CMB is on the one hand iron-rich since the parent material was composed of the last crystallized layers of the initial magma ocean. Additionally this melt resides below the density-inversion depth [17], below which, due to the higher compressibility effect, melt is more dense than the surrounding silicate matrix. Therefore, most likely the melt will remain trapped in the lower mantle. It remains to be clarified if a molten lower layer in the interior of Mars is still compatible with the MOI (moment of inertia) factor of 0.3635 ± 0.0012 [24], which strongly depends on the density structure of the planet. Nevertheless, after the compositional gradient has reached a stable stratification, the upper mantle in the absence of radiogenic heating cools rapidly making it difficult to explain a long lasting volcanic activity [16]. This is also the case if using realistic rheological parameters for the Arrhenius viscosity law. Here, a whole mantle overturn does not take place and the entire amount of radioactive heat sources is trapped in the stagnant lid (Fig. 5a). Below the stagnant lid, a stable chemical gradient is established and the mantle then cools to a nearly conductive profile (Fig. 5c). Moreover, the convection structure is dominated by small-scale features

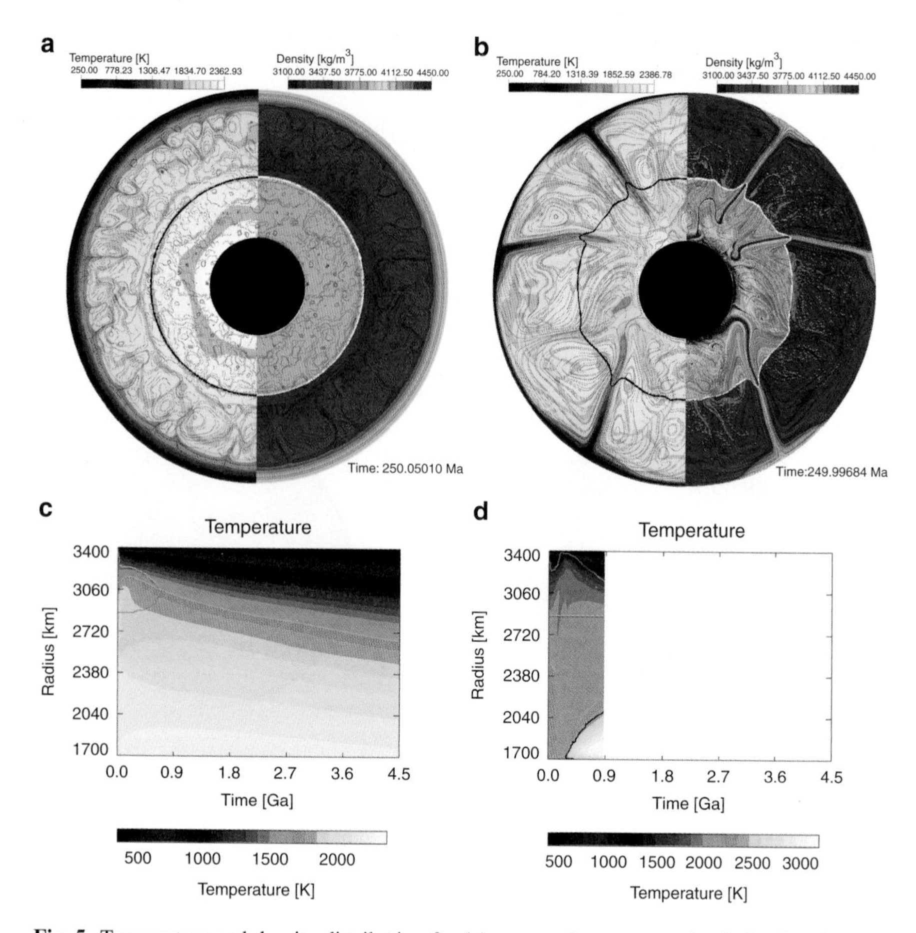

Fig. 5 Temperature and density distribution for (**a**) a case where we use the Arrhenius viscosity law and (**b**) when we use the Frank-Kamenetskii linearization. The temperature evolution for these cases is shown in (**c**) and (**d**) respectively, where the *green line* marks the region where the temperature lies above the *solidus and the black line* the region where the liquidus temperature is exceeded

caused by perturbations in the chemical gradient. This is clearly at odds with the large scale volcanic features of Mars, which are at best explained by strong and localized mantle upwellings e.g. [12, 22].

4 Conclusions and Outlook

In this work we have investigated the implications of a fractionally crystallized global and deep magma ocean overturn in the early stage of planetary evolution for the mantle dynamics. An important aspect is whether compositional convection suppresses or not the onset of thermal convection when the overturn has completed. In fact, thermal and compositional convection are two competing processes. After the compositional gradient has reached a stable stratification (i.e. after the end of the cumulate overturn) it mainly depends on the value of this gradient and on the value of the thermal gradient whether thermal convection can continue or not. Mars presents signs of late volcanism [16], which can hardly be explained if the convection in its mantle stops at a very early stage in the planetary evolution, e.g. after the magma ocean cumulate overturn.

In the first part of the project we have investigated under which conditions (i.e. for what buoyancy ratio) convection becomes dominated by the chemical gradient. We found that for B smaller than 1.0 the mantle undergoes mixing to some extent (the smaller B the faster mixing), however signatures of the chemical structure, that establishes shortly after the overturn may be preserved over 4.5 Ga. A problem of most of these models is that the critical yield stress is rather small and for realistic values the uppermost dense layer will remain trapped in the stagnant lid. This cannot explain the low surface density values from gravity-topography modeling [18]. Increasing B, the critical yield stress reaches more realistic values, however the chemical component starts to dominate suppressing thermal convection. This is at odds with the thermal history of Mars, which presents signs of active volcanism up to the recent past [16]. This study shows that an optimum value of B would most likely lie around 1.0 such that the chemical gradient would not entirely suppress thermal convection but would preserve geochemical reservoirs over the entire 4.5 Ga time-range. It remains to clarify if such a scenario can also explain the isotopic characteristic of the SNC meteorites.

In the second part of the project we have revisited the magma ocean cumulate overturn scenarios proposed in various previous studies to explain the isotopic variation of the SNC meteorites. We have shown that the simplifications applied in previous models highly overestimate the efficiency of the overturn. Although the buoyancy ratio B is higher than 3, the density profile, computed by using geochemical constraints from the fractional crystallization of the magma ocean, prevents a whole mantle overturn. Moreover, the problems revealed in the first part of this project remain. Assuming that a whole mantle overturn is possible considering other mechanisms (e.g. a big impact producing a strong thermal anomaly in the upper mantle layers, or a dense atmosphere lasting for at least 100 Ma), whole mantle overturn will bring the entire amount of heat sources, initially enriched in the uppermost 50 km, to the CMB. An overturn below the stagnant lid, keeps the radiogenic heat producing elements trapped in the 50 km thick surface layer. In both cases, in the absence of radiogenic heating, the upper mantle will rapidly cool, which is difficult to reconcile with the volcanic history of Mars. Since plastic yielding most likely will not take place (at least if no other additional mechanisms are present), additionally, the high density at the surface cannot be explained by gravity-topography modeling similar to the first part of the project for small values of B.

Considering the above presented studies, we can conclude that a fractional crystallized global and deep magma ocean as initial condition for modeling Mars' thermo-chemical history is unlikely to satisfy constraints like long lasting volcanism and maximum crustal density derived from planetary mission data and observations. The geochemical analysis of the SNC meteorites, however, require the existence of a deep magma ocean ($> 1,350$ km [3]) early in the planetary evolution of Mars for the formation of enriched shergottites. The nakhlite isotopic characteristic can at best be explained by the crystallization and segregation of majoritic garnet [4] which implies a fractionated crystallization of the early magma ocean. However, fractional crystallization fails to reproduce the ^{147}Sm/^{144}Nd isotopic ratio for the depleted shergottites [3]. It becomes clear that a more complex crystallization sequence must have taken place in order to explain both nakhlite and shergotite branches of the SNCs. Therefore other scenarios, for example a hemispherical magma ocean or even a mix of equilibrium and fractional crystallization would need to be tested.

Acknowledgements This research has been supported by the Helmholtz Association through the research alliance "Planetary Evolution and Life", by the Deutsche Forschungs Gemeinschaft (grant number TO 704/1-1) and by the High Performance Computing Center Stuttgart (HLRS) through the project Mantle Thermal and Compositional Simulations (MATHECO).

References

1. U. Christensen, Convection with pressure- and temperature-dependent non-Newtonian rheology. Geophys. J. Roy. Astron. Soc. **77**, 343–384 (1984)
2. V. Debaille, A. Brandon, Q. Yin, B. Jacobsen, Coupled ^{142}Nd—^{143}Nd evidence for a protracted magma ocean in Mars. Nature **450**(7169), 525–528 (2007)
3. V. Debaille, Q.-Z. Yin, A.D. Brandon, B. Jacobsen, Martian mantle mineralogy investigated by the ^{176}Lu—^{176}Hf and ^{147}Sm—^{143}Nd systematics of shergottites. Earth Planet. Sci. Lett. **269**, 186–199 (2008)
4. V. Debaille, A. Brandon, C. ONeill, Q. Yin, B. Jacobsen, Early martian mantle overturn inferred from isotopic composition of nakhlite meteorites. Nature Geosci. **2**(8), 548–552 (2009)
5. O. Grasset, E.M. Parmentier, Thermal convection in a volumetrically heated, infinite Prandtl number fluid with strongly temperature-dependent viscosity: Implications for planetary thermal evolution. J. Geophys. Res. **103**, 18,171–18,181 (1998)
6. L.T. Elkins-Tanton, E.M. Parmentier, P.C. Hess, Magma ocean fractional crystallization and cumulate overturn in terrestrial planets: Implications for Mars. Meteoritics Planet. Sci. 38, 12, 1753–1771, (2003).
7. L.T. Elkins-Tanton, S.E. Zaranek, E.M. Parmentier, P.C. Hess, Early magnetic field and magmatic activity on Mars form magma ocean cumulate overturn. Earth Planet. Sci. Lett. **236**, 1–12 (2005)
8. U. Hansen, D. Yuen, Extended-boussinesq thermalchemical convection with moving heat sources and variable viscosity. Earth Planet. Sci. Lett., **176**(3), 401–411 (2000)
9. C. Hüttig, K. Stemmer, Finite volume discretization for dynamic viscosities on Voronoi grids. Phys. Earth Planet. In. (2008). doi:10.1016/j.pepi.2008.07.007
10. C. Hüttig, K. Stemmer, The spiral grid: A new approach to discretize the sphere and its application to mantle convection. Geochem. Geophys. Geosyst. **9**, Q02018 (2008). doi:10.1029/2007GC001581
11. S. Karato, M.S. Paterson, J.D. Fitz Gerald, Rheology of synthetic olivine aggregates: influence of grain size and water. J. Geophys. Res. 8151–8176, **91** (1986)

12. T. Keller, P.J. Tackley, Towards self-consistent modelling of the Martian dichotomy: The influence of low-degree convection on crustal thickness distribution. Icarus **202**(2), 429–443 (2009)
13. S.D. King, C. Lee, P. van Keken, W. Leng, S. Zhong, E. Tan, N. Tosi, M. Kameyama, A community benchmark for 2D Cartesian compressible convection in the Earths mantle. Geophys. J. Int. **180**, 73–87 (2010). doi:10.1111/j.1365–246X.2009.04413.x
14. D.L. Kohlstedt, B. Evans, S.J. Mackwell, Strength of the lithosphere: Constraints imposed by laboratory experiments. J. Geophys. Res. **100**, B9, 17,587–17,602 (1995)
15. T. Lebrun, H. Massol, E. Chassere, A. Davaille, E. Marcq, P. Sarda, F. Leblanc, G. Brandeis, Thermal evolution of an early magma ocean in interaction with the atmosphere. J. Geophys. Res. E Planet (accepted) doi: 10.1002/jgre.20068
16. G. Neukum, R. Jaumann, H. Hoffmann, E. Hauber, J.W. Head, A.T. Basilevsky, B.A. Ivanov, S.C. Werner, S. van Gasselt, J.B. Murray, T. McCord, The HRSC Co-I Team: Recent and episodic volcanic and glacial activity on Mars revealed by the High Resolution Stereo Camera. Nature **432**, 971–979 (2004)
17. E. Ohtani, Y. Nagatab, A. Suzuki, T. Katoa, Melting relations of peridotite and the density crossover in planetary mantles. Chem. Geol. **120**, 207–221 (1995)
18. M. Pauer, D. Breuer, Constraints on the maximum crustal density from gravity-topography modeling: Applications to the southern highlands of Mars. Earth Planet. Sci. Lett. **276**(3–4), 253–261 (2008). doi:10.1016/j.epsl.2008.09.014
19. A.-C. Plesa, Mantle convection in a 2D spherical shell. Proceedings of the First International Conference on Advanced Communications and Computation (INFOCOMP 2011), ed. by C.-P. Rckemann, W. Christmann, S. Saini, M. Pankowska, pp. 167–172, Barcelona, Spain, 23–29 October 2011, ISBN: 978-1-61208-161-8, http://www.thinkmind.org/download.php?articleid=infocomp_2011_2_10_10002. Accessed 3 November 2011
20. A.-C.Plesa, N. Tosi, C. Hüttig, Thermo-chemical convection in planetary mantles: advection methods and magma ocean overturn simulations, in *Integrated Information and Computing Systems for Natural, Spatial, and Social Sciences*, ed. by C.-P. Rueckemann (IGI Global, Hershey, PA, 2013)
21. A.-C. Plesa, N. Tosi, D. Breuer, Can a fractionally crystallized magma ocean explain the thermo-chemical evolution of Mars?. Submitted to Earth and Planet. Sci. Lett.
22. J.H. Roberts, S. Zhong, Degree-1 convection in the Martian mantle and the origin of the hemispheric dichotomy. J. Geophys. Res. E Planet. **111** (2006)
23. G. Schubert, D.L. Turcotte, P. Olson, *Mantle Convection in the Earth and Planets* (Cambridge University Press, Cambridge, 2001)
24. F. Sohl, G. Schubert, T. Spohn, Geophysical constraints on the composition and structure of the Martian interior. J. Geophys. Res. **110**, E12008 (2005). doi:10.1029/2005JE002520
25. V. Solomatov, Fluid dynamics of a terrestrial magma ocean. In *Origin of the Earth and Moon*, ed. by R.M. Canup, K. Righter (University of Arizona Press, Tucson, 2000), p. 555
26. O. Šramek, S. Zhong, Martian crustal dichotomy and Tharsis formation by partial melting coupled to early plume migration. J. Geophys. Res. **117**(E01005) (2012). doi:10.1029/2011JE003867
27. P.J. Tackley, S.D. King, Testing the tracer ratio method for modeling active compositional fields in mantle convection simulations. Geochem. Geophys. Geosyst. **4**(4), 8302 (2003). doi:10.1029/2001GC000214
28. A. Ismail-Zadeh, P.J. Tackley, *Computational Methods for Geodynamics* (Cambridge University Press, New York, 2010)
29. N. Tosi, D.A. Yuen, O. Čadek, Dynamical consequences in the lower mantle with the post-perovskite phase change and strongly depth-dependent thermodynamic and transport properties. Earth Planet. Sci. Lett. **298**, 229–243 (2010). doi:10.1016/j.epsl.2010.08.001
30. N. Tosi, A.-C. Plesa, D. Breuer, Overturn and evolution of a crystallized magma ocean: a numerical parameter study for Mars. J. Geophys. Res. Planet. **118**, 1–17 (2013). doi:10.1002/jgre.20109

31. P.E. van Keken, S.D. King, H. Schmeling, U.R. Christensen, D. Neumeister, M.P. Doin, A comparison of methods for the modeling of thermochemical convection. J. Geophys. Res. **102**, 22477–22495 (1997)
32. S. Zaranek, E. Parmentier, Convective cooling of an initially stably stratied uid with temperature-dependent viscosity: Implications for the role of solid-state convection in planetary evolution. J. Geophys. Res. **109**(B3), B03,409 (2004)

Surface Tension, Large Scale Thermodynamic Data Generation and Vapor-Liquid Equilibria of Real Compounds

Stefan Eckelsbach, Svetlana Miroshnichenko, Gabor Rutkai, and Jadran Vrabec

1 Introduction

The surface tension of oxygen and nitrogen was calculated using molecular dynamics simulation. Due to the inhomogeneity of the system, the long range correction approach of Janeček [1] was used. The results regarding the temperature dependence of the surface tension were compared to simulation data by Neyt et al. [2] and experimental data.

Because of simpler statistical analogs of certain thermodynamic properties in different ensembles, often a particular ensemble is used to sample thermodynamic data. Due to the possibility to measure any thermodynamic property, which can be measured in a given statistical mechanical ensemble, in every other statistical ensemble, this is not a very efficient approach. A simulation framework based on this fact was applied to calculate a comprehensive data set. The results were compared to highly accurate equations of state that were parameterized to basically all available experimental data.

In the third part of this work, vapor-liquid equilibria (VLE) of binary mixtures were simulated, regarding four molecules of the cyanide group. Therefore, two new molecular force field models for cyanogen and cyanogen chloride were developed. The models of hydrogen cyanide and acetonitrile were taken from preceding work. The simulated phase equilibria were compared to experimental VLE data. Furthermore, excess properties were predicted for two liquid mixtures, i.e. dimethyl ether + water and ethylene oxide + water.

S. Eckelsbach · S. Miroshnichenko · G. Rutkai · J. Vrabec (✉)
Lehrstuhl für Thermodynamik und Energietechnik (ThEt), Universität Paderborn, Warburger Str. 100, 33098 Paderborn, Germany
e-mail: jadran.vrabec@upb.de

W.E. Nagel et al. (eds.), *High Performance Computing in Science and Engineering '13*, DOI 10.1007/978-3-319-02165-2_44,

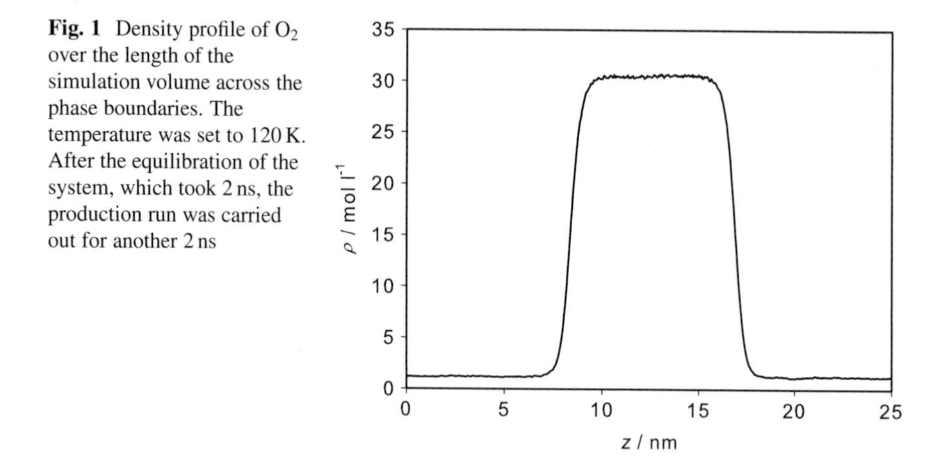

Fig. 1 Density profile of O_2 over the length of the simulation volume across the phase boundaries. The temperature was set to 120 K. After the equilibration of the system, which took 2 ns, the production run was carried out for another 2 ns

2 Surface Tension of Oxygen and Nitrogen

Two fluids, i.e. O_2 and N_2, were studied with respect to their surface tension. The surface tension is of importance for basically every process containing interface interactions, e.g. adhesion or adsorption. Thus it is also needed as an input property for flow simulations based on the continuum approach, e.g. Direct Numerical Simulation or Large Eddy Simulation.

Because of the direct simulation of the interface, a relatively large number of molecules is needed for such simulations, demanding substantial computing power, i.e. parallel computing and a good scalability of the code. To achieve this, the molecular dynamics simulations were carried out with our simulation code *ls*1 [3], which was designed to make efficient use of highly parallel execution.

For the present simulations, the systems were set up with a liquid phase surrounded by a vapor phase. The system thus forms two interfaces, where a direct calculation of the surface tension can be done. After an initial equilibration period, the density profile over the length of the simulation volume remained constant as shown exemplary in Fig. 1 for O_2 at a temperature of 120 K.

Due to the inhomogeneity of the simulated systems, the long range correction of Janeček [1] was used. The surface tension can be calculated by using the virial approach according to Irving and Kirkwood [4]

$$\gamma = \frac{1}{2A}\left(2\Pi_{zz} - \left(\Pi_{xx} + \Pi_{yy}\right)\right), \tag{1}$$

where A is the area of the interface and Π is the virial tensor, which is defined as

$$\Pi_{\alpha\beta} = \left\langle \frac{1}{2}\sum_{i=1}^{N}\sum_{j=1}^{N} r_{ij}^{\alpha} f_{ij}^{\beta} \right\rangle. \tag{2}$$

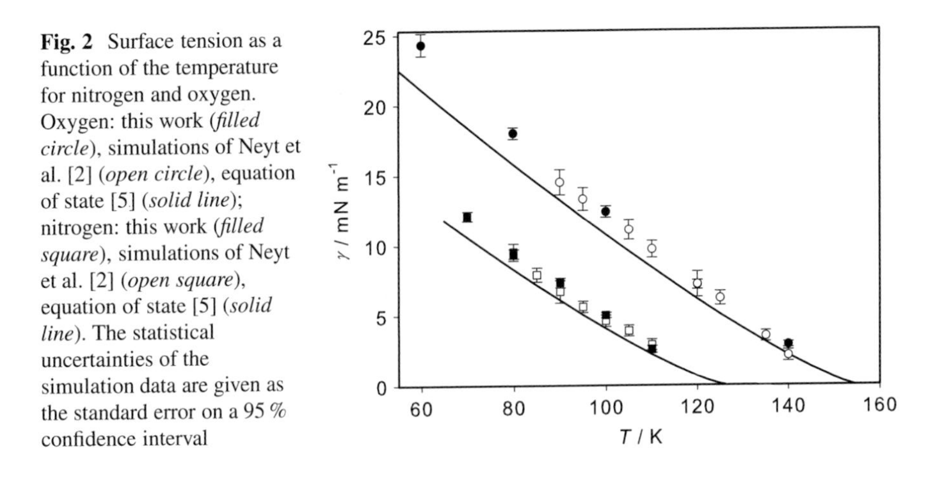

Fig. 2 Surface tension as a function of the temperature for nitrogen and oxygen. Oxygen: this work (*filled circle*), simulations of Neyt et al. [2] (*open circle*), equation of state [5] (*solid line*); nitrogen: this work (*filled square*), simulations of Neyt et al. [2] (*open square*), equation of state [5] (*solid line*). The statistical uncertainties of the simulation data are given as the standard error on a 95 % confidence interval

The upper indices α and β represent the x-, y- or z-directions of the distance vector $\boldsymbol{r}_{ij}$ and the force vector $\boldsymbol{f}_{ij}$, in each case between molecules i and j.

The results of the simulations are shown in Fig. 2. They are compared to the simulations of Neyt et al. [2] and experimental data from an equation of state [5]. The present data show a good agreement with the results of Neyt et al., however both molecular simulations slightly overestimate the surface tension in comparison to the experimental data.

3 Strategies for Large Scale Thermodynamic Data Generation

Almost every effort that aims at technological process design and optimization requires a sufficient amount of reliable thermodynamic data. Despite the extensive effort that was invested in measurements over more than a century, the data availability today is still surprisingly low. Among the about 1,000 chemical pure compounds that are in technological use, a complete thermodynamic knowledge is available for about ten [6], and advanced but still limited knowledge is available for less than 100 [7]. Considering mixtures, where the number of relevant systems is orders of magnitude larger, the data availability is much worse.

Recent progress in molecular simulation has shown that molecular force fields have powerful predictive capabilities with respect to thermodynamic data [8]. Such information can also be straightforwardly accessed for fluids and states which are experimentally too costly or difficult to investigate. Moreover, molecular simulation allows for the generation of comprehensive data sets containing consistent information on arbitrary thermodynamic properties at low cost. At first look, however, the generation of extensive data sets that contain as many different thermodynamic properties as possible may be cumbersome. This is mostly likely caused by the misconception standard textbook approaches in the molecular simulation literature tend to indicate, i.e. certain special techniques or specific statistical mechanical ensembles are required to obtain particular thermodynamic properties. It is true that certain properties have simpler statistical analogs in certain ensembles, nonetheless, any thermodynamic property that can be measured in a given statistical mechanical ensemble can also be measured in any other statistical mechanical ensemble. This is a direct consequence of the physical equivalence of various forms of the thermodynamic fundamental equation, e.g. entropy $S(N, V, E)$, internal energy $E(N, V, S)$, enthalpy $H(N, p, S)$, Helmholtz energy $F(N, V, T)$ or Gibbs energy $G(N, p, T)$, etc., where N is the number of particles, V is the volume, p is the pressure and T is the temperature.

Independent on which form is chosen, any other thermodynamic property can be obtained as a combination of derivatives of the chosen expression with respect to its independent variables. The thermodynamic potential $F/T(N, V, T)$ is a convenient option because it has the advantage that its independent variables can be easily controlled both in the laboratory environment and in molecular simulation. It is therefore a standard choice in empirical fundamental equations of state (FEOS) development [7]. The molecular simulation framework proposed by Lustig [9] offers any A_{mn} to be calculated concurrently from a single *NVT* ensemble simulation at a given state point for $m, n > 0$, where

$$\frac{\partial^{m+n} F/RT}{\partial^m (1/T) \partial^n (N/V)} \equiv A_{mn}. \tag{3}$$

The internal energy, the pressure, the heat capacities, the speed of sound and any other static thermodynamic property can be built up as a combination of these derivatives [7].

Because experimental measurements cannot satisfy the dire need for raw thermodynamic data in process design due to cost and time inefficiency, molecular simulation can be the answer for this problem. With respect to parameter settings, the *NVT* ensemble is particularly convenient, because the simulation does not require human interaction or significant attention to parameters during calculation. Thus, in order to produce an extensive data set consisting of as many state points as possible, the only limiting factor is the computation time. Each state point can be sampled with individual and independent calculations making this problem an ideal scenario for high performance computing. Two examples can be seen in Figs. 3 and 4 for nitrogen and acetone, respectively. It can be concluded that there is a good overall agreement between the FEOS and the prediction by molecular simulation. The full data set contains nine different A_{mn} at each state point, from which five are shown in Figs. 3 and 4, covering the entire homogeneous fluid region of technological interest. Due to the current state of high performance computing, such data set can already be obtained within a single day.

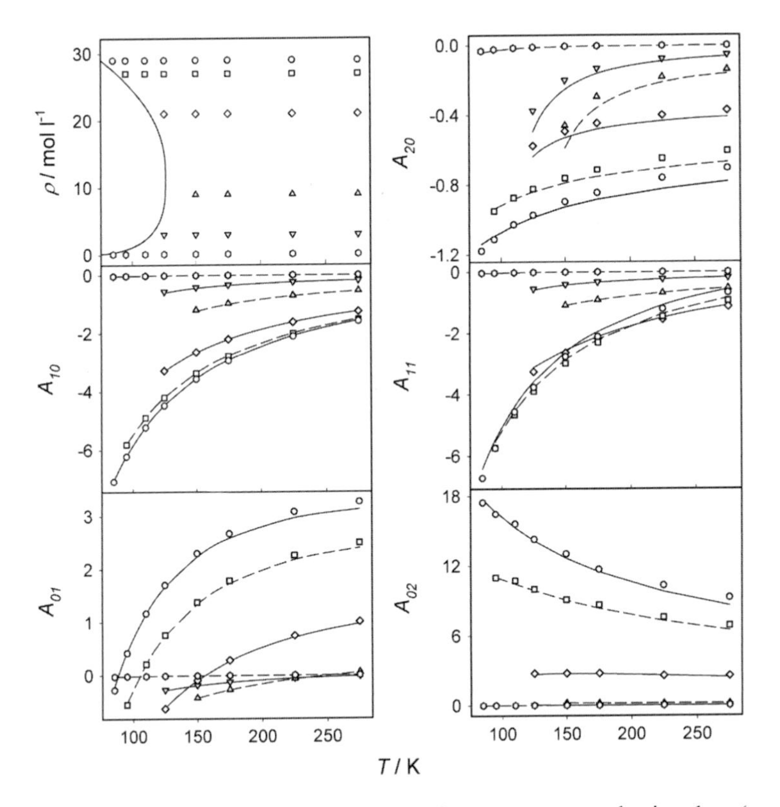

Fig. 3 Distribution of the simulated state points on the temperature vs. density plane (*top left*) and calculated A_{mn} values as a function of temperature from Monte Carlo molecular simulation (remaining subfigures) for nitrogen. State points were selected excluding the two phase region. Each symbol represents an individually simulated state point using the molecular model of Vrabec et al. [10]. The different symbols denote various isochores. The *lines* correspond to the FEOS by Span et al. [11]. The statistical uncertainty of the simulation data does not exceed the symbol size

4 Cyanides

Cyanides are chemical compounds that contain the cyano group N≡C–R. Cyanides, including hydrogen cyanide, cyanogen, cyanogen chloride and acetonitrile, are used industrially in the production of synthetic rubbers and plastics as well as in electroplating, case-hardening of iron and steel and fumigation. Many cyanides are highly toxic and have been used throughout history as a poison. Due to the safety issues associated with the handling of cyanides, molecular modeling and simulation can play a particularly important role for the investigation of the thermodynamic properties of these fluids.

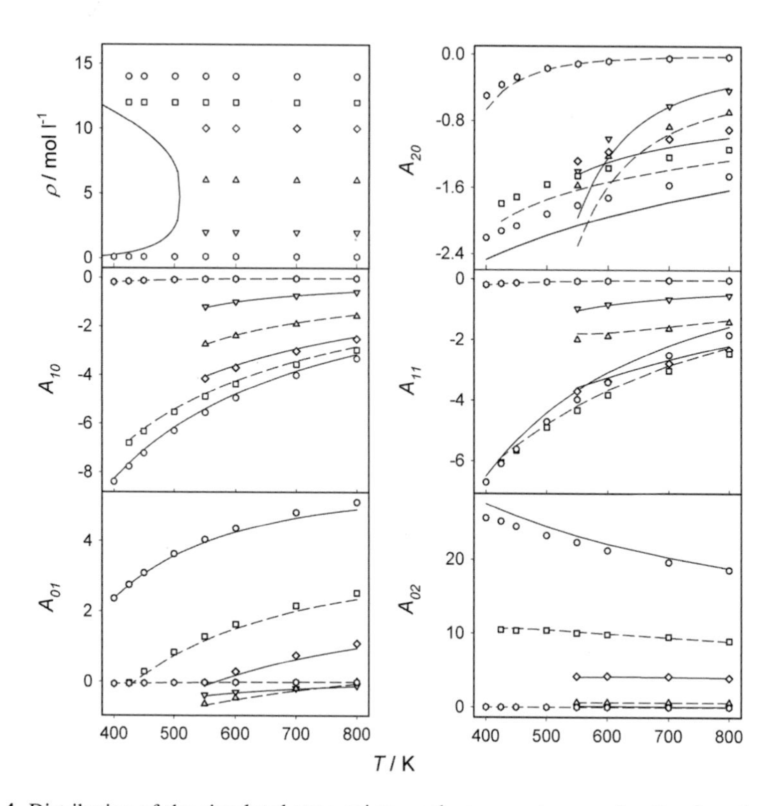

Fig. 4 Distribution of the simulated state points on the temperature vs. density plane (*top left*) and calculated A_{mn} values as a function of temperature from Monte Carlo molecular simulation (remaining subfigures) for acetone. State points were selected excluding the two phase region. Each symbol represents an individually simulated state point using the molecular model of Windmann and Vrabec [12]. The different symbols denote various isochores. The *lines* correspond to the FEOS Lemmon and Span [13]. The statistical uncertainty of the simulation data does not exceed the symbol size

The present molecular models include three groups of potential parameters. These are the geometric parameters, specifying the positions of different interaction sites, the electrostatic parameters, defining the polar interactions in terms of point charges, dipoles or quadrupoles, and the dispersive and repulsive parameters, determining the attraction by London forces and the repulsion by electronic orbital overlaps. Here, the Lennard-Jones (LJ) 12-6 potential [14, 15] was used to describe the dispersive and repulsive interactions. The total intermolecular interaction energy thus writes as

$$U = \sum_{i=1}^{N-1} \sum_{j=i+1}^{N} \left\{ \sum_{a=1}^{S_i^{\mathrm{LJ}}} \sum_{b=1}^{S_j^{\mathrm{LJ}}} 4\varepsilon_{ijab} \left[\left(\frac{\sigma_{ijab}}{r_{ijab}} \right)^{12} - \left(\frac{\sigma_{ijab}}{r_{ijab}} \right)^{6} \right] \right.$$
$$+ \sum_{c=1}^{S_i^{\mathrm{e}}} \sum_{d=1}^{S_j^{\mathrm{e}}} \frac{1}{4\pi\epsilon_0} \left[\frac{q_{ic}q_{jd}}{r_{ijcd}} + \frac{q_{ic}\mu_{jd} + \mu_{ic}q_{jd}}{r_{ijcd}^2} \cdot f_1(\boldsymbol{\omega}_i, \boldsymbol{\omega}_j) + \frac{q_{ic}Q_{jd} + Q_{ic}q_{jd}}{r_{ijcd}^3} \cdot f_2(\boldsymbol{\omega}_i, \boldsymbol{\omega}_j) \right.$$
$$\left.\left. + \frac{\mu_{ic}\mu_{jd}}{r_{ijcd}^3} \cdot f_3(\boldsymbol{\omega}_i, \boldsymbol{\omega}_j) + \frac{\mu_{ic}Q_{jd} + Q_{ic}\mu_{jd}}{r_{ijcd}^4} \cdot f_4(\boldsymbol{\omega}_i, \boldsymbol{\omega}_j) + \frac{Q_{ic}Q_{jd}}{r_{ijcd}^5} \cdot f_5(\boldsymbol{\omega}_i, \boldsymbol{\omega}_j) \right] \right\}, \tag{4}$$

where r_{ijab}, ε_{ijab}, σ_{ijab} are the distance, the LJ energy parameter and the LJ size parameter, respectively, for the pair-wise interaction between LJ site a on molecule i and LJ site b on molecule j. The permittivity of the vacuum is ϵ_0, whereas q_{ic}, μ_{ic} and Q_{ic} denote the point charge magnitude, the dipole moment and the quadrupole moment of the electrostatic interaction site c on molecule i and so forth. The expressions $f_x(\boldsymbol{\omega}_i, \boldsymbol{\omega}_j)$ stand for the dependence of the electrostatic interactions on the orientations $\boldsymbol{\omega}_i$ and $\boldsymbol{\omega}_j$ of the molecules i and j [16,17]. Finally, the summation limits N, S_x^{LJ} and S_x^{e} denote the number of molecules, the number of LJ sites and the number of electrostatic sites, respectively.

It should be noted that a point dipole may, e.g. when a simulation program does not support this interaction site type, be approximated by two point charges $\pm q$ separated by a distance l. Limited to small l, this distance may be chosen freely as long as $\mu = ql$ holds. Analogously, a linear point quadrupole can be approximated by three collinear point charges q, $-2q$ and q separated by l each, where $Q = 2ql^2$ [18].

For a given molecule, i.e. for a pure fluid throughout, the interactions between LJ sites of different type were defined here by applying the standard Lorentz-Berthelot combining rules [19,20].

All molecules studied in the present work do not exhibit significant conformational changes. Hence their internal degrees of freedom were neglected and the molecular models were chosen to be rigid. In a first step, the geometric data of the molecules, i.e. bond lengths, angles and dihedrals, were determined by QC calculations. Therefore, a geometry optimization was performed via an energy minimization using the GAMESS(US) package [21]. The Hartree-Fock level of theory was applied with a relatively small (6-31G) basis set. Intermolecular electrostatic interactions mainly occur due to the static polarity of single molecules that can well be obtained by QC. Here, the Møller-Plesset 2 level of theory was used that considers electron correlation in combination with the polarizable 6-31G basis set.

Fig. 5 shows the considered molecular models.

Two new molecular force field models are presented for cyanogen and cyanogen chloride, molecular models for hydrogen cyanide and acetonitrile were taken from preceding work [22,23]. The results for saturated densities obtained with the present models are compared to the available experimental data [24–28] in Fig. 6.

Fig. 5 Snapshot of hydrogen cyanide, cyanogen, cyanogen chloride and acetonitrile (*left to right*)

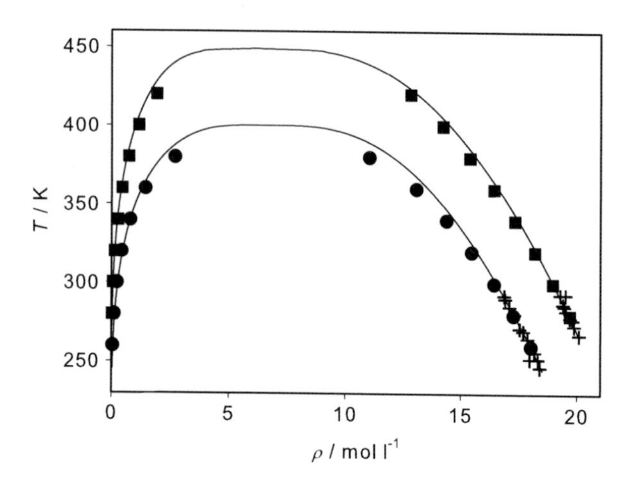

Fig. 6 Saturated densities of cyanogen (*filled circle*) and cyanogen chloride (*filled square*): experimental saturated liquid density [24–28] (*plus*); DIPPR correlation of experimental data [29] (*solid line*). The statistical uncertainties of the present simulation data are within symbol size

Based on the discussed molecular models, VLE data were predicted for the two binary systems cyanogen chloride + hydrogen cyanide and hydrogen cyanide + acetonitrile. This choice was driven by the availability of experimental VLE data for comparison. The vapor-liquid phase behavior of these binary mixtures was predicted by molecular simulation and compared to experimental data as well as the Peng-Robinson EOS [30].

Figure 7 shows the isothermal VLE of cyanogen chloride + hydrogen cyanide at 288.15 K from experiment [31], simulation and Peng-Robinson EOS. The experimental vapor pressure at a liquid mole fraction of $x_{\mathrm{NCCl}} = 0.5$ mol/mol was taken to adjust the binary parameter of the molecular model ($\xi = 1.023$) and of the Peng-Robinson EOS ($k_{ij} = 0.03$). It can be seen that the results obtained by molecular simulation agree well with the experimental results and the Peng-Robinson EOS.

Figure 8 shows the isobaric VLE of hydrogen cyanide + acetonitrile at 0.1013 MPa from the experiments by Jiang et al. [32], present simulations and the Peng-Robinson EOS. The binary parameters $\xi = 1.02$ and $k_{ij} = -0.0365$ were adjusted to the vapor pressure measured by Jiang et al. [32] at 319.25 K and a liquid mole fraction of $x_{\mathrm{HCN}} = 0.5174$ mol/mol. It can be seen in Fig. 8 that the

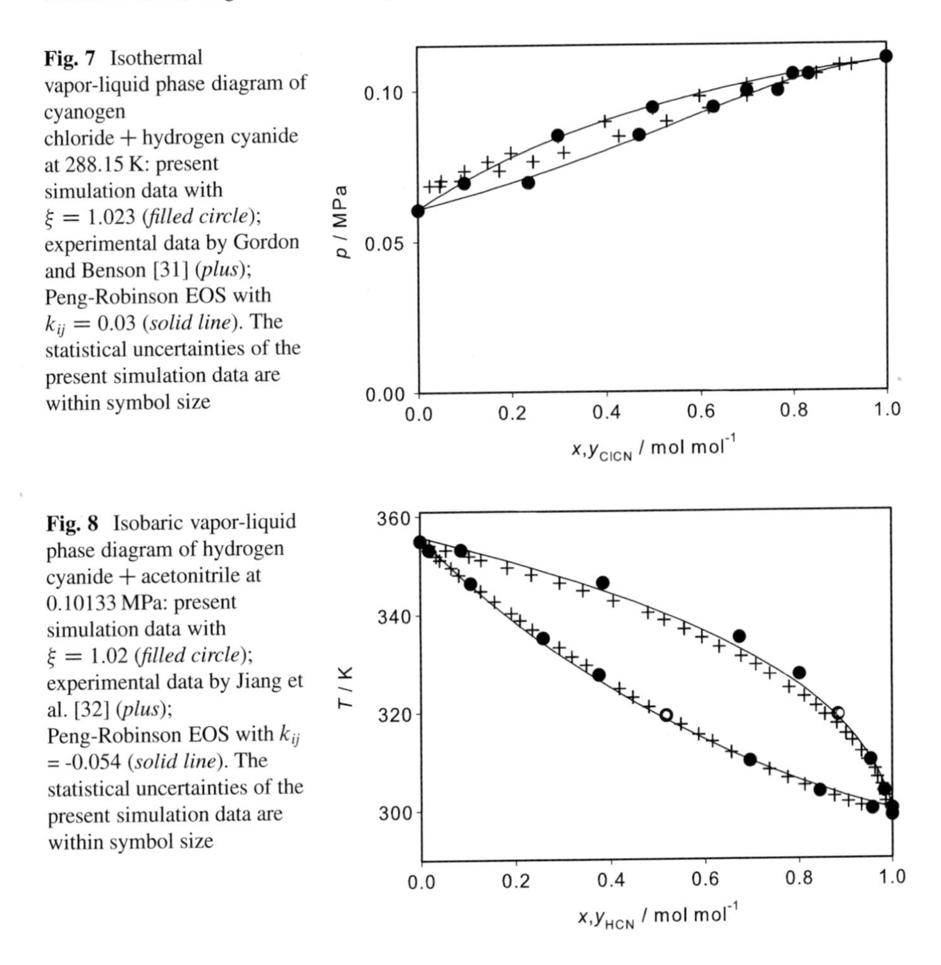

Fig. 7 Isothermal vapor-liquid phase diagram of cyanogen chloride + hydrogen cyanide at 288.15 K: present simulation data with $\xi = 1.023$ (*filled circle*); experimental data by Gordon and Benson [31] (*plus*); Peng-Robinson EOS with $k_{ij} = 0.03$ (*solid line*). The statistical uncertainties of the present simulation data are within symbol size

Fig. 8 Isobaric vapor-liquid phase diagram of hydrogen cyanide + acetonitrile at 0.10133 MPa: present simulation data with $\xi = 1.02$ (*filled circle*); experimental data by Jiang et al. [32] (*plus*); Peng-Robinson EOS with k_{ij} = -0.054 (*solid line*). The statistical uncertainties of the present simulation data are within symbol size

predictions obtained by molecular simulation with the binary interaction parameter $\xi = 1.02$ and those from the Peng-Robinson EOS with $k_{ij} = -0.0365$ agree well with the experimental data.

5 Excess Properties

Excess quantities are properties of mixtures which characterize their non-ideal behavior. It is a common practice to characterize liquid mixtures by means of excess mixing functions

$$y^E = y^{mix} - \sum_i x_i y_i, \tag{5}$$

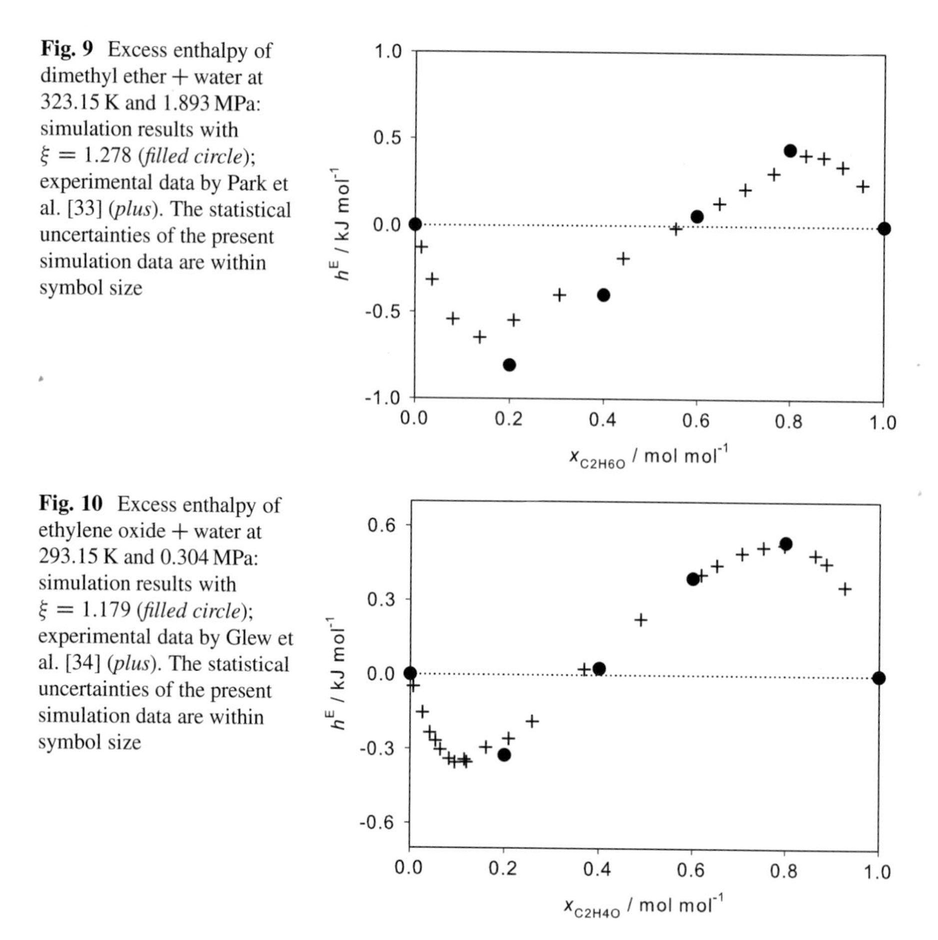

Fig. 9 Excess enthalpy of dimethyl ether + water at 323.15 K and 1.893 MPa: simulation results with $\xi = 1.278$ (*filled circle*); experimental data by Park et al. [33] (*plus*). The statistical uncertainties of the present simulation data are within symbol size

Fig. 10 Excess enthalpy of ethylene oxide + water at 293.15 K and 0.304 MPa: simulation results with $\xi = 1.179$ (*filled circle*); experimental data by Glew et al. [34] (*plus*). The statistical uncertainties of the present simulation data are within symbol size

where y^{mix} is the total measured property of the mixture, y_i the molar property of the pure components at the same temperature and pressure and x_i is the mole fraction of component i.

Molecular simulation was used here to predict excess properties of the liquid binary mixtures dimethyl ether + water and ethylene oxide + water. To determine excess properties, a straightforward approach was used. Three simulations at specified temperature and pressure were carried out, two for the pure substances and one for the mixture at a given composition.

Simulation results for the excess enthalpy are shown in Figs. 9 and 10 where they are compared with experimental results [33, 34]. It can be seen that the results obtained by molecular simulation agree well with the experimental results. The excess enthalpy of both mixtures has a peculiar behavior: at the same temperature and pressure, depending on the concentration, the excess enthalpy takes either a positive (endothermic mixing) or a negative (exothermic mixing) value. For both mixtures exothermal mixing effects are notably strong in the water-rich regions.

6 Conclusion

The surface tension of oxygen and nitrogen was predicted by molecular dynamics simulation. The results show a good agreement with other molecular simulation work and experimental data.

A simulation framework using a single NVT ensemble simulation to obtain as many thermodynamic values as possible was applied. With this approach, a large data set was created, yielding good agreement with FEOS data for acetone and nitrogen.

Molecular modeling and simulation was used to predict the VLE behaviour of the binary mixtures of cyanogen chloride + hydrogen cyanide and hydrogen cyanide + acetonitrile. For this task, two new molecular models were developed in this work. The used models were able to well reproduce the experimental data.

Also excess properties were studied by molecular simulation, namely the excess enthalpy of dimethyl ether + water and of ethylene oxide + water. Again, a good agreement with experimental data was found.

Acknowledgements We gratefully acknowledge support by Deutsche Forschungsgemeinschaft. This work was carried out under the auspices of the Boltzmann-Zuse Society (BZS) of Computational Molecular Engineering. The simulations were performed on the Cray XE6 (Hermit) at the High Performance Computing Center Stuttgart (HLRS).

References

1. J. Janeček, J. Phys. Chem. B **110**, 6264–6269 (2006)
2. J.-C. Neyt, A. Wender, V. Lachet, P. Malfreyt, J. Phys. Chem. B **115**, 9421–9430 (2011)
3. M. Buchholz, H.-J. Bungartz, J. Vrabec, J. Comput. Sci. **2**, 124–129 (2011)
4. J.H. Irving, J.G. Kirkwood, J. Chem. Phys. **18**, 817–829 (1950)
5. E.W. Lemmon, S.G. Penoncello, Adv. Cryog. Eng. **39**, 1927–1934 (1994)
6. K.E. Gubbins, N. Quirke, *Molecular Simulation and Industrial Application* (Gordon and Breach, Amsterdam, 1996)
7. R. Span, *Multiparameter Equations of State: An Accurate Source of Thermodynamic Property Data* (Springer, Berlin, 2000)
8. Industrial Fluid Properties Simulation Collective, http://fluidproperties.org
9. R. Lustig, Mol. Phys. **110**, 3041–3052 (2012)
10. J. Vrabec, J. Stoll, H. Hasse, J. Phys. Chem. B **105**, 12126–12133 (2001)
11. R. Span, E.W. Lemmon, R.T. Jacobsen, W. Wagner, A. Yokozeki, J. Phys. Chem. Ref. Data **29**, 1361–1433 (2000)
12. T. Windmann, J. Vrabec, in preparation (2013)
13. E.W. Lemmon, R. Span, J. Chem. Eng. Data **51**, 785–850 (2006)
14. J.E. Jones, Proc. R. Soc. **106A**, 441–462 (1924)
15. J.E. Jones, Proc. R. Soc. **106A**, 463–477 (1924)
16. M.P. Allen, D.J. Tildesley, *Computer Simulations of Liquids* (Oxford University Press, Oxford, 1987)
17. C.G. Gray, K.E. Gubbins, *Theory of Molecular Fluids. 1. Fundamentals* (Clarendon Press, Oxford, 1984)
18. C. Engin, J. Vrabec, H. Hasse, Mol. Phys. **109**, 1975–1982 (2011)

19. H.A. Lorentz, Ann. Phys. **12**, 127–136 (1881)
20. D. Berthelot, C. R. Acad. Sci. **126**, 1703–1706 (1898)
21. M.W. Schmidt, K.K. Baldridge, J.A. Boatz, S.T. Elbert, M.S. Gordon, J.H. Jensen, S. Koseki, N. Matsunaga, K.A. Nguyen, S. Shujun, T.L. Windus, M. Dupuis, A.M. Montgomery, J. Comput. Chem. **14**, 1347–1363 (1993)
22. B. Eckl, J. Vrabec, H. Hasse, J. Phys. Chem. B **112**, 12710–12721 (2008)
23. S. Deublein, P. Metzler, J. Vrabec, H. Hasse, Mol. Simul. **39**, 109–118 (2012)
24. R.P. Cook, P.L. Robinson, J. Chem. Soc. **19**, 1001–1005 (1935)
25. F. Ullmann, W. Gerhartz, Y.S. Yamamoto, *Ullmann's Encyclopedia of Industrial Chemistry*, 5th edn. (VCH Publishers, Weinheim, 1985)
26. R.C. Weast, *Handbook of Chemistry and Physics*, 62nd edn. (The Chemical Rubber Co., Cleveland, 1981)
27. N.V. Steere, *Handbook of Laboratory Safety*, 2nd edn. (CRC Press, Boca Raton, 1971)
28. N.I. Sax, *Dangerous Properties of Industrial Materials*, 6th edn. (Van Nostrand Reinhold Co., New York, 1984)
29. R.L. Rowley, W.V. Wilding, J.L. Oscarson, Y. Yang, N.A. Zundel, T.E. Daubert, R.P. Danner, *DIPPR Information and Data Evaluation Manager for the Design Institute for Physical Properties* (AIChE, New York, 2011) (Version 5.0.2)
30. D.Y. Peng, D.B. Robinson, Ind. Eng. Chem. Fundam. **15**, 59–64 (1976)
31. A.R. Gordon, G.C. Benson, Can. J. Res. B **24**, 285–291 (1946)
32. W. Jiang, Y. Zhang, M. Cao, S. Han, Zhejiang-Daxue-Xuebao **3**, 25–41 (1983)
33. S.-J. Park, K.-J. Han, J. Gmehling, J. Chem. Eng. Data **52**, 1814–1818 (2007)
34. D.N. Glew, H. Watts, Can. J. Chem. **49**, 1830–1840 (1971)

Molecular Modelling and Simulation of Electrolyte Solutions, Biomolecules, and Wetting of Component Surfaces

M. Horsch, S. Becker, J.M. Castillo, S. Deublein, A. Fröscher, S. Reiser, S. Werth, J. Vrabec, and H. Hasse

Abstract Massively-parallel molecular dynamics simulation is applied to systems containing electrolytes, vapour-liquid interfaces, and biomolecules in contact with water-oil interfaces. Novel molecular models of alkali halide salts are presented and employed for the simulation of electrolytes in aqueous solution. The enzymatically catalysed hydroxylation of oleic acid is investigated by molecular dynamics simulation taking the internal degrees of freedom of the macromolecules into account. Thereby, Ewald summation methods are used to compute the long range electrostatic interactions. In systems with a phase boundary, the dispersive interaction, which is modelled by the Lennard-Jones potential here, has a more significant long range contribution than in homogeneous systems. This effect is accounted for by implementing the Janeček cutoff correction scheme. On this basis, the HPC infrastructure at the Steinbuch Centre for Computing was accessed and efficiently used, yielding new insights on the molecular systems under consideration.

1 Introduction

Molecular simulation provides detailed information on processes on a level which is otherwise inaccessible, on the basis of physically realistic models of the intermolecular interactions. However, despite all efforts to keep these models as simple

M. Horsch (✉) · S. Becker · J.M. Castillo · S. Deublein · A. Fröscher · S. Reiser · S. Werth · H. Hasse
TU Kaiserslautern, Lehrstuhl für Thermodynamik, Erwin-Schrödinger-Str. 44, 67663 Kaiserslautern, Germany
e-mail: martin.horsch@mv.uni-kl.de

J. Vrabec
Universität Paderborn, Lehrstuhl für Thermodynamik und Energietechnik, Warburger Str. 100, 33098 Paderborn, Germany

W.E. Nagel et al. (eds.), *High Performance Computing in Science and Engineering '13*, DOI 10.1007/978-3-319-02165-2_45,

as possible, such simulations are extremely time consuming and belong to the most demanding applications of high performance computing.

The scientific computing project *"Molecular simulation of static and dynamic properties of electrolyte systems, large molecules and wetting of component surfaces"* (MOCOS) aims at advancing the state of the art by developing the molecular simulation codes *ms2* [1] as well as *ls1 mardyn* [2, 3], and by applying advanced simulation algorithms to complex molecular systems, employing the HPC resources at the Steinbuch Centre for Computing (SCC) in Karlsruhe. Various particular topics of relevance to the project are discussed in detail in dedicated publications to which the interested reader is referred here, concerning modelling and simulation of electrolyte solutions [4, 5], hydrogels [6, 7], and mixtures including hydrogen bonding fluids [8–10], as well as interfacial phenomena on small length scales [11, 12] and the role of long-range interactions in heterogeneous systems [12, 13].

The present article gives an overview, addressing the three main subjects of the MOCOS project by presenting recent scientific results which were facilitated by the computing resources at SCC: Sect. 2 discusses molecular model development for ions in aqueous solution and Sect. 3 reports on molecular simulation of large molecules, i.e. of enzymes and fatty acids. Interfacial phenomena such as the wetting behaviour of fluids in contact with a solid surface are discussed in Sect. 4 and a conclusion is given in Sect. 5.

2 Molecular Modelling of Aqueous Electrolyte Solutions

2.1 *Force Field Development for Alkali Cations and Halide Anions*

Aqueous electrolyte solutions play an important role in many industrial applications and natural processes. The general investigation of electrolyte solutions is, hence, of prime interest. Their simulation on the molecular level is computationally expensive as the electrostatic long range interactions in the solution have to be taken into account by time consuming algorithms, such as Ewald summation [14].

The development of ion force fields in aqueous solutions is a challenging task because of the strong electrostatic interactions between the ions and the surrounding water molecules. Previously published parameterization strategies for the adjustment of the ion force fields of all alkali cations and halide anions yield multiple parameter sets for a single ion [15, 16] or employ on additional parameters which have to be adjusted to thermodynamic properties of electrolyte solutions under particular conditions [17].

The recent study of Deublein et al. [18], however, succeeded in obtaining one unique force field set for all alkali and halide ions in aqueous solution. Thereby, the ions are modelled as Lennard-Jones (LJ) spheres with a point charge ($\pm 1\ e$) in their centre of mass. Hence, the ion force fields have two adjustable parameters, namely the LJ size parameter σ and the LJ energy parameter ϵ. The σ parameter of the ions was adjusted to the reduced liquid solution density. The LJ energy parameter showed only a minor influence on the reduced density and was estimated to be $\epsilon = 100$ K for all anions and cations [18].

In the present work, the influence of the LJ energy parameter on the self-diffusion coefficient of the alkali cations and the halide anions in aqueous solutions as well as the position of the first maximum of the radial distribution function (RDF) of water around the ions was investigated systematically. Based on these results, a modified value is proposed for the LJ energy parameter.

The new ϵ_i parameter of the ion force fields is determined by a two step parametrization strategy. First, the LJ energy parameter of the ion force fields is adjusted to the self-diffusion coefficient of the ions in aqueous solution. Subsequently, the dependence of the position of the first maximum in the RDF of water around the ions on ϵ_i is used to restrict the parameter range derived by considering the self-diffusion coefficient.

2.2 *Methods and Simulation Details*

In the present study, the self-diffusion coefficient of the ions in aqueous solutions is determined in equilibrium molecular dynamics by the Green-Kubo formalism. In this formalism, the self-diffusion coefficient is related to the time integral of the velocity autocorrelation function [19]. The radial distribution function $g_{i-\mathrm{O}}(r)$ of water around the ion i is defined by

$$g_{i-\mathrm{O}}(r) = \frac{\rho_\mathrm{O}(r)}{\rho_{\mathrm{O,bulk}}} \tag{1}$$

where $\rho_\mathrm{O}(r)$ is the local density of water as a function of the distance r from the ion i and $\rho_{\mathrm{O,bulk}}$ is the density of water molecules in the bulk phase. In this case, the position of the water molecules is represented by the position of the oxygen atom. The radial distribution functions are evaluated by molecular dynamics (MD) simulation as well.

The calculation of the self-diffusion coefficient by the Green-Kubo formalism is time and memory consuming. The determination of D_i and $g_{i-\mathrm{O}}(r)$ in electrolyte solutions is considerably more expensive due to additional time consuming algorithms, e.g. Ewald summation [14], required for permitting a truncation of the long range electrostatic interactions; however, it should be noted that a completely explicit evaluation of all pairwise interactions would be even much more expensive.

In a first step, the density of the aqueous alkali halide solution was determined in an isobaric-isothermal (*NpT*) MD simulation at the desired temperature and pressure. The resulting density was used in a canonical (*NVT*) MD simulation at the same temperature and composition of the different alkali halide solutions. In this run, the self-diffusion coefficient of the ions and the radial distribution function were determined. In case of the calculation of D_i, the MD unit cell with periodic boundary conditions contained $N = 4{,}500$ molecules, both for the NpT and the NVT simulation run.

For the evaluation of the RDF, there were $N = 1{,}000$ molecules in the simulation volume, i.e. 980 water molecules, 10 alkali cations and 10 halide anions. The radial distribution function was sampled in the NVT simulation within a cutoff radius of 15 Å with 500 bins.

2.3 *Self-Diffusion Coefficients and Radial Distribution Functions*

The self-diffusion coefficient D_i and the position of the first maximum $r_{\mathrm{max},1}$ of the RDF $g_{i-\mathrm{O}}(r)$ of water around the ions was investigated for all alkali cations and halide anions in aqueous solution for $\epsilon = 200$ K. These data were calculated at high dilution so that correlated motions and ion pairing between the cations and anions were avoided. Hence, D_i and $r_{\mathrm{max},1}$ are independent of the counter-ion in solution.

The results for the self-diffusion coefficient D_i are shown in Fig. 1. The overall agreement of the simulation results with the experimental data is excellent. The deviations are below 10 % for all ions, except for the sodium cation, where the deviation is about 20 %.

These simulation results also follow the qualitative trends from experiment, i.e. D_i increases with cation and anion size, respectively. This ion size dependence is directly linked to the electrostatic interaction between the ions and water. In aqueous solution, the cations and anions are surrounded by a shell of electrostatically bonded water molecules (hydration shell). The ions diffuse together with their hydration shell within the bulk water. For small ions, the hydration shell is firmly attached to the ion. Hence, the effective radius, that typically dominates ion motion, is larger for smaller ions than for larger ions, where the hydration shell is less pronounced.

The results of $r_{\mathrm{max},1}$ are shown in Fig. 2. In case of the alkali cations, the simulation results are within the range of the experimental data, except for Na^+ where the deviation from the experimental data is 5.3 %. For the halide anions, only the simulation result for $r_{\mathrm{max},1}$ of the RDF of water around the iodide anion is within the range of the experimental data. The deviations of the simulation results for $r_{\mathrm{max},1}$ from the experimental data are 12.1 % around F^-, 6.5 % around Cl^-, and 2.9 % around Br^-.

Comparing D_i and $r_{\mathrm{max},1}$ of the cesium cation and the fluoride anion, which have almost the same size, it can be seen that Cs^+ diffuses faster in aqueous solution and the water molecules of the hydration shell around F^- are closer to the ion. This can be attributed to the different orientations of the water molecules around the oppositely charged ions. The water molecules are able to build a stronger attached hydration shell around the fluoride ion which is closer to the ion.

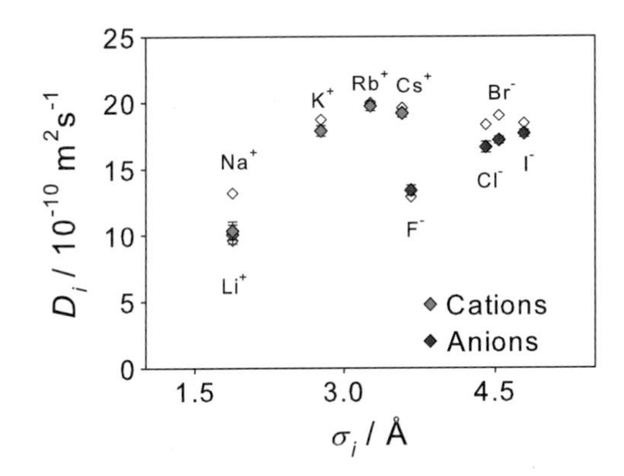

Fig. 1 Self-diffusion coefficient of alkali cations and halide anions in aqueous solutions (x_S = 0.018 mol/mol) at T = 298.15 K and p = 1 bar. Simulation results (*full symbols*) are compared to experimental data [20] (*empty symbols*)

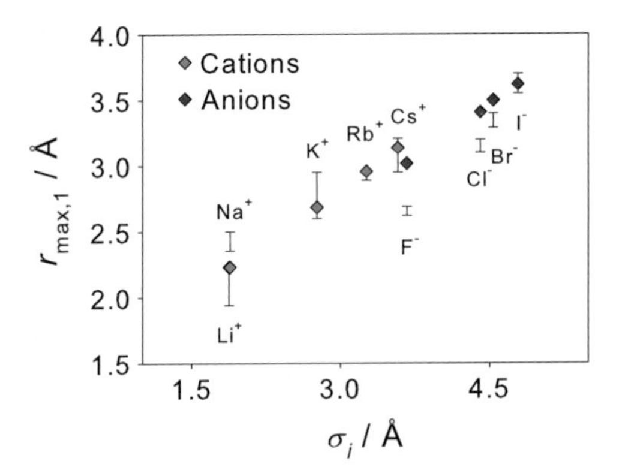

Fig. 2 Position of the first maximum $r_{max,1}$ of the RDF $g_{i-O}(r)$ of water around the alkali cations and halide anions in aqueous solutions (x_S = 0.01 mol/mol) at T = 293.15 K and p = 1 bar. Present simulation data (*full symbols*) are compared to the range of experimental $r_{max,1}$ data [21,22] (*vertical lines*)

2.4 Computational Demands

The molecular simulations in Sect. 2 were carried out with the MPI based molecular simulation program *ms2*, which was developed in our group, cf. Deublein et al. [1]. The total computing time for determining the self-diffusion coefficient of ions in aqueous solutions was 138 h on 36 CPUs (48 h for the *NpT* run and 90 h for the *NVT* run). For these simulations a maximum virtual memory of 1.76 GB was used.

For the evaluation of the radial distribution function of water around the ions a total computing time of 31 h on 32 CPUs (10 h for the *NpT* run and 21 h for the *NVT* run) was required.

3 Simulation of Biomolecules with Internal Degrees of Freedom

3.1 Catalysed Hydroxylation of Unsaturated Fatty Acids

Producing polymers from regenerative feedstock is a highly interesting alternative to polymers made of naphtha. The first step for obtaining biopolymers is the synthesis and study of all possible basic materials that can be used as building blocks. Such materials can for instance be obtained by an enzymatically catalysed reaction of unsaturated fatty acids to dihydroxy-fatty acids.

In nature, there are only mixtures saturated and unsaturated fatty acids rather than the pure compounds. Therefore, separation—before or after the reaction—is necessary to obtain pure products, e.g. by chromatography, using hydrotalcite as adsorbent and a mixture of water and isopropanol as solvent.

Cytochrome P450 monooxygenase, an enzyme well-known for catalysing the hydroxylation of organic molecules, is a suitable catalyst for this process as well. The critical aspect of the catalytic reaction is the contact between the heme group, which contains the active centre of the enzyme, and the double bonds in the carbon chain of the fatty acid. The enzyme is denaturalized in the presence of organic molecules, and is only active in an aqueous phase. Fatty acids have a small dipole moment and are almost insoluble in water. So how does the contact between the active centre of the enzyme and the fatty acid take place?

A series of molecular simulations was conducted to learn more about the distribution of molecules around the enzyme and the behaviour of the system at different conditions. For this purpose, the mixing behaviour of the systems fatty acid + water, fatty acid + water + isopropanol, and fatty acid + water + cytochrome P450 was investigated. In particular, it is relevant to know whether the fatty acid builds micelles in the water-isopropanol solvent, or how the enzyme catalyses the reaction despite the different phase behaviour of the solvent and the fatty acid. The fatty acid of interest in the present work is oleic acid.

3.2 Simulation Details

Molecular simulation of biological systems poses a challenge to scientific computing. The most important limitation in molecular simulation is the system size. As the number of molecules in the simulation box N increases, the computing time required for the simulation increases with $O(N^2)$ if it is implemented in a naive way. The reason for this steep dependence is found in the computation of pair potentials, so that the distance and energy between interacting atoms needs to be computed at every simulation step.

One option for decreasing the simulation time consists in following a coarse-graining approach [23]. By coarse-graining, a group of atoms is modelled as a single interaction centre, reducing the number of pair interactions that have to be calculated. As we are interested in obtaining detailed atomistic information, we prefer to use a full atomistic model. The model we selected is OPLS [24], which has been successfully used to simulate a large variety of biological systems. The model for the heme group, inexistent in the original OPLS force field, was taken from a recent parameterization of this group compatible with the OPLS force field [25]. In these models, repulsion-dispersion interactions are treated with LJ potentials, while electrostatic interactions are taken into account considering point partial charges at the atomic positions. Internal molecular degrees of freedom are modelled via bond, bend, and torsion potentials.

The cytochrome P450 enzyme contains more than 7,000 atoms, and together with the fatty acid and the solvent molecules, we need to simulate up to 180,000 atoms. Molecular simulation of such an enormous system would need months of simulation time unless we use advanced parallel simulation programs. For the present simulations of molecular biosystems, the *GROMACS 4.6* MD program was used [26]. At the beginning of the simulation, the molecules were arranged in a randomized fashion within a periodic simulation box, and the energy was minimized. This procedure for generating a starting configuration attempts to mimic experimental conditions where initially the solutions are vigorously mixed. Then every molecule is assigned a random velocity (corresponding to the temperature of the system), and a short simulation of 2 ps is carried out for an initial equilibration. Subsequently, a long simulation (over a minimum simulation time of 20 ns) is run, where the properties of interest are calculated.

The molecular positions are updated using a leapfrog algorithm to integrate Newton's equations of motion. We truncate the LJ potential at a cut-off radius of $r_c = 1.5$ nm, and use the particle-mesh Ewald summation method [27] to calculate long-range electrostatic interactions. Temperature is controlled by velocity rescaling with a stochastic term, and pressure with a Berendsen barostat. More details about these simulation techniques can be found elsewhere [28].

Our simulations are performed under conditions for which reliable experimental data are available, i.e. at a temperature of 298 K and a pressure of 100 kPa. For the water-oleic acid system, we use a fatty acid mass fraction of 60 % and two different simulation boxes with approximately the same volume: A cubic box with $V = (7.14 \text{ nm})^3$, and a orthorhombic box of the dimension $5\times15\times5$ nm^3. For the isopropanol-water-oleic acid system, we added 10 mg/ml of oleic acid to a mixture water/isopropanol with a mass fraction of 60 % (for isopropanol) in a cubic simulation box of $(12 \text{ nm})^3$. Finally, the oil/water concentration in the cytochrome P450 + water + oil system corresponded to a volume fraction of 30 % (for oleic acid) in a cubic box of $(10 \text{ nm})^3$, wherein a single cytochrome P450 enzyme was placed.

3.3 Simulation Results

Molecular simulations of oil in the presence of water show a swift phase separation, as it is expected from experimental observation. The phase separation takes place

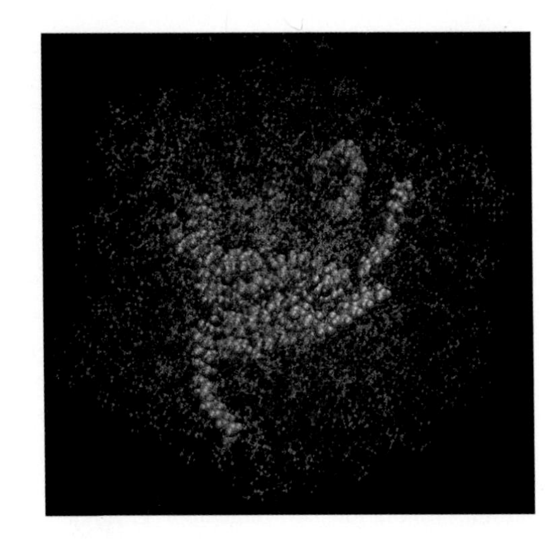

Fig. 3 Cluster of oleic acid molecules in water-isopropanol solution. *Dark blue*, water; *violet*, isopropanol; *light blue*, oleic acid. The acid group of the oil molecules is painted in *red*

only after a few picoseconds. Independently of the shape of the simulation box, the oleic acid forms a bilayer, while other possible structures are not observed. Oil molecules form a ordered phase, where their acid groups point in the direction of the water phase, and their hydrocarbon tails are aligned. This is a consequence of the well known fact that the acid group is hydrophilic while the hydrocarbon tail is hydrophobic.

When we add isopropanol to the water-oil mixture, the situation changes, as shown in Fig. 3. Water and isopropanol are miscible, and their behaviour is similar as in a pure water-isopropanol solution. Oil molecules are mostly surrounded by isopropanol, which can be easily inferred from the radial distribution functions. On the other hand, the double bonds in the oil clusters tend to be in contact with each other.

Oil in the presence of water and cytochrome P450 behaves similarly as in water-oil solutions, cf. Fig. 4. After a few nanoseconds a clear phase separation takes place, where oil forms an ordered, separate phase. Cytochrome P450 stays solvated in water and in contact with the oil phase only at specific points. At the beginning of the simulation there are several oil molecules in the vicinity of cytochrome P450 at favourable contact sites. These molecules maintain their contact with the enzyme during the whole simulation in a position where an interaction with the active centre (which is situated within the heme group) is possible.

3.4 Computational Demands

The computational demands of our simulations were highly dependent on the type of simulation. For the energy minimization simulations, it was sufficient to use a single processor for several minutes. For equilibration, we used 16 CPUs running in

Fig. 4 Cytochrome P450 in water-oleic acid solution. Cytochrome P450 is painted in *orange*; the heme group, in *violet*; the oil molecules, in *light blue*. The acid group of the oil molecules is painted in *red*, the double bond in *green*. For clarity, water has been removed in the picture

parallel for a maximum of 3.5 h. The most demanding simulations during production required 256 CPUs running for 180 h and a virtual RAM of 2.8 GB.

4 Vapour-Liquid Interfaces and Wetting of Component Surfaces

4.1 Planar Vapour-Liquid Interfaces

For simulating vapour-liquid coexistence on the molecular level, a long range correction is needed to obtain accurate results. Thereby, the dispersive contribution to the potential energy U_i of a molecule i

$$U_i = \sum_{r_{ij} < r_c} u_{ij} + U_i^{LRC} \tag{2}$$

is calculated as a sum of the explicitly computed part and the long range correction (LRC), where the latter consists of a summation over N_s slabs, employing a periodic boundary condition. The LRC is applied in the direction normal to the planar interface, corresponding to the y direction here

$$U_i^{LRC} = \sum_{k}^{N_s} \Delta u_{i,k}^{LRC}(y_i, y_k), \tag{3}$$

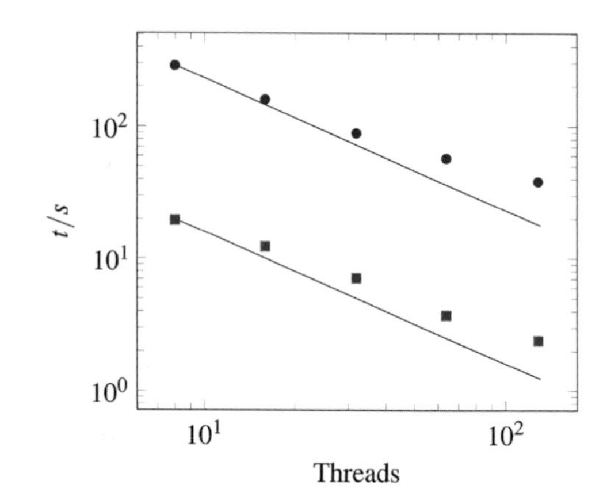

Fig. 5 Strong scaling with $N = 256{,}000$ 2CLJ particles and 512 slabs. The *blue dots* denote the execution time of 100 simulation time steps, while the *red dots* only denote the run time of the long range correction. The *solid lines* represent ideally linear scaling

while the system is homogeneous in the other directions. The correction terms $\Delta u_{i,k}^{LRC}(y_i, y_k)$ for the slabs are calculated as an integral over the slab volume. The resulting term for the potential energy only depends on the distance between the slabs r, the density ρ and the thickness Δy of the slabs [29]. The slabwise interaction is computed in pairs, so that $\frac{1}{2}N_s^2$ individual contributions have to be computed.

Usually, the number of threads is much smaller than the number of slabs. For a scaling test, a system with $N = 256{,}000$ two-centre Lennard-Jones (2CLJ) particles was simulated with different numbers of threads and 512 slabs. Figure 5 shows the results for the strong scaling behaviour.

For small numbers of threads, the program scales almost ideally, i.e. linearly. For larger numbers of threads, the communication between the threads and the decomposition of the particles becomes more time consuming. Eventually, the long range correction requires even more communication between the threads than the rest of the program, since the density profile has to be sent to every thread before the slab interaction begins. After the slab interaction has been calculated, the correction terms for the potential energy, the force and the virial have to be distributed to every thread.

On the basis of the long range correction discussed above, vapour-liquid equilibria were considered by explicit MD simulation of the fluid phase coexistence (i.e. including an interface), so that the vapour-liquid surface tension γ could be computed following the virial route. For the (single centre) LJ fluid, a high precision was obtained by simulating $N = 300{,}000$ particles over roughly a million time steps, resulting in very small statistical uncertainties. From the resulting surface tension values, the regression term

$$\gamma = 2.94 \left(1 - \frac{T}{T_c}\right)^{1.23} \frac{\epsilon}{\sigma^2} \tag{4}$$

was obtained, where the LJ critical temperature is given by Pérez Pellitero et al. [30] as $T_c = 1.3126\epsilon$.

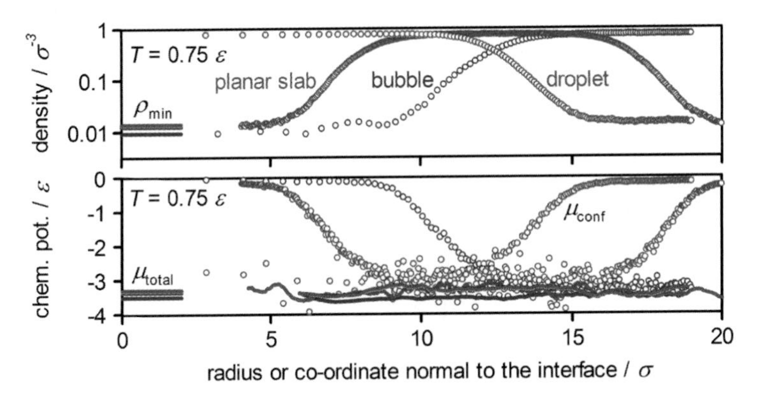

Fig. 6 Density profiles (*top*) and chemical potential profiles (*bottom*) for nanoscopic vapour-liquid interfaces of the TSLJ fluid from canonical ensemble MD simulation of systems containing planar liquid and vapour slabs, a liquid drop (surrounded by a supersaturated vapour), or a gas bubble (surrounded by a subsaturated liquid), respectively. The chemical potential profile includes information on both the configurational contribution (*circles*) and the total chemical potential (*solid lines*), as a function of the characteristic spatial coordinate of the system. The *solid lines* in the *bottom left* part of the two diagrams indicate the density of the vapour phase (*top*) as well as the average value of the total chemical potential (*bottom*)

4.2 *Curved Vapour-Liquid Interfaces*

The influence of curvature on vapour-liquid interfacial properties was examined for the trunacted-shifted Lennard-Jones fluid (TSLJ), with a cutoff radius of $r_c = 2.5\ \sigma$, by simulating both curved (i.e. bubble and droplet) as well as planar interfaces by MD simulation of the canonical ensemble at a reduced temperature of $T = 0.75\ \epsilon$. For the system with planar symmetry, half of the simulation volume was filled with vapour and the other half with liquid, using a simulation box with a total elongation of 17 σ. The density profile from this simulation was compared with that of a bubble with an equimolar radius of $R_\rho = 12.6\ \sigma$ as well as a droplet with $R_\rho = 12.4\ \sigma$, cf. Fig. 6. A system containing a bubble is subsaturated, whereas the fluid phases coexisting at the curved interface of a droplet are supersaturated with respect to the thermodynamic equilibrium condition for the bulk phases coexisting at saturation. In agreement with this qualitative statement from phenomenological thermodynamics, the vapour density (i.e. the minimal value from the density profile) was found to be smaller for the system containing a bubble ($\rho'' = 0.0097\ \sigma^{-3}$) and larger for the system containing a droplet ($0.0140\ \sigma^{-3}$), whereas the vapour density over the planar interface was $0.0127\ \sigma^{-3}$.

Analogous results were obtained for the chemical potential $\mu^{\text{total}} = \mu^{\text{id}} + \mu^{\text{conf}}$, which was determined as the sum of the ideal contribution $\mu^{\text{id}} = T \ln \rho$ and the configurational contribution μ^{conf}. The Widom test particle method [31] was implemented to compute a profile of μ^{conf}, which in equilibrium is complementary to the logarithmic density profile, yielding an approximately constant profile for the total chemical potential, cf. Fig. 6. For the droplet, an average value of $\mu^{\text{total}} = (-3.31 \pm 0.04)\ \epsilon$ was determined, in comparison to $(-3.51 \pm 0.02)\ \epsilon$ for the bubble and $(-3.37 \pm 0.02)\ \epsilon$ for the planar slab. This shows that in the present case, curvature effects are more significant for a gas bubble than for a liquid droplet of the same size, suggesting that the surface tension of the bubble is larger than that of the droplet.

4.3 Temperature Dependence of Contact Angles

The research on the wetting behaviour of surfaces is an active field in materials science. The goal is to reliably predict the wetting properties of component surfaces for design of components with new functional features. In this regard molecular simulations are particularly useful due to their high resolution. Moreover, molecular simulation permits the systematic investigation of the influence of particular parameters on the wetting behaviour.

For the present study, the TSLJ potential was employed and the contact angle dependence on temperature for a non-wetting ($\theta > 90°$) and a partially wetting ($\theta < 90°$) scenario, cf. Fig. 7 was investigated by MD simulation. The solid wall was represented by a face centered cubic lattice with a lattice constant of $a = 1.55\ \sigma$ and the (001) surface exposed to the fluid. For the unlike interaction between the solid and the fluid phase the size and energy parameters $\sigma_{\text{sf}} = \sigma_{\text{ff}}$ and

$$\epsilon_{\text{sf}} = \zeta \epsilon_{\text{ff}}, \tag{5}$$

respectively, were applied.

The simulations were carried out with the massively parallel MD program *ls1 mardyn* [32]. A total number of $N = 15{,}000$ fluid particles was simulated in the *NVT* ensemble. In most cases, 64 processes were employed for parallel computation. The efficient simulation of heterogeneous systems, containing vapour-liquid interfaces and even a solid component, was facilitated here by a dynamic load balancing scheme based on k-dimensional trees [32]. At the beginning of the simulation, the fluid particles were arranged on regular lattice sites in form of a cuboid. The equilibration time of the three phase system was 2 ns, whereas the total simulation time was chosen to be 6 ns. Periodic boundary conditions were applied in all directions, leaving a channel with a distance of at least 35 σ between the wall and its periodic image, avoiding "confinement effects" at near-critical temperatures [33]. The contact angles were determined by the evaluation of density profiles averaged over a time span of 4 ns.

Sessile drops were studied at two different interaction parameters, $\zeta = 0.35$ and 0.65, corresponding to partially wetting and non-wetting conditions. The simulations were carried out in the temperature range between 0.7 and 1 ϵ, covering nearly the entire regime of stable vapour-liquid coexistence for the TSLJ fluid, as

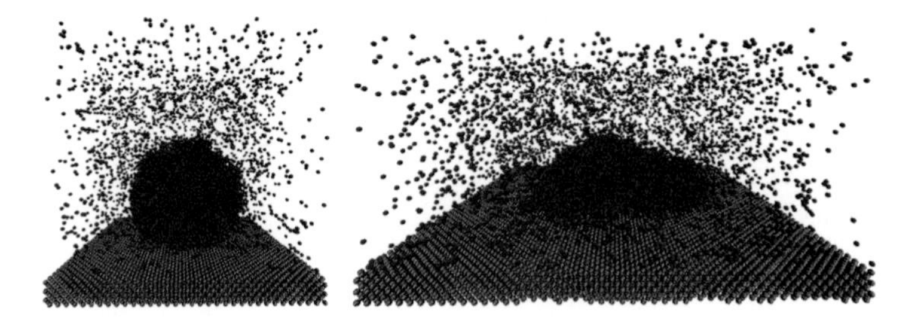

Fig. 7 Snapshots of simulation runs at different fluid-solid interaction parameters ζ of 0.35 (*left*) and 0.65 (*right*) at a temperature of 0.8 ϵ/k

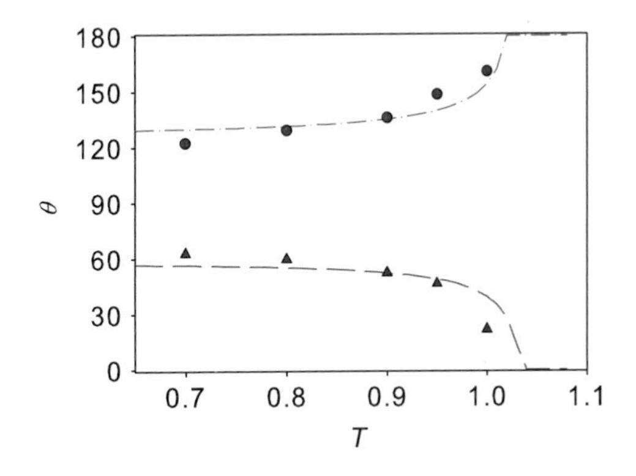

Fig. 8 Simulation results and correlation for the dependence of the contact angle on the temperature for fluid-solid dispersive interaction strengths ζ of 0.35 (*circles, dashed line*) and 0.65 (*triangles, dash-dotted line*)

indicated by the triple point temperature $T_3 \approx 0.65\ \epsilon$, cf. van Meel et al. [34], and the critical temperature $T_c \approx 1.078\ \epsilon$, cf. Vrabec et al. [35]. It is well known that the respective wetting behaviour of the sytem, i.e. wetting or dewetting, is reinforced at higher temperatures, cf. Fig. 8. Close to the critical point of the fluid, the phenomenon of critical point wetting occurs [36], i.e. either the liquid or the vapour phase perfectly wets the solid surface

$$|\cos\theta(T \to T_c)| \to 1. \tag{6}$$

5 Conclusion

Within the MOCOS project, a substantial progress was made regarding software development for massively-parallel molecular simulation. In particular, suitable algorithms for the long range contributions to the intermolecular interaction were implemented in *ms2* (Ewald summation) and *ls1 mardyn* (Janeček correction for long range dispersive effects). In *ls1 mardyn*, a k-dimensional tree-based domain decomposition with dynamic load balancing was implemented to respond to an uneven distribution of the interaction sites over the simulation volume. Furthermore, *GROMACS* [26] was used to simulate large molecules, taking their internal degrees of freedom into account. In this way, systems with up to 300,000 interaction sites were simulated for over 20 ns.

New models were developed for alkali cations and halide anions in order to treat the dispersive interaction accurately in molecular simulations of electrolyte solutions. These models were optimized by adjusting the LJ size parameter to the reduced density of aqueous solutions and the energy parameter to diffusion coefficients as well as pair correlation functions. On this basis, quantitatively reliable results were obtained regarding the structure of the hydration shells formed by the water molecules surrounding these ions. Important qualitative structural information on biomolecules at oil-water interfaces was deduced by analysing MD simulations of the cytochrome P450 enzyme, and the critical wetting behaviour as well as the properties of planar and curved vapour-liquid interfaces were characterized by massively-parallel MD simulation with the *ls1 mardyn* program.

The scenarios discussed above show how molecular modelling can today be applied to practically relevant systems even if they exhibit highly complex structures and intermolecular interactions, including significant long range contributions. In the immediate future, molecular simulation as an application of high performance computing will therefore be able to play a crucial role not only by contributing to scientific progress, but also in its day-to-day use as a research and development tool in mechanical and process engineering as well as biotechnology.

Acknowledgements The present work was carried out under the auspices of the Boltzmann-Zuse Society for Computational Molecular Engineering (BZS), and the molecular simulations were conducted on the XC4000 supercomputer at the Steinbuch Centre for Computing, Karlsruhe. The authors would like to thank Akshay Bedhotiya (Bombay) for his assistance regarding the simulation of gas bubbles and Sergey Lishchuk (Leicester), Martin Buchholz, Wolfgang Eckhardt, and Ekaterina Elts (München) and Gábor Rutkai (Paderborn) as well as Martin Bernreuther, Colin Glass, and Christoph Niethammer (Stuttgart) for contributing to the development of the *ls1 mardyn* and *ms2* molecular simulation programs as well as Deutsche Forschungsgemeinschaft (DFG) for funding the Collaborative Research Centre (SFB) 926. Furthermore, they would like to acknowledge fruitful discussions with Cemal Engin, Michael Kopnarski, Birgit and Rolf Merz as well as Michael Schappals and Rajat Srivastava (Kaiserslautern), George Jackson and Erich Müller (London), Jonathan Walter (Ludwigshafen), Philippe Ungerer and Marianna Yiannourakou (Paris) as well as Nichola McCann (Visp).

References

1. S. Deublein, B. Eckl, J. Stoll, S.V. Lishchuk, G. Guevara Carrión, C.W. Glass, T. Merker, M. Bernreuther, H. Hasse, J. Vrabec, Comput. Phys. Commun. **182**(11), 2350 (2011)
2. W. Eckhardt, A. Heinecke, R. Bader, M. Brehm, N. Hammer, H. Huber, H.G. Kleinhenz, J. Vrabec, H. Hasse, M. Horsch, M. Bernreuther, C.W. Glass, C. Niethammer, A. Bode, J. Bungartz, in *Proc. ISC 2013*. LNCS, vol. 7905 (Springer, Heidelberg, 2013), pp. 1–12

3. M. Horsch, C. Niethammer, J. Vrabec, H. Hasse, Inform. Technol. **55**(3), 97 (2013)
4. S. Deublein, S. Reiser, J. Vrabec, H. Hasse, J. Phys. Chem. B **116**(18), 5448 (2012)
5. S. Deublein, S. Reiser, J. Vrabec, H. Hasse, J. Chem. Phys., accepted (2013)
6. J. Walter, V. Ermatchkov, J. Vrabec, H. Hasse, Fluid Phase Equilib. **296**(2), 164 (2010)
7. J. Walter, J. Sehrt, J. Vrabec, H. Hasse, J. Phys. Chem. B **116**(17), 5251 (2012)
8. Y.L. Huang, T. Merker, M. Heilig, H. Hasse, J. Vrabec, Ind. Eng. Chem. Res. **51**(21), 7428 (2012)
9. S. Reiser, N. McCann, M. Horsch, H. Hasse, J. Supercrit. Fluids **68**, 94 (2012)
10. S. Pařez, G. Guevara Carrión, H. Hasse, J. Vrabec, Phys. Chem. Chem. Phys. **15**(11), 3985 (2013)
11. M. Horsch, H. Hasse, Chem. Eng. Sci., submitted (2013)
12. S. Werth, S.V. Lishchuk, M. Horsch, H. Hasse, Phys. A **392**(10), 2359 (2013)
13. S. Werth, G. Rutkai, J. Vrabec, M. Horsch, H. Hasse, Mol. Phys., accepted (2013)
14. P.P. Ewald, Ann. Phys. (Leipzig) **369**(3), 253 (1921)
15. D. Horinek, S.I. Mamatkulov, R.R. Netz, J. Chem. Phys. **130**, 124507 (2009)
16. M.M. Reif, P.H. Hünenberger, J. Chem. Phys. **134**, 144104 (2011)
17. M.B. Gee, N.R. Cox, Y.F. Jiao, N. Bentenitis, S. Weerasinghe, J. Chem. Theory Comput. **7**(5), 1369 (2011)
18. S. Deublein, J. Vrabec, H. Hasse, J. Chem. Phys. **136**, 084501 (2012)
19. K. Gubbins, *Statistical Mechanics*, vol. 1 (Burlington House, London, 1972)
20. R. Mills, V. Lobo, *Self-Diffusion in Electrolyte Solutions* (Elsevier, New York, 1989)
21. H. Ohtaki, T. Radnai, Chem. Rev. **93**(3), 1157 (1993)
22. J.L. Fulton, D.M. Pfund, S.L. Wallen, M. Newville, E.A. Stern, J. Ma, J. Chem. Phys. **105**(6), 2161 (1996)
23. F. Müller Plathe, ChemPhysChem **3**(9), 754 (2002)
24. W.L. Jørgensen, D.S. Maxwell, J. Tirado Rives, J. Am. Chem. Soc. **110**(45), 11225 (1996)
25. V. Gogonea, J.M. Shy II., P.K. Biswas, J. Phys. Chem. B **110**(45), 22861 (2006)
26. S. Pronk, P. Szilárd, R. Schulz, P. Larsson, P. Bjelkmar, R. Apostolov, M.R. Shirts, J.C. Smith, P.M. Kasson, D. van der Spoel, B. Hess, E. Lindahl, Bioinformatics **29**(7), 845 (2013)
27. M.E. Tuckerman, G.J. Martyna, J. Phys. Chem. B **104**(2), 159 (2000)
28. D. Frenkel, B. Smit, *Understanding Molecular Simulation* (Academic Press, San Diego, 2002)
29. J. Janeček, J. Phys. Chem. B **110**(12), 6264 (2006)
30. J. Pérez Pellitero, P. Ungerer, G. Orkoulas, A.D. Mackie, J. Chem. Phys. **125**, 054515 (2006)
31. B. Widom, J. Chem. Phys. **86**(6), 869 (1982)
32. M. Buchholz, H.J. Bungartz, J. Vrabec, J. Comput. Sci. **2**(2), 124 (2011)
33. A. Oleinikova, I. Brovchenko, A. Geiger, Eur. Phys. J. B **52**(4), 507 (2006)
34. J.A. van Meel, A.J. Page, R.P. Sear, D. Frenkel, J. Chem. Phys. **129**, 204505 (2008)
35. J. Vrabec, G.K. Kedia, G. Fuchs, H. Hasse, Mol. Phys. **104**(9), 1509 (2006)
36. J.W. Cahn, J. Chem. Phys. **66**(8), 3667 (1977)

Focal Adhesion Kinase as a Cellular Mechano-Sensor

Jing Zhou, Agnieszka Bronowska, Bogdan Costescu, and Frauke Graeter

Abstract Focal adhesion kinase (FAK) is a component of focal adhesion sites, which plays a crucial role in cell differentiation and motility. Using molecular dynamic simulations, we observed that binding of the phospho-inositide phosphatidylinositol-4,5-bisphosphate (PIP_2) by FAK can induce major conformational changes of FAK and exposure of its auto-phosphorylation site, but it cannot induce FAK's activation. Our results support the biological experimental data by providing an atomic description of FAK's conformational changes within FERM and kinase domains.

1 Abbreviations

COM	Center of mass
ECM	Extracellular matrix
FAK	Focal adhesion kinase
FAs	Focal adhesions
FDA	Force Distribution Analysis
FK-apo	FAK crystal structure containing FERM and kinase domain
FK-ATP	FK-apo bound to ATP
FK-ATP-PIP	FK-apo bound to ATP and PIP2
FK-PIP	FK-apo bound to PIP2
FPMD	Force Probe Molecular Dynamics

J. Zhou · A. Bronowska · B. Costescu · F. Graeter (✉)
Heidelberg Institute for Theoretical Studies, Heidelberg, Germany

Max-Planck Institute for Intelligent Systems, Stuttgart, Germany
e-mail: Frauke.Graeter@h-its.org

W.E. Nagel et al. (eds.), *High Performance Computing in Science and Engineering '13*, DOI 10.1007/978-3-319-02165-2_46,

MD	Molecular Dynamics
OPLS	Optimized Potential for Liquid Simulations
PIP2	Phosphoinositide phosphatidylinositol-4,5-bisphosphate
PCA	Principal Component Analysis
REDS	RESP charge Derive Server

2 Introduction

In the last decade, it has become evident that cells respond to mechanical signals such as the stiffness of the cellular environment or stretching forces from adjacent cells. These responses are integrated into cellular signaling pathways. Mechanical signals, among others, steer the cell proliferation and motility. Bone renewal occurs only upon physical exercise and gravity forces, since mesenchymal stem cells require mechanical forces for differentiation [15]. However, the molecular sensors which can transduce a mechanical force into a biochemical signals remain poorly understood. We hypothesize that focal adhesion kinase (FAK) acts as a pivotal mechano-sensor in mesenchymal stem cell differentiation and elsewhere, by directly converting force signals into a biochemical response.

Focal adhesion kinase (FAK) is a cytoplasmic non-receptor tyrosine kinase, which exerts its activity at the crossroads of multiple cell signaling pathways. Mechanical deformation of a cell induces the translocation of FAK at the cell periphery towards the direction of cell migration [11]. That leads to the functional activation of FAK by tyrosine phosphorylation. After a sequence of tyrosine phosphorylation events, FAK adheres to downstream signaling proteins leading to the formation of very large multimolecular complexes called focal adhesions (FAs). Through FAs the cytoskeleton of a cell connects to the extracellular matrix (ECM) [13]. In this way, mechanical forces can be transduced from outside to inside the cell, and translated into biochemical signals. Thus, studying the mechanism of FAK activation may be very helpful to understand cell signaling by mechano-sensing. If FAK indeed serves as a direct mechanosensor, i.e. is activated for phosphorylation by mechanical force, it would be a prime candidate for analyzing and re-engineering mesenchymal stem cell differentiation and bone formation in biomedicine.

Despite of multiple experimental studies of FAK's biological function, its activation mechanism remains elusive. Many stimuli have been reported to induce FAK activation, such as integrin signaling and direct engagement of cytoplasmic regions of growth factor receptors [2, 10], but molecular details of such interactions are unclear. Molecular Dynamics (MD) simulation combined with Force Probe Molecular Dynamics (FPMD) simulation, as performed in the present study, provides the direct insight into the underlying molecular mechanism at high resolution, which is inaccessible by experiments.

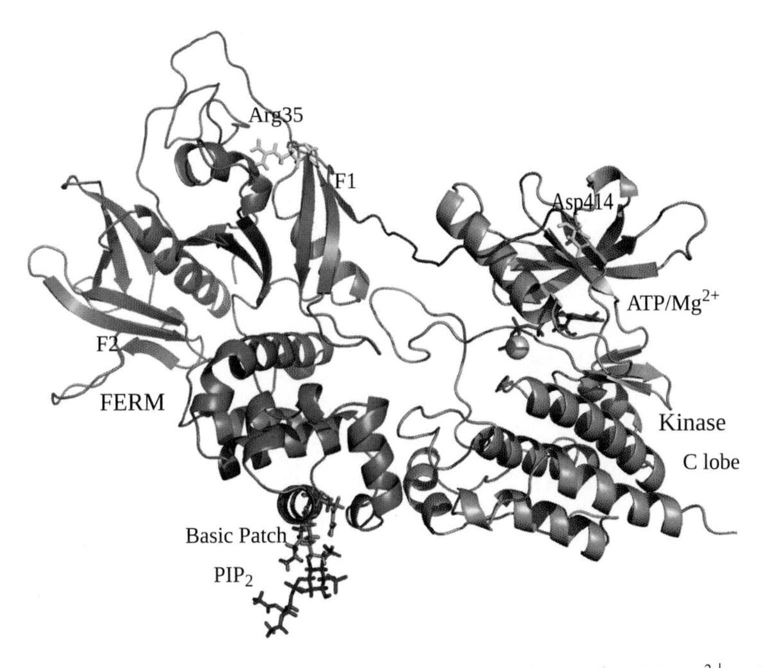

Fig. 1 FAK crystal structure containing FERM and kinase domain bound to ATP/Mg^{2+} and PIP_2

In the current framework, we considered two domains of FAK: namely, the N-terminal FERM domain and the kinase domain (Fig. 1). FERM domain is involved in auto-inhibition of the kinase by blocking a phosphorylation site (Tyr576/577) in kinase domain. The exposure of this phosphorylation site induces the maximum activity of FAK[12]. In addition, we have proposed phosphatidylinositol 4,5-bisphosphate (PIP_2), which is abundant in the cell membrane, as a potential FAK activator[1]. The experimental data recently showed that the negatively charged PIP2 binds to the positively charged basic patch on the FAK FERM domain, which induces conformational rearrangements. This promotes exposure of FAK autophosphorylation site (Tyr397), but cannot induce the exposure of Tyr576/577 phosphorylation site, which is a hallmark of FAK activation [6]. Our MD simulation data provide a detailed molecular insights in the mode of PIP_2-induced FAK phosphorylation. These results are consistent with experimental data available [6]. We demonstrated that PIP_2 binds to a basic region on the FAK FERM domain, close but not over-lapping with an autoinhibitory region on the FERM F2 lobe. ATP binding on the active site lead to partial opening between FERM F2 lobe and kinase C lobe, simultaneously closing the gap between kinase N lobe and kinase C lobe. PIP_2, binding re-opened the gap, enabling the activation of the kinase domain. None of ATP and PIP2 binding could induce an exposure of phosphorylation site (Tyr576/577). These results strongly support our hypothesis that an external mechanical force is required for the FAK activation.

In our previous study, we used FPMD simulation to trace the partial unfolding process of FAK crystal stricture bound to PIP2 (FK-PIP) under the mechanic force. Now, a more realistic simulation setup, which includes PIP2 and the explicit membrane is being studied to validate our preliminary results.

3 Results and Discussion

Our MD simulation results support the experimental observation made by FRET measurement on the allosteric rearrangement of FERM and kinase domains induced by PIP_2 binding [6].

The results presented herein are coming from several studies. Initially, we used protein-ligand docking followed by MD simulation refinement to predict the atomistic structure of FAK binding to PIP_2 phosphoinositide moiety. This was followed by all-atom MD simulations, which provided time-dependent structural and dynamical information about the interaction of FAK with PIP_2. The phosphoinositide head group of PIP_2 was stabilized by the basic patch on the FERM during all the three 300 ns simulations. Since January 2013, we started to perform the MD simulations of (1) FAK crystal structure containing FERM and kinase domain (FK-apo, PDB: 2J0J), (2) FAK bound to ATP (FK-ATP, PDB: 2J0L+2J0J) and (3) FAK bound to ATP and PIP_2 (FK-ATP-PIP) (Fig. 1) by HLRS to study allosteric conformational change of FERM and kinase domains induced by PIP_2 binding and provide the catalytic mechanism of FAK auto-phosphorylation induced by PIP_2 binding in atomic scale.

Because of the protein side-chain interactions, the conformational changes of protein could happen through different pathways. Thus, each of three above mentioned systems has been simulated three times independently. The last 120 ns trajectories of each simulation have been used for the analysis. First, the distances between arginine Arg35 (the N-terminal of the FERM domain) and aspartic acid Asp414 (the N-terminal of the kinase domain), where the FRET sensors bind, were measured (Fig. 2a). As expected, FK-ATP shows the shortest distance between two N-termini. That is consistent with higher FRET signals in experiment. However, in the presence of PIP_2, in the FK-ATP-PIP larger distances between these N-terminal sites were observed than in the apo-FK. This result is consistent with the lower FRET signals in experiment. Then, the distances between center of mass (COM) of FERM F1 and kinase N lobe have been calculated. Surprisingly, we observed that FK-ATP had larger distance between FEMR F1 lobe's center-of-mass (COM) and kinase N lobe than apo-FK (Fig. 2b).

In order to analyze the allosteric rearrangement of FERM and kinase domain, we calculated distances between these domains, as shown in Fig. 2. In the plot showed in this Fig. 2c, the yellow boxes represent the distances between N-termini of FERM and kinase domains. The distances between the COM of FERM and F1

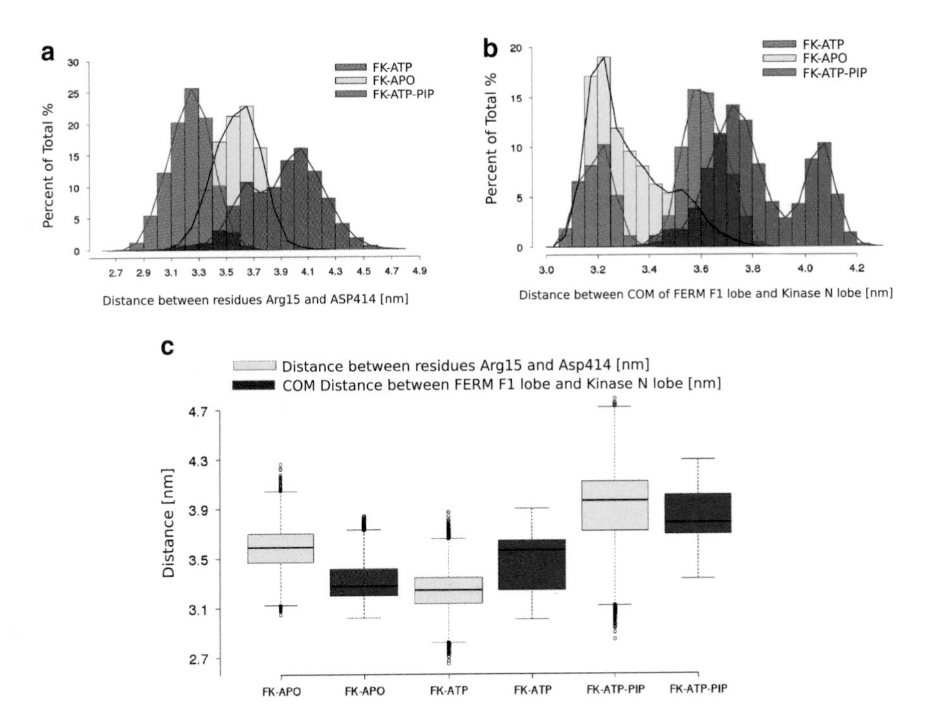

Fig. 2 (**a**) Distances between 2 N-termini of FERM and kinase domain. (**b**) Distances between center of mass of FERM F1 and kinase N lobe. (**c**) All above Distance shown in boxplot

lobe are represented by blue boxes in the plot. As it can be seen, both apo-FK and FK-ATP-PIP have larger distances between two N-termini than distances between COM of FERM F1 and kinase N lobe. In contrast, FAK-ATP has a larger distances between COM of FERM F1 and kinase N lobe. Thus, we suppose that the ATP binding site at the kinase catalytic domain can partially open FERM F2 and kinase N lobe. But the presence of ATP has a large stabilizing effect on the gap between kinase C and N lobes, which results in the inactivation of the kinase domain [6]. The addition of PIP_2 to FK-ATP can induce the switch of an inactive to an active kinase domain [14]. To corroborate this, the Principal Component Analysis (PCA) and Force Distribution Analysis (FDA) [16] of these three systems are being carried out. The calculations are currently in progress.

4 Molecular Docking and Molecular Dynamics Simulations

All simulations were based on the FAK protein crystal structure (PDB: 2J0J) [12] containing FERM and kinase and crystal structure (PDB: 2J0L) containing kinase domain and ANP. Initial geometries of the FK-PIP complex were obtained using a molecular docking procedure. The putative PIP_2 binding site has been assigned to the FERM basic patch (residues 216–222). Atomic charges of the PIP_2 molecule for docking and subsequent MD simulations were assigned by the RESP charge Derive Server (REDS) [4, 18], from Gaussian09 [5] calculations using the charge model and ESP-A1.

Molecular docking was performed using the UCSF DOCK6.5 suite [9] with grid scoring in an implicit solvent. The grid spacing was 0.25 Å, and the grid included 12 Å beyond the FAK basic patch. The energy score was the sum of electrostatic and Van der Waals contributions. In the course of the docking procedure, the PIP_2 molecule was subjected to 2,500 cycles of molecular-mechanical energy minimization. The number of maximum ligand orientations was 5,000. The best-scoring 10 FK-PIP complexes were further analyzed by means of MD simulations. The best structure was chosen according to the Coulomb interaction between the PIP_2 head group and the basic patch in the FERM domain.

All simulations were carried out using the MD software package GROMACS 4.0.5 [17], using the OPLS all-atom force field [7] for the protein and the TIP4P water model [8]. The long-range electrostatic interactions were treated with the Particle Mesh Ewald (PME) [3] method. The molecular systems were simulated under physiological conditions (300 K, 1 atm).

Each of these three structures, FK-apo, FK-ATP, and FK-ATP-PIP_2, was solvated in a triclinic box of volume $13 \times 10 \times 10\,\mathrm{nm}^3$. Sodium and chloride counter-ions were added to neutralize the system with a concentration of 0.1 M. Prior to MD simulations, all systems were minimized using the steepest descent method for 10,000 steps, followed by 2 ns MD simulations during which position restraints were used on all bonds. Finally, the FK-apo and FK-ATP-PIP systems were equilibrated during MD simulations of 200 and 300 ns MD simulation was performed to equilibrate the FK-ATP system. Each system was simulated for three times independently.

GROMACS is a highly scalable Molecular Dynamics (MD) software. MD simulations work by integrating Newton's equations of motion. Empirical potential functions describe the interactions between atoms and allow calculating atomic forces. During each integration step, these forces lead to changes in atomic coordinates, which are communicated to all processors involved in the parallel job. A low-latency interconnect, as found in Hermit, is therefore essential for obtaining a good parallel efficiency. For the simulated molecular systems, the performance on 192 CPU cores (=6 nodes) was around 5 time larger than on 32 CPU cores (=1 node), representing an efficiency of more than 80 %. During approximately 3 months of running jobs on Hermit, we were able to obtain 3 μs of simulated time, which enabled us to have a detailed, molecular-level insight of the protein conformational change.

For the next simulations, we plan to add a biomembrane and enlarge the box of water molecules surrounding the protein-membrane complex (Fig. 3). This will lead to an increase in the number of atoms from currently 220.000 to around 400.000,

Fig. 3 Simulation system of the protein-membrane complex in water box

allowing our simulations to scale to 384 CPU cores (=12 nodes) or more, without a decrease in parallel efficiency.

References

1. X. Cai et al., Mol. Cell. Biol. **28**(1), 201–214 (2008)
2. S.Y. Chen, H.C. Chen, Mol. Cell. Biol. **26**, 5155–5167 (2006)
3. T. Darden, D. York, L. Pedersen, J. Chem. Phys. (1993)
4. F.-Y. Dupradeau et al., Phys. Chem. Chem. Phys. (2010)
5. Gaussian 09
6. M.G. Guillermina et al. (in review)
7. W.L. Jorgensen, J. Tirado-Rives, J. Am. Chem. Soc. (1988)
8. W.L. Jorgensen, J. Chandrasekhar, M.L. Klein, J. Chem. Phys. (1983)
9. P.T. Lang et al., RNA (2009)
10. I. Laza Menacho et al., Biol. Chem. **286**, 17292–17302 (2011)
11. S. Li et al., Proc. Natl. Acad. Sci. USA **99**(6), 3546–3551 (2002)
12. D. Lietha et al., Cell **129**(6), 1177–1187 (2007)
13. S. Lo, L. Chen, Cancer Metastasis Rev. **13**, 9–24 (1994)
14. J. Nowakowski et al., Structure(CAMB) (2002)
15. C.E. Sarraf, W.R. Otto, M. Eastwood, Cell Prolif. **44**(1), 99–108 (2011)
16. W. Stacklies, C. Seifert, F. Graeter, BMC Bioinformatics (2011)
17. D. van der Spoel, E. Lindahl, H.J. Berendsen, J. Comput. Chem. **26**, 1701–1718 (2009)
18. E. Vanquelef et al., Nucleic Acids Res. (2011)

A Highly Scalable Multigrid Method with Parallel Direct Coarse Grid Solver for Maxwell's Equations

Daniel Maurer and Christian Wieners

Abstract We present scalability results on the cluster HERMIT of a parallel direct solver for finite element methods. This is applied in a multigrid iteration to obtain a highly scalable solution method for the computation of reliable and exact approximations of electro-magnetic fields in the cavity problem for the Maxwell's equations, and of electromagnetic eigenfrequencies in Maxwell's eigenvalue problem. Here, we consider in particular the case that several frequencies has to be determined simultaneously. For both problems a Laplace equation has to be solved in addition in order to obtain divergence-free fields.

1 Introduction

Based on the programming model for parallel finite elements with distributed point objects introduced in [4] we realized a direct solver [1, 2]. This solver is a parallel block LU decomposition using the concept of nested dissection to obtain recursively dense Schur complement problems which are then decomposed by cyclic distributions. This allows for the exact solution of the coarse grid problem in multigrid methods in cases that the coarse problem is fairly large, e.g., if the coarse mesh is generated by a mesh generator. Moreover, if the coarse problem is large enough it increases substantially the robustness of the multigrid iteration, e.g., for the case of (nearly) indefinite problems.

Here we present parallel results for electro-magnetic field and eigenfrequency computations. For details on the Krylov-type eigenvalue solver LOBPCG and the hybrid multigrid method for the Maxwell problem we refer to [3], the parallel multigrid method is described in [5].

D. Maurer (✉) · C. Wieners
Institute for Applied and Numerical Mathematics 3, Karlsruhe Institute of Technology, 76128 Karlsruhe, Germany
e-mail: daniel.maurer@kit.edu; christian.wieners@kit.edu

W.E. Nagel et al. (eds.), *High Performance Computing in Science and Engineering '13*, DOI 10.1007/978-3-319-02165-2_47,

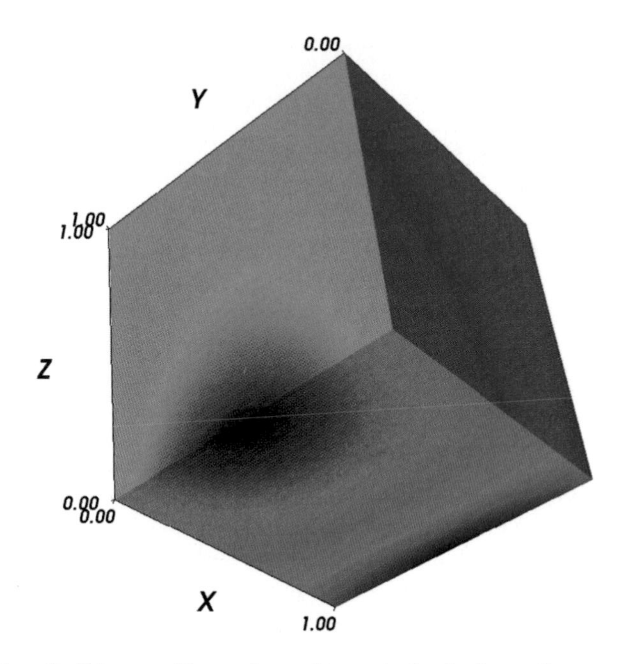

Fig. 1 Solution of a Poisson problem $-\Delta p = f$ in a cube for the first scaling test

2 A Parallel Direct Linear Solver for Finite Elements

In the electro-magnetic applications, the Laplace equation is a sub-problem required for the projection onto divergence-free fields (in the discrete sense). Thus, the scalability of the parallel direct solver is first tested for a three-dimensional Poisson problem (Fig. 1). We consider the unit cube using up to 4,096 cores for 2,146,689 unknowns (cf. Table 1).

We observe that the parallel solver can be used on many cores both for small and large problems.

The applicability for small problems shows a good behavior in the decomposition time when increasing the processor number. A major problem for the parallel decomposition is the strong growth of the communication time. For the smallest problem with 35,937 unknowns the total decomposition time—which includes the communication time—remains within the range of one second. For the largest problem we observe a good scaling up to 2,048 cores.

Hence, a combination of the parallel direct solver with the multigrid method seems to be a consequent concept. Multigrid methods are "optimal" related to the order of convergence when decreasing the mesh width. Nevertheless, on the coarsest grid a preferably exact solver is needed, in particular for ill conditioned problems. Furthermore an increasing problem size requires more cores for the full problem. Since the parallel direct solver has a good scaling behavior it is predesignated as a coarse grid solver in the multigrid method.

Table 1 Execution time for the decomposition $A = LU$ and the solutions $Ly = b$ and $Ux = y$ of the Poisson problem on the unit cube with the parallel direct solver

Unknowns	35,937		274,625		2,146,689	
#cores	Decomposition	Solution	Decomposition	Solution	Decomposition	Solution
1	1:05.58 min	0.20 s	2:03:25 h			
4	27.05 s	0.07 s				
8	6.89 s	0.03 s				
16	1.76 s	0.01 s	5:06.06 min	0.42 s		
32	0.55 s	0.01 s	1:13.94 min	0.23 s		
64	0.83 s	0.01 s	16.06 s	0.21 s		
128	0.77 s	0.01 s	6.87 s	0.17 s		
256	1.11 s	0.01 s	5.05 s	0.17 s	2:45.30 min	2.09 s
512	1.02 s	0.01 s	4.29 s	0.18 s	1:21.68 min	2.04 s
1,024	0.93 s	0.01 s	3.73 s	0.18 s	57.73 s	2.01 s
2,048			3.44 s	0.17 s	52.03 s	2.03 s
4,096					56.55 s	2.00 s

3 The Cavity Problem

One problem in electrodynamics is to find the solution of a cavity problem for a given wave number. For realistic wave numbers the discretization leads to a highly indefinite problem. Therefore, to achieve convergence it is necessary to use a fine coarse grid within the multigrid preconditioner. Furthermore, a reliable coarse grid solver is needed, so that we use the parallel block LU decomposition as direct solver.

The cavity problem in $\Omega \subset \mathbb{R}^3$ with magnetic field $\mathbf{H}$, permittivity ε and frequency ω reads as follows:

$$\begin{aligned} \nabla \times \left(\varepsilon^{-1} \nabla \times \mathbf{H}\right) - \omega^2 \mathbf{H} &= \mathbf{f} \qquad \text{in } \Omega, \\ \nabla \cdot \mathbf{H} &= 0 \qquad \text{in } \Omega, \\ \mathbf{H} \times \mathbf{n} &= 0 \qquad \text{on } \partial\Omega. \end{aligned}$$

If ω is not an eigenfrequency of the cavity, the problem has a unique solution for $\delta = -\omega^2$. The reduction rate of the multigrid preconditioner is depending on δ and the resolution of the coarse grid; if δ is too negative, the multigrid method fails.

We used a test configuration for a resonator with different materials (cf. Fig. 2). The first eigenvalue ($\lambda = \omega^2$) is approx. $\lambda_1 \approx 0.0028$, the thirtieth eigenvalue is $\lambda_{30} \approx 0.014$, such that we compute a highly indefinite problem.

The reduction rates are shown in Table 2. It becomes apparent that δ should not be too negative. The coarse grid for level 0 has about 64,000 unknowns for the Maxwell problem, on level 1 about 500,000 unknowns; the finest problem on level 2 has about four million degrees of freedom. A finer coarse grid improves the multigrid reduction rate.

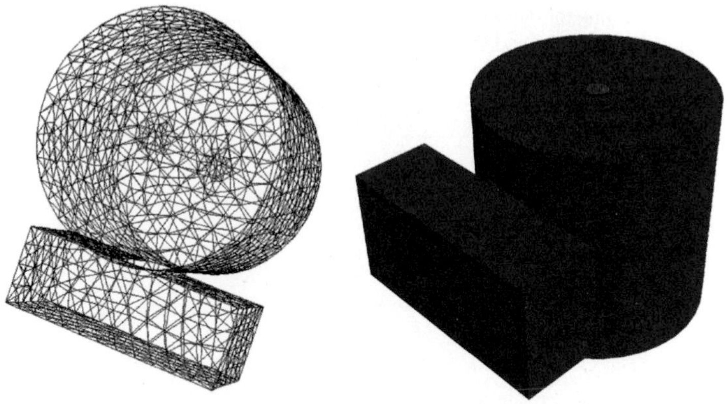

Fig. 2 Tetrahedral mesh and distribution of the permittivity ε of the resonator test configuration

Table 2 Multigrid convergence (average reduction rate) dependent on the shift δ with different coarse grids

δ	0.001	−0.01	−0.014	−0.02	−0.027	−0.038	−0.053	−0.075
Coarse problem size on level 0	0.5	0.64	0.76	0.61	0.75	0.90	0.97	0.96
Coarse problem size on level 1	0.5	0.58	0.68	0.50	0.73	0.78	0.95	0.96

4 Maxwell's Eigenvalue Problems

The eigenvalue problem to find eigenfunctions **H** and eigenfrequencies ω is given as

$$\begin{aligned} \nabla \times \left(\varepsilon^{-1}\nabla \times \mathbf{H}\right) &= \omega^2 \mathbf{H} && \text{in } \Omega, \\ \nabla \cdot \mathbf{H} &= 0 && \text{in } \Omega, \\ \mathbf{H} \times \mathbf{n} &= 0 && \text{on } \partial\Omega\,. \end{aligned}$$

To solve this problem, we use an extended LOBPCG method with projection. A part of the projection is a linear solver for the Laplace problem $-\Delta p = q$ to obtain divergence free approximations. The eigenvalue solver is preconditioned with a linear solver for the Maxwell problem. Both linear solver uses a multigrid preconditioner. The parallel direct solver is used as coarse grid solver. Here, we have an unstructured grid in a cube (cf. Fig. 3). The solutions for the eigenfrequencies are given in Table 3.

To verify our solution methods, we compared the results for a well-known benchmark configuration, the Fichera cube (using two different meshes, see Table 4).

Furthermore, a nested iteration method for the multigrid preconditioner was implemented. The problem is solved on the coarse grid and then prolongated to

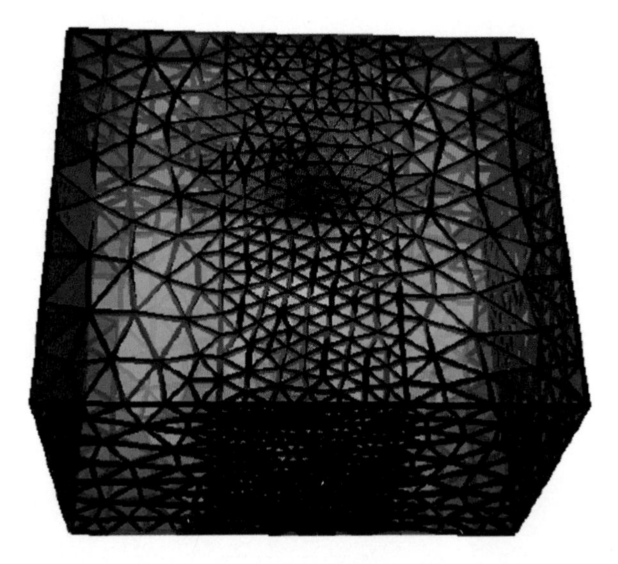

Fig. 3 Test configuration in the unit cube with an unstructured tetrahedral mesh

Table 3 Eigenfrequency results on different discretization levels

Level	Mode 1	Mode 2	Mode 3	Mode 4
0	5.68974259	6.49932331	7.42546687	8.36571135
1	5.68282081	6.48505091	7.41204638	8.02376504
2	5.68089728	6.48377476	7.44170987	8.05350607

Table 4 Finite element results on *tetrahedra* (left) and on *hexahedra* (right)

Dof	13,720	105,008	821,344	6,496,448	12,336	92,256	712,896	5,603,712
λ_1	3.117145	3.181490	3.205313	3.214284	3.192933	3.208178	3.214984	3.217878
λ_2	5.882258	5.880776	5.880493	5.880433	5.904488	5.886762	5.882139	5.880902
λ_3	5.886237	5.881586	5.880650	5.880463	5.904488	5.886762	5.882139	5.880902
λ_4	10.736136	10.702191	10.691251	10.687570	10.821483	10.724291	10.697177	10.689199
λ_5	10.776729	10.725720	10.706467	10.698809	10.838082	10.737451	10.707808	10.698519
λ_6	10.804685	10.731493	10.707554	10.699004	10.838082	10.737451	10.707808	10.698519
λ_7	12.297837	12.305227	12.311102	12.314126	12.443551	12.346881	12.323578	12.318066

the next refinement level. Here, the coarse grid solver has to be constructed only once. The nested approach yields good starting values for the LOBPCG iteration on the next level, so that this is far more efficient than the non-nested method, cf. Table 5.

Finally we compared the different performance of quadratic elements and linear elements. Therefore we used an unstructured grid of a container with a small slot, see Fig. 4. The solutions of linear elements on 512 cores and quadratic elements computed on 2,048 cores are summarized in Table 6.

Table 5 Number of LOBPCG iterations for the nested iteration method (upper part) compared with the non-nested method (lower part)

	Hexahedra			Tetrahedra		
Level	Maxwell	Laplace	# LOBPCG	Maxwell	Laplace	# LOBPCG
2	1,752	665	11	13,720	2,201	11
3	12,336	4401	7	105,008	15,921	7
4	92,256	31,841	6	821,344	120,929	6
5	712,896	241,857	6	6,496,448	942,273	6
6	5,603,712	1,884,545	5			
Total computing time	4:04 min			4:33 min		
5/6	5,603,712	1,884,545	12	6,496,448	942,273	12
Total computing time	9:10 min			7:33 min		

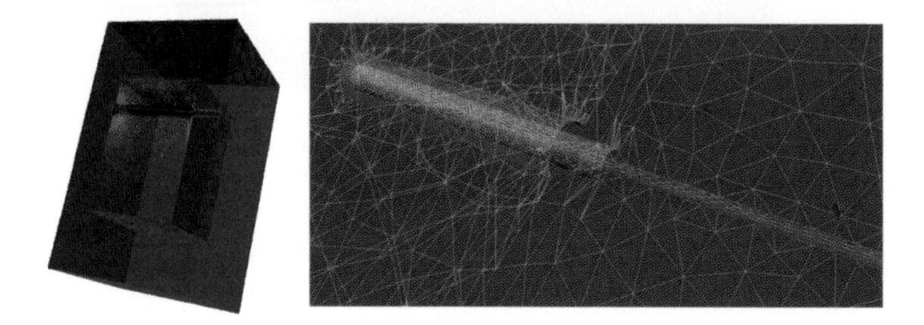

Fig. 4 Geometry of a container configuration (*left*) and geometrical details of the mesh (*right*)

Table 6 Comparison of the results for linear and quadratic elements

	Maxwell	Laplace
Linear elements (512 cores)		
Coarse problem size	698,231	105,629
Decomposition time	7.62 s	1.10 s
Fine level problem size	43,107,006	6,258,687
Quadratic elements (2,048 cores)		
Coarse problem size	3,721,946	803,860
Decomposition time	1:54 min	15.92 s
Fine level problem size	3,721,946	803,860

Mode	Linear (GHz)	Quadratic (GHz)
1	0.0000	0.0000
2	0.5313	0.5314
3	0.8036	0.8036
4	0.8891	0.8893
5	0.8920	0.8921
6	0.8969	0.8972
7	0.9381	0.9382
8	0.9927	0.9927
9	1.0441	1.0445
10	1.0474	1.0474

Acknowledgements We acknowledge the support of the BMBF within the joint research project ASIL (Advanced Solvers Integrated Library—01IH08014A), and we are grateful for the cooperation with CST, DARMSTADT for providing suitable test geometries.

References

1. D. Maurer, Ein hochskalierbarer paralleler direkter linearer Löser für Finite Elemente Diskretisierungen, Ph.D. thesis, KIT, 2013
2. D. Maurer, C. Wieners, A parallel block decomposition method for distributed finite element matrices. Parallel Comput. **37**, 742–758 (2011)
3. D. Maurer, C. Wieners, Parallel multigrid methods and coarse grid solver for Maxwell's eigenvalue problem, in *Proceedings of the CiHPC 2010: Competence in High Performance Computing*, ed. by G. Wittum, Gauss-Allianz (Springer, Berlin, 2012), pp. 205–213
4. C. Wieners, Distributed point objects. A new concept for parallel finite elements, in *Domain Decomposition Methods in Science and Engineering*, vol. 40, ed. by R. Kornhuber, R. Hoppe, J. Périaux, O. Pironneau, O. Widlund, J. Xu. Lecture Notes in Computational Science and Engineering (Springer, 2004), pp. 175–183
5. C. Wieners, A geometric data structure for parallel finite elements and the application to multigrid methods with block smoothing. Comput. Vis. Sci. **13**, 161–175 (2010)

Reduction of Numerical Sensitivities in Crash Simulations on HPC-Computers (HPC-10)

Christiana Eck, Yevgeniya Kovalenko, Oliver Mangold, Raphael Prohl, Anton Tkachuk, and Vladimir Trickov

Abstract For practical application in engineering numerical simulations are required to be reliable and reproducible. Unfortunately crash simulations are highly complex and nonlinear and small changes in the initial state can produce big changes in the results. This is caused partially by physical instabilities and partially by numerical instabilities. Aim of the project is to identify the numerical sensitivities in crash simulations and suggest methods to reduce the scatter of the results.

1 Numerical Sensitivities in Crash Simulation, asc(s

CAE-simulations allow us to recognize and evaluate the characteristics of a vehicle on the basis of a simulation at an early stage without having to build a real prototype. Already at the design engineering stage long before the first prototypes can be tested, we need reliable knowledge about the vehicles characteristics. Advances in applied mechanics, numerical methods and computer technology today permit the simulation of complex phenomena of automotive engineering. In the field of passive

C. Eck · V. Trickov
Automotive Simulation Center Stuttgart e. V., Nobelstrasse 15, 70569 Stuttgart, Germany
e-mail: christiana.eck@asc-s.de

Y. Kovalenko · O. Mangold
High Performance Computing Center Stuttgart, Nobelstrasse 19, 70569 Stuttgart, Germany
e-mail: kovalenko@hlrs.de

R. Prohl (✉)
Steinbeis Center of Innovation Simulation in Technology, Obere Steinbeisstrasse 43/2, 75248 Oelbronn-Duerrn
e-mail: raphael@techsim.org

A. Tkachuk
Institute of Structural Mechanics, Pfaffenwaldring 7, 70550 Stuttgart, Germany
e-mail: tkachuk@ibb.uni-stuttgart.de

W.E. Nagel et al. (eds.), *High Performance Computing in Science and Engineering '13*, DOI 10.1007/978-3-319-02165-2_48,

safety such simulations are used to analyze and optimize the structural behavior of the vehicles. For the realization of these simulations general-purpose-programs like ABAQUS, LS-DYNA, RADIOSS and PAMCRASH are used. Crash simulations are highly sensitive numerical experiments, this means small changes in the input can partly lead to large changes in deformation behavior of the car. On the other hand, physical instabilities due to, for example bifurcation behavior of the material under asymmetric loading, can appear and have to be captured and correctly solved by the numerical procedures in order to obtain reliable results. Achieving this accuracy and reliability is our main goal in this project. The project is build on the knowledge from industry and on the expertise of the participating scientists. We started with the examination of the stability of full-vehicle-models taken from the industry, then we reproduced their main features in reduced models in order to investigate the described instabilities.

Correspondently, one gets in simulations as well as in experiments substantially different results for almost identical input values. Thus the goal of simulation is rather to obtain physically present scatter and its characteristics (e.g. by means of statistics: mean value and standard deviation) than results of unite computations. This requires differentiation between physical and numerical scatter, which is not a trivial task. To investigate the causes of this scatter and to evaluate efficiency of methods that decrease these numerical scatter, parametric studies on reduced models have been done by the ASCS e.V. and IBB. The objective of the studies is to use these findings to develop strategies to avoid the appearance of artificial scatter in the calculated results. For the parametric studies the commercial software packages PAMCRASH, LS-DYNA, ABAQUS and RADIOSS were used. First we do our studies on a very simple model. We use a simple tube and crash it with a plate. The parametric studies show in all software packages similarities [3].

1.1 Parametric Studies on Reduced Model

As a reduced model we were provided with a test-configuration-model from Daimler. A front crash is modelled with 40 % overlap. The car moves with 56 km/h and meets with an obstacle. The mass of the car is 386 kg. The model exists of 222,880 elements and is counted for 55 ms. A crash requires all in all 120 ms. Figure 1 maps the test-configuration-model before the calculation is started. Figure 2 shows the extracted frame rail. We regard this component of the car because it is the most stressed component in a car crash. Figure 3 pictures the extracted frame rail after the computation. This will be the basic model because it is calculated with the standard settings. For the computation the whole front car always has to be computed. Afterwards the crashed component of interest can be extracted. This is necessary because during the calculation the results of each component depend on each other. For the sensitivity studies we have to run 20 calculations for every investigated parameter. In each calculation the initial velocity differs in the third place after the decimal point. We investigated several parameters as

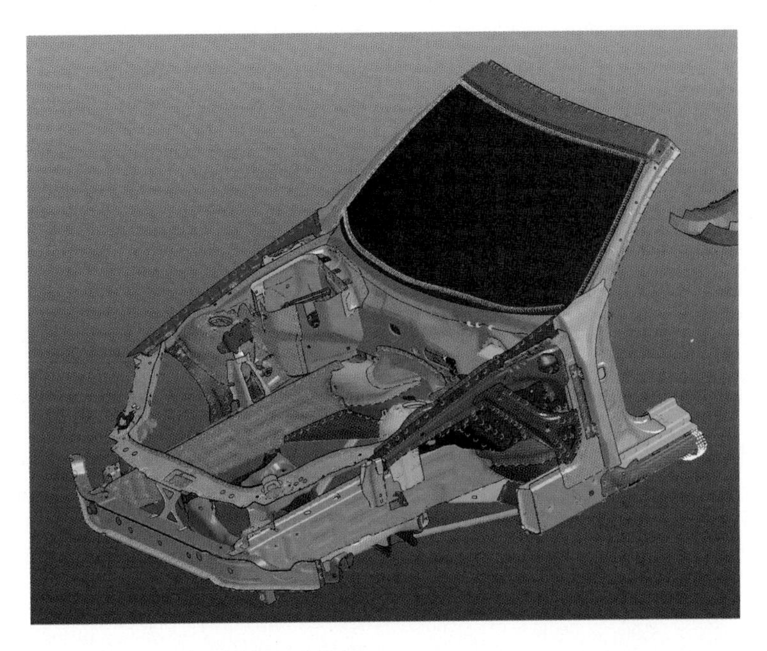

Fig. 1 Test-configuration-model from daimler

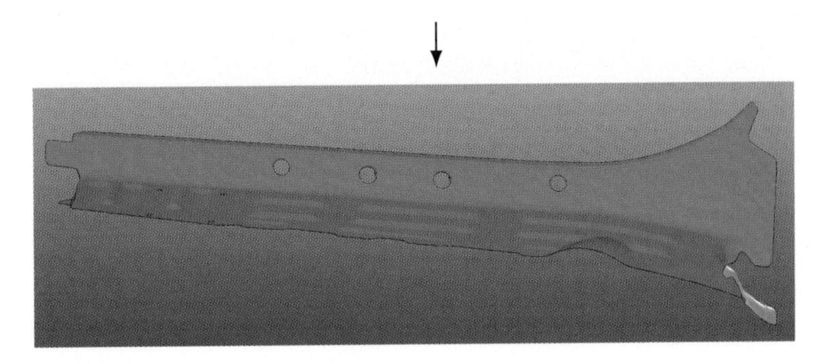

Fig. 2 Extracted frame rail before crashing

timestep, contact, mass-scaling, stiffness, element formulation and damping. For the scatter analysis we use the software DIFFCRASH of Fraunhoferinstitut SCAI. The software computes for the 20 calculation which we use to do for one investigated parameter the maximal scatter of one point in three dimensions.

$$PD3MX(t,p) = max_{1 \le i,j \le n}\left(\left|\left|X_i(t,p) - X_j(t,p)\right|\right|\right)$$

Figure 3 shows the component frame rail after crash which was computed by DIFFCRASH. On the colour scale the displacement in millimetres is visible for

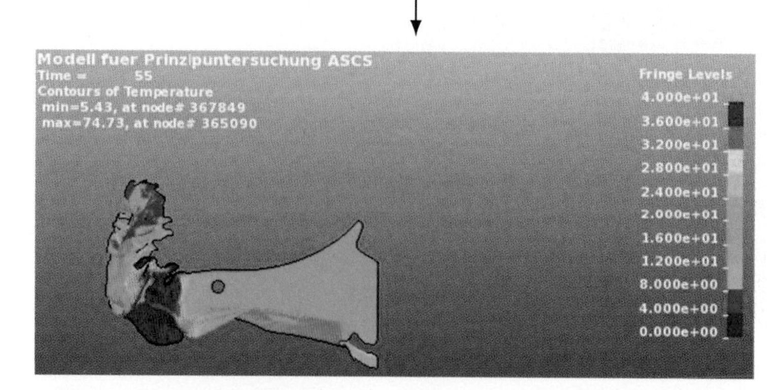

Fig. 3 Extracted frame rail after crashing

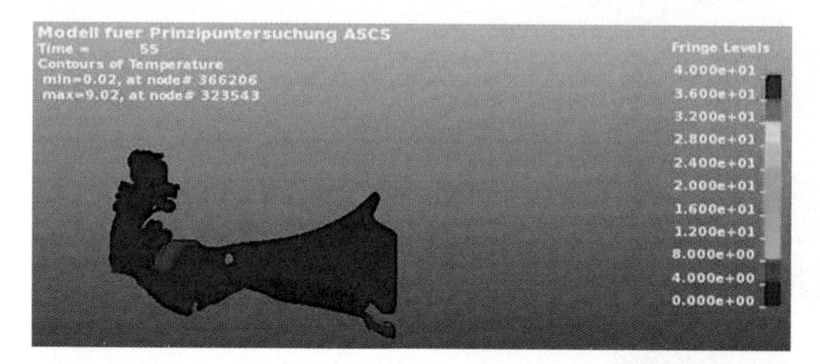

Fig. 4 Frame rail analysed by DIFFCRASH. Damping as the investigated parameter

each point within the 20 calculations. For the basic model the displacement in one point is even more than 4 cm.

Figure 4 shows the displacement of one point after increasing the damping parameter. One can see that the displacement of one point is in the buckling area less than 8 mm and the rest of the component shows displacements of less than 4 mm. Similar results are detected in Fig. 5. This figure maps a result of computation with a very small timestep $\Delta t = 0.1^{-4}$ ms. The basis model is computed by a timestep of $\Delta t = 0.7^{-3}$ ms (mass-scaling timestep). The stable timestep for this model is $\Delta t = 0.3^{-3}$ ms. This means that a much more smaller timestep than the stable one is required in order to obtain less scatter.

1.2 Optimization of the Investigations

The investigations show that if someone wants less scatter and more reliable results it is absolutely essential to calculate with a small timestep. This implies more time

Fig. 5 Frame rail analysed by DIFFCRASH. Small Timestep as the investigated parameter

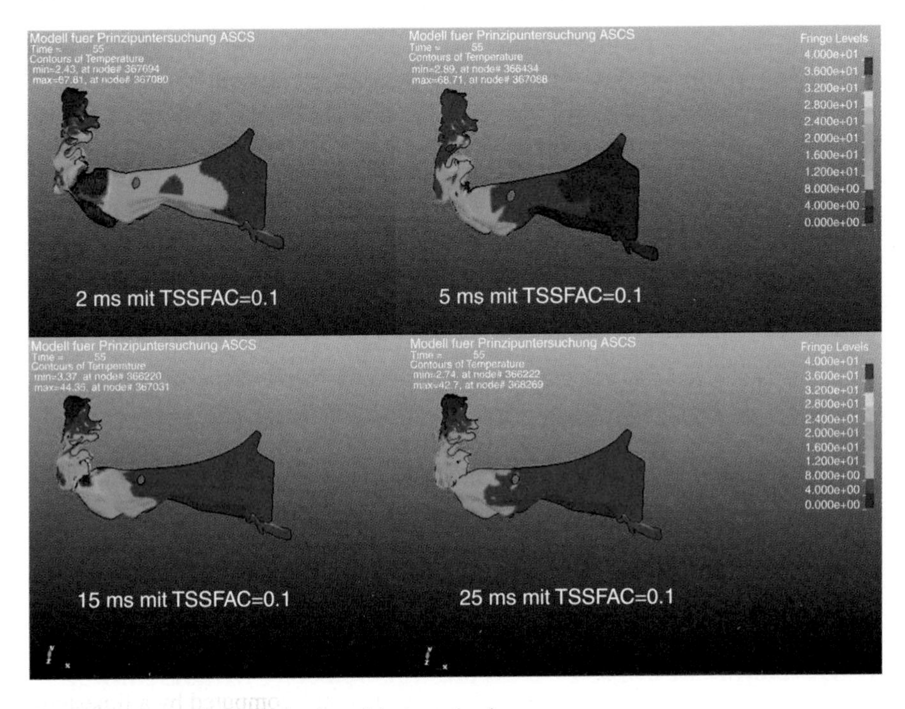

Fig. 6 Case example. Optimization of the investigation

for the computation and therefore more costs which is not realistic to pursue. So we come up with a case example as an optimization. In Fig. 6 the results of this case example are mapped.

The idea is to compute only little time with a small timestep $\Delta t = 0.1^{-4}$ ms. The rest of time until the endtime of 55 ms has been computed with the default timestep $\Delta t = 0.7^{-3}$ ms. The four spans of time in which the calculation runs with a small timestep differ. We choose 2, 5, 15 and 25 ms as you can see in Fig. 6. One can obtain that already with an endurance of 5 ms computation with small timestep the

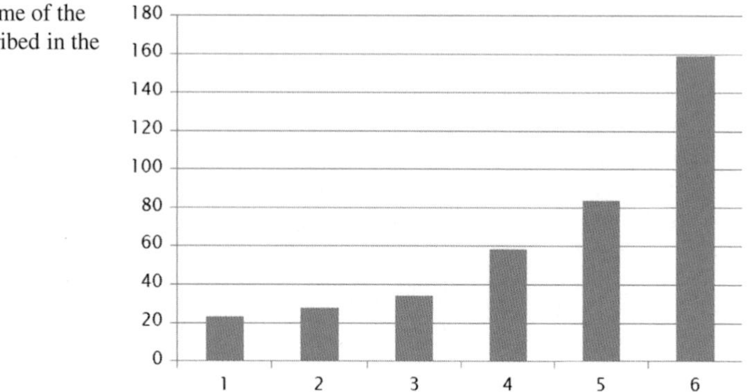

Fig. 7 Compute time of the case example described in the text

result is nearly scatter free. This is not uncommon if you have in mind that the most contact events occur in the first few milliseconds during a car crash.

This case example, of course, is model depending. But if someone wants to optimize the calculation, a possible approach is to first calculate the model and identify the point in time when the most contact events have past. This is the time duration the model should be computed with a small timestep. Subsequently the model can be computed with mass-scaling timestep until the endtime.

In Fig. 7 the computing time in minutes of the case example is shown. Beam number 1 is the computing time of the basic model which was computed with the mass-scaling timestep. Beam number 2 shows the computing time of the model when the first 2 ms have been computed with a small timestep. Beam number 3 shows the computing time of the model where 5 ms have been computed with a small timestep of $\Delta t = 0.1^{-4}$ ms and the rest of runtime until the endtime of 55 ms has been computed with the mass-scaling timestep of $\Delta t = 0.7^{-3}$ ms. In beam number 4 the duration of the calculation with the small timestep was 15 ms and beam number 5 shows the duration of 25 ms. Beam number 6 shows the computing time when the model is entirely computed with a small timestep.

2 Development of New Strategies for Selective Mass Scaling, IBB

The purpose of selective mass scaling (SMS) is to increase the critical time step for an explicit finite element analysis by adding artificial mass. Addition of mass reduces highest frequency, which limits the critical time step. But in order to obtain accurate structural response the lowest modes should not be affected, see Fig. 9. As it was shown in numerical experiment, SMS influences scatter of results. The reason of this may be direct from modification of inertia term and indirect due to the modification of the mass proportional damping force and error of iterative solver (SMS requires solution of linear system, which is usually done iteratively with a given accuracy). Derivation of SMS can be done purely algebraic [8] or using modified weak formulation [1]. The analysis of the approach is done via spectra for linear case or dynamic calculation in non-linear, which is insufficient.

Here we propose an alternative variational formulation that delivers a new family of SMS methods. The SMS method proposed in [11] can be obtained from the new formulation as particular case (see below). The starting point of formulation is modified Hamilton's principle with three penalty terms:

$$\begin{aligned} H^\circ(\mathbf{u},\mathbf{v},\mathbf{p}) &= \int_0^{t_{end}} \left(T^\circ - \Pi^{int} + \Pi^{ext}\right) \mathrm{d}t \to \text{stat}, \\ T^\circ &= \frac{1}{2}(\rho\dot{\mathbf{u}},\dot{\mathbf{u}}) + \frac{C_1}{2}\left(\rho\dot{\mathbf{u}} - \mathbf{p}, \dot{\mathbf{u}} - \frac{\mathbf{p}}{\rho}\right) \\ &+ \frac{C_2}{2}\left(\rho\mathbf{v} - \mathbf{p}, \mathbf{v} - \frac{\mathbf{p}}{\rho}\right) + \frac{C_3}{2}\left(\rho(\dot{\mathbf{u}} - \mathbf{v}), \dot{\mathbf{u}} - \mathbf{v}\right). \end{aligned} \tag{1}$$

where $\mathbf{u}$, $\mathbf{v}$ and $\mathbf{p}$ are displacement, velocity and momentum, respectively, T°, Π^{int} and Π^{ext} are kinetic and potential energy, ρ is density, C_{1-3} are penalty factors. The scalar product over domain $\mathscr{B}$ is denoted $(a,b) = \int_{\mathscr{B}} ab\,\mathrm{d}\mathscr{B}$. Independent discretization of displacement, velocity and momentum yield following matrix expressions

$$\begin{aligned} \mathbf{M} &= \int_{\mathscr{B}} \rho\mathbf{N}^T\mathbf{N}\,\mathrm{d}\mathscr{B} \quad & \mathscr{A} &= \int_{\mathscr{B}} \rho\mathbf{N}^T\boldsymbol{\psi}\,\mathrm{d}\mathscr{B} \quad & \mathscr{C} &= \int_{\mathscr{B}} \mathbf{N}^T\boldsymbol{\chi}\,\mathrm{d}\mathscr{B} \\ \mathscr{Y} &= \int_{\mathscr{B}} \rho\boldsymbol{\psi}^T\boldsymbol{\psi}\,\mathrm{d}\mathscr{B} \quad & \mathscr{G} &= \int_{\mathscr{B}} \boldsymbol{\psi}^T\boldsymbol{\chi}\,\mathrm{d}\mathscr{B} \quad & \mathscr{H} &= \int_{\mathscr{B}} \rho^{-1}\boldsymbol{\chi}^T\boldsymbol{\chi}\,\mathrm{d}\mathscr{B}. \end{aligned} \tag{2}$$

with $\mathbf{M}$ being the consistent mass matrix. Following derivation in [11], local elimination of velocity and momentum parameters is conducted. This leads to equation of motion in form

$$\mathbf{M}^\circ\ddot{\mathbf{U}} + \mathbf{K}\mathbf{U} = \mathbf{F}^{\text{ext}}, \tag{3}$$

where $\mathbf{M}^\circ$ is a modified mass matrix. It can be computed as follows

$$\mathbf{M}^\circ = (1 + C_1 + C_3)\mathbf{M} - \begin{bmatrix} C_3\mathscr{A} \\ C_1\mathscr{C} \end{bmatrix} \begin{bmatrix} (C_2 + C_3)\,\mathscr{Y} & -C_2\mathscr{G} \\ C_2\mathscr{G}^T & (C_1 + C_2)\,\mathscr{H} \end{bmatrix}^{-1} \begin{bmatrix} C_3\mathscr{A}^T \\ C_1\mathscr{C}^T \end{bmatrix}. \tag{4}$$

The template for mass matrix (4) depends on three penalty parameters. Herein case with $C_2 = C_3 = 0$ is used.

This approach is much more general than algebraically constructed SMS [8] and recovers it in special cases, e.g. for simplex elements (2-node line, triangle and tetrahedral) and constant velocity field. The efficiency of the proposed SMS can be

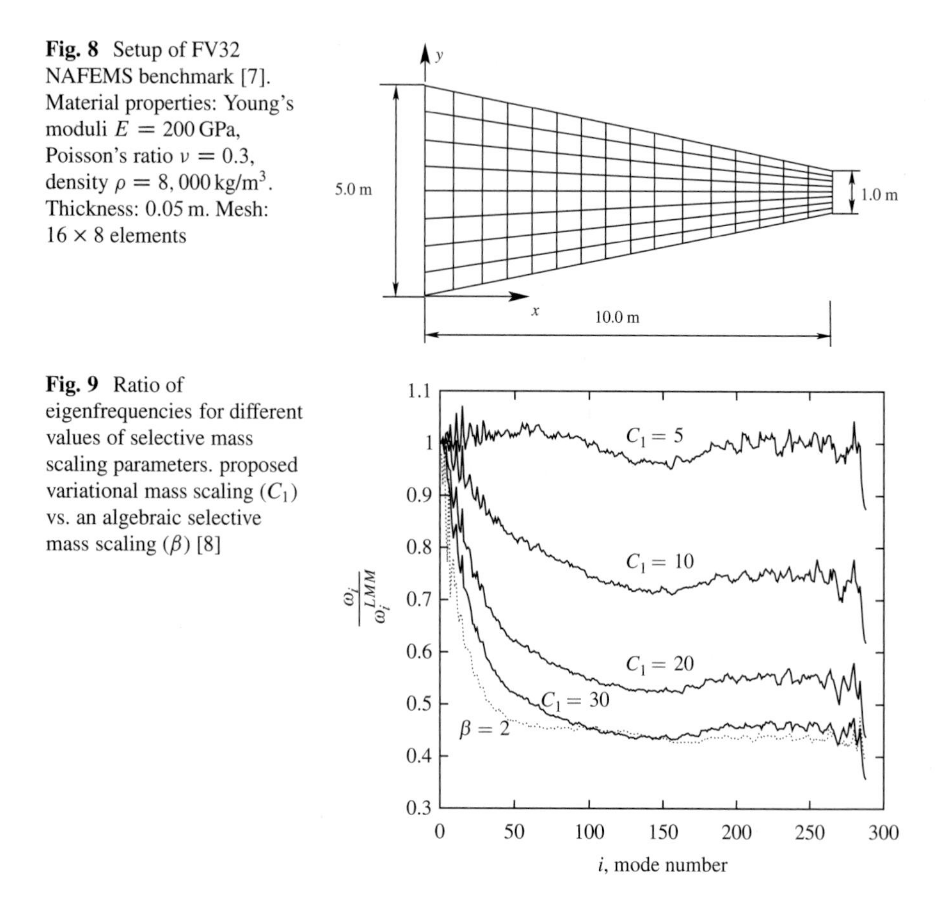

Fig. 8 Setup of FV32 NAFEMS benchmark [7]. Material properties: Young's moduli $E = 200\,\text{GPa}$, Poisson's ratio $\nu = 0.3$, density $\rho = 8,000\,\text{kg/m}^3$. Thickness: 0.05 m. Mesh: 16×8 elements

Fig. 9 Ratio of eigenfrequencies for different values of selective mass scaling parameters. proposed variational mass scaling (C_1) vs. an algebraic selective mass scaling (β) [8]

seen from results of eigenfrequency benchmark Fig. 8. The ratio of eigenfrequencies with scaled mass matrix to eigenfrequencies from consistent mass matrix is shown on the right side of Fig. 9. For $C_1 = 20$ and $\beta = 2$ the maximum frequency is halved with change in lowest 50 modes being less than 5 %. For the same reduction of maximum frequency method [8] requires change in lowest 50 modes by 35 %. This may cause unphysical effects in explicit time integration. Another benchmark considers a transient problem for cantilever beam modeled with 8-node hexahedral elements. Initial zero displacements and velocities are prescribed. The beam is loaded at the tip by an abrupt force F. The quality of solution is observed from the history of the tip displacement w. The deflections obtained with a LMM and the proposed SMS are almost identical even for high value of scaling (Figs. 10 and 11). The method [8] yields larger error for the same reduction of time step. However further studies are required in order to prove efficiency of proposed scheme for distorted meshes and large rotations.

Calculation of the global acceleration vector from Eq. (3) for large systems is difficult with direct solvers because of memory limitation. Therefore preconditioned

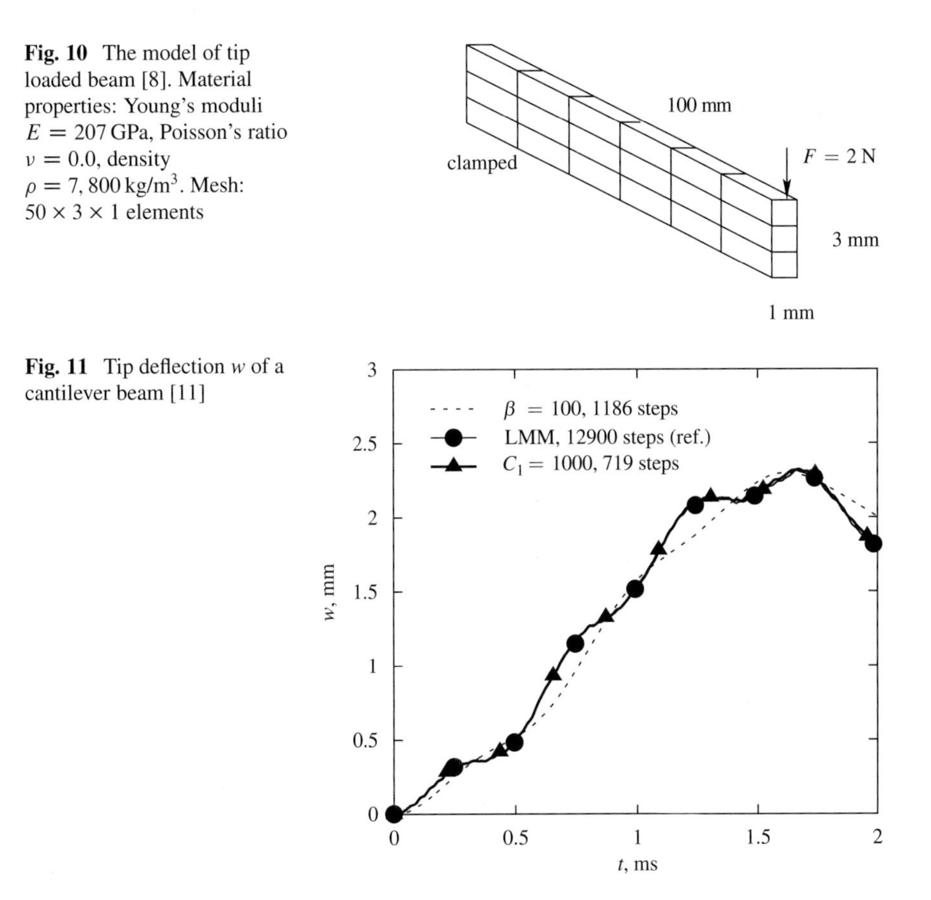

Fig. 10 The model of tip loaded beam [8]. Material properties: Young's moduli $E = 207\,\text{GPa}$, Poisson's ratio $\nu = 0.0$, density $\rho = 7,800\,\text{kg/m}^3$. Mesh: $50 \times 3 \times 1$ elements

Fig. 11 Tip deflection w of a cantilever beam [11]

conjugate gradient (PCG) with Jacobi preconditioner is used [9]. Then the computational cost strongly depends on sparsity and conditioning of matrix $\mathbf{M}^{\circ}$. Large conditioning requires a large number of iterations, which increase the overhead on SMS, see Fig. 12. Small values of mass scaling does not increase the critical time-step a lot. Hence, the speed-up is observed at specific interval of mass scaling factor C_1. Hence, finding these dependencies is crucial for understanding of method.

Consider a tapered plate. Model data is given in Fig. 13. Numerical test are conducted in in-house finite element code *NumPro*. The stiffness matrix is computed using pure displacement formulation with $2 \times 2 \times 2$ and 1 point quadrature for 8-node hexahedral and 4-node tetrahedral elements, respectively. The mass matrix is computed with quadrature rule of 2 order higher, i.e. $3 \times 3 \times 3$ and 4 point quadrature for 8-node hexahedral and 4-node tetrahedral elements, respectively. Velocity field is interpolated with six modes (all rigid body modes). Exact expression for velocity ansatz $\boldsymbol{\psi}$ can be found in [11].

The critical time-step is estimated using iterative algorithm, see for example [2]. The conditioning of a mass matrix is computed using spectral norm. The highest and the lowest eigenfrequency of a mass matrix are computed with subspace

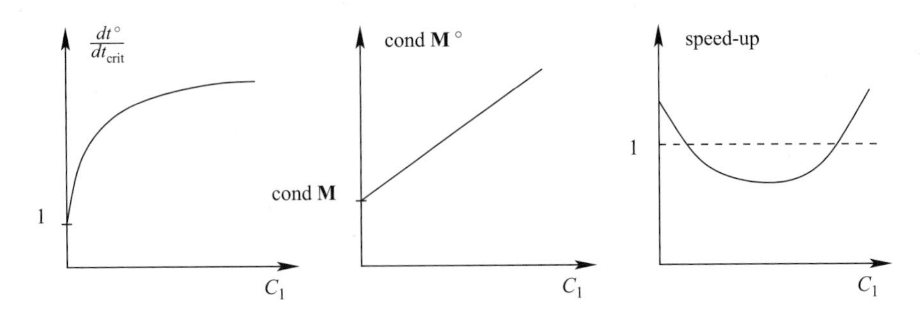

Fig. 12 Schematic dependency of critical time-step (*left*), conditioning (*middle*) and speed up (*right*) on mass scaling factor C_1

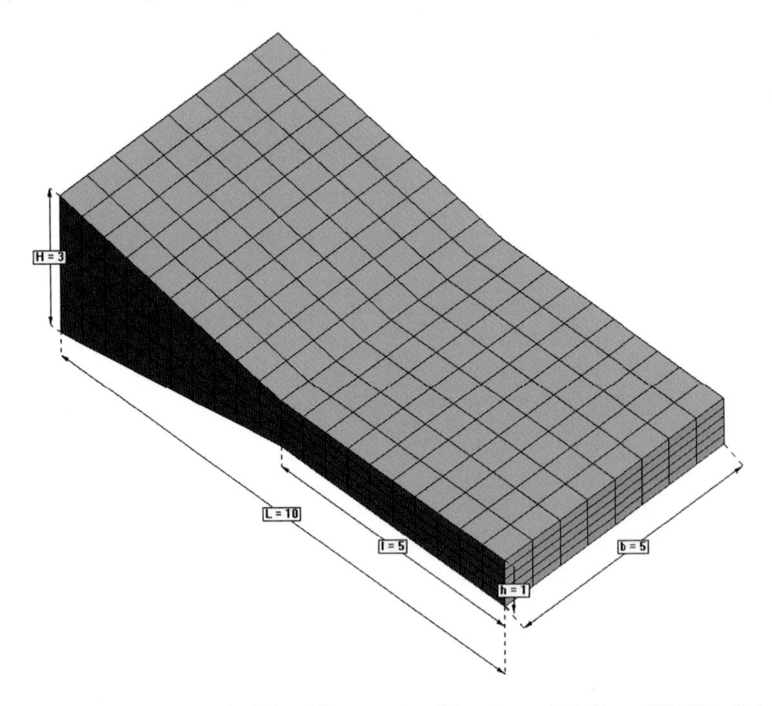

Fig. 13 Tapered plate example. Material properties: Young's moduli $E = 207$ GPa, Poisson's ratio $\nu = 0.3$, density $\rho = 7,800\,\text{kg/m}^3$. Geometry: total length $L = 10$ m, length of thin part $l = 5$ m, height in root $H = 3$ m, height of tip $h = 1$ m, width $b = 5$ m. Boundary conditions: clamped in root $u_x = u_y = u_z = 0$. Mesh: for 8-node hexahedral $20 \times 5 \times 8$ elements (1,134 nodes), for 4-node tetrahedral 19,200 elements (4,634 nodes)

iteration method. Obtained critical time-step and conditioning of mass matrix values are presented at Tables 1 and 2. Least square fit of the data yields following approximated relations for both types of elements

$$\frac{dt^\circ_{\text{crit}}}{dt_{\text{crit}}} \approx \sqrt{1 + \frac{C_1}{2}} \quad \text{cond}\,\mathbf{M}^\circ \approx \text{cond}\,\mathbf{M}_{\text{CMM}} + C_1 \tag{5}$$

Table 1 Critical time-step and conditioning of mass for tapered plate example

	VSMS, C_1							
Mass type	LMM	CMM	10	30	100	300	1,000	2,300
dt_{crit}, (μs)	14.39	7.24	18.75	30.43	53.85	92.30	167.86	254.3
cond **M**	33.73	74.7	50.54	83.70	170.0	380.1	1066.5	2312.3

Computed with 4-node tetrahedral element

Table 2 Critical time-step and conditioning of mass for tapered plate example

	VSMS, C_1								
Mass type	LMM	CMM	2	5	10	30	100	300	1,000
dt_{crit}, (μs)	33.7	19.0	32.84	46.50	63.00	105.9	191.3	330.7	602.7
cond **M**	22.53	140.3	62.21	61.36	168.1	96.62	192.1	436.0	1222

Computed with 8-node hexahedral element

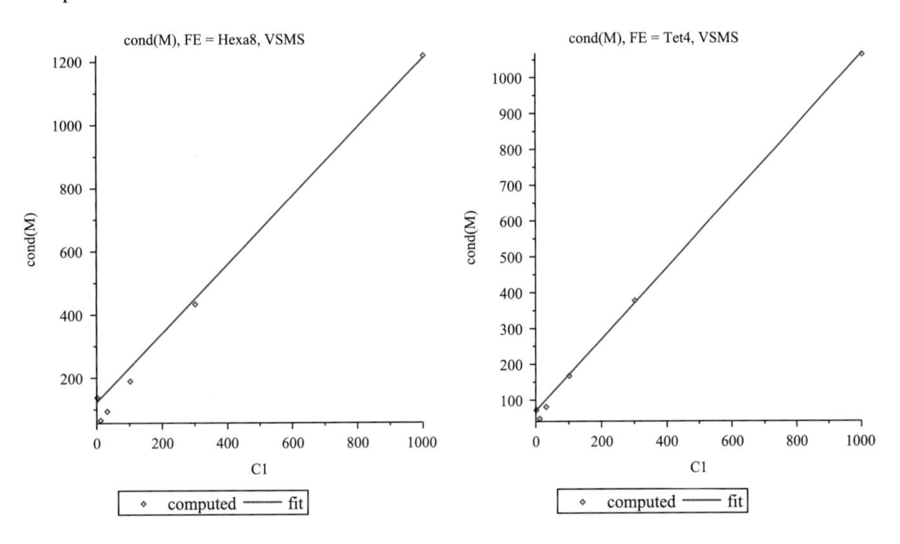

Fig. 14 Conditioning for 8-node hexahedral (*left*) and 4-node tetrahedral (*right*) elements

The quality of such fit is shown in Fig. 14. Growth of the critical time-step and conditioning correspond with one reported in paper [9], where algebraic SMS is used. This means that the proposed method can be potentially used in commercial explicit finite element codes such as *LS-Dyna* and *RADIOSS*. Further relations for conditioning of mass matrix and critical time-step can be found [12].

It is important to state that SMS can be viewed as a non-conventional order reduction scheme (or filter). Increasing inertia of higher modes, their contribution in structural response is significantly reduced. Therefore less mode participate in solution. At the same time the contact is still collocated at all nodes, non-linear material laws are evaluated at each Gauss point of element, which does not reduce the accuracy of results. However, artificial effects may be met during element erosion and material fracture. This question will be studied in following works.

3 Automatic Generation of Efficient Finite Element Codes: Application to Mass Matrix Computations, HLRS

Development of new finite elements includes testing of several competitive formulations. Developer may change shape functions, independent fields, weak forms or quadrature rules, which is often time-consuming and error-prone. Application of automatic generation tool saves time and excludes programming errors, which allows a developer to focus on mathematical formulation rather than implementation details. A recently developed at HLRS tool *Finite Element Generator* can be used for such purpose, see [3]. In addition it generates highly optimized code for common programming languages (C/C++, Fortran 90, CUDA C, CUDA Fortran, OpenCL). This code can be used later for large problems ($> 10M$) on supercomputers.

Efficiency of this tool is demonstrated on the example of mass matrix templates for 10-node tetrahedral element. At the initial stage the goal is to test the element with consistent mass matrix and two types of mass scaling. (variational mass scaling discussed above and so-called enhanced selective mass scaling [5]). Expressions for computation of CMM and VSMS are given in (2). The enhanced selective mass scaling (eSMS) can be formulated as follows

$$\mathbf{M}_{\mathrm{eSMS}} = \int_{\mathscr{B}} \rho \mathbf{N}^T \mathbf{N} \, \mathrm{d}\mathscr{B} + d_1^2 \int_{\mathscr{B}} \rho \nabla \mathbf{N}^T : \nabla \mathbf{N} \, \mathrm{d}\mathscr{B}, \tag{6}$$

with ∇ being a gradient operator and d_1 being a characteristic scaling factor, dimension [m]. The efficient values for d_1 lies in range 0.1 to 0.7 from element size.

The 10-node tetrahedral element can be defined in *Finite Element Generator* by specifying the element type (Tetrahedron), nodes and integration rules. In our example we are dealing with elements, which have different types of nodes. The `SecondOrder` node type consists of 10 nodes and it is used for approximation of displacements. The `FirstOrder` node type is used for approximation of velocities and has only 4 nodes. Each node type is described with its coordinates and base set for the shape functions. At the moment base set can only be polynomials in the element coordinates. The `FirstOrder` and the `SecondOrder` node types use complete linear and quadratic sets of polynomials, respectively. *Finite Element Generator* will automatically generate shape functions accordingly to the base set. All input parameters are checked on correctness otherwise will be produced diagnostic output. The definition of the tetrahedral element can be done as follows

```
Element Tetraeder10 {
  Type = Tetrahedron;
  Nodes SecondOrder {
    Coordinates {
      {0,0,0},
      {0,0,1},
      {0,1,0},
      {1,0,0},
```

```
      {0.5,0,0},
      {0,0,0.5},
      {0,0.5,0},
      {0.5,0,0.5},
      {0.5,0.5,0},
      {0,0.5,0.5}
    }
    ShapeFunctions {
      Polynomial { 1 }, Polynomial { x },
      Polynomial { y }, Polynomial { z },
      Polynomial { x^2}, Polynomial { y^2},
      Polynomial { x y }, Polynomial { x z },
      Polynomial { y z }, Polynomial { z^2 },
    }
  }
  Nodes FirstOrder {
    Coordinates {
      {0,0,0},
      {0,0,1},
      {0,1,0},
      {1,0,0},
     }
    ShapeFunctions {
      Polynomial { 1 }, Polynomial { x },
      Polynomial { y }, Polynomial { z },
     }
  }
  Coordinates = SecondOrder;
  Integration myRule {
    0.0417 -> { 0.5854,0.1381,0.1381 },
    0.0417 -> { 0.1381,0.5854,0.1381 },
    0.0417 -> { 0.1381,0.1381,0.5854 },
    0.0417 -> { 0.1381,0.1381,0.1381},
  }
  Integration Felippa15 {
  0.01199 ->  { 0.09197, 0.09197, 0.09197},
  0.01199 ->  { 0.09197, 0.72409, 0.09197},
  0.01199 ->  { 0.09197, 0.09197, 0.72409},
  0.01199 ->  { 0.72409, 0.09197, 0.09197},
  0.01151 ->  { 0.31979, 0.31979, 0.31979},
  0.01151 ->  { 0.31979, 0.04062, 0.31979},
  0.01151 ->  { 0.31979, 0.31979, 0.04062},
  0.01151 ->  { 0.04062, 0.31979, 0.31979},
  0.00882 ->  { 0.05635, 0.44365, 0.05635},
  0.00882 ->  { 0.05635, 0.44365, 0.44365},
  0.00882 ->  { 0.05635, 0.05635, 0.44365},
  0.00882 ->  { 0.44365, 0.05635, 0.05635},
  0.00882 ->  { 0.44365, 0.05635, 0.44365},
  0.00882 ->  { 0.44365, 0.44365, 0.05635},
  0.01975 ->  { 0.25000, 0.25000, 0.25000},
  }
}
```

Within the element section different integration rules can be provided and then in matrix description can be specified which rule to use. Note, that the decimal representation of quadrature rules is reduced to five digits for readability. In the matrix section first of all must be defined to which elements it belongs to. Constants are simple parameters to the element computation, which do not depend on the location within the element. Coefficients are parameters which vary with the location within the element. For each node in the node set a value for the coefficient variable must be given for computations and the shape function set associated with the node set will be used for interpolation of the variable value [6].

Within a matrix section multiple different operators can be defined, e.g. to allow computation of multiple element matrices in one step, requiring fewer evaluations of shape functions and Jacobians. Only one integration rule can be defined per matrix, so all operators within are integrated using the same set of integration points. A matrix operator has to define the set of variables, test functions and equation to compute. A keyword `Variables` specifies the unknown functions over which a variation will be done in the element matrix computations. A keyword `TestFunctions` declares the weighting functions used for the Galerkin discretization approach. Test functions and variables also need a node set to be specified. Multiple node sets can be given. The equation to compute may depend on any variable, coefficient and constant, with the restriction that it must be linear in the vector of variables. CMM, VSMS and eSMS can be written in language of *Finite Element Generator* as follows

```
 Matrix consistentmass {
  Elements { Tetraeder10 }
  Integration = Felippa15;
  Coefficients { SecondOrder:dens }
  Operator leftSide {
    Variables { SecondOrder:u }
    TestFunctions { SecondOrder:u_t }
    Expression {
        u_t*dens*u
    }
  }
}
Matrix enhancedSelectiveMassScaling {
  Elements { Tetraeder10 }
  Integration = Felippa15;
  Constants { d1 }
  Coefficients { SecondOrder:dens }
  Operator leftSide {
    Variables { SecondOrder:u }
    TestFunctions { SecondOrder:u_t }
    Expression {
        u_t*dens*u+d1*d1*dens*(xx+yy+zz)
    }
  }
  Macros {
    xx=diff(u_t,x)*diff(u,x);
    yy=diff(u_t,y)*diff(u,y);
```

```
      zz=diff(u_t,z)*diff(u,z);
    }
  }
  Matrix VariationalSelectiveMassScaling {
    Elements { Tetraeder10 }
    Integration = Felippa15;
    Constants { C1 }
    Coefficients {  SecondOrder:dens }
    Operator leftSide {
      Variables { SecondOrder:u, FirstOrder:v  }
      TestFunctions { SecondOrder:u_t, FirstOrder:v_t }
      Expression {
          (1+C1)*u_t*dens*u +
          C1*dens*( v_t*(v-u) - u_t*v )
      }
    }
  }
```

The variational mass scaling above use linear ansatz space for velocity. Fifteen point quadrature rule is used for each element [4]. This quadrature results in exact values for matrices even for variable metric element (curved faces and sides).

For those matrices the code, which consist of more than 2,000 lines, will be generated in less than a second. Maybe also benchmarking results...

3.1 Conclusions

Variational justification of selective mass scaling (SMS) given above provides us a basis for research of this technique. Because dynamic contact algorithms and mass proportional damping rely on mass matrix, robustness of the model and scatter of results may depend on the form of SMS. In addition new method for SMS is developed and tested. We obtained dependency of conditioning of mass matrix and critical time-step on mass scaling factor.

Automatic code generation for mass matrices proved to be fast and flexible. It will be employed in future numerical tests.

4 Models and Robust Algorithms for the Finite Elasto-plastic Problem in Crash Test Simulations, SFI

The Steinbeis-Center of Innovation "Simulation in Technology" works on the formulation of appropriate mathematical models and the development of numerical methods for a crash test simulation. To this end the simulation tool UG ("Unstructured Grids") is used which has been developed by the group of Prof. G. Wittum, cf. Sect. 4.3.1. In particular, UG provides a flexible grid manager with finite-element discretizations of arbitrary order, an import-export system which permits the coupling of several partial differential equations in a simple way and a large library of robust, efficient and fast solvers for linear systems, e.g. geometric and algebraic multigrid-methods, [13].

In order to formulate an appropriate mathematical model for a crash test simulation we started with simplified versions of some full-vehicle-models used by automotive industries. Firstly we focused on geometrical nonlinearities (nonlinear kinematic), physical nonlinearities (nonlinear material law) and elasto-plastic material behavior (balance law combined with an evolution law for the plastic distortions).

4.1 Finite Elasto-plastic Model Formulation

In the theory of plasticity exist a lot of model formulations which describe the plastic flow in a different manner. The model should map the material behavior correctly as well as the description needs to be stable and well suited for the mathematical treatment. Classically, plasticity models are based on the assumption that the plastic strains evolute normal to the yield surface which describes the current elastic region of the material. Those flowrules are called associative.

We started with an associative model formulation capturing large elastic and plastic strains, cf. [10], in order to examine the stability of some algorithms applied in the industry.

4.1.1 Balance Law

We consider the balance of linear momentum in local, material form:

$$\begin{aligned} -\,\mathrm{DIV}(\mathbf{FS}) &= \rho_0\mathbf{B} \quad \text{in } \Omega \\ \mathbf{u} &= \mathbf{d} \qquad \text{on } \Gamma_d \end{aligned} \tag{7}$$

In (7), $\mathbf{F}$ denotes the deformation gradient, $\mathbf{S}$ is the second Piola-Kirchhoff stress tensor, ρ_0 is the reference density, $\mathbf{B}$ is the body force per unit mass, and $\mathbf{d}$ represents the prescribed displacement on the Dirichlet boundary of the computational domain. For the first part of our study, inertial terms are not accounted for the force $\rho_0\mathbf{B}$. Apparently, dropping inertial terms in a mathematical model that should be the basis for crash test simulations may sound as a strong contradiction. However, our first goal is to analyze and compare algorithms that should be applied in the numerical simulations of crash tests. In this respect, inertial terms are, at this stage, only temporarily switched off. This saves computational resources when different algorithms are compared, and allows for focusing on the possible numerical instabilities that, hidden behind a given algorithm, may exist independently on the consideration of inertial terms.

We restrict our investigations to a purely mechanical framework. Consequently, thermal phenomena are excluded from the outset and dissipation is expressed in terms of mechanical quantities only.

4.1.2 Constitutive Equations, Associative Flow Rule and Karush-Kuhn-Tucker Conditions

We consider the formulation of the J2 flow theory at finite strains for hyperelastic-plastic materials [10]. The constitutive equations, the flow rule and the KKT-conditions read:

$$\mathbf{S} = 2\frac{\partial W}{\partial \mathbf{C}} \tag{8}$$

$$\frac{\partial \overline{\mathbf{C}}_p^{-1}}{\partial t} = -\frac{2}{3}\gamma \cdot \mathrm{tr}\,[\mathbf{b}^e]\,\mathbf{F}^{-1}\frac{\mathrm{dev}[\boldsymbol{\tau}]}{\|\,\mathrm{dev}[\boldsymbol{\tau}]\|}\mathbf{F}^{-T} \tag{9}$$

$$\gamma \geq 0, \quad f(\boldsymbol{\tau}) \leq 0, \quad \gamma f(\boldsymbol{\tau}) = 0 \tag{10}$$

In (8)–(10), W is a stored energy-function, $\mathbf{C}$ is the Cauchy-Green strain tensor, $\mathbf{b}_e$ is the elastic part of the finger tensor, $\boldsymbol{\tau}$ is the Kirchhoff stress tensor, f is the von-Mises flow-condition and γ is the (Karush-Kuhn-Tucker) plastic multiplier.

4.2 Numerical Methods

The numerical computations of the quasi-static case are performed by an incremental procedure which contains a nonlinear sub-problem in every single incremental step. To treat the governing equations of plasticity, we firstly adopted the well-established return-mapping-algorithm and the consistent-tangent-method for the linearization therein. As remarked in [14], this plasticity-algorithm may be turn instable because it computes stresses that do not necessarily satisfy the global equilibrium equations. Therefore we developed a new class of solution methods with better mathematical properties by reviewing the plasticity problem in the context of nonlinear programming. We achieve higher robustness by additionally linearizing the constitutive equations and the governing equations of plasticity, (8)–(10), with respect to the plastic strains. Hence, our method is an extension of the algorithm proposed in [14] to the case of finite deformations.

Since the new plasticity algorithm is set up in a very general framework, namely the linearization of the KKT-system, the method is capable to be applied to more complex plastic flow behavior like e.g. non-associative flowrules. Classical approaches, e.g. the return-mapping-algorithm, are restricted to models with associative flow rules.

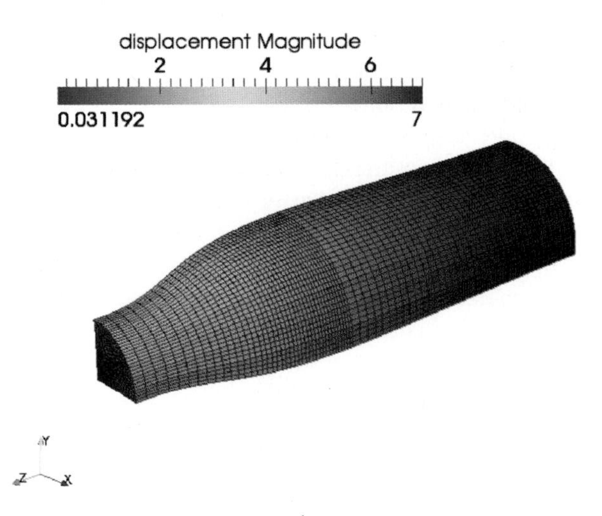

Fig. 15 Deformation behavior of a circular bar ($\frac{1}{8}$ of the original geometry) in a tensile test. This test is enforced by an incrementally applied Dirichlet condition at the back of the domain up to $\mathbf{u}_{back} = 7.0\,\mathrm{mm}$

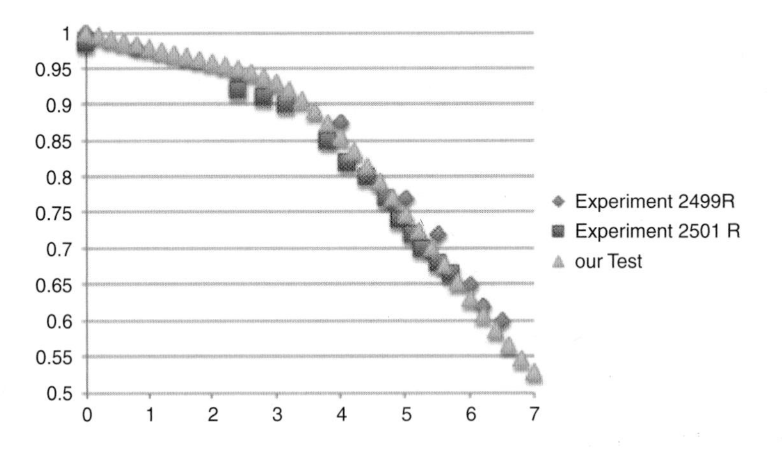

Fig. 16 Change of the sectional area in the necking zone (current radius/initial radius) plotted against the Dirichlet displacement at the back. Our test results are compared to experiments reported in [10]

4.3 Numerical Tests

The implementations of the return-mapping-algorithm and the general plasticity algorithm in UG have been checked by some benchmark problems. As a reference for our numerical tests we used the well-documented necking of a circular bar as an example for exponential hardening behavior [10].

For the model presented in Sect. 4.1 we obtained same results for the return-mapping as well as for the general plasticity-algorithm. These results are plotted by the green line in Fig. 16. The final state of deformation is visualized in Fig. 15.

4.3.1 Software-Framework UG

UG is a general-purpose library for the solution of partial differential equations, which supports parallel adaptive multigrid-methods on high-performance computers. A novel implementation ensures the complete independence of grid and algebra, [13]. Cache aware storage for algebra structures and a parallel communication layer make UG 4 well suited for current and next hardware-architectures.

References

1. H. Askes, D. Nguyen, A. Tyas, Increasing the critical time step: micro-inertia, inertia penalties and mass scaling. Comput. Mech. **47**, 657–667 (2011). doi:10.1007/s00466-010-0568-z
2. D. Benson, Y. Bazilevs, M. Hsu, T. Hughes, Isogeometric shell analysis: the Reissner-Mindlin shell. Comput. Methods Appl. Mech. Eng. **199**, 276–289 (2010). doi:10.1016/j.cma.2009.05.011
3. Ch. Eck, O. Mangold, R. Prohl, A. Tkachuk, Reduction of numerical sensitivities in crash simulations on HPC-computers (HPC-10). in *High Performance Computing in Science and Engineering '12*, ed. by W.E. Nagel, D.H. Krõner, M.M. Resch (Springer, Berlin, 2013), 547–560. doi:10.1007/978-3-642-33374-3_39
4. C.A. Felippa, A compendium of FEM integration formulas for symbolic work, Eng. Comput. **21**, 867–890 (2004). doi:10.1108/02644400410554362
5. S. Gavoille, Enrichissement modal du selective mass scaling (in French), in *Proceedings of 11e Colloque National en Calcul des Structures by Calcul des Structures et Modélisation* (CSMA, 2013)
6. O. Mangold. *Finite Element Generator User's Manual* (2012)
7. National Agency for Finite Element Methods & Standards (Great Britain), *The Standard NAFEMS Benchmarks* (NAFEMS, Hamilton, 1990)
8. L. Olovsson, K. Simonsson, M. Unosson, Selective mass scaling for explicit finite element analyses. Int. J. Numer. Methods Eng. **63**, 1436–1445 (2005). doi:10.1002/nme.1293
9. L. Olovsson, K. Simonsson, Iterative solution technique in selective mass scaling. Commun. Numer. Methods Eng. **22**, 77–82 (2006) DOI:10.1002/cnm.806
10. J.C. Simo, T.J.R. Hughes, *Computational Inelasticity* (Springer, Berlin, 1998)
11. A. Tkachuk, M. Bischoff, Variational methods for selective mass scaling. Comput. Mech. (2013). doi:10.1007/s00466-013-0832-0
12. A. Tkachuk, M. Bischoff, Applications of variationally consistent selective mass scaling in explicit dynamics, in *Proceedings of the Fourth ECCOMAS Thematic Conference on Computational Methods in Structural Dynamics and Earthquake Engineering*, Kos Island, Greece, June 2013, ed. by M. Papadrakakis, V. Papadopoulos, V. Plevris
13. A. Vogel, S. Reiter, M. Rupp, A. Nägel, G. Wittum. UG 4 - a novel flexible software system for simulating PDE based models on high performance computers. Comput. Vis. Sci. (in press)
14. C. Wieners. Nonlinear solution methods for infinitesimal perfect plasticity. Z. Angew. Math. Mech. **87**, 643–660 (2007)

Printing and Binding: Stürtz GmbH, Würzburg